Environmental Division 2014

Core Programming Area at the 2014 AIChE Annual Meeting

Atlanta, Georgia, USA
16-21 November 2014

ISBN: 978-1-5108-1258-1

Printed from e-media with permission by:

Curran Associates, Inc.
57 Morehouse Lane
Red Hook, NY 12571

Some format issues inherent in the e-media version may also appear in this print version.

Printed by Curran Associates, Inc. (2015)

For permission requests, please contact AIChE
at the address below.

AIChE
120 Wall Street, FL 23
New York, NY 10005-4020

Phone: (800) 242-4363
Fax: (203) 775-5177

www.aiche.org

Additional copies of this publication are available from:

Curran Associates, Inc.
57 Morehouse Lane
Red Hook, NY 12571 USA
Phone: 845-758-0400
Fax: 845-758-2634
Email: curran@proceedings.com
Web: www.proceedings.com

TABLE OF CONTENTS

(9a) A Marco-Level Impact Assessment Tool for Emerging Technologies in Chemical Industry 1
Yuan Yao, Diane J. Graziano, Mathew Riddle, Eric Masanet

(9b) Environmental Life Cycle Assessment of OEM Automotive Paint Shop Technologies 2
Rebecca M. Glaspie, Jason Pierce

(9c) Evaluation of the Environmental Impacts of Ethanol Production from Eastern Red Cedar 14
Ife Olukoya, Mark R. Wilkins, Karthikeyan Ramachandriya, Clint P. Aichele

(9d) Life Cycle Assessment of Underground Coal Gasification 15
Sayara Saliyeva, Joseph Anthony Menicucci Jr., Paola Lettieri, Stefaan J. R. Simons

(9e) Life Cycle Analysis for Urban Waste Treatment Optimization 16
Dimosthenis Sarigiannis, Evaggelos Handakas, Spyros Karakitsios, Mariela Antonakopoulou, Alberto Gotti

(9f) Life Cycle Assessment of Green Transportation Fuels Produced from Algal Biomass Via Hydrothermal Liquefaction 24
Sundaravadivelnathan Ponnusamy, Harvind Kumar Reddy, Tapaswy Muppaneni, Peter Lammers, Cara Meghan Downes, Shuguang Deng

(9g) A Life-Cycle Assessment of Greenhouse Gas Emissions for Transportation Fuels from North American Conventional Crudes 25
Amit Kumar, Md. Mustafizur Rahman, Christina Canter

(11b) Specific Energy Consumption in Spiral Wound Reverse Osmosis Water Desalination 26
Mingheng Li

(11c) Enhancement of Oil-Water Separation By Dissolved Air Flotation 38
M. R. Aliff Mohamad Radzi, M. A. Abia-Biteo Belope, R. B. Thorpe

(11d) Direct Integration of RO Desalination and UF Pre-Treatment with RO Concentrate Backwash 48
Larry Gao, Anditya Rahardianto, Han Gu, Panagiotis D. Christofides, Yoram Cohen

(11a) Removal of Organic and Nitrogenous Contaminants of Wastewater By Polyelectrolyte Multilayer Membranes 49
Oishi Sanyal, Anna Sommerfeld, Zhiguo Liu, Rui Chen, Wei Liao, Ilsoon Lee

(42a) Photoactivated SOA Formation: Mechanistic Insight from Modeling and Experiments 50
V. Faye McNeill

(42b) Effect of Humidity and Particle Acidity SOA Derived from Reaction of Beta-Pinene with Nitrate Radical 51
Christopher Boyd, Lu Xu, Javier Sanchez, Xiaoxi Liu, Wing-Yin Tuet, Greg Huey, Nga Lee Ng

(42c) Composition and Oxidative Properties of Particulate Matter Mixtures 52
Wing-Yin Tuet, Vishal Verma, Julie A. Champion, Nga Lee Ng

(42d) Quantify the Impact of Biomass Burning Aerosols on Regional Climate over the Southeastern USA 53
Peng Liu, Yongtao Hu, Armistead Russell, Athanasios Nenes

(42e) Evaluating the Role of Natural Variability in Assessments of Climate Change Impacts on Air Quality 54
Fernando Garcia Menendez, Erwan Monier, Noelle Selin

(42f) Sector-Based Analysis of Deposited BC in the Arctic and Its Impact on Snow Albedo 55
Negin Sobhani, Sarika Kulkarni, Gregory Carmichael

(42g) The Contribution of Different Surfactants within Corexit in Ejection of Oil/Dispersant Material to the Atmosphere 56
Paria Avij, He Huang, Zenghui Zhang, Francisco R. Hung, Kalliat T. Valsaraj

60a) Impacts of CO, H_2 and C_3H_6 on the SCR Reactions over a Cu-Chabazite SCR Catalyst 57
Yang Zheng, Benjamin A Wilhite, Dan Luss

(60b) Development of SSZ-13 Based Sulfur Resistant NH_3-SCR Catalysts 58
Chao Wang, Erdem Sasmaz, Benjamin Galloway, Bihter Padak, Jochen Lauterbach

(60c) Understanding of the Transition Metal Ion Properties and Their Impact on Various Catalytic Functions of Zeolite-Based SCR Catalysts 59
Ashok Kumar, Krishna Kamasamudram, Neal W. Currier, Aleksey Yezerets, Hai-Ying Chen

(60d) Controlled Synthesis of Novel Interweaved Titanium Oxide Nanotubes Confined Metal Oxide Catalytic Formulations: Effect of Morphology on Selective Catalytic Reduction (SCR) of NOx By NH3 60
Thirupathi Boningari, Dimitrios Pappas, Panagiotis Smirniotis

(60e) Bimetallic Pt-Pd Catalyst Kinetics for CO and C_3H_6 Oxidation in Diesel Oxidation Catalysts (DOC) 61
Melanie Hazlett, William Epling

(60f) Characteristics of Pt-BaO/Mgal Mixed Oxides for NOx Storage-Reduction Catalysis 63
Soyeon Jeong, Seunghee Youn, Do Heui Kim

(60g) Hydrogen Purification for Fuel Cells: Preferential CO Oxidation in Excess of Hydrogen 64
Oscar E. Amador Sr., Luis F. Cordoba Sr., Julio C. Vargas

Reverse Electrodialysis: Sustainable Energy from Hydraulic Fracturing Water Recycle 70
Hailey Dunsworth

Modeling of the Aerobic Cometabolic Transformation of Chlorinated Ethenes By the Mycobacterium Elw-1 71
Stephanie Rich

The Enzymatic Hydrolysis of Alfalfa Stalks for Use As a Biofuel Resource 72
Elijah Wade

Synthesis and Characterization of Thin Film Fealcr for High Temperature Corrosion Applications 73
Randy Fang
Crossing Borders and Communities for Healthy Water 82
Joshua Gomez, Meng Zhou, Shuguang Deng
Ferroelectric BTO on Si (001) for High-Efficiency Solar Cell Heterostructures 91
Emma Kaeli
Natural Gas, the Bridge Fuel 92
Sravya Khasnavees
The Federal Role in Fostering an Innovative U.S. Energy Ecosystem 93
Erin Alderink
A Biowall for Improving Indoor Air Quality 94
Caroline Kelemen
(87a) Optimal Performance Management of Clean Water Treatment Processes 95
Eva Sorensen, Folashade Akinmolayan, Nina F. Thornhill
(87b) Using Integrated Controls and Real-Time Weather Forecasting to Drive Zero Discharge 96
Brian Petty, Melissa Fagan
(87c) Water Footprint of Rio Grande River Basin 97
Jonathan Dubinsky, Arunprakash T. Karunanithi
(87d) Water Scarcity Assessment As Part of a Complete Product Life Cycle Assessment, LCA 98
Sergio Galeano
(87f) Choose Appropriate Scenario in Order to Improve Water Quality at EL-Rahawy Drain and the Rosetta Branch, Egypt 105
Mohamed Mostafa, Robert W. Peters
(87g) Dynamic Optimization Strategies for Control of Algae Growth in Eutrophic LAKES with Nonpoint Nutrient Sources 113
Jimena A. Di Maggio, Vanina Estrada, María Soledad Diaz
(87h) Environmental Risk Analysis of Application of Electro-Disinfection on Ballast Water Treatment 114
Yue Ma, Shaj Thayil, J. Paul Chen
(104a) Multi-Objective Optimization for Solid Waste Management Systems 115
José Ezequiel Santibañez-Aguilar, Juan Martínez-Gomez, José María Ponce-Ortega, Fabricio Nápoles-Rivera, Meadardo Serna-González, Mahmoud El-Halwagi
(104b) Global Optimization for Sustainable Design and Synthesis of Algae Processing Network for CO2 Mitigation and Biofuel Production Using Life Cycle Optimization 117
Jian Gong, Fengqi You
(104c) Towards Integration of Life Cycle Assessment within a Multi Objective Supply Chain Optimization Modeling Framework 120
Bahador Mousavi, Arunprakash T. Karunanithi
(104d) Optimal Design of Microgrids with Innovative CHP Systems: Integration of Process Optimisation and Life Cycle Assessment 121
Di Zhang, Sara Evangelisti, Lazaros G. Papageorgiou, Paola Lettieri
(104e) Application of Sustainability Footprint and Multicriteria Optimization for the Choice of Sustainable Chemical Processes 122
Rajib Mukherjee, Debalina Sengupta, Subhas Sikdar
(104f) Sustainable Biorefinery Supply Network Using a Concept of Eco- and Total Profit – Case Study of the European Union 123
Lidija Cucek, Mariano Martín, Ignacio E. Grossmann, Zdravko Kravanja
(104g) Planning and Scheduling Industrial Waste Management Using Knowledge Based Lagrangean Decomposition 125
Elisabet Capón-García, Edrisi Muñoz, José Miguel Laínez, Antonio Espuña, Konrad Hungerbühler, Luis Puigjaner
(104h) Identifying the Preferred Subset of Alternatives for Environmental Improvements Via an MILP Approach Based on the Analytic Hierarchy Process 127
Gonzalo Guillén-Gosálbez, Ruben Ruiz-Femenia, José Antonio Caballero, Laureano Jimenez
(106a) Micro-Kinetics of NO_x Storage and Reduction with $H_2/CO/C_3H_6$ on $Pt/BaO/Al_2O_3$ Monolith Catalysts 128
Arun Kota, Dan Luss, Vemuri Balakotaiah
(106b) Deep SO2 Adsorption at Parts per Billion Level By Alumina-Based Mn/Ce Mixed Oxides for SOFC Cathode Protection 131
Peng Cheng, Bruce Tatarchuk
(106c) A First-Principles Study of the Catalytic Hydrodechlorination of 1,2-Dichloroethane 132
Lang Xu, Eric Stangland, Manos Mavrikakis
(106d) Ultra-Low Loading, Highly Active and Stable Ru/γ-Al_2O_3 Catalyst for Low Temperature Solar Steam Reforming of Methane 133
David Simakov, Yuriy Román-Leshkov
(106e) Spatio-Temporal Features of NOx Reduction on LNT+SCR System: Effect of Hydrocarbon Reductant and Space Velocity 134
Mengmeng Li, Vencon Easterling, Michael P. Harold
(106f) H_2SO_4 Catalysis: Perspective and Opportunities for Reducing SO_2 Emissions Using Particulate and Monolith Catalysts 139
Anuradha Nagaraj, Patrick L. Mills

(106g) Synthesis, Characterization of Vanadia/Titania-Based Catalysts and Their Performance in Ammonia Remediation 143
Hao Chen, Raquel Portela, Wei Han, Pedro Avila, Miguel A. Bañares, King Lun Yeung
(123a) Probing the Active Site Requirements and Mechanistic Details of NOx Selective Catalytic Reduction with NH3 on Cu-SSZ-13 148
Rajamani Gounder, Fabio H. Ribeiro, W. Nicholas Delgass, William F. Schneider, Jeffrey T. Miller, Aleksey Yezerets, Trunojoyo Anggara, Christopher Paolucci, Shane A. Bates, Anuj Verma, Atish Parekh
(123b) A New Method for Exhaust Aftertreatment Design - Integrated Engine and Aftertreatment Analysis 149
Yong Miao, Steve Felix
(123c) Analysis of Ash-Loaded DPF Performance during Drop-to-Idle Test 150
Yi Liu, Changsheng Su, Arvind Harinath
(123d) Transient Kinetic Study of Sulfur Oxide Adsorption/Desorption over Al2O3 and Pt/Al2O3 Catalysts 151
Tayebeh Hamzehlouyan, Chaitanya S. Sampara, William S. Epling
(123e) Effect of Degredation on NOx Storage over Novel Lean NOx Trap Catalysts 154
Travis Wentworth, Susan M. Stagg-Williams, Christopher D. Depcik
(123f) Low Temperature Oxidation of Carbon Monoxide Produced By Diesel-Ignited Methane Dual Fuel Low Temperature Combustion in a Single-Cylinder Diesel Engine 155
Navid Zanganeh, Hossein Toghiani, Kalyan K Srinivasan, Sundar Rajan Krishnan, Taryn Bayles, Mostafa S. Raihan, Jason M. Keith
(163a) The IMPACT of Roofing Material on Building Energy Performance 174
Ali Badiee
(163b) Biowall's Impact on Indoor Air Quality and Energy 175
Dan Newkirk, Bill Hutzel
(163c) Sustainable Buildings: Targets and Measurements 176
Godfried Augenbroe, Jason Brown
(163d) Sustainable Buildings 191
Theresa Weston, Cory Jensen, Yudong Chen, Bhima Sastri
(163e) Avoided GHG Emissions Achieved By Using Architectural Window Film: An LCA Compliant with New WBCSD Avoided Emissions Guidelines 192
Terrie Boguski, Lauren Johnson
(163f) Design of Robust Renewable Energy Supply Systems By Providing Visibility Based upon Improved Neural Network 202
Soo Bin Lee, Jun-Hyung Ryu, In-Beum Lee
(163g) Experimental Investigations on Brick Specimens Made from Deinking Sludge 203
Shilpa Kulkarni, Vivek Kumar, Sanjeev K. Singh, Mukesh C. Bansal
(163h) Multi-Objective Optimization of CHP Systems for Housing Complexes 210
Luis Fabian Fuentes-Cortes, José María Ponce-Ortega, Medardo Serna-González, Carlos Rubio-Maya
(180a) Multi-Layer Monolith Catalysts for Lean NOx Reduction 211
Michael P Harold, Vemuri Balakotaiah, Dan Luss, Sachi Shrestha, Yang Zheng, Bijesh M. Shakya, Pranit S. Metkar, Yi Liu, Krishna Kamasamudram, Aleksey Yezerets
(180b) A Discussion on the Fundamental Aspects of Urea SCR Control for Transient Applications 212
Devesh Upadhyay, M. Van Nieuwstadt
(180c) Investigation of Burning Mode for Diesel Particulate Oxidation: Contrasting O2 and NO2 221
Andrea Strzelec
(180d) TWC Formulation Effects on NH3 Generation for Passive SCR Applications in Lean Gasoline Engine Exhaust 222
Josh A. Pihl, Vitaly Y. Prikhodko, Todd J. Toops, James E. Parks
(180e) Experimental Investigation and Kinetic Modeling of Diesel Oxidation Catalysts for Emissions Aftertreatment from Diesel Engine Exhaust 223
Hom Sharma
(180f) A New Catalyst for the Selective Catalytic Reduction of NO_x By NH_3 over Olivine 224
Yun Shi, Su-Jing Li
(215a) Renewable H2/CO Fuels for FTS and SOFC Applications Via Electrochemical Conversion 226
Wenhua H. Zhu, Bruce J. Tatarchuk
(215c) Maximizing the Recovery of Bio-Commodities from Algal Biomass 227
Ali Teymouri, Edith Martinez-Guerra, Sandeep Kumar, Veera Gnaneswar Gude
(215d) Green Chemical Conversion of Cellulose into Soluble Sugars and 5-Hydroxymethylfurfural (HMF) Using Ionic Liquids 228
Sapna Jain, Alexis Rogers, Kelvin Jones, Jessica Abron
(215e) Liquid-Liquid Extraction of Bio-Oil Components 229
Kyoung Eun Park, Sotira Yiacoumi, Costas Tsouris, Abhijeet P. Borole
Exploiting Polymer-Nanoparticle Interactions to Create a Hydrogel with Biomedical Applications 230
Jessica Greer
Validation of High Throughput Electrochemical Gas Sensing Screening System 231
Zixuan Wang
Long-Range Correlations in Liquid Water 241
Nancy Figueroa
Frictional Study of Polyethylene Glycol Monolayers on Silica Substrate 242
Nadiah Nordin

A Machine-Learning Model to Predict Activation Energies of Hydrogenation Reactions 243
Jack McCullough
Microscopic Modeling of the Self Assembly of Poly(ethylene oxide)-Poly(propylene oxide)- Poly(ethylene oxide) (PEO-PPO-PEO) Block Copolymers: Critical Micelle Concentrations 244
Alexander Colville
(221a) Recommendations for Legislative Actions to Reduce Carbon Emissions in the Electricity Production Sector 245
Sam White
(224a) Life Cycle Assessment of Vaccine Supply Chain in Developing Countries 273
Bahador Mousavi, Joshua Martens, Chris Thai, Arunprakash T. Karunanithi
(224aa) The Study of Metal Oxide Modified Limestone As CO2 Sequestration Materials 274
Jiang-Feng Li, Hui Li, Wen-Bin Yang, Lele Zhang, Binglu Meng, Youhai Yu, Yong Min
(224b) Simple Method of Deposition of CuO Nanoparticles on a Cellulose Paper and Its Antibacterial Activity 275
Amin Yoosefi Booshehri, Rong Xu
(224c) Use of a Commercial-Scale Reverse Osmosis Laboratory Module to Illustrate Interception Technology Applications in Water Pinch Network Design 276
Russell F. Dunn, Jarrid Ristau
(224d) Influence of Stabilizer Size and Chelation Strength on Iron Nanoparticle Oxidation 277
Nikki S. Rentz, Lauren F. Greenlee
(224e) Immobilization Carbonic Anhydrase on Magnetic Polymer Microsphere for Accelerating CO2 Absorption into a Carbonate Solution 278
Zuoming Zhou, Fujun Pan, Guohua Jing
(224g) Optimal Design of Domestic Water-Heating Systems through Solar Collectors 280
Aurora De Fatima Sanchez-Bautista, José Ezequiel Santibañez-Aguilar, José María Ponce-Ortega, Fabricio Nápoles-Rivera, Medardo Serna-González, Mahmoud El-Halwagi
(224h) Simultaneous Design of Water Reusing and Rainwater Harvesting Systems in a Residential Complex 283
Mariana Garcia-Montoya, Andrea Bocanegra-Martinez, José María Ponce-Ortega, Fabricio Nápoles-Rivera, Medardo Serna-González, Mahmoud El-Halwagi
(224i) Mineralization of CO2 with Industrial Solid Waste and Natural Ore 284
Bin Liang, Jiahua Zhu, Chun Li, Hairong Yue, Houfang Lu, Longpo Ye, Chao Wang, Yufei Wang, Heping Xie
(224j) Optimization of Electrode Design for Electrodialysis Reversal 287
Masoume Jaberi, Fattaneh Naderi Behdani, Abbas Ghassemi, Jim Loya
(224k) A Study on Effect of Temperature, COD and Influent Tds on Microbial Desalination Cells' Performance with an Approach to a Unique Predictive Model 288
Mohammad Tanhaemami
(224l) Controlled-Release Antimicrobial for Air Disinfection 289
Yue Tak Lai, Hao Chen, Yan Li, Wei Han, King Lun Yeung
(224m) Impact of Fugitive Emissions on the Greenhouse Gas Emissions of Conventional Crudes 290
Amit Kumar, Christina Canter, Md. Mustafizur Rahman
(224n) Mini Pulsed Electric Field Device for Drinking Water Disinfection 291
Ka Wo Lam, Hao Chen, Pik Shuen Hung, Oi Wa Lee, Siu Ming Kwan, Joseph Kai Cho Kwan, King Lun Yeung
(224o) Antimicrobial and Anti-Adhesion Coating for Water Filtration Membrane 292
Hiu Pang Yu, Hao Chen, King Lun Yeung
(224p) Well to Wheel Life Cycle Assessment of Greenhouse Gas Emissions of Transportation Fuels from Canadian Oil Sands 293
Amit Kumar, Balwinder Nimana, Christina Canter, Md. Mustafizur Rahman
(224q) Advanced Ozone Membrane Reactor for Treatment of Endocrine Disrupting Compounds in Water 294
Tiphaine Corbet, Yakub Fam, Liping Li, Ying Li, Rafael Serra Cuesta, Wei Han, King Lun Yeung
(224r) Greenhouse Gas Calculator for Tracking Combustion and Process Upset Emissions: Methodology and Visual Representation for Simple Ethylene Process Base Case 296
Fahd M. Mohammed, Monzure-Khoda Kazi, Fadwa T. Eljack
(224t) Tower Gardens and Solar Technology 297
Willy Giron Matute
(224u) Investigation of the Influence of Pulsed Corona Discharges on Benzotriazole Degradation 298
Oluwatosin Owoseni
(224v) Modeling of Physical-Chemical Variables Behavior in Rivers Using the Streerter-Phelps Model 299
Lady Andrea Fuertes, Mario Andres Noriega
(224w) Dynamic Model of Stabilization Pond Systems 300
María P. Ochoa, Vanina Estrada, Patricia M. Hoch
(224x) Design and Optimization of the Distillation and Dehydration Steps for Bio-Ethanol Production 301
Mauricio Colombo, Michel Kahwaji Janho, Jorge E. Gatica, Fernando Daniel Mele, María Rosa Hernández
(224y) The Morphology Modification of Limestone By Different Forms of Carbon Additives for CO2 Sequestration 302
Binglu Meng, Hui Li, Wen-Bin Yang, Lele Zhang, Jiang-Feng Li, Youhai Yu, Yong Min
(224z) The Enhancement of CO2 Sequestration of Limestone in Cement Industry By Carbon Black Modification 303
Binglu Meng, Hui Li, Wen-Bin Yang, Lele Zhang, Jiang-Feng Li, Youhai Yu, Yong Min
(710b) Primarily Investigation of Mixed Adsorbents for the Removal of Copper and Methylene Blue from Aqueous Solutions 304
Ahmad Albadarin, Chirangano Mangwandi, Gavin Walker
(256a) Biodegradation of Contaminants in Karst Groundwater, a Dual Continuum Model 310
Roger Painter, Tom Byl, Lonnie Sharpe, Justin Harris

(256b) CFD and Response Surface Modeling of Flare Performance: DRE/CE Vs. Soot ... 311
Daniel H. Chen, Peyton C. Richmond, Helen H. Lou, Xianchang Li, Matthew Johnson
(256c) Development of Detailed Kinetic Mechanisms for Olefins Pyrolysis ... 312
Kun Wang, Stephanie M. Villano, Anthony M. Dean
(256d) Development of a New Low-GWP Refrigerant Composed of HFO-1123 (trifluoroethylene) ... 314
Toshiyuki Tanaka, Hidekazu Okamoto, Katsuya Ueno, Jun Irisawa, Tetsuo Otsuka, Tatsuhiro Nogami, Ritsu Dobashi
(256e) Nitric Oxide Removal By Aqueous Persulfate and Ferrous-EDTA Systems: Effects of Persulfate and EDTA Concentrations and pH ... 322
Yusuf G (Debo) Adewuyi
(256f) Modeling and Analysis of Light-Off Behavior of DOC ... 323
Richa Raj, Michael P Harold, Vemuri Balakotaiah
(256g) Kinetic Study of Anaerobic Digestion of Sewage Sludge ... 325
Pooja Sharma, U K Ghosh, Amiya Kumar Ray
(256h) Impact of Hydrocarbon Fuel Structure on Anaerobic Biodegradation Via the Fumarate Addition Reaction: An Ab-Inito and Kinetic Modelling Study ... 329
Vivek Bharadwaj, C. Mark Maupin, Anthony M. Dean
(288a) Biodiesel Production Via Transesterification in a Mixed Carbon Dioxide-Methanol System with a Heterogeneous Catalyst ... 330
Lindsay Soh, Chun-Chi Chen, Thomas Kwan, Eric J. Beckman, Julie Zimmerman
(288b) Production of Biochar and Combustible Gas from Co-Pyrolysis of Agricultural Plastic Wastes and Animal Manures ... 331
Kyoung S. Ro
(288c) Hydrothermal Liquefaction of Wet Algal Biomass with/without Catalysts ... 332
Tapaswi Muppaneni, Harvind Reddy, Sundaravadivelnathan Ponnusamy, Nagamany Nirmalakhandan, Tanner Schaub, Barry Dungan, Francisco Holguin, Pete Lammers, Wayne Voorhies, Shuguang Deng
(288d) Biorefinery on Biocathodes of Microbial Desalination Cells ... 333
Bahareh Kokabian, Veera Gnaneswar Gude
(288e) Life-Cycle Energy Use and Greenhouse Gas Emissions of Biofuels Production from Sweet Sorghum ... 334
Jihong Li, Shizhong Li
(324a) Improving a Process Site Sustainability Through Waste Heat Recovery ... 335
G. Oluleye, Megan Jobson, Robin Smith
(324b) Multiscale Connectivity for Chemical Mixture Toxicity Assessment ... 358
Dimosthenis Sarigiannis
(324c) Double Containment Piping Systems: A Fail-Safe, Leak-Proof and Environmentally Friendly Solution ... 366
Darin Johnson
(324d) Humic Substances in Treatment of Water Contaminated with Hydrocarbons ... 373
Yair Cruz-Narváez, Hever Honorato-Cervantes, Enrique Rico-Arzate, Jose J. Castro-Arellano, Vanessa Silva-Castro
(324e) Sustainability Assessment and Performance Improvement of Electroplating Systems ... 374
Hao Song, Navdeep Bhadbhade, Yinlun Huang
(324f) Uncertainties of Ozone Increments Caused By Industrial Startup Flaring ... 375
Jian Zhang, Ziyuan Wang, Qiang Xu, Thomas C. Ho
(335a) Protecting our Environment with Catalysis and Reaction Engineering ... 376
Panagiotis Smirniotis
(352a) Using Soybean Derived Crude Glycerol As Co-Digestate in Sewage Sludge Anaerobic Digester to Increase Biogas Production ... 377
Steven Nartker, Michelle Ammerman, Michael Stogsdill, Jennifer Aurandt, Olivia Hayden, Chad Antle
(352b) Kinetic Modeling of Hydrothermal Liquefaction of Algal Biomass ... 378
Harvind Kumar Reddy, Tapaswy Muppaneni, Sundaravadivelnathan Ponnusamy, Thinesh Selvaratnam, Barry Dungan, Nagamany Nirmalakhandan, Tanner Schaub, Francisco Holguin, Peter Lammers, Wayne Voorhies, Shuguang Deng
(352c) Utilization of Solid Residual Wastes Arising from Woody Biomass Gasification ... 379
Thawatchai Maneerung, Zhanyu Yang, Sibudjing Kawi, Chi-Hwa Wang
(352d) Toxicity Assessment of Bottom Ash from Biomass and Sewage Sludge Co-Gasification ... 380
Le Rong, Koon Gee Neoh, Yen Wah Tong, Chi-Hwa Wang
(352e) Single-Step Synthesis of Biodiesel from Crude Jatropha curcas Oil Using Chlorosulfonic Acid Catalyst ... 381
Hanif A Choudhury, Pulkit Srivastava, Vijay Moholkar, Sai Gu
(381a) The Chemistry of the "Herycnite Cycle" Solarthermal Water Splitting Reactions ... 382
Christopher L. Muhich, Kayla Weston, Darwin Arifin, Anthony H. McDaniel, Eric N. Coker, Charles B. Musgrave, Alan W. Weimer
(381b) Hydrogen Production from Biomass Via Microbial Electrolysis ... 383
Alex Lewis, Abhijeet P. Borole
(381c) The Power of Electricity & Chemicals Co-Production ... 384
Abdulrahman Albassam, Vasilios Manousiouthakis
(381d) Low Temperature H_2 Generation from Thermochemical Water-splitting Reaction Using Complex Redox Materials ... 385
Rajesh V. Shende, Vinod S. Amar, J. Puzsynski
(409a) Uncertainty and Sensitivity Analysis in Sustainable Process Design – Environmental Indicators ... 386
Carina Gargalo, Gürkan Sin
(409b) A Techno-Economic, Life-Cycle Modeling Framework for Emerging Technology Assessments in the U.S. Chemical Industry ... 388
Yuan Yao, Diane J. Graziano, Mathew Riddle, Eric Masanet

(409c) Rare Earth Oxide Production: Quantification of Life Cycle Environmental Impacts 389
George G. Zaimes, Vikas Khanna, Berlyn Hubler
(409d) Life Cycle Analysis of Oil Shale Production Using Greet 390
Ghana Paudel, Daniel Steiner, Joseph D. Smith
(409e) Dimethyl Ether Chemical Storage Cycle for Uninterrupted Renewable Power 391
Emre Gençer, Easa I. Al-Musleh, Dharik S. Mallapragada, Rakesh Agrawal
(423a) Plasma-Based Advanced Oxidation: Effects of Reactor Design on the Degradation of Organic Contaminants 392
Selma Mededovic, Gunnar Stratton, Fei Dai, Christopher Bellona, Thomas Holsen
(423b) Effects of the Addition of Catalysts on the Decomposition of Amido Black 10B By Pulsed Corona Discharge 393
Negin Koutahzadeh, Pedro E. Arce
(423c) Novel Spatially and Temporally Confined Microplasmas for Advanced Oxidation 394
Justin Pommerenck, Jordan Pommerenck, Peter Kreider, Yousef Alanazi, Jacob Lum, Alexandre F. T. Yokochi
(423e) Photocatalytic Degradation of Methamphetamine Using UV/TiO2 395
Chin-Sheng Kuo, Cheng-Fang Lin, Pui-Kwan Andy Hong
(423f) Visible Light Mediated Heterogeneous Photo-Fenton Oxidation of Endocrine Disrupting Compounds in Wastewater and Water Using CdS/Carbon Nanotube-Iron Oxide Composite Photocatalyst 396
Jihyun R. Kim, Eunsung Kan
(433a) Technical, Economic and Environmental Viability of Offshore CO2 Reuse from Natural Gas By Dry Reforming 397
Bruna Lima, Ofélia Araújo, José L. Medeiros, Claudia Morgado
(433b) Improved CO2 Capture Process : Rich Vapor Recompression with Split Flow 408
Jaeheum Jung, Yeong Su Jeong, Chonghun Han
(433c) Development of CO2 Capture and Utilization Technologies for Alkaline Wastes for Building Waste-to-Resource Supply Chain: Theoretical Consideration 409
Shu-Yuan Pan, Pen-Chi Chiang, Elisa G. Eleazar, Andrew Chiang, E-E Chang, Yi-Hung Chen
(433d) CO2 Fixation through Carbonation of Waste Cement and Concrete 410
Atsushi Iizuka, Akihiro Yamasaki, Motoki Inoue, Miyuki Noguchi
(433e) Technological Trends in CO2 Capture, Transport and Utilization 411
Rita M. B. Alves, Fernanda M. Fontes, José L. Medeiros, Ofélia Q. F. Araújo
(433f) Effects of Contaminants to Physical Solvents for Pre-Combustion CO2 Capture 426
Fan Shi, Brian Kail, Hunaid Nulwala, Nicholas Siefert, David Luebke
(464a) Self-Emulsification of Alkaline-Dissolved Clove Bud Oil By Whey Protein, Gum Arabic, Lecithin, and Their Combinations 427
Yangchao Luo, Yue Zhang, Kang Pan, Faith Critzer, P. Michael Davidson, Qixin Zhong
(464b) Oil Extraction from Spent Coffee Grounds Using Advanced Techniques 428
Javier Davila Sr., Moshe Rosenberg, Gonzalo Taborda Sr., Carlos A. Cardona
(464c) Microencapsulation of Blackberry Antioxidants. Modeling and Simulation 437
Miguel Rojas, Javier Davila Sr., Moshe Rosenberg, Carlos A. Cardona
(464d) Liquid-Phase Electrical Discharge Plasmas Rapidly Inactivate Pathogenic and Spoilage Microorganisms in Water 449
Tomislava Vukusic, Zoran Herceg, Shane Rogers, Selma Mededovic
(464e) Modeling the Drying Kinetics of Carrot Particles in a Closed Chamber Following Forced Convection and Electric Field Treatment 450
Abdelbasset Bessadok Jemai, L. Khezami, M. Hadjkali, E. Vorobiev
(464f) Antimicrobial Peptide Segments from Soy Protein for Use in Food Safety 451
Ning Xiang, Yuan Lyu, Ganesan Narsimhan
(464g) Making Food Safe in Light of New Demands 452
James Van Wyk
(464h) Low Temperature Batch Conversion of Cellulosic Biomass 457
Mohit Nahata, Galen B Fisher, Johannes W. Schwank
(479a) Influence of the Synthesis Method on Cr Leaching during the Liquid Phase Photocatalysis of Cr-TiO2 458
Siva Nagi Reddy Inturi, Makram Suidan, Panagiotis Smirniotis
(479b) Electrochemical Wastewater Treatment for Industrial Applications 459
Dieter Woisetschläger, Matthaeus Siebenhofer
(479c) Mathematical Model of a Thin Film Slurry Reactor for the Degradation of an Azo-Reactive Dye By Solar Photo-Fenton Process 470
Abdón Parra, Lidia Yokoyama, Fabiana Valéria Da Fonseca
(479d) Performance of Activated Carbon Supported Catalyst during Wet Oxidation of Pulping Effluent 472
Bholu Ram Yadav, Anurag Garg
(479f) Catalytic Wet Oxidation of Phenolic Compounds at Moderate Temperature and Pressure 478
Rajendra Mohite, Anurag Garg
(499a) Establishing Analytical Adsorption Methods for Pore Characteristic Evaluations in Gas Shales 483
Erik C. Rupp, Jennifer Wilcox
(499b) A Generalized Model for Gas-Solid Adsorption Equilibria 484
Austin Ladshaw, Sotira Yiacoumi, Costas Tsouris, David Depaoli, Ronghong Lin, Lawrence L. Tavlarides
(499c) Binding of SO3 to Fly Ash Components: CaO, MgO, Na2O and K2O 485
Benjamin Galloway, Erdem Sasmaz, Bihter Padak
(499d) Metal Organic Frameworks for Selective Adsorption of t-Butyl Mercaptan from Natural Gas 486
Grace Chen, Christopher W. Jones, William J. Koros

(499e) Atomic Layer Deposition Assisted Layer-By-Layer Growth of Metal Organic Frameworks on Polymer Fiber Mats for Hazardous Gas Adsorption 495
Junjie Zhao, Bo Gong, Paul C. Lemaire, William T. Nunn, Eric C. Stevens, Fahim I. Sidi, Matthew A. Browe, Gregory W. Peterson, Mark D. Losego, Gregory N. Parsons
(499f) Amino-Pillared Nanosheet (APN) for High Performance Carbon Dioxide Capture 496
Christopher Cogswell
(499g) Carbon Dioxide Capture Using Elastic Layered Metal-Organic Frameworks: Experimental and Computational Evaluation 497
Francisco Sotomayor, Tran D. Trinh, Christian Lastoskie
(499h) Effects of Flue Gas Contaminants, Regeneration Conditions, and Amine Modification on the Adsorption of Carbon Dioxide By MIL-101(Cr) 498
Qing Liu, Junjie Shi, Shudong Zheng, Yao Shi, Yi He
(506a) Effect of C/N Ratio on Microbial Lipid Production with the Oleaginous Yeast Lipomyces Starkeyi 499
Christopher Giorgio, Stephen Dufreche, William Holmes, Rafael Hernandez, Mark E. Zappi, Rakesh Bajpai, Ramalingam Subramaniam
(506b) Recovery of Polyphenol-Enriched Sugarcane Molasses Solution By Ultrafiltration 500
Michelle Almendrala, Gerald Aaron Ang, Edmar Bernadas, Cedrick Ong
(506c) β-1,3-Glucan Production By Metabolically Engineered Agrobacterium Sp. from Cellulosic Sugar 501
Hyun-Dong Shin, Young-Il Park, Mi-Kyoung Kim, Rachel Ruizhen Chen
(506d) Chemical Imaging of Cheese Microstructure Using ATR Microimaging Spectroscopy 502
Ramazan Kizil
(506e) Study on Swine Manure Foaming to Understand Its Major Reason and Its Relations to Industrial Products 503
Mi Yan, Jing Gan, Yan Yang, Aravindan Rajendran, Hongjian Lin, Qiyang He, Bo Hu
(506f) Economic Analysis of Different Stabilization Methods of Antioxidants 504
Miguel Rojas, Moshe Rosenberg, Carlos A. Cardona
(506g) Preveting Paper Towels from Bacteria Growth Using Selenium Nanoparticles 514
Qi Wang, Thomas J. Webster
(506h) Novel Tools in Genome-Scale Metabolic Flux Modeling to Identify Metabolic Engineering Targets and Predict Microbial Phenotypes 518
Hadi Nazem-Bokaee, Jiun Yen, Ryan S. Senger
(556a) Sustainable Winemaking – Challenges and Opportunities 519
Michael Roland
(556b) The Challenges of Water Use Reductions and Food Safety in Dairy Processing 520
Jill Brigham
(556c) The Challenges of Food Safety and Water Use Reductions in Dairy Processing 521
Joseph Herrud, Montgomery Bohanan
(556d) Maximizing Water and Energy Efficiency in Food Processing While Minimizing Negative Environmental Impact 527
Christopher W. Simmons
(562a) Carbon Footprint of Agrarian Systems 528
Jonathan Dubinsky, Arunprakash T. Karunanithi
(562b) Process Sustainability Evaluation for the Recovery of Metal from Spent Batteries 529
Gerardo J. Ruiz-Mercado, Michael A. Gonzalez, Raymond L. Smith
(562c) Sustainability Control: Theoretical Aspects and Engineering Methods 530
Hao Song, Liwei Yan, Yinlun Huang
(562d) Statistical Analysis of Global Environmental Impact Patterns Using a World Multi-Regional Input Output Database 531
Janire Pascual-González, Gonzalo Guillén-Gosálbez, Laureano Jiménez Esteller, Josep Maria Mateo-Sanz
(562e) Maximizing Sustainability of Ecosystem Model through Socio-Economic Policies Derived from Multivariable Optimal Control Theory 532
Rohan Doshi, Urmila Diwekar, Kirti Maheshkumar Yenkie, Heriberto Cabezas, Pahola T. Benavides
(562f) Understanding Resilience of Metro Systems in Polycentric Megacities: A Case Study of Delhi Metro Rail System 546
Shauhrat Chopra, Vikas Khanna
(565a) Trichloroethylene Oxidation and Hexavalent Chromium Reduction Via Redox Reactions Using KMnO4 and FeSO4 in Aqueous and Soil Systems 548
Jude Ighere, Karina Honjoya, Ramesh Chawla
(565b) Investigation of the Extraction and Recycle Potential of CCA Treated Wood Waste 549
R. Mark Bricka, Amy M. Parker
(565c) Development of a New Paradigm in Biochemical Engineering: Predicting the Genetic Regulation of Aromatic Pollutants Degradation 550
Argyro Tsipa, Michalis Koutinas, Athanasios Mantalaris, Efstratios N. Pistikopoulos
(565d) Studies on Evaluation of Acute Toxicity Factor for Selected Industrial Effluents Using Zebra Fish(Danio Rerio) As a Test Organism 555
Pulipati King, Garika Sheshamma
(565e) Integrated Solar Photocatalytic Oxidation and Biodegradation for Degradation of Antibiotics and Endocrine Disrupting Compounds in Wastewater and Water 556
Jihyun R. Kim, Eunsung Kan

(565f) Enhanced Combustion of Rice Husk for Pollution Abatement and Regional Potential of Silica from Rice Husk Ash (RHA) ... 557
Muhammad Suleman Tahir
(565g) Innovative Design of an Integrated Biotrickling Filter System for Removal of Simulated Waste Gas Containing a Mixture of p-Xylene and Ethyl Mercaptan ... 563
Xiang-Qian Wang, Su-Jing Li, Wei Li
(605b) Land Use Emissions and Environmental Impacts of Agricultural Systems ... 564
Jonathan Dubinsky, Arunprakash T. Karunanith
(608a) Development and Performance of a Novel Cu2+-Imprinted Membrane By Semi-Interpenetrating Polymer Network Technique for Selective Cu2+ Removal ... 565
Jinsong He, J. Paul Chen
(608b) Surfactant-Free Synthesized Mesoporous Carbon Nanocomposites Towards Efficient Pollutants Removal from Water ... 566
Long Chen, Jiahua Zhu
(608c) Evaluation of Usability of Various Real Wastewaters in Microbial Fuel Cells ... 567
Anna Casasus, Luciana Bava, Jaime Lee
(608e) Some Aspects of Hydrodynamics of Continuous Countercurrent Liquid-Solid System ... 575
Krishnan Nagarajan, T. Renganathan, K. Krishnaiah
(660a) Removal of Cadmium Ions from Aqueous Solution Using a New Low-Cost Biosorbent ... 577
Alireza Saraeian, Atefe Hadi, Abbas Ghassemi
(660b) Environmental Remediation of Dense Non-Aqueous-Phase Liquids Using Multifunctional Iron-Carbon Nanocomposites ... 578
Yang Su, Bhanukiran Sunkara, Owoseni Olasehinde, Yueheng Zhang, Jingjing Zhan, Gary McPherson, Vijay T. John
(660c) Adsorption and Oxidative Degradation of Bisphenol a on Surface Modified Iron-Amended Activated Carbon: Effects of Temperature on Adsorption and Fenton Oxidation ... 579
Eunsung Kan, Jihyun R. Kim
(660d) Removal of Metronidazole and Dimetridazole from Aqueous Solution By Adsorption on Multiwalled and Singlewalled Carbon Nanotubes ... 580
Roberto Leyva-Ramos, Damarys H Carrales-Alvarado, Raul Ocampo-Perez
(660e) Synthesis of Zirconium Nanoparticles Doped Activated Carbon Fiber for Optimized Adsorption for Arsenic Removal ... 581
J. Paul Chen
(663b) Integrated Membrane Bioreactor Process for Water Reclamation and Groundwater Recharge Applications ... 584
Woonhoe Kim, Varadarajan Ravindran, Massoud Pirbazari
(663c) Controlling Ammonia Levels with Natural Materials to Preserve Bait and Stored Fish in Fresh- and Seawater ... 585
Wen Zhao, Daniela M. L. Stebbins, Fei Guo, Sarina Ergas, Norma Alcantar
(663d) Production Water Recovery Processing and Reuse ... 596
Christina M. Borgese
(663e) Polyaniline Coated Ethyl Cellulose with Improved Hexavalent Chromium Removal ... 597
Bin Qiu, Cuixia Xu, Dezhi Sun, Suying Wei, Zhanhu Guo
(663f) Study of Fluoride Adsorption Onto Nanostructured Zirconium-Manganese Based Particle ... 598
Alam Akm Khorshed Sr., Jinsong He, Narahari Mahanta, J. Paul Chen
(663g) Removal of Chromium (VI) and Divalent Cations from Flowback Water By Graphene Oxide-Modified Natural Zeolite ... 599
Lucy Mar Camacho, Tanviben Pareshbhai Desai
(688a) Adsorptive Removal of Humic Acids and Microbes By Polyacrylonitrile-Chitosan Composite Membrane ... 600
Swapna Rekha Panda, Munmun Mukherjee, Sirshendu De
(688b) Removal of Gaseous O-Xylene in the Two-Liquid Phase Biotrickling Filters and Airlift Bioreactors ... 608
Chao Wu, Xiang-Qian Wang, Bai-Long Xu, Su-Jing Li, Wei Li
(688c) The Effect of Various Fibrous Filter Media and Newly Designed Packaging Configurations on Sea Salt Particles Loading Performance ... 609
Pengfei Zhao, Bruce J. Tatarchuk
(688d) Removal of Heavy Metals from Wastewater Using Immobilized Biochar ... 618
Edward Trujillo
(695a) UV-Assisted Stabilization of Modified Softwood Kraft Lignin Fibers ... 619
Meng Zhang, Jing Jin, Amod A. Ogale
(695b) Effect of a Natural Cactus Based-Mucilage Dispersant on the Surface Tension and Droplet Size of Dispersed Crude Oil ... 620
Fei Guo, Daniela M. L. Stebbins, Tunan Peng, Rana Falahat, Wen Zhao, Sylvia Thomas, Ryan Toomey, Norma Alcantar
(695c) Artificially Engineered Protein Gels Derived from Nucleoporins ... 621
Minkyu Kim, Wesley Chen, Jeon Woong Kang, Matthew J. Glassman, Katharina Ribbeck, Bradley D. Olsen
(695d) Enhancing Oxygen Permeability in Hydrogel Wound Dressing By Cyanobacterial Gas Vesicles ... 622
Napaporn Vongpanish, Uchechukwu Chamberlin Anozie, Lu-Kwang Ju
(695e) Adhesive Elastin-Based Proteins As Soft Tissue Glues ... 623
M. Jane Brennan, Julie C. Liu, Jessica K. Roman, Julie N. Renner, Renay S.-C. Su, Jonathan J. Wilker
(695f) Simulation and Experimental Investigation of Osteogenic Activity of Hydrogel-Conjugated BMP-2 Peptide ... 624
Seyedsina Moeinzadeh, Esmaiel Jabbari
(695g) Designing Multi-Component Nanostructured Soft Biomaterials Interacting with Charged Nanoparticles ... 626
Fikret Aydin, Meenakshi Dutt

(710a) Novel 3D-Printed Oleophillic Absorbents 627
Duanduan Han, Victor M. Ugaz
(710c) Fate and Removal Behaviour of Antibiotics in an Osmotic Membrane Bioreactor for Municipal Wastewater Treatment 628
Guanglei Qiu, Divya Shankari Srinivasa Ragha, Yen-Peng Ting
(710d) Removal of Trihalomethanes (THMs) By Electro-Coagulation Process Using Aluminium Plate 629
Cheng-Chun He, Ching-Yao Wu, Shang-Lien Lo
(710e) Applications of Sulfur-Containing Radical Treatment in for Water Reuse and Recycling 630
Haizhou Liu, Lucy Li, Lamees Alkhamis
(710f) Copper (II) Removal Using Activated Neem Bark from Waste Water As a Low Cost Adsorbent: Column Studies 631
Suresh Gupta, Utkarsh Maheshwari
(710g) Preparation Characterization and Application of an Affinity Flat Sheet Membrane 632
Ling Yu, Yu Ming Zheng, J. Paul Chen
Author Index

384400 A Marco-Level Impact Assessment Tool for Emerging Technologies in Chemical Industry

Abstract:

In the U.S, chemicals industry is one of major energy consumers and sources of greenhouse gas (GHG) emissions[1-3]. As the economy and population grows, the increasing demand for energy and more severe environmental problems is prompting the development and adoption of emerging technologies in chemicals industry to reduce energy consumption and adverse environmental impacts, which can make contributions to a greener economy. However, not every technology has equal potential for national energy saving and environmental pollution reduction. The assessment of the energy and environmental impacts of these emerging technologies is important because based on these information companies and policy makers make choices on investment and promotion within these technology classes. But current assessment models[4-6] fail to conduct assessments for emerging technologies satisfying the requirements of policy makers for the lack of process data, inflexibility across different products and feedstocks, and ignorance of temporal and spatial dimensions.

In this study, a macro-level modeling framework is developed to serve as a useful tool for policy maker to understand the net energy, emissions and economic impacts of emerging technologies on entire U.S. economy from a life cycle perspective. The results can provide insightful indications on critical factors driving the energy, emissions and resources utilization of chemical production, and offer the answers to which technologies can lead to significant reductions in the national impacts of chemical industry. The modeling framework is generic and flexible enough to be used as a macro-level decision support tool applied across feedstocks, products and technologies and make robust projections and reasonable comparisons over temporal and spatial scales, which is useful for long-term planning of national energy consumption reduction and GHG emissions mitigation.

A preliminary result for the case of ethylene is presented here for demonstration. The results showed that by 2040 a novel catalyst-assisted technology has the potential to reduce 50 million GJ of life-cycle energy consumption of U.S. ethylene industry and avoid 3 million tons of CO_2-equiv GHG emissions. For other emerging technologies applied to ethylene production, the model assesses their net impacts under different economic and resources projections by scenario analysis. Since this model allows for the assessment of economy-wide potential reductions of emerging technologies for a variety of chemical productions, more case studies of different chemicals manufactured by multiple technologies will be presented in oral presentation.

References

[1] ICCA, "Innovations for Greenhouse Gas Reductions " International Council of Chemical Associations 2009.

[2] E. Worrell, D. Phylipsen, D. Einstein, and N. Martin, "Energy Use and Energy Intensity of the U.S. Chemical Industry " Ernest Orlando Lawrence Berkeley National Laboratory, Berkeley, California2000.

[3] EIA, "Annual Energy Outlook 2013 " U.S. Energy Information Administration Washington, DC 2013.

[4] EIA. (01/02). *The National Energy Modeling System: An Overview* Available: http://www.eia.gov/oiaf/aeo/overview/

[5] ETSAP. (12/11). *MARKAL*. Available: tp://www.iea-etsap.org/web/Markal.asp

[6] M. D. Tabone, J. J. Cregg, E. J. Beckman, and A. E. Landis, "Sustainability Metrics: Life Cycle Assessment and Green Design in Polymers," *Environmental Science & Technology,* vol. 44, pp. 8264-8269, 2010/11/01 2010.

EASTMAN

Environmental Life Cycle Assessment of Automotive OEM Paint Shop Technologies

Jason Pierce, PE
Rebecca Glaspie

Outline

- About Eastman and LCA
- Auto Coatings 101
- The Assessment
- Conclusions

EASTMAN

Who we are

- A global specialty chemical company headquartered in Kingsport, Tennessee
- Approximately 14,000 employees and over 40 manufacturing sites around the globe
- Serving customers in approximately 100 countries
- A company dedicated to environmental stewardship, social responsibility and economic growth
- 2014 ENERGY STAR® Partner of the Year Sustained Excellence
- 2014 Ethisphere's World's Most Ethical ® Companies
- 2014 Glassdoor Employees' Choice Best Places to Work (# 4)
- 2013 revenue of $9.4 billion

EASTMAN

Diversified product lines & technologies

2013 sales revenue by market

EASTMAN

What is LCA?

A tool to quantify environmental impacts associated with a product, process, or activity by:

Compiling a Life Cycle Inventory (**LCI**) of relevant energy and material inputs and environmental releases

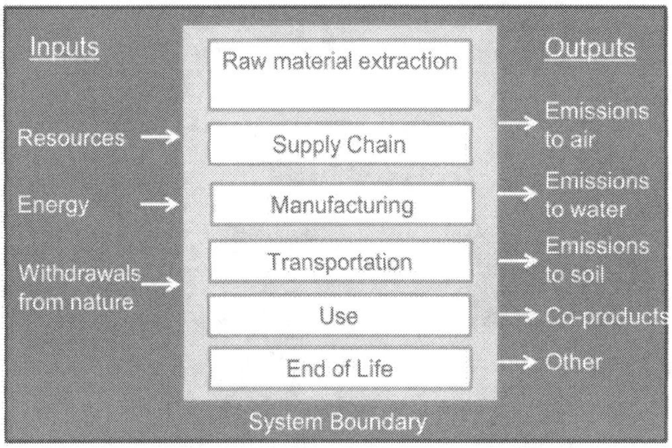

Impact Assessment
the potential environmental impacts associated with identified inputs and releases

Interpreting Results
To enable innovations and informed business decisions.

EASTMAN

Why do an LCA?

- Comparing the sustainability position of two alternatives may provide surprising results
- Example to discuss today: Waterborne vs. Solventborne automotive coatings

What do you think is greener?

EASTMAN

Understand the process

Paint Shop Processes

Traditional 3 coat , 2 bake (3C2B) Process:

1. Primer with full cure at 285°F (140°C)
2. Base Coat + Clear Coat with full cure at 285°F (140°C)

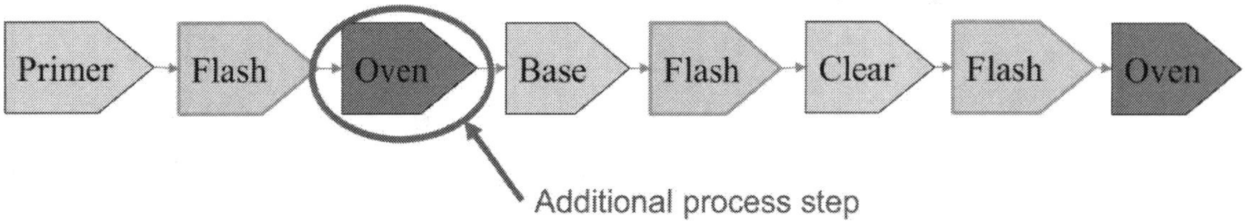

Compact 3 coat, 2 bake (3C1B) Process:

1. Primer + Base Coat + Clear Coat with full cure at 285°F (140°C)

In all processes flash zones are ambient for SB and **heated** for WB layers

EASTMAN

The Life Cycle Assessment Eastman & IHS

Study Scope

EASTMAN

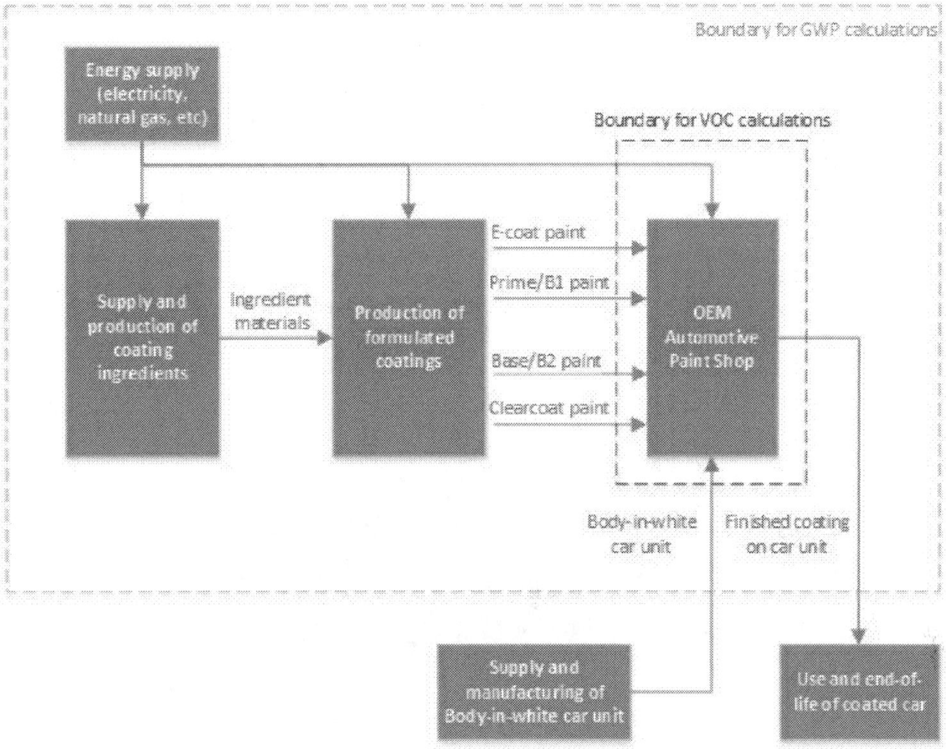

Scenarios Evaluated

EASTMAN

Abbreviation	Description
3C2B-WB-1K	Conventional WB basecoat with 1K clearcoat
3C2B-WB-2K	Conventional WB basecoat with 2K clearcoat
3C2B-LSSB-1K	Conventional low-solids SB basecoat with 1K clearcoat
3C1B-SB-1K	Compact high-solids SB basecoat with 1K clearcoat
3C1B-WB-1K	Compact wet-on-wet WB basecoat with 1K clearcoat

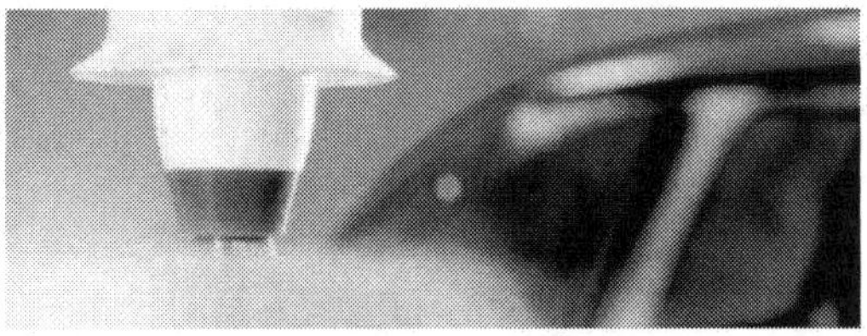

Scenarios Evaluated

EASTMAN

Abbreviation	Description
3C2B-WB-1K	Conventional WB basecoat with 1K clearcoat
3C2B-WB-2K	Conventional WB basecoat with 2K clearcoat
3C2B-LSSB-1K	Conventional low-solids SB basecoat with 1K clearcoat
3C1B-SB-1K	Compact high-solids SB basecoat with 1K clearcoat
3C1B-WB-1K	Compact wet-on-wet WB basecoat with 1K clearcoat

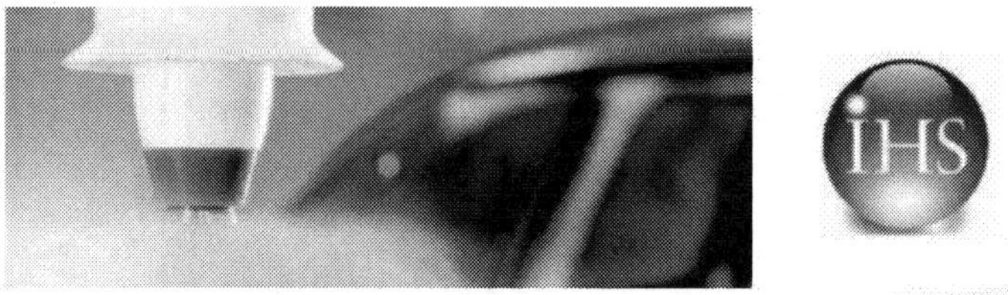

Life Cycle Emissions

EASTMAN

Life Cycle GWP for paint shop and materials

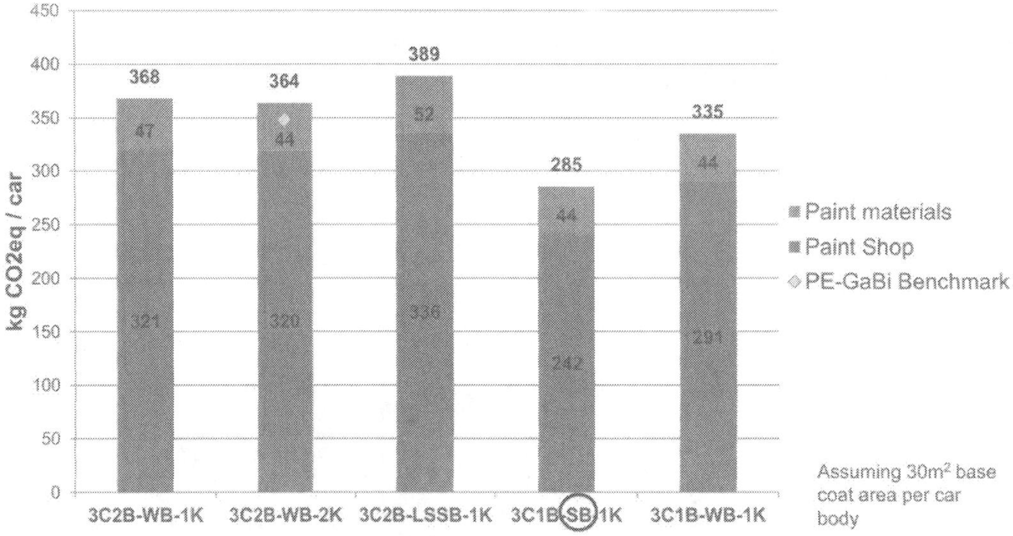

Assuming 30m² base coat area per car body

Modern compact high solids SB processes demonstrates reduced Global Warming Potential (GWP)compared to WB processes

Paint Shop GHG emissions

EASTMAN

Breakdown of Life Cycle GWP for paint shop only

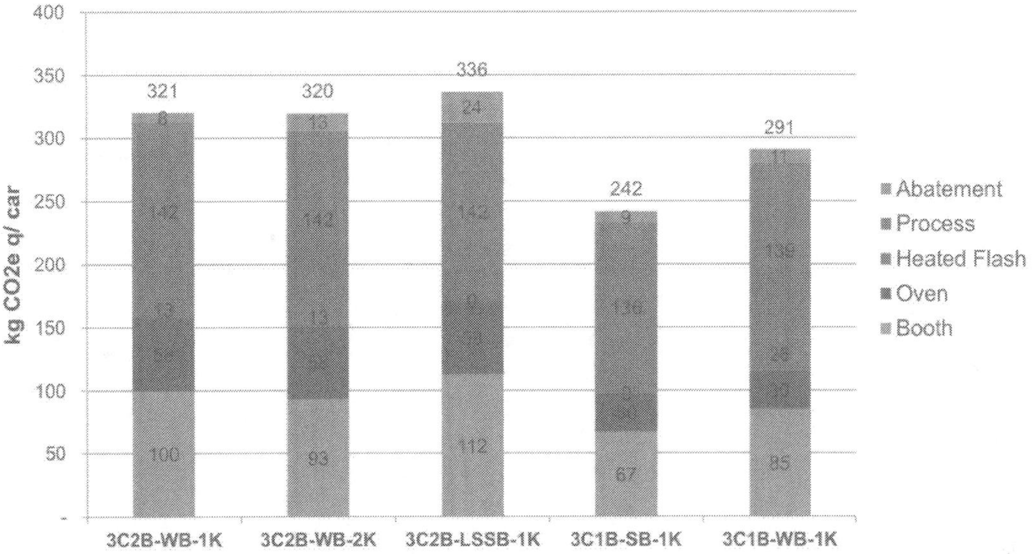

The compact solventborne process has a lower energy demand because it contains higher solids and does not require heated flashes

GWP benchmarking

EASTMAN

Normalized comparison with Ford and BASF/Dürr Studies

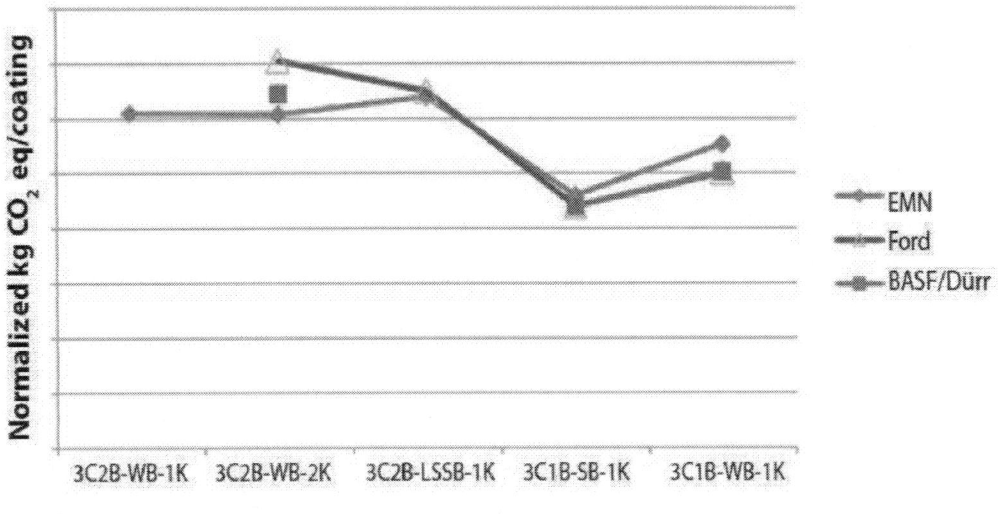

(1) Ford Motor Company, "Sustainable Painting Processes: *Global Alternatives.*" NCMS Sustainable Designs and Manufacturing Roundtable, June 2012

(2) Durr and BASF. "Eco-Efficiency Analysis of Global Coating Processes" Presented at Strategies in Car Body Painting 2012

VOC Emissions at the paint shop

EASTMAN

Paint shop sensitivity study showing minimum and maximum VOC emissions by varying film build

The redline indicates the EU emission limit of 35 g/m² of VOC.

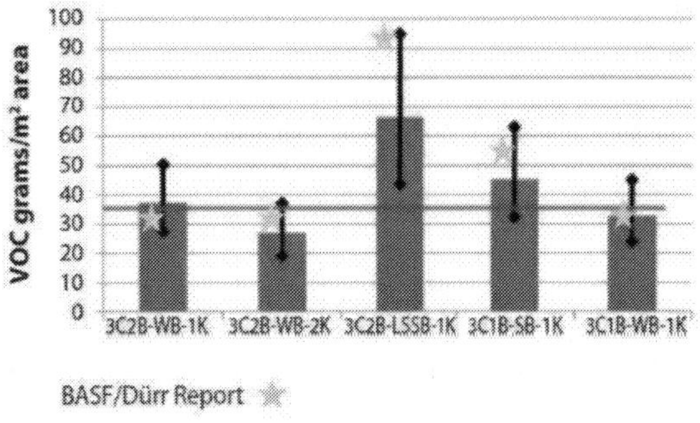

Unabated VOC emissions

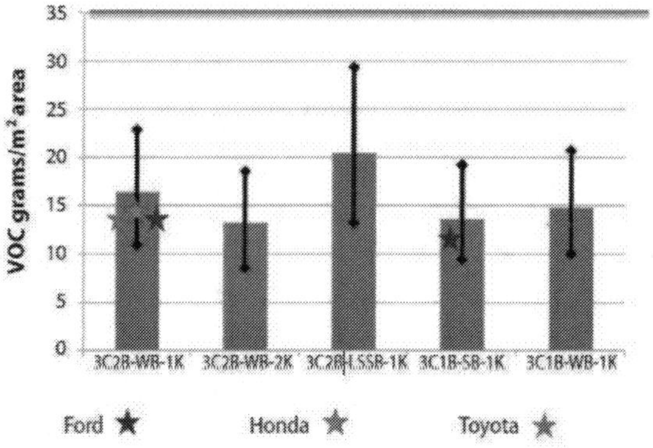

Abated VOC emissions

Upper: High film build with all high VOC materials
Lower: Low film build with all low VOC material

Study Conclusions

EASTMAN

- Compact Solventborne process can achieve the lowest global warming potential (GWP) while meeting European VOC limits
- The Compact Solventborne process has lower GWP compared to waterborne compact process due to elimination of heated flash and lower spray booth energy
- This study supports and attractive sustainability position for the Compact Solventborne process

Key Acknowledgements

- The Eastman team
  - Primary authors

Jason Pierce and Josh Seeley

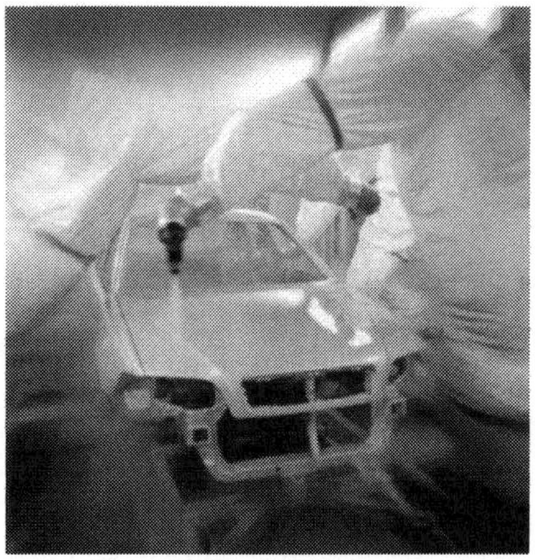

- IHS for data collection and collaboration on the study

IHS

EASTMAN

EASTMAN

Want to learn more?

- Attend Eastman's architectural window film LCA presentation Monday 4:35pm in room M304
- Read the full article presented today at PCIMAG.COM

July 2014
issue

Appendix

References:

1. Weingartz, T. et al. (2012, June). "Sustainable Paint Processes: Global Alternatives". Automotive Circle International. Berlin, Germany.
Also available a
http://www.ncms.org/wp-content/NCMS_files/sustainability/2012Sustainability/amPresentations/1130_Weingartz_NCMS_Roundtable_2012.pdf

2. http://corporate.honda.com/images/banners/environment/Honda_2012_North_American_Environmental_Report.pdf

3. http://www.toyota.com/about/environmentreport2012/data_center.html

4. Schumacher, H & Haunschild, A. (June, 2012). "Strategies in Car Body Painting 2012", Automotive Circle International. Berlin, Germany. BASF/Dürr: Eco-Efficiency Analysis of Global Coating Process

Additional uncited references:

http://www.nissan-global.com/EN/DOCUMENT/PDF/ENVIRONMENT/EFB/2012/FB2012_p05-19.pdf

http://www.nissan-global.com/EN/ENVIRONMENT/LIBRARY/DATA/OVERSEAS_SUBSUDUARIES/

http://gmsustainability.com

http://corporate.ford.com/microsites/sustainability-report-2011-12/environment-data-emissions

EASTMAN

Paint basics

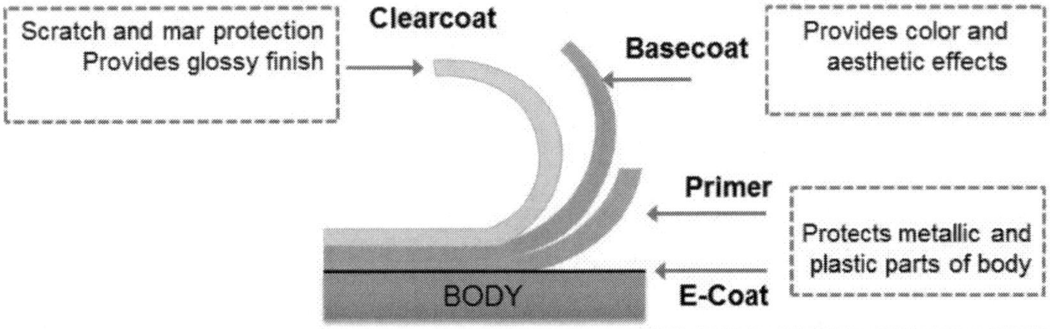

Conventional Solvent borne (SB) Basecoat "Paint Can"

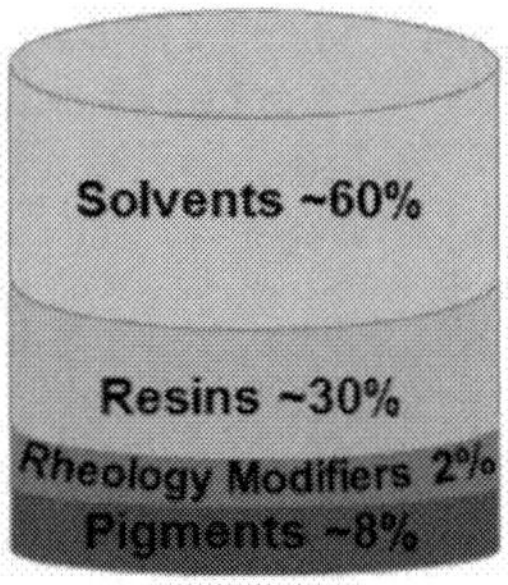

GHG sources

EASTMAN

Breakdown of GHG emission by source

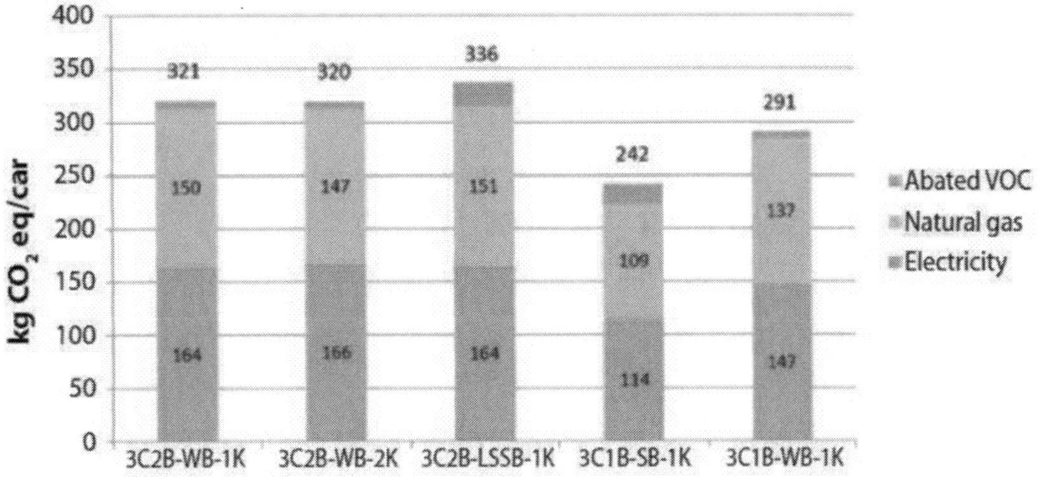

374634 Evaluation of the Environmental Impacts of Ethanol Production from Eastern Red Cedar

Abstract:

Based on the global interest in renewable fuel options, there has been significant research directed towards investigating lignocellulosic feedstocks that are suitable for ethanol production. These include: corn stover, wheat straw, eucalyptus, and switchgrass. This study evaluates the environmental impacts of an ethanol production system using eastern red cedar trees as the biomass feedstock using life cycle assessment (LCA) methodology. A process simulation was developed using Aspen Plus® to model the acid bisulfite pretreatment, enzymatic hydrolysis, fermentation, and distillation steps. A cradle to gate LCA will be presented that evaluates the environmental impacts from the cutting down of the trees to the production of anhydrous ethanol. The functional unit for this cradle to gate analysis is 1 MJ of equivalent energy and the relative mass, energy and economic (RMEE) system boundary selection method is used. The environmental impacts of this process are compared to that of the production of ethanol from corn. Inventory data for the system are collected and used to calculate a life cycle impact assessment (LCIA) using the IMPACT 2002+ framework in simapro 8. Three impact categories will be evaluated: water use, greenhouse gas (GHG) emissions, and non-renewable energy use. Preliminary results indicate that a significant percentage of environmental impacts are from the pretreatment step.

374443 Life Cycle Assessment of Underground Coal Gasification

Abstract:

Increasing energy demand, impending oil and natural gas resource depletion, and relatively low cost of energy production, have allowed coal to remain an important source of energy. However, increasing awareness of climate change has led to public resistance against the operation and new construction of coal-fired power plants. As a result, there is a growing interest in coal gasification as a more environmentally friendly alternative to traditional coal combustion.

Underground coal gasification (UCG) is a promising technology where unmined coal is converted, in-situ, into valuable synthetic gas (syngas) via the injection of oxidants into the coal seam. The syngas can then be collected above ground for further processing via numerous routes. For instance, via the Fischer-Tropsch process, it could be processed into synthetic fuels, or the hydrogen and carbon monoxide can simply be combusted to provide power to electricity generating turbines.

Whilst there are obvious advantages to not being required to extract the coal, there are potential impacts from UCG that need to be considered when assessing the viability and sustainability of the process. These in turn must be considered alongside other issues, such as energy security and human health and well-being.
UCG employs the techniques developed for oil and gas industry, it not only has the potential of reducing the cost of energy generation from coal, but also opens the possibility of generating energy from coal previously too deep to be mined in an economically feasible manner.

This study uses a life cycle assessment methodology to investigate the environmental impacts, to assess possible process improvements, including costs, and ascertain the competitiveness of the technology via comparison with alternative process routes. It also investigates how the environmental impacts and resource flows change with different potential end products of UCG.

LIFE CYCLE ANALYSIS FOR URBAN WASTE TREATMENT OPTIMIZATION

D.A. Sarigiannis[1], E.J. Handakas[1], S.P. Karakitsios[1], M.P. Antonakopoulou[1], A. Gotti[1]
[1]Chemical Engineering, Aristotle University of Thessaloniki, Thessaloniki, Greece

Introduction

Global consumption has rapidly increased and the amount of waste generated by mankind has raised significant concerns over the associated burden on human and environmental health. In a sustainable society waste should be thought of as raw materials, which can be utilized for use in other economic sectors. Also, waste disposal methods should be transformed to ensure an ecological balance with the intention that this vital resource does not harm the environment or humans' health. If all waste streams could have multiple options at the end of their life then this would be a giant step towards zero waste.

Greece currently deals with a number of changes regarding environmental management. The most significant one is management of municipal solid waste (MSW). Here, the country has not kept at pace with its European or other Western world counterparts. In contrast to what happens in the rest of Europe and much of the Western world, landfilling remains the dominant waste management option. The main factors contributing to these changes are European and national waste legislation, continuous public pressure and the increasing awareness of adverse health and environmental problems [1]. The above lead to the need for more advanced treatment processes, which will reinforce recycling, pre-selection of waste, energy and material recovery. In this paper five alternative scenarios aimed at minimizing the material fraction disposed of in landfills are analyzed. The methodological framework followed is life cycle impact assessment (LCIA), a tool commonly used to analyze environmental interactions during the life of a product from raw material acquisition to production, use and, finally, disposal (ISO 14040, 1997). LCIA was applied to assess municipal solid waste management options in the two largest cities of Greece, Athens and Thessaloniki, paying special attention to energy and material balances. SimaPro 7.1 was used to assess abiotic deletion accounting, cumulative energy and exergy requirements and airborne emissions.

Methodology

Introduction of Analysis Methods

Life cycle assessment is a system analysis tool to describe environmental and health impacts of products and processes over their lifetime [2]. An integrated LCA has been realized using using SimaPro 7.1 software including the production, collection, treatment and final disposal of the MSW as well as the matter and energy demand and the pollutant emissions. The environmental impact categories were assessed using CML 2001 method, the Cumulative Energy Demand method using Ecoinvent 1.01 and the Cumulative Exergy Demand method using Ecoinvent 2.0. In addition, the analysis includes potential adverse health impact of incineration and landfilling on the basis of the latest relevant epidemiological information.

The total amount of municipal solid waste produced in Athens is equal to 1350 ktn/yr [3]. The MSW generated in Thessaloniki is 600 ktn/yr [4]. Table 1 presents the MSW composition in the two cities.

Table 1. Composition of municipal solid waste in Athens and Thessaloniki

Elements of MSW	Athens[a]	Thessaloniki[b]
Organics	42%	36%
Paper	29%	25%
Plastic mixture	14%	19%
Metal	3%	3%
Glass	3%	3%
Others	6%	11%

a: information taken by HSWMA [5] , b: information taken by Chatziangelou [6]

Description of Scenarios

The waste management systems studied are described below.

Scenario 1: In this case all the waste generated from the cities directly go to landfill

Scenario 2: Only the fraction of the waste which consists of plastic, paper and ferrous material is recycled. The rest fraction goes to landfills (Figure 1).

Scenario 3: Municipal solid waste is pre-treated and pre-sorted into biodegradable and non-biodegradable fraction. The biodegradable fraction goes through anaerobic digestion process. The digestate product undergoes a composting process. The non-biodegradable fraction is recycled. The residues of the previous processes are disposed of to landfills (Figure 1).

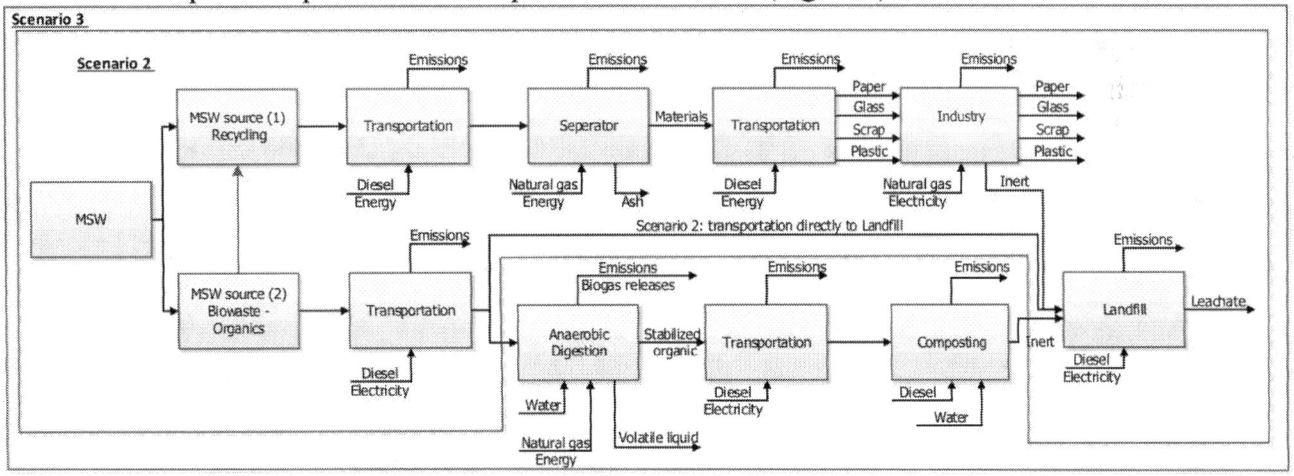

Figure 1. Scenario 2 Recycling and Scenario 3 Recycling & Anaerobic Digestion & Composting

Scenario 4: Raw waste without any pre-treatment or pre-sorting is used as feedstock for thermal waste treatment process. The solid residue state is disposed of in landfills (Figure 2).

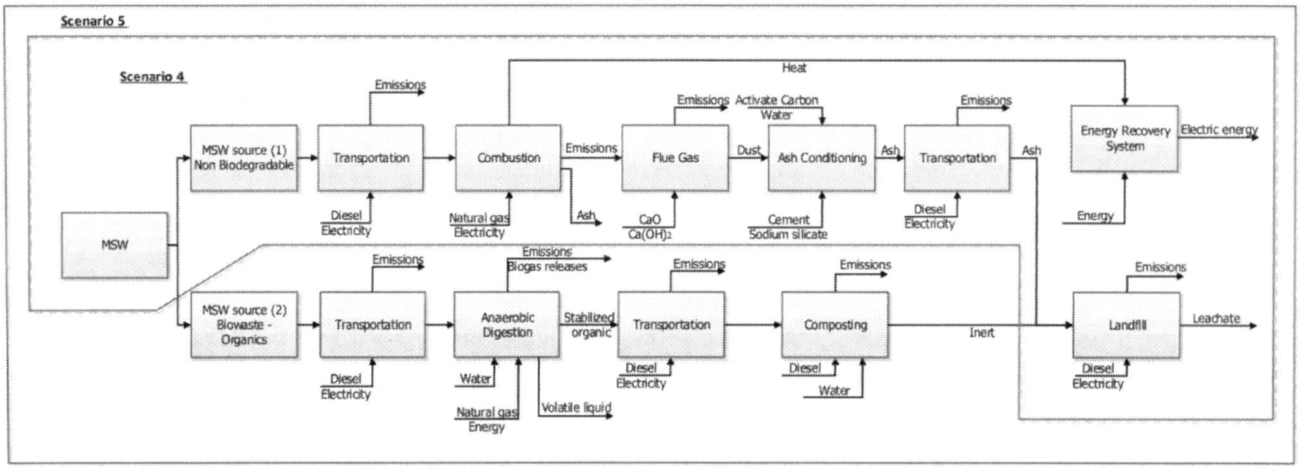

Figure 2. Scenario 4 Incineration and 5 Incineration & Anaerobic Digestion & Composting

Scenario 5: The organic fraction of the waste is composted and undergoes anaerobic digestion process whereas all the other material are combusted into incinerators (Figure 2).

Clearly, more scenarios integrating different technological options are available in theory, or based on international practice. However, the choice of these scenarios was based on pragmatic solutions that could be practical in Greece at this point in time.

Environmental Impact Assessments

The waste streams of both cities were simulated in the SimaPro database and were analyzed with CML 2001 methodology. CML '01 focuses on a series of environmental impact categories expressed in terms of emissions to the environment or resource use. The large number of elements make this a preferable method to assess environmental damages calculated from the disposal inventories of this study. The CML '01 impact categories are the following. Abiotic depletion is related to the extraction of minerals and fossil fuels due to inputs in the system. Acidification Potentials (AP) for emissions to air are calculated describing the fate and deposition of acidifying substances such as NH_3, NO_x, SO_X, and are expressed as kg SO_2 equivalents/ kg emission. Eutrophication is caused by various substances, the most significant ones are nitrate, phosphate, phosphorus, phosphorus pentoxide, they are expressed in kg $PO^{-3}{}_4$ equiv. and cause damage of terrestrial and aquatic ecosystems. Global warming (kg CO_2 equiv.) has adverse effects on ecosystem health, human health and material welfare and is related to emissions of greenhouse gases to air. Photo-oxidant formation is the formation of reactive substances (mainly ozone) which are hazardous to human health and ecosystems especially by damaging crops. The impact potentials are expressed as the equivalent emission of the reference substance ethylene. [7].

Cumulative Energy and Exergy Demand (CED, CExD)

Cumulative Energy Requirement evaluates energy resources, foregoing any burden associated with emissions from waste material. It is an indicator, which calculates the total energy through the life cycle of a good or service. The method to calculate Cumulative Energy Demand is based on the method published by Ecoinvent version 1.01 in the SimaPro 7.1 database. The energy categories included in the methodology are the non-renewable fossil and nuclear, and renewables such as biomass, wind, solar, water and geothermal energy. Cumulative Exergy Demand calculates total exergy removal from nature so as to provide a new product or service. In this method exergy is used as a measure of the potential loss of useful resources. The indicator is distinguished into non-renewable and renewable resources. The exergy categories investigated are non-renewable fossil and nuclear, renewable kinetic, solar, biomass, non-renewable metals and minerals, other renewables. The weighting factor applied across the board is unity to avoid computational bias in the presence of lack of additional information.

Health Impact due to Incinerators and Landfills

The assessment of the health effects due to the operation of landfills and incinerators, was based on the study previously carried out by Forastiere et al. [8]. Due to multiple environmental pressures related to the operation of these SWM options [9], a hybrid methodology was followed. For landfills, the health impact indicators investigated were congenital malformations and low birth weights [10, 11]. Excess risk values reported by Elliott et al. [12] regarding cancer incidence associated to proximity of residence to incinerator plants, as well as the effect of long term mortality due to PM and NO_2 inhalation exposure were employed for assessing the health risks imposed by the operation of the incinerators. In the study carried out by Elliott et al. [12], cancer incidence between 1974 and 1987 among over 14 million people living near 72 solid waste incinerator plants in Great Britain were studied. The excess risk estimate for living within 3 km of an incinerator for all cancers combined was 3.5%. Linear no-threshold exposure-response functions related to the long-term effects on mortality from PM_{10} and NO_2 have been derived from the extensive existing reviews of epidemiological and toxicological data [13]. The following ERFs were used:

RR = 1.06 (95% CI = 1.03 – 1.09) increase in mortality for 10 $\mu g/m^3$ PM_{10}

RR = 1.06 (95% CI = 1.04 – 1.08) increase in mortality for 10 $\mu g/m^3$ NO_2

A critical component for the estimation of attributable mortality is population exposure, which in turn is based on the calculation of ambient air concentrations at a radius of 3 km around the incinerator. For this purpose, the Industrial Source Complex - Short Term regulatory air dispersion model (ISCST3) was employed [14]. Incinerator plume and stack parameters were obtained by the study of Morselli et al. [15], while meteorological parameters for the two sites of interest were retrieved from Hellas Weather [16]. In the national study conducted by Elliott et al. [11] on 9,565 landfill sites in Great Britain, operational at some time between 1982 and 1997, statistically significant increased risks were found for all congenital malformations and low and very low birth weight in residents within 2 km of the sites, relative risk being for congenital anomalies equal to 1.02 (99% CI = 1.01-1.03) and 1.06 (99% CI = 1.052-1.062) for low birth weight. Prevalence of congenital malformations at birth in Greece, was given by Brilakis et al. [17], found equal to 5.5%, while background rate of low birth weight was obtained by Tsimbos and Verropoulou [18], found equal to 10.5%. For landfills it was assumed that the emissions will last up to 30 (an assumption supported by the available knowledge that landfilled biodegradable waste starts to emit biogas a few years after deposit and continues to do so for several decades) and the health effects, in terms of congenital anomalies and low birth weight, are constant throughout this period. Overall population data regarding population age stratification nearby the sites of interest, birth and death rates were retrieved from the Hellenic Statistical Authority [19].

Results and discussion

Environmental Impacts of the Different Waste Management Options

The results of the different waste management options for each city are shown below in Figure 3 (left for Athens – right for Thessaloniki).

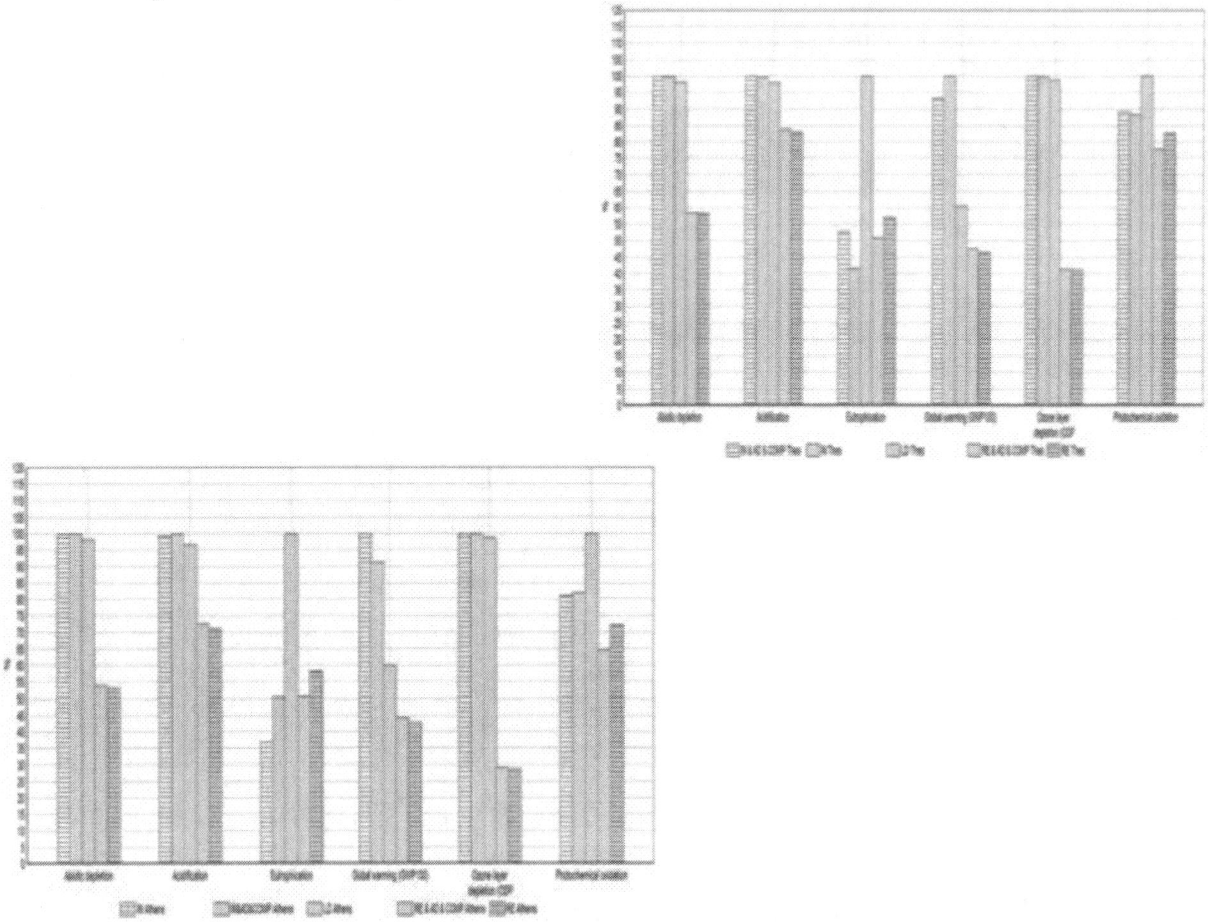

Figure 3. Environmental impact categories assessment for Athens and Thessaloniki

The results indicate that the scenarios of recycling consume less abiotic matter. As far as acidification is concerned all scenarios show approximately the same behavior. However, due to big amounts of emissions of ammonia and nitrogen dioxide incineration processes have higher acidification values among all scenarios. Landfill process is strongly correlated to eutrophication effect as shown by the results of the study whereas incineration treatment proves to be the most environmentally sound option. The results show that global warming effect of thermal treatment is much higher than all other scenarios. Also landfilling has an important impact to Green House Effect due to excessive untreated methane emissions and to the residues of all other waste management systems which end to landfills. Recycling and composting together have a considerable low global warming effect. Moreover recycling is significantly more favorable to the ozone layer depletion among all treatment methods. Landfill treatment of solid waste has the highest photochemical potential among all other treatments as it has very high CH4 emissions.

The following tables indicate the comparative assessment of cumulative energy requirement and cumulative exergy demand in Athens and Thessaloniki for the waste management options. Waste collection as mentioned above is included in process in order to account for the energy and exergy demand. It is clear from the tables that there are no significant differences between all the scenarios. An important reason for which this happens is the multiple and heavy infrastructure of the plants for each waste treatment system as mentioned by [1]. It is necessary to take into account that all residues from all other systems cannot avoid landfilling. This also indicates the reason for which landfilling has a higher cumulative energy demand. It's important to point out that landfill reaches the same energy demand level as thermal treatment system. However, thermal treatment systems can generate significant amounts of energy substituting the energy lost during the system process. Meanwhile, thermal treatment methods have the highest cumulative exergy demand which verifies that the degradation of the original stored energy in the municipal solid waste stream is significant. Landfilling process has also high levels of exergy demand as the waste stream materials are not further sustainably used. In addition, the results show that the biological treatment methods combined with recycling treatment have the least energy and exergy requirement and seem to be the best waste management scenario. This is the only scenario where both organic and inorganic materials are taken into consideration for a more sustainable use of MSW.

Table 2. CED and CExD for waste management systems in Athens &Thessaloniki

Impact category	**CED Athens**	**CED Thessaloniki**	**CExD Athens**	**CExD Thesslaoniki**
Scenarios /Unit	MJ eq	MJ-Eq	MJ eq	MJ eq
IN	4.56E+10	2.43E+10	1.26E+ 12	2.89E+ 11
IN&AD&CO MP	4.58E+10	2.44E+10	1.26E+ 12	2.91E+ 11
LD	4.49E+10	2.39E+10	1.24E+ 12	2.89E+ 11
RE&AD&CO MP	3.20E+10	1.78E+10	1.24E+ 12	2.84E+ 11
RE	3.18E+10	1.77E+10	1.23E+ 12	2.82E+ 11

Health Impacts due to Incinerators and Landfills

The concentrated results on health impact assessment of the relevant waste management options are presented in Figure 4. The number of additional cancer cases is minimal (less than one case expected), due to the very low population density (especially in Thessaloniki) around the incineration site. The same occurs for the total number of Years of Life Lost, in the two sites, attributable to exposure to PM_{10} and NO_2 from incinerators, corresponding to 5.22 years for Athens and 4.71 for Thessaloniki. Finally, regarding the health effects of landfills in the two sites, less than a case of birth defects for all the entire life cycle of the landfill is expected. This is due to the low population density around the selected sites.

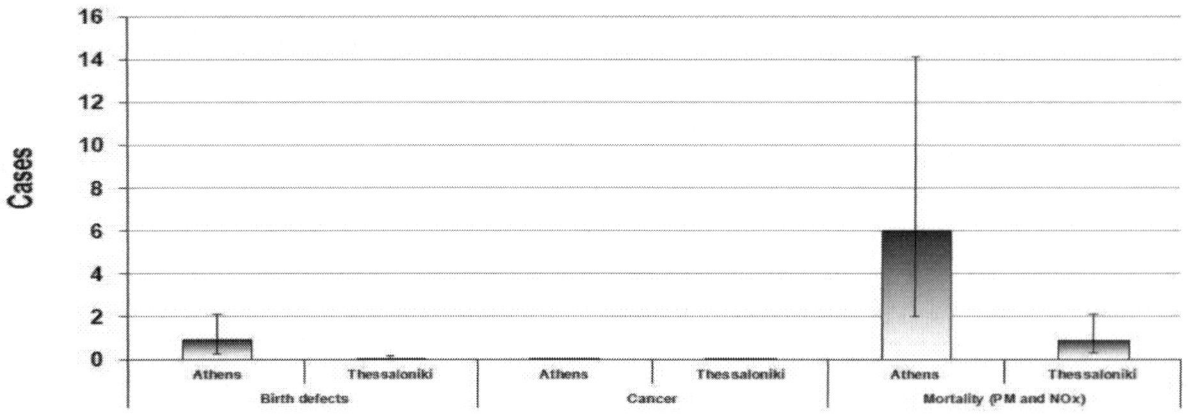

Figure 4. Health impact assessment of the relevant waste management options

Conclusions

The life cycle impact of collection and different waste disposal strategies in Athens and Thessaloniki including landfilling, recycling, biological treatment (both aerobic and anaerobic) and incineration and their potential combinations, was performed by means of a multi-method and multi-scale approach. The results of the assessment based on selected impact indicators lead to the following conclusions.

Biological methods place the smallest burden regarding abiotic matter, acidification potential, greenhouse gas effect, ozone depletion and photochemical oxidation among the other waste management systems. In terms of energy and exergy demand biological methods prove to be more sustainable options. These results underline the need of exploring further the biological treatment of waste in Greece. In order to have a more comprehensive picture at different scales of the MSW management, further studies are essential for more developed LCA-approaches which will integrate socio-economic and environmental points. Our analysis has shown that landfilling is the worst waste management strategy and that other options for waste treatment coupled with energy and material recovery would result in very important benefits such as reduction of greenhouse emissions. However, not all options are benign to the local environment and to the health of the local population, since both can be influenced by non-negligible local emissions.

References

1. Cherubini F., Bargigli S., Ulgiati S. (2008). Life cycle assessment of urban waste management: Energy performances and environmental impacts. The case of Rome, Italy. Waste Management. 28(12), pp. 2552-2564.
2. G.Finnveden J.J., P.Lind, A. Moberg, 2000. *Life cycle assessment of Energy from solid waste,* Stockholm University: Stockholm.
3. Greek_Government, 2012. http://government.gov.gr/.
4. Koroneos C.J., Nanaki E.A. (2012). Integrated solid waste management and energy production - a life cycle assessment approach: the case study of the city of Thessaloniki. Journal of Cleaner Production. 27(0), pp. 141-150.

5. HSWMA. http://www.eedsa.gr. 2011 [cited 2011 December].
6. Chatziaggelou H., 2007. *Qualitative and quantitative analysis of municipal solid waste in Thessaloniki.*
7. Banar M., Cokaygil Z., Ozkan A. (2009). Life cycle assessment of solid waste management options for Eskisehir, Turkey. Waste Management. 29(1), pp. 54-62.
8. Forastiere F., Badaloni C., De Hoogh K., Von Kraus M.K., Martuzzi M., Mitis F., Palkovicova L., Porta D., Preiss P., Ranzi A., Perucci C.A., Briggs D. (2011). Health impact assessment of waste management facilities in three European countries. Environmental Health: A Global Access Science Source. 10(1), pp.
9. Porta D., Milani S., Lazzarino A.I., Perucci C.A., Forastiere F. (2009). Systematic review of epidemiological studies on health effects associated with management of solid waste. Environmental Health: A Global Access Science Source. 8(1), pp.
10. Dolk H., Vrijheid M. (2003). The impact of environmental pollution on congenital anomalies. British Medical Bulletin. 68, pp. 25-45.
11. Elliott P., Briggs D., Morris S., De Hoogh C., Hurt C., Jensen T.K., Maitland I., Richardson S., Wakefield J., Jarup L. (2001). Risk of adverse birth outcomes in populations living near landfill sites. British Medical Journal. 323(7309), pp. 363-368.
12. Elliott P., Shaddick G., Kleinschmidt I., Jolley D., Walls P., Beresford J., Grundy C. (1996). Cancer incidence near municipal solid waste incinerators in Great Britain. British journal of cancer. 73(5), pp. 702-710.
13. WHO, 2005. *Air quality guidelines for particulate matter, ozone, nitrogen dioxide and sulfur dioxide. Global update 2005*, WHO, Editor: Copenhagen, Denmark.
14. USEPA, 1995. *User's Guide for the Industrial Source Complex (ICS3) dispersion model.*
15. Morselli L., De Robertis C., Luzi J., Passarini F., Vassura I. (2008). Environmental impacts of waste incineration in a regional system (Emilia Romagna, Italy) evaluated from a life cycle perspective. Journal of Hazardous Materials. 159(2-3), pp. 505-511.
16. Hellas-Weather. *Network of Greek private meteorological stations. Retrieved 2012, from* http://www.meteorologia.gr/. 2012.
17. Brilakis E., Fousteris E., Papadopulos J., *Congenital Abnormalities in Greece: Functional Evaluation of Statistical Data 1981 – 1995 Congenital Diseases and the Environment*, P. Nicolopoulou-Stamati, L. Hens, and C.V. Howard, Editors. 2007, Springer Netherlands. p. 323-338.
18. Tsimbos C., Verropoulou G. (2011). Demographic and socioeconomic determinants of low birth weight and preterm births among natives and immigrants in greece: An analysis using nationwide vital registration micro-data. Journal of Biosocial Science. 43(3), pp. 271-283.
19. EL.STAT, 2008. http://www.statistics.gr.

376636 Life Cycle Assessment of Green Transportation Fuels Produced from Algal Biomass Via Hydrothermal Liquefaction

Abstract:

A life cycle assessment (LCA) study is performed for the energy requirements and greenhouse gas emissions in the production of transportation fuels from algal biomass. Different fuel production pathways have been studied to produce different types of transportation fuels. This study uses hydrothermal liquefaction (HTL) as the primary method to produce biocrude oil from algal biomass. Fuel production pathways studied in this work include the production of biodiesel, green diesel, green gasoline, and green jet fuel. A sensitivity analysis is conducted to identify the key factors affecting the total energy consumption in the production of green transportation fuels from algal biomass. Net energy demand requirements are calculated with/without coproducts to identify a suitable production pathway for sustainable transportation fuel production. Importance of co-product management to achieve sustainable transportation fuel is discussed. A well-to-wheel analysis has been conducted to assess the greenhouse gas emissions in all production processes.

389764 A Life-Cycle Assessment of Greenhouse Gas Emissions for Transportation Fuels from North American Conventional Crudes

Abstract:

The life-cycle analysis for transportation fuels includes the well-to-tank (WTT) and tank-to-wheel (TTW) stages. The WTT stage includes crude recovery, transportation of crudes to the refinery, the refining of crudes to produce transportation fuels, and the transportation of these fuels to local refueling stations. The crude recovery part of the WTT stage consists of oil well drilling, crude extraction from the reservoir, processing of crude, associated gas and water, and oil field venting and flaring. Fuel consumption in engines is considered as TTW stage. Energy is consumed, and a significant amount of greenhouse gases (GHGs) are emitted in all of these stages. The purpose of this study was to quantify the energy used and GHGs emitted in all the stages in transforming different North American conventional crudes to transportation fuels, mainly gasoline and diesel. A spreadsheet-based model was developed based on fundamental scientific equations to calculate the WTW energy use and GHG emissions. Emissions from crude recovery depend on crude and reservoir, depth of well, extraction methodology, gas-to-oil ratio, water-to-oil ratio, and oilfield venting, flaring, and fugitives. Recovery emissions vary from 3.94 g-CO_{2eq}/MJ of crude to 23.85 g-CO_{2eq}/MJ of crude. Transportation from the feedstock to the refinery and the refinery to refueling stations contributes only 0.51%-1.87% of the total WTW life-cycle emissions, depending on the transportation methods and total distance transported. The HYSYS refinery model was used to calculate energy used during refining. Energy consumption in a refinery depends on the crude quality (e.g., API gravity and sulfur content), distillation curves, and refinery configuration. GHG emissions in refining were calculated from the amount of energy used in the refining of crudes for conversion into gasoline and diesel. Refining emissions vary from 12.40 g-CO_{2eq}/MJ of gasoline to 14.83 g-CO_{2eq}/MJ of gasoline and 5.37 g-CO_{2eq}/MJ of diesel to 8.75 g-CO_{2eq}/MJ of diesel. Emissions from fuel consumption in vehicles (passenger cars) were taken from the literature. This stage contributes 66%-79% of total life-cycle GHG emissions in the case of gasoline. Total WTW GHG emissions range from 91.03 g-CO_{2eq}/MJ of gasoline to 109.37g-CO_{2eq}/MJ of gasoline and 86.63 g-CO_{2eq}/MJ of diesel to 107.92g-CO_{2eq}/MJ of diesel.

Specific Energy Consumption in Spiral Wound Reverse Osmosis Water Desalination

Mingheng Li

minghengli@cpp.edu

Nov. 16, 2014

OUTLINE

1. Motivation
2. Introduction
    - Definition of Normalized Specific Energy Consumption (NSEC)
    - Mathematical model
    - Differences between SWRO and BWRO
3. NSEC in RO
    - Review of NSEC in thermodynamically reversible RO
    - Theorems of NSEC in SWRO
    - Analysis of NSEC in BWRO
4. Concluding Remarks

MOTIVATION OF THIS WORK

- Reverse osmosis is the most common method of water desalination.
- Energy consumption in reverse osmosis membrane module accounts for a major portion (up to 45%) of the total cost of water desalination (Manth et al., 2003; Busch and Mickols, 2004; Wilf and Bartels, 2005; Zhu et al., 2009).
- There is a huge gap between specific energy consumption in RO plants and its theoretical minimum.
- Mathematical models helps elucidate the understanding of energy issues in RO.

DEFINITION OF SEC AND NSEC IN RO

- Schematic of RO

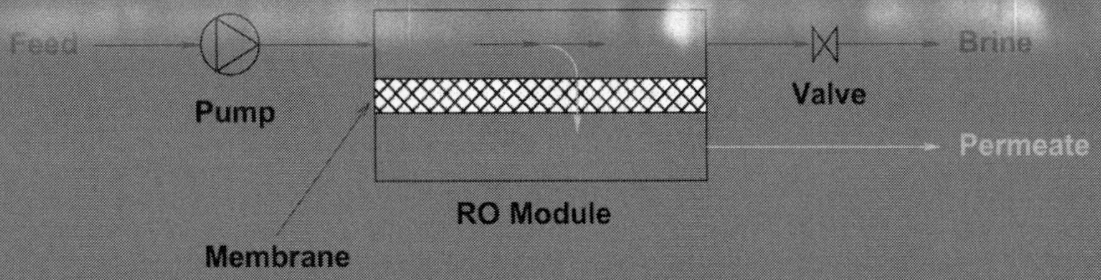

- Specific Energy Consumption (SEC)
- Normalized Specific Energy Consumption (NSEC)

MATHEMATICAL MODEL FOR RO

- Spiral wound reverse osmosis module

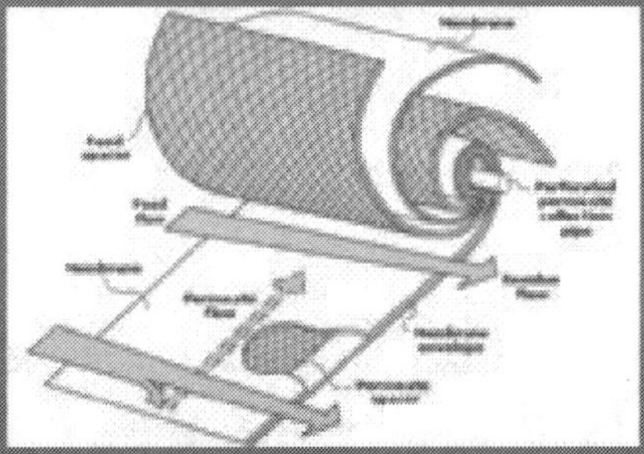

- Mathematical model (Li, Desalination, 2012)

Mingheng Li | AIChE Annual Meeting | Nov. 16, 2014 | 5 / 23

DIMENSIONLESS RO MODEL

- Dimensionless parameters

α: Osmotic hydraulic pressure ratio
γ: Membrane demand capacity ratio
κ: Retentate pressure drop ratio

- Dimensionless RO model

Mingheng Li | AIChE Annual Meeting | Nov. 16, 2014 | 6 / 23

DIFFERENCES BETWEEN SWRO AND BWRO

- Plant data and optimization (Li, IECR, 2013)

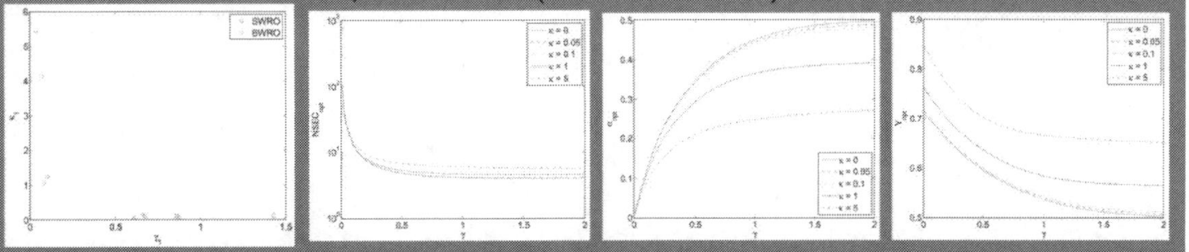

- Difference is due to $\Delta\pi_0$ (390 psi for seawater and 10-20 psi for brackish water)
- SWRO is operated near thermodynamic limit
- BWRO is operated far away from thermodynamic limit

NSEC OF FULLY REVERSIBLE RO

- Schematic of an ideal, thermodynamically reversible RO

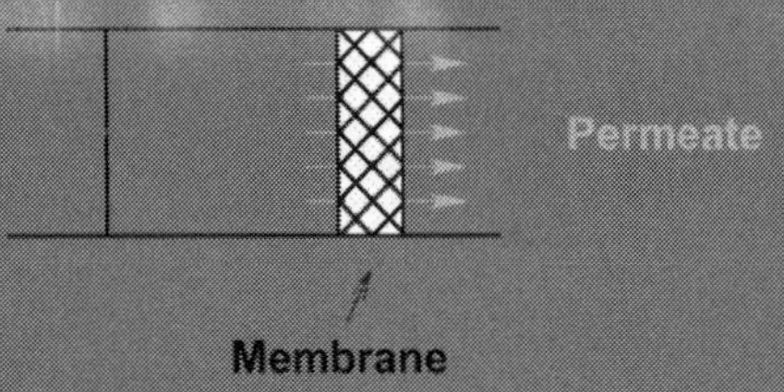

- NSEC of a fully reversible RO

(Spiegler, 1977)

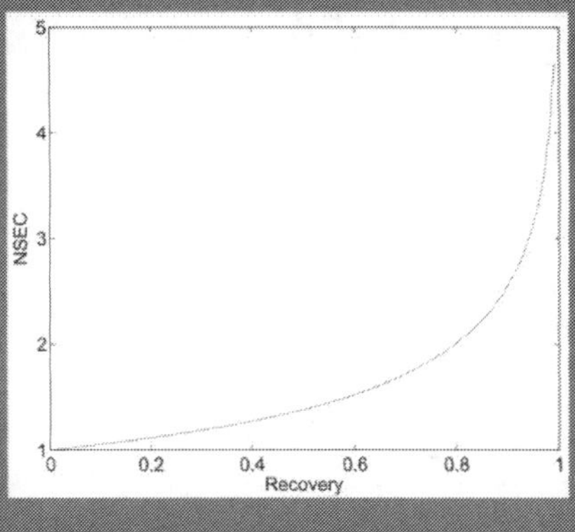

NSEC OF MULTI-STAGE SWRO WITH ERD

- Schematic of a generic multi-stage RO with interstage booster pumps and ERD

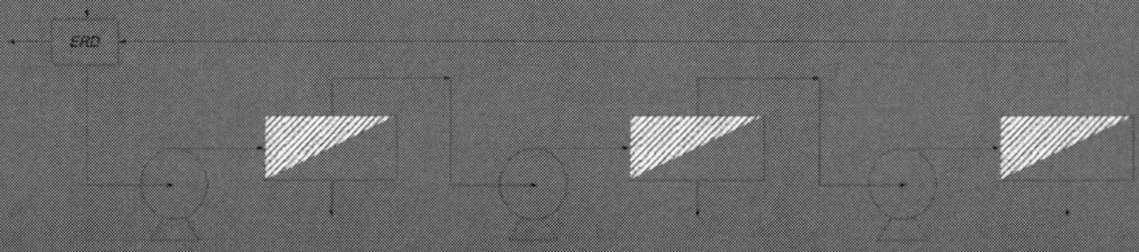

- NSEC (Li, Desalination, 2011)
  - Pressure drop is ignored
  - α_j, Y_j are based on each stage

NSEC IN SWRO AT THERMODYNAMIC LIMIT

Theorem 1

The minimum of NSEC for a single- or multi-stage RO with/without ERD at the thermodynamic limit is:
$NSEC = \left\{N\left[(1-Y_o)^{-1/N}-1\right]+(1-\eta_{erd})\right\}/Y_o$, where Y_o is system level water recovery, N is number of stages ($N \geq 1$) and η_{erd} is energy recovery efficiency of ERD ($0 \leq \eta_{erd} \leq 1$).

Proof

$$\min_{\alpha_j} NSEC$$

$$\begin{aligned} Y_o &= 1-\prod_{j=1}^{N}(1-Y_j) \\ \alpha_j + Y_j &= 1 \end{aligned}$$

Solution:

$$\begin{aligned} \alpha_1 = \alpha_2 = \ldots = \alpha_N = \alpha &= (1-Y_o)^{1/N} \\ NSEC &= \left\{N\left[(1-Y_o)^{-1/N}-1\right]+(1-\eta_{erd})\right\}/Y_o \end{aligned}$$

NSEC IN SWRO AT THERMODYNAMIC LIMIT

Theorem 2

For single- or multi-stage RO, the reduction in NSEC due to an ERD at the thermodynamic limit is η_{erd}/Y_o. The reduction in NSEC increases as η_{erd} increases. The reduction in NSEC is more significant at low recoveries.

Proof

$$\begin{aligned}\Delta_{erd} &= \frac{N\left[(1-Y_o)^{-1/N}-1\right]+(1-\eta_{erd})}{Y_o} - \\ &\quad \frac{N\left[(1-Y_o)^{-1/N}-1\right]+1}{Y_o} \\ &= -\frac{\eta_{erd}}{Y_o}\end{aligned}$$

or,

$$\Delta_{erd} = -\frac{\eta_{erd}[Q_0(1-Y_o)][\Delta\pi_0/(1-Y_o)]}{(Q_0 Y_o)\Delta\pi_0} = -\frac{\eta_{erd}}{Y_o}$$

NSEC IN SWRO AT THERMODYNAMIC LIMIT

Theorem 3

For infinite number of stages, $NSEC = \frac{-\ln(1-Y_o)+(1-\eta_{erd})}{Y_o}$ at the thermodynamic limit. If $\eta_{erd} = 100\%$, it is equivalent to a fully reversible RO process.

Proof

$$\begin{aligned}\lim_{N\to\infty} NSEC &= \lim_{N\to\infty} \frac{N\left[(1-Y_o)^{-1/N}-1\right]+(1-\eta_{erd})}{Y_o} \\ &= \frac{1}{Y_o}\lim_{x\to 0}\frac{(1-Y_o)^{-x}-1+(1-\eta_{erd})x}{x} \\ &= \frac{1}{Y_o}\lim_{x\to 0}\frac{e^{-x\ln(1-Y_o)}-1+(1-\eta_{erd})x}{x} \\ &= \frac{-\ln(1-Y_o)+(1-\eta_{erd})}{Y_o}\end{aligned}$$

EFFECT OF ERD AND MULTI-STAGE IN SWRO

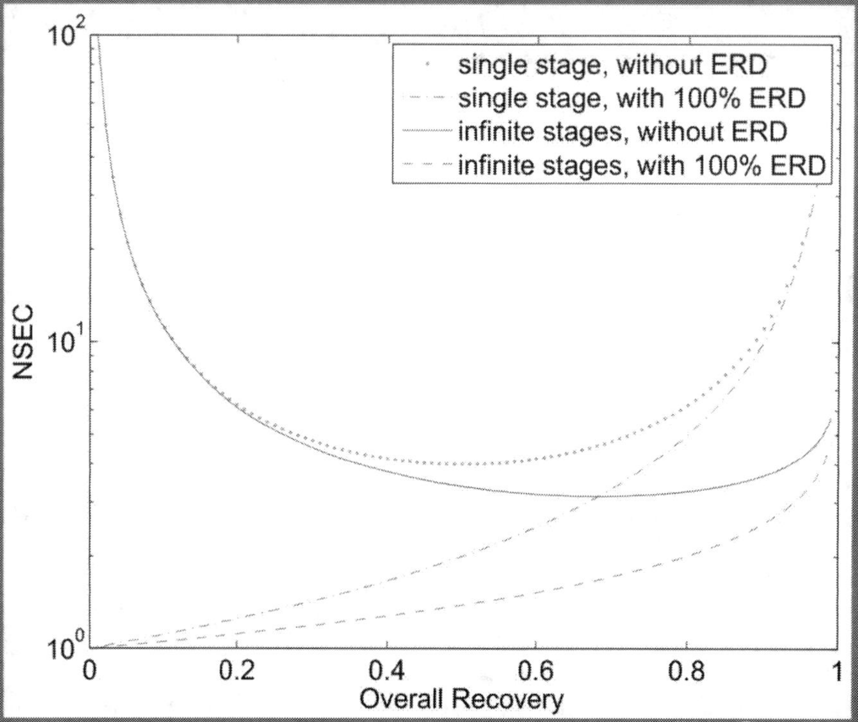

NSEC IN SWRO AT THERMODYNAMIC LIMIT

Theorem 4

With/without ERD, the NSEC at the thermodynamic limit reduces monotonically as the number of stages N increases. However, the reduction in NSEC due to N is less significant as N becomes larger. Multi-stage design works better at high recoveries.

Proof (see Li, IECR, 2014 for details)

$$\Delta_N = NSEC_{N+1} - NSEC_N < 0$$

$$\Delta_{N-1} < \Delta_N < 0$$

$$\Delta'_N(Y_o) < 0$$

- For high-recovery design, it is motivated to employ multi-stage configuration.

NSEC IN SWRO AT THERMODYNAMIC LIMIT

Theorem 5

Without ERD, as the number of stages increases from 1 to ∞, the corresponding optimal Y_o monotonically increases from 0.5 to 0.6822, and the minimum of NSEC monotonically reduces from 4 to 3.1462.

Proof (see Li, IECR, 2014 for details)

$$(1-N)(1-Y_o)^{(N+1)/N} + (N+1)(1-Y_o) - 1 \;=\; 0$$

$$NSEC_{N+1}(Y_{o_{N+1}}) \;<\; NSEC_{N+1}(Y_{o_N}) \;<\; NSEC_N(Y_{o_N})$$

- $N = 1$
  - $2(1-Y_o) = 1$
  - $Y_o = 0.5$
  - $NSEC = 4$
- $N = \infty$
  - $\dfrac{Y_o}{1-Y_o} - 1 + \ln(1-Y_o) = 0$
  - $Y_o = 0.6822$
  - $NSEC = 3.1462$

NSEC IN SWRO AT THERMODYNAMIC LIMIT

Theorem 6

At the thermodynamic limit, when $Y_o = 0.6822$, the NSEC of a single-stage RO with 100% ERD is equivalent to the one of an infinite-stage RO without ERD. Below this recovery, a single-stage RO with 100% ERD has a lower NSEC. Above this recovery, an infinite-stage RO has a lower NSEC.

Proof (see Li, IECR, 2014 for details)

$$\Delta \;=\; NSEC_{erd} - NSEC_{\infty} \;=\; \frac{1}{1-Y_o} - \frac{1-\ln(1-Y_o)}{Y_o}$$

$$\Delta'(Y_o) \;>\; 0$$

$$\Delta(Y_o = 0.6822) \;=\; 0$$

NSEC IN SWRO AT THERMODYNAMIC LIMIT

Theorem 7

A RO using brine recirculation is equivalent to a RO with ERD in which feed osmotic pressure difference $= \Delta\pi_0/(1 - fY_o)$, RO recovery $= Y_o(1-f)/(1-fY_o)$, and $\eta_{erd} = f$, where f is brine recycle ratio. Using brine recirculation would not reduce NSEC.

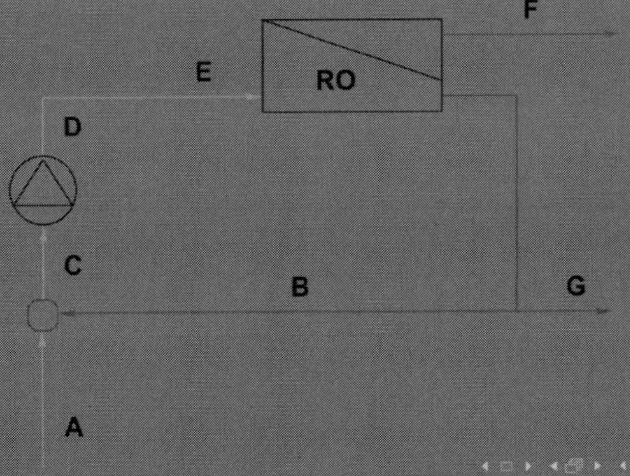

NSEC IN SWRO AT THERMODYNAMIC LIMIT

Proof of Theorem 7

Location	no recirculation		with recirculation	
Fresh Feed	Q_0	$\Delta\pi_0$	Q_0	$\Delta\pi_0$
Recycled Brine	0	-	$(1-Y_o)\dfrac{f}{1-f}Q_0$	$\dfrac{\Delta\pi_0}{1-Y_o}$
RO Inlet	Q_0	$\Delta\pi_0$	$\left[1+(1-Y_o)\dfrac{f}{1-f}\right]Q_0$	$\dfrac{\Delta\pi_0}{(1-f)+f(1-Y_o)}$
RO Outlet	$(1-Y_o)Q_0$	$\dfrac{\Delta\pi_0}{1-Y_o}$	$\dfrac{1-Y_o}{1-f}Q_0$	$\dfrac{\Delta\pi_0}{1-Y_o}$
Product	Y_oQ_0	0	Y_oQ_0	0
Rejected Brine	$(1-Y_o)Q_0$	$\dfrac{\Delta\pi_0}{1-Y_o}$	$(1-Y_o)Q_0$	$\dfrac{\Delta\pi_0}{1-Y_o}$

$$NSEC^{br} = \frac{N\left[(1-Y_o^{br})^{-1/N}-1\right]+(1-f)}{Y_o^{br}}\frac{\Delta\pi_0^{br}}{\Delta\pi_0}$$

- $N = 1$, $NSEC^{br} = NSEC$
- $N \geq 2$, $NSEC^{br} > NSEC$ (see Li, IECR, 2014 for details)

NSEC IN BWRO FAR AWAY FROM THERMODYNAMIC LIMIT

- Small γ, high pressure drop and recovery (Li, IECR, 2013)
- ERD may not be so helpful as in SWRO

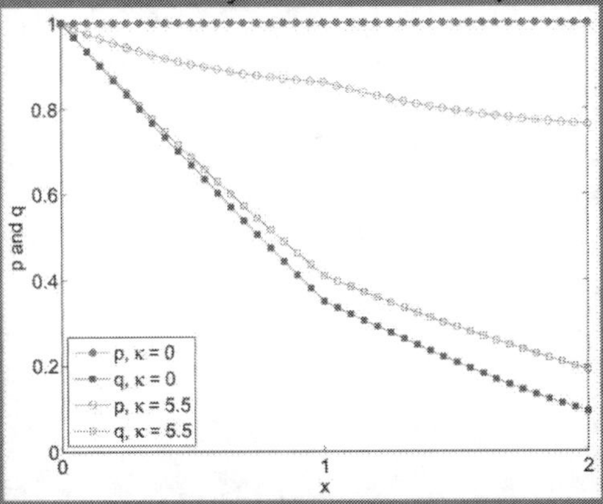

Brine energy for ERD:

- Seawater: $(0.5Q_0)(\Delta P_0) = 0.50(Q_0 \Delta P_0)$
- Brackish water: $(0.2Q_0)(0.75\Delta P_0) = 0.15(Q_0 \Delta P_0)$

- Brine recirculation will increase NSEC because of increased flow resistance.

NSEC IN BWRO FAR AWAY FROM THERMODYNAMIC LIMIT

- One-stage is more energy-efficient than two-stage design. However, two-stage design with 2:1 array are used in industry due to other reasons.
- For a two-stage design, an interstage booster pump does not reduce NSEC because of small γ.

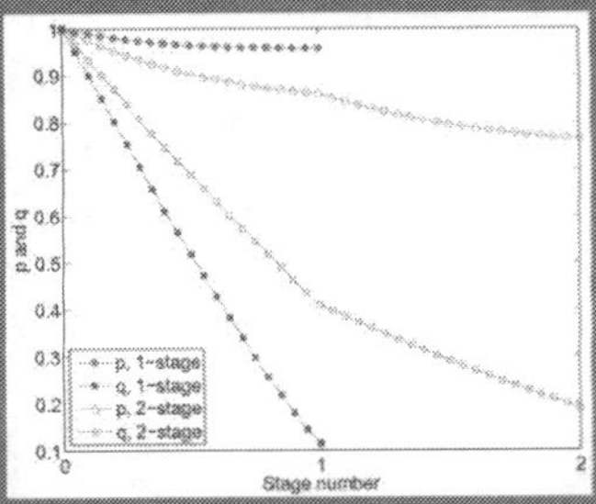

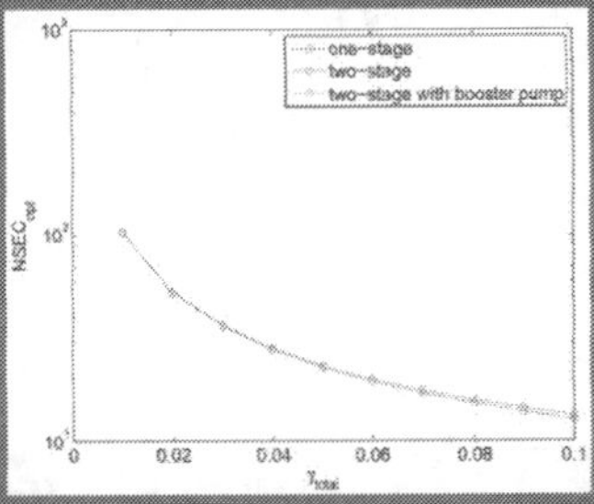

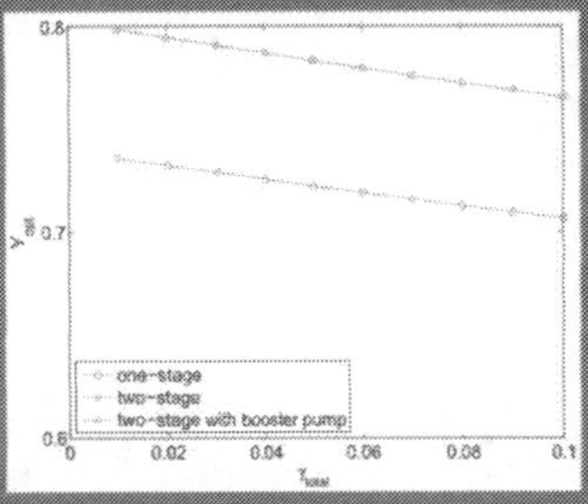

CONCLUSIONS

- How to approach the theoretical value of NSEC?
  - Very large $\gamma = AL_p\Delta\pi_0/Q_0$
  - Very small $\kappa = kQ_0^2/\Delta\pi_0$
  - Multiple stages (especially at high recoveries)
  - ERD
  - Efficient pump
- SWRO (near thermodynamic limit)
  - Multiple stages and ERD are beneficial
- BWRO (far away from thermodynamic limit)
  - One stage is the best in terms of NSEC
  - Other factors (e.g., to achieve a more uniform flow) lead to two-stage designs
  - Interstage pump is not necessary from a viewpoint of energy saving
  - ERD in BWRO is not so beneficial as in SWRO
- Brine circulation does not help NSEC in SWRO or BWRO

RELATED PUBLICATIONS

- Li, M. "Energy Consumption in Spiral Wound Seawater Reverse Osmosis at the Thermodynamic Limit," *Ind. Eng. Chem. Res.*, **53**, 3293-3299, 2014.
- Li, M. "A Unified Model-Based Analysis and Optimization of Specific Energy Consumption in BWRO and SWRO," *Ind. Eng. Chem. Res.*, **52**, 17241-17248, 2013.
- Li, M.; Noh, B. "Validation of Model-Based Optimization of Reverse Osmosis (RO) Plant Operation," *Desalination*, **304**, 20-24, 2012.
- Li, M. "Optimization of Multitrain Brackish Water Reverse Osmosis (BWRO) Desalination," *Ind. Eng. Chem. Res.*, **51**, 3732-3739, 2012.
- Li, M. "Optimal Plant Operation of Brackish Water Reverse Osmosis Water Desalination," *Desalination*, **293**, 61-68, 2012.
- Li, M. "Reducing Specific Energy Consumption in Reverse Osmosis Water Desalination: An Analysis from First Principles," *Desalination*, **276**, 128-135, 2011.
- Li, M. "Minimization of Energy in Reverse Osmosis Water Desalination using Constrained Nonlinear Optimization," *Ind. Eng. Chem. Res.*, **49**, 1822-1831, 2010.

The End

Mingheng Li AIChE Annual Meeting Nov. 16, 2014 23 / 23

Enhancement of Oil-Water Separation by Dissolved Air Flotation

M.R. Aliff Radzuan[1], *M.A. Abia-Biteo Belope, R.B. Thorpe*
Department of Chemical & Process Engineering, Faculty of Engineering and Physical Sciences, University of Surrey, Guildford, GU2 7XH, UK

Abstract

Dissolved air flotation (DAF) is one of many separation techniques that can be used to remove oil droplets from an aqueous stream by using air bubbles. Most of the droplets are attached to the bubbles to form agglomerates that are easier to float onto the surface of the DAF tank. A DAF tank consists of two main zones. The collision and attachment of oil droplets-air bubbles take place in the contact zone before they flow into the separation zone where the separation occurs. The main aim of the experiments reported in this paper is to remove the oil droplets in the range 15-80 μm from an oil-in-water mixture. Two kinds of oil, vegetable and mineral oil were used and several operating parameters were varied. An oil-in-water measuring technique and a droplet counting method were used to calculate the removal efficiency. The results from DAF experiments reported in this paper reveal that the technique can achieve up to 95% removal efficiency. The best separation efficiency was obtained from the experiment of 4 barg, Veg. oil + salty water at 35 °C and at the lowest inlet oil concentration.

Keywords: Dissolved air flotation, oil droplet, air bubble

1. Introduction

Dissolved Air Flotation (DAF) aims to remove oil droplets from an aqueous continuous phase often in a range 15-80 μm. The application of this technique in the oil-based industry to separate oil droplets carried with the produced water after a primary separation is not entirely new, but to date, there are very few scientific studies that have discussed its removal efficiency and the impact of different design parameters (*1–7*).

The air bubbles are introduced from the nozzles located at the bottom of the contact zone. The air is dissolved into the water in a packed column under pressure of 1 to 4 barg. Several researchers have suggested different ranges of saturator pressure for the best operation, which are 4-5 barg (*8*), 4-6 barg (Al-shamrani et al. 2002) and 5 barg (*9*). This water passes through the nozzles and because of the rapid pressure drop to the atmosphere, small air bubbles form within the range of 10-120 μm. However, most of the bubbles produced in the DAF rig used in this study are within 50-55 μm (*10*). The attachment of the air bubbles and the oil droplets increases the average density difference between agglomerate and continuous phase, which increases the rise velocity of oil to the surface and therefore, provides a better separation from the aqueous phase.

2. DAF Analysis

2.1 Oil-In-Water Method (FastHEX)

An ultraviolet-visible spectrophotometry (UV-Vis) was used to analyse the amount of oil extracted into hexane from the samples of influent and effluent from the DAF tank. This method is

[1] Corresponding author: Tel.: +441483686943
E-mail address: a.mohamadradzi@surrey.ac.uk

widely used in the oil based industry and was used for the Alba Field oil in water measurement (*11*). The absorbance of the mixtures (Veg. oil/Lamp oil with Hexane) was measured at the optimum wavelength of 400 nm. Finally, the removal efficiency from the FastHEX method E_{FH} was calculated by using equation **(1)**. The value of α is based on oil/Hexane calibration curve at 400 nm, which are 0.045 for lamp oil and 0.060 for vegetable oil.

$$E_{FH} = \frac{Abs_{in} - Abs_{out}}{Abs_{in} - \alpha_i} \times 100\% \qquad (1)$$

2.2 Droplet Counting Method (Coulter Counter)

A Coulter Counter was used to count the number of oil droplets passing through the aperture based on the size range. The aperture was immersed in the cuvette filled with the sample mixed with the isotonic solution supplied by Beckman Coulter (*12*). The analysis was started quickly to reduce the chances of droplet coalescence. The measurement was done for 5 μm bands starting from 15 μm. Data obtained from the Coulter Counter was analysed to produce the inlet and outlet droplet size distributions. The volume of the oil droplets at a particular range was calculated from the number/count obtained by the coulter counter. The removal efficiency from the coulter counter method E_{CC} was calculated using equation **(2)**.

$$E_{CC} = \frac{\sum_{in} n_i V_{o,i} - \sum_{out} n_i V_{o,i}}{\sum_{in} n_i V_{o,i}} \times 100\% \qquad (2)$$

3. Results and Discussions

3.1 Results from the Oil-In-Water Method (FastHEX)

3.1.1 The effect of inlet oil concentration

The effects of the inlet oil concentration entering the flotation tank were measured by varying the oil flow rate going into the static mixer (See **Fig. A** 1), while other variables were kept constant. Three different inlet oil concentrations were used for each set of conditions. As the concentration of the inlet oil increased, either the quality of the attachment seems to be lowered, or the larger number of oil droplets in the flotation tank decreases the probability of collision and attachment of any droplet, has resulted in lower removal efficiency, or both. As can be observed in **Fig. 1**, the lowest inlet oil concentration gave the highest removal efficiency, regardless of any parameters involved in the experiments.

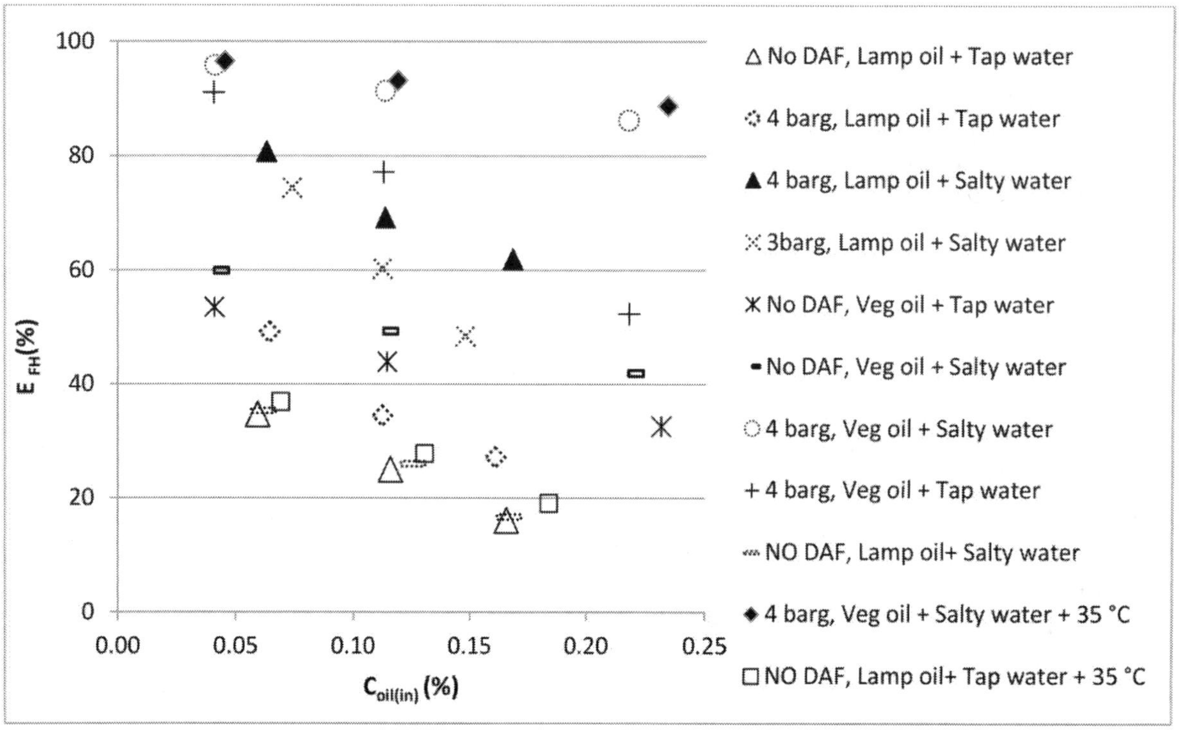

Fig. 1. Removal efficiency of oil from oil-in-water mixture

3.1.2 The Effect of Air Saturator Pressure

Due to the pressure limitation from the air supply, which could only achieve up to 4 barg, only the pressure 3, 4, and 0 barg (No DAF) were applied to the flotation process. It has been identified that the bubbles size stays the same when the pressure is increased to more than 3.5 barg (Han et al. 2002). The results at 4 barg may be partially reflective of higher pressures. However, the amount of gas dissolved or released is proportional to the pressure. This may give a better matching between numbers of bubbles and number of oil droplets, especially at higher inlet oil concentration. This is a hypothesis that seems to be supported by the graph shown in **Fig. 1.** The results show that the best oil droplet separation efficiency was obtained at an air saturator pressure of 4 barg. The creation of bubbles distribution at any pressure in the DAF tank resulted in a better improvement in the separation of oil droplets compared to using gravity separation (0 barg / No DAF). The results of applying DAF at 4 barg resulted in separation efficiency of about 81 %, when lamp oil was used as the dispersed phase and salty water as the continuous phase. 3 barg of DAF presented slightly lower removal efficiency and this became even lower when DAF was not used, which at 35 %. This emphasizes the importance of the DAF in separating dispersed oil droplets from the continuous phase.

3.1.3 The Effect of Temperature

The temperature of the continuous phase in the flotation tank was varied from room temperature to 35 °C. **Fig. 1** presents the removal efficiency increasing with increasing temperature. This maybe expected as the increase in temperature reduces the viscosity of the fluids being separated; therefore, the rise velocity of the oil droplets is increased (*14*). The effect of the temperature on kinematic viscosity was also measured. A low viscosity increases the separation efficiency. However, for both type of oil tested in this experiment, the improvement in the removal efficiency seems relatively small to the change in temperature.

3.1.4 The Effect of Salinity

The investigation was carried out using two different types of continuous phase: Guilford, Surrey tap water and salty water. For the experiment of lamp oil + salty water, 4 barg and at the lowest inlet oil concentration, the removal efficiency obtained was at 80.8% instead of 49.1 % for the experiment with tap water. Similar with the experiment involved with vegetable oil, which 95.8 % and 91.1 % for the runs with salty and tap water respectively. The possible reasons for this behaviour are because salt has altered the repulsive charges of the oil droplets and encouraged the attachment of the air bubbles to the oil droplets (*5*). Also, adding salts to water supress the coalescence of bubbles (*15*). Hence, the bubble size did not grow and so captured more of the desired size of oil droplets. To keep the collision efficiency high, coalescence of the bubbles should be avoided to maintain them below 100 μm. Another possible reason is because the addition of salt enhanced the spreading coefficient, resulting in a better attachment of the air bubbles to the oil droplets (*7*). **Fig. 1** shows that the experiments conducted with salty water obtained a higher separation efficiency compared to tap water.

3.1.5 The Effect of Type of Oil

Based on the terminal rise velocity, the removal efficiency of lamp oil was expected to be higher than vegetable oil. However, the experiment conducted showed a contrasting result, i.e. the experiments with the vegetable oil presented higher removal efficiency (See **Fig. 1**). The results may be explained as a combination of factors one of which is the fact that vegetable oil has a positive and a greater spreading coefficient than lamp oil (*16, 17*). At 20 °C, the surface tension of Veg. oil, 14.6 x 10^{-3} N/m is higher than n-Tetradecane -7.11 x 10^{-3} N/m. n-Tetradecane is closely similar to lamp oil and sunflower oil is to be read as been closely similar to the veg. oil used in the experiments.

3.2 Results from the Droplet Counting Method (Coulter Counter)

3.2.1 Droplet Size Distribution of Lamp Oil at 0.2 % inlet oil concentration

A static mixer was used to produce the oil droplets below 100 μm. The lamp oil droplet size distributions of the inlet and the outlet of the DAF tank are shown in **Fig. 2**. Each experiment was run twice and the average of the volume fractions were used to plot the droplet size distribution. The average Sauter mean diameter of (a) and (b) are 29.1 μm. Most of the inlet oil droplets were produced in a range of 30-35 μm. No oil droplet was detected at 65-70 μm onwards. The used of DAF was proved to enhance the separation of oil from oil-in-water mixture especially within the range of 50-55 μm. Coalescence of droplets was obviously occurring for the lamp oil. That can be seen especially above the range 50-55 μm on **Fig. 3**. The removal efficiency differences were not obvious until at 50-55 μm, which approximately 70 %. The average removal efficiency of run (a) with 4 barg and run (b) No DAF were found at 27.7 % and 12.1 %

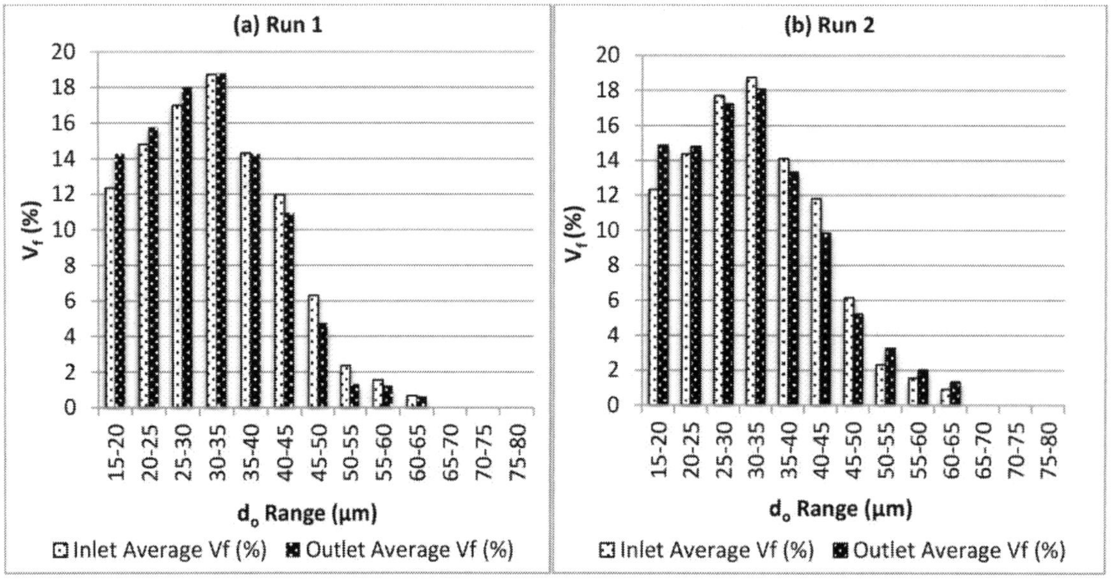

Fig. 2. Average droplet size distribution by volume of the Lamp oil at (a) 4 barg (b) No DAF, $C_{oil(in)}$ at 0.2%

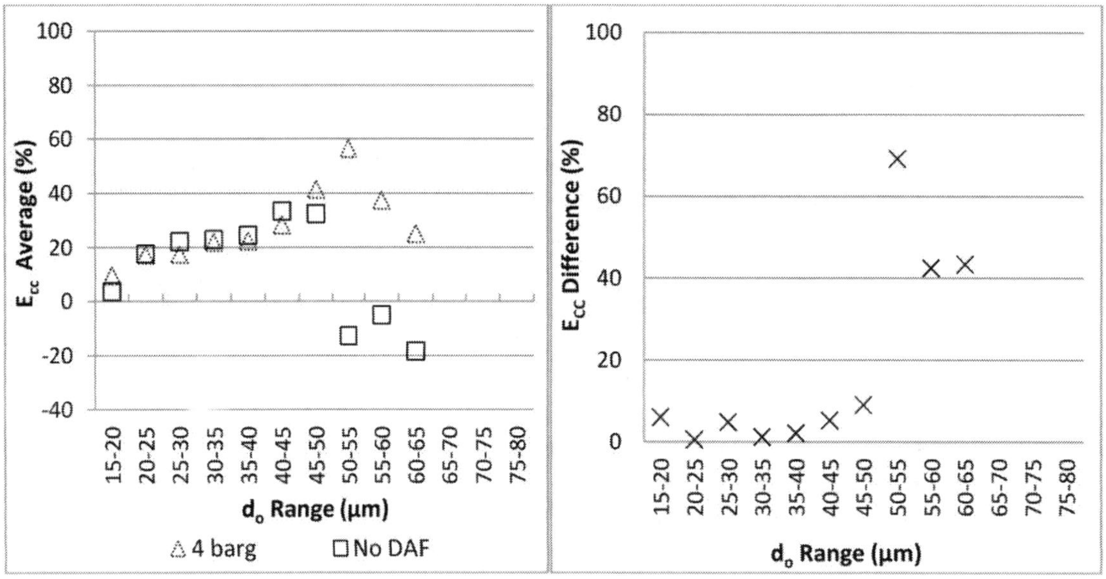

Fig. 3. Average and difference of removal efficiency of Lamp oil $C_{oil(in)}$ at 0.2%

3.2.2 Droplet Size Distribution of Veg. Oil at 0.2 % Inlet Oil Concentration

Run 5 and 6 were conducted twice to obtain the average V_f. These runs were conducted with the veg. oil at 0.2 % inlet oil concentration. **Fig. 4** shows that the runs with veg. oil produced wider droplet size distribution compared to the lamp oil. The Sauter mean diameter of (a) run 5 and (b) run 6 were 31.0 and 31.4 µm.

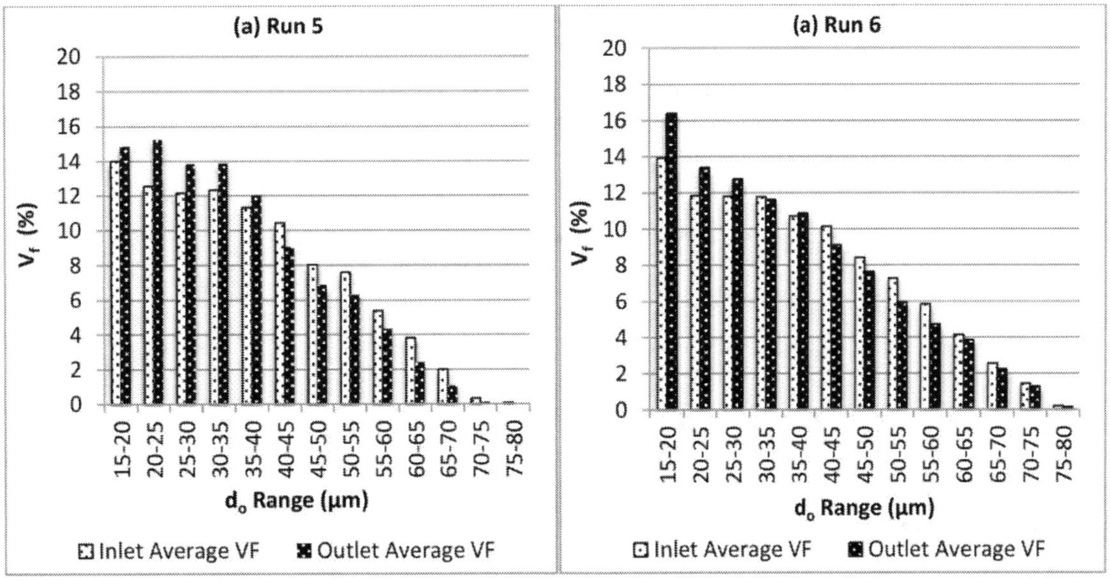

Fig. 4. Average droplet size distribution of the Veg. oil at (a) 4 barg (b) No DAF, $C_{oil(in)}$ at 0.2%

No coalescence was noticed on the veg. oil experiments. The removal efficiency was up to 10 % for the biggest droplet range (75-80 µm). Most of the removal efficiency for the experiment with No DAF was laid below 40 %. According to **Fig. 5** the difference became larger started at 60-65 µm. The average removal efficiency of (a) run 5 with 4 barg of DAF and (b) run 6 with no DAF were found at 43.7 % and 25.5 %.

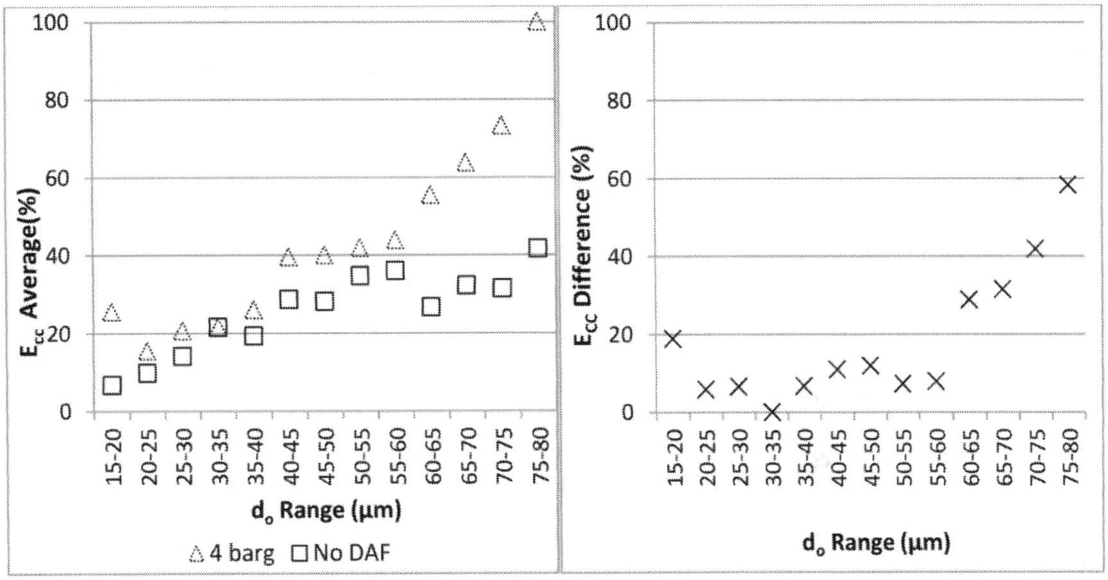

Fig. 5. Average and difference of removal efficiency of Veg. oil $C_{oil(in)}$ at 0.2%

3.3 Comparison of FastHEX and Droplet Counting Method

These experiments were investigated to confirm the accuracy and the reproducibility of the removal efficiency of oil from the oil-in-water mixture by using FastHEX and Droplet Counting methods. This investigation revealed that both of the methods show agreements. According to **Fig. 6**, the highest difference between these methods was around 8.9 %. The difference was due to the different inlet oil concentration when conducted the experiment.

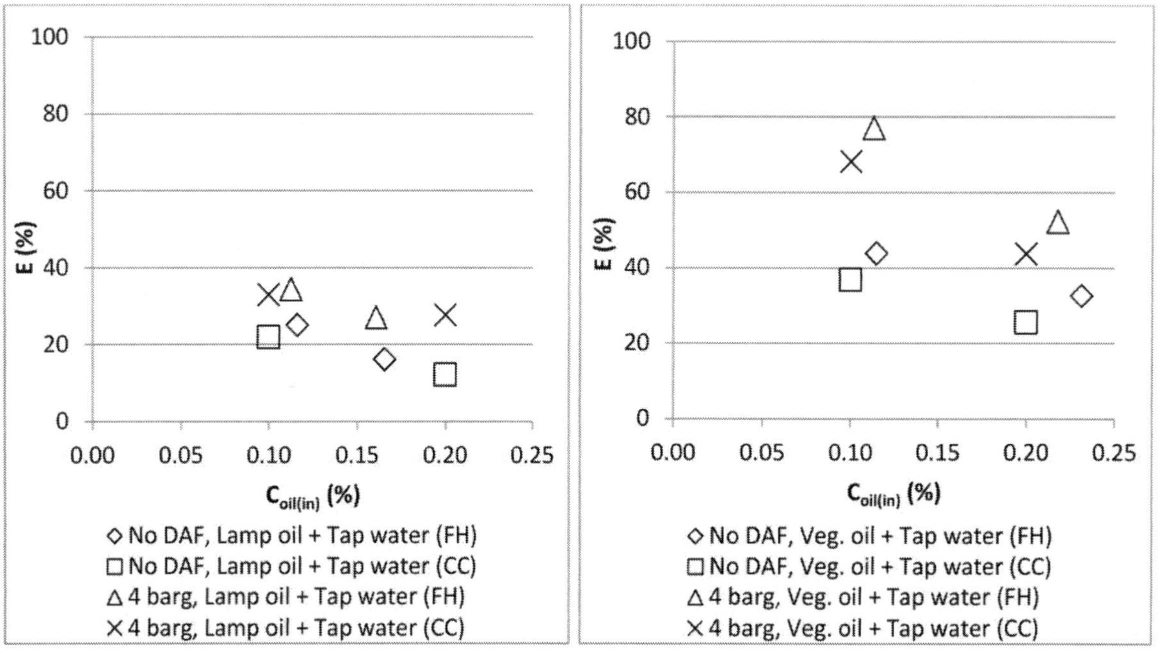

Fig. 6. The comparison of the FastHEX and the Droplet Counting method for the lamp oil

4. Conclusions

Several operational parameters were investigated to determine the removal efficiency of oil from oil-in-water mixture by using DAF. All the parameters tested proved that DAF could provide a significant intensification to the separation performance. As a conclusion, the experiments revealed the following:

a) The inlet oil concentration affected the removal efficiency. It was found to increase with decreasing inlet oil concentration. This happened due to the increase of the collisions frequency and attachment of bubble-oil droplet.

b) Increasing the pressure had increased the removal efficiency, but limited up to 4barg, due to unaffected bubble size beyond that.

c) Temperature has not showing a great effect on the removal efficiency, although the viscosity of the oil has changed drastically at higher temperature.

d) Addition of salt into the process tends to improve the removal efficiency, where experiment with saline water produced better performance.

e) Experiment with the lamp oil produced awful removal efficiency for bigger size of oil droplets. Coalescence happened after 50-55 µm of diameter size.

f) According to droplet size distribution, better removal efficiency obtained in the ranges of 45-50 µm onwards for vegetable oil.

The best removal efficiency from the DAF experiment was obtained in the condition of 4 barg, Vegetable oil + salty water at the temperature of 35 °C, at the lowest inlet oil concentration. The 30% difference of separation efficiency compared to gravity separation showed its impacts to various operations, especially in the oil and gas industry, where the environmental legislation is getting tighter.

5. References

(1) Al-Shamrani, A. A.; James, A.; Xiao, H. Separation of Oil from Water by Dissolved Air Flotation. *Colloids and Surfaces* 2002, *209*, 15–26.

(2) Galil, N. I.; Wolf, D. Removal of Hydrocarbon from Petrochemical Wastewater by Dissolved Air Flotation. *Water Sci. Technol.* 2001, *43*, 107–113.

(3) Hanafy, M.; Nabih, H. I. Treatment of Oily Wastewater Using Dissolved Air Flotation Technique. *Energy Sources* 2007, *29*, 143–159.

(4) Karhu, M.; Leiviskä, T.; Tanskanen, J. Enhanced DAF in Breaking Up Oil-in-Water Emulsions. *Sep. Purif. Technol.* 2014, *122*, 231–241.

(5) Moosai, R.; Dawe, R. A. Gas Attachment of Oil Droplets for Gas Flotation for Oily Wastewater Cleanup. *Sep. Purif. Technol.* 2003, *33*, 303–314.

(6) Moursy, A. S.; Abo El-Ela, S. E. Treatment of Oily Refinery Wastes Using A Dissolved Air Flotation Process. *Environmnetal Int. a J. Sci. Technol. Heal. Monit. policy* 1982, *7*, 267–270.

(7) Oliveira, R. C. G.; Gonzalez, G.; Oliveira, J. F. Interfacial Studies on Dissolved Gas Flotation of Oil Droplets for Water Purification. *Colloids and Surfaces* 1999, *154*, 127–135.

(8) Zouboulis, A. I.; Avranas, A. Treatment of Oil-In-Water Emulsions by Coagulation and Dissolved-Air Flotation. *Colloids and Surfaces* 2000, *172*, 153–161.

(9) Edzwald, J. K. J. . Dissolved Air Flotation and Me. *Water Res.* 2010, *44*, 2077–2106.

(10) Hague, J. Experimental and Modeling Study of Two-Phase (Air/Water) Flow in Dissolved Air Floatation (DAF) Tanks., Eng.D. Thesis, University of Surrey, 2002, pp. 25–56.

(11) Abia-Biteo Belope, M. A. The Design and Performance of Offshore Gas/Oil/Water Separation Process, PhD Thesis, University of Surrey, 2010, pp. 1–315.

(12) Beckman-Coulter Cell Counting and Sizing. Z2 Series Coulter Counter User's Manual.

(13) Han, M.; Park, Y.; Lee, J.; Shim, J. Effect of Pressure on Bubble Size in Dissolved Air Flotation. *Water Sci. Technol. Water Supply* 2002, *2*, 41–46.

(14) Abdel-Aal, H. K.; Aggour, M.; Fahim, M. A. *Petroleum and Gas Field Processing*; Heinemann, H., Ed.; Marcel Dekker, Inc.: Kuwait, 2003; pp. 243–245.

(15) Thoma, G. J.; Bowen, M. L.; Hollensworth, D. Dissolved Air Precipitation/Solvent Sublation for Oil-Field Produced Water Treatment. *Sep. Purif. Technol.* 1999, *16*, 101–107.

(16) Bassam, Z. S. Surface Spreading and Surface Tension. In *Chemical Demonstrations, A Handbook for Teachers of Chemistry*; The University of Wisconsin Press, 1989; pp. 4–13.

(17) Hotrum, N. E.; Cohen Stuart, M. A.; Van Vliet, T.; Van Aken, G. A. Spreading of Partially Crystallized Oil Droplets on an Air/Water Interface. *Colloids and Surfaces* 2004, *240*, 83–92.

Nomenclature

Abbreviation	Description
barg	bar gauge
DAF	Dissolved Air Flotation
NaCl	Sodium Chloride
SMV-02	Static Mixer
UV-Vis	Ultraviolet-Visible Spectrophotometry
Veg. oil	Vegetable Oil

Symbol	Description
α	Constant value from the oil/Hexane calibration curve at 400 nm
Abs_{in}	Absorbance value as measured by UV-Vis for the inlet mixture sample
Abs_{out}	Absorbance value as measured by UV-Vis for the inlet mixture sample
D_{pipe}	Pipe diameter
d_{32}	Sauter mean diameter of oil droplets
E_{CC}	Removal efficiency as measured by Coulter Counter method
E_{FH}	Removal efficiency as measured by FastHEX method
n	Number of droplet as measured by Coulter Counter
V_f	Volume fraction
V_o	Volume of oil droplet

Appendix

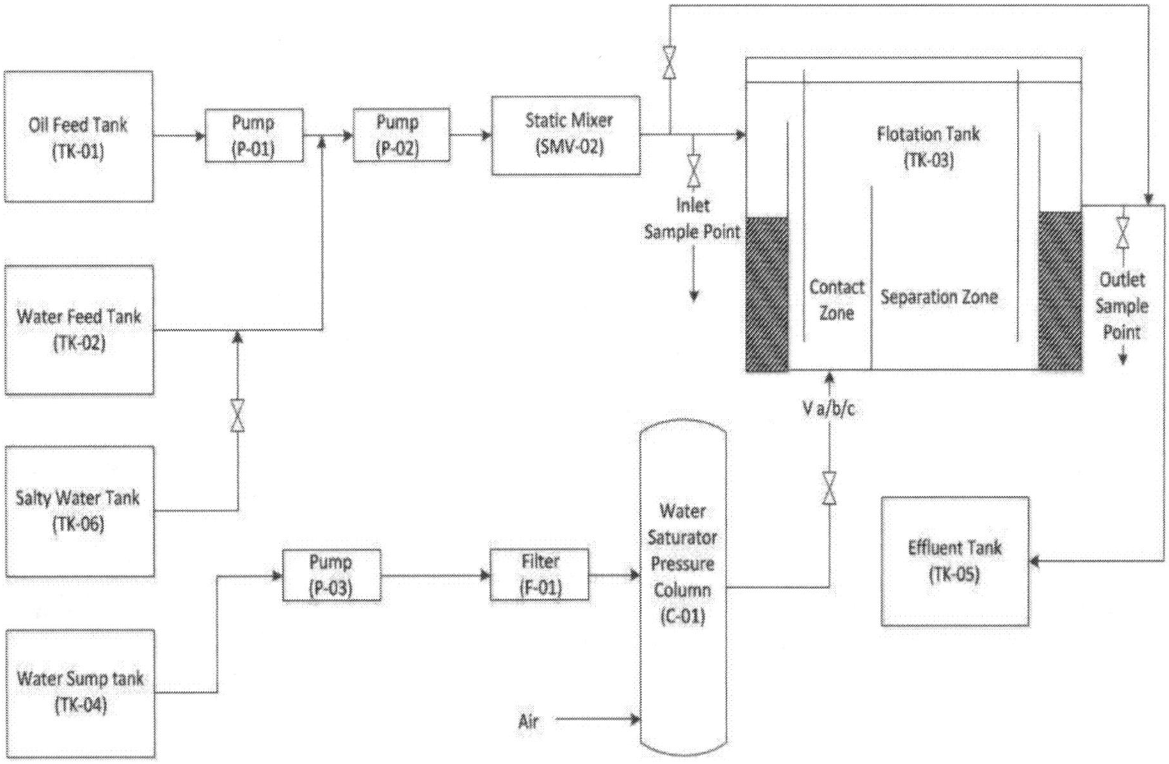

Fig. A 1. Flow diagram of DAF at University of Surrey

386619 Direct Integration of RO Desalination and UF Pre-Treatment with RO Concentrate Backwash

Abstract:

A design for a reverse osmosis (RO) desalination system directly integrated with an ultrafiltration (UF) pre-treatment unit was developed. The integration involves direct RO feed from the UF filtrate and UF backwash using the RO concentrate. This alignment removes the necessity of intermediate tanks for a more compact design which is vital in areas where space is at a premium. The use of RO concentrate for UF backwash allows the UF to operate at 100% recovery since all UF filtrate is fed directly to the RO. Also, due to the constant production of RO concentrate, the length of backwash is not constrained and the system is capable to carrying out a variable UF backwash strategy where the backwash duration is dynamically adjusted to adapt to temporal variability of feed water quality. Since the UF pre-treatment system is a "slave" of the RO desalination system due to its task of constantly providing the RO feed, a control system is necessary to ensure the RO inlet retains a specified level of flow and pressure. The UF pre-treatment system is equipped with UF membrane modules capable of operating independently of each other so that the UF pre-treatment system can still produce RO feed if some of its membrane modules are being backwashed. This design was used in the construction of an 18,000 gallons/day seawater desalination plant. Through field testing, the advantages of direct integration of UF and RO, and RO concentrate adaptive (in terms of both frequency and duration) backwash of the UF system, were demonstrated.

376478 Removal of Organic and Nitrogenous Contaminants of Wastewater By Polyelectrolyte Multilayer Membranes

Abstract:

The alarming depletion of fresh water resources all over the world has led to widespread water shortages. As the gap between demand and supply is growing with the increase in population the occurrence of several water-borne diseases is also on the rise especially in the developing countries. Designing cost-effective water purification technologies to recycle and reuse wastewater is therefore the need of the hour. Reduction in the chemical oxygen demand (COD) and total Nitrogen (TN) values of wastewater are two very important parameters for wastewater purification. Reverse osmosis (RO) which is one of the most efficient membrane-based separation processes offers very high reduction of these parameters. However the RO membranes inherently offer very low water fluxes across them making the process energy-intensive. Nanofiltration (NF) membranes on the other hand allow higher water fluxes but lower solute rejections as compared to RO membranes. In our work we modify the surfaces of existing NF membranes using the Layer-by-layer (LbL) assembly technique in order to render them higher perm-selectivity as compared to commercial RO membranes. LbL is as aqueous-based process involving the sequential deposition of alternately charged polyelectrolytes on a substrate leading to the formation of polyelectrolyte multilayer (PEM) films. The tremendous versatility associated with this method in terms of its design parameters helps optimizing the membrane performance with extreme precision. The PEM membranes owing to their high perm-selectivity can potentially replace the traditional RO membranes. The wastewater sample under consideration is the effluent from an anaerobic digestion (AD) process which was further treated in an electrocoagulation (EC) unit prior to being used as the feed for the membrane processes. The membranes are tested in a stirred dead end membrane module and the COD and TN quantifications are done using commercial HACH kits.

391675 Photoactivated SOA Formation: Mechanistic Insight from Modeling and Experiments

Abstract:

Reactions in or on aerosol particles involving the participation of light-absorbing organic aerosol species as photoactivators have been suggested as pathways for the photochemical processing of volatile organic species (Yu and Keutsch, 2012; Monge et al., 2012; Aregahegn et al., 2013, Rossignol et al. 2014), leading to the efficient formation of secondary organic aerosol (SOA) material. Additional insight into the mechanisms and kinetics of these processes is required so that we may evaluate their importance on the regional and global scales via atmospheric chemistry modeling. We have used a combination of bulk phase and surface-specific spectroscopy, density functional theory, and box modeling to this end. Bulk flash photolysis kinetics studies, along with the surface-active nature of many likely photoactivator species, show that this chemistry most likely occurs at the gas-aerosol interface. Analysis of published aerosol chamber data using our photochemical box model, GAMMA, shows evidence for SOA formation from glyoxal via photoenhanced fulvic acid and, to a lesser degree, particle-phase products of glyoxal SOA formation. Finally, density functional theory provides insight to candidate photoactivator-VOC pairs, demonstrating the general nature of this phenomenon.

379273 Effect of Humidity and Particle Acidity SOA Derived from Reaction of Beta-Pinene with Nitrate Radical

Abstract:

Secondary organic aerosol (SOA) is a major component of atmospheric aerosol, which has important implications on health, climate, and visibility. SOA is formed when gas-phase products oxidize in the atmosphere to form lower volatility products that can partition onto pre-existing particles. Studies have shown that nighttime chemistry can account for about 20-50% of biogenic precursor oxidation in polluted urban areas. The primary nighttime oxidant is the nitrate radical (NO_3) formed by reaction of ozone and anthropogenic NO_2. The interactions of biogenic VOCs with NO_3 radical represent a direct way for positively linking anthropogenic and biogenic emissions. Since reaction with nitrate radical is predominant at night, it primarily reacts when the relative humidity is high due to the cooler nighttime temperatures. Beta-pinene is an important biogenic precursor in nighttime aerosol chemistry because of its relatively high abundance and high SOA yields when oxidized by NO_3. In the ambient, the products of this reaction will condense onto aerosol of high anthropogenic sulfate content. The sulfate in this pre-existing aerosol increases the particle acidity. In this study, chamber experiments with beta-pinene are performed at the Georgia Institute of Technology Environmental Chamber facility. Reactions are carried out at high humidity to simulate elevated night-time humilities in the ambient environment. Seed aerosol acidity is controlled by the addition of either neutral ammonium sulfate ($(NH_4)_2SO_4$) or an acidic mixture of $(NH_4)_2SO_4/H_2SO_4$. N_2O_5 is injected into the chamber by pre-reacting NO_2 and O_3 in a flow tube to initiate beta-pinene oxidation. Aerosol chemical composition is characterized by a High Resolution-Time of Flight-Aerosol Mass Spectrometer (HR-TOF-AMS) with corresponding gas-phase oxidation products measured by a Chemical Ionization Mass Spectrometer (CIMS). The hydrolysis of organic nitrate products from the beta-pinene/NO_3 reaction under humid conditions represents an important organic nitrate sink in the night-time environment. More detailed analysis will discuss the influence of relative humidity and particle acidity on gas-phase and aerosol composition.

Composition and Oxidative Properties of Particulate Matter Mixtures

Wing Y. Tuet[1], Vishal Verma[2], Julie A. Champion[1], Nga L. Ng[1,2]
1 School of Chemical and Biomolecular Engineering - Georgia Institute of Technology;
2 School of Earth and Atmospheric Sciences - Georgia Institute of Technology;

Particulate matter (PM) is classified as a human carcinogen and associated with increases in respiratory and cardiovascular diseases. However, the mechanisms by which PM exposure causes adverse health effects are not well characterized. Toxicology studies have suggested oxidative stress induced by reactive oxygen species (ROS) may be the missing link. ROS production resulting from PM exposure, which includes oxygen ions, free radicals, and peroxides, can trigger inflammatory cascades, and prolonged stimulation of these cascades leads to cellular damage and disease. Previous health studies either involved particles not representative of ambient aerosols, such as titanium oxides, or focused on primary organic aerosols (POA). Field studies have shown that secondary organic aerosols (SOA) mass loadings often dominate over POA implying that health effects may be grossly underestimated in studies that only consider POA.

Here we present modifications to an existing ROS assay in which macrophages are exposed to suspensions of PM, and the ability of PM to indirectly induce cellular ROS production is measured. We use murine alveolar macrophages to represent the first line of defense against environmental insults to the respiratory system and vary several parameters to maximize the assay's signal to noise ratio for our samples. SOA from various precursors will then be generated in the Georgia Tech Environmental chamber facility, and the oxidative stress of these particles will be measured using the macrophage assay. The oxidative properties of SOA generated under different experimental conditions will be evaluated and discussed.

383206 Quantify the Impact of Biomass Burning Aerosols on Regional Climate over the Southeastern USA

Abstract:

Biomass burning has been and will continue to be a major contributor to the particulate matter over the southeastern USA. In order to quantify the direct effects of biomass burning aerosols on the regional climate, we employ the WRF and coupled WRF-CMAQ (Wong et al., 2012), both of which are driven by NARR data and downscale to 12km resolution covering the southeastern USA.

By comparing WRF and coupled WRF-CMAQ (control run with no fire emissions), we will be able to quantify when feedback being considered the background of the impact of emissions other than fires on regional climate. In addition, we will separate the impacts into two parts. The first part is due to the aerosol direct effect and the second part is due to bulk property change of clouds. By comparing the results from control and emission scenarios, we are interested in first, how the regional climate (especially, surface temperature, cloud coverage and precipitation) would change due to the extra forcing from biomass burning, and second, to what extent the relative contributions from aerosol direct effect and bulk cloud change would alter due to extreme conditions.

References:

Wong D, Pleim J, Kang D, et al, (2012): WRF-CMAQ two-way coupled system with aerosol feedback: software development and preliminary results. *Geoscientific Model Development* , 5,299-312,doi: 10.5194/gmd-5-299-2012

386174 Evaluating the Role of Natural Variability in Assessments of Climate Change Impacts on Air Quality

Abstract:

Climate change is expected to significantly influence meteorology and may therefore impact future air quality. Several studies have attempted to estimate the effect of climate change on air pollution by using meteorological fields derived from general circulation model simulations to drive atmospheric chemistry and transport models and project future pollutant concentrations. However, large uncertainties are associated with climate simulations and may propagate into predictions of future air quality. Beyond uncertainty in emissions and model response, climate projections are significantly influenced by natural variability. However, little attention has been given to the role of natural climate fluctuations in modeling analyses aimed at quantifying the effect of climate change on air pollution. As internal variability intrinsically limits the ability of models to predict climate on time scales smaller than a decade, multiyear or multidecadal mean data may be necessary to adequately capture the climate signal and should be considered for air quality assessments.

In this study we investigate the effect of natural variability inherent to climate projections on future estimates of U.S. air quality. Future ozone and particulate matter concentrations are simulated with the Community Earth System Model driven by meteorological fields derived from an ensemble simulation of 21st century climate change carried out with the MIT Integrated Global System Model–Community Atmosphere Model. The influence of natural variability is explored by carrying out a series of multidecadal atmospheric chemistry simulations under present and future climates, using fixed anthropogenic emissions levels and multiple initial conditions. By modeling the changes in surface O_3 and $PM_{2.5}$ concentrations across the climate ensemble, we identify the magnitude of unforced interannual variations in air quality and the adequate time scales for climate-related impacts assessments. In addition, the analysis further examines how uncertainty in air quality projections due to internal variability may propagate into health effects estimates and influence the findings of climate impacts studies.

387701 Sector-Based Analysis of Deposited BC in the Arctic and Its Impact on Snow Albedo

Abstract:

Long-range transport of atmospheric particles from mid-latitudes to the Arctic is the main contributor to the Arctic aerosol distribution. Black Carbon (BC) is considered of great climatic importance and is the main absorber of sunlight in the atmosphere. Besides, wet and dry deposition of BC on snow cause reduction of snow and ice albedo. This will cause sea ice and snow to warm and melt more quickly. WRF-STEM model is used to predict the transport of aerosols from different geographical sources and sectors (i.e. transportation, residential, industry, biomass burning and power) to the Arctic. The model is evaluated and interpreted with the observations from ARCTAS campaign flights in spring and summer 2008. Furthermore, the amount of particles deposited (by wet and dry deposition) is calculated. The results show that the major sector contributor to BC concentration is residential and the main source region contribution is from Europe annually. Both wet and dry deposition shows seasonality with the maximum occurring during spring for BC dry deposition and during summer for BC wet deposition.

388173 The Contribution of Different Surfactants within Corexit in Ejection of Oil/Dispersant Material to the Atmosphere

Abstract:

The Contribution of Different Surfactants within Corexit in Ejection of Oil/Dispersant Material into the Atmosphere

Paria Avij, He Huang, Zenghui Zhang, Francisco R. Hung, Kalliat T. Valsaraj

Cain Department of Chemical Engineering, Louisiana State University

The Deepwater Horizon (DWH) oil spill was one of the largest marine oil spills in U.S. history, during which vast quantities of crude oil were released into the sea-surface environment of the Gulf of Mexico. Ample research has been focused on the transportation of oil into the different environmental sections which includes air, water, sediment, and biota. Nevertheless, the only considered transportation pathway of oil spill matter into atmosphere has been evaporation. In our work, we study the aerosolization of oil and dispersant matter by whitecaps, which are simulated by bursting bubbles in a laboratory aerosolization reactor. The reactor incorporates an annular shear sparger which produces abundant, small sized bubbles.

In our experiments, different organic material including oil, mixtures of oil and Corexit (dispersant), and mixtures of oil and surfactants (DOSS, Tween 80, Tween85, Span 80) have been injected into the reactor. Since the reactor is capable of simulating both evaporation and surface adsorption transport processes, we collect the gas and particle phases at the reactor effluent and analyze these phases by GC-MS/GC-FID and LC-MS-MS for their alkane and surfactant content. We have clearly shown the air-water interface driven ejection of oil/dispersant material to the atmosphere via bursting bubbles.

Moreover, our experiments show that the ejection of oil matter via bursting bubbles occurs both in the presence and the absence of Corexit and this ejection is enhanced (especially C20-C30) when a dispersant is present. Also, through experiments of mixtures of oil and the individual components (surfactants) of Corexit, different ejection rates and dispersion behavior of oil matter were observed. Investigation and knowledge of these experiments will help us to have a better understanding about the surfactant behavior when they are exposed to the oil at the air-water interface. Likewise, this will help us implement possible corrections to the current dispersant formulation, which includes a lower ejection of oil matter into the atmosphere and a higher dispersion level within the water column.

Effects of CO, H_2 and C_3H_6 on the NH_3-SCR reactions over a Cu-chabazite SCR catalyst

Yang Zheng, Michael P. Harold*, Dan Luss*

Department of Chemical & Biomolecular Engineering
University of Houston
Houston, TX 77204

Abstract

The effects of reducing agents of CO, H_2 and C_3H_6 on the performance of a Cu-chabazite catalyst were investigated, including NO_2 decomposition, NH_3-SCR reactions, NH_3 storage, as well as N_2O production. Different reductants affect NH_3-SCR reactions to different extents by means of different pathways. Propylene is most effective in promoting NO_2 decomposition to NO via organic intermediates. CO effectively reduces nitrates to nitrites that can then react with NO_2 to release NO, while H_2 has a slightly promoting effect when NO_2 becomes thermodynamically unstable at high temperatures. As for SCR reactions, propylene competes with NH_3 for adsorbed NO_2 or nitrates to release NO and thus increase NO/NOx ratio. This leads to the dominance of either fast (NH_3 + NO + NO_2) or standard SCR (NH_3 + NO + O_2) reaction under slow SCR (NH_3 + NO_2)condition if C_3H_6 present. The competition of CO with NO for nitrates reduction to nitrites, has a negligible effect on both standard and fast SCR reactions but a promoting effect on slow SCR reaction. In contrast to steady-state operation, lean/rich cycling enhances cycle-averaged NOx conversion of all NH_3-SCR reactions when adding either C_3H_6 or CO + H_2 in rich phase. Both high NO_2/NOx ratio and long lean-phase purge can mitigate propene poisoning, enabling the combined contribution of both HC- and NH_3-SCR pathways. In addition, decreased N_2O emission from slow SCR reaction is observed when CO or C_3H_6 present partly due to each reaction with ammonium nitrates. All three reductants do not compete with NH_3 for storage sites and only C_3H_6 inhibits NH_3 oxidation. This study underscores the importance of accounting for the impact of different reducing agents on conventional NH_3-SCR reaction mechanism during the rich regeneration for integrated systems (LNT + SCR, SCR on filter) or cold-start period.

Development of SSZ-13 based sulfur resistant NH_3-SCR catalysts

Chao Wang, Erdem Sasmaz, Benjamin Galloway, Bihter Padak, Jochen Lauterbach*
Department of Chemical Engineering
University of South Carolina
Columbia, SC, 29208, U.S.A.
*lauteraj@cec.sc.edu

The ammonia-selective catalytic reduction (NH_3-SCR) of NO_x is a widely used technology for the reduction of NO_x emissions in stationary sources (gas-, oil-, and coal-power plants) and mobile sources (e.g., diesel engines). Previous studies have shown that the activity of SCR catalysts depends on a wide parameter space including the catalyst type (i.e., V_2O_5/TiO_2 vs. zeolites), metal type, synthesis method, and reaction parameters (flue gas composition, reaction temperature, and space velocity). Moreover, the presence of excess O_2, SO_2 and H_2O vapor in combustion exhaust affects both catalyst selectivity and stability at high and low temperatures.

Many studies have shown that zeolites are promising catalysts to be operated under coal combustion flue gas conditions. Previous work indicates that SSZ-13 based zeolite catalyst own higher thermal stability and NO_x conversion than other zeolite catalysts under ideal flue gas condition (NO, NH_3, O_2 and N_2) [1]. In this work, SSZ-13 catalysts loaded with Cu, Fe, Ce are synthesized and characterized to test NH_3-SCR activity under SO_2 and H_2O feed conditions. All catalysts are found to deactivate at 250°C with the addition of SO_2 and H_2O vapor. Among them, Cu-SSZ-13 shows the best poisoning resistance and keeps more than 90% NO conversion at high temperature. At low temperature, catalysts deactivate more than at high temperature, which might be due to the formation of $(NH_4)_2SO_4$ on the catalyst [2]. XPS results support the possible formation of $(NH_4)_2SO_4$, indicating that both NH_4^+ and SO_4^{2-} present on the surface of the catalysts. Detailed deactivation mechanism for different metal loading SSZ-13 catalysts will be explored.

References

1. D. W. Fickel, E. D. Addioa, J. A. Lauterbach, R. F. Lobo, Appl. Catal. B Environmental 102, 441 (2011)
2. G. S. Qi, R. T, Yang, R. Chang, Appl. Catal. B Environmental, 51, 93 (2004).

387796 Understanding of the Transition Metal Ion Properties and Their Impact on Various Catalytic Functions of Zeolite-Based SCR Catalysts

Abstract:

Copper and iron-zeolite SCR catalysts are widely used in US and European diesel after-treatment system to achieve drastic reduction in NOx emissions. The type of transition metal ion in zeolite has a significant impact on their key performance and durability characteristics including the response to sulfur containing exhaust gas. For example, it has been shown that at low temperatures, under standard SCR conditions, Cu-zeolites show higher steady state NOx conversion than Fe-zeolites and reverse is true at high temperatures. These differences have been reported in literature, but the underlying mechanisms were not very well understood mainly due to lack of comparable reference catalysts, which potentially introduces confounding effects for various reasons. In this article, we made an attempt to address the impact of only the transition metal ion on various SCR catalytic functions, by using a single source of zeolite BEA in the preparation of H-, Fe- and Cu-BEA samples.

Similar molar amounts of copper and iron were ion-exchanged into H-BEA and washcoated on the cordierite support by Johnson Matthey Inc. NOx conversion with NH_3 as a reductant and NO and NH_3 oxidation, and their storage were among the catalytic properties evaluated. The above catalytic properties were measured before and after sulfation of the SCR catalysts. Sulfation of the catalysts is achieved by exposing them to a mixture of SO_2 and SO_3 containing gas feed. The cumulative amount of sulfur stored on the catalyst and the rate of SOx release from sulfated SCR catalysts were determined using temperature programmed desorption under oxidizing conditions. Detailed experimental set up and conditions were described from our previous publications on SCR topic [1].

Below 300C, Fe-BEA showed lower NOx conversion than Cu-BEA despite the former catalyst more active for NO oxidation to NO_2, a potential reaction intermediate. Upon sulfation, the NOx reduction activity of Cu-BEA degraded much more than Fe-BEA SCR catalysts. Adsorption and desorption of various probe molecules such as NO_2, SOx, and NH_3 indicated that the Fe-BEA is relatively more acidic than Cu-BEA therefore, the former catalyst is prone to strong NH_3 adsorption on Fe active species inhibiting the NOx conversion at lower temperatures. On the other hand, more acidic Fe-BEA shows lower affinity to acidic species such as sulfur oxides. This study reconciles a number of observed differences in catalytic properties of Fe- and Cu-zeolite SCR catalysts, mainly based on their acid-base characteristics. A systematic and comprehensive analysis, which yielded practically relevant insights into the catalyst performance and sulfation and desulfation behaviors, will be presented.

[1] Ashok Kumar, Michael A. Smith, Krishna Kamasamudram, Neal W. Currier, Hongmei An, Aleksey Yezerets, "Impact of different forms of feed sulfur on small-pore Cu-zeolite SCR catalyst", Catalysis Today, 231 (2014) 75-82

374792 Controlled Synthesis of Novel Interweaved Titanium Oxide Nanotubes Confined Metal Oxide Catalytic Formulations: Effect of Morphology on Selective Catalytic Reduction (SCR) of NOx By NH3

Abstract:

The textural properties of various metal oxide confined titania nanotubes and their influence on surface active sites for NH_3−SCR were studied. Remarkably, the surface texture and tubular morphology of Me/TNT-H catalysts greatly promoted the NO_x conversions and temperature window compared to their nanoparticles. It can be ascribed to the preferentially located active species in the tubular framework. The impregnation of MnO_x over TNT-SA catalyst occur bare minimum because of the low specific surface area and low porevolume compared to other titania natotube supports, and this leads to poor deNO_x activity and narrow temperature window. Raman and X-ray diffraction patterns indicated that tubular structure in Mn/TNT samples significantly increases the concentration of oxygen vacancies and thus facilitate NO oxidation to NO_2. The relative atomic percentage value of Mn^{4+}/Mn^{3+} characterized by deconvoluted XPS spectra was considerably high for the Mn(0.25)/TNT-H catalyst ($Mn^{4+}/Mn^{3+} = 2.15$) whereas, surface Mn^{3+} species are in equivalence with MnO_2 phase in other catalysts ($Mn^{4+}/Mn^{3+} = 0.36$-1.34). In H_2-TPR, the disappearance of high-temperature reduction peak at 482 °C suggests that the Mn^{4+} species are dominant in Mn/TNT-H sample compared to Mn/TiO_2 nanoparticles. Our Raman results illustrate the redshift of 14 cm^{-1} from 640 cm^{-1} (TiO_2-Hombikat) to 626 cm^{-1} in Mn/TNTs-H sample which indicates the formation of a new phase or structural difference. In comparison with the Mn supported on TiO_2 (Hombikat) nanoparticles, the Mn species on the developed tubular Mn/TNT-H were less pronounced to coagulate under the reaction temperatures (100-300 °C) and demonstrated an impressive promotion in the deNO_x performance and SCR activity window. The intrinsic activity of the prepared catalysts measured under differential reaction conditions and the effect of morphological features will be discussed in the presentation.

Bimetallic Pt-Pd Catalyst kinetics for CO and C_3H_6 Oxidation in Diesel Oxidation Catalysts (DOC)

Melanie J. Hazlett and William S. Epling
University of Houston, Houston, TX (USA)
*wsepling@uh.edu

Introduction

Today's diesel engines are more fuel efficient than gasoline engines, but even so increases in fuel economy are needed. With this in mind and due to increasingly stringent environmental policies on NO_X, hydrocarbons (HC), CO, CO_2 and particulate matter (PM) emissions from diesel engines, low temperature combustion (LTC) engines have been developed. The emissions from these LTC engines have significantly lower concentrations of emitted NO_X (by a factor of 6) and PM (by a factor of 8), however the HC and CO concentrations are much higher, by factors of 20 and 7 respectively. Also the exhaust temperatures are 40-70°C lower. With these changes, significantly more emphasis needs to be put on the activity of the oxidation catalyst function and low temperature NO_X reduction performance.

Diesel oxidation catalysts (DOCs) are designed to oxidize CO, NO, and HCs, and typically contain Pt and Pd. It has been shown that adding Pd to Pt based DOCs leads to less Pt sintering relative to monometallic Pt catalysts [1]. Pt is a much better NO oxidation catalyst than Pd, and Pd is less sensitive to CO poisoning than Pt [2]. Non-uniformly distributed oxidation catalysts have been studied [3], where the premise for DOCs is axially distributing active sites will minimize the amount of metal used to meet conversion targets. A non-linear relationship between oxidation performance and Pt or Pd content has been observed, with different minimum Pt:Pd ratios achieving low light-off temperatures for different compounds [4]. With these features, there is a unique possibility that DOCs can be designed with gradients in Pt and Pd amounts, as well as ratios, that could lead to improved low temperature performance.

Methodology

Pt, Pd, and different ratios of Pt-Pd bimetallic catalysts, supported on Al_2O_3, were prepared via incipient wetness impregnation. For the data to be described, all catalysts are based on 1 wt% Pt catalyst loading, with a molar equivalent of Pd considered (0.55 wt%). Catalysts were dried overnight and calcined for 4 hours at 550°C. Catalysts were pressed and sieved to 40-60 mesh, and placed in a quartz tube packed bed reactor with SiO_2 pellets of same size to minimize temperature gradients. Temperature programmed oxidation (TPO) experiments were conducted with CO (1000-4000ppm), and C_3H_6 (500-2000 ppm) with excess O_2 (6-10 vol%) from 100-300°C. Outlet concentrations were measured with a MKS MultiGas2030 FTIR gas analyzer. Results were modeled as a PFR, with low conversion (1-5%) data used to calculate kinetic parameters (activation energies and frequency factors).

Results and Discussion

The CO and C_3H_6 kinetics for monometallic Pt and Pd have been tested and modeled, and we are currently characterizing the kinetics for various bimetallic Pt-Pd catalysts with varying Pt:Pd ratios. We started with the Voltz kinetics [5] as a basis. Using these, CO oxidation kinetics calculated for the Pt catalyst using LHHW kinetics ($r = kC_{CO}C_{O2}^{1/2}/(1+KC_{CO})^2$) show good agreement with experimental data, although we did modify the mechanism to include dissociative oxygen adsorption instead of molecular adsorption. Oxidation kinetics for Pd were also modified, since PdO is the assumed active state rather than metallic Pd. This also led to good fits between experimental and modeling results. At this stage we are characterizing kinetic and performance data from the bimetallic catalysts, and these will be included in the presentation, with insight into designs for improved CO and HC oxidation included.

In terms of performance, a 1:1 ratio of Pt:Pd typically proved to have the best T_{50} value (temperature where 50% conversion was achieved). This was true for CO oxidation, C_3H_6 oxidation and the oxidation of both when added together. However, different ratios led to the best T_{90} values. Again, this suggests a possible optimized design to attain the best performance with minimum precious metal content.

Overall, the goal of this research is designing a better DOC (reduction in precious metal content or improved performance), by taking advantage of the effects of different Pt-Pd ratios. Kinetic parameters calculated can be used in a monolith reactor model to determine optimal axial distribution of PGM.

REFERENCES

1. J.C. Summers, and L.L. Hegedus, *Journal of Catalysis* 51(1978)185.
2. R. Chen, Z. Chen, B. Ma, X. Hao, N. Kapur, J. Hyun, K. Cho, and B. Shan, *Computational and Theoretical Chemistry* 987(2012)77.
3. Y.D. Kim, S.J. Jeong, and W.S. Kim, *Chemical Engineering Science* 64(2009)1373.
4. M. Skoglundh, L.O. Lowendahl, and J.E. Ottersted, J.E. *Applied Catalysis* 77(1991)9.
5. S. E. Voltz, C. R. Morgan, D. Liederman and S. M. Jacob, *Industrial & Engineering Chemistry Product Research and Development,* 12(1973)294.

377116 Characteristics of Pt-BaO/Mgal Mixed Oxides for NOx Storage-Reduction Catalysis

Abstract:

Efficient NOx abatement catalyst operated under lean condition must be developed to promote the use of more fuel-efficient diesel and/or lean-burn engines since conventional three-way catalyst does not work properly. For this purpose, selective catalytic reduction (SCR) catalyst or NOx storage-reduction (NSR) catalyst were commercially being used. NSR technology, which is more appropriate for the small-sized engine, primarily consisted of noble metals such as Pt, alkali earth metal (e.g. Ba) and support (e.g. Al_2O_3 or $MgAl_2O_4$). It was demonstrated that $MgAl_2O_4$ support provided the improved NOx storage activity at higher temperature compared with Al_2O_3 support, implying that support material can play a crucial role in affecting the stability of barium nitrate species.

Hydrotalcite (HT) material, which can be precursors to $MgAl_2O_4$ spinel structure, has received more attention due to its various advantages, such as large surface area, controllable acid-base properties, and the improved thermal stability. In this work, HT materials with different Mg/Al ratios were prepared, resulting in the different amount of $MgAl_2O_4$ and Al_2O_3 in the support after calcination. Then, Pt-BaO active components were loaded on these HT-derived supports, and the catalysts were extensively investigated by using NOx storage measurement and various analytical techniques. In addition, the effect of the promoter on the NOx storage activity in Pt-BaO/MgAl mixed oxides was also investigated.

Combination of activity measurement and characterization results indicates that MgAl mixed oxides supports give rise to the higher thermal stability of barium nitrates and Pt species as well as superior activity than one supported on Al_2O_3, moreover, can be optimized further for the high temperature NSR application by changing Mg/Al ratios in HT support. This work provides fundamental understanding for optimizing HT-derived $MgAl_2O_4$ support, which would be a strong candidate for the high temperature NSR application.

HYDROGEN PURIFICATION FOR FUEL CELLS:

Preferential CO Oxidation in Excess of Hydrogen

Oscar Eduardo Amador [a,*], Luis Fernando Córdoba [b], Julio Cesar Vargas [c]

[a] Department of Chemistry and Environmental Engineering, Faculty of Engineering, National University of Colombia, Cr 30 Cll 45, Bogotá, Colombia.
[b] Department of Chemistry and Environmental Engineering, Faculty of Engineering, National University of Colombia, Cr 30 Cll 45, Bogotá, Colombia.
[c] Department of Chemistry and Environmental Engineering, Faculty of Engineering, National University of Colombia, Cr 30 Cll 45, Bogotá, Colombia.

Abstract

In the present work we have studied the effects of the gold load, as well as the O2/CO ratio, using gold catalyst supported on cerium and zirconium oxides, on the catalytic performances in PROX reaction. The catalysts of Au-CeO2-ZrO2 were prepared by the pseudo sol-gel method from the thermal decomposition of a mixed resin precursor metal propionates. The synthesized catalysts were characterized by various techniques in order to determine the main physic-chemical and structural properties and then related with the results of selective CO oxidation reactions. The specific surface area of the catalysts was determined by N2 adsorption desorption at 77K and interpreted using the Brunauer - Emmet -Teller (BET). The crystalline phases and the determination of the degree of crystallinity of the synthesized materials were identified by X-ray diffraction (XRD). The texture, shape and size of the materials was studied by scanning electron microscopy - SEM and analyzed by elemental microcomposition energy dispersive spectroscopy X-ray - EDXS coupled to SEM.
The catalytic activity and selectivity measurements in PROX over the studied gold catalysts were carried out using a quartz glass U-shaped reactor, equipped with a temperature programmed controller. The reactor effluent was analyzed on-line by an assembly of two, in series, gas chromatographs. The reaction was evaluated in a temperature range of 70-300 ° C, the catalytic tests were performed at ensuring higher GHSV 40000 h-1 hourly space velocity, the activity, hydrogen yield and products distribution was analyzed.

Keywords
ProxCO, Preferential CO Oxidation, Direct Anionic Exchange Method, Gold Catalyst.

1. Introduction

Most of the energy used today comes from fossil fuels that cause serious environmental problems, one of the most attractive options that is being considered is the use of H_2 as fuel, both for issues of power generation (fuel cells) and for direct use in internal combustion engines (heat engines H_2. The production of electrical energy from H_2 feed directly electric motor engines, either by direct combustion or use in fuel cells. The H_2 is a clean way to generate power, as the only product obtained in the process is H_2O [1].

Combustion Engine.
$H_2 + \frac{1}{2} O_2 \rightarrow H_2O$ + Heat generation

* Corresponding Author.
oeamadorc@unal.edu.co (O. Amador)

Fuel Cell
$H_2 + \frac{1}{2} O_2 \rightarrow H_2O$ + Electric Power

The reforming of methane with water vapor is one of the most important used for the production of H_2 with 95% usage. There is other methods, but in all these processes always involve the simultaneous CO formation [2]. Although the water gas shift reaction is effective to oxidize CO, efficient elimination of trace amounts of CO is required due to small amounts of CO present in the H_2 streams poison the Pt electrocatalysts of PEM fuel cells: an objective which can be achieved by the preferential oxidation reaction of CO (PROX), because its low-cost, effective, available method for reducing CO to a desired level (less than 100 ppm of CO) without excessive hydrogen consumption [3-4].

In this paper we present the influence of the active phase integration on the catalytic behavior of impregnated gold at two concentrations over a bimetallic mixed oxide support with formula $Ce_{0.65}Zr_{0.35}O_2$ in processes of selective oxidation of CO. Further, a trimetallic Nickel doped (Ce–Zr) mixed oxide ($Ce_{0.65}Zr_{0.3}Ni_{0.05}O_2$), were prepared and studied as catalysts. In order to understand the influence of the active phase and the role of this concentration on the selective oxidation of CO, the synthesized fresh mixed oxides were characterized by XRD and TPR and carbonaceous deposits produced during the reaction was quantified by TPO.

2. Experimental

2.1.Preparation of Catalysts

Ce–Zr support and Ce–Zr –Ni mixed oxide was synthesized using the pseudosol–gel method based on the thermal decomposition of metallic propionates[5]. The salts used were cerium(III) acetate hydrate, zirconium(IV) acetylacetonate and nickel(II).The support and mixed oxide prepared could be represented with next the formulas:

$Ce_{0.65}Zr_{0.35}O_2$ *(CeZr)*

$Ce_{0.65}Zr_{0.35}Ni_{0.05}O_2$ *(Ni/CeZr)*

The starting salts were dissolved separately in boiling propionic acid (0.12 mol/L). The solutions were mixed and the solvent was evaporated until obtaining a resin, which was calcined in air at 600 °C over 3 h (2 °C /min).

The impregnation of gold on the support of CeZr was conducted by direct ion exchange method, since according to reported results [6] through this method of preparation predominant presence of gold metal on the surface.

Because this method does not change the pH, the deposition of gold species is a function of the concentration of absorbable species. Analysis of mass spectrometry and UV-Visible (UV-VIS) spectroscopy reported [7] show that is favored and favorable deposition prevails species [8] in solutions of concentration 10^{-4} M and 10^{-3} M temperatures near 70 ° C, under these conditions we have obtained the best results for this support.

The crystalline structure of mixed oxides was determined by X-ray diffraction with Brucker D8 Advanced apparatus, Cu Kα radiation, in an interval 2θ = 10° up to 2θ = 100°.

The reducibility was studied by thermo-programmed reduction (TPR) of a 0.10 g sample, in a hydrogen/argon mixture (50 mL/min, 3.85% (v/v)). The temperature was increased at 15 °C/min from room temperature to 900 °C.

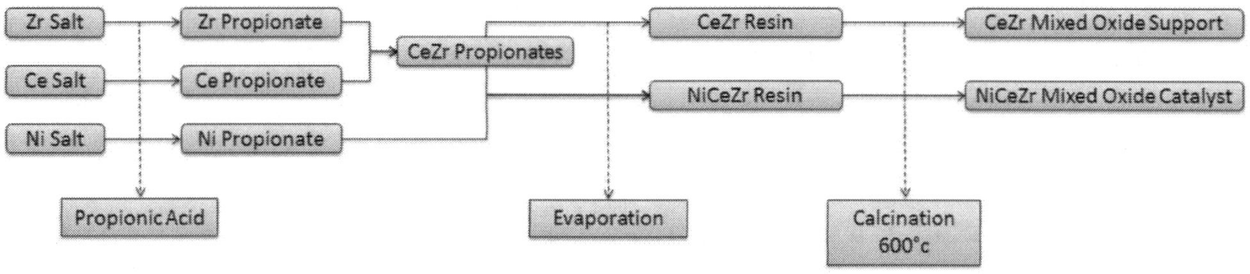

Fig1: Global catalyst preparation procedure following the sol-gel like method.

Fig2: Global catalyst preparation procedure following the DAE like method.

2.2.Catalyst Test

The CO selective oxidation was conducted in a tubular quartz reactor of fixed bed with 6 mm inner diameter, at atmospheric pressure in the temperature range 25 ° C - 250 ° C with 0.100 g of catalyst, the heating system consists of a cylindrical electric furnace. The reactant feed was a mixture of helium (purity 99.9995%), hydrogen, CO and synthetic air (80% N2 + 20% O2). In the experiment a total gas flow of 100 ml / min was handled. In the first test a concentration of 1.5% O2 in flux and for the second test of 3.0% was handled. The catalyst was heated to 300 ° C under a flow of 60 ml / min of H2 for one hour, this in order to simulate a reducing atmosphere and generate the activation of the catalyst, then with this reducing atmosphere the temperature was initial catalytic test, once obtained this temperature, the mixture of gases (He, CO, H2 and air) was introduced and the temperature was increased to that of the catalytic test. The temperature was increased in stages 25, 50, 100, 150, 200 and 250 ° C. At the end of the catalytic test gas flow was stopped and the catalyst was cooled under a stream of He, and stored for further characterization.

The gas reaction products were analyzed on two chromatograph connected in series; the first was a HP 5890 series II GC equipped with a HP-PLOT / Q column (0.53 mm ID and 30 m in length, Agilent Technologies, USA) in which CO2 is analyzed, and the second column GC equipped with HP 5890 A 5 A Molecular grid (Agilent Technologies, USA) for H2, CH4, and CO.

The catalytic tests were performed at ensuring higher GHSV 40000 h-1 hourly space velocity. To analyze the catalytic activity, selectivity and yield to hydrogen production and distribution of products, considering the effect of the charge on the metal (Au) and the effect of the concentration of O2 in the feed mixture.

2.3.Catalytic activity and selectivity

The analysis of the effluent gas was performed by online gas chromatography.

The conversion of CO to CO2 is calculated from the following equation:

$$X_{CO}(\%) = \left(1 - \frac{[CO]_{out}}{[CO]_{in}}\right) * 100$$

Where CO_{in} and CO_{out} are the initial and outlet CO concentration, respectively. The selectivity of oxygen to oxidize CO is defined as follows:

$$S(\%) = 0.5 * \frac{[CO]_{in} - [CO]_{out}}{[O_2]_{in} - [O_2]_{out}} * 100$$

3. Analysis and Discussion

The structural and morphological properties of catalysts Gold supported on Cerium-zirconium oxides were investigated and discussed in detail in a previous work by Laguna et al. Characterization was carried by several techniques (N2 adsorption,

XRF, XRD, Raman Spectroscopy, Techniques SEM and H2-TPR. [11] also found that the DAE method is more suitable than those methods as DP, IWI for the preparation of gold catalysts supported on oxides of cerium-zirconium, allowing a higher fraction of the species of nano-sized gold to be exposed on the catalyst surface [8].

In the case of the Au / CeZr10, there is a high-temperature reduction peak with maximum at 746 ° C. This signal has a very broad shoulder at lower temperature, which extends until 300 ° C. Finally, a low-temperature reduction process, around 145 ° C, Which is not observed for the CeZr support, Appears [11]. The resulting behavior is related to the method of preparation, which leads to the different sizes of the gold particles and improving the capacity reduction of the mixed oxide.

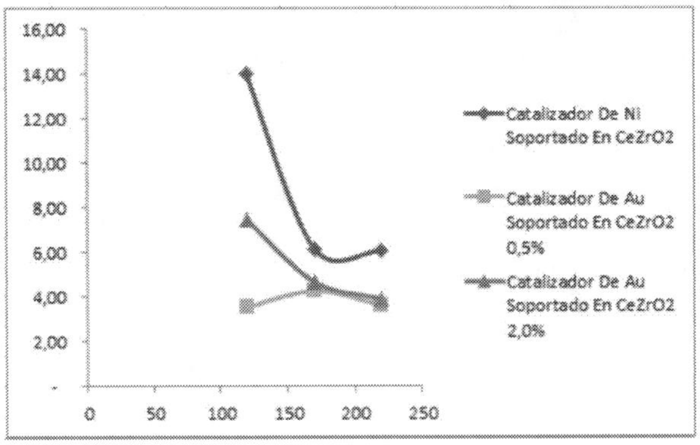

Fig 3. Selectivity (%) vs. Temperature (°C). Feed: 1.0% CO, 1.5% O2, 50.0% H2 and He of Balance..

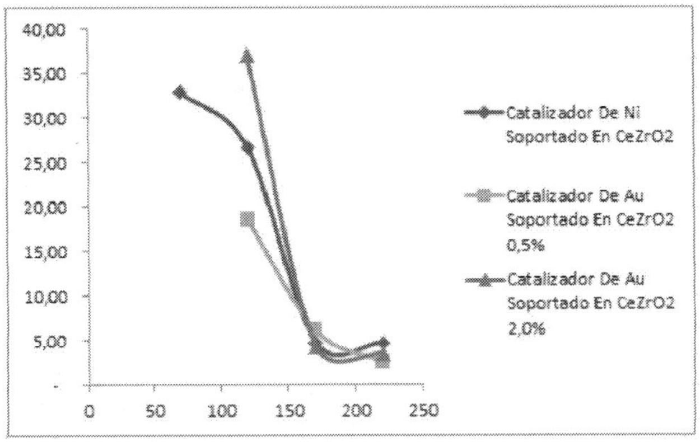

Fig 4. Selectivity (%) vs. Temperature (°C). Feed: 1.0% CO, 3.0% O2, 50.0% H2 and He of Balance.

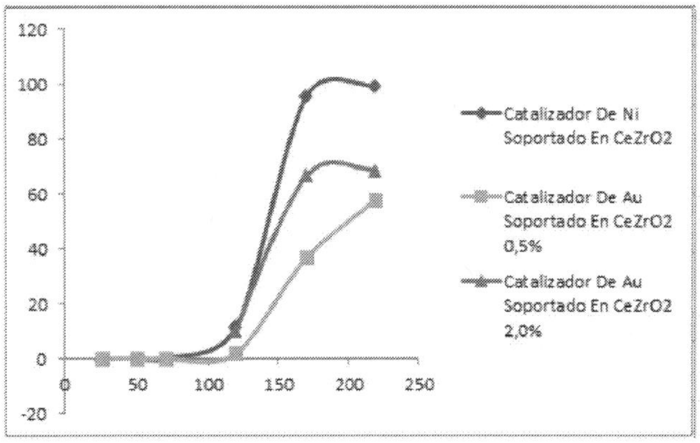

Fig 5. Conversion (%) vs. Temperature (°C). Feed: 1.0% CO, 1.5% O2, 50.0% H2 and He of Balance..

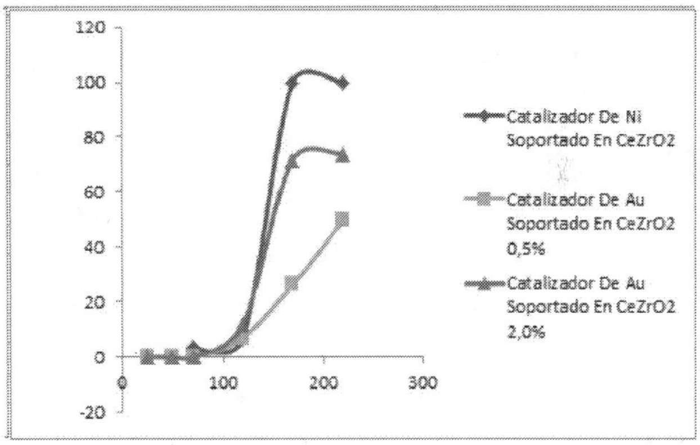

Fig 6. Conversion (%) vs. Temperature (°C). Feed: 1.0% CO, 3.0% O2, 50.0% H2 and He of Balance.

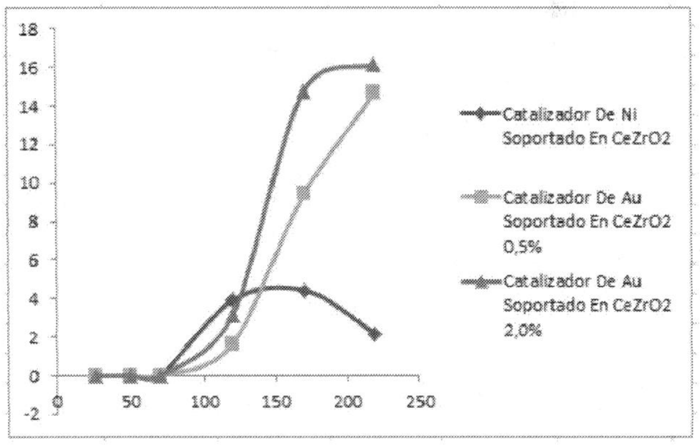

Fig 7. Conversion (ppm) vs. Temperature (°C) of CO_2 Catalyzed. Feed: 1.0% CO, 1.5% O2, 50.0% H2 and He of Balance.

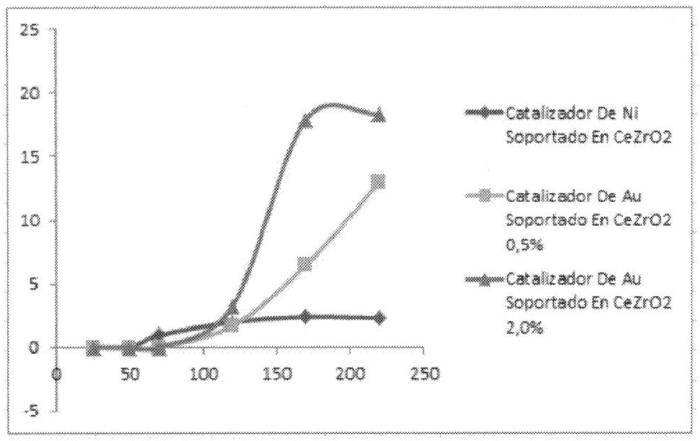

Fig 8. Conversion (ppm) vs. Temperature (°C) of CO_2 Catalyzed. Feed: 1.0% CO, 3.0% O2, 50.0% H2 and He of Balance.

Figures 3 and 4 show selective activity toward the production of CO_2 0.5% Au/CeZr, 2.0% Au/CeZr and NiCeZr catalyst prepared by the method DAE for the selective oxidation of CO. Figures 5 and 6 show the obtained conversion to CO_2 production of 0.5% Au/CeZr, 2.0% Au/CeZr and NiCeZr catalyst prepared by the DAE method for the selective oxidation of CO. Likewise by Figures 7 and 8 we can observe the behavior of the oxidation of CO to CO_2, and this varies as a function of increasing temperature and the change in oxygen concentration in the feed streams.

It can be observed that the conversion and selectivity depend strongly on the nature of the catalyst. In fact, none of the catalysts produces measurable catalytic activity at temperatures below 120 °C. It was observed that the selectivity decreased progressively from 14% for the first cup of food stabilizing close to 4% and for the second cup had to be decreased from 40% to re-stabilize around 4%, this phenomenon presented with increasing progressive temperature.

Among the three catalysts tested, the NiCeZr is the most active, while the Au/CeZr is the most selective for the oxidation of CO and that the largest concentration of tested gold. The method of preparation strongly influences the catalytic performance and greater efficiency to the production of CO_2 was obtained with the DAE method.

The maximum conversion of CO achieved on the sample 2.0 Au / CeZr was 74% with 4% selectivity, 220 ◦C, the highest selectivity of 37% was achieved for a CO conversion of 12% and the test with the highest concentration of O2 (3%). For the sample of 0.5 Au / CeZr the highest conversion obtained was 58% with a selectivity of 4%, this for a feed of 1.5% in oxygen flow and greater selectivity was 19% at a conversion of 7% for run with higher oxygen concentration (3.0%). As the catalyst NiCeZr obtained the highest conversion was 100% at a selectivity of 4.75% and the maximum selectivity obtained was 33% to 3% conversion results for both oxygen supply flow of 3%.

4. Conclusions

Considering that high concentrations of hydrogen show no significant effects on the selective oxidation of CO [12], we can assign the significant increase in selectivity (passing from 7% to 37%) to increase the oxygen concentration in the feed stream.

The temperature increase during the tests developed to catalysts where the active phase is gold causes a significant increase in the selective oxidation of CO to CO2 increase, likewise the increase is also benefited by the increased concentration of O2 in the feed stream. (Figure 7 and 8).

As the catalyst of NiCeZr which has 5% of active phase (Ni) very low selectivity was found in the oxidation of CO to CO2, because due to the high concentrations of H2 this catalyst produced the oxidation of hydrogen to water and the methanation of carbon monoxide to methane. Although this catalyst showed the best results in terms of conversion is concerned, the results of selectivity to CO2 is very low.

Acknowledgements

Thanks to the Department of Chemical and Environmental Engineering of the National University of Colombia by the flanged financial support for the project, also thanks to Mr. Ricardo Cortez, for their cooperation in conducting the investigation.

References

[1] Linares, J. Moratilla, B. Ins. de Ing. de España (2007) 25-86.

[2] A. Iulianellia, G. Manzolinib, M. De Falcoc, S. Campanarib, T. Longoa, S. Liguoria, d, A. Basilea. International Journal of Hydrogen Energy. 20 (2009) 11514-11524

[3] Jenny E. Hu, Zhufang Liu, Bryan W. Eichhorn, Gregory S. Jackson, International Journal of Hydrogen Energy. 15, (2012), 11268-11275

[4] Seyyed M. Mousavi, Siew H. Chan, Electrochimica Acta. 93, (2013) 334-345

[5] F. Romero-Sarria, Julio C. Vargas, Anne C. Roger, Alain Kiennemann , Catalysis Today. (2008) 149-153.

[6] Ivanova, S., Petit, C., Pitchon, V., Applied Catalysis A: General, 2004. 267: p. 191-201.

[7] Ivanova, S., Formation de nanoparticles d'or supportèes : De la préparation à la réactivité catalytique.Tesis Doctoral, 2004, Université Louis Pasteur de Strasbourg. Strasbourg. p. 221.

[8] Aguirre, L. Estudio de la influencia del método de síntesis en las propiedades catalíticas del oro soportado en la activación del glicerol. Tesis de Maestria, 2010, Universidad Nacional de Colombia. P 35-36.

[9] T.R. Reina, E. Papadopoulou, S. Palma, S. Ivanova, M.A. Centeno, T. Ioannides, J.A. Odriozola. Applied Catalysis B: Environmental, 150–151 (2014) 554-563.

[10] Xuemei Liao, Wei Chu, Xiaoyan Dai, Véronique Pitchon. Applied Catalysis A: General 449 (2012) 131-138.

[11] O.H. Laguna, F. Romero Sarria, M.A. Centeno, J.A. Odriozola. Journal of Catalysis. 2 (2010) 360-370

[12] Cécile Rossignol, Sandrine Arrii, Franck Morfin, Laurent Piccolo, Valérie Caps, Jean-Luc Rousset. Journal of Catalysis. 2 (2005) 476-483.

[13] G. Avgouropoulos, T. Applied Catalyst. A:Gen. 244 (2003) 155-167.

[14] D.L. Trimm, Z.I. Onsan, Catalyst Rev. Sci. Eng 43 (2001) 31-84.

394219 Reverse Electrodialysis: Sustainable Energy from Hydraulic Fracturing Water Recycle

Abstract:

Due to the growing limitations in dependence on fossil fuels, attention has been turned towards searching for alternative means for harvesting energy. One emerging source is energy produced from natural salinity-gradients between freshwater and seawater. Two different processes used to accomplish this are pressure-retarded osmosis and reverse electrodialysis. Significant amounts of research are currently under way regarding pressure-retarded osmosis, however very few studies have been conducted on reverse electrodialysis due to heavy resistance in the diluate chamber of a reverse electrodialysis cell. The focus of this research is to reduce this resistance by using wafers and ion-exchange beads in order to maximize net power gain. Surrogate brackish/saltwater feed streams were used to measure overall power density.

By reducing resistance to levels previously deemed impossible, we are exploring the fundamental limits of separation technology. This research proposes an innovative resolution to high resistance levels through the use of ion exchange wafers to promote ionic movement. Here we show that such wafers have a great impact on the reverse electrodialysis process, and succeeded in achieving a sizeable gain in power density.

Traditional applications include harnessing energy from areas where freshwater bodies meet saltwater, as in an estuary. The future of reverse electrodialysis, however, lies in fracking. Produced water from a fracking well site in Oklahoma was used for power density calculations under realistic conditions. Preliminary results show that our system is capable of producing power densities of up to $10W/m^2$, which are five times higher than the current state-of-the-art. Environmental concerns will be all but eliminated if fracking can evolve to be a sort of cyclic process, where brackish fracking water is converted to energy to, in turn, power the overall operation. This poster presentation will model the power increase of electrode ionization and comment on future potential for improvement, especially in the realm of hydraulic fracturing applications.

394215 Modeling of the Aerobic Cometabolic Transformation of Chlorinated Ethenes By the Mycobacterium Elw-1

Abstract:

ELW1 is a recently isolated aerobic bacterium, in the mycobacterium family, that uses isobutene as a growth substrate. ELW1 was also discovered to have the ability to cometabolically transform substances with chemical structures similar to isobutene, such as chlorinated ethenes, into chloride ion and water. Chlorinated ethenes are notorious carcinogens, and groundwater pollutants, whose remediation has been a worldwide problem for over 40 years. Bioremediation of chlorinated ethenes by the augmentation of contaminated soils with microorganisms such as ELW1 is a proposed method of reducing contaminant concentrations in groundwater. To determine the viability of using ELW1 for groundwater remediation, an initial kinetic analysis of transformation rates must be performed. The focus of this research was to determine the rates of contaminant transformation when this organism was exposed to high concentrations of chlorinated ethenes such as trichloroethylene (TCE), *cis*-dichloroethene (c-DCE), and vinyl chloride (VC).

The ELW1 culture was grown and sustained in both batch culture and steady-state chemostat environments with isobutene as a growth substrate on a minimal growth media. Cells that were used for experiments were concentrated by centrifugation, washed with excess media and placed in a new media solution containing high concentrations of the compound to be tested. Initial rates of transformation were determined using gas chromatography to analyze 100μL samples of headspace gas over a period of one to three days to determine the rate of compound loss. The rates at which ELW1 can utilize isobutene were compared with rates of chlorinated ethene transformation, which overall quantifies the practicality of using this organism bioremediation purposes in the field.

Kinetic parameters of these cometabolic processes were determined using an analytical equation based on a modified Monod equation in order to account for decrease in cell activity over time. Results of this investigation indicate that the ELW-1 culture would be effective for the bioremediation of c-DCE and VC, but would be less effective for treated TCE.

394214 The Enzymatic Hydrolysis of Alfalfa Stalks for Use As a Biofuel Resource

Abstract:

Most biofuel resources, especially plant sources, have two key issues: farming space and slow growth. After overexpressing the sucrose phosphate synthase gene (SPS) in Medicago sativa (alfalfa) plants, large differences in phenotype and growth rate were seen in the plant. While the leaves of the alfalfa plants can be used for animal forage, the stems of the alfalfa can be fractionated into sugars that can then be fermented into bioethanols. This technique could solve the growth problem faced by bioethanol producers.

The purpose of the project was to fractionate and analyze the extractability of fermentable sugars from a control and several transgenic species that show overexpression in the SPS gene. Using a strong acid pre-treatment and enzymatic hydrolysis to break down the cellulose into glucose, the alfalfa stalks were fractionated then analyzed using ultraviolet spectroscopy with anthrone as an indicator for glucose content. Two of the three tested transgenic
species were shown to have more glucose content per gram than the control species (2-4% more glucose/gram). A further development of the overexpression technique can lead to faster growth, increased concentration of fermentable carbohydrates, and larger crop yields.

UNIVERSITY OF SOUTH CAROLINA
College of Engineering and Computing

SAGE
Power through Engineering

Synthesis and Characterization of FeAlCr via High Throughput Techniques

Randy Fang
Jonathan Kenneth Bunn and Dr. Jason Hattrick-Simpers
Department of Chemical Engineering
University of South Carolina
AICHE National Conference 2014
11/17/2014

UNIVERSITY OF SOUTH CAROLINA
Honors College

Fukushima Daiichi Nuclear Disaster

https://coto2.files.wordpress.com/2011/03/2-fuel-pellet-assembly.jpg

http://digginginthe clay.files.wordpress.com/2011/03/5525847859_e1934af238_o_d1.jpg

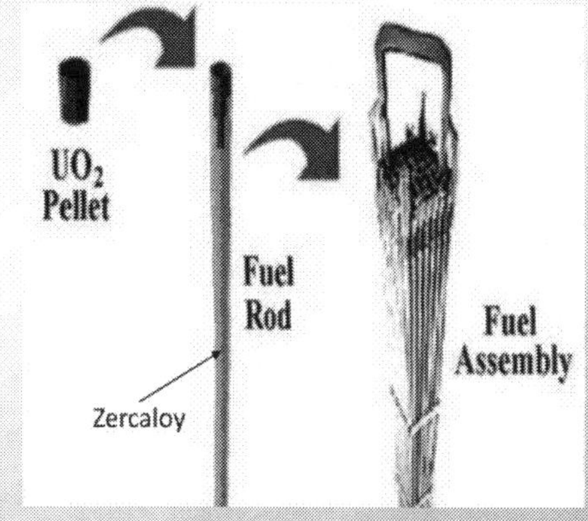

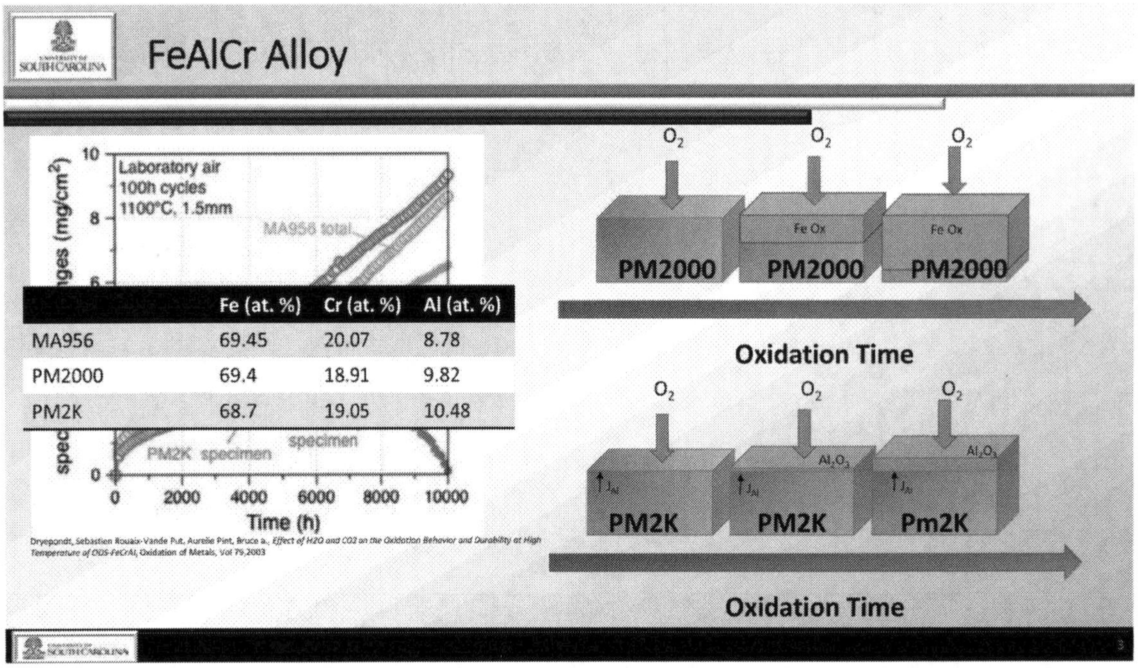

	Fe (at. %)	Cr (at. %)	Al (at. %)
MA956	69.45	20.07	8.78
PM2000	69.4	18.91	9.82
PM2K	68.7	19.05	10.48

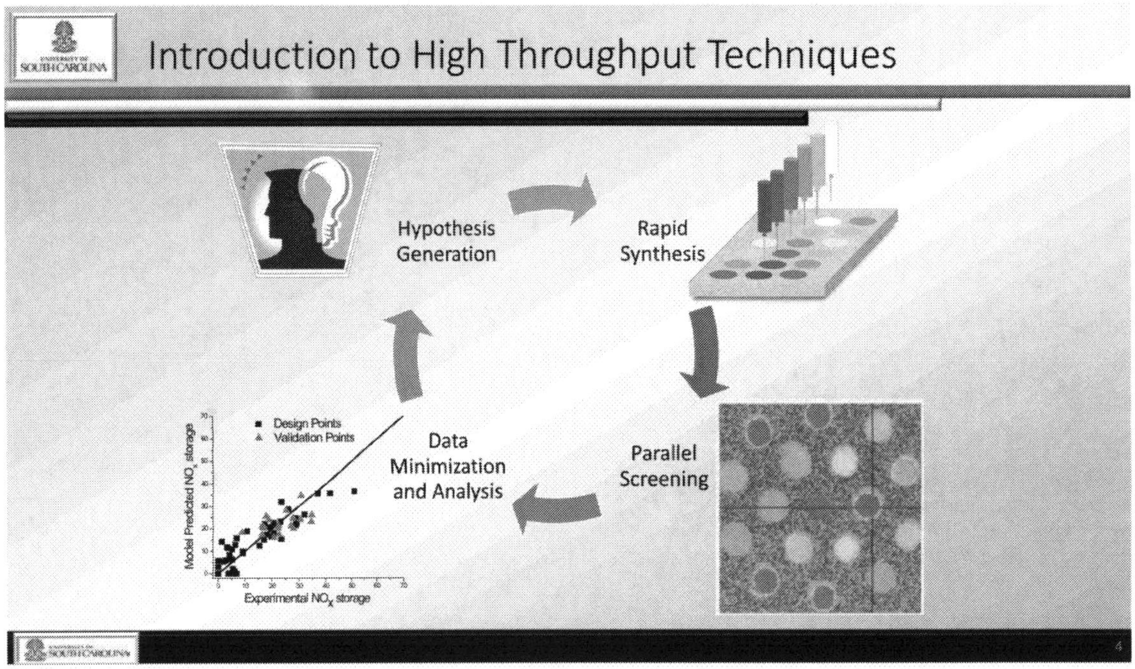

Research Objective

- **Implementing established high-throughput methods for FeAlCr high temperature oxidation resistance screening**
- Non-destructive characterization
- Gain comprehensive understanding of:
  - Time dynamics of oxide formation at high temperature
  - Role of metallurgical phase in promoting growth and continued growth of passivating oxide
- Identification of promising compositions for passivating oxide growth

5

High Throughput Approach to Explore the Ternary Phase Diagram

Qu Materials Science and Engineering A Vol 562 (2013)

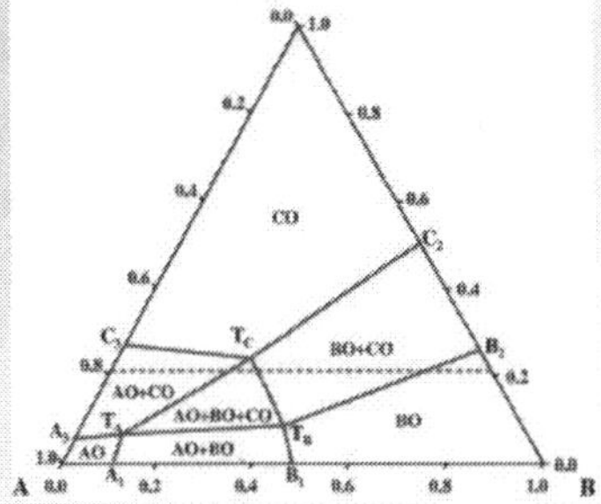

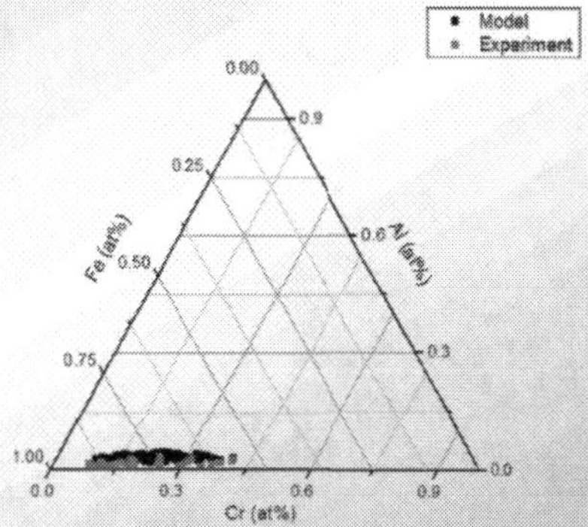

J. K. Bunn, C. J. Metting, and J. Hattrick-Simpers, "A Semi-Empirical Model for Tilted-Gun Planar Magnetron Sputtering Accounting for Chimney Shadowing", JOM, invited paper, Accepted January 2014 Special Issue

6

Sputtering Geometry and Sample Locations

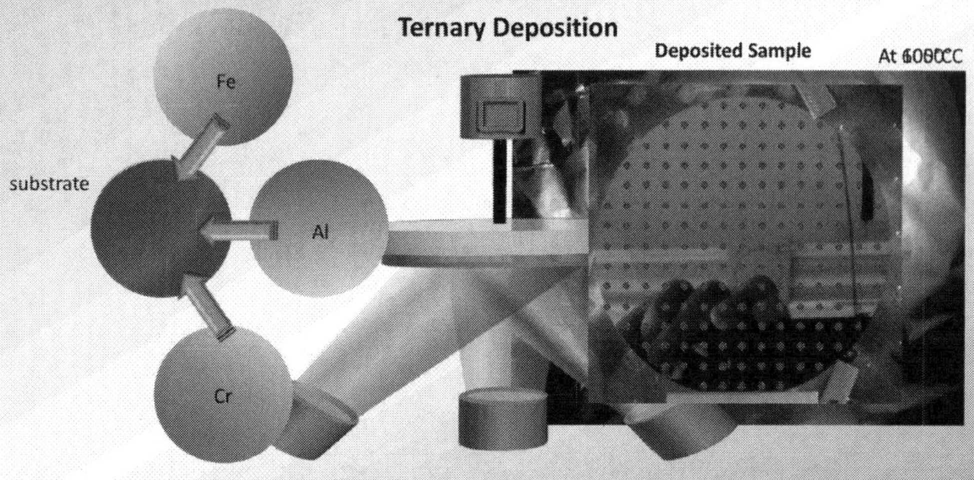

Magnetron Sputtering

- **5 Guns (only used 3)**
- **DC/RF power sources**
- **Ultra high vacuum**
- **Deposition temperature at room temperature**

Experimental Design for High Temperature

Sample fabrication
-12 FeAlCr samples prepared on Al_2O_3 substrate and Ti with adhesion layer
-1 micron nominal thickness

Heat treatment
- Tube furnace treated
- Treated at 1050˚C
- Air quenched

After deposition

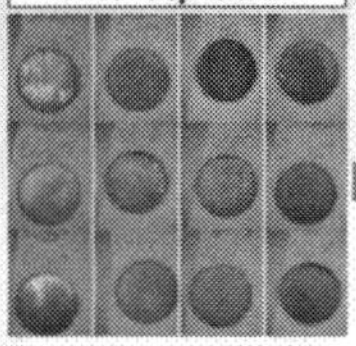

After heat treatments

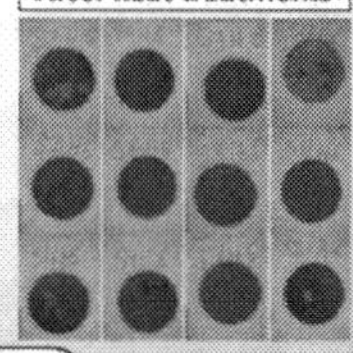

Cycled 7 times for cumulative heat treatments at 1050°C for:

-5 min -10 min
-20 min -40 min
-60 min -120 min
-240 min -360 min

Characterization
-SEM/EDS
-XRD

Characterization Techniques
- **Raman**: Fe_2O_3, Fe_3O_4, and Al_2O_3
- **XRD**: FeCrAl, Surface Oxides, and the substrate
-**Optical Microscopy**

9

Oxide Formation at 5 minutes, 1050˚C

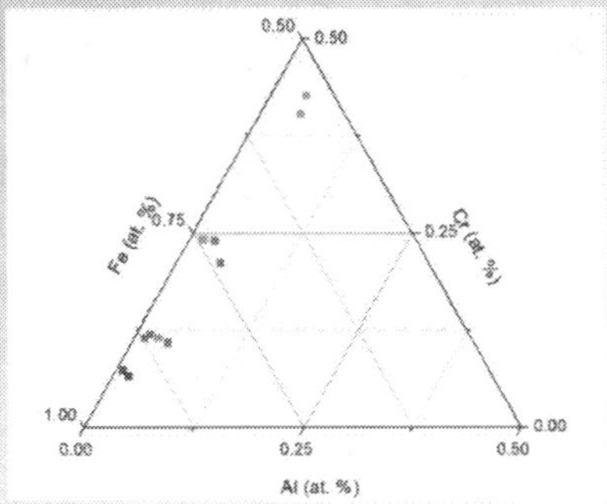

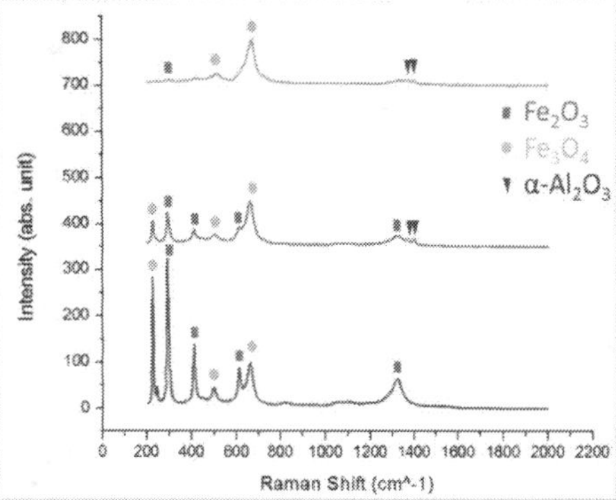

Blue: mostly Fe_2O_3, some Fe_3O_4
Red: mostly Fe_3O_4, some Fe_2O_3, some α-Al_2O_3
Green: very little Fe_2O_3, some α-Al_2O_3, some Fe_3O_4

- No α-Al_2O_3 growth was observed when the composition is Fe heavy

10

Oxide Formation at 360 minutes, 1050˚C

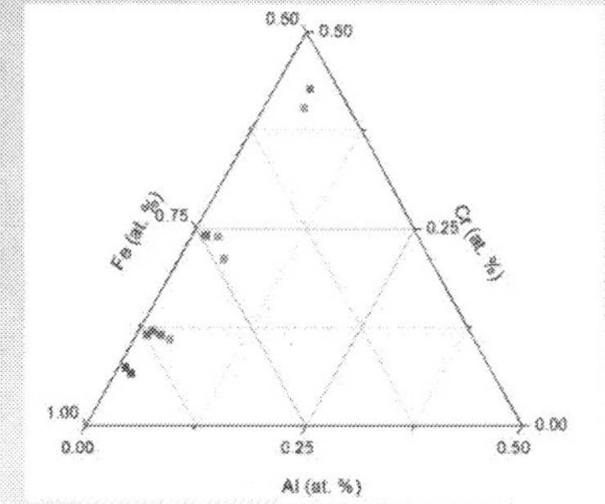

Blue: some Fe_2O_3 and Fe_3O_4, very little α-Al_2O_3
Red: some α-Al_2O_3, some Fe_3O_4
Green: majority α-Al_2O_3, some Fe_3O_4

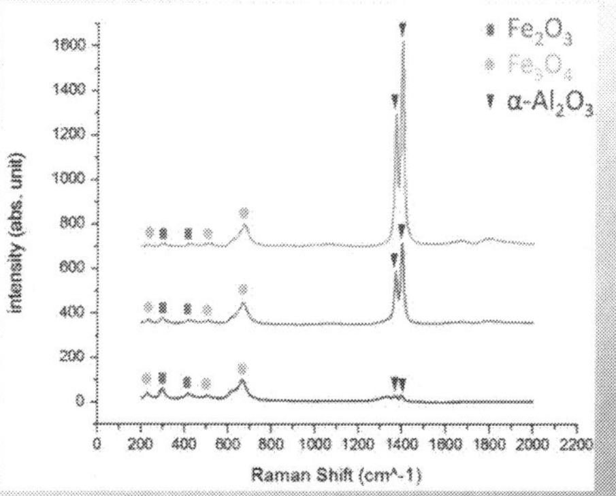

- Large amount of α-Al_2O_3 growth was observed at the Cr heavy region, indicating that Cr promotes the growth of the alumina oxides

Experimental Design at Low Temperature

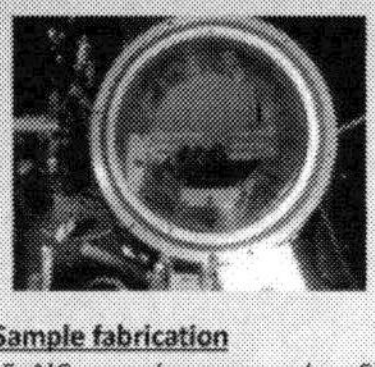

Sample fabrication
-FeAlCr samples prepared on Si substrate with Ti diffusion barrier
-1 micron nominal thickness

Cycled 6 times for cumulative heat treatments 1 hour each (with the exception of 600°C):

-room temperature	-200°C
-300°C	-400°C
-500°C	-600°C (for 2hr)

Characterization
-*in-situ* GIXRD

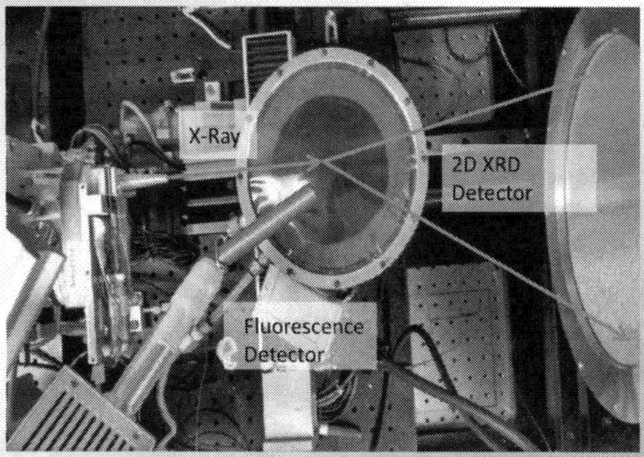

Oxide Growth Study through *in-situ* XRD

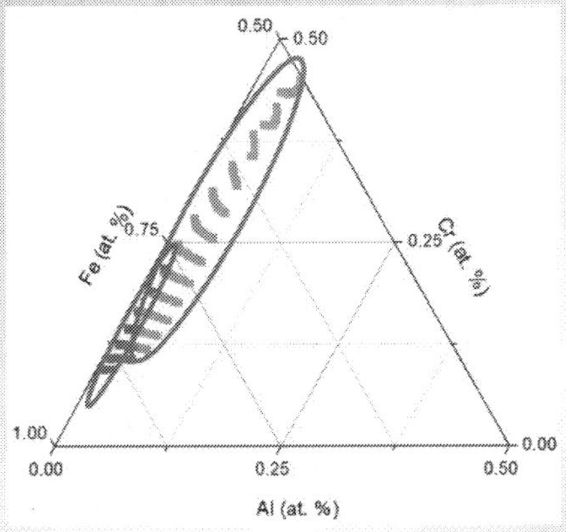

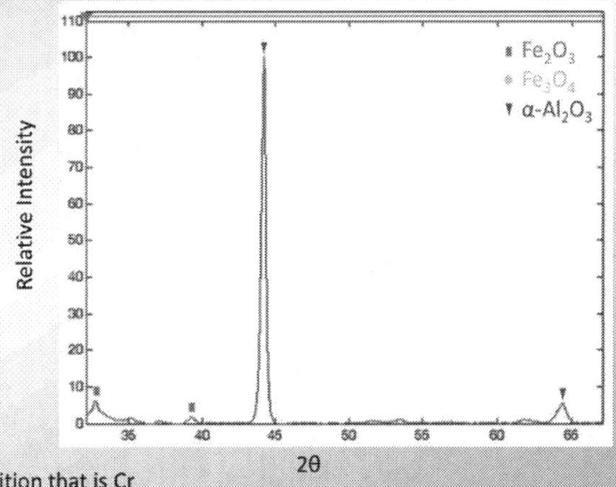

- Iron oxides are suppressed when the composition that is Cr rich (small amount are observed)

Comparison of Trends Between Low and High Temp Oxidation

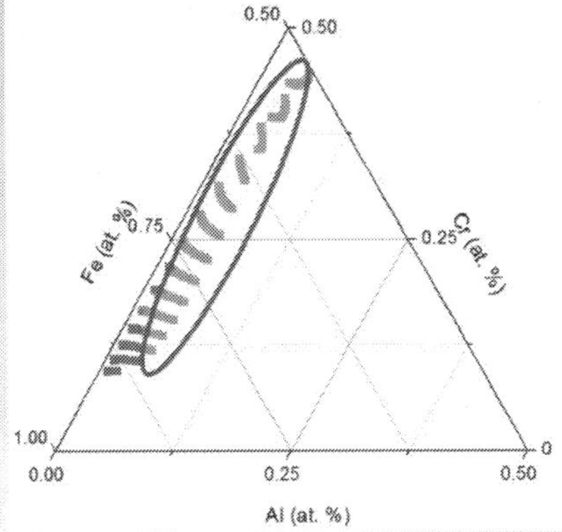

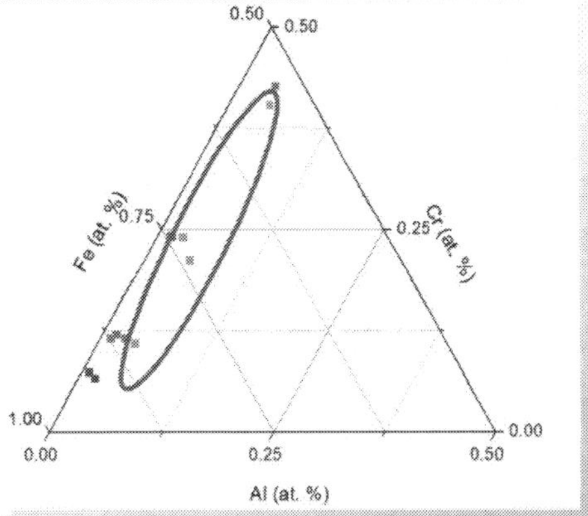

- XRD and Raman matched up where the more Al and Cr rich region are a lot easier to form α-Al_2O_3

Summary

- Probed the temporal effect of temperature on the metallurgic and oxide phase transitions
- A method for the **initial high-throughput and non-destructive screening** of FeCrAl thin film was developed
- **Cr promotes** α-Al_2O_3 growth on the surface of the samples to form a protective layer preventing further development of other oxides
  - This indicates that Cr rich FeCrAl are more favored to operate under high temperature conditions
- Oxidation studies indicate that severe oxide growth was not observed until 400°C
- The methodology can be applied to more complicated systems for high temperature oxidation resistant materials.
- Detailed studies of **oxidation and phase transitions of single compositions at different temperatures** will refine the results

15

Acknowledgement

- **Dr. Jason Hattrick-Simpers**
- **Jonathan Kenneth Bunn**
- **SmartState Center for Strategic Approaches to the Generation of Electricity**
- **USC Honors College (Science Undergraduate Research Fellowship Program)**
- **University of South Carolina**
  - **College of Engineering and Computing Science**

This project was funded in part by the South Carolina Honors College

16

UNIVERSITY OF SOUTH CAROLINA
College of Engineering and Computing

SAGE
Power through Engineering

Thank you for your Attention

Are there any questions?

17

Crossing Borders and Communities for Healthy Water

Joshua Gomez, Meng Zhou, Dr. Shuguang Deng

Chemical and Materials Engineering Department, New Mexico State University, Las Cruces, New Mexico, 88003

In Partnership with:

NM Water Resources Research Institute

NMSU Chemical Engineering Department

The New Mexico Alliance for Minority Participation

New Mexico Environment Department

Border Partners (Non-profit)

EPA 2020

National Science Foundation

New Mexico State University NM STATE

Puerto Palomas, Chihuahua, MX

- Located just south of Columbus, NM.
- Population of 4,688.
- Arsenic contamination (0.030-0.1 mg/L).
  - Horrible Carcinogen.
- Fluoride contamination (5-40 mg/L).
  - Bone Fluorosis and neurological damage.
- Concentration depends entirely on well water.
  - Naturally occurring concentrations of arsenic and fluoride exceed the permissible limits of Mexico and the U.S.

New Mexico State University NM STATE

New Mexico Transboundary Aquifers

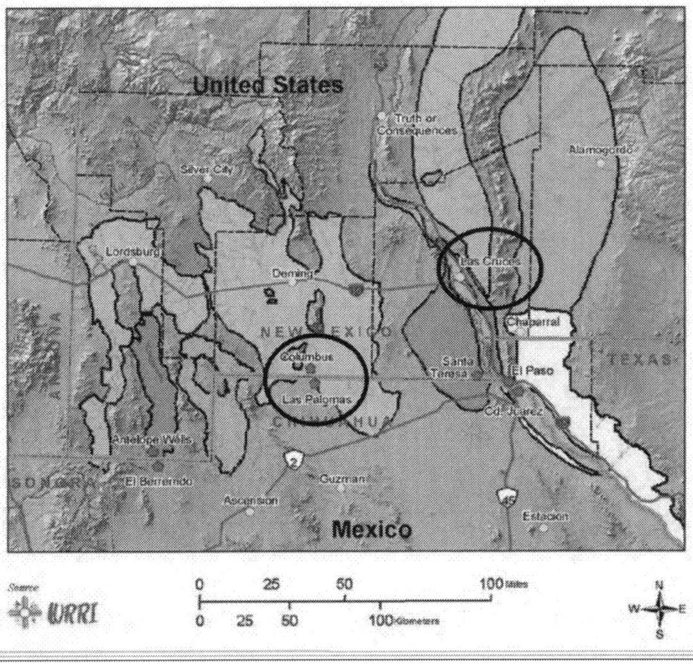

New Mexico State University

Proposed Solution

- Homemade water filter
  - Low cost <$15.00
  - Easy to assemble
  - Removes arsenic
  - Removes fluoride
  - Utilizing activated carbon (AC)
  - Utilizing activated alumina(AA)
    - Iron coated
  - Improves the taste

New Mexico State University

Filter Designs

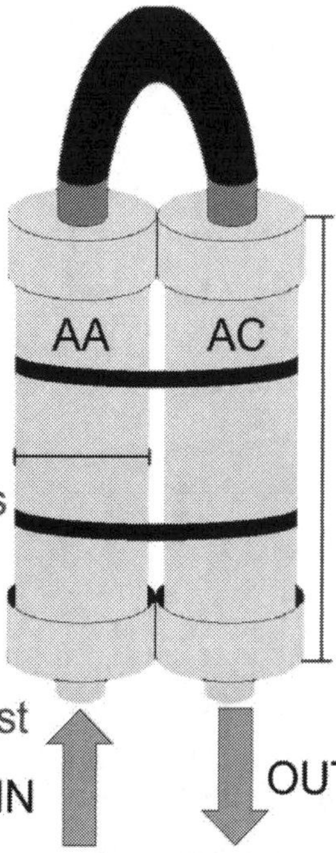

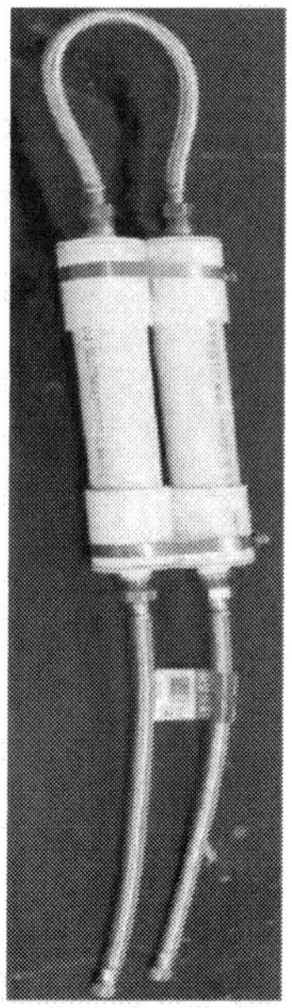

- 1 filter of the final design costs $5.00 to make.
- The Activated Alumina costs $1/kg.
- The Activated carbon used cost $35/lb.

New Mexico State University

Our Approach

NMSU demonstrated to Palomas science teachers how to assemble a simple water filter.

Science teachers Juan Carlos Huerta and Joel Carreon pose with the filter they assembled for installation at the Palomas primary school.

New Mexico State University

What's Working

Hall Environmental Laboratory, Albuquerque

CONTAMINANT	RESULTS	STANDARDS					
		U.S.	Exceeded	%	Mexico	Exceeded	%
Fluoride							
Unfiltered	4.90 mg/L	4.00 mg/L	Y	22.5	1.50 mg/L	Y	226.7
Filtered	0.240 mg/L	4.00 mg/L	N	-	1.50 mg/L	N	-
Arsenic							
Unfiltered	0.030 mg/L	0.01 mg/L	Y	200.0	0.02 mg/L	Y	20.0
Filtered	ND	0.01 mg/L	N	-	0.02 mg/L	N	-

ND = Not Detected

New Mexico State University

Optimization of Algae Derived Activated Carbon (AC)

- Low cost alternative to commercial activated carbon.
- More effective adsorbent than commercial carbon.
- Chlorella algae is used as a carbon source.
- Potassium Hydroxide and Iron (II) Chloride
  - Activating agent.
- Carbonization times of 500° C, 600° C, and 700° C.
- Shaker experiments to determine most effective adsorbent for Arsenic(III).

New Mexico State University

Carbon Experimental data

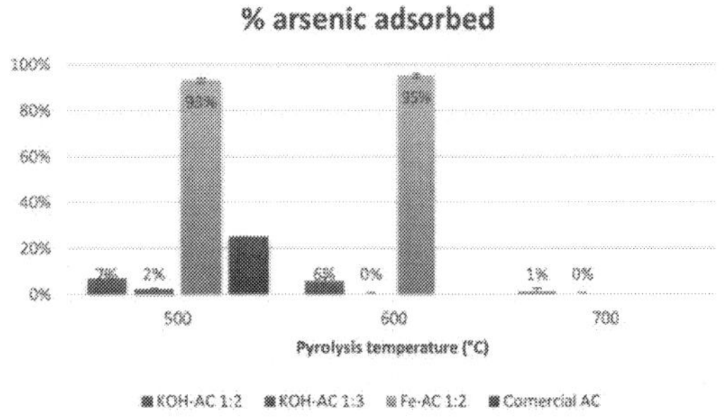

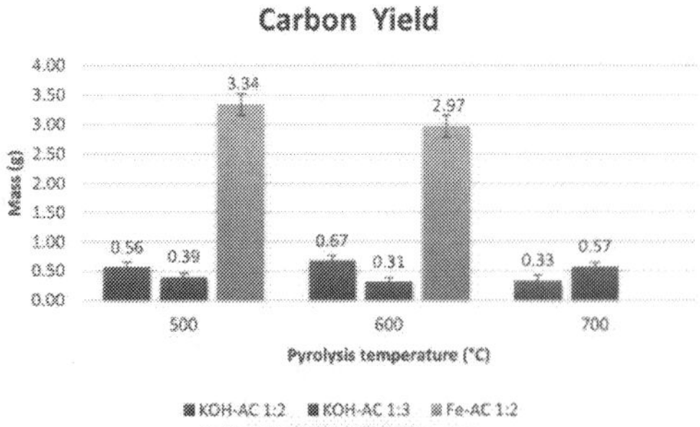

adsorption capacity (mg/g)			
	500°C	600°C	700°C
KOH-AC 1:2	0.082	0.067	0.010
KOH-AC 1:3	0.024	0.000	0.000
Fe-AC 1:2	1.143	1.168	-
Comercial AC	0.275	-	-

$$capacity = \frac{(InitialConcentration - FinalConcentration)Volume}{Mass\ of\ adsorbent}$$

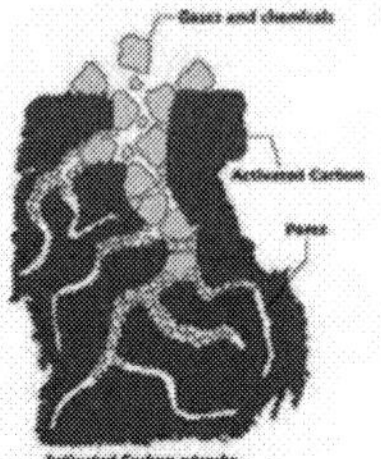

Faust, M.
http://www.carbonanswers.com/carbonadsorptionadsorption.html

New Mexico State University

X-Ray Difraction Pattern

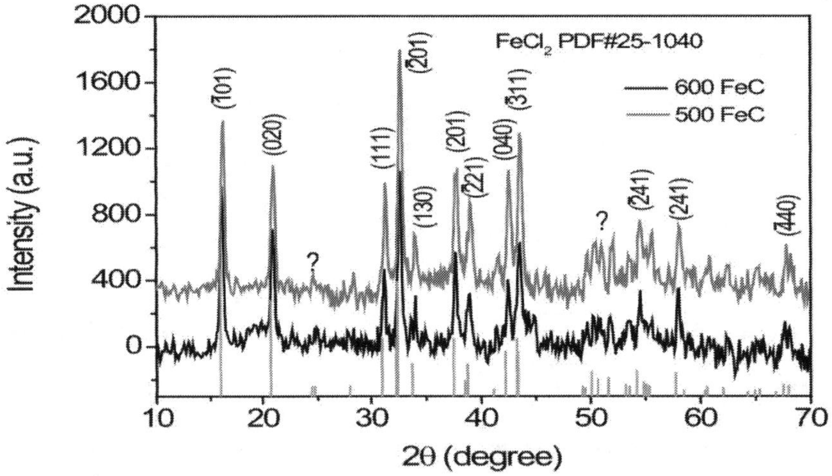

New Mexico State University

Scanning Electron Microscope Imaging

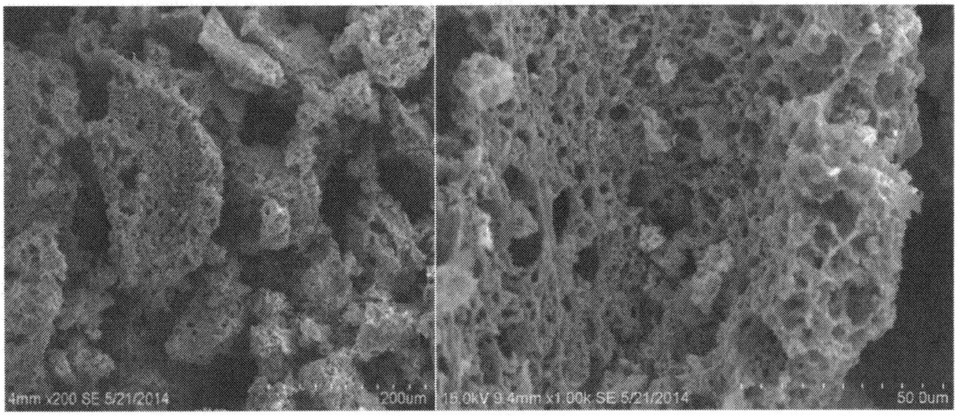

New Mexico State University

Unwashed FeCl(II) Activated Carbon Textural Properties

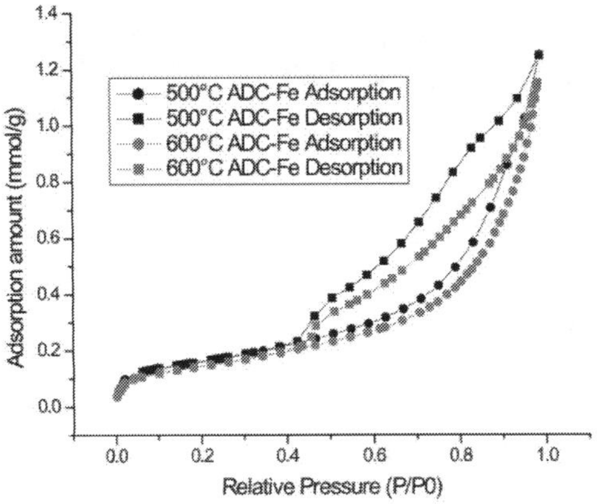

N_2 adsorption and desorption isothermal @ 77K for unwashed samples.

	500 ° C FeCl2/ADC	600 ° C FeCl2/ADC
BET surface area (m^2/g)	12.9893	11.6336
Average pore size (nm)	8.6408	6.5769
Pore volume (cm^3/g)	0.0394	0.0476

New Mexico State University

Washed FeCl(II) Activated Carbon Textural Properties

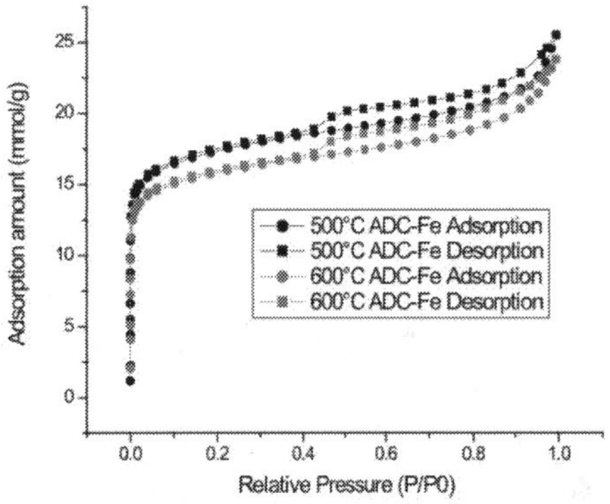

	500 ° C ADC (after washing)	600 ° C ADC (after washing
BET surface area (m^2/g)	1326.0368	1208.1831
Average pore size (nm)	7.8704	7.9340
Pore volume (cm^3/g)	0.3362	0.3305

N2 adsorption and desorption isothermal @ 77K for washed samples.

New Mexico State University

Shaker Experiments

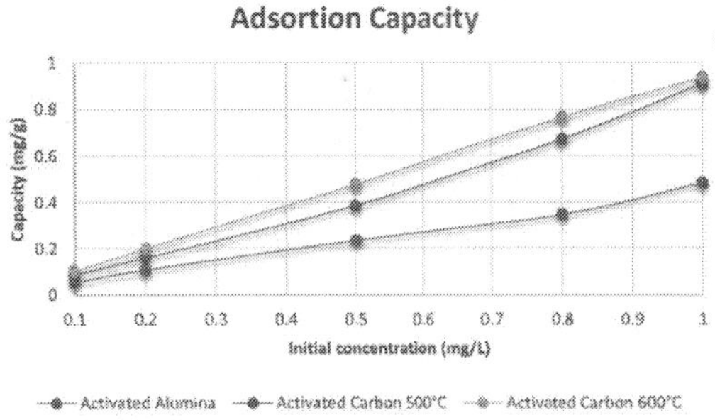

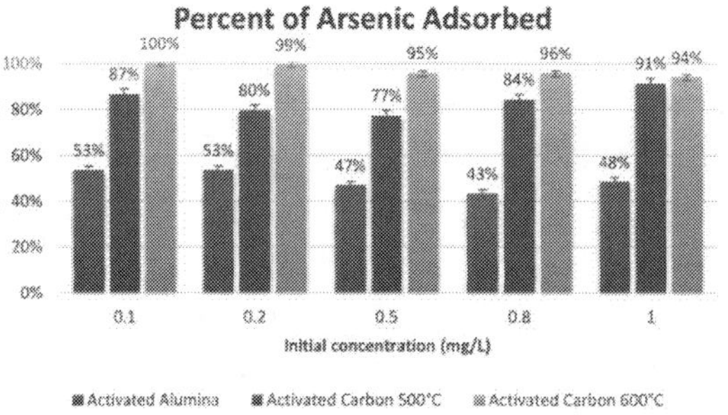

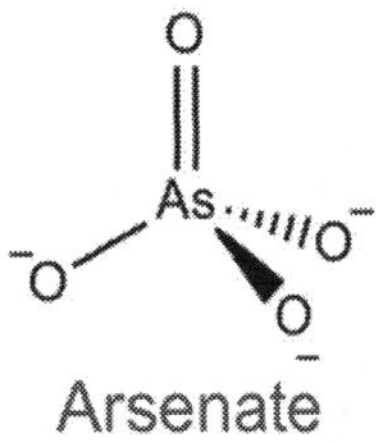

Arsenate

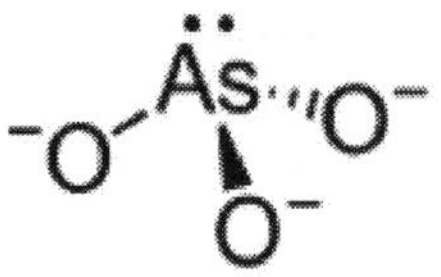

Arsenite

New Mexico State University

Conclusions

- Filters are still being made in Palomas by the teachers today and will continue to be made.
- The results of the shaker experiments and textural property data shows that the carbon activated with Iron (II) Chloride is a viable replacement.

New Mexico State University

Next Steps

- Further Development of the filter.
- Reach out to other communities.
- Advance research with optimized carbon.

New Mexico State University

Acknowledgments

- Dr. Shuguang Deng, Chemical & Materials Engineering dept.
- Meng Zhou, Chemical & Materials Engineering dept.
- Dr. David A. Rockstraw, Chemical & Materials Engineering dept.
- Erin Ward, Water Resources Research Institute.
- Mark Chidester, Chemical & Materials Engineering dept.
- Peter Edmunds, Border Partners.

NEW MEXICO AMP

NSF HRD # 1305011

Deng's Renewable Energy and Advanced Materials

New Mexico State University

Questions?

Email: jgomez38@nmsu.edu

New Mexico State University

394186 Ferroelectric BTO on Si (001) for High-Efficiency Solar Cell Heterostructures

Abstract:

Ferroelectric oxides such as barium titanate (BTO) have recently become a topic of interest in relation to thin-film solar cell production due to the material's stability at a wide range of temperatures, its spontaneous polarization, and the potentially low-cost of production. Integration of BTO on silicon could enable the creation of highly efficient solar cells, as BTO's ferroelectric properties encourage the detachment of photo excited electrons. We have previously demonstrated successful integration of BTO on silicon carbide (SiC) substrates, using a magnesium oxide (MgO) heteroepitaxial layer to enable single-crystal orientation of the BTO film deposited on SiC.[1] BTO has also been successfully grown on silicon substrates, but with the use of strontium titanate (STO) and MgO buffer layers to resist formation of SiO_2.[2] Integration of BTO on Si with a single MgO interface layer would decrease interfacial losses in a final solar cell device. Based on previous work with SiC, we believe a single interface can be possible by controlling the surface interactions during film growth.

In this research, MgO is grown on Si (001) substrates in an ultrahigh vacuum (UHV) environment using a number of growth methods including molecular beam epitaxy (MBE) and sputtering. In comparing differences in surface structure using reflection high-energy electron diffraction (RHEED) and surface chemistry using X-ray photoelectron spectroscopy (XPS), we can understand surface interactions between Si, Mg, and O. By understanding the relationship between atomic level interactions and ultimate film characteristics, we can engineer the most effective process for BTO integration on Si. The substrates are cleaned using wet chemicals to create a hydrogen-terminated surface that resists contamination from laboratory exposure. XPS and RHEED are used to verify the lack of contamination on the surface and proper crystallographic orientation. Contaminant levels of 7.61 % oxygen (0.15 O/Si ratio) and 18.0% carbon (0.31 C/Si ratio) are the most successful results that have thus far been attained. XPS and RHEED will serve to confirm proper growth of MgO (001) on Si and determine thickness of the MgO film. BTO is then grown on the correctly-oriented MgO film using MBE, as has already been demonstrated on MgO/SiC substrates.[1] This presentation will discuss cleaning and growth conditions for successful production of high quality BTO on Si using the single MgO interface layer.

393873 Natural Gas, the Bridge Fuel

Abstract:

Medium and heavy duty trucks and buses produce almost 18 percent of all carbon emissions from transportation. Alternative energy sources need to be used on a larger scale to reduce these emissions, as well as to reduce dependence on foreign oil. Natural gas is the future of fuel in large vehicles. This paper tackles different issues surrounding the usage of natural gas as the main fuel for natural such as public awareness, public policy, existing infrastructure, consumer choice and controversies surrounding hydraulic fracturing. Current pilot programs, in which public city buses are using natural gas as the fuel of their choice, will be analyzed to determine the best way to implement these programs in other areas of the United States as well.

397811 The Federal Role in Fostering an Innovative U.S. Energy Ecosystem

Abstract:

Although renewable sources of energy – solar, wind, hydro, and geothermal – have been around for over one hundred years, they only generate 12.5 percent of electricity in the United States today. Renewable technology has advanced in cost, reliability, performance, and efficiency, but not enough to allow renewables to be a large contributor to electricity generation and widely used in the U.S. Renewables need continued innovation and progression to become more efficient, more reliable, and more cost competitive with fossil fuels, as well as expanding the possibilities of their use. Increased R&D funding for renewable sources is the key to advancing the technology and giving renewables the tools it needs to provide a larger portion of electricity in the U.S.

With renewables generating a larger portion of electricity, the U.S. can reach environmental sustainability, energy security, and a new economic opportunity. In reference to reaching environmental sustainability, renewables do not emit carbon and greenhouse gases, which is important for a cleaner future and mitigating climate change. Because the “fuel” for renewables includes the sun and wind, electricity generation from these sources does not depend on obtaining fossil fuels inside or outside of our country. They also extend the lifetime of these resources to be used for base load power, providing energy security to the U.S. Finally, if the U.S. is able to quickly and effectively develop competitive renewable technologies and deploy them into the electricity sector, it can provide competitive renewable technologies to the global community, boosting the economy and maintaining world leadership.

Although renewables have the capability to provide many benefits for the U.S., they face obstacles in receiving the funding they need to be widely used. These obstacles include climate change skeptics, the fossil fuel industry, the polarized political environment, and having to cut funding for other areas such as education and defense in order to allocate more money to R&D. The demand for renewables has also been relatively weak because electricity generated by renewables is no different than electricity generated by fossil fuels when the consumer uses it. The main way for renewables to be competitive with fossil fuels is through the cost and since wind is the only renewable that is cost competitive with fossil fuels, a strong customer demand has not been present to spur large private sector investment in renewable technologies.

In order for renewables to accomplish necessary advancements and contribute a larger portion of electricity to the U.S., financing to R&D for the technologies needs to be increased through government funding, capital generated through a carbon tax, public-private partnerships, and incentives for investment in the technology and its R&D. A commitment to innovating renewables in order to reach environmental sustainability, energy security, and a new economic opportunity is necessary from governmental officials to make the tough decisions advocating for widespread use of renewables in the U.S.

397856 A Biowall for Improving Indoor Air Quality

Abstract:

The Biowall is a living plant wall that filters air in the HVAC system of a home or small commercial building. This new technology improves indoor air quality be removing Volatile Organic Compounds and has the potential for energy savings by reducing the need for heating and cooling large amounts of outside air in a building. The presentation provides an overview of biofiltration for buildings and discusses the new research being conducted in this emerging field.

384797 Optimal Performance Management of Clean Water Treatment Processes

Abstract:

A major challenge currently facing a sustainable future is the ever increasing demand for clean water of adequate quality and quantity. With growing populations and industrial growth, the need for cost-effective water purification processes has become more urgent than ever. The processes used for potable water treatment can vary greatly depending on the sources of raw water and the associated water quality. One of the biggest operational risks to water companies arise from their ability to control the day-to-day management and optimisation of their water treatment systems; and with emerging pressures to remain competitive within the global market, companies are looking for better solutions to be able to make predictions on how their treatment processes should be managed and operated.

Work in literature has so far focused solely on performance optimisation of *individual* water treatment unit operations, due both to the complexity of the overall process and to the interactions between unit operations. Such an approach will inevitably lead to sub-optimal performance as the operation of one unit operation has an impact on the next, and their operations can therefore not be considered in isolation. A plant-wide approach to performance optimisation will therefore be of significant benefit to the water industry by increasing the *overall* process efficiency and thus decreasing overall plant costs. By considering the plant as a whole, the interactions between individual processing units can be taken into account and properly balanced. A design and operating procedure that considers an optimal series of several unit operations, including units which may run in parallel, whilst at the same time reducing the number of steps required to achieve a given product quality, will therefore improve the *overall*plant efficiency.

Optimization based on plant superstructures has proven to be an effective tool for the synthesis of chemical engineering process flowsheets and for overall plant performance optimization. The design of the process, and the operating parameters for each individual piece of equipment, can thus all be determined optimally, but more importantly, simultaneously.

This work addresses the current gap in clean water performance research by developing an approach based on a superstructure concept for the synthesis of clean water treatment works through the application of mixed integer optimisation techniques. A systematic framework is presented for the representation of superstructures and derivation of optimisation models for process synthesis. The state task network (STN) and state equipment network (SEN) are proposed as the two fundamental representations of superstructures used to describe the overall plant operation including multiple sources of raw water and water treatment processes. The proposed framework is illustrated by several examples to demonstrate that the approach can provide valuable guidance in clean water treatment process design and operation, thus providing a tool to achieve better day-to-day performance management.

390744 Using Integrated Controls and Real-Time Weather Forecasting to Drive Zero Discharge

Abstract:

New technologies are needed to better manage our environmental and industrial infrastructure. One emerging technology couples real-time monitoring with advanced cloud computing tools to create "smart" infrastructure such as cisterns, environmental remediation systems, and dust suppression systems that use weather forecasts and other predictive tools. The result is an integrated approach to driving zero discharge at facilities that are under increasing pressure to better manage water use, treatment, and discharge. A recent example project included the integration of a real-time weather forecasting control system into a rainwater harvesting system currently being constructed at the CSX Curtis Bay Piers. The real-time weather forecasting control system improves on-site capture and use of storm and gray waters at the facility and although the primary objective of the project was to help the Piers achieve its zero discharge goals, the system also improves the efficiency of on-site water usage for coal dust suppression and will substantially reduce the use of potable water at the facility. The system, designed and installed by Geosyntec Consultants, interfaces with the current dust suppression system (DSS) used at the site, and contains automated control components as well as indicators to advise manual control of the DSS based on forecasted meteorological conditions.

Better management of environmental infrastructure has numerous benefits. The cost of water, which is often perceived as a relatively inexpensive resource, increases considerably when accounting for the tangible costs related to environmental compliance, treatment, and discharge. Water costs increase even more when accounting for the intangible costs related to regulatory scrutiny, negative public perception, and increased risk of lawsuits. This presentation will use the Curtis Bay Piers case study to illustrate the benefits of integrated infrastructure. Additional applications and benefits will also be summarized.

391462 Water Footprint of Rio Grande River Basin

Abstract:

Water resource management at the scale of river basin is extremely critical for the sustainability of rural agrarian systems. Most agrarian systems are characterized by water stress and depleting water table. This is primarily due to unsustainable extraction and use of ground water and surface water for irrigation. In this presnetation we outline our work related to the development of water footprint metric for the San Luis Basin in south central Colorado. We will present data collection, modeling and water footprint calculations for the regional system. In addition, we will discuss long term implications related to water sustainability in this region through scenario analysis. Finally, we will review a systems framework where water footprint along with other sustainability metrics is utilized to analyze the overall sustainability of regional systems at the scale of river basin.

Water Scarcity Footprint as Part of a Complete Product Life Cycle Assessment. 87d. Sergio F. Galeano. 09B02. Water Sustainability and Integrated Water Resource Management

One issue not properly addressed in product LCA methodology and studies is the one related to water scarcity. Water scarcity represents the risk in the magnitude of additional water withdrawals in the life cycle of a product system in terms of time and the specific basin. However, water scarcity is not included in existing impact indicator sets such as TRACI 2 used by LCA practitioners, and most LCA studies just mentioned an inventory item for water use, a volumetric figure devoid of an impact element. The globally accepted ISO 14044 for LCA of products albeit not selecting specific impact indicators, provides guidance on requirements to develop valid impact indicators. With the advent of the recently approved ISO 14046: 2014, dedicated to water footprint based on ISO 14044, efforts to further complete the classical LCIA profile with the additional water availability indicator are emerging. This paper shows how to develop a new water scarcity impact indicator and how it can be integrated in new or already completed LCA studies. . The teachings of this paper add value to the LCIA methodology by adding a very desirable impact indicator for a resource of importance. Also, value is added by making its integration in LCA studies possible for new and existing LCA studies. Finally, the use of data from the Aqueduct Water Risk Atlas to develop the scarcity impact indicator adds context to the results by assessing hydrological considerations of the basin and sub-basin from which withdrawals and replenishments take place. It addresses the business case about the sustainability of the production level in a given plant or mill according to its water supply demands. .

.

INTRODUCTION

Globally, water use has been growing as a result of population increases and the economic development needed for higher levels of quality of life. As one consequence, fresh water scarcity is already affecting many regions in the world and these effects will be increasing in intensity and number with corresponding environmental and sustainability concerns. Although the magnitude and number of affected regions scattered globally makes it a global concern, water scarcity has a unique regional and basin perspectives. It is a resource sustainability issue rather than a pollutant emission or discharge issue. Among the different predictive analytical initiatives developed, and in development, to address these environmental and sustainability concerns, the "footprint" has gained some popularity because a belief of easier understanding by the populace. Carbon footprint, CFP, for products and corporations, were among the first and more popular "footprint" approaches based on different adaptations of the life cycle assessment, LCA, of the ISO LCA methodology. More recently, ISO attempted to follow similar steps of CFP ISO/TS 14067 in a recently approved water foot print standard, WFP, ISO 14046:2014. It is a complex standard whereby both WFP from inventory (volumes) and impact indicators are addressed with subtle terminology differentiation, say; water scarcity foot print or water foot print. A dedicated Technical Report ISO 14073, with examples, is being finalized. The author on behalf of USA had presented an example in the draft TR 14073 dealing with this application to develop a water scarcity impact indicator.

The author have applied the pertinent guidance and requirements of both ISO 14044:2006 and ISO 14046:2014 to the development of an additional water scarcity impact indicator to be integrated to the traditional LCIA profile of a LCA study, either existing or new. The mechanism and characterization factor are based on data from the Aqueduct Water Risk Atlas [1] developed by the World Resources Institute, WRI, available free in the website and explained in more general details in the following of this paper. During the preparation of this paper the author has been in close consultation with experts from WRI.

BACKGROUND

Typically, product Life Cycle Assessment (LCA) addresses water use and/or consumption as an inventory item in the LCI without the proper characterization to develop an impact indicator in the LCIA profile. In spite of the ISO 14046 WFP standard, there still is not a recognized method to establish preference for a suitable water impact indicator. Berger and Finkbeiner [2] and the called WULCA (water use LCA) working group of the SETAC/UNEP Life Cycle Initiative [3] have surveyed and evaluated different methods to characterize water use and water scarcity in terms of LCIA. Albeit helpful, the results of these two most recent survey studies are not in agreement in the selection of assessment criteria and the most acceptable methods.

LCIA considerations in ISO 14046 and in this application

ISO 14046 [4] allows for either stand-alone water footprint assessment or water footprint assessment as part of an LCA. If the assessment meets the comprehensiveness principle requirements of ISO 14046 the assessment can be named a water footprint, WFP. Otherwise, the term WFP needs more specificity, say "water scarcity footprint ". In this paper, the author present an application of the WFP standard ISO 14046:2014 for a new or existent LCA study which add a new water scarcity impact indicator to the selected LCIA profile. It does so following ISO 14044 requirements for impact indicator categories and categorization factors. It can be considered also a novel application of ISO 14044:2006. It will allow for a more complete LCA report with a new impact indicator of great interest for producers and consumers. The water scarcity impact indicator in isolation could be considered the WFP metric for the product system in question. Additionally, this application uses data from the Aqueduct Water Risk Atlas developed by the World Resources Institute, WRI, for characterization purposes. It will provide context to the water scarcity impact indicator and thus added real value to business planners, customers and LCA practitioners in answering with less uncertainty the question: Is the present or planned use of freshwater at this plant or mill really sustainable? Finally, the integration of the new water scarcity impact indicator is smooth and according to the already understood and practiced LCA ISO methodology.

METHODOLOGICAL ASPECTS OF THE ISO 14046 APPLICATION AND CURRENT LCA PRACTICE

This application entails certain differentiating aspects from most current practice in LCA and "footprint". The application's description and its understanding will benefit with the following review of some methodological aspects.

1- In water footprint (WFP) assessment of freshwater scarcity the issue is the balance of a resource at specific points in the supply chain. In other "footprints" like carbon footprint (CFP) the issue is the emission of GHG at different points of the supply chain considered as global impact.

2- WFP assessment for water scarcity is a regional and basin-driven analysis and it is not a global impact albeit it is already scattered in so many regions and countries that the problem is of global concern.

3- There is a gap of information and clarity in addressing databases. Inventory databases such as Eco-invent, PE, and Quantis,. provide freshwater water use information for agricultural and manufacturing sectors. Databases for basins and sub-basins, either surface or groundwater, are scarce and appear neglected in LCA studies. LCA is mainly about eco-efficiency of potential and relative impacts in a product system. The intrinsic lack of context in LCA methodology could be a reason for the apparent neglect in basin, sub-basin database for LCA. As McElroy and van Engelen [5] explain the profile of impact indicators in LCIA "do not measure or express environmental impacts relative to norms, standards or thresholds for what such impacts ought to be".

4- This application using data from the Aqueduct Water Risk Atlas as a means to develop the characterization factor for the water scarcity impact indicator enhances the value of LCA by providing the answer to the question of interest; is your present or planned use of fresh water at that facility or point in the supply chain sustainable? For a business case it is very important to have an answer for the above question and evidence about water supply. This application with Aqueduct provides context thus fully answering the question. We see this type of question more frequently requested by customers, consumers and internal planners.

5- Since the methodological approach of the application adheres closely to the requirements of ISO 14044, the introduction of a new additional water scarcity impact indicator is not a "shock" but rather a smooth addition adhering to a practice and procedures the stakeholders are already familiar with.

6- The water scarcity metric can be adapted to the functional unit of the traditional LCA methodology previously selected in an existing or planned LCA study. Proper algorithms depending on the basic LCA into which the integration will be performed can be developed with ease. This case study provides one such example. A clear and consistent format for the LCIA profile of indicators is thus available as well as a single metric in case the WFP be expressed as a standalone metric.

ABOUT AQUEDUCT METHODOLOGY AND ITS USE IN THIS APPLICATION

As explained in the above there is a demand for comprehensive and comparable data. To satisfy this demand, the World Resources Institute (WRI) developed the Aqueduct Water Risk Atlas which provides a set of twelve indicators (not LCA impact indicators) that capture a wide range of variables, and aggregates them into comprehensive risk scores using the Aqueduct Water Risk Framework (AWRF). The AWRF is fully described in the website at http://www.wri.org/our-work/project/aqueduct .

The 12 indicators are grouped in 3 water risk to business categories. These are; Physical Risk: Quantity, Physical Risk: Quality and Reputational and Regulatory Risk. In this case study focusing on water scarcity impact indicator the applicable category is Physical Risk: Quantity and its seven indicators. The seven indicators for this category are: baseline water stress, inter-annual variability, seasonal variability, flood occurrence, drought severity, upstream storage and groundwater stress. To develop many of the 7 indicators in this category of the AWRF, there is need to measure total blue water and available blue water. Physical Risks related to quantity are defined as the exposure to changes in water quantity that may affect a company, plant, supply chain and logistics. Physical Risks related to quantity may also include disruptions of needed electric power due to water issues.

Data selection is a rigorous process with proper transparency. The seven indicators are calculated according

to the methodology described in full detail in the cited website. Once calculated, the results can be used individually or aggregated into composite risk scores. The first step in aggregation is to normalize the indicators over a set of thresholds, chosen to divide the calculated indicators into 5 levels say from "low" to "extremely high". Thresholds are determined, for each indicator, using existing literature, governmental or intergovernmental guidelines and expert judgment. Indicator values are normalized to a score between 0 and 5; 0-1 to the lowest level and 4-5 to the highest. This method of threshold indicator normalization has several advantages and one disadvantage; the scoring of normalized indicators is subjective. Only extreme transparency, as it is provided, permits assess the effects of this subjectivity.
Once scores have been established, weights are assigned based on level of importance and relevance. There are five distinct weights for each indicator. AWRF prefers an exponential scale over a linear scale to better reflect human tendency to categorize by orders of magnitude differences. Utmost transparency allows the user to view aggregated scores or use, as appropriate, default preset weighting profiles from 10 different industrial sectors. AWRF explains how aggregated risk scores are calculated via weighted averages and rescaled to compensate the pull of weighted averages towards the mean. A final score $\propto$ is then computed.

RESULTS AND DISCUSSION

A CASE EXAMPLE ILLUSTRATING THE APPLICATION OF A WATER SCARCITY FOOTPRINT ASSESSMENT AS PART OF A COMPLETE LCA STUDY.

Options provided by ISO 14046 and the selected one

ISO 14046 considers two main options of water footprint assessment; Option1, stand-alone assessment where only impact related to water are assessed, and Option 2, as part of a life cycle assessment (LCA) where consideration is given to a comprehensive set of environmental impacts including water scarcity. This latter Option 2 is addressed in this illustrative case example. Two main cases are identified in this option 2; (a) a new LCA in which the water footprint assessment is selected together with other impact categories and (b) when a product LCA exists with different impact indicators but the water footprint scarcity impact indicator does not exist.

The Original LCA study – Background and methods

The reference paper of Galeano et. al. [6] provides whole information about the original LCA study conducted in a conventional corrugated packaging system with manufacturing in different linerboard mills and converting plants of the same manufacturer and supply (the Company) to different assembling and packaging plants of a consumer product customer (the Customer). The original LCA study was an environmental optimization analysis to satisfy the Customer and Company desire to optimize the existing production and distribution system factoring environmental impacts.

In the following we will explain briefly first, the fundamentals of the existing LCA study, simplified in the LCIA profile to the linerboard and not the final corrugated box. Linerboard, material wise, is about 2/3 of the corrugated box and in the integrated pulp and linerboard mills the use of water is very significant. The simplification allows for more clarity. Figure 1 provides a visual of the typical corrugated board with two linerboards and the corrugated medium.

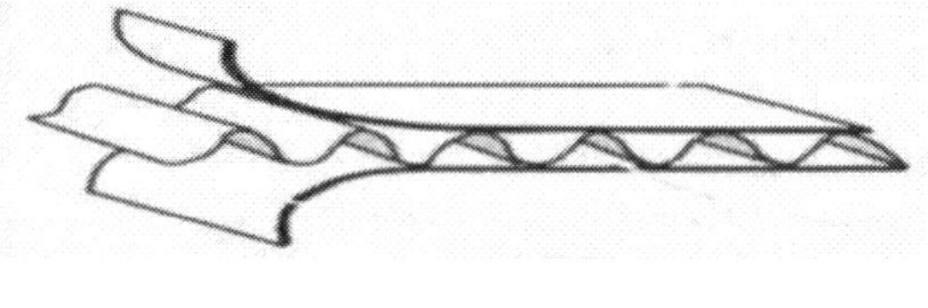

Figure 1—Components of the generic corrugated fibreboard: 2 Linerboards and 1 Medium

The existing LCA study relied in a set of indicators from TRACI. 2 [6] which is a well-known and used set of impact indicators in LCA studies for forest product in the United States. TRACI 2 does not include a water scarcity impact indicator. It was considered that the goal and objectives for the existing LCA study will not need modification for the planned integration of a water scarcity impact indicator. Simply, the original optimization assessment will include the additional water scarcity impact indicator. The boundaries of the original LCA will need to expand as to include the demarcation of the involved basins without influence on prior LCA inventory and LCIA impact indicator results. For simplification in this case example we have considered as functional unit a tonne of linerboard instead of a tonne of corrugated container.

The existing LCA study report, adapted to the linerboard simplification explained above, results in a LCIA profile described in Table 1

Table 1— LCIA profile results for linerboard in original LCA

Impact category	Units (per tonne of Linerboard)	Results LCIA profile
Climate change fossil	T Fossil fuel CO_2 eq.	1,0
Energy non-renewable	GJ	1,05
Eutrophication	T N equiv.	$7{,}3 \times 10^{-4}$
Ecotoxicity	T 2,4-D equiv.	0,24
Acidification	T H+ equiv.	$5{,}7 \times 10^{-4}$

Methodological details for the water scarcity impact indicator.

To address the water scarcity category indicator in the water footprint assessment we considered the applicable provisions of ISO 14044 in regards to indicators in general and the new impact indicator result of water scarcity not considered in conventional LCA. We submit that the provisions of ISO 14044 in clauses 4.4.2.5 and 4.4.2.2.1, the latter for new indicators, are met in the context and content of this example. For this case example in developing the water scarcity footprint we selected the Physical Risk: Quantity indicator from the Aqueduct Water Risk Atlas which provides the mechanisms justifying the transformation factor that permits converting the inventory volume per tonne into the water scarcity impact indicator. For the purposes of this example and according with the referenced paper on the original LCA study, we considered two linerboard manufacturing locations, A and B. This score will be reflected in the final profile of the integrated LCA including the water scarcity footprint indicator. Although a weighting of all six impact indicators could be possible and permissible under ISO 14044, the author does not have a reliable way or method to calculate such.

For the determination of the water scarcity footprint the following relationship was used.

$$W_{SC} = \propto \times \; C_W \; \times \; Cx/Cw$$

W_{SC}	is the water scarcity footprint indicator
$\propto$	is the Aqueduct score for the location
C_W	is the water use per functional unit for this case example
C_x	is the water use per functional unit for another situation

In this example, $\propto$ was identified as 1(range 0 to 1) and C_W was determined as 113 m^3, the average of the 2 linerboard mills. An algorithm is useful because it records progress or drawbacks in water use influencing the water scarcity footprint indicator. When applying the data from the Aqueduct Water Risk Atlas, the Physical Risk: Quantity risk score is considered low for both locations and lies between 0 and 1 in these locations. We selected 1 to provide a value to the indicator. The final integrated LCIA profile is shown in Table 2.

Table 2 —New LCIA profile for linerboard with water scarcity indicator

Impact category	Units	Linerboard
Climate change fossil	T Fossil fuel CO_2 eq.	1,0
Energy non-renewable	GJ	1,05
Eutrophication	T N eq.	$7{,}3 \times 10^{-4}$
Ecotoxicity	T 2,4-D eq.	0,24
Acidification	T H+ eq.	$5{,}7 \times 10^{-4}$
Water scarcity footprint	m^3-eq	113

CONCLUSIONS

The water scarcity impact indicator as part of a comprehensive product LCA study illustrated in this example, allows for a sensible examination of the business risks or impacts of water withdrawal for product manufacturing in its scarcity and other aspects, both present and in the future. This examination was done in the context of a simplified product system of one published LCA study as example. It complements, in a logical manner, the most classical LCA methodology that typically does not focus on the water scarcity impact category.

1- Besides its use in new LCA studies, this application would allow in many cases, as in the LCA study used in this example, the enhancement of many existing LCA studies with the new water scarcity impact category. The LCA practice could benefit by incorporating this additional impact indicator
2- The example also redirects the conversation to a more holistic sustainability-inspired analysis whereby not only volumetric-related impacts are examined but other factors both in space and time. By providing context to this impact indicator the facility owner or operator as well as customers would have better answer to the water scarcity risk associated with the product or location. The business case is properly satisfied.
3- The application narrows the gap in unabridged areas of LCA by providing guidance on addressing the water scarcity impact category.
4- Besides its use in a more complete LCIA profile, the water scarcity impact indicator can be used alone as a water scarcity footprint providing more flexibility to users and designers in satisfying marketplace demands.

5- There is a good fit in the methodology and procedures applied in the example and the routine ones utilized in conventional LCA of product systems in LCA studies.

REFERENCES

[1] World Resources Institute (WRI) Markets and Enterprise Program. Aqueduct Water Risk Atlas. Aqueduct Water Risk Framework. http://www.wri.org/publications/aqueduct-water-risk-framework.

[2} Berger M, Finkbeiner M (2010), Water footprinting: how to assess water use in life cycle assessment? Sustainabiity 2:919-944

[3] WULCA Group, SETAC/UNEP LCA Program. Review of methods addressing freshwater use in life cycle inventory and impact assessment. JILCA (2013), 18 (707-721).

[4] ISO 14046:2014. Environmental management – Water footprint – Principles, requirements and guidelines. Geneva,

[5] McElroy M and van Engelen Jo M.L., Corporate Sustainability Management, Earthscan, Routledge, 2012

|6] S. F. Galeano, P. M Smorch, M. S. Richardson, "Application of Life Cycle Assessment to Supply Optimization". ISIE 2011 Conference, 6th International Conference on Industrial Ecology. Berkeley CA, June 9, 2011

ACKNOWLEDGMENT. To Georgia-Pacific LLC for its foresight and support in the development of environmental management tools and ISO environmental standard-making activities.

CHOOSE APPROPRIATE SCENARIO IN ORDER TO IMPROVE WATER QUALITY AT EL-RAHAWY DRAIN AND THE ROSETTA BRANCH, EGYPT

Mohamed Mostafa[1] and Robert W. Peters[2]
[1]Doctoral Student of Environmental Engineering and [2]Professor of Environmental Engineering
Department of Civil, Construction and Environmental Engineering
University of Alabama at Birmingham
1075 13th Street South
Birmingham, Alabama 35294-4440

Abstract

El-Rahawy drain is the major cause of water pollution at the Rosetta branch of the Nile River, where approximately 1.45 million m^3 of primary treated wastewater are discarded daily into its water from the Abu-Rawash Waste Water Treatment Plant (WWTP), which is characterized by its high biological oxygen demand (BOD), chemical oxygen demand (COD), and total suspended solids (TSS). Two scenarios were simulated using river pollutant (RP) modeling in order to effectively manage water quality at the Rosetta branch. The first scenario is to discontinue discharging the effluent from the Abu-Rawash WWTP into the El-Rahawy drain and change the effluent path to the desert to be used in the cultivation of woody trees. The second scenario is to minimize contamination of surface water at the El-Rahawy drain by discharging clean water from the Al-Buhairi Water Canal, and improving effluent quality at the Abu-Rawash WWTP with the application of aluminum chloride ($AlCl_3$) as a coagulant at a lower pH value. By comparing the two scenarios, it was concluded that the second scenario would be more effective and less costly than the first scenario, where the negative influence of the El-Rahawy drain on the Rosetta branch water quality is temporary and will end once the biological treatment units at the Abu-Rawash WWTP begin operation.

Introduction and Background

The Rosetta branch flows from Delta Barrage into the Mediterranean for about 239 km (148.50 miles) [1]. It was estimated that the surface water of the Rosetta branch receives more than 3.0 million cubic meters daily of drainage water from domestic, industrial, and agricultural activities [2]. Five drains outfalls located along the Rosetta branch which include the El-Rahawy, Sabal, the El-Tahreer, the Zawyet El-Bahr, and the Tala [3]. The El-Rahawy is the major source of pollution along the Rosetta branch because it receives daily around 1,450,000 m^3 of untreated and primary treated wastewater from the Abu-Rawash WWTP [4]. Therefore, Abu-Rawash

WWTP considers the major source of water quality degradation along the El-Rahawy drain [5, 6]. Typical effluent water quality from the Abu-Rawash WWTP has been 174 mg/L biological oxygen demand (BOD), 0.95 mg/L dissolved oxygen (DO), 154 mg/L total suspended solids (TSS), 258 mg/L chemical oxygen demand (COD), and pH 7.1 [7, 8].

Researchers have conducted several studies of water pollution in the El Rahawy drain. El-Bouraie *et al.* (2011) conducted a study to assess the Rosetta branch water quality at the outlet of the El-Rahawy drain [9]. Low levels of DO, as well as high concentrations of COD, TDS, BOD, and ammonia (NH_3) were recorded at the discharge point of El-Rahawy Drain. Ezzat *et al.* (2012) conducted a study to assess water quality at the Rosetta branch [2]. A Water Quality Index (WQI) was calculated at different points along the Rosetta branch. The mean value of the Water Quality Index at the El-Rahawy drain outlet and downstream of the El-Rahawy drain was found to be 24.5 and 34.5, respectively. Consequently, a very bad water quality was recorded at El-Rahawy drain outlet and a bad water quality was recorded downstream of the El-Rahawy drain. Badr *et al.* (2006) conducted research to study the effect of the El Rahawy drain on the Rosetta branch water quality; the results indicated the presence of high concentrations of trace metals (lead (Pb); zinc (Zn); iron (Fe); and cadmium (Cd)) downstream of the El Rahawy drain [10]. El-Gammal and Hesham (2008) conducted a study to manage water quality at the El-Rahawy drain and at the Rosetta branch [11]. Two different scenarios were simulated using the Q2K model. Q2K is a river modeling used to calculate the following parameters, chlorophyll a, coliforms, temperature, nitrate, dissolved phosphorus, organic N, nitrite, DO, and BOD [12]. The first scenario addresses the mitigation of organic loads by dilution. The second scenario includes changing the path of the Abu-Rawash WWTP effluent to the desert for reuse in the cultivation of wooden trees. The results showed that the second scenario is the best solution, where the concentrations of organic-nitrogen and ammonia at the Rosetta branch are reduced to very low levels.

Materials and Methods

The aim of this research investigation was to manage water quality at the Rosetta branch. The length of the study area was 228.5 km (141.9 miles), starting from the El-Rahawy drain and ending downstream of the Tala drain. Two scenarios were simulated using river pollutant (RP) modeling. RP modeling is a river modeling used to calculate the following parameters: chlorides

(Cl^-), total organic carbon (TOC), total dissolved solids (TDS), TSS, COD, BOD, pH, temperature, and DO [4]. The first scenario investigated was to improve water quality at the Abu-Rawash WWTP with the application of coagulant at a lower pH value and consequently improving water quality at the El-Rahawy drain. Carbon dioxide (CO_2) was injected into the wastewater in order to reduce the pH value. The first scenario also included discharging clean water from the Al-Buhairi Water Canal to the El-Rahawy drain. The second scenario was to divert 1.45 million cubic meters of primary treated wastewater daily from the Abu-Rawash WWTP to the desert, and thus reduce pollution in the El-Rahawy drain. The total cost to apply the first and second scenarios is \$2,655,000 and \$57,143,000, respectively [4]. By 2024, the biological treatment units will begin working at the Abu-Rawash WWTP that will significantly improve the effluent quality [4].

Mass balance and exponential equations were used to estimate the flow and concentrations of the selected parameters at the El-Rahawy drain for the proposed scenarios. In order to run the RP modeling, water samples were collected upstream of the El-Rahawy drain and from five drains located along the study area, which included the El-Rahawy drain, the Sabal drain, the El-Tahreer drain, the Zawyet El-Bahr drain, and the Tala drain. Water samples were analyzed for TOC, TDS, TSS, BOD, COD, pH, and DO. The flow rate at the El-Rahawy drain is expected to increase to 2.9×10^6 m^3/day after applying the first scenario, while it is expected to decrease to 450,000 m^3/day after applying the second scenario. Table 1 below lists the flow and water characterization at the El-Rahawy drain for the existing situation and the two proposed scenarios. Then, The RP modeling was used to calculate the concentration of the studied parameters at different points along the study area for the existing situation and the proposed scenarios.

Table 1. The flow and water characterization at the El-Rahawy drain.

Term	Current situation	First scenario (improve water quality at El-Rahawy drain)	Second scenario (change the effluent path of Abu-Rawash WWTP)
Flow, (m^3/day)	1,900,000	2,900,000	450,000
pH, (dimensionless)	7.4	6.43	7.4
TSS, (mg/L)	150	32.25	30
COD, (mg/L)	270	65.6	70
TDS, (mg/L)	820	296.6	412
BOD, (mg/L)	140	34.48	34.5
TOC, (mg/L)	12.0	3.17	1.5
DO, (mg/L)	1.45	3.22	4.0

Results and Discussion

Applying the first scenario is estimated to reduce the TSS concentration at the El-Rahawy drain from 150.0 mg/L to 32.25 mg/L. This modification will increase the DO concentration at the El-Rahawy drain by 54.9%, and will also reduce the levels of BOD, TDS, TOC, and COD by 75.37%, 63.8%, 73.5%, and 75.70%, respectively. Applying the first scenario will also decrease the pH value at the El-Rahawy drain from 7.4 to 6.43 due to using CO_2 in treating the wastewater. Applying the second scenario will reduce the TSS concentration at the El-Rahawy drain from 150 mg/L to 30.0 mg/L. This modification will increase the DO concentration at the El-Rahawy drain by 63.75%, and will also reduce the levels of BOD, TDS, TOC, and COD by 75.35%, 49.75%, 87.5%, and 74.0%, respectively.

The following discussion provides an example of using the RP modeling in the prediction of BOD, DO, and COD concentration along the study area. The BOD concentration specified in EPA standards and Egyptian law 48/1982 is $\leq$ 6.0 mg/L [13, 14]. The analysis of water samples showed that the BOD concentration upstream the El-Rahawy drain was within the permissible limits. Figure 1 provides a simulation of the BOD concentration along the study area for the existing situation and the two proposed scenarios. After receiving discharge from the El-Rahawy drain, the BOD concentration directly downstream of the El-Rahawy drain increased to 5.6, 8.4, and 16.0 mg/L for the second scenario, the first scenario, and the existing situation, respectively. For the existing situation, the El-Rahawy drain is negatively affecting the water quality at the Rosetta branch. In comparison with the existing situation, applying the first and second scenarios reduced BOD concentration downstream the El-Rahawy drain by 46.25 and 65.0%, respectively. Consequently, applying the proposed solutions, especially the second scenario, will significantly reduce the BOD concentration in the Rosetta branch.

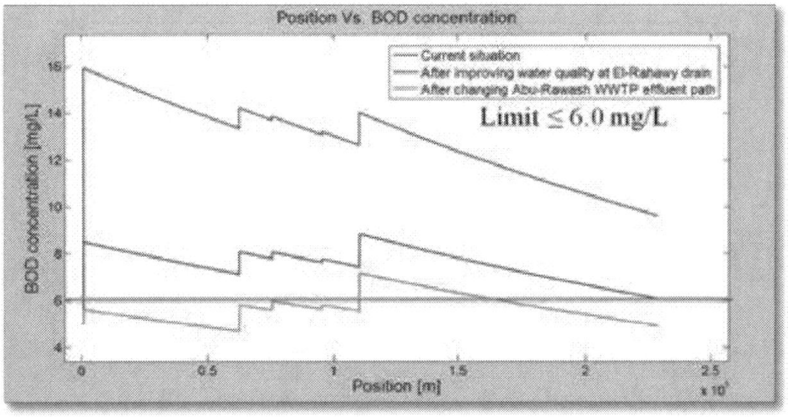

Figure 1. BOD concentrations along the Rosetta branch for the existing situation and the proposed scenarios.

The water quality standard specified in EPA and Egyptian law 48/1982 for DO is ≥ 4.0 mg/L [13, 14]. The analysis of water samples showed that the DO concentration upstream of the El-Rahawy drain was within the permissible limits. Figure 2 simulates the DO concentration along the study area for the existing situation and the two proposed scenarios. After receiving discharge from the El-Rahawy drain, the DO concentration directly downstream the El-Rahawy drain reduced slightly to 5.94, 5.65, and 5.61 for the second scenario, the first scenario, and the existing situation, respectively. Consequently, the El-Rahawy drain is not negatively affecting the DO concentration at the Rosetta branch.

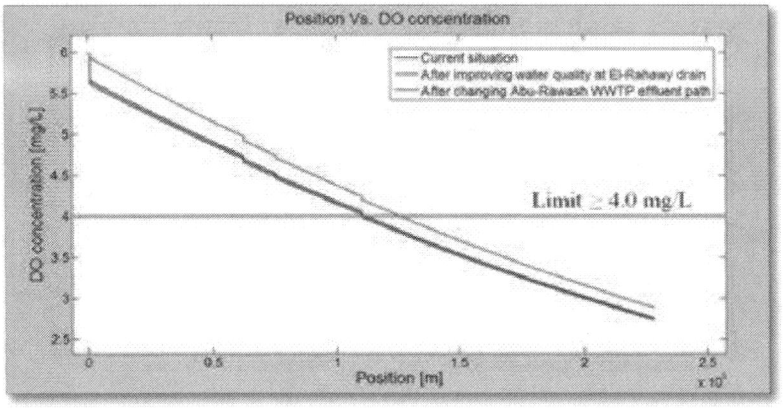

Figure 2. DO concentrations along the Rosetta branch for the existing situation and the proposed scenarios.

The water quality standard specified in EPA and Egyptian law 48/1982 for COD is ≤ 10.0 mg/L [13, 14]. The analysis of water samples showed that the COD concentration upstream the

El-Rahawy drain was within the permissible limits. After receiving discharge from El-Rahawy drain, the COD concentration at the Rosetta branch increased to 11.2, 16.5, and 31.1 for the second scenario, the first scenario, and the existing situation, respectively (see Figure 3). For the existing situation, the El-Rahawy drain is negatively affecting the water quality at the Rosetta branch. In comparison with the existing situation, applying the first and second scenarios reduced the COD concentration downstream of the El-Rahawy drain by 46.45 and 63.87%, respectively. Consequently, applying the proposed solutions will significantly reduce the COD concentration at the Rosetta branch, with knowing that the second scenario is more effective in improving water quality at the branch.

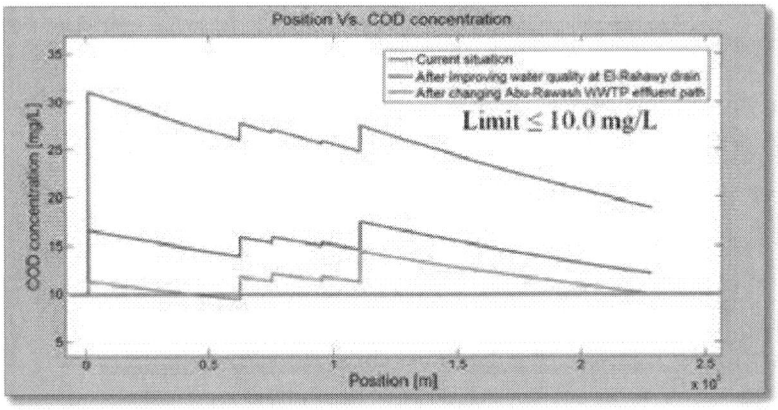

Figure 3. COD concentrations along the Rosetta branch for the existing situation and the proposed scenarios.

Summary and Conclusions

For the existing situation, the RP modeling showed that the TDS and pH values in the Rosetta branch are not negatively affected by receiving discharge from the El-Rahawy drain, while the TOC, TSS, COD, DO, and BOD concentrations are negatively affected by this discharge. The second scenario is more effective in reducing TOC, BOD, and COD concentrations in the Rosetta branch than the first scenario, while there is no significance difference was recorded between the two scenarios in reducing TDS and TSS concentrations or increasing DO concentration. Since, applying the first scenario is more cost-effective than the second scenario, it is recommended to apply the first scenario as a temporary solution, where the negative impact of the El-Rahawy drain on the Rosetta branch water quality will end once the biological treatment units in Abu-Rawash WWTP begin working.

LIST OF REFERENCES

1. Milliman, J., and B.U. Haq, 1996. ***Sea-Level Rise and Coastal Subsidence: Causes, Consequences, and Strategies***, Kluwer Academic Publishers, Norwell, MA.

2. Ezzat, S.M., M.M. Hesham, A.A. Mervat, H.A. Essam, and A.E. Mostafa, 2012. "Water Quality Assessment of River Nile at Rosetta Branch: Impact of Drains Discharge", ***Middle-East Journal of Scientific Research, 12*** (4): 413–423.

3. Lasheen, M.R., K.A. Fagr, A.A. Aly, and M.H.A. Hassan, 2012. "Fish as Bio Indicators in Aquatic Environmental Pollution Assessment: A Case Study in Abu-Rawash Area, Egypt", ***World Applied Sciences Journal, 19*** (2): 265–275.

4. Mostafa, M., 2014. "Modeling of Pollutant Transport in the Nile Delta, Egypt", Ph.D. Dissertation, University of Alabama at Birmingham, Department of Civil, Construction and Environmental Engineering, Birmingham, AL.

5. Moustafa, M.M., H.H.A. Mohamed, M.A. Amal, Y.M. Talat, and M.M. Siliem, 2010. "Water Quality Assessment of Rosetta and Damietta Branches, River Nile, Egypt", ***African J. Biol. Sci., 6*** (2): 127–142.

6. Gaber, H.S., A.E. Midhat, A.I. Seham, and M.N.A. Mohammad, 2013. "Effect of Water Pollution in El-Rahawy Drainage Canal on Hematology and Organs of Freshwater Fish *Clarias gariepinus*", ***World Applied Sciences Journal, 21*** (3): 329–341.

7. Ashmawy, M.A., M.S. Moussa, A.K. Ghoneim, and A. Tammam, 2012. "Enhancing the Efficiency of Primary Sedimentation in Wastewater Treatment Plants with the Application of *Moringa oliefera* Seeds and Quicklime", ***Journal of American Science, 8*** (2): 494–502.

8. Misheloff, R., 2010. "Integrated Water Resource Management II: Feasibility of Wastewater Reuse", International Resources Group, Report No.14.

9. El-Bouraie, M.M., A.M. Eman, G.M. Gehad, H.A. Essam, and M.Y. Mohamed, 2011. "Water Quality of Rosetta Branch in Nile Delta, Egypt", ***Suoseura, Helsinki, 62*** (1): 31–37.

10. Badr, M.H., E. Abdelsamie, B.S. Mohammed, F.M. Laila, and S.A. Ghada, 2006. "Studies on the Effect of El-Rahawy Drain on the River Nile Water Pollution by Trace Metals and Major Cations at El-Kanater-El Khyria Area Under the Effect of Seasonal Variation", ***Ass. Univ. Bull. Environ. Res., 9*** (2): 205–218.

11. El-Gammal, H.A., and S.E. Hesham, 2008. "Water Quality Management Scenarios in Rosetta River Nile Branch, Egypt", Paper presented at the Twelfth International Water Technology Conference, Alexandria, Egypt.

12. Frey, M., and H. Evan, 2002. "Modeling and Total Maximum Daily Loads", Clean Water Network. Boise, ID.

13. National Water Research Center (NWRC), 1995. "River Nile Protection and Development Project Phase II", Environmental Pollution and Legislative Regulations (Law 48. 1982 and

Decree 8, 1993), Ministry of Public Works and Water Resources, Report No. A3010901–9508–01 – FR.

14. U.S. Environmental Protection Agency, 2007. "Drinking Water Standards and Health Advisories Table". San Francisco, CA. Accessed July 14, 2014 http://www.epa.gov/region9/water/drinking/files/dwsha_0607.pdf.

389535 Dynamic Optimization Strategies for Control of Algae Growth in Eutrophic LAKES with Nonpoint Nutrient Sources

Abstract:

We address the optimal planning of alternative restoration strategies of eutrophic water bodies, mainly affected by nonpoint nutrient sources, through advanced dynamic optimization techniques. Restoration strategies that involve chemical and physical processes and biological manipulation have been studied and implemented to reduce algal blooms and their consequences (Paerl and Otten, 2013; Jeppesen et al., 2012; Estrada et al., 2011) as eutrophication is the most serious environmental problem in many lakes and reservoirs.

Artificial wetlands have been largely used to decrease external nutrient loading from nonpoint sources and, in this way, carry out bottom up control on phytoplankton growth. However, biogeochemical processes that take place within water bodies delay restoration (phosphorus and nitrogen recycles, as well as nutrient release from sediments). Additional inlake strategies have been proposed and applied, such as hypolimnetic oxygenation and biomanipulation (Søndergaard et al., 2007). Moreover, macrophytes play an important role on biological processes of water bodies, since they capture nutrients from the water column and sediments, act as zooplankton refuge and, in some cases, produce substances that inhibit algal growth (Jeppesen et al., 2012; Asaeda and Bon, 1997).

We have implemented some of these strategies as different optimal control problems in previous work, by developing ecological water models integrated to optimization strategies to evaluate management strategies in both the short and long term (Estrada et al., 2011). In this work, we include mass balances for the macrophyte population, modeling roots and leaves biomass as different submodels. The net growth of macrophytes results from photosynthesis, respiration, dead, reserves and reallocation of dead biomass (Asaeda and Bon, 1997). The resulting dynamic optimization problems are subject to complex partial differential algebraic equations (PDAE) systems representing the main biogeochemical processes that take place within these water bodies. The PDAE systems result from dynamic mass balances for the different submodels for macrophytes, three phytoplankton groups (cyanobacteria, diatomea, chlorophyta); two zooplankton groups (cladocera, copepoda) and three size classes of local zooplanktivorous fish, as well as dissolved oxygen and main nutrients. Algebraic equations stand for forcing functions profiles, such as temperature, solar radiation, river inflows and concentrations, etc. Optimization variables (time dependent degrees of freedom) are associated to the reduction of nutrient loading by deviation to an artificial wetland (tributary flowrate to wetland profile) and in-lake restoration through fish removal rate for each of the class sizes. The dynamic optimization problem is formulated within a control vector parameterization framework (PSEnterprise, 2013). The present study has been performed on Paso de las Piedras Reservoir (38° 22′ S and 61° 12′ W), which is the drinking water source for two cities in Argentina. Numerical results provide optimal profiles for planning of the restoration actions, as well as a quantitative estimation of restoration effects on the water body, along a middle term time horizon.

References

Asaeda, T. and T. Van Bon (1997), Modeling the effects of macrophytes on algal blooming in eutrophic shallow lakes, Ecological Modeling, 104, 262-267.

Estrada V., J. Di Maggio, M.S. Diaz (2011), Water Sustainability: A Process Systems Engineering approach to the restoration of eutrophic lakes, Computers and Chemical Engineering, 35, 8, 1598-1613

Jeppesen, E., Søndergaard, M., Lauridsen, T., Davidson, T., Liu, Z., Mazzeo, N., Trochine, C., Özkan, K., Jensen, H., Tolle, D., Starling, F, Lazzaro, X., Johansson, L., Bjerring, R., Liboriussen, L., Larsen, S., Landkildehus, F., Egemose, S. & Meerhoff, M. (2012). Biomanipulation as a Restoration Tool to Combat Eutrophication: Recent Advances and Future Challenges. Advances in Ecological Research, 47.

Paerl, H. and Otten, T. (2013) Harmful Cyanobacterial Blooms: Causes, Consequences, and Controls. Environmental microbiology, 65, 995-1010.

Process Systems Enterprise (2013), gPROMS, www.psenterprise.com/gproms

Søndergaard, M., Jeppesen, E., Lauridsen, T. L., Skov, C., van Nes, E. H., Roijackers, R., Lammens, E. & Portielje, R. (2007). Lake restoration: successes, failures and long-term effects. Journal of Applied Ecology, 44, 1095–1105.

377028 Environmental Risk Analysis of Application of Electro-Disinfection on Ballast Water Treatment

Abstract:

Ballasting and de-ballasting are essential in balancing stability of ship during loading and unloading operation. As a result, biological invasion becomes a serious global issue to the fishing industry as well as eco-environmental system. It is there important to reduce the organisms prior to the water discharge to sea. It is observed that electro-disinfection is a cost-effective technology in the ballast water treatment. Due to the operation, a series of disinfection by-products (DBPs) is created, which may cause risk to the waters. This study addresses the formation and speciation of DBPs, including four trihalomethanes (THMs) and nine haloacetic acids (HAAs), resulting from the chlorination in ballast water treatment. The impact of several variables that influence DBPs formation, such as chlorine dosage, water quality, organic content, temperature, and chlorine contact time or residence time was investigated. The chlorine decay in the chlorination process was monitored and modeled. The potential risk of the formed DBPs to human being and marine environment was evaluated. Higher water temperature and/or total organic carbon (TOC) in the water increased the total residual chlorine decay rate. The rate of decay is different in the samples that had different represent organic compounds. The effect of salinity however less was obvious. Water quality and operation conditions affected the different classes of DBPs at different magnitudes. Four compounds, namely Ethanol, acetic acid, glucose and humic acid were used to represent the organic matters in seawater. More DBPs were produced as the TOC concentration was increased. In the predication of the DBPs, both TOC concentration and the distribution of organic species should be involved. Higher chlorine dose caused a higher DBPs formation. The dominant constitutes of THMs and HAAs were the bromo-substitutions, namely tribromomethane (TBM), dibromoacetic acid (DBAA) and tribromoacetic acid (TBAA). The concentrations of chloro-dervatives were at much lower levels in comparison with bromo-derivatives. The rate and total yield of THMs increased as the temperature was increased. For HAAs, more HAAs were generated as the temperature was increased from 4 to 25oC. When the temperature reached above 35 oC, HAAs concentration initially increased and then decreased after reaching its peak. DBAA and TBAA were the predominant HAAs species at the temperature ranging from 4 to 25oC, which accounted for more than 94% of total HAAs. Less TBAA content was observed when the temperature was increased. At 45 oC, no TBAA could be detected at all after 5 days reaction. The formation and speciation of THMs and HAAs were further investigated under various water quality and operational conditions typically used in land-based and shipboard tests. Higher salinity led to more THMs generation; more THMs species with chlorine components could be found at low salinity. Conversely, at same experimental temperature, it was observed that more HAAs were found in lower salinity seawater. Five separate seawaters were collected from different locations in Singapore to simulate ballast water taken from different sources/locations in tropical marine environment. The THMFP and HAAFP of the five tested natural seawater samples were different from each other. Among the five samples, the predominant THMs compound formed was TBM. The next of the THMs found was due to the DBCM generation; other two THMs compounds, namely BDCM and TCM, were not formed during the testing. TBAA was the major HAAs species found in all the natural water samples studied. DBAA appeared in all samples but its concentrations were at lower levels than TBAA. Generally speaking, the levels of DBPs in the chlorinated seawater at the points of discharges were found to be far higher than the drinking water regulated values set by USEPA and WHO. However, they would be well diluted when they are released to the sea. Modeling simulation by MAM-PEC was conducted based on various situations, with the consideration of several physico-chemical and biological processes and interactions. It was demonstrated that the PEC values calculated by the MAM-PEC model were at a safe level in most cases. It indicates that there is no risk to the marine environment caused by the discharged chlorinated ballast water.

Multi-Objective Optimization for Solid Waste Management Systems

José Ezequiel Santibañez-Aguilar,[1] Juan Martínez-Gómez,[1] José María Ponce-Ortega,[1] Fabricio Nápoles-Rivera,[1] Medardo Serna-González[1], and Mahmoud M. El-Hawagi[2, 3]

[1]*Chemical Engineering Department, Universidad Michocana de San Nicolás de Hidalgo, Morelia, Michoacán, México, 58060*

[2]*Chemical Engineering Department, Texas A&M University, College Station, TX, USA, 77843*

[3]*Adjunct Faculty at the Chemical and Materials Engineering Department, King Abdulaziz University, Jeddah, Saudi Arabia.*

Several options for reusing the municipal solid waste have been reported; however, integral approaches for optimizing the entire supply chains associated to the municipal solid waste have not been considered. Supply chains associated to waste management might help to reduce the waste generation in an environmentally and economically sustainable way. Particularly, multi-objective optimization approaches accounting for economic, environmental, social and safety issues in the entire life cycle have not been reported. Therefore, this work presents a mathematical programming model to the optimal planning for reusing municipal solid waste taking into account economic, environmental, social and safety issues simultaneously in a distributed framework. The optimization model is able to select the processing technologies, consumers, cities producing wastes, amount of recycled waste, products and location of processing facilities.

The optimization problem is formulated as a multi-objective problem that considers three different objectives: the net annual profit, the amount of reused wastes and the total fatalities generated with the considered risks. The addressed problem considers several cities with a given municipal solid waste production, these wastes can be utilized to obtain different useful products like energy, recycled material, olefins, fuels, etc. The waste is stored in several disposition centers or dumps; these dumps can affect the population's safety in different ways. The first one is the fatalities associated for consuming polluted water, since the pollutants in dumps can be transported through leaching from the surface to shallow well that can serve as water source to the population. In addition, the risk is associated the intoxication by the gas emissions caused by the burning of a given dump; this risk is taken into account by the implementation of a gas dispersion model. One of the main considerations is that each city is divided in several grids named sites, each one of these sites have different waste production, population and may have processing plants and dumps, this division is accomplished because each dump can affect in different magnitude each one of the sites, since each site is located at different distance from the dump considered.

The economic objective function is the net annual profit and considers all the costs of the entire supply chain; in this context, the net annual profit is the revenue from sales minus the costs for operation, capital, transportation, disposing and separating.

$$\begin{aligned} NETPROFIT = {} & REVENUE - OPCOST - CAPCOST^{total} \\ & - TRANSPCOST - DISPCOST - SEPCOST \end{aligned} \qquad (1)$$

The safety objective function is based on the total fatalities generated by intoxication and leaching. For the fatalities for intoxication, a frequency factor ($f^{frequency}_{factor}$) to obtain the number of times that the landfill is burned yearly is considered. The total fatalities are equal to the sum of the total fatalities ($F_{k,s}^{\substack{leaching\\fatalities}}$) for leaching plus the total fatalities for intoxication ($F_{g,d,a}^{\substack{intoxication\\fatalities}}$).

$$F^{\substack{total\\fatalities}} = f^{\substack{frequency\\factor}} \sum_{g}\sum_{d}\sum_{a} F_{g,d,a}^{\substack{intoxication\\fatalities}} + \sum_{k}\sum_{s} F_{k,s}^{\substack{leaching\\fatalities}} \quad (2)$$

The total consumed waste is a way to take into account the environmental impact although there are many forms to assess the environmental impact; this is an adequate way for the case of the municipal solid waste because one of the main impacts is for the huge amount of waste that is disposed in landfills. In this way, the total consumed waste is equal to the total waste that is sent to the processing facilities from dumps divided by the total waste that is produced in the each site of all cities.

$$WASTECONS = \sum_{wn\in WASTES}\sum_{d\in DUMPS} W_{wn,d}^{\substack{out\\dump}} \Big/ \sum_{wn\in WASTES}\sum_{s\in SITES} W_{wn,s}^{produced} \quad (3)$$

A case study for planning the waste management in the city of Morelia in Mexico was considered. Figure 1 shows the tradeoffs between the number of fatalities and the net annual profit; notice the number of fatalities can decrease until 21 % from the maximum number of fatalities while the change in the profit is about 19 %. It is important to note that the fatalities for leaching are the most important in the proposed case study. Furthermore, Figure 1 shows that the consumed waste is almost constant when the Pareto curve is carried out with the net annual profit and the number of fatalities, and this way the number of fatalities is not function of the consumed waste. Results show that it is possible to implement a distributed processing system to reuse municipal solid waste in an economically attractive way. In addition, results can be used for governments to make decisions about the waste disposal and define the amount of waste that must be reused to obtain several products. It should be noted that results include the supply chain configuration.

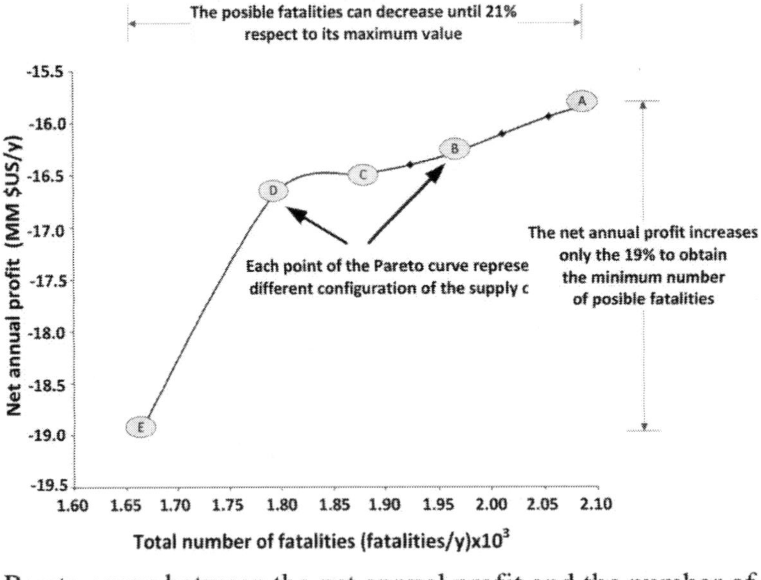

Figure 1. Pareto curve between the net annual profit and the number of fatalities.

384303 Global Optimization for Sustainable Design and Synthesis of Algae Processing Network for CO2 Mitigation and Biofuel Production Using Life Cycle Optimization

Abstract:

Renewable biofuel is considered as a promising fossil fuel substitute due to its potential for large-scale production and carbon mitigation. Among various biomass feedstocks with increasing attention in the recent decades, algal biomass offers significant advantages over conventional energy crops [1]. Algae have high area productivity and accumulate lipid materials up to 80% of its dry cell weight. They are able to utilize non-arable land and waste water for cultivation. Additionally, the hydrocarbon algal biofuels converted via hydroprocessing accommodate the specifications of drop-in fuels, which can be directly incorporated into the existing infrastructure [2]. In order to reduce environmental impacts and improve the overall economic performance, systematic modeling and optimization frameworks can be utilized to identify and assess the sustainable design and synthesis of a processing network [3-5]. Therefore, applying such a framework to the algae processing network provides valuable insights into the technology pathways in the context of sustainability. The goal of this research is to identify and assess the optimal sustainable design and synthesis of an algae processing network with both economic and environmental concerns.

In this work, we propose a superstructure of an algae processing network including 7,800 processing routes and 11 processing sections, namely cultivation, harvesting, primary dewatering, secondary dewatering, storage, cell disruption & hydrothermal liquefaction, lipid extraction, upgrading, remnant treatment, electricity generation, and steam generation. Feed gas, off-gas, makeup nutrients, and recycled materials are fed into the cultivation system and mature algae product from cultivation is concentrated by several technologies. First, 10% of the dilute slurry is harvested by auto flocculation and dissolved air flotation. The water content is further reduced by primary and secondary dewatering. Most of the reduced materials are recycled to the cultivation facility. In order to keep downstream facilities operating continuously, we employ a storage tank to temporarily store part of the enriched algae biomass for the production at night [6]. Next, the enriched microalgae paste undergoes cell disruption and lipid extraction or HTL so that most of the lipid materials are separated from algal cells. The lipid materials are then upgraded to biodiesel or renewable diesel depending on the specific conversion technologies. The rest of the algal body or remnant is fed to an anaerobic digester and decomposed into biogas, aqueous nutrients, and unreacted solids. Additionally, we couple electricity and steam generation into the network to satisfy the onsite energy consumption and reuse the resulting off-gas.

Based on this superstructure, a multi-objective mixed-integer nonlinear programming (MINLP) model for the sustainable design and synthesis is developed following the life cycle optimization frame work that integrates the multi-objective superstructure optimization scheme with life cycle assessment and techno-economic analysis [3, 4, 7-10]. The model consists of two objective functions and simultaneously minimize the unit annualized cost and the unit Global Warming Potential (GWP). As the final biofuel product can be biodiesel or renewable diesel depending on the specific upgrading technology, the unit objective functions are associated with the unit gasoline gallon equivalent (GGE) of the specific biofuel produced. To efficiently solve the problem, we propose a global optimization strategy that integrates a branch-and-refine algorithm based on successive piecewise linear approximations and an exact parametric algorithm based on Newton's method [11-15]. The multi-objective optimization problem results in two Pareto-optimal curves for biofuel production and biological carbon sequestration, respectively. Three optimal processing routes are identified. From the biofuel production perspective, the most environmentally sustainable processing route achieved a unit GWP of 16.520 kg CO_2-eq/GGE and a unit annualized cost of $9.712/GGE, while the most economical processing route resulted in a unit annualized cost of $7.017/GGE and a unit GWP of 26.791 kg CO_2-eq/GGE. If the algae processing network is utilized for biological carbon sequestration, the unit carbon sequestration and utilization cost ranges from $10.43 /t of CO_2 to $1.64 /t of CO_2, corresponding to unit GWP's from 352.41 kg CO_2-eq/t of CO_2 to 412.90 kg CO_2-eq/t of CO_2, respectively.

References

[1] J. Ferrell and V. Sarisky-Reed, "National algal biofuels technology roadmap," US Department of Energy College Park, MD2010.

[2] D. Yue, F. You, and S. W. Snyder, "Biomass-to-bioenergy and biofuel supply chain optimization: Overview, key issues and challenges," *Computers & Chemical Engineering,* 2013.

[3] D. J. Yue, M. A. Kim, and F. Q. You, "Design of Sustainable Product Systems and Supply Chains with Life Cycle Optimization Based on Functional Unit: General Modeling Framework, Mixed-Integer Nonlinear Programming Algorithms and Case Study on Hydrocarbon Biofuels," *Acs Sustainable Chemistry & Engineering,* vol. 1, pp. 1003-1014, Aug 2013.

[4] B. Wang, B. H. Gebreslassie, and F. Q. You, "Sustainable design and synthesis of hydrocarbon biorefinery via gasification pathway: Integrated life cycle assessment and technoeconomic analysis with multiobjective superstructure optimization," *Computers & Chemical Engineering,* vol. 52, pp. 55-76, May 10 2013.

[5] B. H. Gebreslassie, R. Waymire, and F. You, "Sustainable design and synthesis of algae-based biorefinery for simultaneous hydrocarbon biofuel production and carbon sequestration," *AIChE Journal,* vol. 59, pp. 1599-1621, 2013.

[6] J. Gong and F. You, "Optimal Design and Synthesis of Algal Biorefinery Processes for Biological Carbon Sequestration and Utilization with Zero Direct Greenhouse Gas Emissions: MINLP Model and Global Optimization Algorithm," *Industrial & Engineering Chemistry Research,* vol. 53, pp. 1563-1579, 2014/01/29 2014.

[7] F. Q. You, L. Tao, D. J. Graziano, and S. W. Snyder, "Optimal design of sustainable cellulosic biofuel supply chains: Multiobjective optimization coupled with life cycle assessment and input-output analysis," *Aiche Journal,* vol. 58, pp. 1157-1180, Apr 2012.

[8] F. Q. You and B. Wang, "Life Cycle Optimization of Biomass-to-Liquid Supply Chains with Distributed-Centralized Processing Networks," *Industrial & Engineering Chemistry Research,* vol. 50, pp. 10102-10127, Sep 7

2011.

[9] B. H. Gebreslassie, M. Slivinsky, B. L. Wang, and F. Q. You, "Life cycle optimization for sustainable design and operations of hydrocarbon biorefinery via fast pyrolysis, hydrotreating and hydrocracking," *Computers & Chemical Engineering,* vol. 50, pp. 71-91, Mar 5 2013.

[10] Q. Zhang, J. Gong, M. Skwarczek, D. Yue, and F. You, "Sustainable process design and synthesis of hydrocarbon biorefinery through fast pyrolysis and hydroprocessing," *AIChE Journal,* vol. 60, pp. 980-994, 2014.

[11] F. Q. You and I. E. Grossmann, "Stochastic Inventory Management for Tactical Process Planning Under Uncertainties: MINLP Models and Algorithms," *Aiche Journal,* vol. 57, pp. 1250-1277, May 2011.

[12] Z. X. Zhong and F. Q. You, "Globally convergent exact and inexact parametric algorithms for solving large-scale mixed-integer fractional programs and applications in process systems engineering," *Computers & Chemical Engineering,* vol. 61, pp. 90-101, Feb 11 2014.

[13] F. Q. You, J. M. Pinto, I. E. Grossmann, and L. Megan, "Optimal Distribution-Inventory Planning of Industrial Gases. II. MINLP Models and Algorithms for Stochastic Cases," *Industrial & Engineering Chemistry Research,* vol. 50, pp. 2928-2945, Mar 2 2011.

[14] D. J. Yue and F. Q. You, "Planning and Scheduling of Flexible Process Networks Under Uncertainty with Stochastic Inventory: MINLP Models and Algorithm," *Aiche Journal,* vol. 59, pp. 1511-1532, May 2013.

[15] F. Q. You, P. M. Castro, and I. E. Grossmann, "Dinkelbach's algorithm as an efficient method to solve a class of MINLP models for large-scale cyclic scheduling problems," *Computers & Chemical Engineering,* vol. 33, pp. 1879-1889, Nov 12 2009.

391695 Towards Integration of Life Cycle Assessment within a Multi Objective Supply Chain Optimization Modeling Framework

Abstract:

This paper presents advancements in integration of multiple environmental indicators (CO_2e, Human health, Ecosystem health) based on life cycle assessment methods within the framework of supply chain modeling and optimization. We present results related to the optimal design of supply chain networks with consideration of environmental and economic objectives. The environmental objective is measured through life cycle assessment indicators. IMPACT 2002+ methodology is used to evaluate the environmental scores. Global warming, human health damages and ecotoxicity impacts are considered for evaulating the environmental impacts. On the other hand, minimization of total production cost and maximization of net present value of the supply chain is another goal. A multi objective mixed integer linear programing (MILP) model is developed for optimal design of green supply chain. Multi objective optimization is accomplished through non dominated sorted genetic algorithm. The resulting Pareto-optimal curves demonstrate the tradeoff between economic and environmental pathways. This method considers life cycle emissions from cradle to gate for each process and suggests alternatives technologies for the reduction of these pollutants while considering financial implications. The method can be used to derive optimal solutions for scheduling and planning or could be used to design future green supply chains. The methodology is demonstrated via several case studies.

383858 Optimal Design of Microgrids with Innovative CHP Systems: Integration of Process Optimisation and Life Cycle Assessment

Abstract:

Optimal design of microgrids with innovative CHP systems: integration of process optimisation and life cycle assessment

Di Zhang, Sara Evangelisti, Paola Lettieri, Lazaros G. Papageorgiou

Department of Chemical Engineering, University College London, Torrington Place, London WC1E 7JE, UK

As an alternative to current centralised energy generation, microgrids are adopted to provide local energy with lower energy expenses and gas emissions by utilising distributed energy resources (DER). DER is defined as local resources related to the energy system, such as generators, boilers and energy storage devices. Micro Combined Heat and Power (micro CHP) units simultaneously produce heat and power for a single building. In order to meet renewable energy target in the UK, the installation of CHP units is considered because of their flexibility, reliability and safer operating conditions with high overall efficiencies. Several micro CHP technologies have been developed recently for applications at domestic scale, including fuel cell, micro turbine, internal combustion engine and sterling engine. The optimal design of DERs within microgrid plays an important role in promoting the penetration of microgrid. It attracts money research attentions from the economic view.

Different tools exist to evaluate the environmental impacts of a process or a service. Life Cycle Assessment (LCA) is recognised as one of the best tools available. Moreover, LCA is one of the most developed and widely used environmental assessment tools for comparing alternative technologies when the location of the activity is defined. Studies have been published on carbon footprint of micro – CHP units rather than a holistic environmental appraisal. A few studies have focused on the environmental impacts of different micro-CHP technologies analysing the functioning of single units rather than microgrids.

Limited studies in the literature have been focused on coupling environmental and economic sustainability in a multi-objective optimisation model: the majority of them focused on the operation of a single unit. To our knowledge, no studies have been published on the optimal design of microgrids considering environmental and economic metrics. In this work, the optimal design of microgrids with innovative CHP units is addressed while involving both concerns. A multi-objective modelling framework for the optimal design is presented. The trade-off between the two conflicting objectives is obtained by solving the proposed model using the ε-constraint method.

This work considers a general microgrid, which involves several participant sites of different building types. Macrogrid is available to provide electricity to the participant in the microgrid and extra electricity can also be sold back to the macrogrid when it benefits. The candidate technologies include CHP generators (with different capacities and heat-to-power ratios), boilers, thermal storage and a macrogrid power connection. Micro CHP technologies considered in this work are Proton exchange membrane (PEM) Fuel Cell, Solid Oxide Fuel Cell (SOFC), Internal Combustion Engine (ICE) and Stirling Engine (SE). Turn-key costs of CHP generators are based on the CHP types as well as the capacity range. The microgrid and the macrogrid are interacted and constrained through exporting or importing electricity.

The goal of the LCA in this work is to estimate the environmental impacts of four different CHP systems with additional thermal energy storage (TES) working in a distributed generation network. The life span considered is 15 years. The impact categories considered in this study are global warming potential (GWP) and acidification potential (AP). The output of LCA is used as the input for the optimisation model. A multi-objective mixed-integer linear programming (MILP) approach is developed for this problem. The key decision variables include unit capacities and resources utilised. They are determined by minimise the total equivalent annualised cost (EAC), GWP and AP of all participants, subject to equipment capacity constraints, CHP ramp limit constraints, energy demand constraints, CHP selection constraints, thermal storage constraints and transfer price level constraints.

The proposed model is implemented for different microgrid modes. For each mode, optimal design of the microgrid is obtained with (a) the candidate technologies selected and their capacities, (b) energy resources consumed, (c) energy production plan, (d) thermal energy storage plan. The trade-off between the economic and environmental objectives is analysed with a set of Pareto-optimal solutions.

Keywords: microgrids; life cycle assessment; multi-objective optimisation; ε-constraint method

Application of sustainability footprint and multicriteria optimization for the choice of sustainable chemical processes

AIChE Annual Meeting, Atlanta, GA, November 16-21, 2014
For submission to "Advances in Life Cycle Optimization for Process Development"
Rajib Mukherjee[1], Debalina Sengupta[2], Subhas K. Sikdar[3]

1. *Vishwamitra Research Institute, Chicago, IL*
2. *Texas A &M University, College Station, TX*
3. *NRMRL, US EPA, Cincinnati, OH,*

Abstract

A chemical product can be obtained by different process pathways involving varying amounts and types of resources, utilities, product, and byproduct formation. Evidently, the sustainability in terms of economic, environmental and social impacts of these processes will vary relatively. When such competing process options are presented, and compared to an existing process pathway, it is necessary to evaluate and identify the most sustainable process among these competing options. Sustainability of a chemical process is generally evaluated with indicators that need process and chemical properties data. These indicators individually reflect the impacts of the process on the respective areas of sustainability. In order to choose among several alternative processes an overall comparative analysis is essential.

To make a decision on the most sustainable process, the indicators need to be rationally aggregated into a single index. In this research, the suite of sustainability indicators includes traditional economic indicators to represent the economic dimension, environmental indicators, and human health indicators to represent the social dimension. The environmental and human health indicators were computed using the TRACI impact characterization method. Process flow and economic data were used to compute the indicators in these categories. These indicators were then aggregated to compute the sustainability footprint (D_e) of each process and evaluate the relative sustainability of the competing processes.

This method of process selection is applied to five competing processes for methanol production from carbon dioxide. These are novel chemical processes which involve carbon dioxide as raw material, and a uniform industrial scale capacity for production of methanol. The processes differ in the amount of raw material required, byproducts formed, and utilities used. Based on process data, these processes are compared using the sustainability footprint method. This alone should give us information to evaluate the best process as the one which has the minimum sustainability footprint, D_e.

However, we wanted to see if the minimization of the sustainability footprint, D_e, can be used as an objective function for a mixed integer optimization problem. The aim in solving the multi-criteria problem will be to select the most sustainable option among the five competing processes, and also determine the optimum values of process parameters which will enable the same.

Based on our results, this research will demonstrate how an aggregate sustainability footprint index can be utilized to compare between competing processes, as well as how it can be integrated into a mixed integer problem formulation to determine the optimum process conditions which help in determining the most sustainable process among competing options.

Sustainable Biorefinery Supply Network using a Concept of Eco- and Total Profit – Case Study of the European Union

Lidija Čuček,[a] Mariano Martín,[b] Ignacio E. Grossmann,[c] Zdravko Kravanja[a]

[a]*Faculty of Chemistry and Chemical Engineering, University of Maribor, Smetanova ul. 17, Maribor, Slovenia, lidija.cucek@um.si, zdravko.kravanja@um.si*

[b]*Department of Chemical Engineering, University of Salamanca, Salamanca, Spain, mariano.m3@usal.es*

[c]*Department of Chemical Engineering, Carnegie Mellon University, Pittsburgh, USA, grossmann@cmu.edu*

Abstract

Nowadays the shortages of energy and water resources, the increasing prices, the demand for and dependence on fossil fuels and a growing environmental consciousness, have become societal priorities. These concerns have supported the development of alternative transportation fuels and the use of renewable sources of energy. Thus, governments across the globe have set targets on fossil fuels' substitutions over the next decade. In order to be able to meet these targets, not only process designs for the productions of alternative fuels are needed, but also the development of the networks which account for all the stages from the energy source to the final product. Biomass, in particular, is a challenging resource because of its seasonality, its degradation over time, its low density, regional availability, growth yield, and the large area required to grow it. However, biomass can provide a reasonable alternative to crude-based gasoline and diesel in the short and medium terms, although the competitiveness of biomass as an energy source is highly dependent on its supply chain. Lately, several studies have extended the supply-chain studies to actual regions across the US, but only for a limited number of raw materials and fuels, focusing mainly on lignocellulosic resources and in most of the cases only for bioethanol production, although FT-products have also been accounted for [1,2].

This contribution deals with a sustainable large-scale biorefinery supply network by considering both economic and environmental objectives. A biorefinery's supply network accounts for different biomass sources (algae, lignocellulosic materials, corn, wheat, cooking oil), conversion technologies (biochemical, thermochemical, thermo-biochemical, oil extraction, transesterification) and includes first, second, and third generations of biofuels, such as bioethanol, biodiesel, hydrogen, Fischer-Tropsch diesel, and green gasoline [3] accounting for water consumption and availability and energy integration within the network. Multi-period synthesis of an optimally-integrated biorefinery supply network [4] was performed through maximization of the economic objective, and applied on the case study of the European Union [5].

In this contribution the synthesis has now been extended to simultaneously accounting for environmental impact. The concept of Eco- and Total profit [6] was used for this purpose, which offers two important advantages over conventional multi-objective approaches: i) the concept accounts for both, burdening and unburdening effects [6] on the environment, as Eco-Profit is defined as the difference between Eco-Benefit (unburdening) and Eco-Cost (burdening), expressed in monetary terms. ii) by applying the concept where multi-objective solutions are reduced into single-objective ones as Total profit is composed of both, economic and Eco-profits. This enables the obtaining of sustainable solutions directly, which is especially important when solving supply network problems at the continental level when the models are huge. The solutions obtained for the case study of the European Union are those alternatives that would unburden the environment the most while having significant profit.

Acknowledgments

The authors acknowledge the financial support from the Slovenian Research Agency (Programs P2-0032 and P2-0377) and Centre for Advanced Process Decision-making.

References

[1] Elia, J.A., Baliban, R.C., Floudas, C.A. Nationwide energy supply chain analysis for hybrid feedstock processes with significant CO_2 emissions reduction. AIChE Journal. 2012; 58(7), 2142-2154.

[2] Marvin, A. W., Schmidt, L. D., Benjaafar, S., Tiffany, D. G., & Daoutidis, P. Economic Optimization of a Lignocellulosic Biomass-to-Ethanol Supply Chain. Chemical Engineering Science. 2012;67, 68-79.

[3] Martín M, Grossmann IE. On the systematic synthesis of sustainable biorefineries. Industrial & Engineering Chemistry Research. 2013;52(9):3044-64.

[4] Čuček L, Martín M, Grossmann IE, Kravanja Z. Multi-period Synthesis of Optimally-Integrated Biomass and Bioenergy Supply Network. Computers and Chemical Engineering. 2014:doi: 10.1016/j.compchemeng.2014.02.020.

[5] Čuček L, Martín M, Grossmann IE, Kravanja Z. Large-Scale Biorefinery Supply Network – Case Study of the European Union. In: J. J. Klemeš, P. S. Varbanov, P. Y. Liew, editor. Computer Aided Chemical Engineering: Elsevier; 2014, accepted for publication.

[6] Kravanja Z, Čuček L. Multi-objective optimisation for generating sustainable solutions considering total effects on the environment. Applied Energy. 2013;101:67-80.

Planning and Scheduling Industrial Waste Management Using Knowledge Based Lagrangean Decomposition

Elisabet Capón-García[a], Edrisi Muñoz[b], José Miguel Laínez[c], Antonio Espuña[d], Konrad Hungerbühler[a], and Luis Puigjaner[d]

[a]*Chemistry and Applied Biosciences, Institute for Chemical- and Bioengineering, Swiss Federal Institute of Technology, Zurich (ETHZ)*
[b]*Information Technologies, Centro de Investigacion en Matematicas A.C., Guanaguato, Mexico,*
[c]*Department of Industrial & Systems Engineering, University at Buffalo, Buffalo, NY,*
[d]*Chemical Engineering Department, Universitat Politècnica de Catalunya - ETSEIB, Barcelona, Spain*

Waste liquid effluents originated from industrial activities contain contaminants and certain species, such as organic solvents, with material and energy recovery potential. Therefore, not only does industrial waste management stand as an end-of-pipe problem, but also as a potential origin of resources, which can result insignificant benefits if material and energy integration strategies are used. The simultaneous planning and scheduling of industrial waste management allows improved resource utilization and reduced environmental impact. However, rigorous mathematical models usually result in large scale problems. In order to tackle the problem complexity, decomposition techniques based on information flows between a master and a set of sub-problems are widely applied (Maravelias and Sung, 2009, Calfa et al., 2013).

In this work, we exploit this capability of ontologies to address the optimal integration of planning and scheduling using a Lagrangean decomposition approach. Thus, ontologies improve information sharing and communication in the enterprise and can even represent holistic mathematical models facilitating the use of analytic tools and providing higher flexibility for model building. Therefore, the scheduling/planning sub-problems are created for the waste management facility and their dual solution information is shared by means of the ontological framework. Furthermore, this work presents a multi-period waste management multi-objective optimization, considering economic and environmental issues. The behavior of waste treatment units is included in the optimization problem as black-box models based on industrial practice.

A case study of the planning and scheduling of industrial waste facilities is presented to show the advantages and limitations of the proposed approach in a multi-objective environment considering economic and environmental issues using the decomposition strategy. The proposed framework leads to reduced computation effort and better solutions in terms of solution quality, since waste stream scheduling is integrated in decision-making.

Acknowledgements
Financial support received from the Spanish "Ministerio de Economía y Competitividad" and the European Regional Development Fund, both funding the research Project

SIGERA (DPI2012-37154-C02-01), and the "Agència de Gestió d'Ajuts Universitaris i de Recerca (2014SGR1092)" is thankfully acknowledged.

References

Calfa, B. A., Agarwal, A., Grossmann, I. E., Wassick, J. M., 2013. Hybrid bilevel-lagrangean decomposition scheme for the integration of planning and scheduling of a network of batch plants. Industrial & Engineering Chemistry Research 52 (5), 2152-2167.

Maravelias, C. T., Sung, C., 2009. Integration of production planning and scheduling: Overview, challenges and opportunities. Computers & Chemical Engineering 33 (12), 1919-1930.

391026 Identifying the Preferred Subset of Alternatives for Environmental Improvements Via an MILP Approach Based on the Analytic Hierarchy Process

Abstract:

The combined use of multi-objective optimization (MOO) and LCA has recently gained wider interest in process systems engineering. This approach provides as output a set of Pareto solutions that represent the optimal trade-off between the economic and environmental concerns considered in the analysis. From this set of optimal alternatives, decision-makers should identify the ones that better fulfill their preferences. Generating a large and representative enough subset of Pareto points to aid decision-making is challenging. This task is particularly difficult in problems with a large number of (environmental) objectives, as is the case when incorporating LCA principles in MOO.

In this work we present a mixed-integer linear programming method that simplifies MOO problems by concentrating on determining only a reduced number of Pareto points that are particularly appealing. Our approach, which relies on the analytic hierarchy process (AHP), identifies a set of weights to be assigned to the environmental objectives so as to translate them into a single aggregated indicator that reflects to the maximum extent possible the decision-makers' preferences. For every such combination of weights, we solve a single objective problem that optimizes the corresponding weighted sum of objectives, thereby generating only a subset of alternatives (Pareto solutions) reflecting preferences with a large degree of consistency. We illustrate the capabilities of our approach through its application to the design of supply chains for bioethanol production.

Micro-kinetics of NO_x Storage and Reduction with $H_2/CO/C_3H_6$ on $Pt/BaO/Al_2O_3$ Monolith Catalysts

Arun Kota, Vemuri Balakotaiah and Dan Luss

Department of Chemical and Biomolecular Engineering,

University of Houston, 4800 Calhoun Rd, Houston, TX - 77204

Abstract:

Steady state NO_x storage and reduction on $Pt/BaO/Al_2O_3$ by $H_2/CO/C_3H_6$ reactant mixtures is studied using an isothermal, 1+1 dimensional two-phase model that accounts for washcoat diffusional limitations. It predicts the storage and reduction performance over a wide range of inlet feed concentrations and temperatures. Global kinetic models have a major limitation of not being able to predict accurately the performance of the catalyst with changing feed composition. To account for such variations, micro-kinetic models elucidating the role of individual reductants have been developed. The temperature dependence of the rate constants has been quantified to reflect the surface chemistry that is consistent with experimental observations. Figure 1 shows the comparison between the experimental and model predicted data for NO/H_2, NO/CO and NO/C_3H_6 systems. The model predicts the experimentally observed trends in individual systems reasonably well and is further used to predict the conversions and selectivities for a combined $NO/H_2/CO/C_3H_6$ system.

Inclusion of washcoat diffusion enables the model to accurately predict the NO_x conversion and scale-up performance of the catalyst over a wide temperature range. Figure 2 shows that the model accurately represents different controlling regimes during NO/H_2 reaction. It can be seen that the system goes from being kinetic controlled to washcoat diffusion controlled as temperature increases. At higher temperatures most of the reaction is restricted to a thin layer of the washcoat close to

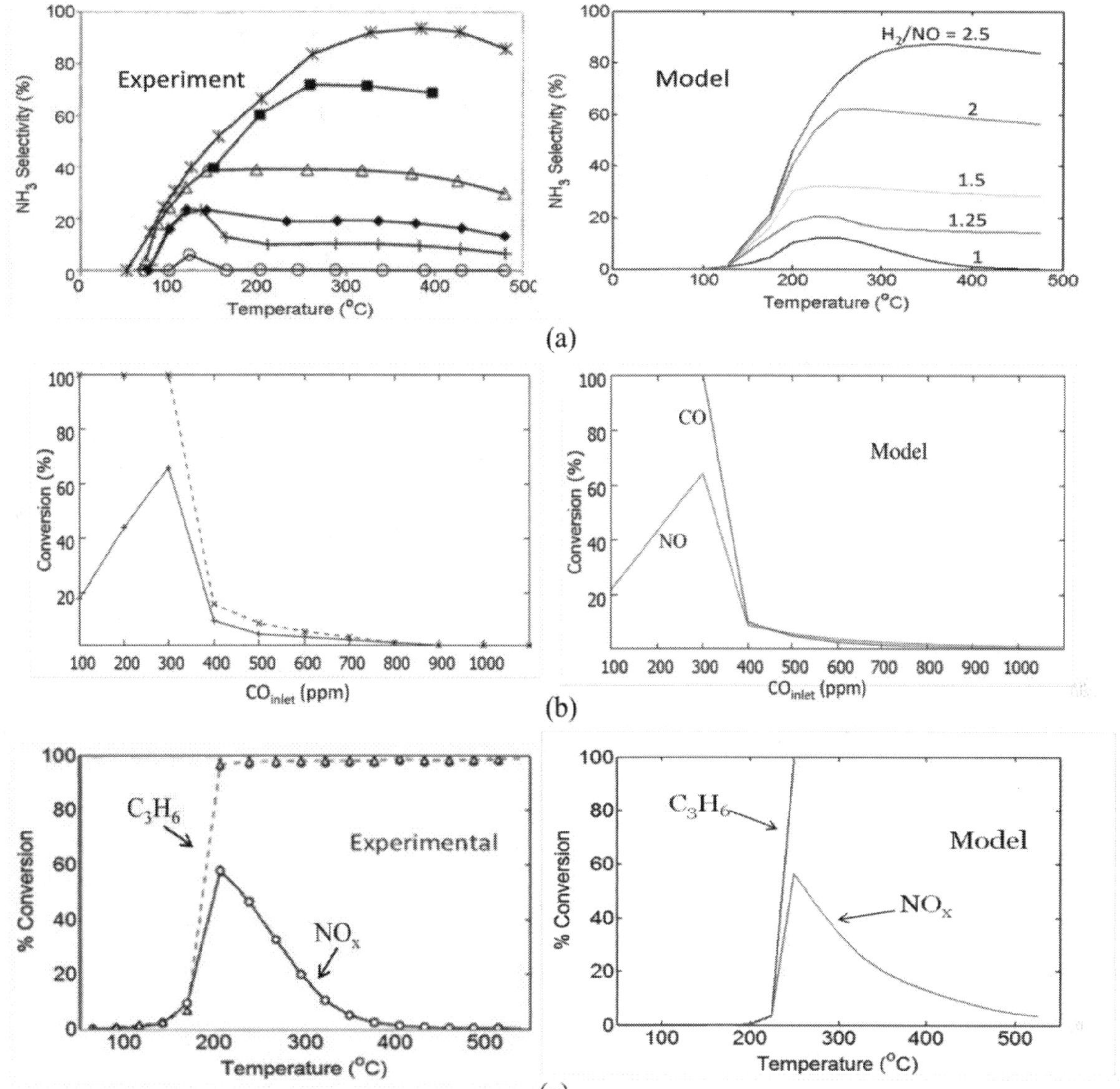

Figure 1: Comparison of the experimental and model predicted conversions and selectivities (a) NH_3 selectivity in NO/H_2 system (b) NO and CO conversions in NO/CO system and (c) NO and C_3H_6 conversions in NO/C_3H_6 system.

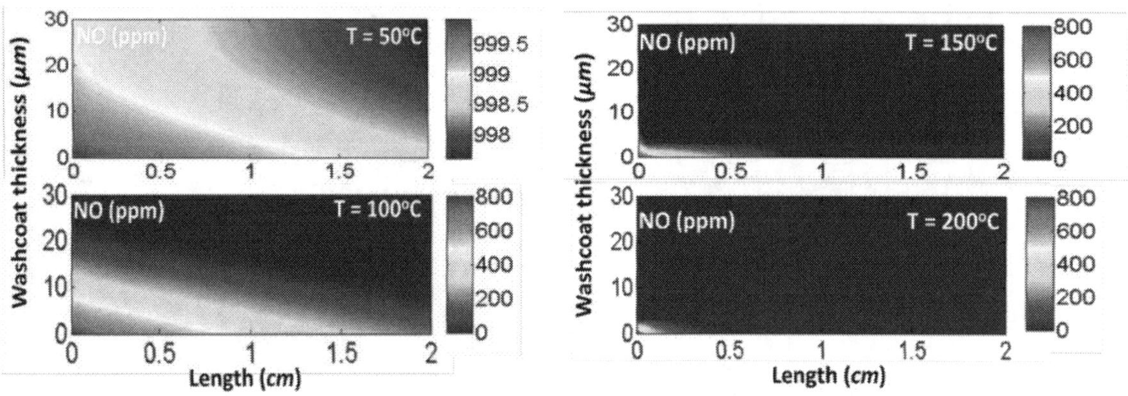

Figure 2: NO concentration inside the washcoat at different temperatures for NO/H_2 system [NO_{inlet}=1000ppm, H_{2inlet}=2500ppm; Pt loading = 2.2 wt%].

the fluid-washcoat interface. The sensitivity of the estimated kinetic parameters to the effective diffusivities in the washcoat is also analyzed by pore scale comparison between the time constants of adsorption ($C_0/R_{ads}(C_0)$) for H_2/CO/C_3H_6 and pore scale diffusion time constants ($\frac{r_{eff}^2}{D_{eff}}$). Results indicate that the estimated micro-kinetic parameters are influenced by the washcoat pore structure and the properties of the precious metal and storage components.

385847 Deep SO2 Adsorption at Parts per Billion Level By Alumina-Based Mn/Ce Mixed Oxides for SOFC Cathode Protection

Abstract:

Gas phase contaminant removal is a critical process for current and future power supply systems including fuel cells, gas turbines (GT), or a combination on ships. Because of high energy efficiency, low emissions, and fuel flexibility, Solid Oxide Fuel Cells (SOFCs) are promising energy converting devices. Recently, however, researchers have shown a significant degradation caused by SO_2 reacts with cathode side materials and thus hinders the active sites available for oxygen reduction. The major SO_2 source is the emission from marine fuels such as JP-5 and JP-8 whose sulfur content can be up to 1200 ppmw. The cell exposed to as low as 100 ppbv of SO_2 can result in a current loss during the test. Therefore, it is necessary to remove SO_2 from cathode side. In this work, alumina-supported/alumina-ceria supported manganese oxides were used as adsorbents operated at high space velocity. Breakthrough capacity and saturation capacity were evaluated by varying Mn loadings, namely 2.5, 5.0, 7.5, 10, and 15 wt %. The structural and surface properties of the adsorbents were further characterized by XRD, N_2 physisorption, H_2-TPR, CO_2-TPD, and TEM. The SO_2 adsorption capacity of the mixed oxides correlated closely with their metallic dispersion, surface basicity, and average oxidation state. Mn (5 wt %)/CeO_2-Al_2O_3 was found to be an optimum adsorbent in a wide temperature range of 20-500°C. H_2-TPR technique revealed that the average oxidation state for 5 wt % Mn was around +3.8; while Mn loading (<2.5 wt % or >15 wt %) upon impregnation of nitrate precursor solution was likely to lower the oxidation state to around +3.3, which was related to catalytic activity for S^{4+} oxidation. Furthermore, the reducibility of mixed oxides can be enhanced by introducing Ce into Al_2O_3, facilitating the regeneration of spent Mn/CeO_2-Al_2O_3.

A first-principles study of the catalytic hydrodechlorination of 1,2-dichloroethane

L. Xu, E. Stangland, M. Mavrikakis

[1]Department of Chemical & Biological Engineering, University of Wisconsin-Madison

[2]Core Research and Development, the Dow Chemical Company, Midland MI

Chlorinated hydrocarbons are common intermediates and by-products from various chemical processes. Industrial chlorocarbon wastes are serious environmental pollutants, and an efficient treatment method is highly desired [1]. The catalytic hydrodechlorination, which utilizes hydrogen to convert chlorinated species into hydrogen chloride and valuable hydrocarbon products, is an attractive approach from both the economic and environmental standpoints. Catalytic hydrodechlorination of 1,2-dichlorethane produces both ethylene and ethane, and thus poses an additional challenge of selectivity, where ethylene, due to its higher economic value, is the desired product over ethane.

Here we aim to investigate the hydrodechlorination of 1,2-dichloroethane over transition metal surfaces. Past studies have shown that bimetallic catalysts yield good catalytic performance in terms of both the dechlorination activity and the ethylene selectivity [2]. The lack of fundamental understanding of the reaction mechanism and the catalytic behavior, however, hinders catalyst design and optimization. Therefore, we adopted a first-principles approach in our study. Through density functional theory calculations, we evaluate the energetics of adsorbates and reactions over metal surfaces. The results offer valuable insights into the surface reaction mechanism and will potentially facilitate the rational design of optimal catalysts. This presentation will demonstrate the trend of hydrodechlorination reactivity over a series of monometallic surfaces, followed by a detailed mechanistic study on platinum catalysts.

[1] E. D. Goldberg, Halogenated hydrocarbons: past, present and near-future problems, *Science of the Total Environment* **100**, 17-28 (1991).
[2] V. I. Kovalchuk and J. L. d'Itri, Catalytic chemistry of chloro- and chlorofluorocarbon dehalogenation: from macroscopic observations to molecular level understanding, *Applied Catalysis A: General* **271**, 13-25 (2004).

Ultra-low loading, highly active and stable Ru/γ-Al_2O_3 catalyst for low temperature solar steam reforming of methane

David S. A. Simakov and Yuriy Román-Leshkov

Department of Chemical Engineering, Massachusetts Institute of Technology, Cambridge MA 02139, USA

Converting solar irradiation into chemical energy via solar thermal steam reforming of natural gas is an attractive method to increase the energy content of methane, consequently reducing both natural gas consumption and greenhouse gases emission. Of particular interest is collecting solar radiation through the parabolic trough technology, which is mature and relatively inexpensive, but limited to temperatures ranging from 400-600 °C. One of the major challenges is developing catalysts that are highly active and stable in this low temperature range. Commercial nickel-based catalysts do not have sufficient activity, are prone to fast oxidation deactivation and coke excessively. Catalysts based on platinum group metals do not suffer from these drawbacks, but their cost is a prohibiting factor. In the present work, we demonstrate that Ru/γ-Al_2O_3 catalyst prepared by sonication-assisted wet impregnation with the ultra-low loading of Ru of 0.15 wt% has superior activity (Fig. 1). The catalyst outperforms commercial Ni catalyst (12 wt% of Ni) by about two orders of magnitude in terms of methane conversion rate per amount of active metal (Fig. 2), while also providing excellent resistance to sintering, oxidation and coking. The systematic evaluation of catalytic activity over a range of Ru nanoparticles loading and under various operating conditions (pressure, temperature and space velocity) will be presented, along with a comprehensive characterization study using a variety of analytical techniques to investigate particle size, dispersion, and metal-support interaction effects.

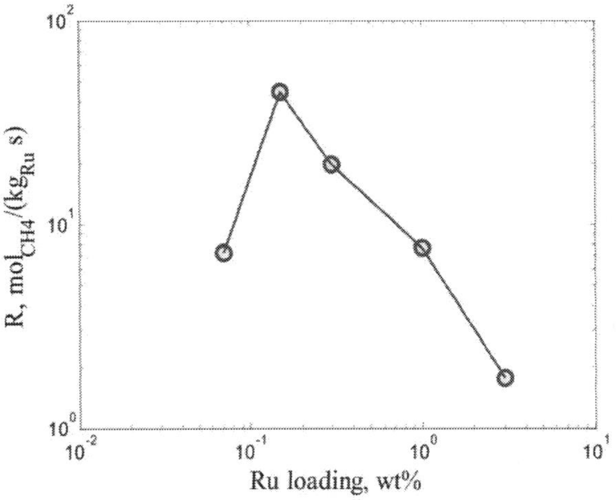

Figure 1: Steady state rate of methane conversion per amount of active metal as a function of Ru loading. P=1 bar, T=600 °C, H_2O/CH_4=3 in the feed, CH_4 GHSV=7500 mL/(g h)

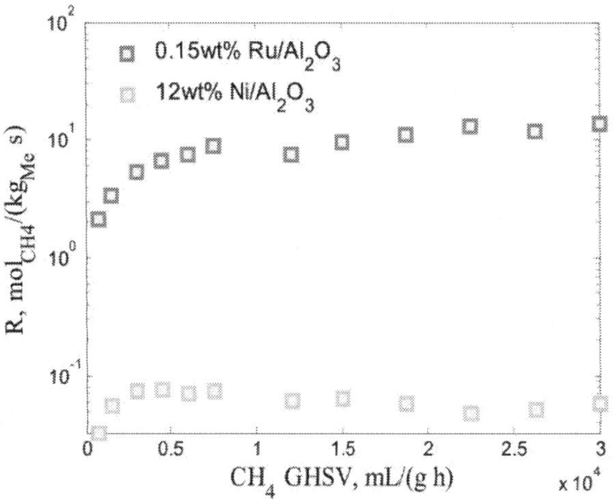

Figure 2: Steady state rate of methane conversion per amount of active metal as a function of GHSV. P=1 bar, T=500 °C, H_2O/CH_4=3 in the feed

Spatio-temporal Features of NOx reduction on LNT+SCR System:

Effect of Hydrocarbon Reductant and Space Velocity

Mengmeng Li, Vencon G. Easterling, Michael P.Harold

Previous studies have shown, in order of effectiveness, that H_2, CO, and hydrocarbons (HCs) are capable of reducing NOx present in diesel exhaust across a lean NOx trap (LNT) catalyst. Additionally, HCs and NH_3 have been utilized to reduce NOx to N_2 using selective catalytic reduction (SCR) catalysts. Recently, the synergy of combining LNT with SCR has attained NOx reduction levels that equal to or greater than stand-alone LNT or SCR while avoiding the need for the urea feed and reducing the loading of the expensive precious metal over LNT catalyst. Although simpler reductants such as H_2 and CO are capable of attaining higher NOx reduction levels, the study of more complex reductants in a combined LNT+SCR system is warranted for real world application on a vehicle.

In this study, NOx reduction using hydrocarbons (HCs) during the rich was investigated over a lab made LNT-SCR dual brick catalyst system. The effects of three reductants were studied in separate and blended combinations of CO, C_3H_6, C_3H_8, and CO+C_3H_6+C_3H_8. Figure 1 shows the overall performance of using the belended gas as reductant. At temperatures below 300°C, NO_2 was found re-reduce back to NO during the rich phase across the SCR catalyst. NOx reduction with HCs becomes active above 300°C, especially for C_3H_6, which could achieve 90% NOx conversion across the dual brick catalyst system. While C_3H_8 is not as active as C_3H_6 when fed in a mixture with C_3H_6 it starts to react at lower temperature than when C_3H_8 is the sole reductant. Figure 2 shows the comparison of three different reductants. The yield of NH_3 was ~40% greater when a C_3H_6 was used as compared to CO at higher temperature.

The gas hourly space velocity (GHSV) effect on NOx reduction was also investigated in this study. Figure 3 shows the GHSV was increased in order to achieve a lower NOx conversion to N_2 over the LNT which could provide a more desirable NH_3 to NOx ratio (ANR) across the SCR. It was found out at high temperature, GHSV in the range of 105,000 h^{-1} to 120,000 h^{-1} could achieve nearly 50% NOx conversion and an ANR of 1.0 from LNT catalyst which is the ideal gas mixture to the downstream SCR catalyst. The recent research shows that the LNT-SCR dual brick catalyst system could also benefit from the HC-SCR mechanism which not only reduced the HC slip from the system but also improved NOx reduction over the combined catalysts. The HC steam reforming and partial oxidation were carried out by spatially resolved mass spectrometry (SpaciMS), which was utilized to decouple the regeneration reaction from NOx reduction. This method enabled a detailed study of NH_3 formation over LNT catalyst and utilization over SCR catalyst while using a reductant feed that approach real world conditions.

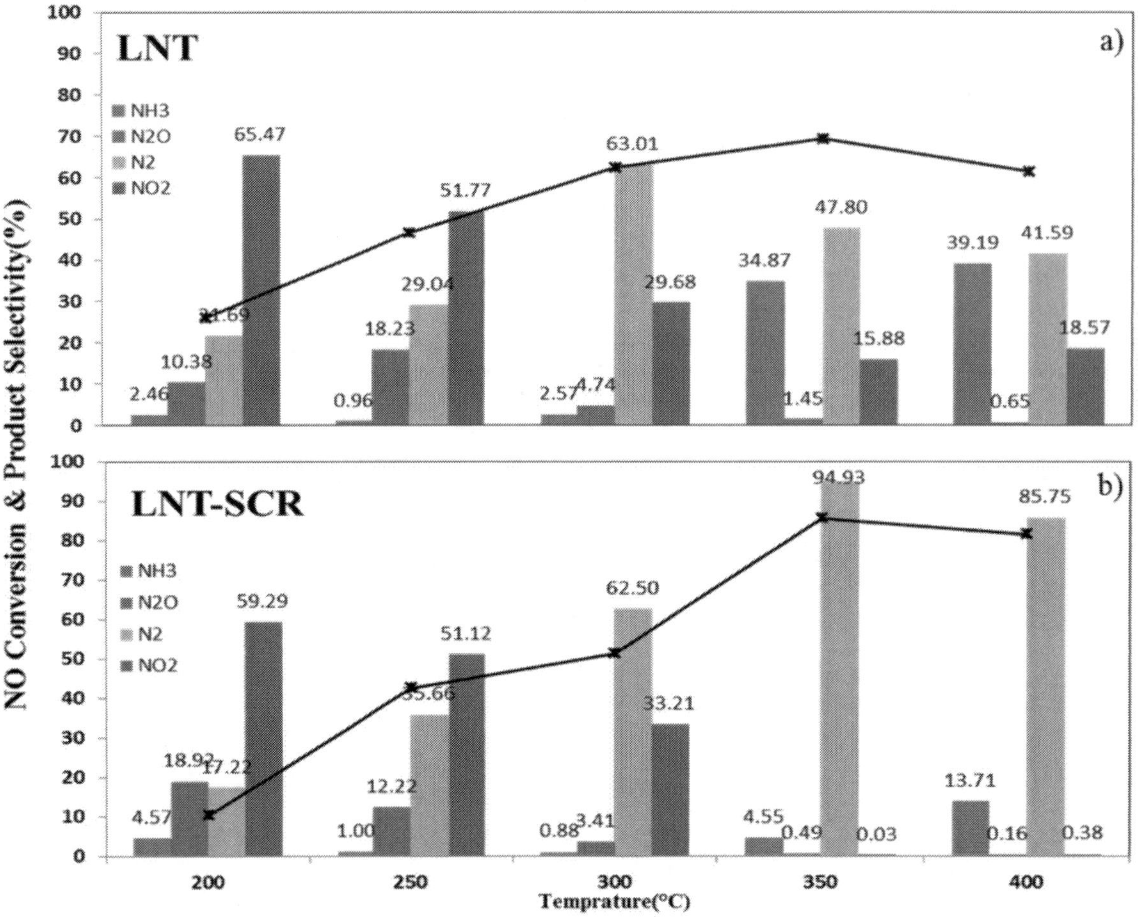

Figure 1. NO conversion and product selectivity using 1.5% CO, 500ppm C_3H_6, 3500ppm C_3H_8 and 1% O_2 as reductant over a) LNT A and b) LNT A-SCR A catalyst (NO conversion in black line).

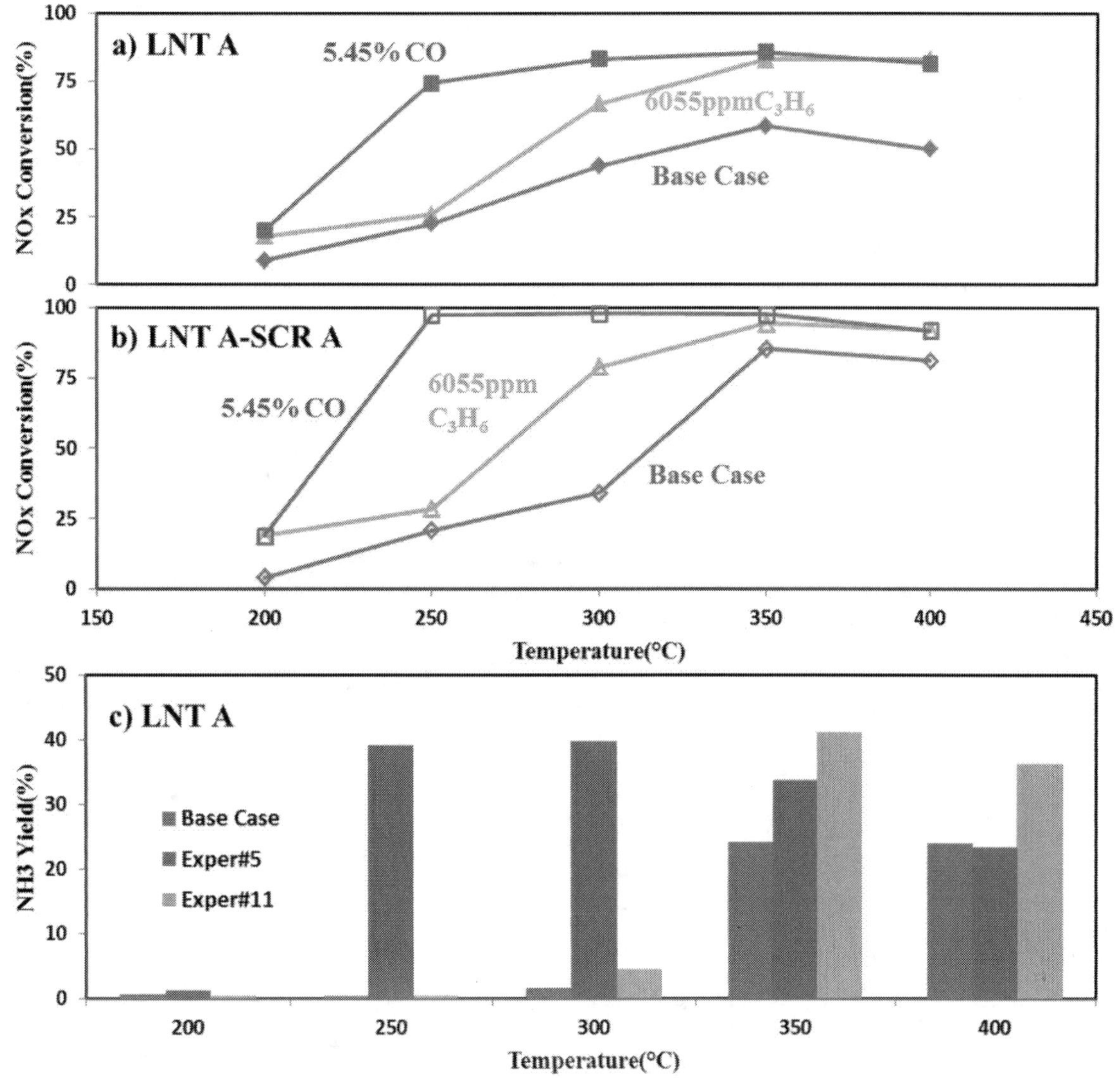

Figure 2: Cycle averaged NOx conversion and NH_3 Yield for Base case (500ppm NO, 1.5% CO, 500ppm C_3H_6, 3500ppm C_3H_8 and 1% O_2) , Exper #5 (500ppm NO, 5.45%CO and 1% O_2), and Exper #11 (500ppm NO, 6055ppm C_3H_6 and 1% O_2). a) NOx conversion for LNT A b) NOx conversion for LNT A-SCR A c) NH_3 Yield for LNT A. GHSV=90,000 h^{-1}.

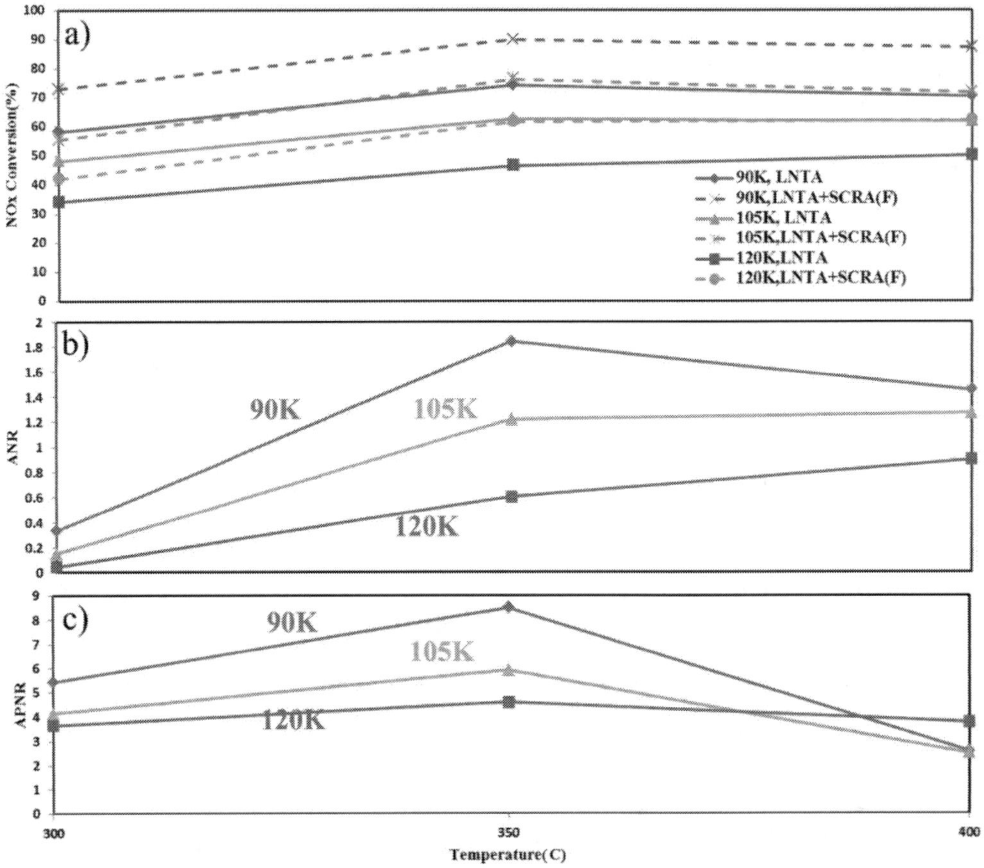

Figure 3: Cycle-averaged a) NO conversion and b) ANR c) APNR using 9300ppm C_3H_6 and 1% O_2 as reductant over LNT A by varying different GHSV (90,000 - 120,000 h^{-1}).

H_2SO_4 Catalysis: Perspective and Opportunities for Reducing SO_2 Emissions Using Particulate and Monolith Catalysts

Anuradha Nagaraj[1] and Patrick L. Mills[2*]
[1]Department of Environmental Engineering
[2]Department of Chemical & Natural Gas Engineering
Texas A&M University-Kingsville TX-78363-8202
*Corresponding author: patrick.mills@tamuk.edu

Abstract: Development of next-generation chemical processes that have zero emissions is a key environmental objective for sustainable development. The manufacture of H_2SO_4 by the air oxidation of SO_2 to SO_3 is an important technology where an opportunity exists for new catalyst development and process innovation by reducing emissions of unconverted SO_2 in process reactor tail gases owing to the sheer number (>1500) and scale (*ca.* 500 to 4500 metric tons/day) of typical plants. Development of new catalysts having higher activity, lower pressure drop, low dusting, and sufficient mechanical crush strength to meet the anticipated reduction in SO_2 emissions from H_2SO_4 manufacturing plants will potentially benefit by using this more realistic approach for particle-scale shape modeling. Transport-kinetic interactions in commercial porous catalyst shapes used for SO_2 oxidation are analyzed using the Wilke, Wilke-Bosanquet, Maxwell-Stefan, and Dusty-Gas flux models.

Keywords: catalyst particle shape, SO_2 oxidation, monolith, flux models.

1. Introduction

The manufacture of H_2SO_4 by the air oxidation of SO_2 to SO_3 is an important technology where an opportunity exists to develop new innovations in environmental catalysis and reaction engineering for an important technology that has a rich and long history with positive economic growth. Best Available Control Technologies (BACT) for controlling SO_2 emissions from stack gases of H_2SO_4 plant involves additional processing resulting are cost intensive and well-studied. However, emissions control technologies using front-of-pipe technology, such as new catalyst technology, improved reactor designs, and process operational strategies have greater potential to be more economical versus end-of-pipe scrubbing technologies.

The primary objectives of this paper are to develop a rigorous modeling framework that accounts for diffusion and non-isothermal reaction in various realistic particulate catalyst shapes and monolith catalysts, and to employ this framework to compare the performance of these catalysts for typical multi-pass SO_2 convertor operation.

2. Catalyst Shapes

From a catalyst design perspective, specification of catalyst composition, catalyst size, catalyst shape and internal pore structure are interlinked parameters that have received notable attention by catalyst scientists and engineers since these properties affect the catalyst effectiveness factor, catalyst bed pressure drop, rate of catalyst attrition, catalyst mechanical integrity, catalyst life, amongst other key measures of performance. It is evident that specification of preferred catalyst design parameters requires careful optimization since tradeoffs exist in how the observed reaction rate and other measures of performance may be affected by altering one or both of these parameters.

2.1 Particulate Catalysts

Commercial particulate catalysts for the air oxidation of SO_2 to SO_3 are available in various shapes, such as solid cylindrical pellets, hollow rings, and multi-lobed rings. A typical composition of these catalysts is based on alkali metals, such as potassium, sodium or cesium, vanadium pentoxide salts, and non-reactive silica support.

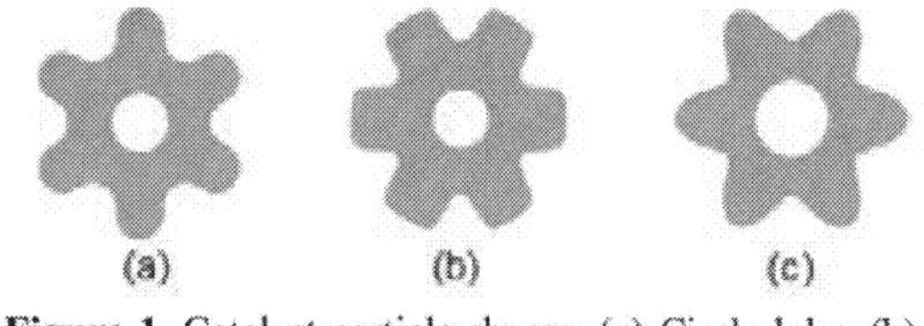

Figure 1. Catalyst particle shapes. (a) Circle lobe, (b) Rounded step, and (c) Trigonometric lobe.

Various complex catalyst shapes based on multi-lobed shape technology used in this study are illustrated in Figure 1.

2.2 Monolith Catalysts

Limited open literature is available on monolith synthesis and modeling for SO_2 oxidation in H_2SO_4 manufacturing process.

In early 1990s, two patents were issued to Felthouse and coworkers at Monsanto Enviro-Chem Systems (MECS) (Felthouse, 1992; Felthouse and Vavere, 1993) for synthesizing monolith catalysts for SO_2 oxidation. In 1991, Bespalov and coworkers at Moscow Chemical Engineering Institute developed a model for SO_2 oxidation in a monolith catalyst (Bespalov *et al.*, 1991). No experimental data was provided, however.

The monolith catalyst channel used in this study is shown in Figure 2. A single channel with 75 mm length, 1.5 mm channel width and washcoat of 0.625 mm thickness is used.

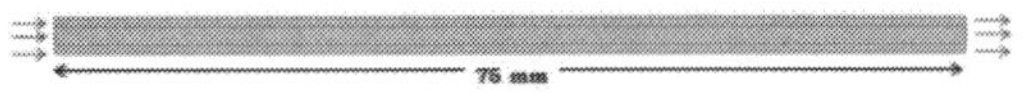

Figure 2. Washcoated single monolith channel

3. Transport-Kinetics Model

The modeling framework used to describe transport-kinetic interactions in both particulate and monolith catalyst is presented in this section.

The reaction kinetics for SO_2 oxidation are described by the following equation which is based on a comprehensive set of experimental measurements (Collina *et al.*, 1970).

$$-r_{SO_2} = \frac{k_1 p_{O_2} p_{SO_2}\left(1 - \dfrac{p_{SO_3}}{p_{SO_2}\sqrt{p_{O_2}}\, K_P}\right)}{22.414\,(1 + K_2 p_{SO_2} + K_3 p_{SO_3})^2}$$

3.1 Particulate transport model

The specie mass and energy balance equations for particulate shape are:

Mass balance

$$\nabla \bullet N_i = \upsilon_i \, r \, \rho_p$$

where i = SO_2, O_2, SO_3 and N_2 and υ_i = -1, -0.5, 1 and 0 for SO_2, O_2, SO_3 and N_2 respectively.

Energy balance

$$\nabla \bullet \bar{q} = -(\Delta H_{rxn})\, r \, \rho_P$$

3.2 Monolith transport model

The flow through monolith channel is modeled as free and porous media flow. The transport equations used for this monolith channel include balances for species, energy and momentum.

Mass balance

$$\nabla \bullet \bar{N} - u \bullet \nabla \bar{C} = \upsilon \, r \, \rho_p$$

Energy balance

$$\nabla \bullet \bar{q} - \rho C_p u \bullet \nabla T = -(\Delta H_{rxn})\, r \, \rho_P$$

Momentum balance

For channel flow:

$$\rho(u \bullet \nabla)u = \nabla \bullet [-\rho I + \eta(\nabla u + (\nabla u)^T)] + F$$

$$\nabla \bullet u = 0$$

For flow through the catalyst layer:

$$(\eta / \kappa + Q)u = \nabla \bullet [-\rho I + (1/\varepsilon_p)\{\mu(\nabla u + (\nabla u)^T) - (2\eta/3 - \kappa_{dv})(\nabla \bullet u)I\}] + F$$

$$\nabla \bullet u = Q/\rho$$

4. Diffusion flux models

Transport-kinetic interactions are analyzed using the Wilke, Wilke-Bosanquet, Maxwell-Stefan, and Dusty-Gas flux models. Particle effectiveness factors derived from the various flux models can differ for otherwise identical values for kinetic and transport parameters.

Wilke model

$N_i = -D_{ei,m} \nabla C_i$ where

$$D_{ei,m} = \frac{1}{\sum_{j=1, j\neq i}^{n} x_j / D_{ij}^e}$$

Wilke-Bosanquet model

$N_i = -D_{i,eff} \nabla C_i$ where

$$\frac{1}{D_{i,eff}} = \frac{1}{D_{ei,m}} + \frac{1}{D_{ei,k}}$$

Maxwell-Stefan model

$$N_i = \frac{-\nabla C_i + \sum_{j=1, j\neq i}^{n} \dfrac{x_i N_j}{D_{ij}^e}}{\sum_{j=1, j\neq i}^{n} \dfrac{x_j}{D_{ij}^e}}$$

Dusty-Gas model

$$N_i = \frac{\sum_{j=1, j\neq i}^{n} \frac{x_i N_j}{D_{ij}^e} - \frac{C_i v^*}{D_{ei,k}} - \nabla C_i}{\sum_{j=1, j\neq i}^{n} \frac{x_j}{D_{ij}^e} + \frac{1}{D_{ei,k}}} \quad \text{where}$$

$$v^* = -\frac{\varepsilon d_{pore}^2}{32 \tau \mu} \nabla P$$

5. Results

5.1 Particulate catalyst

The 1-D diffusion-reaction system was first solved for a solid cylindrical pellet with a diameter 6 mm. The specie and energy equations are solved using two of the COMSOL Multiphysics™ modules: (1) diffusion, and (2) heat transfer by conduction. Typical molar composition profiles for all four flux models are compared in Figure 3. The mean pore radius used in this study is 319 nm with a pellet porosity of 0.44 (Davis *et al.*, 1982).The profiles suggest that the Knudsen diffusion is not significant.

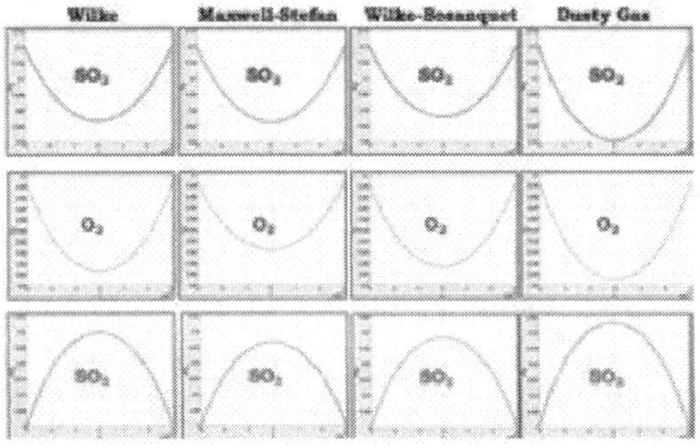

Figure 3. Comparison of the steady-state molar composition profiles for different flux models (T_B = 420°C, P = 1 atm, % SO_2 = 11%; % O_2 = 9%).

Figure 4 shows the 3-D concentration and temperature profiles for the rounded-step catalyst illustrated in Figure 1b for the Wilke and Dusty gas flux models. The boundary conditions correspond to specified bulk concentrations and temperatures so no external transport resistances are present. The effectiveness factor was obtained by integration of the reaction rate at the local values for the particle temperature and specie concentrations over the particle volume and then dividing this result by the reaction rate evaluated at the external bulk conditions.

The profiles for both flux models show that the highest particle temperature and the lowest concentrations for both SO_2 and O_2 are reached in the center of the lobes. This is expected owing to the longer diffusion path length and hence longer characteristic time for transport of mass and energy in the lobe regions. The effectiveness factor for the Wilke model is lower (η = 0.89) than the one obtained when the Dusty gas model is used (η = 0.92). This difference is attributed the presence of a viscous flux term in the latter model that likely occurs owing to the non-equimolar stoichiometry that is not accounted for in the Wilke model. The contribution of Knudsen diffusion is not important since the mean pore radius is 319 nm.

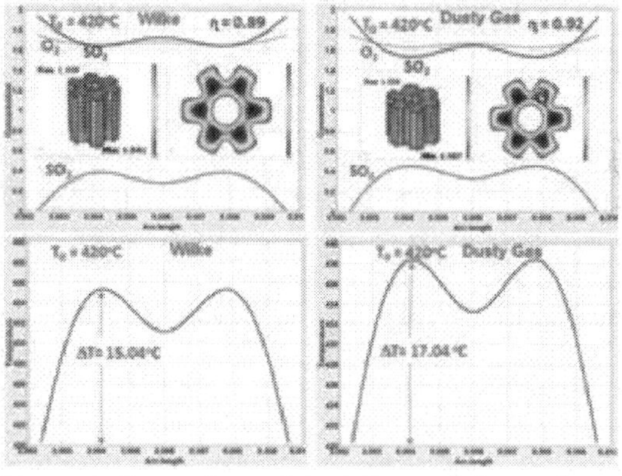

Figure 4. Comparison of the 3-D rounded step shape concentration profiles for Wilke and Dusty Gas flux models.

5.2 Monolith catalyst

The 2-D convective diffusion-reaction system was solved for the single monolith channel illustrated in Figure 2. The momentum, specie and energy equations are solved using the following COMSOL Multiphysics™ modules: (1) Navier-Stokes (channel flow) and Brinkman (catalyst flow); (2) convection and diffusion; and (3) convection and conduction.

Figure 5 shows the 2-D temperature profiles for the monolith channel using different flux models.

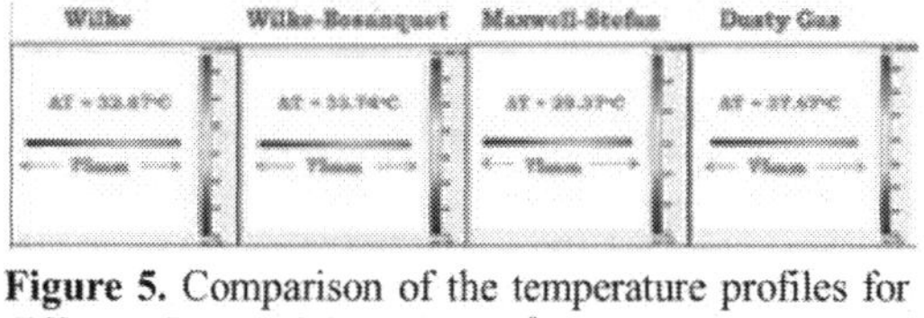

Figure 5. Comparison of the temperature profiles for different flux models (T_B = 420°C, p = 1 atm, % SO_2 = 11%; % O_2 = 9%).

The overall temperature gradients are largest for the Dusty gas model while the smallest occurs for the Maxwell-Stefan model. Since the latter model does not account for convective transport or Knudsen diffusion, these must account for the difference in the predicted gradient. Additional analysis that provides more insight into the differences between the various models will be provided in a future manuscript (Nagaraj and Mills, 2014).

6. Conclusions

COMSOL Multiphysics™ provides an efficient platform for modeling transport-kinetic interactions in particulate and monolith structures. Wilke specie flux model is sufficient when molecular diffusion is controlling, otherwise more complex models are needed. Synthesis of tailored porous supports could lead to more active catalysts for SO_2 oxidation to SO_3. The initial modeling results suggest monoliths are a potential catalyst platform SO_2 oxidation if the structure can be tailored to provide high activity with attractive economics.

7. Nomenclature

C_i	Concentration of species 'i' (mol/m^3)
C_p	Specific heat capacity (J/kg-K)
$D_{ei,m}$	Wilke molecular diffusivity (m^2/s)
$D_{ei,k}$	Knudsen diffusion coefficient for species i (m^2/s)
$D_{i,eff}$	Bosanquet diffusivity (m^2/s)
D_{ij}^{e}	Effective binary diffusivity of species 'i' (m^2/s)
d_{pore}	Mean pore diameter (m)
k	Rate constant (kmol/kg cat-hr)
K_p	Equilibrium constant (atm$^{1/2}$)
K	Adsorption Constant (atm)
N_i	Molecular diffusion flux (mol/m^2-s)
q	Heat flux (W/m^2)
Q	Heat source (W/m^3)
P	Pressure (atm)
p_i	Partial pressure of component i (atm)
r	Reaction rate (mol/m^3-s)
r_{so2}	Reaction rate of SO_2 (kmol/kg cat-hr)
T	Temperature (^{o}C)
u	Velocity (m/s)
v^*	Dimensionless velocity
x_i	mole fraction of component i

Greek Letters

η	Effectiveness factor
κ	Permeability (m^2)
μ	Viscosity (Pa-s)
ρ_p	Density of catalyst pellet (kg/m^3)
ε	Catalyst pellet porosity
ν_i	Stoichiometric coefficient of component i
τ	Catalyst tortuosity

8. References

Bespalov, A. V., E. M. Chechetkina, et al. (1991). "Numerical modeling of sulfur dioxide reaction in through-channel of a block catalyst with a honeycomb structure." Zhurnal Prikladnoi Kimii 64(10): 2048-2053.

Collina, A., D. Corbetta, et al. (1970). Europeam Symposium on "Use of computers in the design of chemical plants", Firenze.

Davis, M. E., G. Fairweather, et al. (1982). "Analysis of SO2 oxidation in non-isothermal catalyst pellets using the dusty-gas model." Chemical Engineering Science **37(3): 447-452.**

Felthouse, T. R. (1992). Monolithic catalysts for conversion of sulfur dioxide to sulfur trioxide. US, Monsanto Company.

Felthouse, T. R. and A. Vavere (1993). Monolithic catalysts for conversion of sulfur dioxide to sulfur trioxide. US, Monsanto Company.

Nagaraj, A. and Mills, P. L. (2014). Comparison between particulate and monolith catalyst performance for the oxidation of SO_2 using various specie transport flux models," Chemical Engineering Science (in preparation).

Synthesis, Characterization of Vanadia/titania-based Catalysts and their Performance in Ammonia Remediation

Hao Chen[1,3], Raquel Portela[4], Wei Han[1], Pedro Avila[4], Miguel A. Bañares[4], King Lun Yeung[1,2*]

[1]*Department of Chemical and Biomolecular Engineering,* [2]*Division of Environment,* [3]*Nano Science and Technology Program, The Hong Kong University of Science and Technology, Clear Water Bay, Kowloon, Hong Kong, P.R. China*
[4] *Instituto de Catalisis y Petroleoquimica, CSIC, Marie Curie 2, 28049, Madrid Spain*

*Corresponding author: kekyeung@ust.hk

Division: 20 Catalysis and Reaction Engineering Division

Subdivision: 20037 Applied Environmental Catalysis

Format: Oral Presentation

Abstract:

Malodor is an important class of air pollution that poses a serious threat to personal health and quality of life. Ammonia gas is a major cause of malodor complaints and is produced by microbial decompositions of organic wastes. A catalyst capable of attaining better than 99 % conversion of 50 ppm NH_3 in air at room temperature was successfully prepared. The E_A of the reaction is dramatically decreased by using the catalysts when the ratio of VO_x/TiO_x increases. The NH_3 was oxidized to nitrogen and water during the reaction without producing NO_x. The catalyst was characterized by BET, micro-Raman and X-ray photoelectron spectroscopy to identify the active sites to be monomeric VO_x on TiO_x.

Introduction:

Indoor air quality has captured much attention of many researchers recently because of its effect on human health. Ammonia is one of the most odor origin gases in the world that can be easily emitted in wash room, fish markets, meat treatment industries, piggery, landfill and agricultural activities. They have a sharp pungent smell which can be detected by human nose at even several ppm levels. The damages caused by Ammonia mainly role in eye, respiratory and skin.[1] Selective catalytic oxidation gives a new way to remove ammonia in oxygen or air in the recent years, ammonia reacts with air can produce nitrogen, nitrous oxide and nitrogen monoxide. This technology has reported in literature that different kinds of catalysts have been used for selective catalytic oxidation of ammonia into nitrogen, such as noble metals (Pt, Rh and Pd) iron exchanged into zeolite ZEM-5[2] or supported on gamma alumina[3], Ni oxides[4], copper oxides and Ag oxides supported on γ-Al_2O_3[5], V_2O_5/TiO_2 and CuO/TiO_2[6], Fe exchanged ZSM-5[7], Fe_2O_3/TiO_2.[8]

Experimental:

2.5g Ammonia Vanadate Meta (NH_4VO_3, 98%, Nacalai Tesque INC) is dissolved in 500 ml DDI water, then 38.87g UV100 (TiO_2, Sachtleben) is added into 100 ml the above solution to prepare 1wt% of V_2O_5/TiO_2 catalyst. The mixed suspension is dried via rotary evaporator under 65°C with 120 r/min. After drying, the powder is put inside oven and further dried at 115°C for 24 hours. The obtained powder is then calcined at 200°C for 4 hours. A serial of 0.1wt%, 5wt% and 15wt% catalysts are also prepared with this method.

Results and Discussion:

The catalysts used in the experiments in this report are 1wt%, 5wt% and 15wt% V_2O_5/TiO_2 which are calcined in air at 200°C for 4 hours. To obtain the morphology of catalyst, SEM is measured. Figure 1 shows SEM images of 15wt% V_2O_5-TiO_2 catalyst. Figure 2 shows the Micro-Raman Spectra of 1wt%, 5wt% and 15wt% V_2O_5/TiO_2. The four main peaks of 160cm^{-1}, 390cm^{-1}, 520cm^{-1} and 640cm^{-1} are assigned to anatase phase of TiO_2. The intensity of peaks at 810cm^{-1} and 940cm^{-1} increase as the V_2O_5/TiO_2 loading increases from 1wt% to 5wt% which are associated to polymeric form of V_2O_5, when the loading increases to 15wt%, the peak of 940cm^{-1} shifts to 1020cm^{-1}, which is associated to monomeric form of surface of V_2O_5. In the four catalysts, the peak of crystal V_2O_5 isn't seen in the Micro-Raman spectra at 995cm^{-1}. We can run the experiments at different temperature, to observe the relation between T and conversion shown in Figure 3. To design the experiment condition, the reaction flow rate is 400 SCCM containing 10 SCCM 8.73% NH_3 in N_2 and 390 SCCM N_2, so the concentration of the inlet NH_3 is 2175ppm. The reaction temperature is ranging from 25°C to 250°C. Then the Arrhenius equations of three V_2O_5-TiO_2 catalysts are obtained, shown in Table 1. It is found that 15wt% one has the lowest Ea, that means the reaction barrier decreases a lot as the V_2O_5/TiO_2 ratio increases.

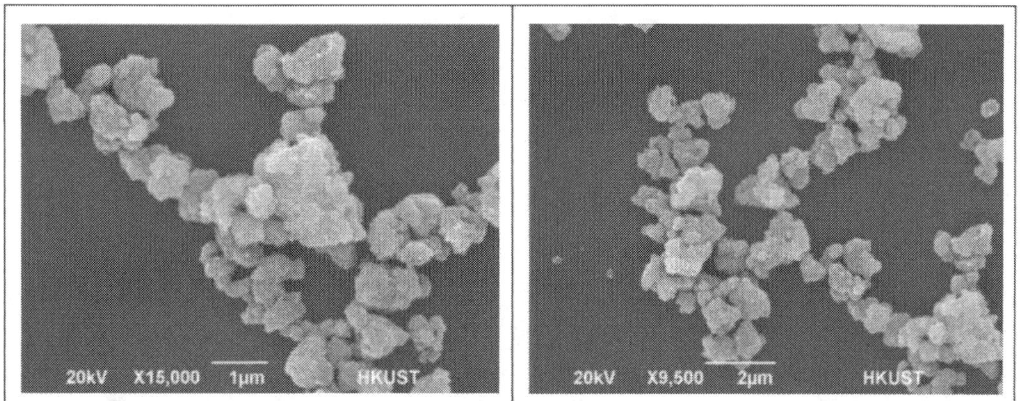

Figure 1 SEM images of 15wt% V_2O_5/TiO_2

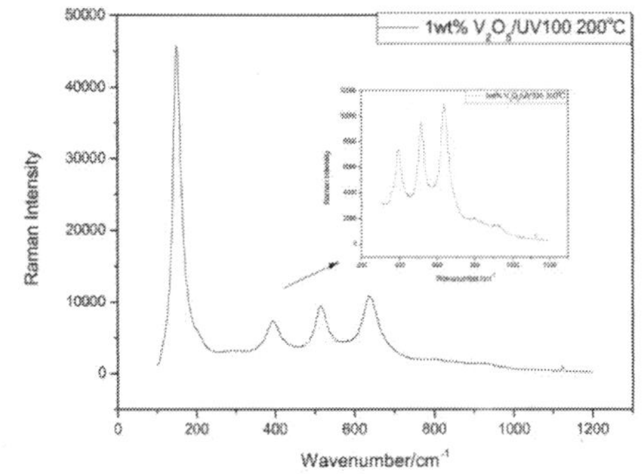

(b)

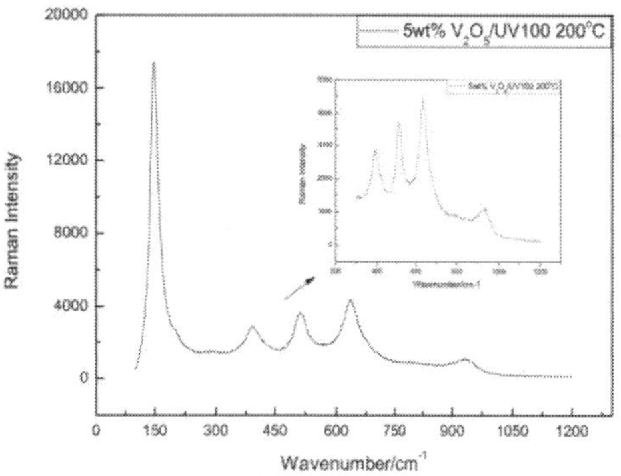

(c)

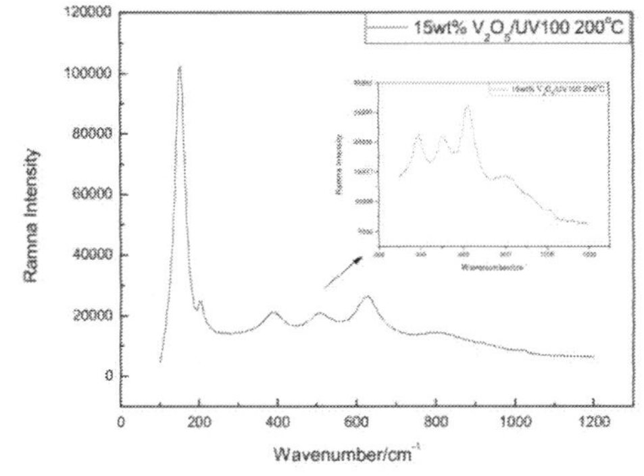

(d)

Figure 2 Micro-Raman Spectra of 1wt%, 5wt% and 15wt% V_2O_5/TiO_2.

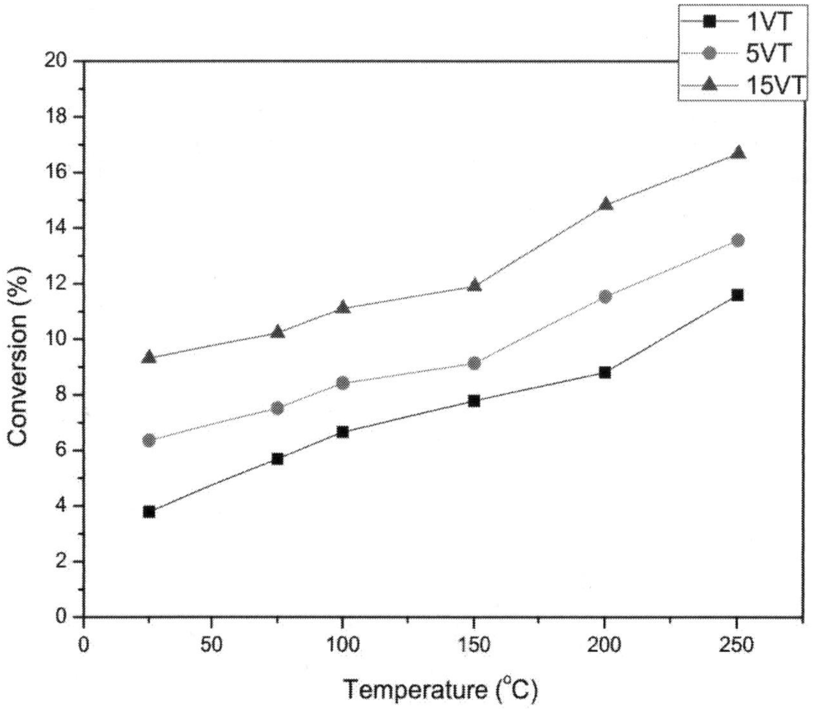

Figure 3. Figure 5 NH_3 conversion using 1wt%, 5wt% and 15wt% V2O5-TiO2 catalyst

Table 1. Arrhenius equations and Ea of 1wt%, 5wt% and 15wt% V2O5-TiO2

Catalyst	Arrhenius equation	Ea/kJ*mol^{-1}
V_2O_5-TiO_2-1wt%	lnk=-1.081-722/T	6.00
V_2O_5-TiO_2-5wt%	lnk=-1.087-513/T	4.27
V_2O_5-TiO_2-15wt%	lnk=-0.827-406/T	3.37

Conclusion:

The obtained V_2O_5-TiO_2 catalysts dramatically decrease the reaction barrier of NH_3 oxidation, where the catalysts achieve a 99% conversion for 50ppm NH_3 gas in air at room temperature.

Acknowledgement:

The authors gratefully acknowledge financial supports from the Hong Kong Research Grant Council and Innovation & Technology Fund.

Reference:

1. ***Agency for Toxic Substances and Disease Registry***, toxicological profile for NH_3

2. Li, Y., Armor, J. N. ***Appl. Cata. B*** 1997(13) 131-139

3. Long, R. Q., Yang, R. T. ***Catal. Lett.*** 2002(78) 353-357

4. Amblard, M. Burch, R. Southward, B. W. L. ***Appl. Catal. B*** 1999(22) L159-L166

5. Gang, L., van Grondelle, J., Anderson, B. G., van Santen, R. A. ***J. Catal.*** 1999(186) 100-109

6. Sazonova, N. N., Simakov, A. V., Nikoro, T. A., Barannik, G. B., Lyakhova, V. F., Zheivot, V. I., Ismagilov, Z. R., Veringa, H. ***React. Kinet. Catal. Lett.*** 1996(57) 71-79

7. Qi, G. S., Gatt, J. E., Yang, R. T. ***J. Catal.*** 2004(226) 120-128

8. Long, R. Q., Yang, R. T. ***J. Catal.*** 2002(207) 158-165

380527 Probing the Active Site Requirements and Mechanistic Details of NOx Selective Catalytic Reduction with NH3 on Cu-SSZ-13

Abstract:

Cu-exchanged SSZ-13 (CHA) zeolites have been adopted commercially in automotive exhaust deNO_X applications for selective catalytic reduction (SCR) with NH_3 because they show preferred characteristics regarding performance at low temperatures and stability under hydrothermal conditions. Here, we use an integrated experimental and theoretical modeling approach to probe the active site requirements and mechanistic details of standard NO SCR with NH_3 on Cu-SSZ-13 zeolites. The number of isolated Cu ions (per total Cu) in 6-MR locations correlate linearly with standard NO SCR rates (473 K, per total Cu), while the number of Cu species located in clusters with Cu-O-Cu connectivites (per total Cu) in 8-MR cages correlate linearly with dry NO oxidation rates (573 K, per total Cu). The distribution of isolated and clustered Cu species in SSZ-13 depends on the density of paired Al atoms in the framework and, in turn, the framework Si/Al ratio. *Operando* X–Ray absorption spectroscopy (XAS) of Cu-SSZ-13 samples under working conditions indicates that standard SCR proceeds via a redox mechanism involving isolated Cu(II) and Cu(I) species, with oxidation and reduction half-reactions predominating under different gaseous mixtures that enable isolation and probing of Cu in either the Cu(II) or Cu(I) oxidation state. We discuss a plausible redox mechanism consistent with experiment and density functional theory (DFT), and the different mechanistic roles of Brønsted acid sites also present in Cu-SSZ-13 samples that mediate standard SCR with NH_3.

392233 A New Method for Exhaust Aftertreatment Design - Integrated Engine and Aftertreatment Analysis

Abstract:

Currently, the engine and aftertreatment (catalytical converter) design are separate process. This might lead to several empirical interactions to finalize the engine calibration and aftertreatment layout for a specific project (tailpipe emission target together with engine performance target).

This report describes a 'back of the envelope" methodology to quickly assess the tailpipe emission of a given engine and aftertreatment system. This allows to do combination system (engine and aftertreatment) selection at the early stage of development to reduce time and cost.

There are three main benefits from this methodology:

- It can define the optimal operating range for a given powertrain, with the consideration of engine out emission, fuel economy targets and aftertreatment conversion efficiency.
- It can estimate the catalyst light-off time with provided engine out emission data, or vice versa.
- For a given engine out exhaust gas temperature, species concentration and tailpipe emission target, this methodology can provide the maximum allowable aftertreatment system thermal inertia with defined catalyst light-off time, or vice versa, to meet a specific tail-pipe emission target.

Analysis of ash-loaded DPF performance during drop-to-idle test

Diesel particle filter (DPF) has been widely used in the automobile industry to reduce particle matter emission. The soot reduction process in DPF requires the cycling of two operation modes. During the collection phase, particle matters, including soot and ash, are filtered and accumulated in the wall-flow DPF. The accumulated soot leads to a slow increase in DPF backpressure and triggers the O2-based soot regeneration after reaching the DPF backpressure limit. DPF inlet temperature is then increased to enable the soot oxidation in a short time and the DPF backpressure falls back into the normal range. This process has two innate deficiencies: 1. DPF collects the incombustible ash, which cannot be removed during soot regeneration and increases the DPF pressure drop; 2. High temperature excursion during soot regeneration could damage DPF. Based on the DPF failure observed in the field, these two failure modes could interact and compound with each other, making a more complex situation for analysis.

In this work, we evaluate the exotherm from an ash-loaded DPF using a 1D transient model. Drop-to-idle condition was used as a simulated high temperature event. During this test, flow rate drops during the soot regeneration, while inlet temperature and oxygen concentration increase. A variety of ash-loaded DPFs with different sizes, materials and cell structures were assessed in this test. Simulation results showed that DPF design could change the temperature profile along the channel. In some cases, temperature could be high enough to cause the ash sintering inside DPF channel and thus reduce the effective length, leading to a higher DPF backpressure than prediction.

390165 Transient Kinetic Study of Sulfur Oxide Adsorption/Desorption over Al2O3 and Pt/Al2O3 Catalysts

Abstract:

Background

Sulfur oxides in the engine exhaust interact with diesel aftertreatment catalysts and deactivate them. SO_2 is the main sulfur species exiting the engine, which in turn can be further oxidized to SO_3 over the diesel oxidation catalyst (DOC). SO_2 and SO_3 may have different impacts on the DOC activity, and downstream catalysts, with SO_3 leading to more severe deactivation. In the present work, in order to study the activity recovery of a Pt/Al_2O_3 DOC after sulfur poisoning, sulfur storage/release was studied using temperature-programmed desorption (TPD) experiments and a transient kinetic model was developed based on the experimental data. Diffuse reflectance infrared spectroscopy (DRIFTS) experiments were also utilized to identify the surface species.

Methodology

To separate out the support effect, TPD experiments were performed in a flow reactor system using Al_2O_3 and Pt/Al_2O_3 samples. An Al_2O_3-coated monolith with a loading of 1.60 g/in^3 and a Pt/Al_2O_3-coated monolith with a Pt loading of 50 g/ft^3 were used in the experiments. During the TPD experiments, the catalyst was saturated with a specific concentration of either SO_2 or SO_3, and a temperature ramp from the saturation temperature up to ~900°C was used to desorb the surface species. For the SO_3 TPD experiments, a separate reactor upstream of the main reactor was used where SO_2 was completely oxidized to generate SO_3 required in the feed. SO_2 and SO_3 TPD experiments were carried out under four different sets of operating conditions where saturation temperatures of 120°C and 160°C and approximate concentrations of 50 and 100 ppm for the adsorbing gas species were employed. Temperature ramps of 20 and 10°C/min were used in the SO_2 and SO_3 TPD experiments, respectively. For all flow reactor experiments, the outlet gas concentrations were measured using a MKS MultiGas MG-2030 FT-IR analyzer. In the DRIFTS experiments, pellets of Al_2O_3 and Pt/Al_2O_3 were prepared from the monolith washcoat and IR studies were performed with SO_2 and SO_3 concentrations similar to those employed in the flow reactor experiments.

Results and Discussion

According to the DRIFTS and TPD results, three different sulfur species were identified during SO_2 adsorption on Al_2O_3. A small high temperature peak in the TPD spectra, i.e. above 850°C, was assigned to bulk aluminum sulfate, whereas the other species with less stability were identified as weakly adsorbed SO_2 (with its desorption peak centered at 250°C) and surface sulfites or sulfates (with a peak maximum in the range of 400-500°C). Similar surface species were identified in the case of SO_2 TPD on Pt/Al_2O_3, however, with a higher contribution of the high temperature peaks indicating the promoting effect of Pt on the sulfate formation as well as on the spillover of the surface sulfates to the bulk. 1.45×10^{-4} mol SO_2 adsorbed was calculated based on the SO_2 TPD for Pt/Al_2O

sulfur oxides significantly increased, to 1.50×10^{-3} mol, when exposed to SO_3 instead, suggesting a higher level of sulfur storage upon exposure to SO_3. Also according to these results, SO_3 exposure resulted in the formation of stable sulfate species on the catalyst. The SO_3 TPD results obtained with the Al_2O_3 and Pt/Al_2O_3 samples indicated that SO_2 and SO_3 release patterns were strongly affected by thermodynamic equilibrium. The important role of the Al_2O_3 support in the catalyst sulfation, as previously reported in the literature [1-3], was also confirmed in the experimental study.

A three site adsorption mechanism, with S1, S2 and S3 representing the adsorption sites with an increasing order of strength, was assumed to develop a transient kinetic model for the SO_2 storage/release over Al_2O_3 and Pt/Al_2O_3. First order kinetics were assumed for the SO_2 adsorption on the S1 (weak) and S3 (strong) sites, whereas higher order kinetics, i.e. second or third order depending on the catalyst, was assumed for SO_2 adsorption on the S2 site. The S2 kinetics are explained by the structure of the surface sulfates being formed as a result of SO_2 adsorption on the under-saturated oxygen anions on the surface of alumina. These oxygen species were recently proposed as possible sites for the strong adsorption of SO_2 in the form of surface sulfates [4].

A uniform temperature profile was assumed for the reactor and mass transfer limitations were neglected in the radial direction of the monolith channel. A set of governing equations were solved to obtain the concentration and surface coverages as a function of time and axial position. Some of the kinetic parameters were optimized to describe the experimental data with thermodynamic constraints imposed upon the optimized parameters. A similar methodology was also applied in the development of the SO_3 TPD model over Al_2O_3 and Pt/Al_2O_3. The kinetic model was validated using other sets of experimental data which were not included in the fitting process. The transient models were able to accurately predict the experimental behavior of the catalyst. According to the modeling results, the binding energy of the adsorbed species on the S1 and S2 sites are coverage dependent, indicating surface interactions between the adsorbed species as well as the heterogeneous nature of the alumina surface sites. Coverages of different sulfur species as well as their variation as a function of temperature were also predicted by the model indicating differences in the SO_2 and SO_3 storage mechanism.

REFERENCES

1. O. Saur M. Bensitle, S. Mohammed, J.C. Lavalley, C.P. Tripp and B.A. Morrow, *Journal of Catalysis* 110(1986)104.
2. F. Gracia, S. Guerrero, E. Wolf, J. Miller and A. Kropf, *Journal of Catalysis* 233(2005)372.
3. J.H. Pazmiño, C. Bai, J.T. Miller, F.H. Ribeiro and W.N. Delgass, *Catalysis Letters* 143(2013)1098.
4. M.Y. Smirnov, A.V. Kalinkin, A.V. Pashis, A.M. Sorokin, A.S. Noskov, K.C. Kharas and V.I. Bukhtiyarov *The*

Journal of Physical Chemistry B 109(2005)11712.

Lean NO_x Trap catalysts are a promising potential strategy for reducing NO and NO_2 emissions from lean burn engine exhausts. Operating in the lean burn regime generates higher NO_x and oxygen concentrations than are found in traditional stoicheometric engines, while virtually eliminating reducing components in the exhaust gas. Under these conditions traditional three-way catalysts (TWC) are rendered virtually ineffective. The traditional LNT catalysts formulation consists of a noble metal and alkali earth metal on a high surface area support. The addition of cerium to the traditional support has been shown to increase the reduction of stored NO_x during rich phase operation [1]. For novel catalysts to be effectively implemented it is important to understand the effect of new catalyst formulations on stability, and degradation, specifically the effect of degradation on lean phase oxidation and NO_x storage.

In the current study LNT catalysts have been produced by incipient wetness impregnation of gamma alumina, to yield the desired weight percentage loading. The catalysts investigated are: Pt/Ba/Al, Pt/Ba/Ce/Al with weight ratio loadings of 1/20/100 and 1/20/5/100 respectively. The effect of ceria addition has been characterized by long time NO_x storage experiments, and steady state NO oxidation measurements. Surface characterization has been completed via XRD, EDAX, TEM and FTIR. Further characterization has been conducted through CO chemisorption, BET analysis and H_2-TPR. Ceria addition shows a significant effect on surface morphology, storage component surface species and overall NO_x storage with thermal degradation when compared to the standard alumina supported catalysts. As a result degradation and stability of other novel LNT catalysts will be evaluated in the future.

1. Infantes-Molina, A., et al., *Characterization and reactivity of Ce-promoted PtBa lean NOx trap catalysts.* Catalysis Today, 2012. **197**(1): p. 178-189.

Preliminary Study of Low Temperature Oxidation of Carbon Monoxide Produced By Diesel-Ignited Methane Dual Fuel Low Temperature Combustion in a Single-Cylinder Diesel Engine

Navid Zanganeh[1], Mostafa S. Raihan,[2], Kalyan K Srinivasan[2], Sundar Rajan Krishnan[2], Hossein Toghiani[1] and Jason M. Keith[1]
(1)Dave C. Swalm School of Chemical Engineering, Mississippi State University, Mississippi State, MS
(2)Department of Mechanical Engineering, Mississippi State University, Mississippi State, MS

14AIChE
Annual Meeting, Atlanta, GA

MISSISSIPPI STATE
UNIVERSITY

Presentation Outline

- Introduction
- Diesel-Ignited Dual Fuel Combustion
- Experimental Section- Engine
- Result and discussion
- Catalyst preparation and characterization
- Summary
- Future work

MISSISSIPPI STATE UNIVERSITY

2

14AIChE Annual Meeting, Atlanta, GA

Introduction

- ❑ Diesel engine exhaust is a mixture of thousands of different gases and tiny particles.

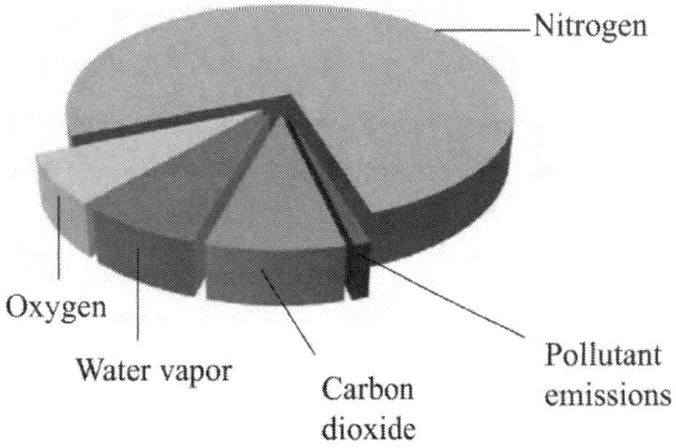

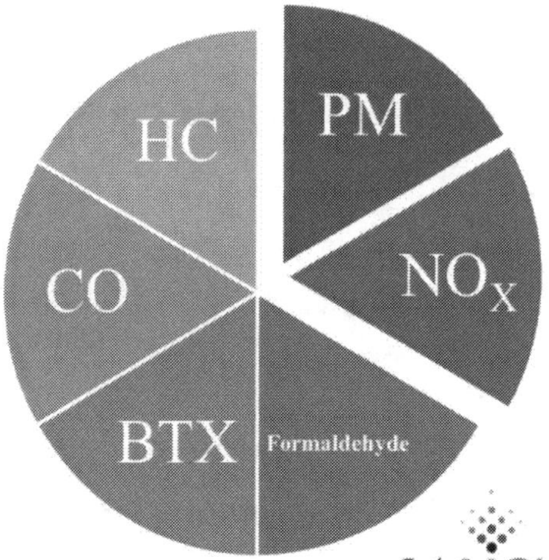

Introduction

- ❑ Diesel Exhaust Health Effect
- ❑ Diesel Exhaust Environmental Effect
- ❑ US Environmental Protection Agency (EPA) Emission Standard

Year	HC (g/bhp-hr)	CO (g/bhp-hr)	NO_x (g/bhp-hr)	PM (g/bhp-hr)
1990	1.30	15.5	6.0	0.60
1991	1.30	15.5	5.0	0.25
1994	1.30	15.5	5.0	0.10
1998	1.30	15.5	4.0	0.10
2002	0.50	15.5	2.4	0.10
2007+	0.14†	15.5	0.2	0.01

† 0.14 is given for non-methane hydrocarbons
Regulation for heavy duty trucks www.epa.gov

Introduction

❑ Reducing Diesel Exhaust Emissions

❖Engine treatment
- ❑ Diesel engines with exhaust gas recirculation
- ❑ Low temperature combustion strategies

❖Change the fuel
- ❑Reformulated diesel fuels
- ❑Natural gas
- ❑Electricity

❖Install vehicle accessories that reduce diesel exhaust
- ❑Diesel Oxidation Catalyst
- ❑Diesel Particulate Filter

Diesel-Ignited Dual Fuel Combustion

Video Presentation in Original e-media

www.youtube.com

Center for advanced vehicular systems (CAVS)

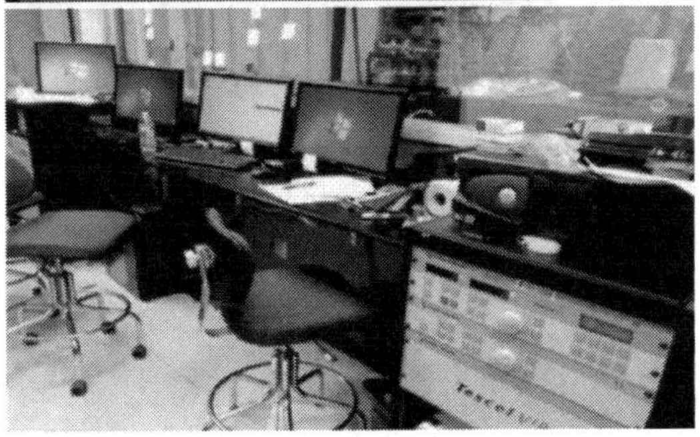

MISSISSIPPI STATE UNIVERSITY | CAVS | Advanced Combustion Engines Laboratory | 7 | 14AIChE Annual Meeting, Atlanta, GA

Experimental Setup

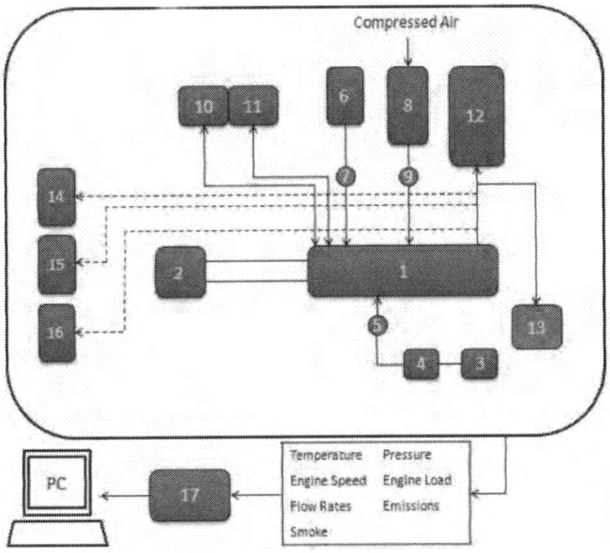

1-Single cylinder engine
2-AC regenerative dynamometer
3-Diesel supply tank
4-Diesel level tank
5-Coriolis flow meter
6-Methane supply tank
7-Coriolis flow meter
8-Intake tank
9-Sonic flow meter
10-Engine water temperature conditioner
11-Engine oil temperature conditioner
12-Exhaust tank
13-Smoke meter
14-Engine exhaust particles sizer (EEPS)
15-AVL FTIR
16-Six-gas emission bench
17-Data acquisition system

MISSISSIPPI STATE UNIVERSITY | CAVS | Advanced Combustion Engines Laboratory | 8 | 14AIChE Annual Meeting, Atlanta, GA

Experimental Setup

Six-gas emission bench

FTIR

Smoke meter

MISSISSIPPI STATE UNIVERSITY

CAVS

Advanced Combustion Engines Laboratory

14AIChE Annual Meeting, Atlanta, GA

Test Matrix

Engine Load (IMEP)	Varied Parameter	Constant Parameters
5.1 bar IMEP at 1500 RPM	PES : 0% to 90%	SOI = 355 CAD Boost pressure = 1.5 bar Rail pressure = 500 bar
	SOI : 355 CAD to 280 CAD	PES = 80% Boost pressure = 1.5 bar Rail pressure = 500 bar
	Rail pressure : 200 bar to 1300 bar	PES = 80% Boost pressure = 1.5 bar SOI = 310 CAD
	Boost pressure : 1.1 bar to 1.8 bar	PES = 80% Rail pressure = 500 bar SOI = 310 CAD

PES: Percentage Energy Substitution
SOI: Start of Injection
CAD: Crank Angle Degree
IMEP: Indicated Mean Effective Pressure

MISSISSIPPI STATE UNIVERSITY

Mostafa S. Raihan. A comparative study of diesel ignited methane and propane dual fuel low temperature combustion in a single cylinder research engine. Diss. MISSISSIPPI STATE UNIVERSITY, 2014.

14AIChE Annual Meeting, Atlanta, GA

PES Sweep Results

Apparent Heat Release Rate (AHRR)

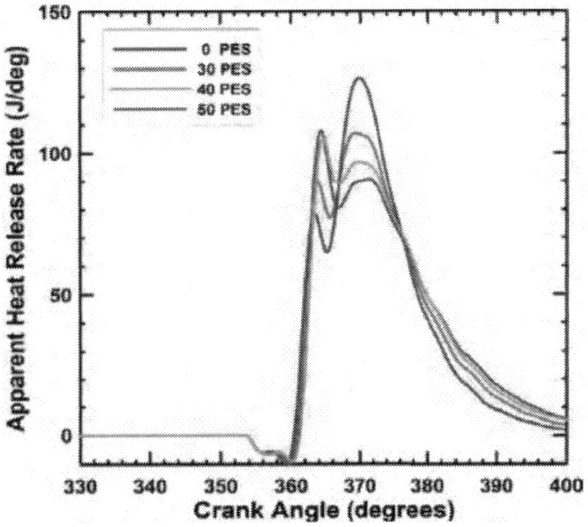

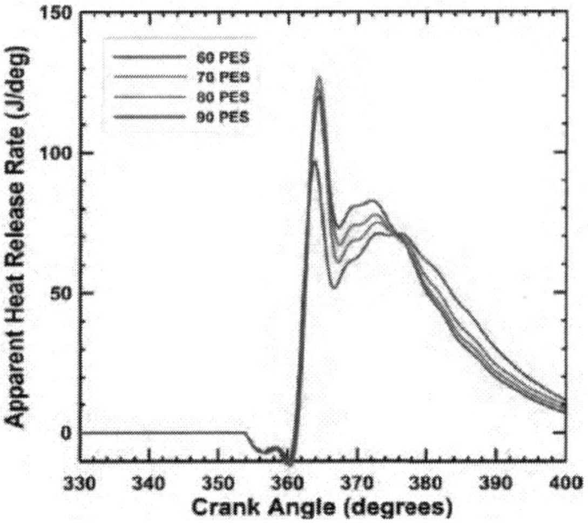

AHRR for 0 to 90 PES at 5.1 bar net IMEP, 1500 RPM, SOI = 355 CAD, Prail = 500 bar.

MISSISSIPPI STATE UNIVERSITY. CAVS Advanced Combustion Engines Laboratory 11 14AIChE Annual Meeting, Atlanta, GA

PES Sweep Results

In-cylinder Pressure

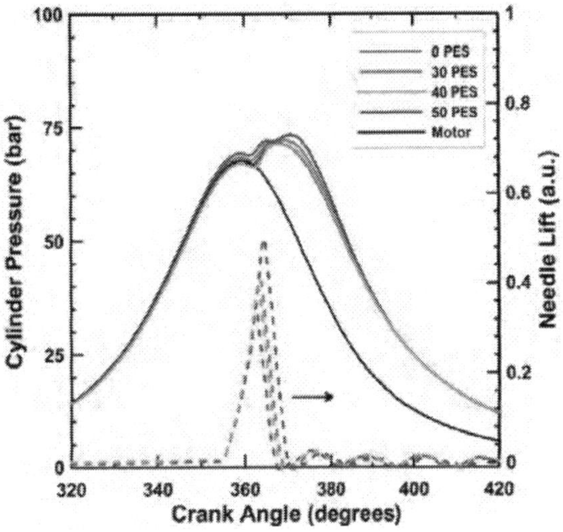

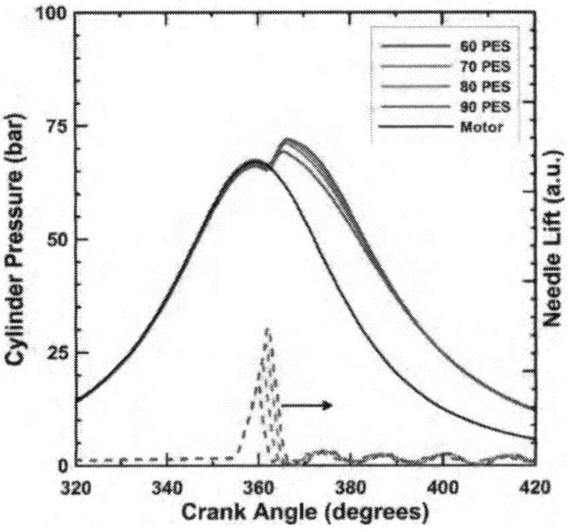

Cylinder pressure for 0 to 90 PES at 5.1 bar net IMEP, 1500 RPM, SOI = 355 CAD, P_{rail} = 500 bar.

MISSISSIPPI STATE UNIVERSITY. CAVS Advanced Combustion Engines Laboratory 12 14AIChE Annual Meeting, Atlanta, GA

PES Sweep Results

Global temperature

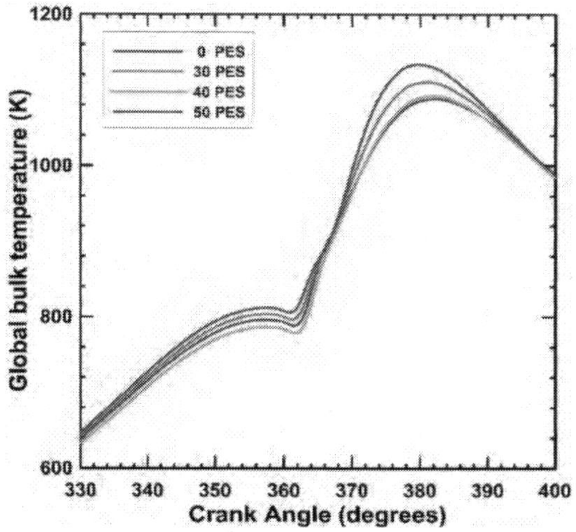

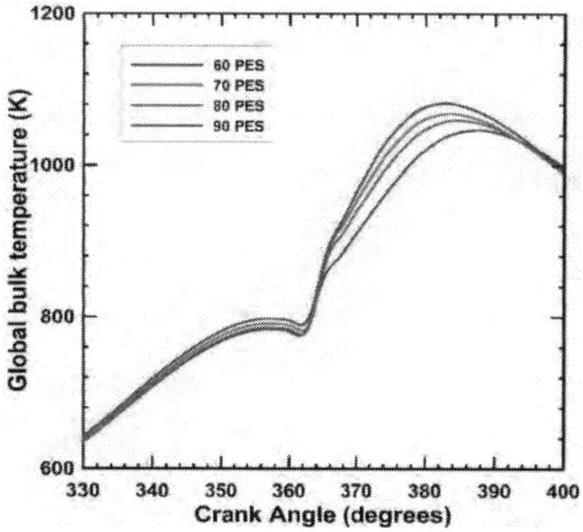

Global temperature profiles for various PES at 5.1 bar net IMEP, 1500 RPM, SOI = 355 CAD, P_{rail} = 500 bar.

MISSISSIPPI STATE UNIVERSITY CAVS Advanced Combustion Engines Laboratory 13 14AIChE Annual Meeting, Atlanta, GA

PES Sweep Results

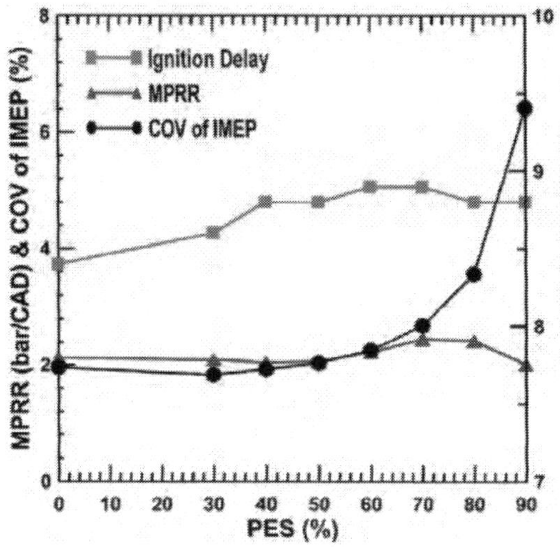

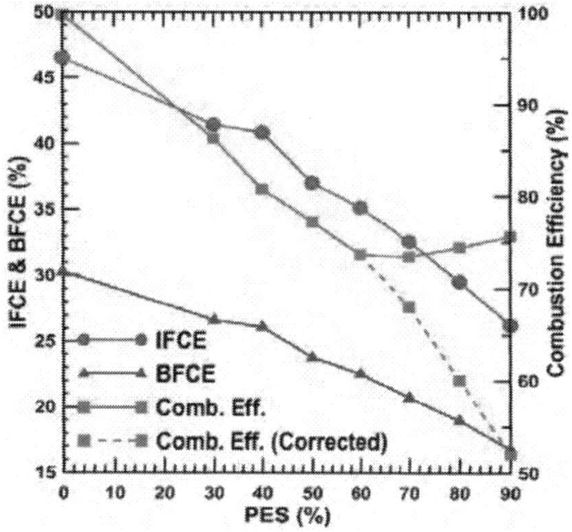

Coefficient of Variance (COV) of IMEP = Combustion instability

MISSISSIPPI STATE UNIVERSITY CAVS Advanced Combustion Engines Laboratory 14 14AIChE Annual Meeting, Atlanta, GA

PES Sweep Results

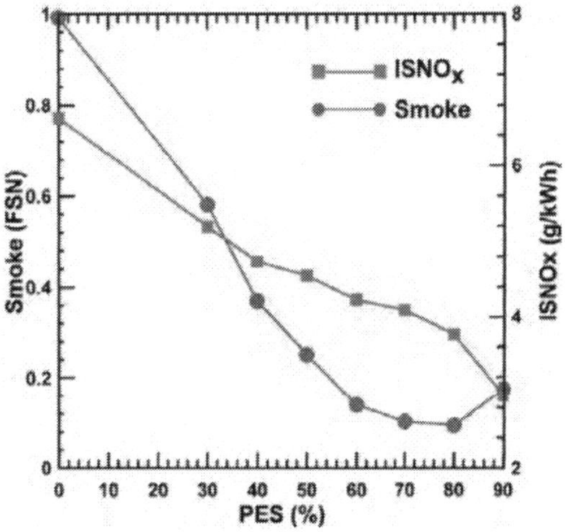

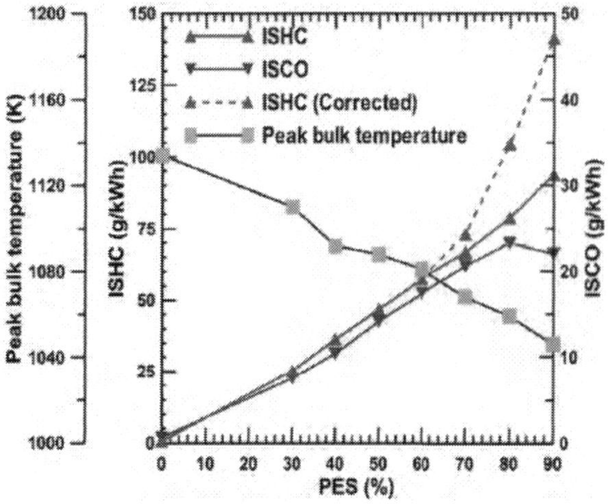

NOx: form by thermal mechanism; extremely dependent on **high local temperature**

HC : Combustion chamber crevices, Partial or bulk quenching, Exhaust valve leakage

CO: **Intermediate combustion species** formed during the oxidation of hydrocarbon fuel

- CO oxidation reaction (CO + **OH** → CO_2 + H) depends on **bulk temperature**

MISSISSIPPI STATE UNIVERSITY CAVS Advanced Combustion Engines Laboratory

Summary

- ❑ Diesel ignited dual fuel low temperature combustion is successful to reduce PM and NO_X emissions simultaneously.
- ❑ Diesel ignited dual fuel low temperature combustion results higher CO and HC emissions.

Exhaust Temperature Variation (°C)			
PES Sweep Experiment	SOI Sweep Experiment	Boost Pressure Experiment	Rail Pressure Experiment
324 - 348	231 - 272	186 - 311	234 - 247

MISSISSIPPI STATE UNIVERSITY

Gold Catalysis

- In the late 1980's Haruta et al. demonstrated that Au nanoparticles (less than 5 nm) are highly active for the oxidation of CO at low temperatures (200 K)*

- Gold Precursor

$HAuCl_4$ | $Au(en)_2Cl_3$

$[Au(OH)_XCl_{4-X}]^-$ | $[Au(en)_2]^{3+}$

*M. Haruta, N. Yamada, T. Kobayashi and S. Iijima, J. Catal.,1989, 115, 301

Catalysis Laboratory

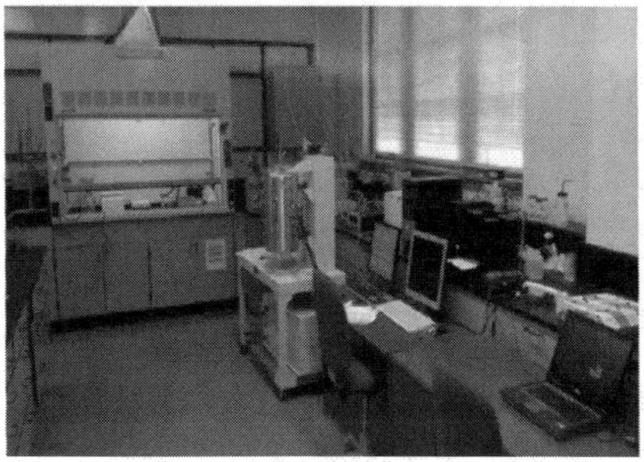

Reactor bench and monitoring

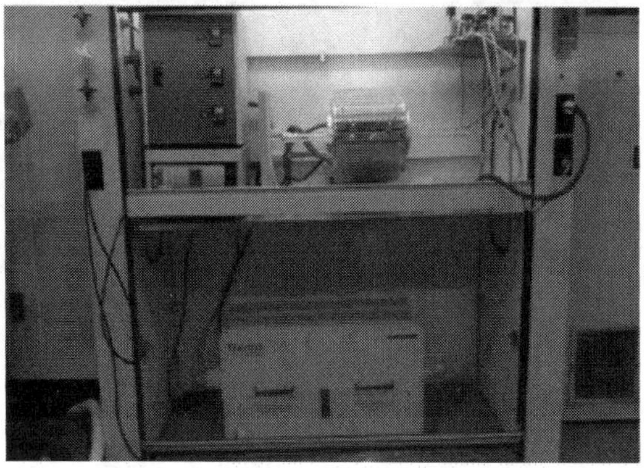

Furnaces which are able to increase the temperature to 1200 °C

Catalysis Laboratory

Continuous flow tubular reactor

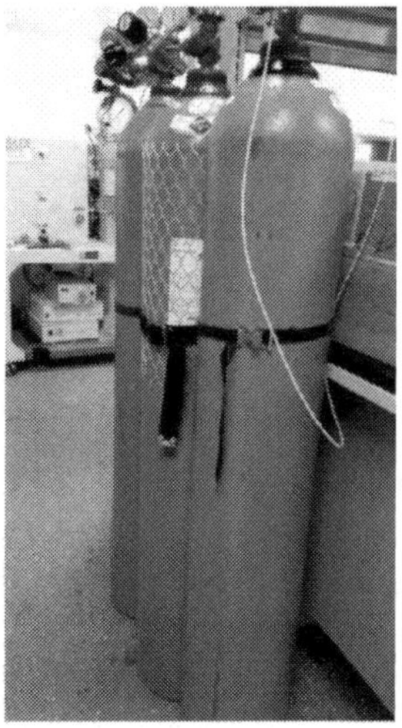

Calibration and feed gas cylinders

Catalysis Laboratory

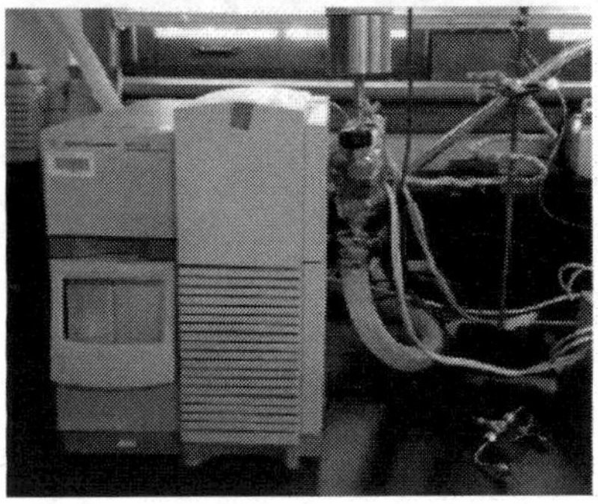

Micro GC and Mass spec

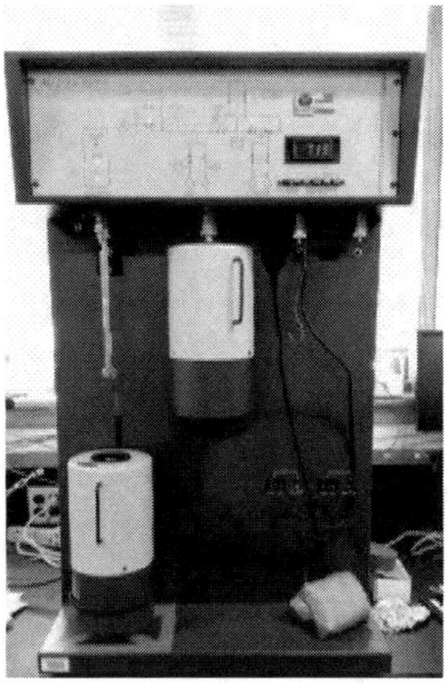

BET

Gold Nanoparticles

Turkevich method*

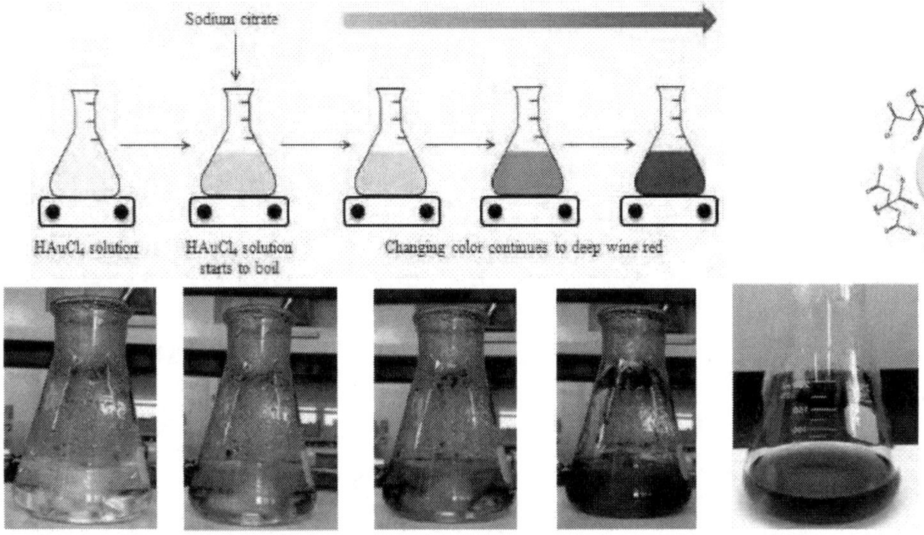

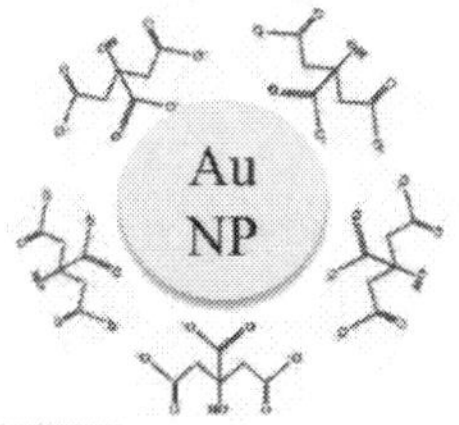

*Turkevich, J., P. C. Stevenson and J. Hillier (1951). "A study of the nucleation and growth processes in the synthesis of colloidal gold." Discussions of the Faraday Society **11**(0): 55-75

Gold Nanoparticles

Turkevich method

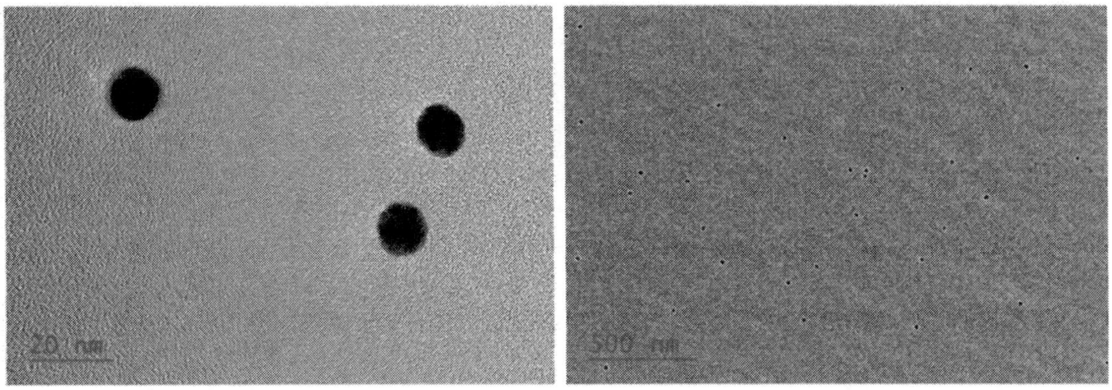

TEM images for sodium citrate stabilized gold nanoparticles

Au/SiO_2 Catalyst Preparation

Deposition-Precipitation

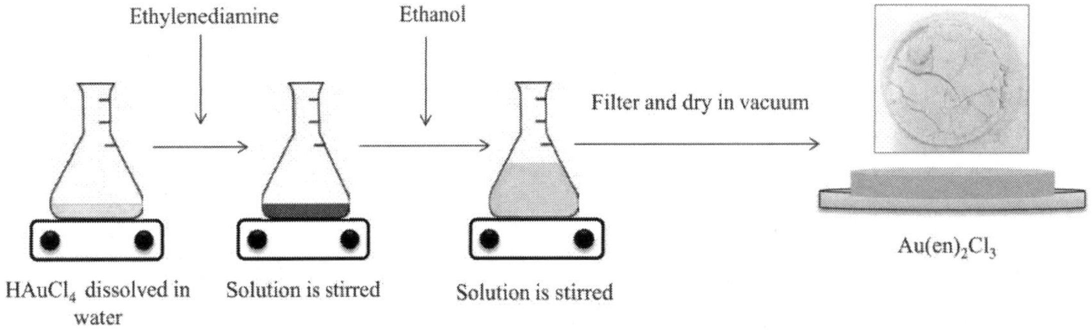

$$HAuCl_4.3H_2O + C_2H_4(NH_2)_2 \rightarrow Au(en)_2Cl_3 + HCl$$

MISSISSIPPI STATE UNIVERSITY

Zhu, H., et al., *Low-temperature CO oxidation on Au/fumed SiO2-based catalysts prepared from Au(en)2Cl3 precursor.* Applied Catalysis A: General, 2007. **326**(1): p. 89-99

23

14AIChE Annual Meeting, Atlanta, GA

Au/SiO_2 Catalyst Preparation

Deposition-Precipitation

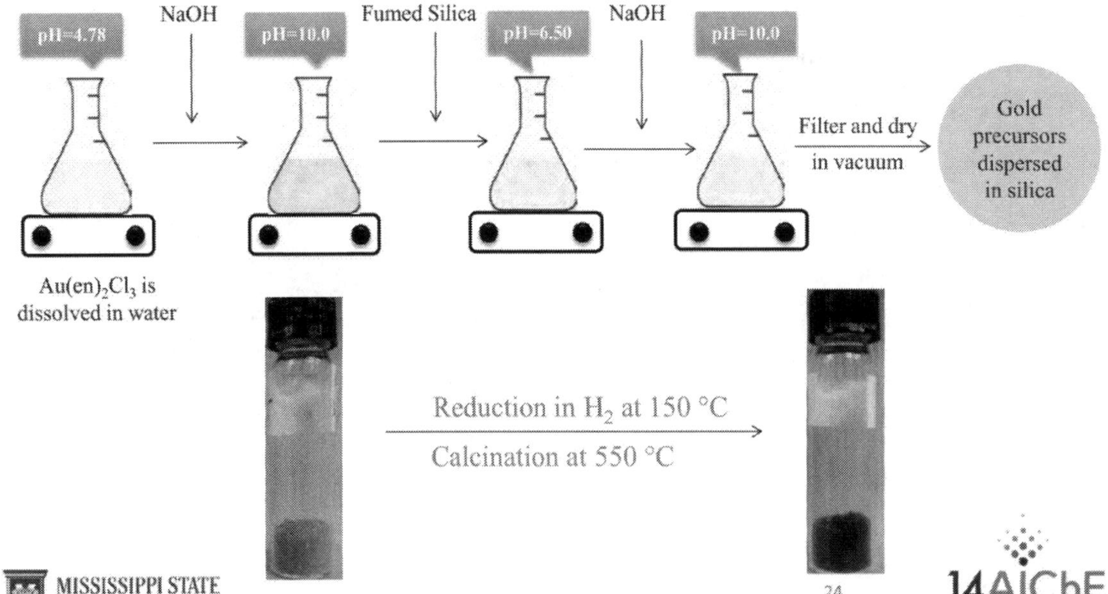

MISSISSIPPI STATE UNIVERSITY

24

14AIChE Annual Meeting, Atlanta, GA

Au/SiO_2 Catalyst Characterization

X-ray Diffraction Spectroscopy (XRD)

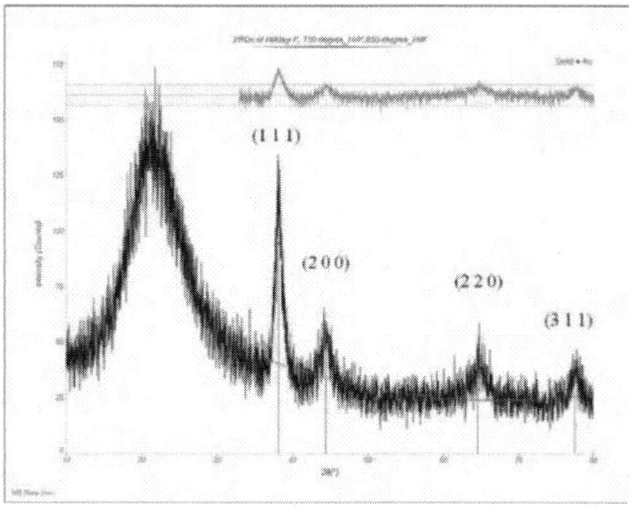

MISSISSIPPI STATE UNIVERSITY

14AIChE Annual Meeting, Atlanta, GA

Au/SiO_2 Catalyst Characterization

Energy Dispersive Spectroscopy (EDS)

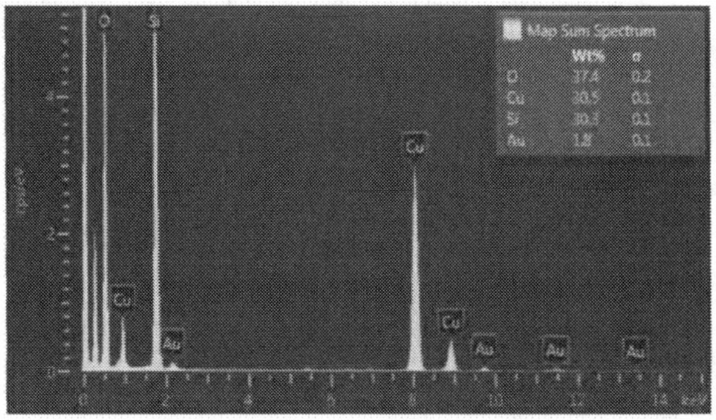

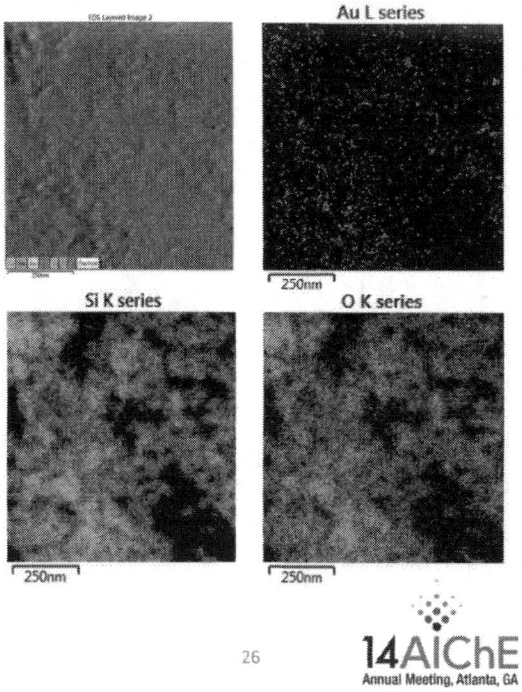

EDS results confirms the formation of Au/SiO_2 Catalyst.

MISSISSIPPI STATE UNIVERSITY

14AIChE Annual Meeting, Atlanta, GA

Au/SiO_2 Catalyst Characterization

Transmission Electron Microscopy (TEM)

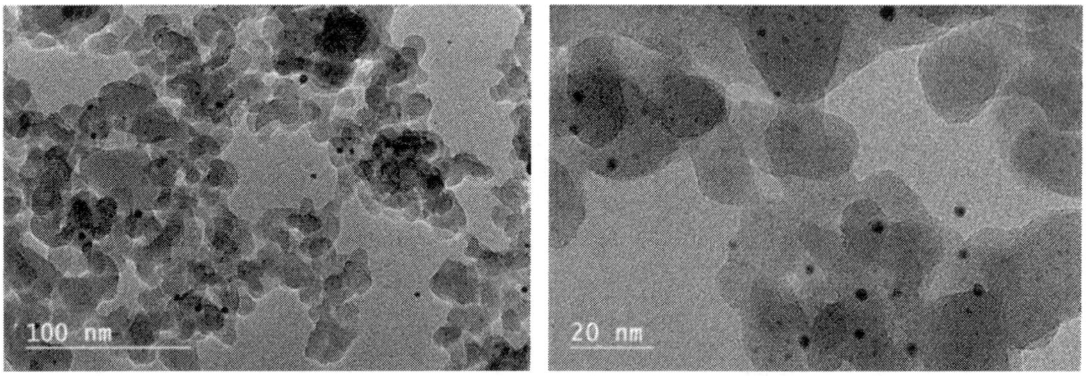

TEM images for Silica supported gold NPs

Thiolate-protected Gold Clusters

- Compounds with a core of gold atoms and organic groups covalently bound to the surface gold atoms.

- Thiolate groups are widely used as stabilization of gold cluster In solution.

- $Au_{25}(SHC2H4)_{18}$

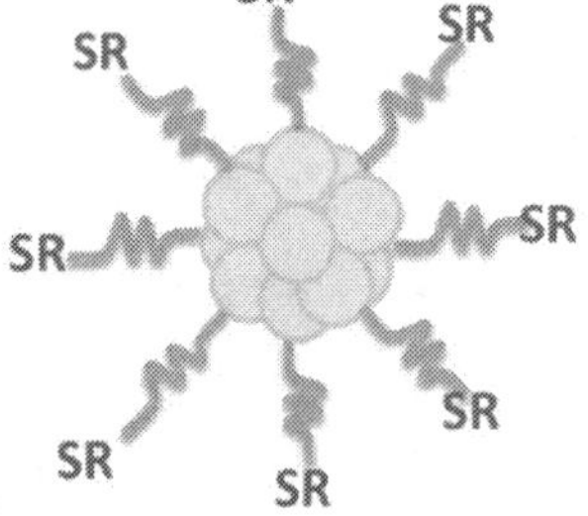

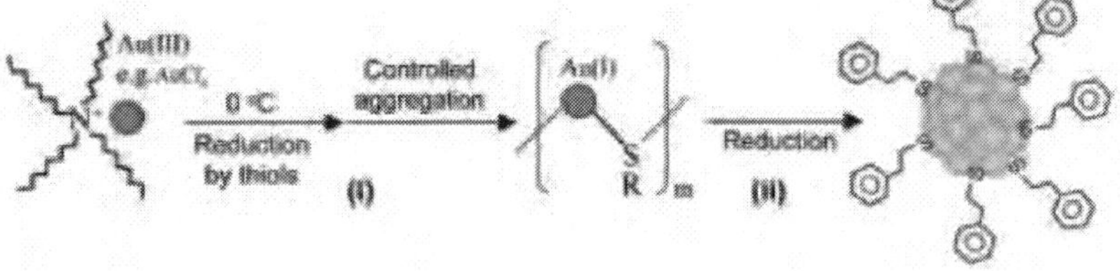

Zhu, Manzhou, et al. "Kinetically controlled, high-yield synthesis of Au25 clusters." *Journal of the American Chemical Society* 130.4 (2008): 1138-1139.

$Au_{25}(SHC_2H_4)_{18}$ Clusters

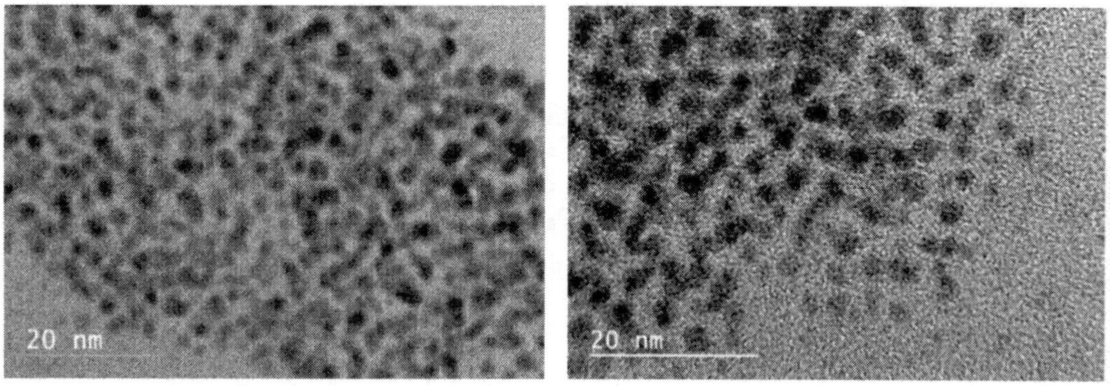

TEM images for gold clusters

$Au_{25-n}Cu_n(SHC_2H_4)_{18}$ Clusters*

$HAuCl_4$ $CuCl_2$ | Tetrabutylammonium bromide | phenylethanethiol | Sodium borohydride

Negishi, Yuichi, et al. "Effect of copper doping on electronic structure, geometric structure, and stability of thiolate-protected Au25 nanoclusters." *The Journal of Physical Chemistry Letters* 3.16 (2012): 2209-2214.

$Au_{25-n}Cu_n(SHC_2H_4)_{18}$ Clusters

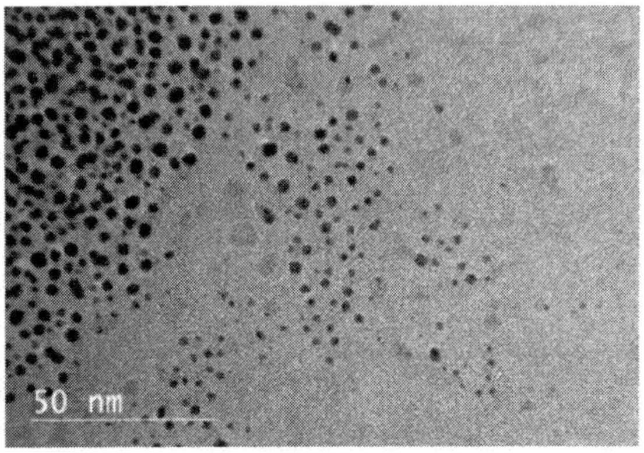

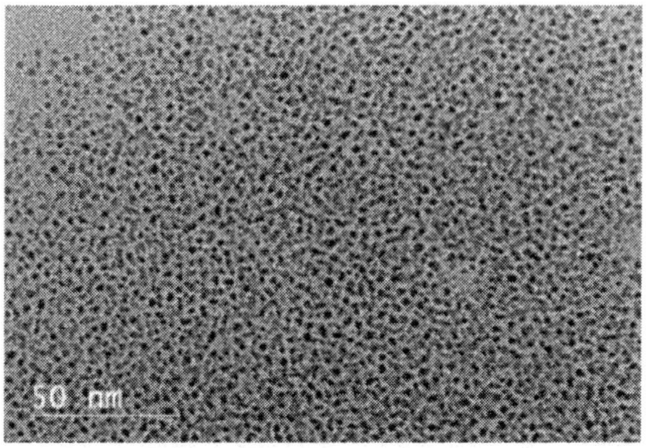

TEM images for gold-copper clusters

MISSISSIPPI STATE UNIVERSITY

31

14AIChE Annual Meeting, Atlanta, GA

$Au_{23}Cu_2(SHC_2H_4)_{18}/SiO_2$

- ❑ $Au_{23}Cu_2(SHC_2H_4)_{18}/SiO_2$ was produced by stirring certain amount of Au-Cu clusters with Cab-O-Sil EH5 dispersed into toluene.
- ❑ Product was centrifuged and dried.
- ❑ Final catalyst obtained after calcination at 300°C for 3h.

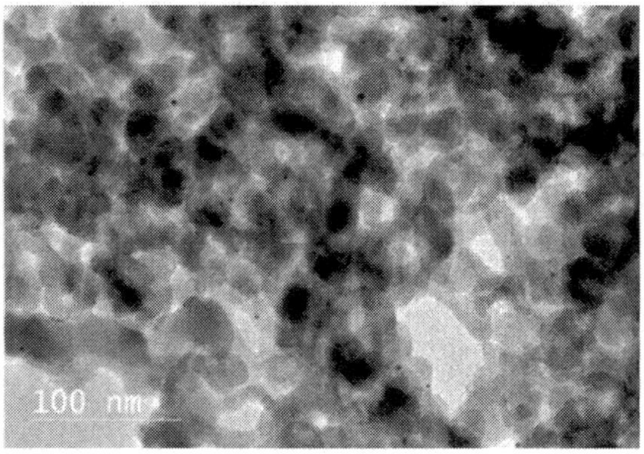

TEM images silica supported $Au_{23}Cu_2(SHC_2H_4)_{18}$ thiolate-protected nano alloy

MISSISSIPPI STATE UNIVERSITY

32

14AIChE Annual Meeting, Atlanta, GA

$Au_{23}Cu_2(SHC_2H_4)_{18}/SiO_2$

EDS results confirms the formation of $Au_{23}Cu_2(SHC_2H_4)_{18}/SiO_2$ Catalyst.

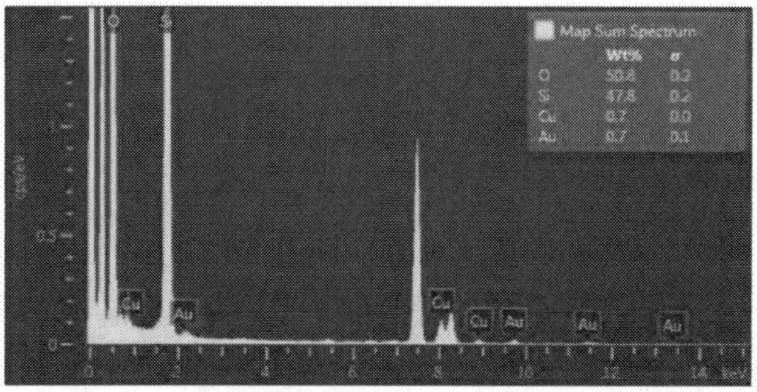

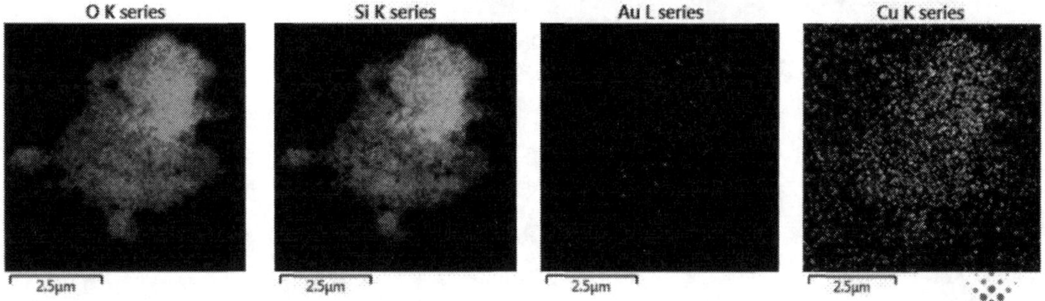

Summary

- ❑ Using $Au(en)_2Cl_3$ salt produces stable and active Au/SiO_2 catalyst via DP method which gold NPs diameter are less than 5nm.
- ❑ Gold NPs with diameter less than 2 nm can be produced by $HAuCl_4$ salt assisted by thiol groups as stabilizers.
- ❑ Silica supported gold-copper nano alloy with diameter less than 2 nm was successfully produced by using thiol ligands.

Future Work

- Low temperature oxidation of CO using $Au_{23}Cu_2(SHC_2H_4)_{18}/SiO_2$
- Investigating the effect of thiol ligands and support on activity of catalyst
- Investigating the CO oxidation in presence of exhaust species (e.g. NO and Hydrocarbons)
- In-situ oxidation of CO produced by diesel-ignited methane (Propane, Gasoline) dual fuel low temperature combustion

MISSISSIPPI STATE UNIVERSITY

35

14AIChE Annual Meeting, Atlanta, GA

Acknowledgement

The authors gratefully acknowledge the financial support from

Bagley College of Engineering
Earnest W. Deavenport, Jr. Chair
Thomas B. Nusz Professorship
Sustainable Energy Research Center (US DOE Award # DE-FG36-06GO86025)
Facilities support from the Center for Advanced Vehicular Systems at Mississippi State University.

JAMES WORTH BAGLEY COLLEGE OF ENGINEERING MISSISSIPPI STATE UNIVERSITY

CAVS Center for Advanced Vehicular Systems

U.S. DEPARTMENT OF ENERGY

MISSISSIPPI STATE UNIVERSITY

36

14AIChE Annual Meeting, Atlanta, GA

Thanks for your attention

MISSISSIPPI STATE UNIVERSITY

37

14AIChE
Annual Meeting, Atlanta, GA

THE IMPACT OF ROOFING MATERIAL ON BUILDING ENERGY PERFORMANCE

Ali Badiee

CIVIL ENGINEERING

ABSTRACT

The way of using the water, electricity, natural gas, materials or in general "energy" more efficiently in all phases of residential and commercial building constructions have grown significantly in the last decade since the times when heating and cooling prices were a reasonably minor line item on a facility manager's budget are long gone. Oil costs, nevertheless lower than they were last year, remain high and tremendously unstable. Coal prices and natural gas are similarly on the growth. All of these instability and rises have gone to higher heating and cooling costs, and engineers are doing all of their efforts to keep them under control by using energy efficient building materials. An argument can be made that the emphasis on building energy efficiency has influenced the roofing manufacturing more than the other building industries. Several energy codes such as ASHRAE 90.1 (2010) and IECC (2012) have been developed, organizations, administrations created and guidelines established, all in the interest of demonstrating the matter of energy efficient roofing. The topic of this thesis will be to address the concern of roofing energy productivity and ranks different roofing materials in terms of energy efficiency. The yield of various roofing materials installed on five different UAB campus buildings have been survived, analyzed and evaluated in this study. Two parameters will be considered in this evaluation, primarily the energy consumption, and energy bills as a second factor. These data have been provided by the UAB Department of Energy and Facilities Management and monitored from 2007 to 2013, creating several spreadsheets. The energy utilities examined in this study are consisted of electricity, natural gas, domestic water and chilled water (where it was applicable) which measured separately in four different seasons during this particular period. It has been expected to observe a rooftop classification indicating green roof, white roof, river rock roof, concrete paver roof and black traditional roof respectively at the conclusion of this research.

Keywords: ASHRAE 90.1, IECC (2012), Green roof, White roof, Black roof, Concrete paver roof, River rock roof

392682 Biowall's Impact on Indoor Air Quality and Energy

Abstract:

The average American spends more than 90% of his or her time indoors where the air is 2-5 times more polluted than outdoors. A principal cause of indoor pollution and its associated health concerns are volatile organic compounds (VOCs), which are emitted by a wide array of household products. Taken as a whole, poor indoor air quality (IAQ) is estimated to cost the United States at least $40 billion annually due to related illness and lost productivity. The elderly and the young are especially at risk to illness due to lifestyle habits and weaker immune systems.

Energy efficient buildings typically seal the leakage paths for outside air infiltration but this can lead to poor IAQ as an unintended side effect. The Biowall is a new concept for maintaining high levels of indoor air quality and has the potential to reduce HVAC energy use by reducing outside air requirements for buildings. The device consists of a living plant filter that is integrated into the heating and cooling system of a residence or a small commercial building. This technology leverages the natural ability of plants to reduce CO2 and remove VOCs.

To evaluate this concept, a living plant filter was designed and grown in a simulated building environment. The plant filter was integrated with a heating and cooling system and fully instrumented so precise measurements of air temperature, air humidity, air quality, and component-level energy consumption could be made. Various growth media, plant varieties, and watering strategies were evaluated to investigate options for improving a Biowall's performance. The experimental work involved introducing toluene, a common VOC in buildings, into a test chamber where precise measurements of the VOC decay rate could be made. Comparisons were made with the toluene removal rate, both with and without the Biowall.

Sustainable Buildings: Targets and Measurements

Godfried Augenbroe
Jason Brown
Georgia Institute of Technology School of Architecture

AIChE Annual Meeting
November 16-21, 2014
Atlanta, GA

Energy and emissions

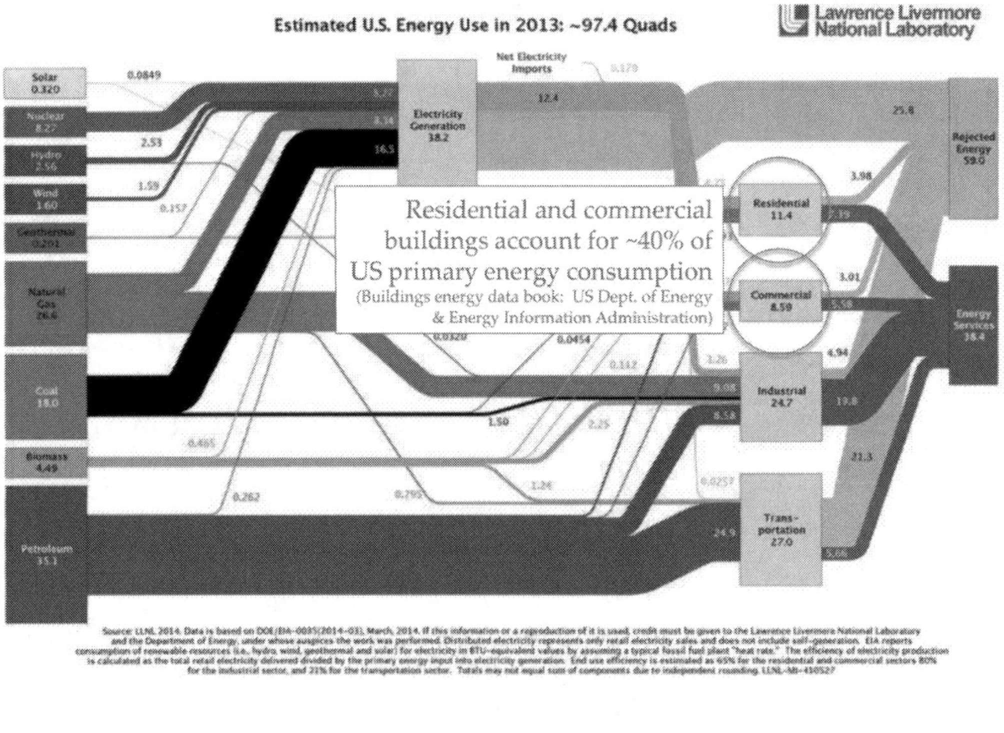

Energy and emissions

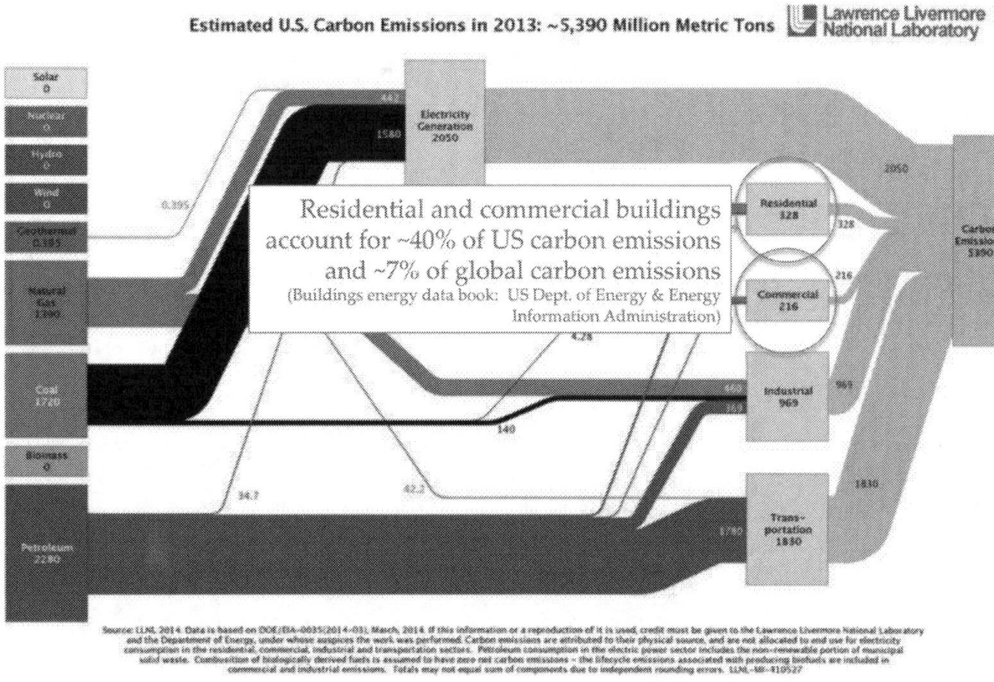

Energy and emissions

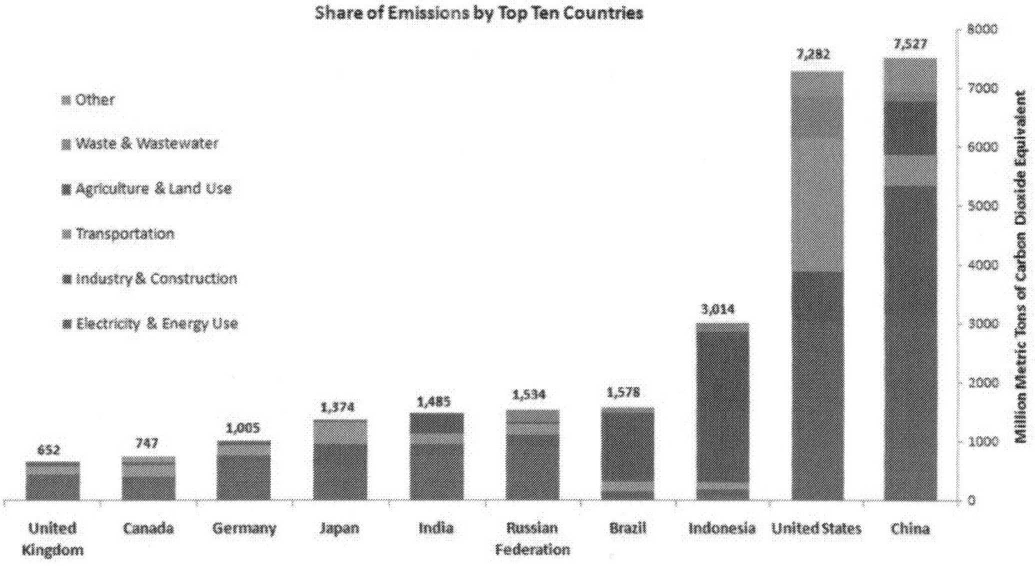

Source: Brown and Sovacool. 2011. *Climate Change and Global Energy Security*, Figure 2.1

Water

In 2005, US buildings used roughly 40,000 million gallons of water per day

This is roughly 10% of total US water consumption

Conservation measures

Commercial buildings end-uses, 2010

Residential buildings end-uses, 2010

Site energy

Primary energy

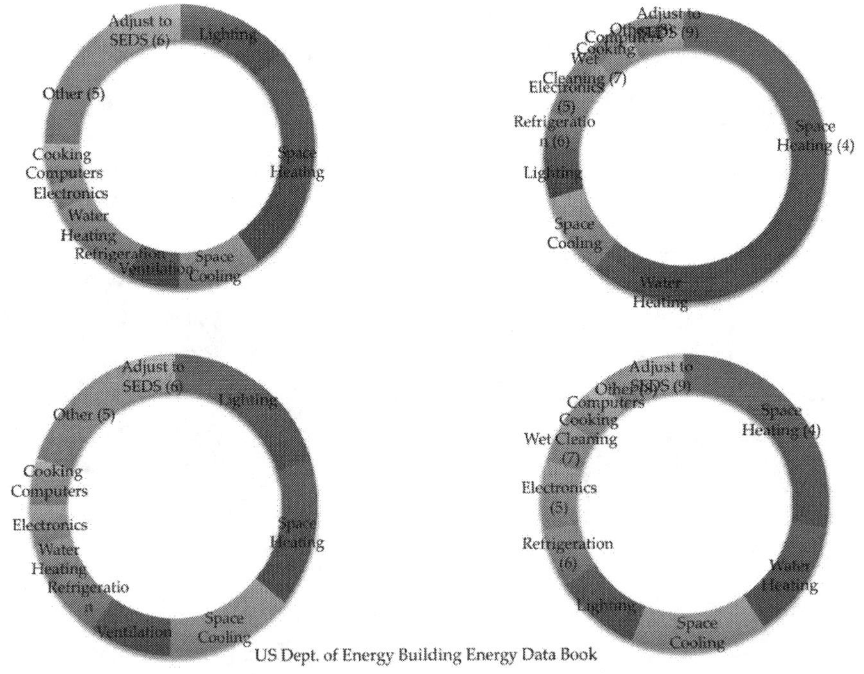

US Dept. of Energy Building Energy Data Book

Conservation measures

First things:

- Improvements to envelope: insulation, low U-value and low SHGC windows, etc.
- More efficient equipment: lighting, chillers, boilers

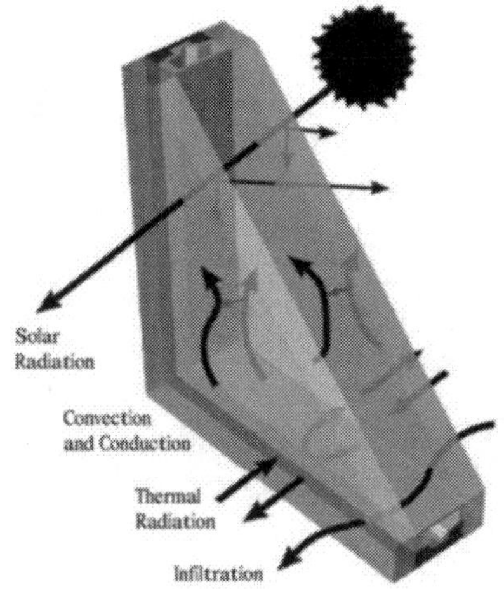

Conservation measures

Second: an integrated system

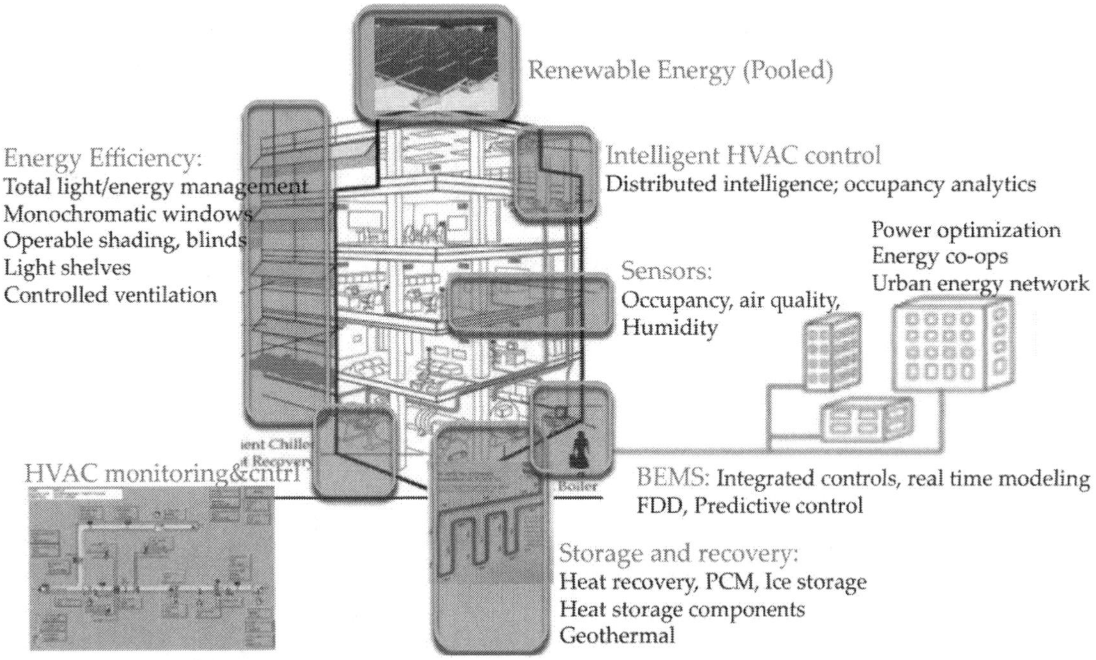

Targets

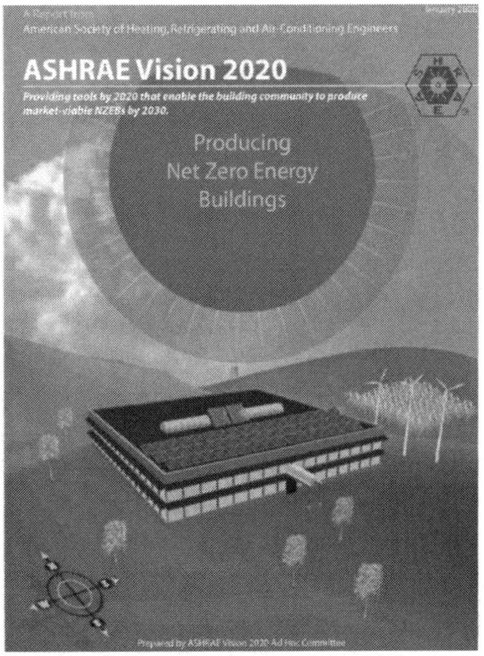

Goal: market-viable net-zero energy at building site by 2030

Goal: carbon-neutral buildings by 2030

Sustainability

Bruntland definition: sustainable development

> "...meets the needs of the present without compromising the ability of future generations to meet their own needs."

(World Commission On Environment and Development. (1987). Our common future. Oxford University Press, Oxford, United Kingdom.)

Daly's definition:

1. "harvesting rates [of renewable resources] do not exceed regeneration rates"
2. "nonrenewable resources are depleted at a rate equal to the rate of creation of renewable substitutes"
3. emission rates of substances harmful to life are less than the rate at which those substances are rendered harmless or are naturally recycled.

Daly, H.E. (1993) "Sustainable Growth: An Impossibility Theorem." In Daly, H.E. and Townsend, K.N., eds. (1993) *Valuing the Earth: Economics, Ecology, Ethics.* The MIT Press, Cambridge, Massachusetts.

Sustainability

Something is either sustainable or it is not

How do we assess the sustainability of buildings?

Building environmental assessment (BEA) systems

Building environmental assessment

In the spirit of, but more expansive than, life-cycle assessment

- Includes not just ecological impact, but social issues as well
- Ideally would be a full cradle-to-grave assessment of environmental and social effects

Actual practice deviates significantly from the ideal, however, due to:

- Difficulty in properly defining system boundaries
- Lack of data
- Time-consuming calculations even if the above are overcome
- AND: not all impacts can be verified

Building environmental assessment

As a result:

- A building's sustainability is not fully assessed
- Proxy or "feature-based" assessments stand in for difficult-to-measure outcomes

  e.g. giving points for locating a building next to public transit or providing showers and bike racks, etc.

- Such surrogates co-exist with more measurable, verifiable (at least partially) and a priori calculable estimates of outcomes

  e.g. estimates of energy consumption and ensuing emissions

- Scores are weighted and combined (sometimes in an ad-hoc way)
- A high "sustainability" score may not indicate good energy performance

  (e.g. Scofield, J.H. (2009) "Do LEED-certified buildings save energy? Not really…" Energy and Buildings 14, pp. 1386-1390)

- We only assess how less unsustainable a building is compared to a baseline

Building environmental assessment

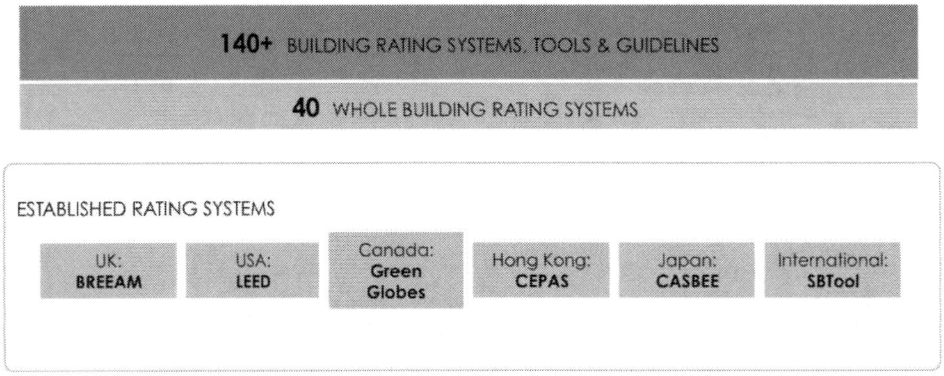

A great success of BEA systems is the creation of a sustainability or "green" marketplace

Building environmental assessment

LEED Green Building Rating System

LEED for New Construction 2009

EA Credit: Energy & Atmosphere

- **EA Prerequisite 2:** Minimum Energy Performance (Prerequisite)
  Need to show at least **10%** improvement (beyond that, earn extra points)
- **EA Credit 1:** Optimize Energy Performance (1-19 Points)
  More than **12%, up to 48%** (1-19 Points)

ANSI/**ASHRAE**/IESNA **Standard 90.1-2007**
"Energy Standard for Buildings Except Low-Rise Residential Buildings"

Use **Appendix G (Performance Rating Method)-Computer Simulation**

Envelope | **HVAC** | **Water Heating** | **Power** | **Other Equip.** | **Lighting**

Other options: Prescriptive Compliance Path
(ASHRAE Advanced Energy Design Guide or Advanced Buildings™ Core Performance™ Guide)

Building environmental assessment

LEED Green Building Rating System

ANSI/**ASHRAE**/IESNA **Standard 90.1-2007**
"Energy Standard for Buildings Except Low-Rise Residential Buildings"

1. Comply with the mandatory Provisions
(a prescriptive path to compliance)

If a building cannot satisfy the prescriptive provisions (1),

2. Energy Cost Budget
- An alternative to the prescriptive provisions
- Evaluating the compliance of all proposed designs

If you need to rate the energy efficiency of building designs that exceed the requirements of this standard (i.e. LEED EA Scores),

3. APPENDIX G: Performance Rating Method (for LEED EA)
- **Using a computer simulation model for the whole building project**
- **Energy Cost Savings Percentage compared with the baseline building performance rating**

Envelope | **HVAC** | **Water Heating** | **Power** | **Other Equip.** | **Lighting**

Building environmental assessment

LEED Green Building Rating System

ANSI/**ASHRAE**/IESNA **Standard 90.1-2007**
"Energy Standard for Buildings Except Low-Rise Residential Buildings"

3. APPENDIX G: Performance Rating Method

- **Using a computer simulation model for the whole building project**
- **Energy Cost Savings Percentage compared with the baseline building performance rating**

$$\text{Percentage improvement (\%)} = 100 \times \frac{\text{(Baseline building performance − Proposed building performance)}}{\text{Baseline building performance}}$$

Building environmental assessment

LEED Green Building Rating System

- **EA Credit 1: Optimize Energy Performance (1-19 Points)**

New Buildings	Existing Building Renovations	Points
12%	8%	1
14%	10%	2
16%	12%	3
18%	14%	4
20%	16%	5
22%	18%	6
24%	20%	7
26%	22%	8
28%	24%	9
30%	26%	10
32%	28%	11
34%	30%	12
36%	32%	13
38%	34%	14
40%	36%	15
42%	38%	16
44%	40%	17
46%	42%	18
48%	44%	19

*The minimum energy cost savings percentage for each point threshold

Building environmental assessment

Changes in LEEDv4 Green Building Rating System (coming in 2015)

- **Energy and atmosphere**: Energy consumption metering will be required; track and report performance for 5 years
- **Materials and resources**:
    - Includes verified LCA rather than proxies like % recycled content; early indications suggest LCA may become more business as usual, improving sustainability assessments
    - Incorporates hazard-based (chemical) ingredients reporting by a variety of pathways (e.g. Health Product Declaration)

Building environmental assessment

QSAS

QATAR SUSTAINABILITY ASSESSMENT SYSTEM

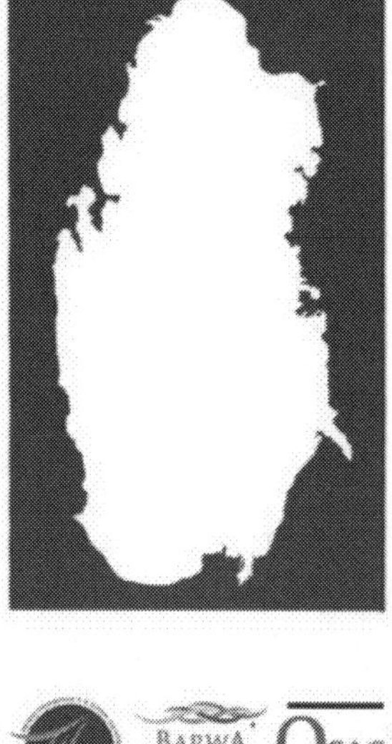

Building environmental assessment

QSAS Goals

[UC] The building's planning shall incorporate **urban** considerations.

[S] The building's development in relation to the existing **site** conditions shall be controlled.

[E] The building's depletion of fossil **energy** over its service life shall be controlled.

[W] The building's impact on the overall **water** resource shall be controlled.

[M] The impact of the building's use of **materials** on the environment shall be controlled.

[IE] The building's **indoor environment** shall be controlled.

[CE] The building's **cultural and economic value** shall be maintained or enhanced.

[MO] The building's **management and operations** plan shall be defined.

Building environmental assessment

QSAS Scoring

No	Category / Criteria	Weight
UC	Urban Connectivity	8.00%
UC.1	Load on Local Traffic Conditions	1.80%
UC.2	Pedestrian Pathways	1.08%
UC.3	Proximity to Amenities	0.69%
UC.4	Light Pollution	0.58%
UC.5	Noise Pollution	0.43%
UC.6	Public Transportation	1.30%
UC.7	Private Transportation	0.39%
UC.8	Sewer & Waterway Contamination	1.08%
UC.9	Shading of Adjacent Properties	0.65%
S	Site	9.00%
S.1	Ecological Value of Land	2.34%
S.2	Vegetation & Shading	1.05%
S.3	Desertification	1.75%
S.4	Rainwater Runoff	1.17%
S.5	Mixed Use	0.88%
S.6	Heat Island Effect	0.58%
S.7	Adverse Wind Conditions	0.88%
S.8	Acoustic Conditions	0.35%
E	Energy	24.00%
E.1	Energy Demand Performance	5.20%
E.2	Energy Delivery Performance	5.20%
E.3	Fossil Fuel Conservation	3.64%
E.4	CO2 Emissions	4.55%
E.5	NOx, SOx, & Particulate Matter	5.41%
W	Water	16.00%
W.1	Water Consumption	16.00%

M	Materials	8.00%
M.1	Regional Materials	1.85%
M.2	Responsible Sourcing of Materials	0.00%
M.3	Structure Reuse: On-site	0.91%
M.4	Materials Reuse: Off-site	1.54%
M.5	Recycled Materials	1.85%
M.6	Design for Disassembly	1.85%
M.7	Life Cycle Assessment (LCA)	0.00%
IE	Indoor Environment	14.00%
IE.1	Thermal Comfort	1.20%
IE.2	Low-Emitting Materials	1.60%
IE.3	Natural Ventilation	1.60%
IE.4	Mechanical Ventilation	1.60%
IE.5	Indoor Chemical & Pollutant Source Control	1.60%
IE.6	Views	1.20%
IE.7	Glare Control	1.20%
IE.8	Illumination Levels	1.20%
IE.9	Acoustic Quality	1.20%
IE.10	Daylight	1.60%
CE	Cultural & Economic Value	13.00%
CE.1	Heritage & Cultural Identity	8.67%
CE.2	Support of National Economy	4.33%
MO	Management & Operations	8.00%
MO.1	Commissioning Plan	2.67%
MO.2	Energy Use Sub-metering	0.89%
MO.3	Leak Detection	1.78%
MO.4	Organic Waste Management	0.00%
MO.5	Recycling Management	0.00%
MO.6	Intelligent Building Control System	2.67%
	Total	100.00%

Building environmental assessment

QSAS Scoring

- QSAS consists of six certification levels to denote a project' s environmental impact
- Certification can only be achieved when the final score exceeds 0, earning 1, 2, 3, 4, 5, or 6 stars
- The highest certification level a building can reach is 6 stars

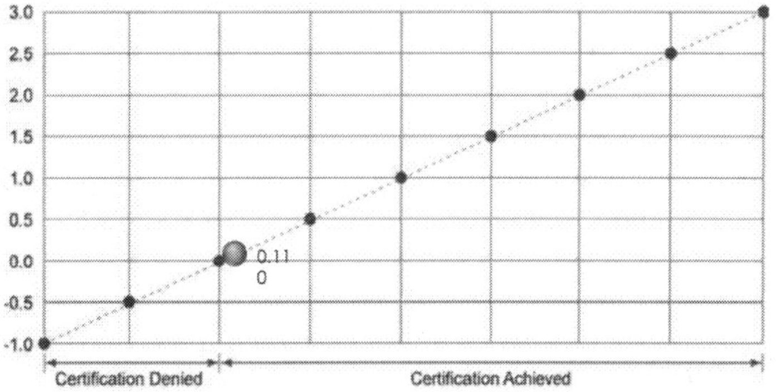

Score	Certification Level	QSAS Certification
X < 0	-	Certification denied
0.0 ≤ X ≤ 0.5	★	Certification achieved
0.5 < X ≤ 1.0	★★	
1.0 < X ≤ 1.5	★★★	
1.5 < X ≤ 2.0	★★★★	
2.0 < X ≤ 2.5	★★★★★	
2.5 < X ≤ 3.0	★★★★★★	

Building environmental assessment

QSAS Scoring: uses targets as baseline rather than % improvement over an ad-hoc baseline

Building environmental assessment

QSAS Scoring: energy consumption assessment based on an international, *normative* standard (ISO 13790)

E_{design}: Calculated energy use

Calculated based on energy standard

E_{ref}: Reference Energy use

Energy consumption of a basket of buildings that represent local building practices.

$$EPC = \frac{E_{design}}{E_{ref}}$$

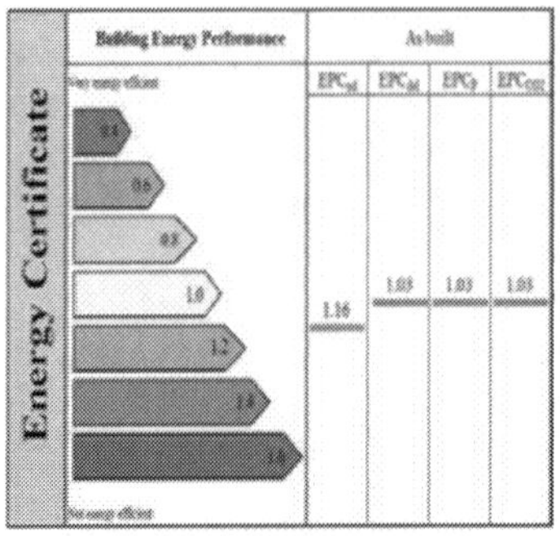

An example of how The EPC is used to "rank" a building

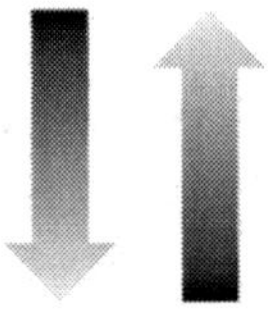

EPC **Energy Performance**

Building environmental assessment

[E.1] Q_{nd} , EPC_{nd}

-Performance evaluation of thermal energy demand, based on calculation of heating & cooling demand
-Includes heat transmission, ventilation, solar heat gain, internal heat gain

Thermal Energy Need

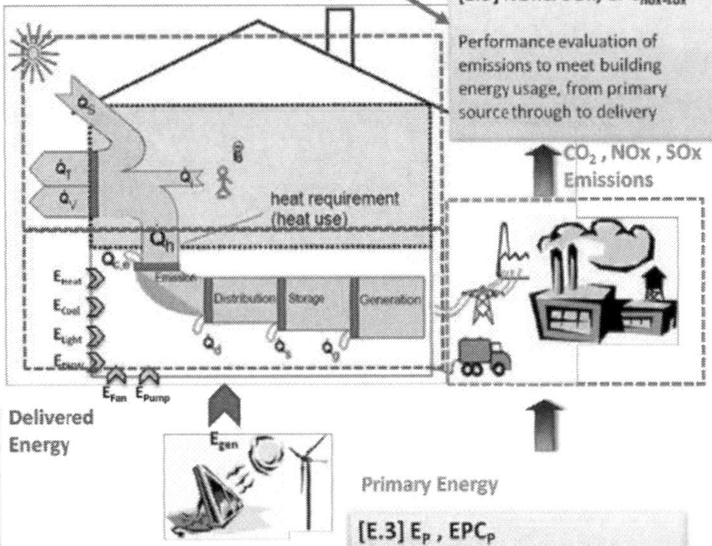

[E.4] CO_2, EPC_{CO2}
[E.5] NOx& SOx, $EPC_{nox\text{-}sox}$

Performance evaluation of emissions to meet building energy usage, from primary source through to delivery

CO_2 , NOx , SOx Emissions

[E.2] E_{del} , EPC_{del}

-Performance evaluation of building energy consumption based on calculation of heating, cooling, lighting, DHW, receptacles, auxiliary energy consumption
-Consideration of all on site systems, including local energy generation and renewable energy systems

Delivered Energy

Primary Energy

[E.3] E_P , EPC_P

-Performance evaluation of primary energy use at the source, based on calculation of generation, distribution, district systems, delivery on site

Summary

Buildings are important components of the sustainability agenda

If sustainability is to be achieved, aggressive targets need to be met

Current assessments of sustainability vary in approach and are imperfect, but improving

One success of BEA systems is the creation of a sustainability or "green" marketplace

Backup: do LEED buildings save energy?

J.H. Scofield / Energy and Buildings 41 (2009) 1386–1390 1387

Table 1
Comparison of building-weighted mean site and source (case B only) energy intensities for LEED office sets with matched CBECS building sets (see text).

Case	LEED			CBECS			EUI reduction	LEED > CBECS	*T*-statistic	*p* value
	N	Mean	sdm	*N*	Mean	sdm				
Site energy intensity										
A	35	67.8	4.2	27	97.0	8.7	30%	12 (34%)	3.36	0.0019
B	32	67.6	4.4	26	98.9	9.0	32%	10 (31%)	3.53	0.0013
Source energy intensity										
B	32	178.3	13.2	26	227.5	18.9	22%	10 (31%)	2.53	0.0167

J.H. Scofield / Energy and Buildings 41 (2009) 1386–1390 1389

Table 2
Comparison of gsf-weighted mean site and source energy intensity (for case B) for LEED office sets with matched CBECS building sets (see text).

Case	LEED			CBECS			EUI reduce	LEED > CBECS	*T*-statistic	*p*
	gsf	Mean	sdm	gsf	Mean	sdm				
Site energy										
A	5,303,460	77.1	3.4	5,196,100	80.7	8.4	4%	68%	−0.40	0.69
B	3,835,620	77.6	3.5	3,776,100	86.6	9.3	10%	63%	−0.97	0.37
Source energy										
B	3,835,620	205.0	9.6	5,196,100	211.9	18.7	3%	64%	−0.35	0.73

Backup: abstract

The presentation starts with a discussion of the role of buildings in the global energy and water conservation agenda, well-established design measures in new buildings, and retrofits in existing buildings are reviewed, going from the low hanging fruit of energy and water efficiency measures, to improved management and control, and ultimately to the installation of renewables and recyclables. We compare measures especially on their cost per saving effectiveness, and discuss the feasibility of (net) zero-impact as the ultimate target of sustainable buildings.

The second part of the presentation focuses on building environmental assessment (BEA) systems. BEA systems measure the outcomes of buildings through a broad range of criteria, ranging from occupant health and productivity, energy and water conservation, material impacts, transport needs, etc. In the material impact criterion, they have recently started using LCA methods, or some reduced version of it when appropriate data is not available.

The primary aim of any BEA system is to encourage the design of green buildings and stimulate the market for sustainable construction components and materials. A properly managed set of guidelines, supported by an adequate BEA system, is the best instrument for local governments to regulate industries and to improve the sustainability of the built environment.

The current state of sustainability targets and assessment systems is critically analyzed.

373642 Sustainable Buildings

Abstract:

This unique session combines such a broad perspective of engineering practices in the face of contemporary society that we like to make time to promote audience interaction. If time allows or a speaker cannot make the session, a chair will prompt appropriate discussion in order to get feedback from the audience. We greatly appreciate your attendance and look forward to any comments you may have.

Avoided GHG Emissions Achieved By Using Architectural Window Film:

An LCA Compliant With New WBCSD Avoided Emissions Guidelines

Terrie Boguski, Harmony Environmental, LLC
Lauren Johnson, Eastman Chemical Company

Agenda

- Avoided Emissions Guidelines
- Architectural Window Film LCA
  - Product description
  - Goal and Scope
  - Results
  - Compliance with Avoided Emissions Guidelines
- Conclusions
- Questions

Avoided Emissions Guidelines

Avoided Emissions Guidelines

Image credit: ICCA and WBCSD, 2013

Avoided Emissions Guidelines

- Elements
  - Level in the Value Chain
  - Comparison Criteria
  - Calculation Methodology
  - Significance of Contribution

Architectural Window Film

Architectual Window Film

- Daylighting – +70% light transmittance
- Energy Savings – SHGC 0.41
- Transparent Appearance – 8% visible reflectance
- Fade Reduction – Blocks 99% of harmful UV radiation
- Uniform Appearance
- Noise Reduction

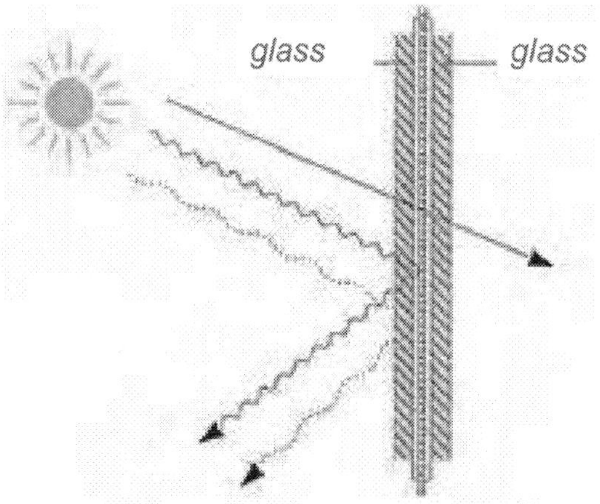

Architectural Window Film LCA

Architectural Window Film LCA

- Goal
  - Understand and document the environmental impact of using window film
  - Evaluate energy savings and avoided GHG emissions

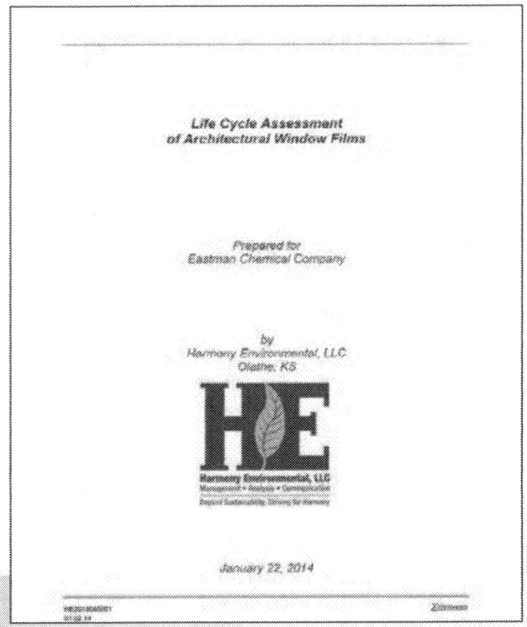

Life Cycle Assessment
of Architectural Window Films

Prepared for
Eastman Chemical Company

by
Harmony Environmental, LLC
Olathe, KS

January 22, 2014

Architectural Window Film LCA

- Scope
  - Cradle-to-Grave LCA of Window Film
    - 15 years of use
    - Follow ISO 14044
    - Follow the WBCSD Guidelines for Avoided Emissions

Architectural Window Film LCA

- LCA Boundaries

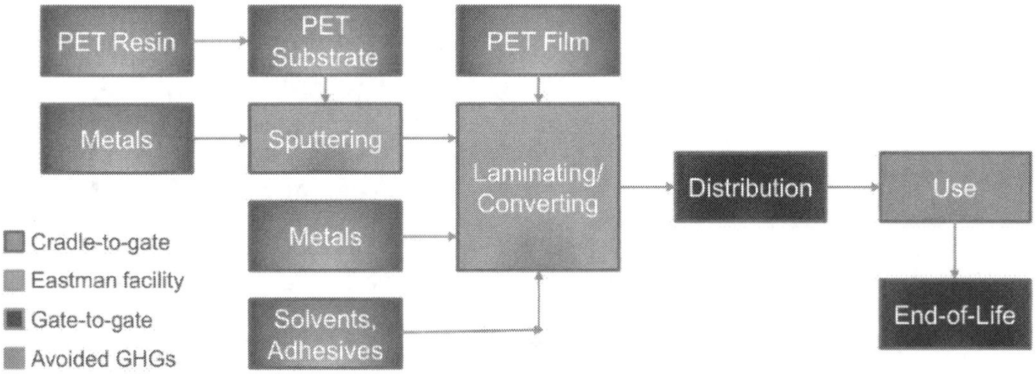

Architectural Window Film LCA

- Geographic coverage

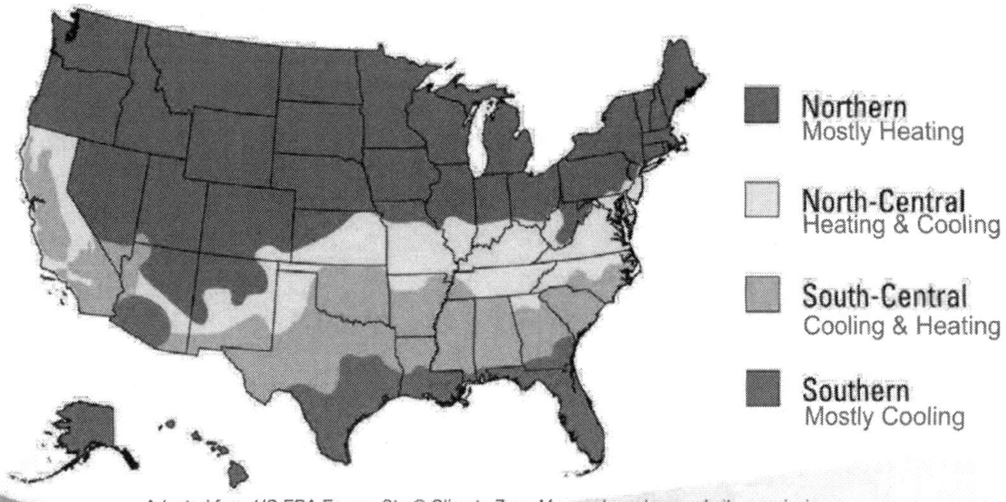

Adapted from US EPA Energy Star® Climate Zone Map and used per website permissions
http://www.energystar.gov/ia/products/windows_doors/Promotional_Map.pdf?a8d7-b26d

Architectural Window Film LCA

- Surrogates and Excluded Materials

Component	% of Mass of Product	Notes
Proprietary Adhesives and Coatings	13% to 15% depending on specific film	Estimated from MSDS component lists
Metals	<0.3% total metals	Surrogate data used for some metals
Silicone	<0.1%	Excluded

Architectural Window Film LCA

- Results

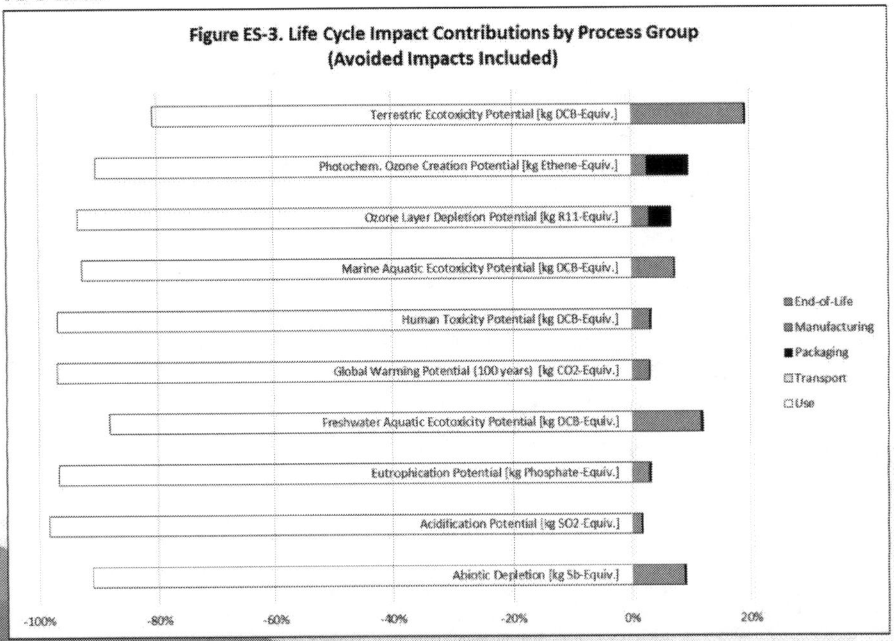

Figure ES-3. Life Cycle Impact Contributions by Process Group (Avoided Impacts Included)

Architectural Window Film LCA

- Results for GHG emissions

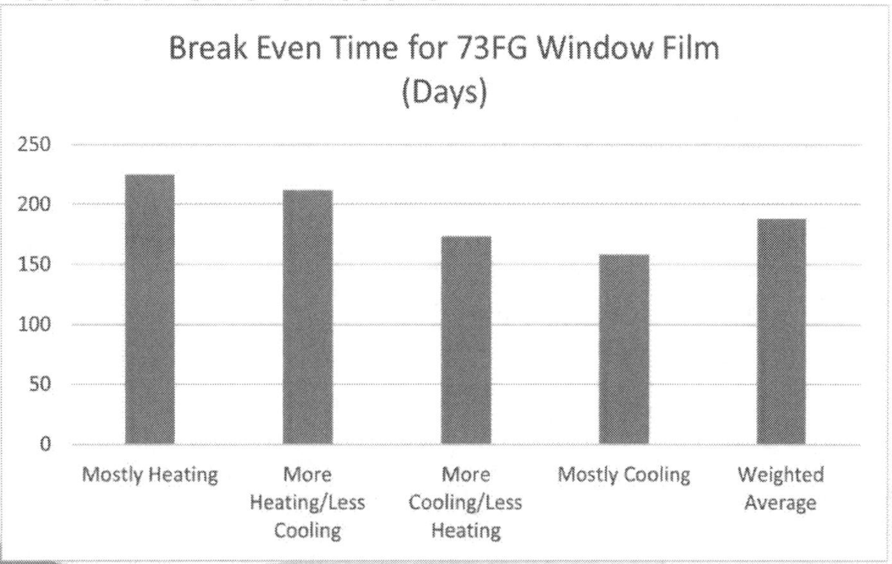

Architectural Window Film LCA

- Compliance with Avoided Emissions Guidelines
  - Level in the Value Chain
    - end-use level
  - Comparison Criteria
    - 15 years of window film use
    - compared fuel use for buildings with and without the window film applied to the windows.
    - all other building aspects are the same

Architectural Window Film LCA

- Compliance with Avoided Emissions Guidelines
  - Calculation Methodology
    - GHG reductions accumulated
      - over lifetime of 1 square meter of window film (0.87 square meters of window space)
  - Significance of Contribution
    - "fundamental"
      - provides the key technology
      - directly manufactures the end-use product.

Architectural Window Film LCA

- Conclusions
  - Benefits outweigh impacts
    - for all impact categories considered
  - Improvement Potential
    - film performance
    - sputtering process
    - greener electricity
    - metals/coatings with lower impacts
  - Provided meaningful results for consumers
    - Following the ICCA/WBCSD Avoided Emissions Guidelines

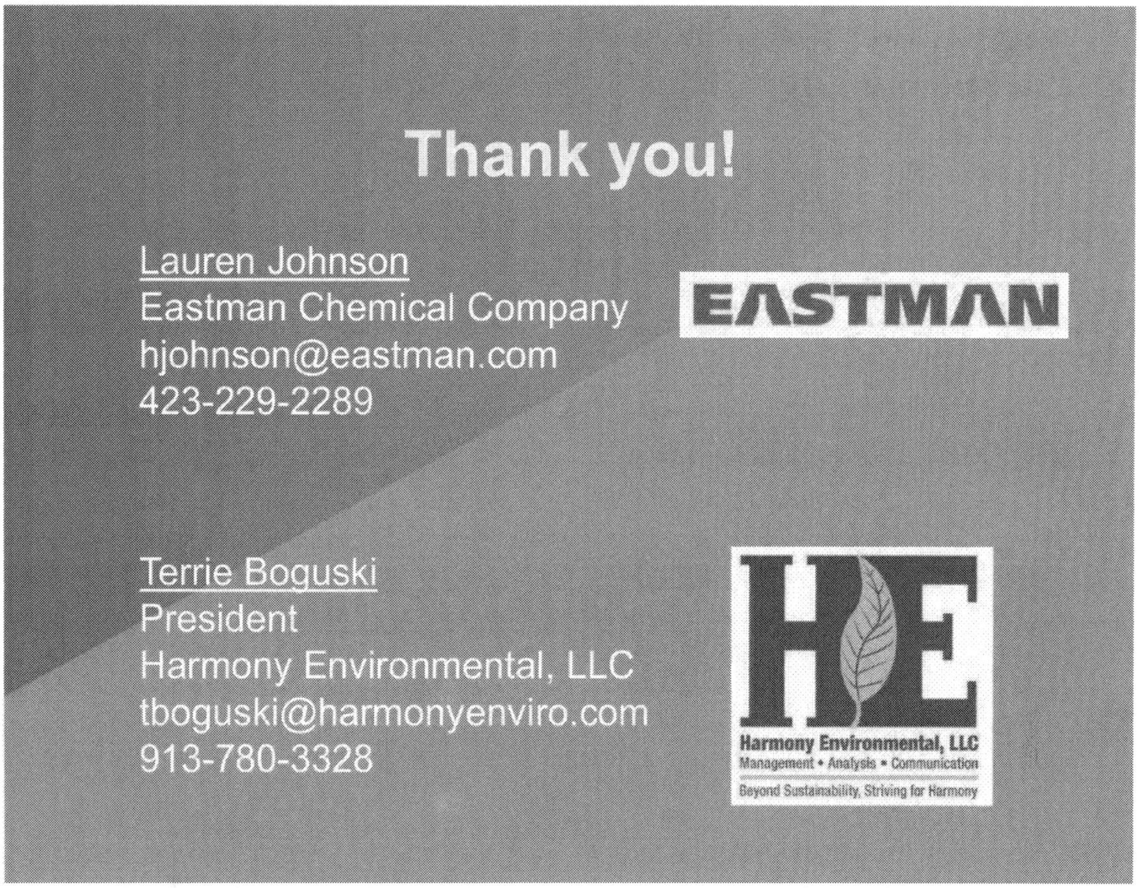

Avoided GHG Emissions Achieved By Using Architectural Window Film: An LCA Compliant With New ICCA/WBCSD Avoided Emissions Guidelines

This presentation will include an overview of the International Council of Chemical Associations (ICCA) and World Business Council for Sustainable Development (WBSCD) avoided emissions guidelines "Addressing the Avoided Emissions Challenge" and a compliant cradle-to-grave life cycle assessment study performed for architectural window film.

The goal of the life cycle assessment study was twofold: 1) to document the life cycle environmental impacts of architectural window films, and 2) to calculate the potential energy savings and avoided GHG emissions during use in compliance with new ICCA/WBSCD avoided emissions guidelines.

A full cradle-to-grave study was conducted covering the entire value chain of window film manufacture. Calculation of potential energy savings and avoided GHG emissions from use of the window film products is a key component of the project. Avoided emissions are the calculated differences in the life cycle emissions between two alternative solutions. For architectural films, this study compares the full life cycle of window films in buildings versus the alternative of the same windows in the same buildings without window films. Architectural window films are an optional upgrade to windows which may be factory installed or a retrofit. The impacts due to manufacturing, use and disposal of the building structure and window panes are identical between the two cases and are excluded from the scope. Potential energy savings and avoided GHG emissions calculations are made at the end-use level solely on whether or not window films are employed.

The results of the LCA show that the architectural window films studied have the potential to greatly improve building environmental performance by saving energy and avoiding emissions from the combustion of fuels used to provide heating and cooling. The environmental benefits from using architectural window films far outweigh the environmental impacts of making the film.

378735 Design of Robust Renewable Energy Supply Systems By Providing Visibility Based upon Improved Neural Network

Abstract:

Climate change and rising supply costs make us agree alternative energy sources should be introduced in replace of fossil fuels. Various renewable energy sources are thus investigated. It is clear that renewable energies become a practical contributor to energy supply system when we can manipulate their performances. However unpredictable nature of renewable energy outputs make it difficult to incorporate them into the main energy supply system directly. The wide implementation of renewable energies makes it even more necessary.

Alternatively the estimated output of renewable energy supply should be available before they are realized. When such estimation is available, the visibility of energy output would be greatly improved. In the end the portion of fossil fuels can be reduced.

This paper investigates existing forecasting methodologies such as neural network and the up-to-date linear / nonlinear time series methodologies are with the comparison of their performances. We propose a new optimized neural network based forecasting methodology for the case of photovoltaic power generation. The applicability of the proposed methodology is illustrated in 10KW real case in Korea.

Experimental Investigations on Brick Specimens made from Deinking Sludge

Shilpa Kulkarni [a 1], Vivek Kumar [b 1], Sanjeev K. Singh[2]., Mukesh C. Bansal [3]

[1]*Department of Paper Technology, Indian Institute of Technology- Roorkee, India,*
[2]*CSIR- Central Building Research Institute, Roorkee, India,*
[3]*Bharat Institute of Technology-Meerut, India*
[a]*shilpa.ips@gmail.com* [b]*vivekfpt@gmail.com* [2]sksingh_cbri@yahoo.co.in

---

Abstract:

The growing population, increasing pollution and decreasing land filling sites for the disposal of solid wastes possess potential problem. One such type of solid waste is the waste from recycled paper industries. It consists of organic and inorganic materials and also some heavy metals because of the writing or printing on the paper. However manufacturing process and raw material used plays a vital role in deciding the chemical composition of residue. Worldwide researchers are trying to use these wastes effectively and also economically into construction materials like bricks and titles. In this paper, deinking sludge, a recycled paper industry waste product, has been chemically and thermally characterized, in order to evaluate the possibility of its use as a construction materials. Characterization of materials is performed by using X- ray Fluorescence, X- ray Diffractometry, and Thermo gravimetric analysis for the determination of its composition, presence of crystalline material and to identify percentage of weight loss with change in temperature. Besides, the investigation also covers the brick making by utilizing the deinking sludge. The results indicate that the brick prepared from the combination of (paper sludge -cement) are light in weight. This novel construction material serves the objectives of reducing solid waste.

Keywords: Deinking Sludge *Compressive Strength *X- ray Fluorescence *Thermo gravimetric analysis *X- ray Diffractometry

---

1. Introduction

Nowadays there is an urgent need to manage the solid wastes as these are produced in huge quantities from various industries and agricultural sectors. The growing population, increasing pollution and decreasing land filling sites for the disposal of solid wastes possess potential problem. One such type of solid waste about which the present study is carried out is the waste from recycle paper industries. Deinking sludge is the by-product of paper industries which mainly consists of ink, fillers, fines, cellulose fibers *[1]*. The moisture content in deinking sludge normally varies from 60-75% (3, 4). Moreover manufacturing process, waste water treatment applied, and raw material used also play a vital role in deciding the chemical composition of residue *[2]*. It also possess problem of disposal because of its high volume production and decreasing land filling sites. Therefore utilizing these waste generated in huge quantities from recycle paper industry into some construction materials like bricks can address the problem of solid waste management. Moreover it can generate revenue also. Additionally, bricks manufactured from these wastes contain air pockets which make them light in weight *[3]*.

Worldwide researchers are trying to use these wastes effectively and also economically into some other developmental works like construction of homes, buildings, stadiums, dams and schools etc. Moreover it has been investigated that these wastes can be used for making bricks and can fulfill the ever growing demand of construction materials. It will not only save natural resources but also contribute to solid waste management. [*4*].

The main aim of present research work is to use recycle paper industry waste for making bricks with required properties. Solve the problem of disposal of solid waste and can revolutionize solid waste management.

2. Objectives of the study

This research aims is to evaluate the suitable mixing proportion in term of percentage of deinking sludge, Flyash, and Ordinary Portland Cement to achieve the required compressive strength. To achieve the above mentioned objectives, the investigation covers the following characterization: particle size analysis, specific gravity, moisture content and comparison of physical properties in terms of compressive strength, TGA analysis, FE-SEM image analysis.

3. Materials and methods

3.1 Materials used

Recycle paper industry waste called deinking sludge was obtained from one of the Recycled Paper Industry Ltd., Punjab, India. This sludge was received in grey color. Ordinary Portland cement (43-grade, Vikram Brand of Ultratech) conforming to IS: 12269 and Flyash (grade F) were also used.

3.2 Characterization of raw materials

Chemical analysis of recycled paper industry sludge and flyash were done by using Energy Dispersive X-ray Fluorescence Spectrometer (XRF, S4, Germany). X-ray diffraction pattern was recorded by model XRD- Philips X'Pert Pro with a scan rate of 2 degree/min. Particle size analysis was carried out by using Horiba's laser scattering diffraction Particle size distribution analyzer Partica LA-950. Thermogravimetry –differential analysis (ESTAR 6500 SII) was carried out to determine the thermal stability. Scanning electron microscope image analysis was recorded by using (Perking, Germany).

3.3 Experimental Investigation

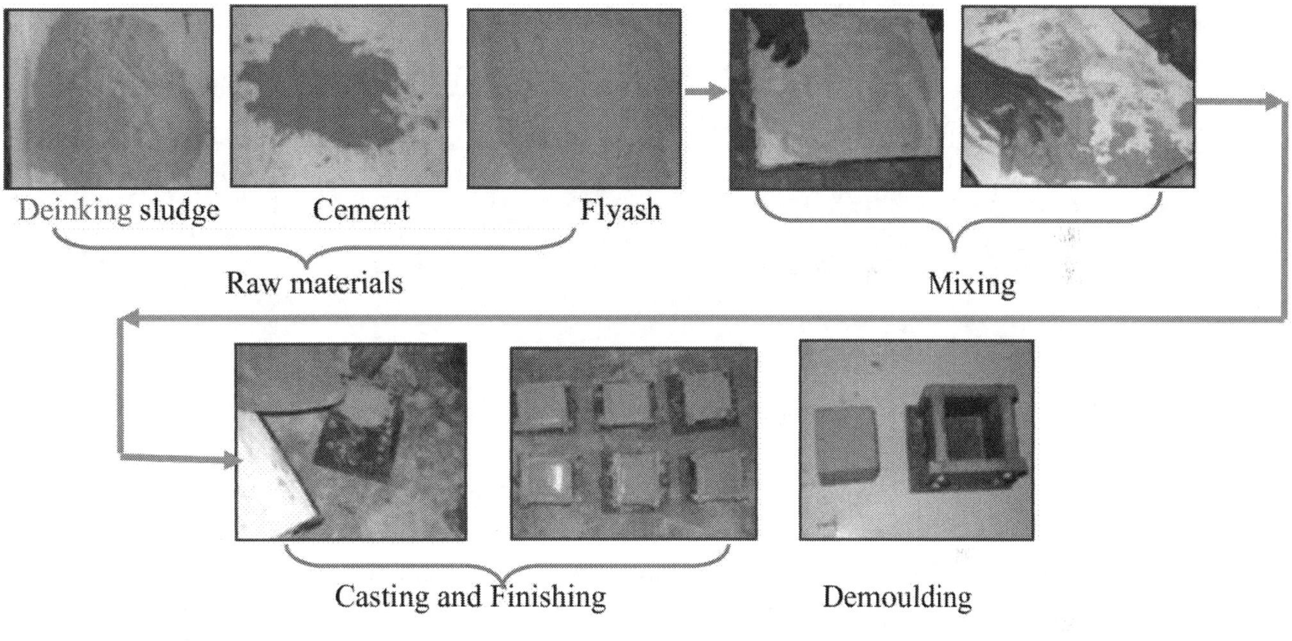

Figure 1: Experimental sequence for making specimens with deinking sludge

Deinking sludge was dried in oven at 60°C. After that, it was grinded and passed through the mesh sieve of 240 microns. The deinking sludge and flyash were mixed properly for 10 mins and agglomerates were left after that cement was sprayed uniformly over whole mixture and mixed it again for 10 mins. After that, water was added to the mixture and mixed well for 20 mins. When mixture was properly mixed then it was poured into the mould with three layers having better compaction. The mould size of 2 inches was used. Thereafter, finishing was carried out by using small trowel. Similarly all the different mix proportions were cast. After 48hrs, samples were demoulded carefully, because of its light weight and transfer into the tray. Moist curing was done for these specimens because deinking sludge is fibrous in nature. So it requires moist curing. Six numbers of cubes were caste for the determination of compressive strength testing. Different mix proportion represented in Table 2 and experimental sequence for fabricating the specimen is given in Figure 1.

Table 1: Chemical composition of Deinking Sludge, Flyash and Cement										
Oxides (%)	SiO_2	Al_2O_3	Fe_2O_3	CaO	K_2O	Na_2O	SO_2	MgO	TiO_2	Loss of ignition
Flyash	55.27	26.69	8.14	1.61	1.76	0.13	0.25	0.39	3.19	1.33
Deinking Sludge	15.50	8.90	1.17	62.43	0.17	0.38	1.62	3.44	1.64	-
Cement	19.32	5.84	5.02	64.29	-	-	2.37	0.83	-	4.0

4. Results and discussion

Chemical analysis of deinking was sludge shown in Table 1. According to the XRF analysis, the residue includes a large fraction of Ca in addition to Si, Al and Mg. This calcium most probably exists in the form of calcite ($CaCO_3$). The moisture content of deinking sludge is found to be 54 %.

Table 2: Mixing proportion of cubes specimen made with cement, flyash and deinking sludge with their compressive strength

Mixing designation	Cement (g)	Flyash (g)	Deinking sludge (g)	Water (g)	Total	Compressive strength (KN)	
						28 days	56 days
M1	50	0	450	568	1068	1.2	1.8
M2	100	0	900	1330	2330	0.5	0.8
M3	100	900	0	308	1308	10.0	12.9
M4	150	1350	0	500	2000	13.4	15.3
M5	75	450	225	300	1050	1.1	1.8
M6	88	528	264	454	1334	1.4	2.1
M7	100	600	300	500	1500	7.4	10.8

From the above mentioned results, it was observed that these specimens required moist curing because deinking sludge was light weight material. M1 and M2 cubes that were made from deinking sludge only showed least compressive strengths. It was observed that it required some reactive material/adhesives/fillers/ binders for improving strength. While M3 and M4 made with flyash only showed better compressive strength as compared to other mixtures. However M5 and M6 showed lower compressive strength then M3, M4 and M7. The results showed that it required some increment in mixing proportion and additions of some reactive materials/binders/fillers/adhesives. Moreover M7 showed better strength then M1, M2, M5 and M6. From the overall results, it has been concluded that deinking sludge is a useful material for construction of light weighted bricks. Some reactive materials/binders/fillers/adhesives are required to improve the strength.

From the above mentioned different mixing proportion shown in Table 2, we made specimen cubes of 2 inches which is shown in Figure 2 (2a and 2b). While casting the cube specimens made with deinking sludge, flyash and cement and after demoulding, it became grey in colour. It also became hard and showed good compressive strength. While cubes specimens made with deinking sludge and cement only, the colour obtained after demoulding was yellow and cubes show poor compressive strength. During compressive strength testing of cubes specimen made with deinking sludge and cement only it was found that these specimens instead of breaking became compressed and also shows less compressive strength. While cubes specimens made with deinking sludge, flyash and cement broke easily and shows brittle nature. The result indicates that it requires some binder/reactive and filler materials for providing good strength.

Figure 2a: Deinking sludge + Flyash + Cement
(Cubes Specimens 2 inches)

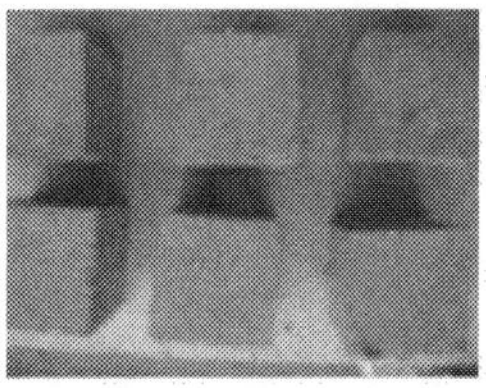

Figure 2b: Only Deinking sludge + Cement

(Cubes specimens, 2 inches)

Figure 2: Specimens made with addition of flyash shown in figure 2a and without addition of flyash shown in figure 2b

X-ray diffraction pattern was recorded with a scan rate of 2°/min and (0 to 100°) at 2θ angle with CuKa radiation (l = 1.542 A °) at 40 kV as given in Figure. 3. In this figure, it shows the intensities of the major and minor peaks for the minerals present. The major mineral phases in the specimens were calcium oxide , quartz (SiO_2), alumina (Al_2O_3) and hematite (Fe_2O_3).

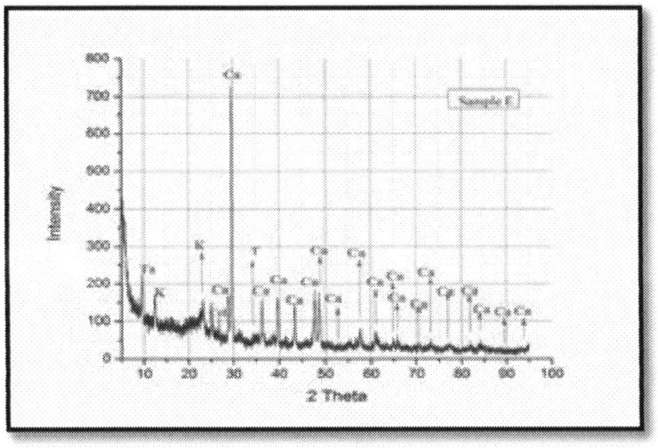

Figure 3: X-ray diffraction analysis of deinking sludge sample with continuous scanning from (0-95 °) at 2θ with intensity having (0-800 counts)

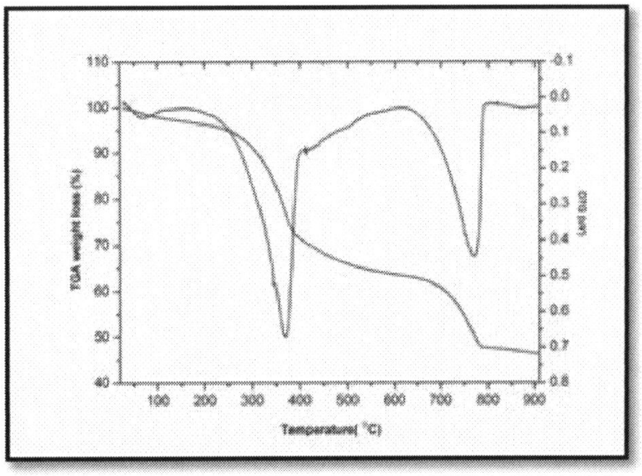

Figure 4: Thermal behavior of deinking sludge sample

Thermogravimetric Analysis (TGA) of deinking sludge was carried out to measure the amount and rate of change in the weight loss of a material as a function of temperature or time in a controlled atmosphere. Thermal behavior of deinking sludge for which the heating rate is 10 °C/min with temperature from ambient to 900 °C in the presence of nitrogen supply of 200 ml/min **as** shown in Figure 4. The curve shows the loss in weight that occurred at different temperatures. This curve reveals the appearance of three distinct mass loss regions. The first loss (7.5%) occurred between 30°C and 280°C which is premature loss and could be attributed to the removal of superficial water molecules that may be present in the solid pores. The second loss occurred beyond 280 ° C where the material gets thermally degraded and the third mass loss beyond 600 ° C is due to combustion of solid organic matter present in deinking sludge.

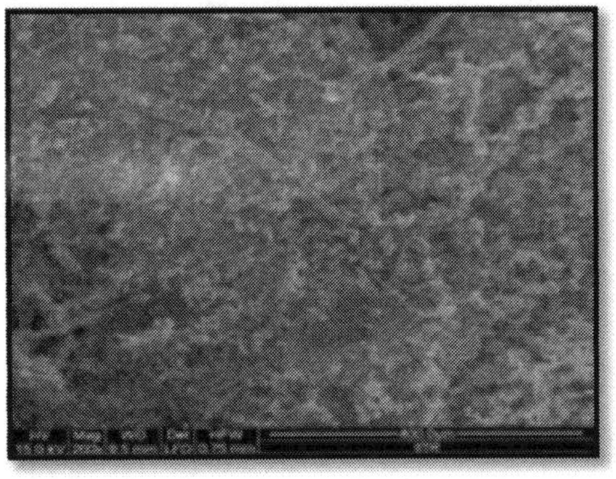

Figure 5: FE-SEM analysis of deinking sludge sample

Scanning Electron Microscopy (SEM) images shown in Figure 5 for deinking sludge sample. In this figure, it indicates the presence of irregular pores and fibrous nature.

5 Conclusion

Deinking sludge brick's specimens, measuring cubes (2 inches) was molded in laboratory. The following conclusions can be drawn about specimens fabricates with deinking sludge:

- The specimen fabricates with paper sludge only **was** brittle in nature. Some binder is required.
- Recycling of such wastes by incorporating them into useful materials is a practical solution to reduce the solid waste.

- Innovative way for using waste material into useful products
- Further investigation is required in this area to make good quality/ useful product.
- Mix stuff, extra energy was needed for compaction
- Color of mix proportion obtained grey
- Proportion needed to be modified
- As the time of curing increases, specimen made with only paper sludge without flyash and cement it shrinks.

Acknowledgement

The financial assistance to this experimental study received from Ministry of Human Resource Department, Govt. of India, New Delhi.

References

1. Mohammad Ismail C, M.A. Ismail[1], S.K. Lau, Bala Muhammad, and Zaiton Majid' "*Fabrication of Bricks from Paper Sludge and Palm Oil Fuel Ash*", (1), 2010
2. Mucahit Sutcu, Sedat Akkurt "*Use of recycled paper processing residues in making porous brick with reduced thermal conductivity*", 35, 2625-2631, 2009
3. S. P. Raut, Rohant Sedmake, Sunil Dhunde, R.V. Ralegaonkar, Sachin A. Mandavganec," *Reuse of recycle paper mill waste in energy absorbing light weight bricks*", 27, 247-251, 2012.
4. S. Raut, R. Ralegaonkar, S.Mandavgane , "*Utilization of recycle paper mill residue and rice husk ash in production of lightweight bricks*", 4-7, 2012

388436 Multi-Objective Optimization of CHP Systems for Housing Complexes

Abstract:

Energy demand in the residential sector of Mexico is defined by the requirements of electricity and hot water for domestic use (HWS). These requirements may be satisfied by a cogeneration system (CHP). However, the implementation of such systems presents a particular problem.

On one hand, due to the nature of the domestic user, the energy demand is not uniform throughout the day and the season of the year, presenting differences between electricity demand and hot water demand. On the other hand, the costs of investment in such systems can become elevated to the domestic consumer. In addition, although the CHP systems have good efficiency compared to conventional power generation schemes, due to the use of fossil fuels continue to generate greenhouse gases emissions (GHG).

The above mentioned problems can be solved by using optimization models in the design stage. In this paper, the design of a CHP system, which consists of a diesel internal combustion and thermal storage tank, is presented. The thermal storage tank is used to smooth the gaps between electricity demand and hot water demand. The system interacts with the network of the electric company through a scheme in which the surplus power is sold or bought to cover missing energy demands.

The optimization model uses as objective functions to maximize the net present value (NPV) of the CHP system and minimize the generation of GHG emissions. As independent variables the schedule scheme operating internal combustion engine and the volume of thermal storage tank used. The partial load of the internal combustion engine and the temperature inside the thermal storage tank is restricted. The model has been solved by NLP models and genetic algorithms. Additionally, in order to make a comparison over conventional generation, the rate of primary energy savings (IPES) and avoided GHG emissions are calculated. Furthermore, the relationship between the temperature of hot water supply, the operation scheme of the primary drive and the volume of the thermal storage tank are shown.

A case study from the state of Michoacan in Mexico was considered to prove the proposed approach, this case study consists of a residential l complex of 1,110 houses; and the results show that there economically and environmentally attractive solutions with the implementation of the proposed approach.

382884 Multi-Layer Monolith Catalysts for Lean NOx Reduction

Abstract:

The challenge of achieving high NOx conversion and N_2 selectivity and minimal release of unreacted NOx and NH_3 in lean burn vehicle emission aftertreatment calls for structured catalysts with multiple functions. We describe experimental and modeling studies of three different multi-layer applications. The first application combines Fe- and Cu-exchanged zeolitic layers to expand the high NOx conversion temperature range. The second application combines periodic NOx storage and reduction with in situ NH_3 generation and selective catalytic reduction. The third application combines a Pt/Al_2O_3 bottom layer and a zeolitic toplayer to oxidize NH_3 with high selectivity to N_2.

A discussion on the fundamental aspects of Urea SCR control for Transient (Automotive) Applications.

Annual AIChE meeting Atlanta GA, Nov. 17, 2014

Session: Development and Applications of Automotive Catalysts

Devesh Upadhyay

Michiel Van Nieuwstadt

Acknowledgements:
Emil Serban
Giovanni Cavataio

Overview

1. Evolution timeline for SCR hardware and control
2. Typical aftertreatment system configuration/s for automotive applications.
3. Storage dynamics and its implications
4. Control
   - Objectives
   - Constraints
   - Architecture
   - Adaptation
   - Stability
5. Storage Observers
6. NH_3 NO_x Signal Separation, need and method.
7. New Directions in Control
8. Diagnostics
   - Threshold selection
   - SCR Aging
9. Conclusions/Outlook

Urea SCR Aftertreatment System time line with Control implications

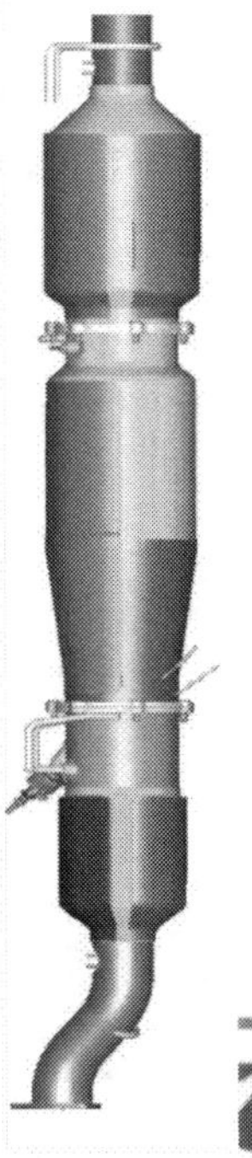

2000-2004: 1st Generation Urea SCR control

- FF control only → FG NO_x following. NH_3 slip not a performance criterion.
- Urea injection with air assist. No spray mixers.
- Cu, F_e Zeolites →US, T_iO_2/V_2O_5 →EU
- DOE sponsored research at Ford.

2004 -- 2006: 2nd Generation Urea SCR Control

- FF + FBK → FG (model/sensor) and TP NO_x sensor
- Embedded SCR models developed and in use.
- Adaptive control laws considered / in use.
- Direct Urea Injection → Air assist less prevalent.
- Mixers → high γ values >90%
- Urea management for freeze and thaw.
- SCR OBD on the horizon for 2010.
- ZSM-5 in active use, SSZ-13 being assessed.

2006 -- 2010: 3rd Generation Urea SCR Control

- Focus on output signal: NH_3, NO_x signal separation.
- NH_3 slip management is of interest, sensors available for use.
- AMOX catalysts and Hydrolysis catalysts.
- Predictive Control techniques: IMC, GPC explored.
- Complexity of storage mechanisms becoming clear, number and role of various storage sites ? Brøensted, Lewis, strong , weak ....
- SCR OBD is in use → age vs system drift confounding.
- DEF warning standardized
- SSZ-13 zeolites (Cu-CHA) dominant, SAPO studied.

2010 – 2014+ : 4th Generation Urea SCR Control

- Transition to Multi Slice storage models (n x 0D).
- Multi Brick configurations under consideration.
- Investigations into active control via kinetic manipulation.
- Urea Quality sensors available.

Research

Diesel Controls and Diagnostics
Powertrain Controls R&A

Aftertreatment system configuration/s for automotive applications

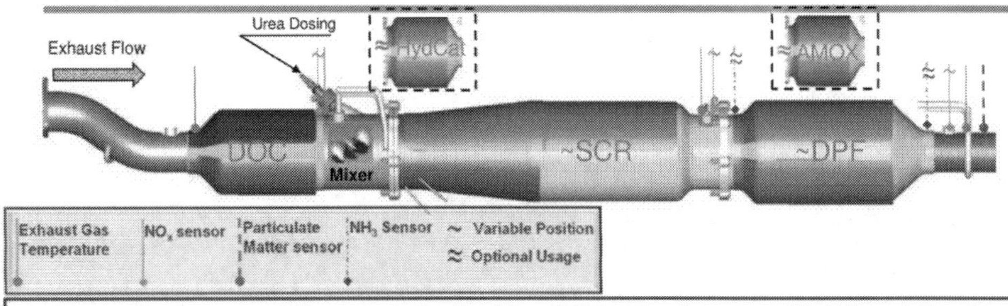

- Most applications use a NO_x sensor in the Feed Gas (FG) location, however, a FG NO_x model has been tried (Ford) and shown to work for control purposes. OBD placer harder constraints on accuracy.
- FG NO_x sensor location can vary between pre or post DOC with varying implications.
- Tail-Pipe (TP) NO_x sensor is necessary for all applications and may be placed either directly after the SCR or after the DPF when the DPF is located after the SCR.
- Due to confounding of the output NO_x signal (TP NO_x sensor) with NH_3 (slip) many applications are resorting to an Ammonia sensor for output signal disambiguation. Others may use a filtering algorithm.
- Ammonia Oxidation or "clean up catalysts" are also becoming an option to reduce TP NH_3 → costly but allows for less constrained control objectives, including relaxed urea cost.
- Some applications may also use Hydrolysis catalysts to enhance the rate and efficiency of NH_3 availability from urea dosage, although this is less prevalent.

The Minimal AT system is the most constrained (cost effective) and uses only the following components: 1. FG NO_x (sensor or model) 2.TP NO_x sensor, 3. Urea Injector 4. mixer 5.DOC, 6. SCR, 7. DPF.

Research

Diesel Controls and Diagnostics
Powertrain Controls R&A

The dynamics of NH_3 storage: Adsorption-Desorption

Critical Storage Identifiers for SCR performance control

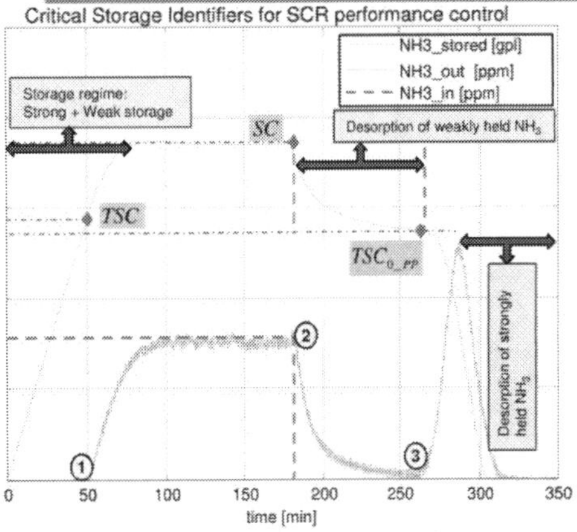

$$TSC = \left(\max\left(m^{stored}_{NH_3}\right)_{C^{in}_{NH_3}>0} \middle| C^{out}_{NH_3} = 0 \right) \qquad TSC_{0\_PP} = \left(\max\left(m^{stored}_{NH_3}\right)_{C^{in}_{NH_3}>0} \middle| C^{out}_{NH_3} = 0 \right)$$

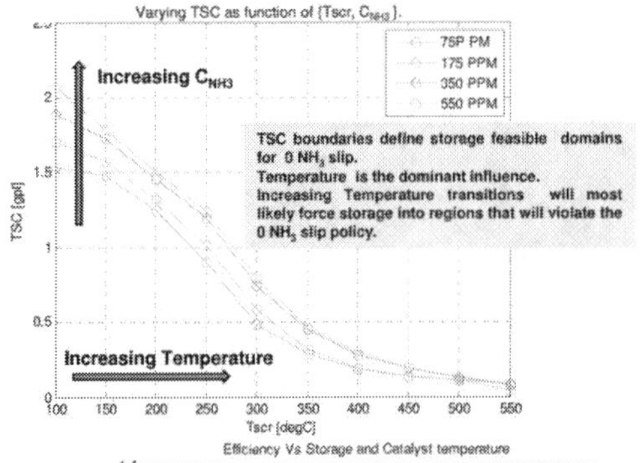

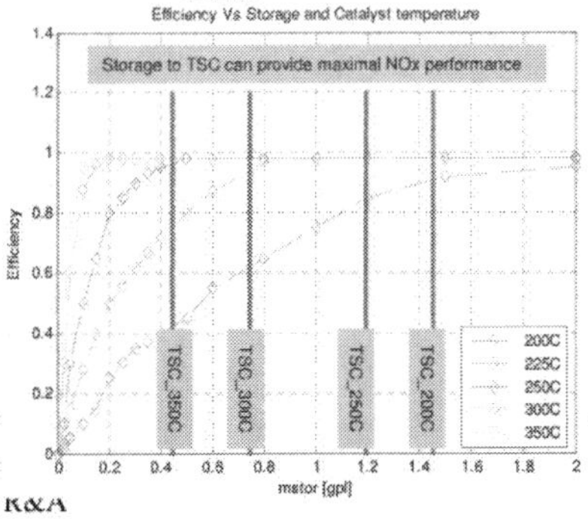

- Threshold Storage Capacity (TSC), for the operating pair {Tscr, C_{NH3}}, is the maximum storage possible without any NH_3 slip. For most Conditions the TSC also provides a storage level that allows maximal or near maximal NO_x conversion.
- TSC_{0_pp} is the TSC with 0 NH_3 partial pressure (C_{NH3} = 0) in the gas phase and TSC ≥ TSC_{0_pp} for all operating regimes.
- Markers 1,2 and 3 define the 3 basic storage and NH_3 slip transients that impact all SCR's during continuous operation.

Powertrain Controls R&A

Understanding NH_3 storage on the SCR

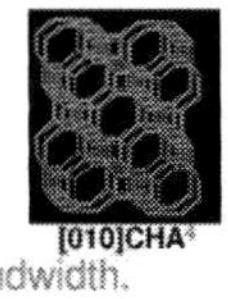

[010]CHA[4]

- Exact nature of storage remains an active area of research with regard to :
  - The nature and distribution of storage sites → Brøensted, Lewis, I.E. metal-Lewis.
  - Participation of the various stored species in NO_x conversion.
  - Migration of NH_3 within the zeolite pore structure → Strong to weak transition.
  - Role and behavior of Cu.
- For Control design storage behavior has implications on system response → control bandwidth.

<u>TSC and its correlation with storage: a workable control interpretation</u>

- In traditional terms , TSC may be understood to account for ALL the strongly held NH_3 as well as one or more minimal layer/s of the strongest of the weakly held NH_3. This weakly held layer/s is/are hypothesized to be a result of hydrogen bonded[1] NH_3 as, $NH_4^+.nNH_3$. This layer is a strong function of the gas phase partial pressure of NH_3, and the surface adherence is strong enough to not result in any measurable NH_3 slip.
- TSC_{0_PP} can therefore be logically concluded to account for all the weakly held layer/s that will be released when the partial pressure of NH_3 is reduced to 0 → state transition from 2 to 3 in the previous figure.
- For control purposes it is critical to understand the contributing dynamics of the different types of NH_3 storages with respect to NOx activity and influence on NH_3 slip.
- *<u>Q. Is there a benefit for embedded, control oriented, storage models to account for these various sites ?</u>* Up to 5 storage sites have been explored and reported in literature[2,3]
- We find that the quality of prediction of multi-site storage models does not improve significantly when compared to a single-site model especially when considered in the context of model based control.

1. Rodriguez-Gonzalez, L., Rodriguez-Castellon, E., Jimenez-Lopez, A., and Simon, U., "Correlation of TPD and impedance measurements on the desorption of NH3 from zeolite H-ZSM-5," *Solid State Ionics* 179:1968–1973, 2008, doi:10.1016/j.ssi.2008.06.007.
2. Sjövall, H., Fridell E., Blint, R., and Olsson, L., "Identification of adsorbed species on Cu-ZSM-5 under NH3 SCR conditions," *Top Catal* 42-43(1-4):113–117, 2007, doi:10.1007/s11244-007-0162-6.
3. Skarlis, S., Berthout, D., Nicolle, A., Dujardin, C. et al., "Modeling NH3 storage over Fe- and Cu-zeolite based, Urea-SCR catalysts for mobile diesel engines," *Procedia - Social and Behavioral Sciences* 48:1672–1682, 2012, doi:10.1016/j.sbspro.2012.06.1142.
4. Data base of zeolite structures.

Control Objectives and desired storage

Control Objective

- The ideal control objective for the Minimal AT system may be defined by the following hard constraints:
  - $C_{TP\_NOx} \equiv 0 \forall t$ → achieve 100% NO_x conversion for all time (full control horizon).
  - $C_{TP\_NH3} \equiv 0_{\forall} t$ → achieve 0 NH_3 slip for all time (full control horizon).

- ***It is obviously impossible to continually maintain NH_3 storage such that both of these objectives are met for all operating conditions.***

- This can be easily verified from the TSC vs η plot shown earlier. As an example consider the following:
  - For a given catalyst temperature, 0 NO_x slip → NH_3 storage $m_{NH3\_k} \approx TSC_k$.
  - Any positive temperature transition will create a storage condition where: $TSC_{k+1} < TSC_{k}, \rightarrow m_{NH3_{k+1}} \geq TSC_{k+1}$ such that the NH_3 slip criterion is violated.
  - Or both criteria can be simultaneously violated when the SCR catalyst and/or the AT system has deviated from nominal (biased system) or for unrealistically high conversion demands from a kinetically limited SCR (cold).

- Relaxed control objectives are typically defined requiring maximal NO_x conversion and minimal NH_3 slip.
- Urea injection control objectives translate to storage manipulation as shown below.

NH_3 slip minimizing criterion — NO_x slip minimizing criterion

$$m^*_{NH3} \in \left(\left\{\max\left(m^{stor}_{NH3}\right)\middle|\left(C^{TP}_{NH_3} < C^{\max}_{NH_3}\right)\right\} \cap \left\{\min\left(m^{stor}_{NH3}\right)\middle|\left(\eta_{NOx} = \eta^*_{NOx}\right)\right\}\right) \forall t, T_{SCR}.$$

Note: minimal NH_3 slip → minimal urea consumption.

Pareto point is an intersection set between two objectives.

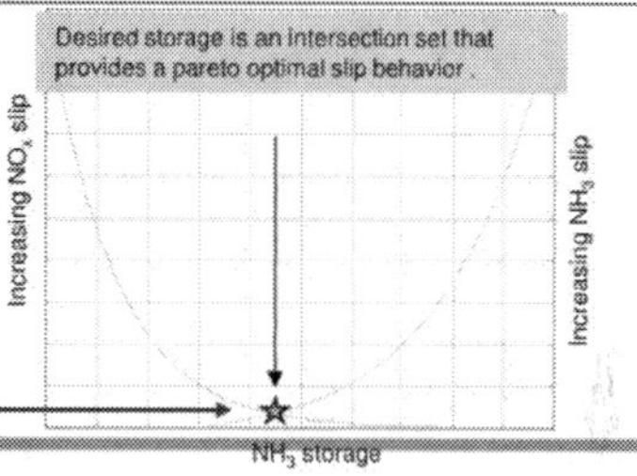

Control Constraints

- The feedback signal from the TP NO_x sensor, is confounded → $Y_{TP\_mes} = C_{NOx} + \alpha C_{NH3}$
  - Unless this confounding is managed ALL closed loop control designs will be prone to instability.
  - Since the system is naturally dissipative the instability will typically emerge as an unstable limit cycle.
- Control authority with respect to storage is single sided → can only add urea, storage depletion is determined by NO_x flow and/or thermal desorptions both of which are un-manipulated events.
- What if we could partition the control authority $U = \begin{bmatrix} U^+ \\ --- \\ U^- \end{bmatrix} \overset{\text{def}}{\equiv} \begin{bmatrix} urea \\ ---- \\ NO_x \end{bmatrix}$, such that storage depletion was manipulated via NO_x, ... there are several issues
  - FG NO_x must be increased without a thermal event → EGR intervention → "Defeat Device" management
  - Since $urea \equiv 0$ during this mode it is expected that additional NH_3 slip from weakly adsorbed NH_3 will occur, additionally to achieve even moderate levels of depletion a large NO_x flux may be needed making the Defeat Device management task challenging.
- If temperature (fast) is used as U^- → translates to fuel cost and storage depletion will be from NH_3 slip.
- *It is therefore clear that any active method of fast storage depletion will result in NH_3 slip, defeating the very objective being pursued → deplete storage to avoid potential NH_3 slip.*
- These restrictions also make any preview information (if available) ineffective.
- We have found that maintaining storage via a stable limit cycle, within a narrow boundary of the TSC surface, allows us to nominally maintain storage at desired levels

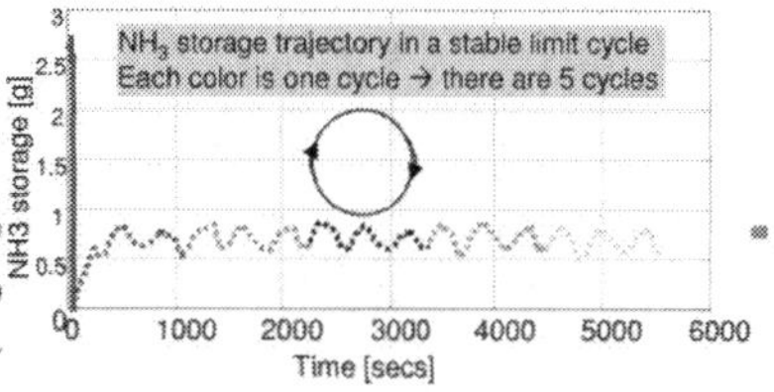

Typical Control Architecture

- Typical control architecture includes Feed Forward (FF), Feed Back(FB) and Adaptive corrections
- Closed loop control adds robustness against modeling error and system and measurement noise.
- Closed loop Control should ideally be some combination of storage control and slip control.
- Storage Control does not have state measurement → hence control to a reference state for ex: TSC as discussed earlier.
- Slip control should be based on disambiguated output signal to influence the correct control direction.
- This may lead to a controller with a structure similar to an IMC structure.

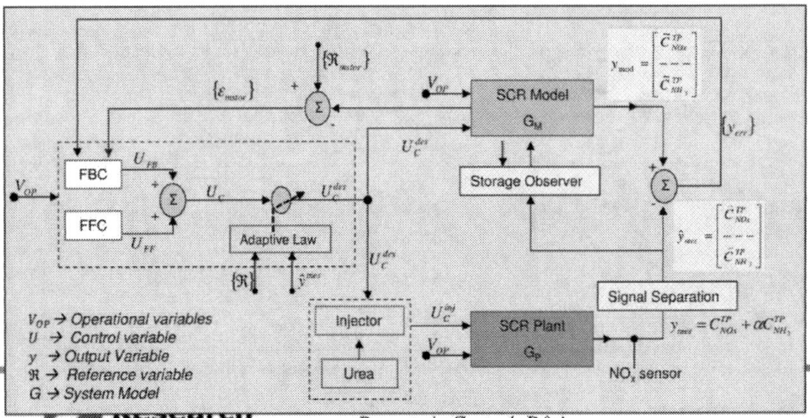

Adaptive correction

- Robustness against diverse set of expected system drifts and uncertainties cannot be addressed by Closed Loop control alone.
- A suitable adaptive algorithm must be used, we use a gradient based update law with a slow update (to avoid instabilities from fast corrections especially for large biases).
- Two options were exercised to isolate the source of efficiency loss:
  1. A game theoretic approach that assess nature of slip from the gradient of performance
  2. A direct approach to distinguish between NH_3 and NO_x as the dominant slip species.

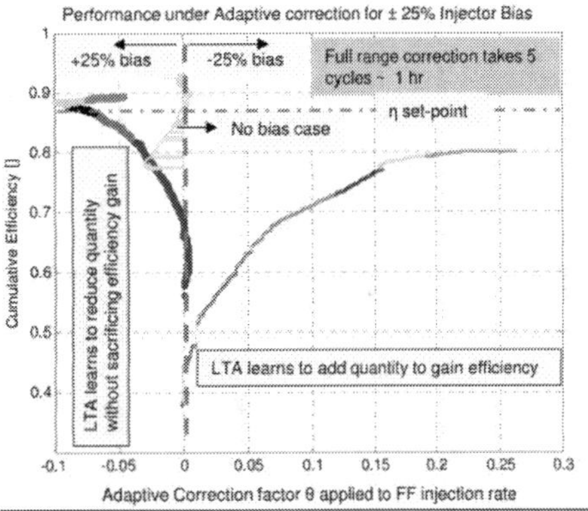

Adaptive law learns out bias errors, although learning is not symmetric to achieve desired target performance. The no Bias case dithers along the 0 correction line to achieve desired performance.

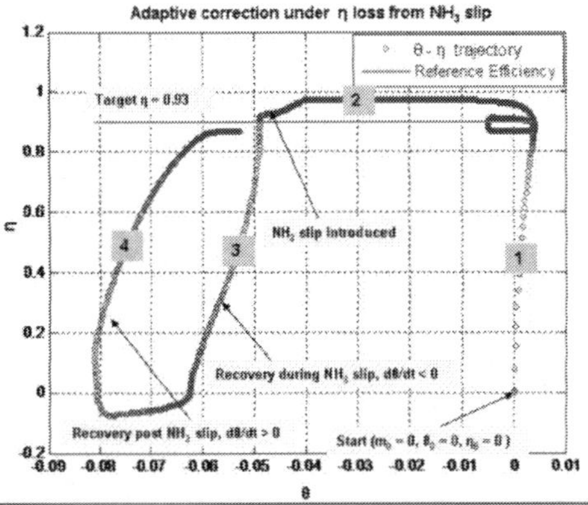

Adaptive law tried in situation of sudden NH_3 release. Update law reduces NH_3 fast after onset of NH_3 slip. Note there was already some –ve correction prior to the onset of slip due to the system over achieving.

What does stability look like for SCR control

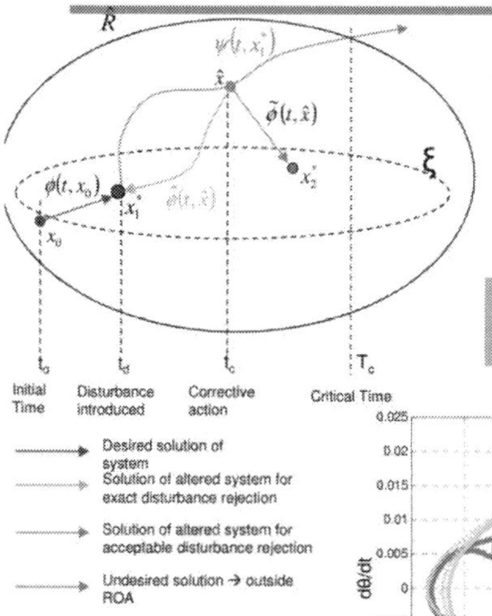

R = Region of attraction
$\hat{R}$ = Relaxed R
ξ = domain of acceptable equilibria
$\phi(t, x_0)$ = solution with initial state, x_0
$\psi(t, x_1^*)$ = solution after disturbance

$$R = \{x \in \Omega \,|\, \phi(t, x_o) \to x^* \text{ as } t \to \infty\}$$

$$\hat{R} = \{x \in \Omega \,|\, \|\phi(t, x_o) - x^*\| < \xi \text{ as } t \to \infty\}$$

- SCR catalysts are dissipative systems with finite ranges for states.
- Instability in SCR's catalysts is observed via sustained (t_∞) oscillation in some control variable.

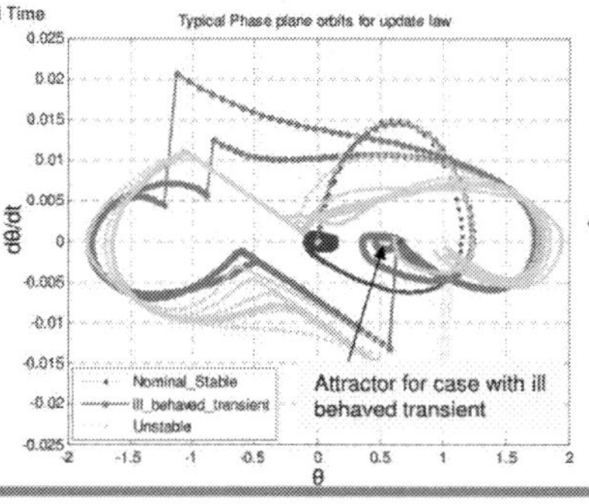

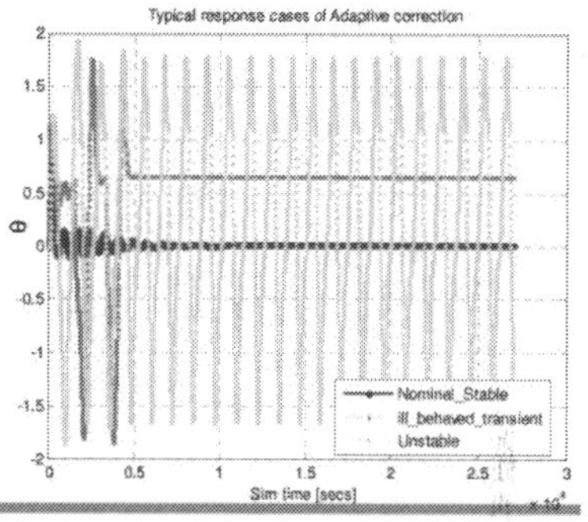

Research

Diesel Controls and Diagnostics
Powertrain Controls R&A

Storage Observer

- All NH_3 storage models are prone to storage error creep over time, hence storage must be periodically adjusted.
- Direct storage measurement is currently not a possible (will probably remain so in the near future).
- A traditional observer may be designed based on measured signal from TP NO_x sensor, provided signal separation into constituent species is available, subject to observability criterion[1].
- A direct reference is available for storage adjustment in the form of TSC. A standard Luenberger observer may be designed with TSC as the reference storage state whenever a deterministic evaluation of NH_3 slip is made. Under this condition it is clear that the current NH_3 storage is ≥ TSC and the observer can make at least a partial correction.

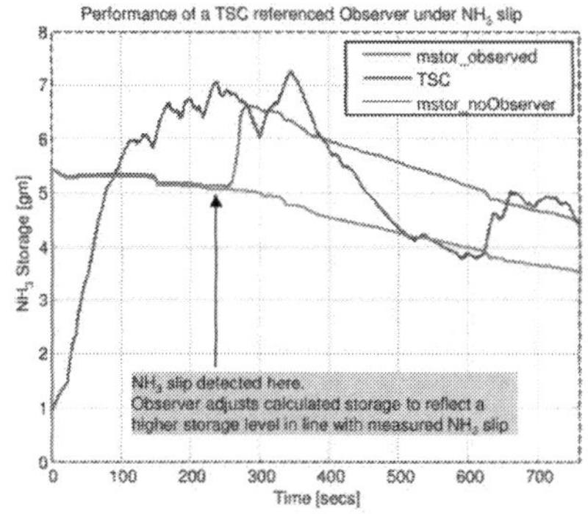

An example of the Observer performance under NH_3 slip is illustrated. The actual calculated storage value was biased low (-2gm) for this test. The observer resets the storage to TSC which recovers most of the storage error.

The same observer is also used to adjust storage downward, when slipping NO_x , by using a reference storage obtained by inversion of a NO_x conversion model, $\eta_{NOx} = f(m_{stor}, T_{SCR})$.

1. Model Based Analysis and Control Design of a Urea-SCR deNOx Aftertreatment System, Devesh Upadhyay and Michiel Van Nieuwstadt, J. Dyn. Sys., Meas., Control 128, 737 (2006)

Research

Diesel Controls and Diagnostics
Powertrain Controls R&A

Signal separation

- Deterministic feedback is critical for robustness of control and OBD performance as well as control stability.
- In the absence of an NH_3 sensor the TP NO_x signal must be separated into its constituent species.
- This has remained an open problem in the Urea SCR community until recently. At Ford we developed a purely Signal Processing based algorithm to accomplish this task → avoids uncertainty form modeling error.
- The fundamental approach relies on convolving the forward difference of the FG and TP NO_x signals over a window defined by the decreasing phase of the FG NO_x signal[1] .

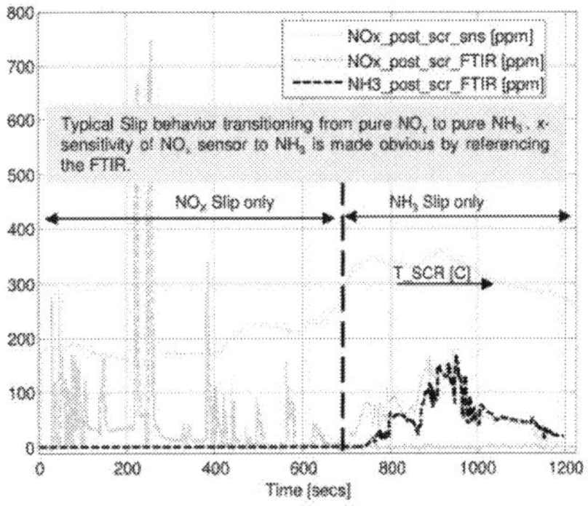

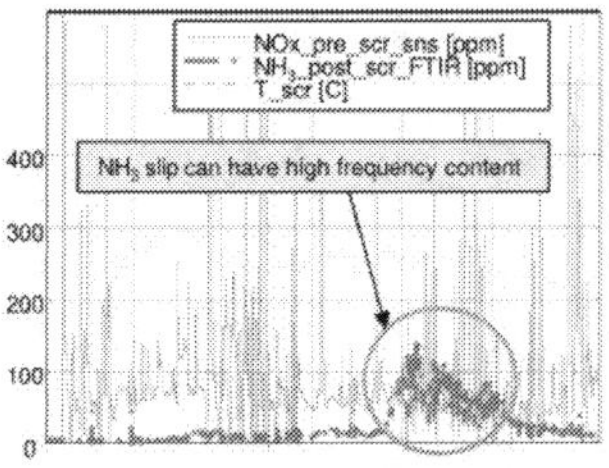

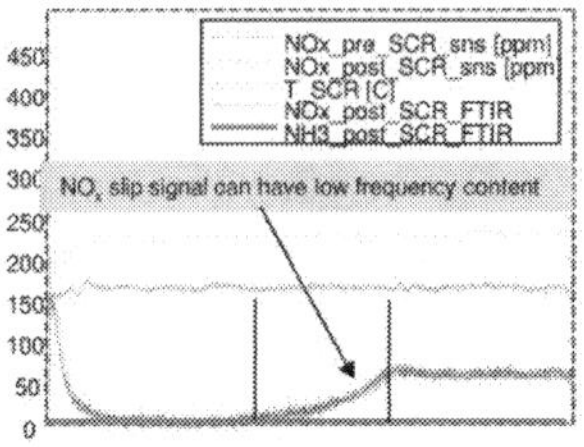

Myths about NH_3 and NO_x slip:

1. NH_3 slip signal will always have a low frequency spectrum
2. NO_x slip will always have a high frequency spectrum

1. Devesh Upadhyay, M van Nieuwstadt, Robust Separation of Signal Domain From Single Channel Mixed Signal Output of Automotive Urea Based Selective Catalytic Reduction Systems, J. Dyn. Sys., Meas., Control 136(1), 011012 (Oct 07, 2013) (9 pages)

Signal slip domain separation Performance

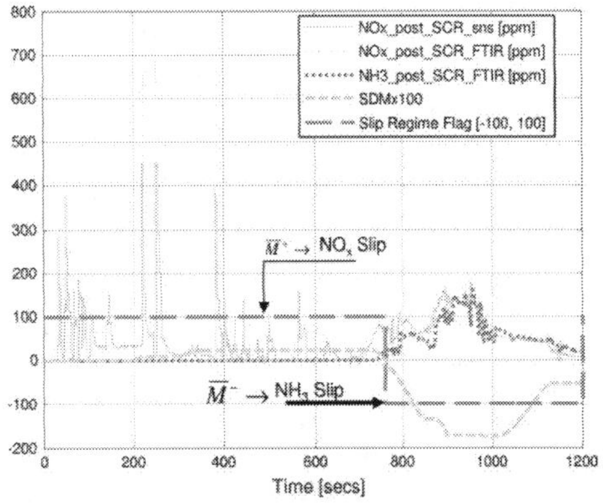

Slip Detection Metric (SDM) identifies NO_x and NH_3 slip domains with correct transition

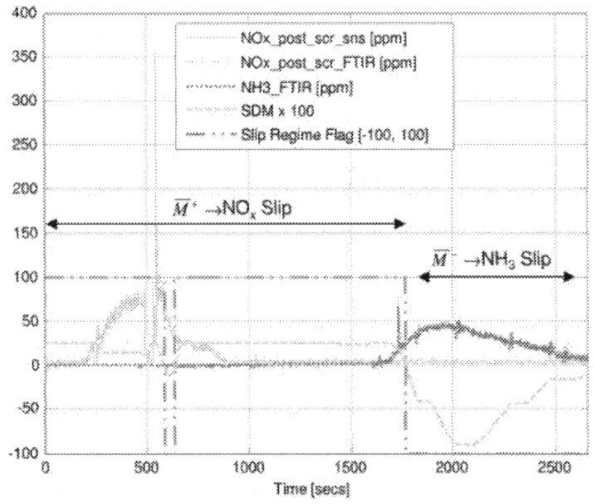

Slip Detection Metric identifies NO_x and NH_3 slip domains even though both signals have similar low frequency spectra.

New Directions in controls – storage models

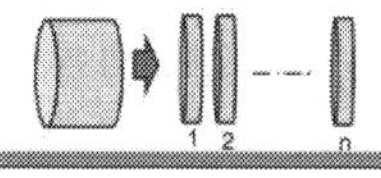

- Trend is to transition from Lumped parameter (0-D) models to n x 0-D models → pseudo 1-D model.
- Thermal Hypothesis: Single node 0-D models will incur inaccuracies from axial temperature gradient. This error maybe mitigated by adopting a n x 0-D model coupled with an axial thermal model of the AT system.
- Scale-Up Hypothesis: Full scale catalyst models, scaled up from SGB reactor core data, may have scale up ratios > 500x. The error from such large scaling factors can also be reduced by using several nodes each with a smaller scale up ratio relative to the SGB core.
- However, It is critical to perform a critical analysis of these assumptions prior to adopting any particular structure for the embedded model. Some points to note are:
  - **Typically only a fraction (~20%) of the catalyst is actively engaged** in catalytic activity, hence the 1x0D → nx0D reconstruction must be done with full understanding of the reactor sizing and usage.
  - Under most conditions axial temperature gradients are small across bricks of nominal dimensions, 6" –12" long → nx0-D benefits will be limited to conditions generating axial temperature gradients → cold starts, very long catalyst bricks, 2 or more separate bricks operating in kinetically distinct thermal regimes.
  - Under certain operating conditions and/or ambient influences the **radial temperature gradients may be more significant** compared to axial gradients. The nx0D approach will **NOT** address this.
  - Observer correction of storage becomes more involved for a n x 0-D model since the **storage adjustment implies a redistribution of the storage correction** between the n-nodes. This is an under-determined problem forcing heuristic schemes for storage adjustment.

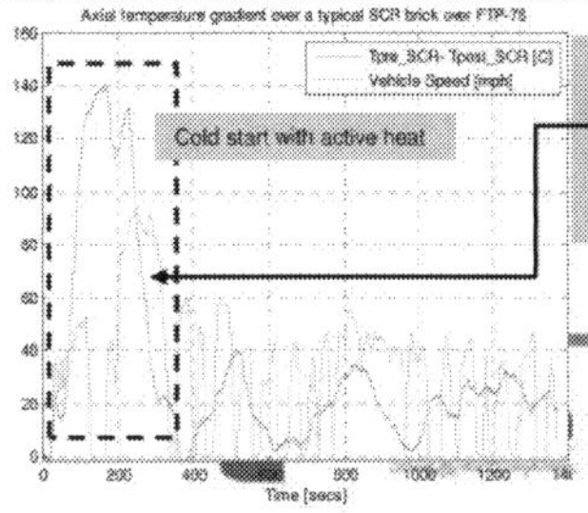

Over FTP cycle (Tamb ~= 25C):
Axially: Front is hotter than back (expected)
Radially: Outer edge can be considerably cooler

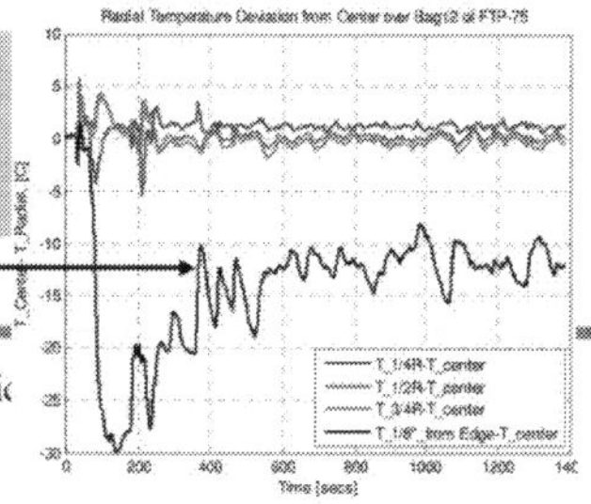

Diagnostics

- At current OBD levels of **0.4 g/mi** TP NOx for CC U10K HD trucks the SCR catalyst must (typically) be flagged as deteriorated at activity levels in the range 70% to 75% over the FTP-75.
- A **70 ~75%** SCR catalyst can return very high efficiencies O(90%+) on a variety of drive cycles/operating conditions.
- Conversely cold ambient (< 5C) can make a healthy SCR return poor performance.
- These behaviors make selection of a robust performance threshold for minimizing Type-1 and Type-2 errors a non-trivial task.
- Robust OBD in the face of **Noise factors** only increases the challenge .
- The primary path is to exploit the characteristic behavior of a deteriorated SCR → propensity to incur dual mode slip → NOx + NH_3. NH_3 slip will lead to a lower measured performance due to the cross sensitivity of the NO_x sensor (discussed before).
- Given the possibility of a healthy SCR to slip NH_3 under poor control or transients, for example, robust detection of a threshold SCR dictates a two step process whenever a failing result is returned. ***This is to ensure that the efficiency loss is persistent and not transient***. And that a healthy catalyst is not flagged as deteriorated (cost implications).

Aging

- OBD testing also entails a catalyst aging procedure. Current generation SCR's Cu SSZ-13 are hydrothermally quite stable → Defining an aging protocol that allows consistent aging to the OBD threshold level involves some degree of trial and error.

Conflict with control

- Since control will attempt to drive the catalyst in the direction of improved performance, a synergistic approach or "decoupling" may become necessary to avoid premature/ delayed detection of an SCR that has deteriorated to the OBD threshold level.

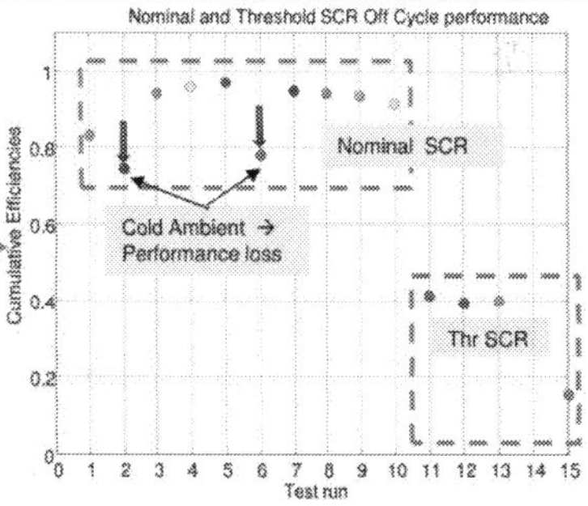

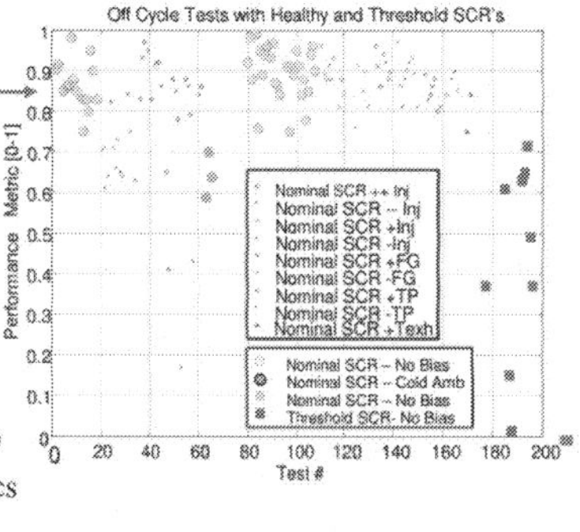

Conclusions/Outlook

- C_u, F_e, Zeolite based urea SCR catalyst will most likely continue to be the dominant diesel aftertreatment system.
- Investigations into chemistry/physics of these catalysts will continue to remain an active are of research until all fundamental behaviors are understood.
- Control of these systems, for on-road transient applications has achieved some degree of maturity with (industry) consensus on the need for:
  - Model based control (storage model)
  - Control architecture → FF+FB+Adaptation
  - Feedback signal separation (or NH_3 sensor)
- Potential areas of continued work in Control/Systems:
  - AT system arrangement variants : 2 brick systems, SCRDPF-SCR.
  - Higher fidelity (distributed), yet reduced order models.
  - Control via active manipulation of kinetics through large gain sensitivities such as substrate temperature.
  - Active management of reactor volume via manipulation of storage distribution.

Research

Diesel Controls and Diagnostics
Powertrain Controls R&A

Thank you for your attention

Research

Diesel Controls and Diagnostics
Powertrain Controls R&A

392229 Investigation of burning mode for diesel particulate oxidation: contrasting O2 and NO2

Abstract:

The physical characteristics of diesel particulates from conventional diesel, biodiesel and blends of the two were investigated at several points along the O_2 and NO_2 oxidation reactions. BET measurements elucidated the surface area evolution with extent of oxidation for the two oxidizers and revealed differences in surface area evolution in the samples when oxidized with O_2. Particulate matter from biodiesel initially has greater surface area than particulate from ULSD, an advantage, which is maintained under O_2 oxidation until approximately 40% of the sample has been consumed. HR -TEM and fringe analysis confirm that initial differences in the lamella lengths have disappeared by 40% burnout of the samples. TEM images suggest that surface area evolves mostly internally for O_2 oxidation as the primary particle diameters do not follow the shrinking core trajectory. However, for NO_2 there are no sample-specific dependencies in either reactivity and surface area. TEM images suggest that NO_2 oxidation occurs by a mostly external trajectory.

392232 TWC formulation effects on NH3 generation for passive SCR applications in lean gasoline engine exhaust

Abstract:

Due to the dominance of gasoline vehicles in the U.S. light duty fleet, improvements in gasoline engine fuel efficiency would significantly reduce U.S. petroleum consumption and greenhouse gas emissions. Operating gasoline engines with more air than is required to burn the fuel (known as lean conditions) can improve fuel efficiency by 10% or more. However, lean gasoline vehicles have been unable to meet U.S. emissions regulations for oxides of nitrogen (NOx). The two lean NOx reduction technologies that have been successfully deployed on diesel engines, LNTs (lean NOx traps) and urea SCR (selective catalytic reduction), are both challenged by lean gasoline applications. The high NOx concentrations and exhaust temperatures from gasoline engines result in excessive NOx slip during LNT regeneration and a high fuel penalty due to frequent regeneration events. With SCR systems, high NOx concentrations require increased urea dosing rates, necessitating a large onboard storage tank and/or easily accessible refueling infrastructure. The cost of the urea storage and dosing system may also be prohibitive in gasoline applications. An alternate strategy, termed passive SCR, generates NH_3 from engine-out NOx over a close-coupled catalyst during periodic excursions to slightly fuel-rich conditions. The NH_3 is stored on an underfloor SCR catalyst, where it is used to reduce NOx upon returning to lean conditions. By generating NH_3 over the three-way catalyst (TWC) already present on gasoline vehicles, the passive SCR approach enables use of proven commercial SCR catalysts while avoiding the cost and packaging issues associated with a urea dosing system. Oak Ridge National Laboratory, in collaboration with partners at General Motors, Umicore Autocat USA, and the University of South Carolina, is currently working to identify passive SCR catalyst formulations, system architectures, and operating strategies that would enable lean gasoline vehicles to meet emissions regulations while minimizing the associated fuel penalty and costs. This presentation will describe the results of flow reactor and engine experiments that evaluated the effects of TWC formulation on the NH_3 formation process under both steady state and dynamic operating conditions. The TWC formulations used in the experiments included variations in precious metal composition, oxygen storage capacity, and NOx storage capacity. Correlations between TWC formulation and expected passive SCR fuel penalty will be discussed.

Experimental Investigation and Kinetic Modeling of Diesel Oxidation Catalysts for Emissions Aftertreatment from Diesel Engine Exhaust

Hom N. Sharma
Department of Chemical and Biomolecular Engineering, University of Connecticut, Storrs, CT

Keywords
Diesel engine exhaust, Emission oxidation, Diesel oxidation Catalyst (DOC), microkinetic modeling

Emissions from the diesel engine is one of the major concerns for global warming, human health and environmental pollution, due to many toxic and harmful exhaust gasses like carbon monoxide (CO), carbon dioxide (CO_2), nitrogen oxides (NO_x), sulfur oxides (SO_x) and many other volatile organic compounds (VOCs) [1] and poly aromatic hydrocarbons (PAHs) [2]. Diesel Oxidation Catalyst (DOC) is a predominant component in emissions aftertreatment to meet such regulations [3]. As shown in Figure 1a, In Urea- or aqueous ammonia-based Selective Catalytic Reduction (SCR) of NO_x, the Pt/Pd based DOC is typically placed upstream of the SCR to achieve an optimum NO: NO_2 ratio for improved SCR efficiency. On the other hand, as shown in Figure 1b, in HC-based SCR (e.g., using Diesel/ethanol/mixtures), a downstream DOC would be required to oxidize the toxic byproducts, such as CO, ammonia (NH_3), formaldehyde (HCHO), hydrogen cyanide (HCN), acetonitrile (CH_3CN), and acetaldehyde (CH_3CHO), to relatively 'less' harmful CO_2 and NO_x.

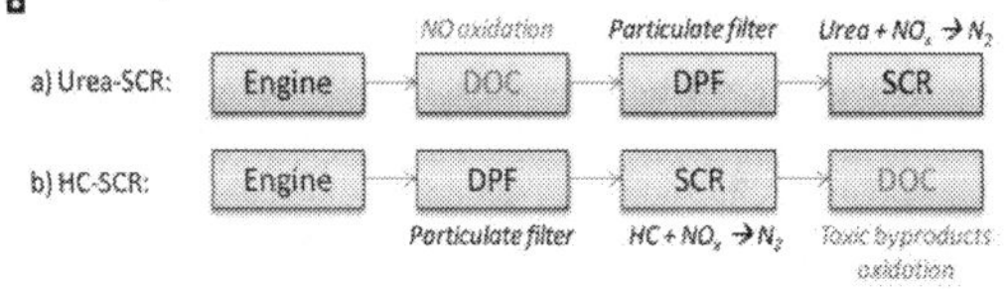

***Figure1** Role of DOC in toxic emissions oxidation.*

The noble metal based DOC is one of the most expensive components of the emissions aftertreatment system, and is typically overdesigned to some extent to ensure complete oxidation of toxic byproducts. Despite the extensive research conducted on DOC performance, comprehensive and predictive kinetic models for simultaneous catalytic oxidation of multiple emissions are limited due to the complex multi-species interactions.

In this work, we have developed a detailed elementary-step microkinetic model for DOC that accounts for (i) CO oxidation, (ii) NO oxidation, (iii) oxidation of other N-containing emissions, such as NH_3 and HCN, (iv) oxidation of toxic aldehydes, such as HCHO, (v) oxidation of hydrocarbons, such as C_2H_4, and (vi) oxidation of toxic sulfur oxides, such as SO_2 [4-6]. Furthermore, A state-of-the-art BenchCAT reactor was be used for the experimental investigation of CO and C_2H_4 oxidation experiments on Pt/Al_2O_3 powder and monolith scale DOC catalysts. Catalyst activity for oxidation of various toxic components was measured as a function of operating conditions, such as temperature, flow rate, and inlet composition. The experimental data was used to rigorously validate the microkinetic model and to probe the ability of the reaction mechanism to capture the interactions among various species during emissions oxidation.

The detailed reaction mechanism development is carried out using various parameter estimation techniques, such as semi-empirical Unity Bond Index-Quadratic Exponential Potential (UBI-QEP) method [7], Transition State Theory (TST), and quantum mechanical Density Functional Theory (DFT). The microkinetic model performs well against various experiments conducted in practically relevant conditions (atmospheric pressure, ppm level concentrations of emission components) and in fixed bed and monoliths reactors. The elementary-step surface reaction mechanism could be utilized for improved design and control of the aftertreatment system as well as appropriate DOC sizing.

References

(1) Kim, Y.; Kim, W. *Ind. Eng. Chem. Res.* **2009,** 48 *(14)*, 6579-6590.
(2) Lim, M. C. H.; Ayoko, G. A.; Morawska, L.; Ristovski, Z. D.; Rohan Jayaratne, E. *Atmos. Environ.* **2005,** 39 *(40)*, 7836-7848.
(3) Environmental Protection Agency *Federal Register* **2006,**Vol. 71, No. 83*(Air Docket No. EPA-HQ-OAR- 2006-0224)*.
(4) Sharma, H. N.; Mhadeshwar, A. B. *Appl. Catal. B-Environ.* **2012,** 127 , 190-204.
(5) Sharma, H. N.; Suib, S. L.; Mhadeshwar, A. B. Interactions of Sulfur Oxides With Diesel Oxidation Catalysts (DOCs). In *Novel Materials for Catalysis and Fuels Processing;* Bravo-Suarez, J. J., Schwartz, V. and Kidder, M. K., Eds.; ACS: 2013; Vol. 1132, pp 117-155.
(6) Sharma, H. N.; Sharma, V.; Hamzehlouyan, T.; Epling, W.; Mhadeshwar, A. B.; Ramprasad, R. *J. Phys. Chem. C* **2014,**.
(7) Shustorovich, E.; Sellers, H. *Surface Science Reports* **1998,** 31 *(1-3)*, 1-119.

A new catalyst for the selective catalytic reduction of NO_x by NH_3 over olivine

Yun Shi, Su-jing Li[1]

Key Laboratory of Biomass Chemical Engineering of Ministry of Education, Institute of Industrial Ecology and Environment; Institute of Environmental Engineering, Zhejiang University (Yuquan Campus), Hangzhou 310027, China.

Abstract:

Selective catalytic reduction (SCR) of NO_x by urea (Urea-SCR), using urea as the ammonia source, is a potential method to remove NO_x from automobile exhausts. Various catalysts, including metal-supported Al_2O_3 and ion-exchanged ZSM-5, have been widely investigated. However, the hydrothermal deactivation, narrow active temperature window, and high cost limited the application of these catalysts. The main challenge for the NO_x removal for automobile exhausts is the low-temperature activity (<300 °C). Researchers reported that transition metal oxides like FeO_x, MnO_x and NiO_x exhibit good activity for Low-temperature (LT) SCR. Olivine, a magnesium iron silicate with the formula $(Mg^{2+}, Fe^{2+})_2SiO_4$, and containing minor amounts of manganese and nickel, is a promising SCR catalysts.

This work reports for the first time the performance of olivine for the selective reduction of NO_x by NH_3. The catalytic activity and parameters that may affect olivine catalytic performance were investigated. It was hoped that this work could provide theoretical foundation for the application of Olivine-SCR.

Catalytic activity tests were carried out at atmospheric pressure in a fixed bed quartz tube reactor (inner diameter 6 mm). A reaction mixture containing 500 ppm NO, 500 ppm NH_3, 5% O_2, and N_2 as the balance gas was fed to the catalyst bed (GHSV = 11,000 h^{-1}). The reaction temperature was increased from 100-450 °C with 10 °C min^{-1}. Each step lasted 20 min to obtain steady state conditions. Without introducing any other elements, olivine itself displays 45-80% for NO_x conversion in the temperature range 100-350 °C, and with a maximum conversion of about 80% at 300 °C. The results also show that very low concentration of N_2O was detected, and N_2 is the main products during olivine-SCR.

The effects of gas hourly space velocity (GHSV) and inlet concentration of O_2 were investigated. The increase of GHSV decreased the NO_x conversion. The inlet concentration of O_2 had an obvious effect on the NO_x conversion. The NO_x conversion was below 30% without O_2 at 300 °C due to lack of active catalytic center that could be regenerated with O_2. The maximum NO_x conversion was obtained when the O_2 concentration was 2%. Further increase of O_2 could decrease NO_x conversion due to competitive adsorption on the active catalytic center between O_2 and NO_x.

As a result, some natural minerals, such as olivine, perovskite and clinoptilolite, are potential catalysts for NO_x reduction due to good activity and stability, low cost and flexible composition.

Key words: NO_x, NH_3-SCR, Olivine, FeO_x

1

[1] Corresponding author: E-mail: sujing-li@zju.edu.cn

385080 Renewable H2/CO Fuels for FTS and SOFC Applications Via Electrochemical Conversion

Abstract:

Liquid fuel is in high demand because it has high energy density due to its inherent chemical bonds, but unfortunately, fossil fuel supplies and related liquid fuel products are quickly diminishing. One solution is to explore non-fossil renewable H_2/CO fuels or even sustainable hydrogen fuel. The technical progress of CO_2 capturing makes CO_2 a potential feedstock for renewable syngas fuel (a gas mixture of H_2/CO). This alternative solution is to develop more efficient and durable electrochemical reduction conversion system such as a solid-oxide electrolytic cell to generate the H_2/CO syngas fuel by co-electrolysis of CO_2 and H_2O. The clean fuel can directly generate electricity using a SOFC power system at relatively high energy efficiency. To further combat the energy issue, especially for onboard energy storage requiring of significant high energy density, one proposition is the direct synthesis of liquid fuels using the syngas via the FTS technique. As a whole, the solid oxide electrolysis process at high temperature can use the CO_2 feedstock to subsequently produce non-fossil synthetic fuels. More R&D efforts potentially reduce CO_2 emissions and decrease environmental concerns. The utilization of CO_2 as the starting material for fuel production is highly desirable, gradually reduces CO_2 contribution to the environment, and creates a positive impact on the global CO_2 emission levels.

Maximizing the recovery of bio-commodities from algal biomass

Ali Teymouri[1], Edith Martinez-Guerra[2], Sandeep Kumar[1], Veera Gnaneswar Gude[2*]

1. Department of Civil and Environmental Engineering, Old Dominion University, Norfolk, VA 23529,

2. Department of Civil and Environmental Engineering, Mississippi State University, Mississippi State, MS 39762

Abstract

Algae have the potential to provide the most promising long-term solution for our energy and food supplement needs due to their high lipid and protein content. To maximize the recovery of these bio-commodities (lipids and proteins), a suitable processing scheme needs to be developed. Here, we developed a two-step process in which algal biomass is subjected to flash hydrolysis (FH) in the first step to extract the algal proteins followed by extractive-transesterification of lipids into crude biodiesel in the second step. Recovering proteins prior to the biofuel production from lipids would enhance the quality of both bio-commodities.

The novel FH process developed under this study applies short residence time under near-subcritical water conditions to selectively hydrolyze the proteins and preserve the other valuable components (lipids) of microalgae. A continuous flow process was specifically designed to work with *Scenedesmus sp.* slurry to extract proteins. More than 60 wt % of the total nitrogen content (dry basis) in *Scenedesmus sp.* was extracted within 10 s of residence time above 240°C in the form of hydrolyzate leaving lipid-rich biofuel intermediate product behind. The percentage of lipids in biofuels intermediate was measured gravimetrically using chloroform:methanol (2:1) extraction method. The results showed the lipid contents increased from 18 wt% to 70 wt% on dry weight basis.

Followed by the FH process, this biofuel intermediate product was subjected to extractive-transesterification under microwave irradiation using methanol as solvent to yield crude biodiesel. Simultaneous extraction and transesterification (also known as *in situ* transesterification) method simplifies the conversion process, reducing the number of unit operations and consequently the overall process costs and final biodiesel product costs. By extractive-transesterification, more than 90% of the algal lipid content was converted to crude biodiesel in a very short reaction time of less than 10 minutes.

This presentation will discuss the merits of the novel integrated two-step process for maximizing selective lipid and protein recovery from algal biomass supported by the results from the experimental studies along with a comparison of other existing process alternatives.

375968 Green Chemical Conversion of Cellulose into Soluble Sugars and 5-Hydroxymethylfurfural (HMF) Using Ionic Liquids

Abstract:

An ionic liquid-based chemical hydrolysis strategy was developed to obtain high-yielding soluble sugars and 5-hydroxymethylfurfural (HMF), a promising platform chemical derived from cellulose. The efficiency of two biomass conversion technologies, dilute acid hydrolysis and dissolution in ionic liquids (ILs), are compared in terms of delignification, saccrification efficiency and saccharide yields with crystalline cellulose serving as a model biomass. Several ILs such as 1-ethyl-3-methylimidazolium chloride [EMIM], 1-butyl-3-methylimidazolium chloride [BMIM], 1-ethyl-3-methylimidazolium chloride aluminum chloride [EMIM] $AlCl_3$ were tried and analyzed for their conversion efficiency and selectivity for various products such as glucose, fructose, HMF, and formic acid etc. Initial ILs dissolution and subsequent acid hydrolysis improved the dissolution and release of total sugars from cellulose. The amount of mineral acid load was also reduced considerably by initial ILs dissolution of cellulose. [EMIM] treatment on the long fibers of cellulose had no significant effect on the cellulose morphology as evident by scanning electron microscopy (SEM), and optical microscopy analysis (OM). Acid catalyzed hydrolysis and dissolution in ILs resulted in higher conversion of sugars which was evident by Benedict's test', analysis for reducing sugars. Conversion of various products was also analyzed by Fourier Transform-infrared spectroscopy (FT-IR). The condition including the ratio of Ils to cellulose, temperature and time for cellulose conversion was optimized. This ionic liquid-based hydrolysis strategy showed the great potential to produce soluble sugars and HMF from cellulose.

387238 Liquid-Liquid Extraction of Bio-Oil Components

Abstract:

Increases in the world population and energy demand have raised the importance of developing sustainable energy resources. Utilizing bio-oil (pyrolysis oil) from biomass may alleviate global energy problems; however, wide applications of bio-oil are hampered by difficulties in attaining usable, high-quality bio-oil. Particularly, separating the aqueous phase and water-soluble compounds from crude bio-oil has not been effective. Bio-oil has high moisture content, low pH, high viscosity, and low heating value. Liquid-liquid extraction has been investigated in this work as a method to effectively separate the aqueous and organic phases of bio-oil. The effects of pH, ionic strength, and volume ratio of aqueous solutions used in liquid extraction on the separation of water and water-soluble species from bio-oil are being investigated. Results thus far show that, as the ionic strength of the aqueous solution is increased, more acidic compounds are selectively extracted from the bio-oil. This is desirable, as acidic components are utilized downstream for hydrogen production in bio-electrochemical cells. Results from a wide range of experimental variables will be reported. Furthermore, liquid-liquid extraction experiments using a centrifugal contactor for selected experimental conditions of volume ratio, pH, and ionic strength of the aqueous phase will be presented.

394223 Exploiting Polymer-Nanoparticle Interactions to Create a Hydrogel with Biomedical Applications

Abstract:

Injectable hydrogels have recently become a relevant material platform in applications such as in vivo drug delivery or as cell carriers because they can be implanted in a minimally invasive manner and closely mimic biological tissue. Attractive properties for such injectable hydrogels are that they easily flow under applied stress (*shear-thinning*) and rapidly heal upon relaxation of the applied stress (*self-healing*). Previously, we developed a shear-thinning and self-healing injectable hydrogel driven by non-covalent interactions between hydrophobically modified cellulose and PEG-PLA nanoparticles. [1] The nature of this gel enables facile loading with a variety of therapeutics as well as tunable release profiles of multiple therapeutics. Specifically, the PEG-PLA nanoparticles allow for loading with hydrophobic therapeutics while a hydrophilic therapeutic can be entrapped within the aqueous phase of the gel.[2] In this work, we have quantified the factors that affect the mechanical properties and release profiles of the gels, demonstrating the range of their tunability. Additionally, we explored other nanoparticles and biopolymers for the fabrication of similar materials. Lastly, we studied the use of this injectable hydrogel as a cell carrier with osteogenic properties. Human mesenchymal stem cells (hMSCs), which are capable of differentiating into bone, cartilage, and fat cells, were encapsulated in the gel and survived injection through a 28 gauge needle into three-dimensional collagen hydrogels. Additionally, PEG-PLA nanoparticles were loaded with dexamethasone to stimulate osteogenic differentiation of the encapsulated hMSCs. We are currently exploring the use of this drug-loaded cell carrier as an injectable system for bone healing. Overall, we present an injectable hydrogel with tunable properties for the controlled delivery of therapeutics with facile synthesis and minimally invasive application *in vivo* that is capable of serving as a stimulatory cell carrier.

---

[1] Appel, E.A., Tibbitt, M.W., Langer, R., Nature Communications., **2014**, *in revision*.

[2] Greer, J.M., Appel, E.A., Tibbitt, M.W., Langer, R., *manuscript in preparation*.

Validation of High Throughput Electrochemical Gas Sensing Screening System

Zixuan Wang

November 17, 2014

UNIVERSITY OF SOUTH CAROLINA

Background

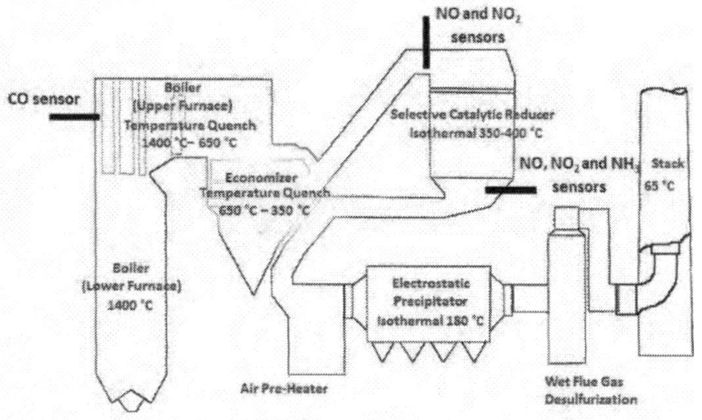

Harsh combustion environment - 800 ° C

Accurate and selective to target gas

UNIVERSITY OF SOUTH CAROLINA

High Throughput Methodology

Subatomic difference affect large scale

↓

Not feasible to individually test material

↓

High throughput synthesis and characterization methods

UNIVERSITY OF SOUTH CAROLINA

High Throughput Methodology

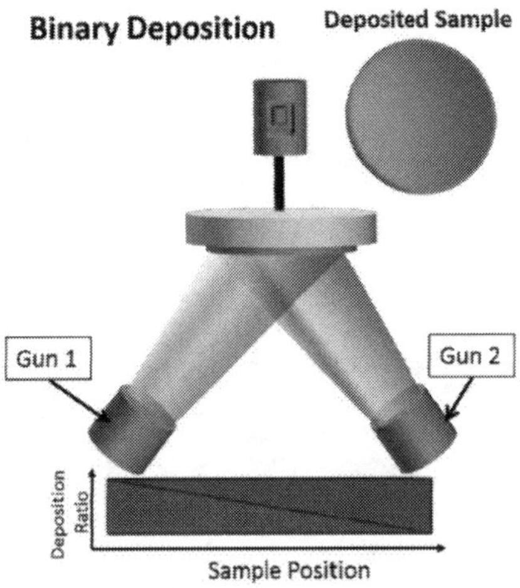

- Normalize affect of distance on the potential

UNIVERSITY OF SOUTH CAROLINA

Development of Uniform Standard Sample

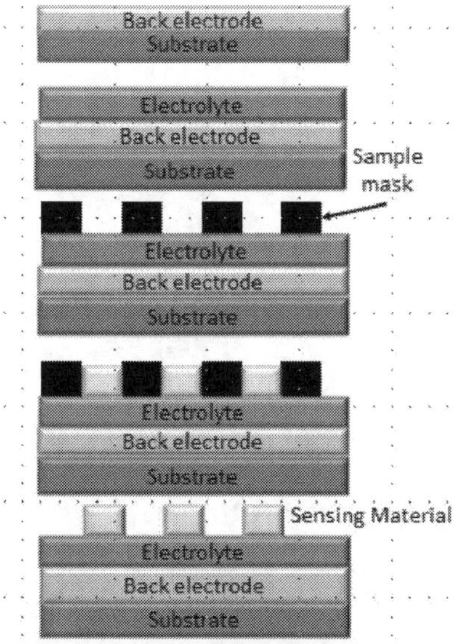

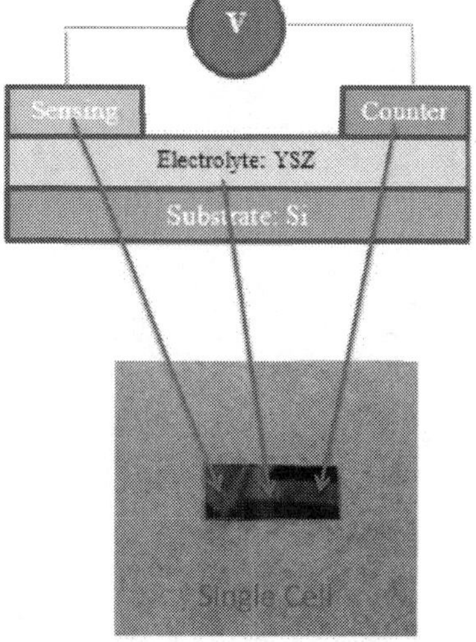

Surface Mechanisms

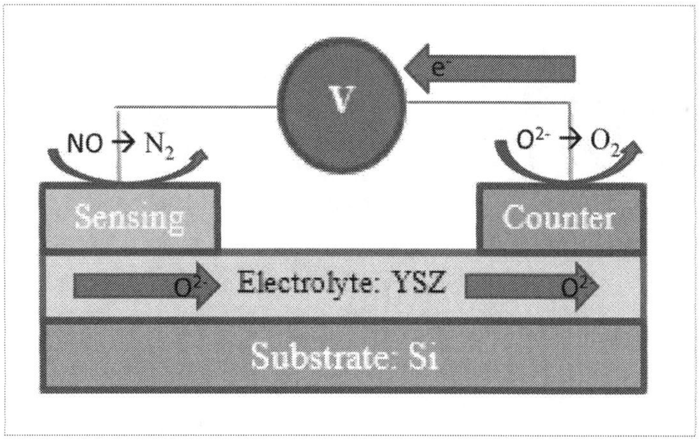

Sensors and Actuators B 121 (2007) 652-663

Reflectometry Results - YSZ

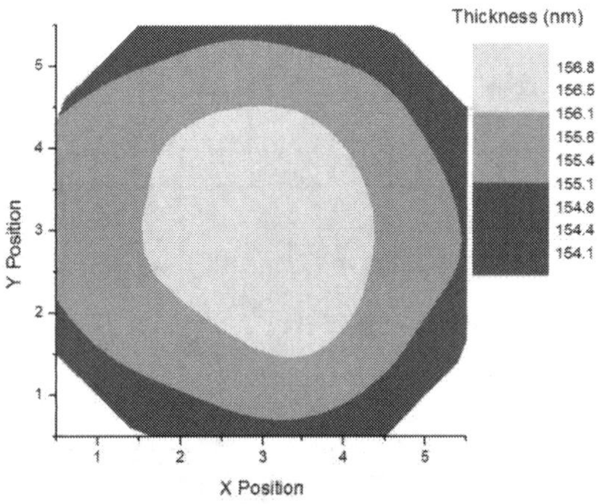

- Results confirm relative uniformity of sample (1.8% difference)

UNIVERSITY OF SOUTH CAROLINA

XRD Results - YSZ

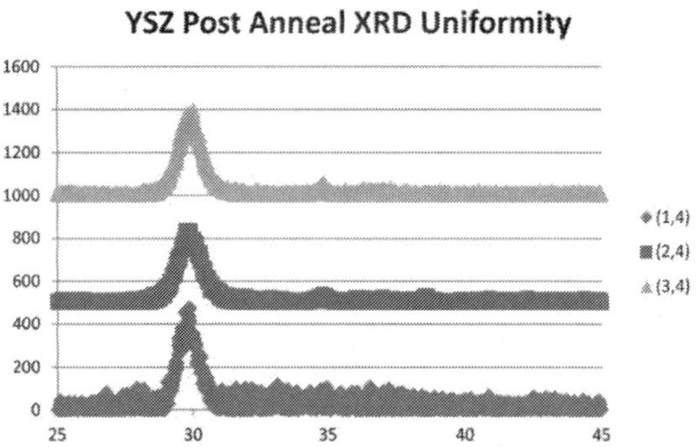

(1,4)	29.745	
(2,4)	29.836	34.848
(3,4)	29.876	34.859

- XRD analysis using Jade
- YSZ Cubic:
  - (111): 29.675°
  - (200): 34.398°
- Confirm identical phases

UNIVERSITY OF SOUTH CAROLINA

Reflectometry Results – WO_3

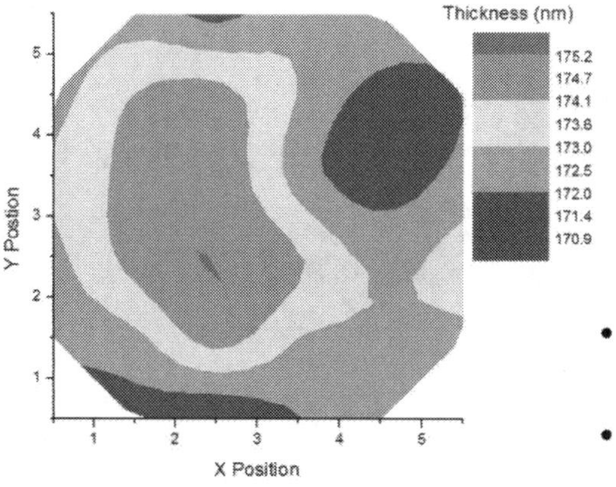

- Confirms relative uniformity (2.5% difference)
- Evidence some parts of the plate that are slightly thicker or thinner

XRD Results – Post Annealed WO_3

(1,4)	23.143	23.653	24.362
(2,4)	23.140	23.671	24.397
(3,4)	23.127	23.641	24.377

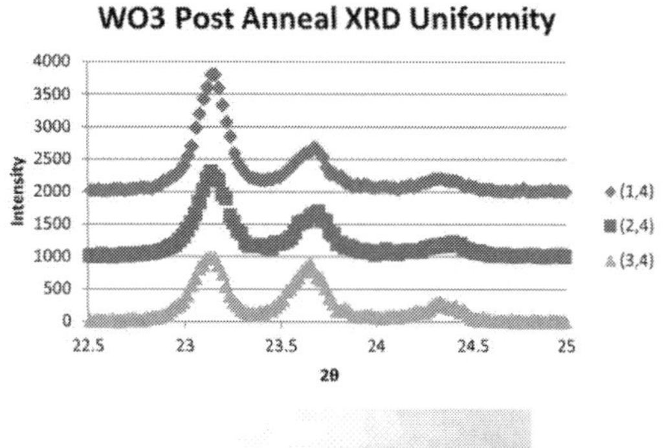

- XRD analysis using Jade
- WO_3 monoclinic:
  - (002): 23.118°
  - (020): 23.582°
  - (200): 24.365°
- Confirm identical phases with phase shifts present

Reflectometry Results - WO_3-YSZ

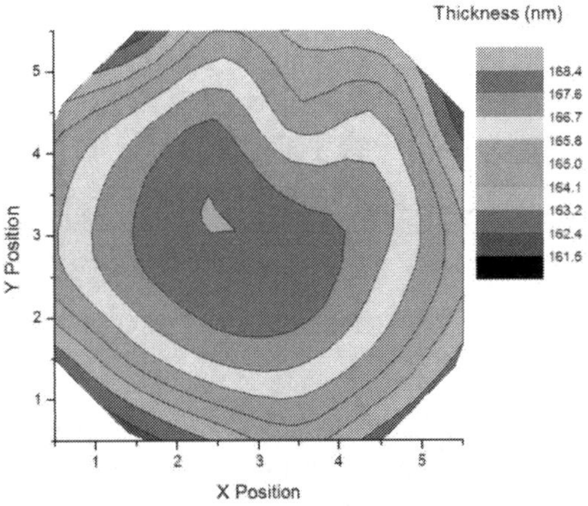

- Sputtering conditions: same
- YSZ held constant
- Confirms relative uniformity (4.3% difference)

UNIVERSITY OF SOUTH CAROLINA

XRD Results – WO_3-YSZ Post-annealed

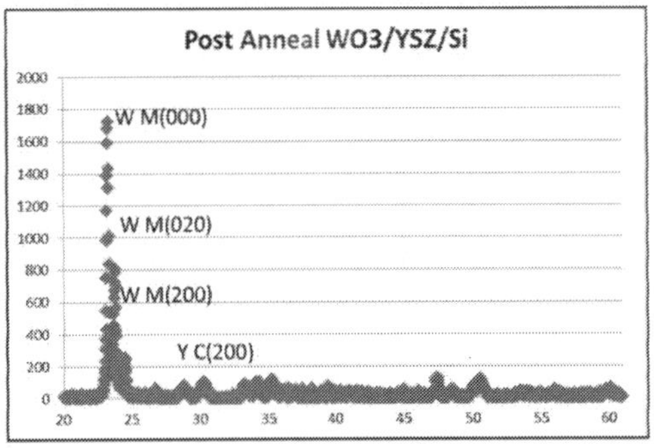

WO_3 monoclinic:
- (002): 23.118°
- (020): 23.582°
- (200): 24.365°

YSZ Cubic:
- (111): 29.675°
- (200): 34.398°

UNIVERSITY OF SOUTH CAROLINA

Electrochemical Cell Setup

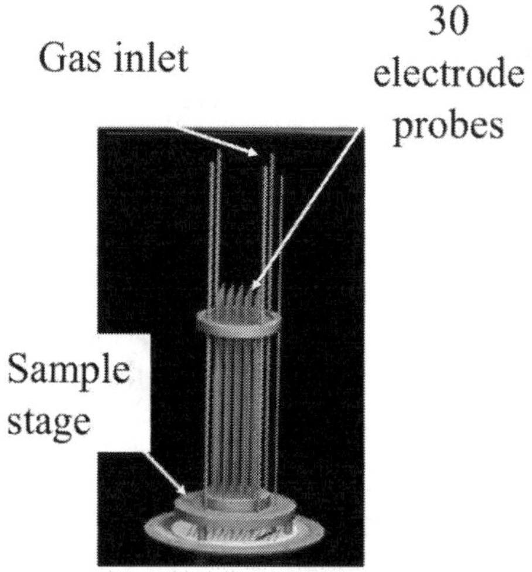

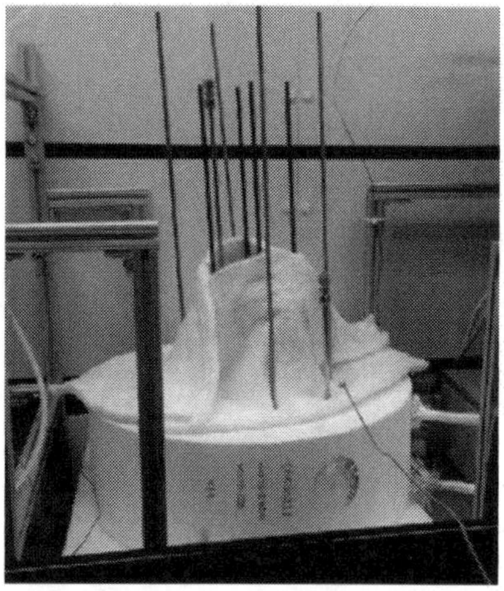

Single Electrochemical Cell Tests

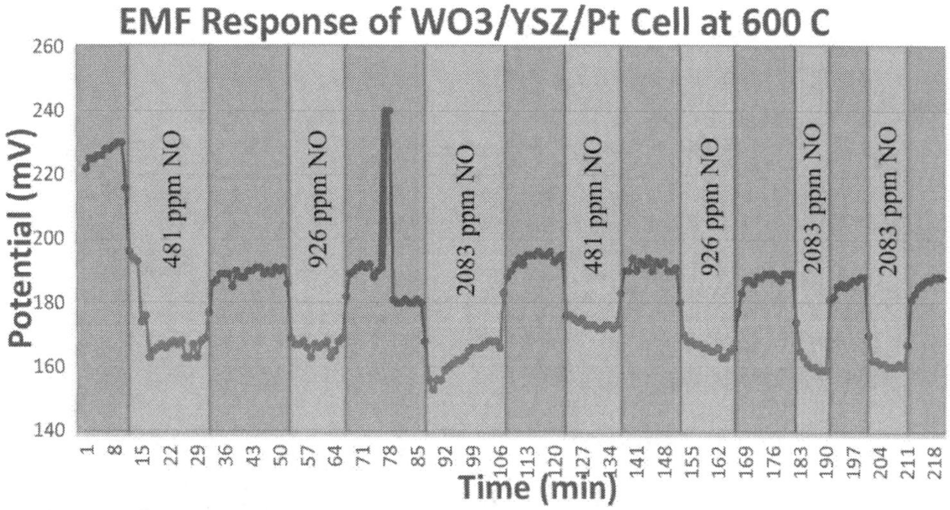

Electrochemical Cell Tests

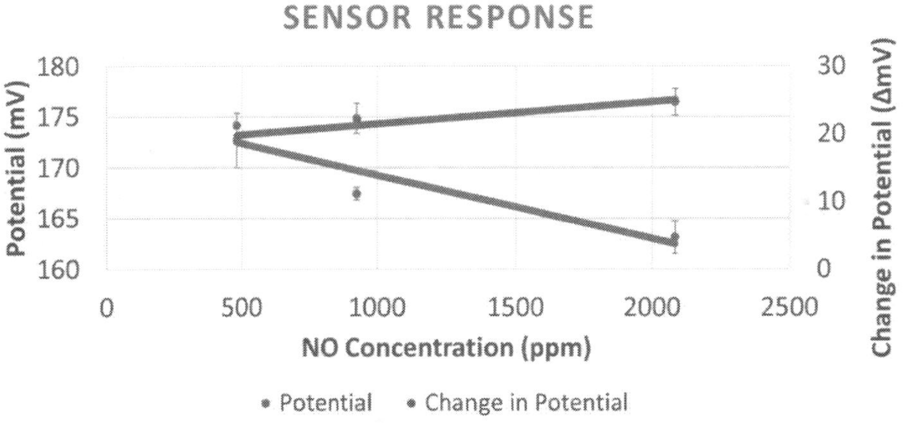

Multielectrode Cell Tests

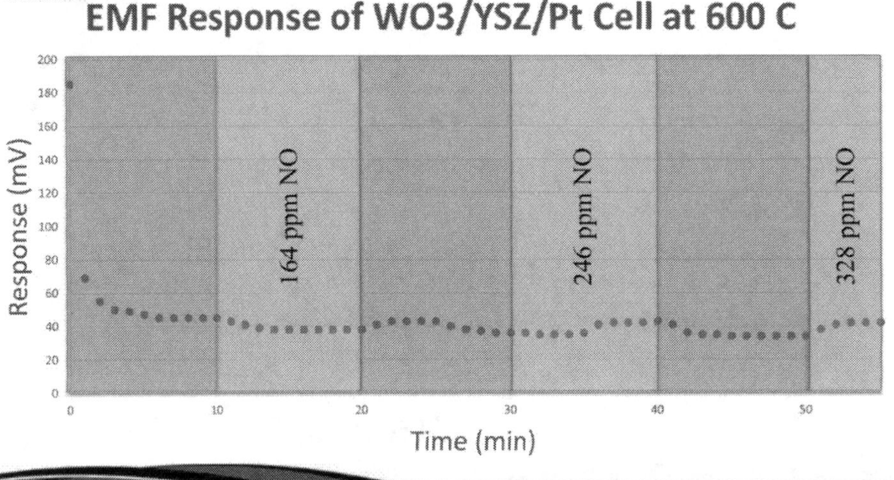

Conclusions

- Designed planar sensor array
  - Use for combinatorial screening of electrochemical gas sensors
- Confirm deposition uniformity
  - Reflectometry: thickness
  - XRD: phases
- Cell potential
  - Changes valid use of planar sensor design

UNIVERSITY OF SOUTH CAROLINA

Future Work

- Create multielectrode with increased surface area/perimeter of the electrodes
- Normalize potential to distance of electrodes
- Create multielectrode with different composition gradient

UNIVERSITY OF SOUTH CAROLINA

Acknowledgements

- National Science Foundation (NSF)
- University of South Carolina – College of Engineering and Computing
- Strategic Approaches to the Generation of Electricity (SAGE)
- Principal investigators: Dr. Jochen Lauterbach and Dr. Jason Hattrick-Simpers
- Graduate mentor: Ben Ruiz-Yi

394216 Long-Range Correlations in Liquid Water

Abstract:

The purpose of this research project was to observe water molecules' environments to determine whether or not the system forms identifiable clusters of similar environments. The different clusters aid in proving the hypothesis that different water patches at differing high and low densities values form in a size of 1nm in a system at room and lower temperatures. Molecular dynamics computer simulations were done using the dl_poly classic program with the water model potential TIP4P/2005. The simulations were run at various configurations for density, temperature, and pressure. The results were read by a program, which took the file from the simulations and determined the coordinate positions of the water molecules. Different clusters were found based on the results. These results are indicative of distinctive structures formed in supercooled water that may lead to the water patches of varying densities.

394155 Frictional Study of Polyethylene Glycol Monolayers on Silica Substrate

Abstract:

Lubricating micro- and nano-electromechanical systems (MEMS and NEMS) has become increasingly challenging as devices are continuously miniaturized. As the dimensions of these systems decrease, their surface area to volume ratios increase significantly and frictional forces become dominant where moving parts exist. Solid-state lubrication via self-assembled monolayers presents a promising alternative to traditional macro-scale lubricants which are too bulky to use in such small systems.[1,2] Previous studies have shown that self-assembled monolayers containing alkylsilanes and mixed alkylsilane and perfluoroalkylsilane chains reduce friction coefficients compared to equal length systems. Although, hydrocarbon-based monolayers have been widely studied and shown to improve frictional properties and device life, mixed monolayers impart the beneficial properties of perfluoroalkylsilanes (low critical surface tension and high thermal and mechanical stability) to the film. Furthermore, a mixed length configuration with longer alkylsilane chains relative to the perfluoroalkylsilane chains leads to a liquid-like layer formed by the taller chains that fold over and align with the shearing direction.[3]

In this study, we investigate polyethylene glycol (PEG) monolayers using molecular dynamics simulations as a potential solid-state lubricant for NEMS. The presence of oxygen atoms within the chains render these chains more flexible than their alkyl- and perfluoroalkyl-counterparts. Thus, even sparsely functionalizing a surface leads to a uniform film. We have investigated the tribological properties of such films across a range of surface coverages and chain lengths. Preliminary results show the effects of surface coverage on friction coefficients is diminished as chain length is increased. This suggests that equivalent friction coefficients could be achieved while reducing separation distances by using longer chains on sparser surfaces.

394001 A Machine-Learning Model to Predict Activation Energies of Hydrogenation Reactions

Abstract:

The goal of this research is to develop a machine-learning model to predict activation energies associated to the hydrogenation of organic compounds in the presence of specific catalytic substrates. The current method of predicting activation energies involves a linear relationship with the change in Gibbs free energy of the reaction, which can be very inaccurate. Other methods include experimentation and simulation, both of which have significant cost and/or time drawbacks. This research seeks to improve on these methods to develop a means to predict activation energies quickly, accurately, and with minimal cost.

393764 Microscopic Modeling of the Self Assembly of Poly(ethylene oxide)-Poly(propylene oxide)- Poly(ethylene oxide) (PEO-PPO-PEO) Block Copolymers: Critical Micelle Concentrations

Abstract:

Single chain mean field theory (SCMFT) is applied as a coarse-grain simulation model for Pluronics, a trademarked branch of triblock copolymers. The microscopic model is used to quantify the micellization process of the nonionic class of surfactants at 37 degrees Celsius for drug delivery applications. The simple model with only three interaction parameters is shown to be able to reproduce with high accuracy experimental data for the critical micelle concentrations of different chain lengths of the polymers. These parameters are physically significant and are close to the experimental values from Flory models. In addition, aggregation number of the surfactants and volume fraction profiles are explored. The effects of the chain lengths of hydrophobic tails and hydrophilic heads on CMC values are discussed. SCMFT is transferable in surfactants of different ABA block conformations and other more complex polymer structures where previous simulations have not excelled.

Recommendations for Legislative Actions to Reduce Carbon Emissions in the Electricity Production Sector

Sam White

American Institute of Chemical Engineers
University of Michigan
Washington Internships for Students in Engineering

August 2014

Executive Summary

There is currently a movement in America pushing for environmental effects to be taken into consideration when assessing new and existing business ventures. As emphasized by President Obama's 2014 State of the Union Address, a major area for this movement to make progress is within the energy sector [1]. Electricity production currently produces the most greenhouse gas emissions of any energy sector in the US, making it the energy sector where the largest environmental impact can be made with proper policy action [2].

Luckily, currently-available, low-emission renewable energy technologies are most well suited for electricity production over any other energy sector application. As a result of recent government regulations combined with high coal prices, coal plants, which currently dominate the electricity production landscape, are being retired at an accelerated rate [8]. This, combined with increased electricity demand, has opened up the door for new electricity production technology to be implemented. However, due to the low price of natural gas resulting mainly from the onset of fracking, efforts to implement new renewable-sourced electricity production facilities have been stalled in favor of natural gas power plants. While natural gas plants produce only half the carbon emissions of coal, and do not produce harmful byproducts such as sulfur dioxide or mercury, they are not as environmentally friendly as the other available alternatives, such as nuclear, solar, or wind [11] [12] [13] [14].

The fact that natural gas dominates the shift away from coal is not surprising, however, given the current economic landscape. The figure below shows the levelized cost of electricity (LCOE) for select available electricity production technologies, giving a $/Mwh cost of production by factoring in all applicable costs (construction, fuel, maintenance, operation, etc.) [23].

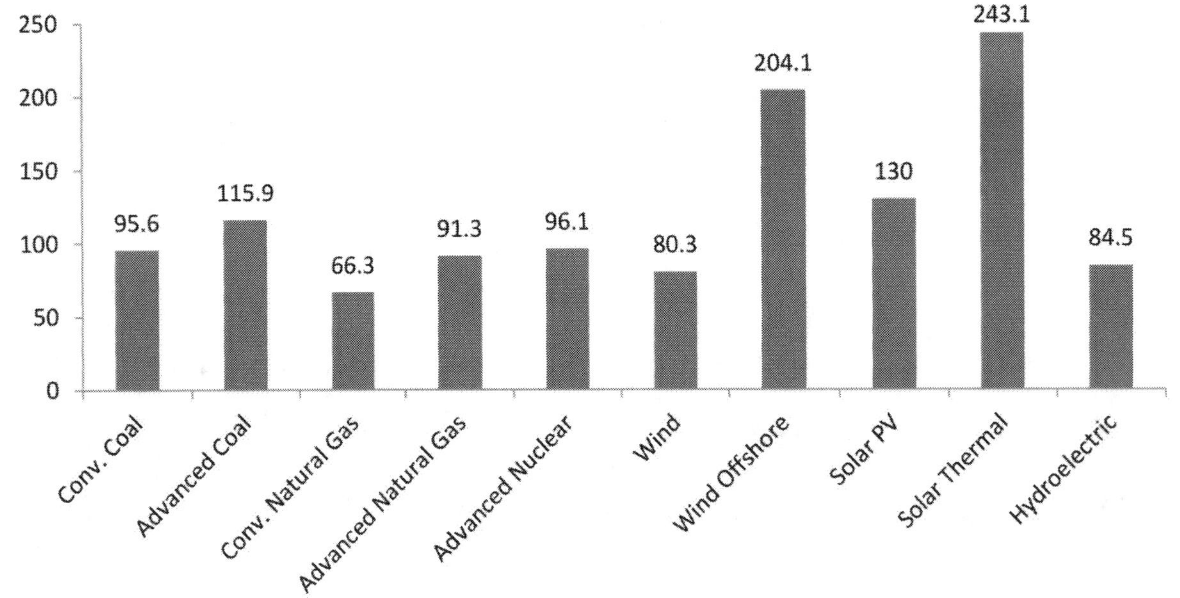

US Average LCOE for Plants Entering Service in 2019, in 2012 $/Mwh [23 EIA LCOE]

As seen in figure above, conventional natural gas currently offers the cheapest form of electricity production. It is important to note, though, that some alternatives like nuclear or wind are already competitive with other conventional electricity production methods like coal. Other alternatives, like solar or offshore wind, are still too expensive to be competitive at the moment, mainly due to the fact that electricity production from renewable sources is still too variable. With the advent of advanced electricity storage technology this issue would be eliminated, but such technology is currently unavailable or too expensive to be implemented economically [19].

Because of their positive environmental impacts, alternative electricity production methods are given some assistance by the federal government in order to make them more economically competitive. These economic aids mainly come in two forms: tax credits and loan programs. On the tax side, an installment tax credit allows solar production facilities a credit equal to 30% of the cost of implementation, while a production tax credit allows wind, hydro, and geothermal plants to receive a .23 cents/kwh credit. However, the installment tax credit is only available until the end of 2016, and the production tax credit already expired at the beginning of 2014 [26]. On the loan program side, the government has extended loans to help both nuclear and renewable power plants, though the renewable plant loan program expired in 2011 [27] [28].

On top of aid being given to alternative electricity production methods, the federal government can also level the playing field by putting a burden on conventional electricity sources. This mainly comes in one of two forms, a carbon tax or a cap-and-trade program, both of which seek to charge electricity producers for the amount of carbon emissions they produce. On top of producing revenue for the federal government, these programs help incentivize electricity producers to innovate new ways to reduce their carbon emissions. Though no such program has ever been implemented, a cap-and-trade program was able to at least pass the House in 2009 [24] [25].

Even small actions by the federal government can have noticeable effects in shifting the production methods used in the electricity production sector. Just a relatively small tax such as $10/metric ton of carbon emissions would cause significant carbon emission reductions compared a scenario in which no government actions are taken [33]. Additional analysis specific to natural gas and renewables shows that any carbon tax would actually increase the amount of natural gas-based electricity production for the next couple of decades, and increase the amount of renewable-based electricity production indefinitely. Additionally, if current favorable tax policies for renewables are renewed long-term, the use of renewables is significantly higher than those projected given current tax policy deadlines, especially after 2030 [34] [35].

As a result, it is recommended that a two-pronged approach be taken to limit the amount of economic disturbance while ensuring the electricity production sector makes a greater shift towards renewables given the reduction of coal use. First, a carbon tax equivalent of $10/metric ton of carbon emissions should be enacted to reflect the social cost of emissions. Each year this tax would increase 5% per year after inflation in order to push companies to innovate new carbon reduction technologies. Second, the production tax credit should be extended for all renewable electricity production methods using the current credit of .23 cents/kwh. This credit would decrease 5% per year after inflation to mirror the carbon tax in order to account for the increased rate of improvement in renewable technologies and the relative advantage given by the increasing carbon tax.

It is important that such a program be long-term in order to allow industry to accurately model the economic landscape their projects will enter. Given the current gridlock in Washington and the long-term nature of implementing new electricity production facilities, a long-term tax plan is the only way to ensure economic stability. Factoring in only revenue from taxes on natural gas use and costs from credits extended to renewables use, such a program would pull in an average surplus of $5.5 billion per year between 2015 and 2040; factoring in coal would increase this figure significantly, particularly in the near future. Feeding at least part of this surplus into loan programs for nuclear plants would ensure that an increased amount of low carbon electricity production facilities are able to come to market.

Table of Contents

Executive Summary

Table of Contents

Table of Figures

Table of Tables

1.0 Introduction

2.0 Overview of Electricity Production Methods and Their Environmental Impacts

- 2.1 Coal and Natural Gas
- 2.2 Nuclear
- 2.3 Solar and Wind
- 2.4 Other Renewables
- 2.5 Environmental Impact Summary Table

3.0 Economics in the Electricity Production Sector

- 3.1 Comparing Costs of Electricity Production Technologies
- 3.2 Capacity Factor

4.0 Recent Policy Strategies

- 4.1 Tax Credits for the Electricity Production Sector
- 4.2 Loan Programs
- 4.3 Carbon Tax and Cap-and-Trade Program

5.0 Projected Electricity Sector Growth: Electricity Sector Capacity Additions for 2013

- 5.1 Policy Action Effects: Overall Carbon Emissions
- 5.2 Policy Action Effects: Conventional Fuels
- 5.3 Policy Action Effects: Renewables

6.0 Policy Recommendations

7.0 References

Table of Figures

Figure 1. Total Greenhouse Gas Emissions by Economic Sector in 2012 [2]

Figure 2. Share of Total US Electricity Production by Source [Adapted from 8]

Figure 3. US Average LCOE for Plants Enetering Service in 2019, in 2012$/MWh [23]

Figure 4. Capacity Factor (%) of Various Electricity Production Technologies [23]

Figure 5. Electricity Generation Capacity Addition Projections by Fuel Type, 2008 (Gw) [30]

Figure 6. Electricity Generation Capacity Addition by Fuel Type 2013 (GW) [Adapted from 31]

Figure 7. Electricity Generation Addition Projections by Fuel Type 2014 (GW) [32]

Figure 8. Electricity-Related CO_2 Emissions in Five Cases, 2000-40 [33]

Figure 9. Natural Gas Electricity Generation in Five Cases, 2000-40 [34]

Figure 10. Renewable Electricity Generation in Four Cases, 2000-40 (billion kWh) [Adapted from 35] .

Table of Tables

Table 1. Life-Cycle Environmental Impact Summary of Electricity Production Technologies [Adapted from 11]

1.0 Introduction

In his 2014 State of the Union address, President Obama stated that "[O]ne of the biggest factors in bringing more jobs back is our commitment to American energy"[1]. He stressed the importance of "a cleaner energy economy" and acknowledged that "[it] won't happen overnight, and it will require tough choices along the way." Though he touted the success of his "all-of-the-above energy strategy", and highlighted the advantages of natural gas and solar, the President laid out no definitive plans for how the US would approach energy production in the future. However, with the release of recent carbon emission regulations, the executive branch seems to be taking at least some action to direct the country towards a cleaner energy economy. However, if a cleaner energy economy is going to happen in America, the legislative branch needs to get more involved by developing and implementing a long-term energy policy plan.

Of all the areas of energy production in the America, the most promising area for change is the electricity production sector, since a vast majority of alternative energy sources are most viable for electricity production instead of applications like transportation or residential heating. Luckily, this is also the energy sector in which the most impactful changes can be made: in 2012, electricity production was the largest single contributor to greenhouse gas emissions in the US, accounting for 32% of greenhouse gas emissions as seen in Figure 1 below [2].

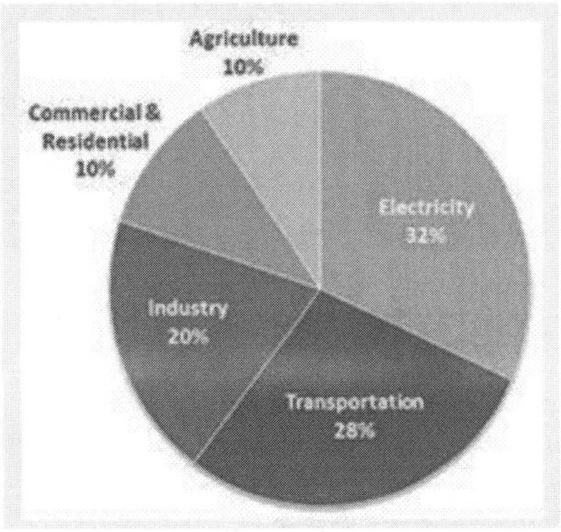

Figure 1. Total Greenhouse Gas Emissions by Economic Sector in 2012 [2]

The implementation of a cleaner energy economy has shown to be a popular policy among voters from both parties. According to a Washington Post – ABC News poll from June 2014, 70 percent of voters say that the federal government should require limits to greenhouse gases from existing power plants. 57 percent of Republicans, 76 percent of Independents, and 79 percent of Democrats support state-level limits on greenhouse gas emissions. Most significantly, voters are willing to pay extra for energy if it means a reduction in greenhouse gas emissions. "Asked whether Washington should still go forward with limits [on energy-sector greenhouse gas emissions] if „they significantly lowered greenhouse gases but raised your monthly energy bill by 20 dollars a month," 63 percent of respondents say yes, including 51 percent of

Republicans, 64 percent of Independents and 71 percent of Democrats" [3]. These results suggest that Americans on both sides of the aisle are in support of implementing a cleaner energy economy.

In addition to the popularity of a cleaner energy economy, it also makes sense from an environmental and health stand point. On top of the fact that the reduction of greenhouse gas emissions reduces the rate of climate change, it brings along health benefits as well. With the release of its Clean Power Plan in June 2014, which seeks to cut carbon pollution from the power sector by 30% by 2030 by imposing stricter emission limits on existing plants, the EPA projects that "annual public health benefits will total somewhere between $55 billion and $93 billion by avoiding up to 100,000 asthma attacks and 2,100 heart attacks each year." By contrast, they predict compliance cost of $7.3 to $8.8 billion per year, significantly lower than the expected public health benefits [4]

However, despite the ongoing executive branch actions regarding energy policy, the legislative branch has taken relatively little action regarding energy policy lately, mainly due to party-politics-driven gridlock. At the beginning of 2014, the Production Tax Credit for renewables was allowed to expire for the first time since 2004, and only the fourth time since its enactment in 1992 [5]. In May of 2014, the Senate allowed a bipartisan energy efficiency bill to die due to disagreement over other energy related issues, such as the fate of the Keystone XL pipeline and possible legislative response to President Obama"s climate change regulations. In fact, it has been 7 years since Congress passed a major energy bill, which means there has been no major legislative action in response to the changing energy landscape as a result of the shale fracking boom [6].

A plan for implementing a cleaner energy economy becomes even more important as energy demands increase, particularly when it comes to electricity production. As electricity demand increases, more electricity production facilities are built, and how these facilities produce electricity greatly affects their environmental impact. According to the EIA"s Annual Energy Outlook 2014 report, US electricity demand will increase from anywhere between 14% and 41% between 2012 and 2040, though their best estimates suggest that electricity demand will increase by 29% over this time period. Additionally, the growth of electricity demand is projected to increase sometime around 2025, marking the beginning of an upwards slope in electricity demand growth for the first time in 75 years [7].

In order to meet this demand and fill the gap left behind by plant retirements, new electricity production facilities need to be built. As a result of recent emission regulations like the Clean Power Plan, coal power plants are being retired at an accelerated rate. However, due to the low price of natural gas in recent years thanks to fracking, much of this gap has been filled by natural gas fired plants instead of alternatives, like nuclear or renewables. This is mainly due to economics: even the cheapest alternative energy sources, which are on the verge of being if not already cost competitive to coal, simply cannot compete with the low price of natural gas. While natural gas is a lower-emissions alternative to coal, it has emissions nowhere near as low as alternative electricity sources.

The onset of natural gas is not a bad thing, however: using natural gas as a bridge fuel to reduce emissions in the immediate future while hoping that renewables become more cost competitive in the future is not unreasonable given the rate of development and investment in renewable technologies. Natural gas also offers the unique advantage of "spinning reserve", a source that can quickly be powered up to meet rapid fluctuations in demand. This means natural gas can offer a flexibility that base load electricity sources like coal or nuclear cannot, while acting as a buffer against the variability of renewables like solar or wind.

Embracing natural gas in the here and now is also economically responsible, as it keeps electricity prices low by offering a cheap alternative to coal and helps support the growing American natural gas production market. However, without government action, this shift towards natural gas will dwarf any efforts to make renewables a larger share of our electricity production sector. If the government is serious about using natural gas as a bridge fuel to a cleaner energy economy in the electricity sector, a long-term policy needs to be enacted to slowly guide the market in the right direction.

2.0 Overview of Electricity Production Methods and Their Environmental Impacts

In order to discuss energy policy, it is first important to establish the electricity needs of the country, as well as the current capabilities of various electricity production methods and their effects of the environment. As seen in Figure 2 below, coal and natural gas currently dominate the electricity production sector, with nuclear and renewables making up the bulk of the remaining production.

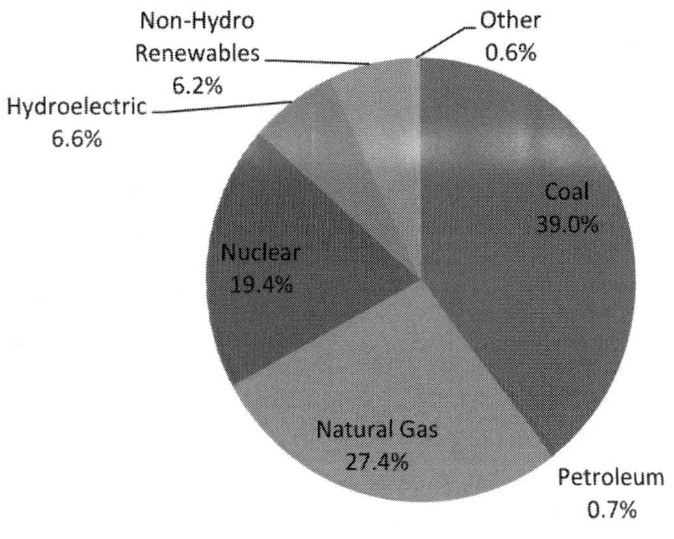

Figure 2. Share of Total US Electricity Production by Source [Adapted from 8]

2.1 Coal and Natural Gas

The two main fossil fuel based / conventional energy producing sources, coal and natural gas power plants are currently the largest source of electricity in America, accounting for nearly two-thirds of the country"s total electricity production [8]. Coal plants are relatively easy to keep in operation over a long time scale, but it is difficult for them to make relatively quick changes to production capacity in response to the changing needs of the grid, particularly during peak-usage hours. As a result, coal plants are very useful in providing a baseline power supply somewhere between peak and off-peak requirements.

Natural gas, on the other hand, allows for a very flexible energy production method. Gas flow rates into a furnace can be controlled with relative precision, unlike solid input streams found in coal plants. As a result, natural gas plants can be very responsive to the changing needs of the energy grid. In fact, natural gas plants are typically used to augment other energy production facilities for this very reason.

From an environmental perspective, the EIA considers conventional electricity sources like coal and natural gas as being the only contributors to CO_2 emissions in the electricity production sector. In 2010, the EIA reported that 1969 million metric tons of CO_2 were released by coal plants nation-wide, while 1285 million metric tons of CO_2 were release by natural gas plants [9].

However, coal tends to have other unique environmental impacts. First, burning coal tends to release nitrogen oxides, which tend to have larger warming potentials than CO_2. For example, N_2O has a SAR 100 year warming potential of 310, meaning that one ton of N_2O has 310 times the impact on global warming compared to one ton of CO_2 over the course of 100 years [10]. From a life cycle assessment standpoint, coal plants produce 974 metric tons of CO_2 equivalent per gigawatt-hour, which accounts for this increased warming potential as well as CO_2 emissions due to non-production factors, such as plant construction or fuel transportation [11]. Coal also tends to produce SO_2, which, along with nitrogen oxides, is the major man-made cause of acid rain. According to the EPA, "In the United States, roughly two-thirds of all SO_2 and one-quarter of all nitrogen oxides come from electric power generation that relies on burning fossil fuels, like coal" [12]. Additionally, coal plants are a major source of mercury, a highly toxic element. According to the EPA, "Coal-burning power plants are the largest human-caused source of mercury emissions to the air in the United States, accounting for over 50 percent of all domestic human-caused mercury emissions" [13].

In the case of natural gas, environmental impacts are lower than that of coal. According to the EPA, "... the burning of natural gas produces nitrogen oxides and carbon dioxides, but in lower quantities than burning coal". In fact, from a life cycle assessment standpoint, natural gas plants produce 469 metric tons of CO_2 equivalent per gigawatt-hour, less than half that of coal [11]. Additionally, the EPA states that "[e]missions of SO_2 and mercury compounds from burning natural gas are negligible". [14]. While the EPA also mentions that unburned natural gas being released either from a power plant or during transportation is a concern, though in 2009 the

EIA stated that "[i]ncreases in energy-related methane emissions [were] largely from underground coal mining", suggesting that methane releases are more prevalent in the procurement of coal than in the use and transport of natural gas [14] [15].

2.2 Nuclear

In the US, nuclear energy makes up nearly one-fifth of all electricity production [8]. According to the World Nuclear Association, "it [nuclear power] is especially suitable for large-scale, continuous electricity demands, which requires reliability (i.e. base load)" [16]. In this way, nuclear power plants are analogous to coal plants in the sense that they are very useful in meeting baseline power needs, but have trouble making quick changes in response to fluctuating peak-hour demand.

From an environmental standpoint, "nuclear power plants do not emit CO_2, SO_2, or nitrogen oxides as part of the power generation process" according to the EPA [17]. However, as with all power production facilities, the construction of nuclear power plants leads to non-negligible greenhouse gas emissions. Based off a compilation of life-cycle emissions studies gathered by the Nuclear Energy Institute (NEI), nuclear energy"s life-cycle emissions are significantly lower than those of coal or natural gas plants, and are in fact comparable to those of renewable like solar, wind, or hydro [18]. From a life cycle assessment standpoint, nuclear produces 15 metric tons of CO_2 equivalent per gigawatt-hour, which is the second-lowest behind wind, and only about 1.6% that of coal plants [11]. Much of these emissions come from the construction of plants, which are significantly larger than other electricity production facilities, and the transportation of fuel.

2.3 Solar and Wind

As shown in Figure 2, only about 6% of electricity production in the US comes from non-hydro renewable sources. Of this 6%, a majority comes from solar (in the form of both photovoltaic cells and solar thermal) and wind [8]. Of all of the available technologies, solar and wind are the most challenging to integrate into the grid due to the fact that they are considered variable sources: if the sun doesn"t shine or the wind doesn"t blow, these technologies are not actively putting energy into the grid, meaning demand has to be met by some other means [19] [20]. In order to offset this, large-scale production facilities are looking into using some form of energy storage system. This would allow excess energy collected during times of sunshine or wind gusts to be used when the sun doesn"t shine or the wind doesn"t blow [19]. However, the storage capacity of current electricity storage technology is limited. As a result, any useful amount of electricity storage is currently too expensive, which limits the viability of renewables.

From an environmental standpoint, neither solar nor wind produce carbon emissions during their operating lifetime, since their energy is derived from the sun and not from burning fossil fuels [19] [20]. However, like nuclear, solar and wind do have some emission impacts due to the processes used to manufacture materials and construct plants: wind is responsible for 14 metrics tons of CO_2-equivalent per gigawatt-hour, while solar photovoltaics are responsible for

39 metric tons of CO_2 equivalent per gigawatt-hour [11]. As with nuclear plants, many of the emissions related to renewables come from construction of production facilities. In the case of solar photovoltaics, higher life-cycle emissions are seen as a result of the emissions that result from the manufacturing process of the solar cells themselves.

2.4 Other Renewables

Solar and wind are not the only renewable sources of energy available to the US. As seen in Figure 2, nearly 6.6% of electricity production comes from hydro plants. The main issue with hydro is that most of the available sites for dams in the US have already been dammed, meaning there is little room for expansion. As a result, the Obama administration has identified a collection of dams that currently do not produce electricity, but outfitting these dams with turbines would only increase the current US hydro capacity by roughly 15% [21]. Another alternative energy source is geothermal, which utilizes naturally-occurring underground sources of heat to power turbines. While this energy is clean, it is only available on a regional basis: Hawaii and the western US have a large geothermal production potential, while areas like New England do not [22]. Since hydropower does not have a lot of potential for growth, and geothermal is regionally specific, they will not be explored as in-depth in this paper.

2.5 Environmental Impact Summary Table

Table 1 below summarizes the life-cycle environmental impact per gigawatt-hour of the main electricity production methods highlighted in this paper.

Technology	Metric Tons of CO_2 released / GWh
Coal	974
Natural Gas	469
Nuclear	15
Solar	39
Wind	14

Table 1. Life-Cycle Environmental Impact Summary of Electricity Production Technologies [Adapted from 11]

3.0 Economics in the Electricity Production Sector

On top of technological limitations and environmental concerns, economics plays a major role in determining which technologies get used for energy production. On the production side, economic factors tend to play a major role in determining which electricity production technologies are utilized by producers. On the consumer side, the electricity production technology used by electricity providers plays a role in the price of electricity sold to consumers.

3.1 Comparing Costs of Electricity Production Technologies

Due the wide range of variables present in energy production, it can be difficult to compare the cost of one electricity production technology to another. However, a useful metric, levelized cost of electricity (LCOE), lays out the cost of given technologies on a per-MWh basis, factoring in such costs as construction, fuel, operating costs, maintenance, and financing costs [23]. Figure 3 below shows the US average LCOE for plants entering service in 2019, in 2012 dollars/MWh, and does not factor in any public financial support (tax credits, subsidies, etc.).

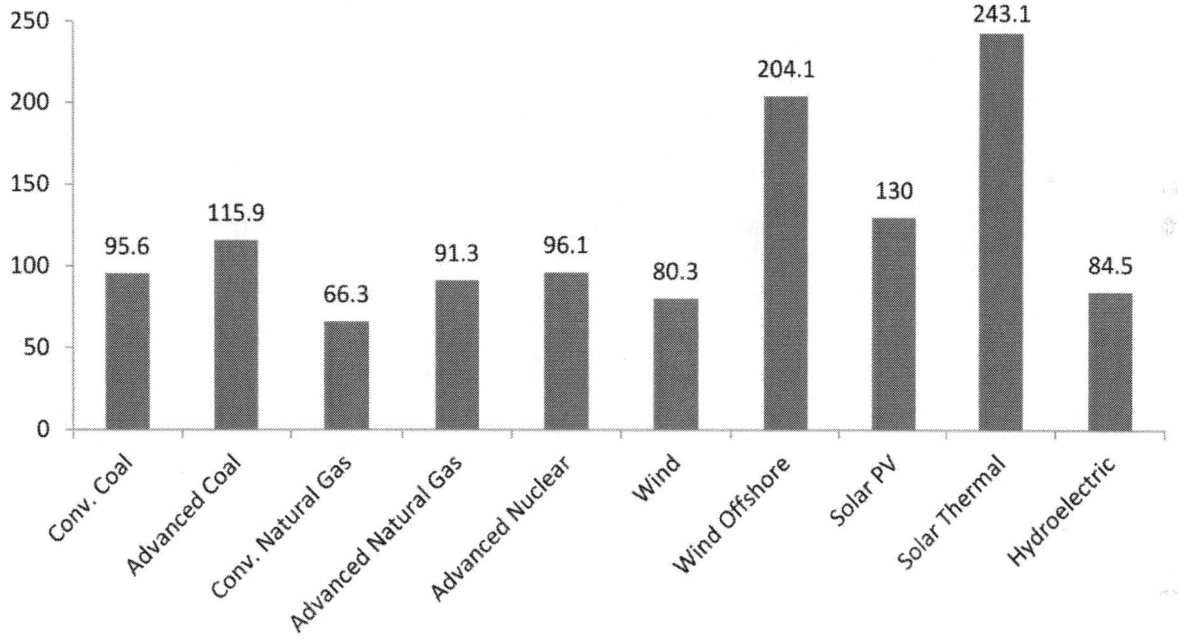

Figure 3. US Average LCOE for Plants Enetering Service in 2019, in 2012$/MWh [23]

As Figure 3 shows, the conventional energy production sources, coal and natural gas, are currently the cheapest available electricity production technologies, which explains why they currently account for such a large portion of US electricity production. Much of this can be contributed to the low price of fuel. Natural gas in particular is benefitting greatly from low fuel prices as a result of the recent American shale fracking boom. However, advanced coal and natural gas facilities, which incorporate carbon capture and sequestration technologies, are significantly more expensive than their conventional counterparts.

While coal and natural gas remain the cheapest electricity production technologies, it is important to note that there are alternatives that are already competitive. For example, the LCOE of nuclear energy is nearly the same as conventional coal and advanced natural gas, and is in fact less expensive than advanced coal. This is particularly noteworthy since nuclear fills the same base load electricity production role as coal, and produces no greenhouse gas emissions during its operational lifetime. It is also important to note that, while many of the renewable energy sources are significantly more expensive than conventional sources or nuclear, wind built on land has the second lowest LCOE, higher than only conventional natural gas.

3.2 Capacity Factor

One factor contributing to the relative expense of renewable sources, such as offshore wind or solar, is their variable nature. Since they are not able to produce electricity at all times of the day, renewable sources suffer from prolonged periods of not operating at peak capacity. The easiest way of quantifying this is through capacity factors, the ratio of actual output to the theoretical output if the plant could operate at full capacity indefinitely. A comparison of capacity factors for each electricity production technology can be seen in Figure 4 below.

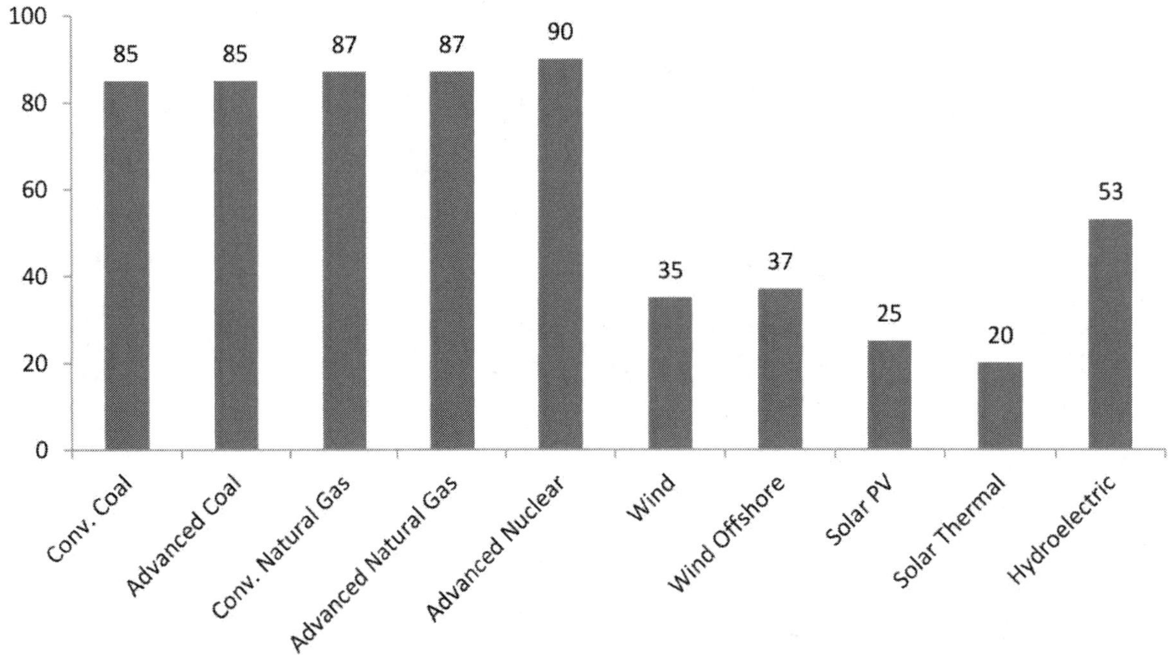

Figure 4. Capacity Factor (%) of Various Electricity Production Technologies [23]

It is important to note that there is no electricity production technology with a capacity factor of 100%: plants do not always run at maximum capacity, and require downtime for things like routine maintenance or system upgrades. However, it is easy to see that renewable sources have significantly lower capacity factors, mainly due to the fact that they cannot produce electricity at all times of the day. Because of this, capital and operating costs of renewable energy plants factor significantly more into their LCOE.

However, with advancements in technology, particularly in the energy storage field, the capacity of renewable sources could significantly increase, and possibly even become comparable with conventional sources. This would cut LCOE by anywhere from 40-70% (depending on current capacity factor). If this electricity storage is cheap enough, even the most currently expensive renewable energy technologies would be competitive with conventional sources.

4.0 Recent Policy Strategies

In recent years, legislators have used a wide variety of strategies to help shape the electricity production sector, most notably by offering some form of economic incentive to desirable technologies. These economic incentives typically come in the form of tax credits, though they may also come in the form of subsidies or guaranteed loan programs. Congress has also made efforts in the past to enact a carbon tax program, but such a program has never been able to make it out of Congress [24] [25].

4.1 Tax Credits for the Electricity Production Sector

The major economic incentive congress has implemented is by extending tax credits to electricity producers, which reduces the amount of taxes they have to pay to the federal government and thus improving their bottom line[1]. Recently, these tax credits have been targeted at renewable electricity resources, most notably solar and wind, though to a lesser extent hydroelectric and geothermal as well [26]. These tax credits have taken on two different forms: an investment tax credit and a production tax credit.

The investment tax credit allows qualifying electricity producers to receive a tax credit equal to a certain percentage of their initial cost of implementation. Solar energy installments are eligible for a 30% credit, while geothermal plants are eligible for a 10% credit. These credits have no limit. As of right now, these tax credits extend to any qualifying facility put in service before January 1, 2017. Wind power facilities put in service between 2009 and 2012 were also eligible for a 30% investment credit, but that has not been renewed [26].

The production tax credit, on the other hand, allows qualifying electricity producers to receive a tax credit based off the amount of electricity produced. As of the end of 2013, this tax credit was 2.3 cents / kWh, and extended to wind, geothermal, and hydropower facilities [5] [26]. If a facility qualifies for both the investment and production tax credit, they are only permitted to claim one of them, at their choosing. The production tax credit extended to any qualifying facility under construction by January 1, 2014, and is available for qualifying facilities for the first 10 years of operation. However, since January 1, 2014, the production tax credit for renewable electricity production facilities has not been extended [26].

4.2 Loan Programs

The federal government also offers loan programs for electricity producers. These loan programs differ from tax cuts mainly in the fact that producers are required to pay back their loan; while there is a cost to the federal government, it is only in the risk of borrowers defaulting,

[1] It is important to note that non-profits, such as municipal utilities and utility cooperatives, do not pay taxes, meaning that federal tax cuts do not have an effect on their bottom line. While alternative programs to incentivize non-profits do exist, they are not covered in the scope of this paper.

which has historically been low. Under Title XVII of the Energy Policy Act of 2005, two loan programs were outlined that are still used today: section 1703 and section 1705 [27] [28].

Section 1703 provides loans for "a new or significantly improved technology that is NOT a commercial technology" [27]. Qualifying facilities must also be located in the US. In the current state of the electricity production sector, the only facilities that really meet these requirements are related to nuclear energy; the most recent loans are an $8.3 billion loan for two new nuclear power plants in Georgia and a $2 billion loan for a uranium enrichment plant in Idaho [27]. This program is essential to nuclear power plants, as the most prohibitive cost to entering the market for nuclear plants is upfront capital costs. Due to relatively low operating costs once the plant is built, loan repayment is usually not an issue.

Section 1705, on the other hand, focuses on loan programs for renewable energy production facilities. Unlike section 1703 rules, section 1705 has a time limit, and was only available to projects that had begun construction before September 30, 2011. Since then, this project has not been renewed. Section 1705 loans tend to be lower than their section 1703 counterparts: one of the more expensive section 1705 loans was for $1.3 billion to an Oregon-based wind farm [28].

The major significant difference between section 1703 and 1705 loans is how the Credit Subsidy Cost (CSC) is handled. By law, the CSC, which is "the expected long-term liability to the Federal Government by issuing the loan guarantee", must be paid either by budget appropriations or by the borrower directly. For section 1703, which tend to have more expensive loans and thus larger CSCs, the CSC is paid by the borrower. For section 1705, the CSC was paid for by appropriations [27] [28]. In the American Recovery and Reinvestment Act of 2009, $6 billion was appropriated towards section 1705 CSCs. [29]

4.3 Carbon Tax and Cap-and-Trade Program

On top of offering economic benefits to desirable electricity production methods, Congress can also place economic burdens on undesirable electricity production methods. In the electricity production sector, this can be practically done in one of two ways: either a carbon tax or a cap-trade-program.

A carbon tax works very simply: electricity producers pay an additional tax for each unit of carbon produced during electricity production. This means that carbon-heavy electricity production facilities like coal (and to a lesser extent, natural gas) power plants would pay an additional amount for their electricity production, thus increasing their LCOE. In the same way that tax credits and subsidies seek to level the economic playing field by making renewables cheaper, a carbon tax would seek to level the playing field by increasing the cost of conventional electricity production methods. However, there has never been a carbon tax approved by either chamber of Congress.

An alternative method to a carbon tax is a cap-and-trade program, which seeks to place an artificial limit on the amount of carbon that can be emitted. Once this limit has been determined, electricity producers can buy permits to emit certain amounts of carbon dioxide. The government only sells as many permits as it takes to reach the cap, and then monitors the emissions of each producer to make sure they do not exceed their permitted limit. Producers are able to buy or sell permits to or from other producers as needed.

Cap-and-trade programs create economic incentive to reduce carbon emissions in two ways: first, reducing carbon emissions means that electricity producers need to buy fewer permits, saving them money. Second, if producers reach their permitted cap, their competitors determine the cost of additional permits (since each company now control its own excess permits, not the government). This means exceeding the permitted carbon emission limit can be disproportionately costly, either through the purchase of competitor permits or paying government fines.

In order to ensure long-term carbon emission reduction, cap-and-trade programs also incorporate a permit allowance reduction system. This means that, each year, the amount of permits allowed is reduced, effectively lowering the acceptable amount of carbon dioxide that can be emitted by electricity producers. While a cap-and-trade program did make it out of the House in 2009, the bill was never heard by the Senate, and died when that session of Congress ended [24] [25].

5.0 Projected Electricity Sector Growth: Electricity Sector Capacity Additions for 2013

In order to establish a reference case, it is useful to look at how the electricity production sector has changed recently. Figure 5 below shows the EIA projections for electricity sector capacity additions from 2008.

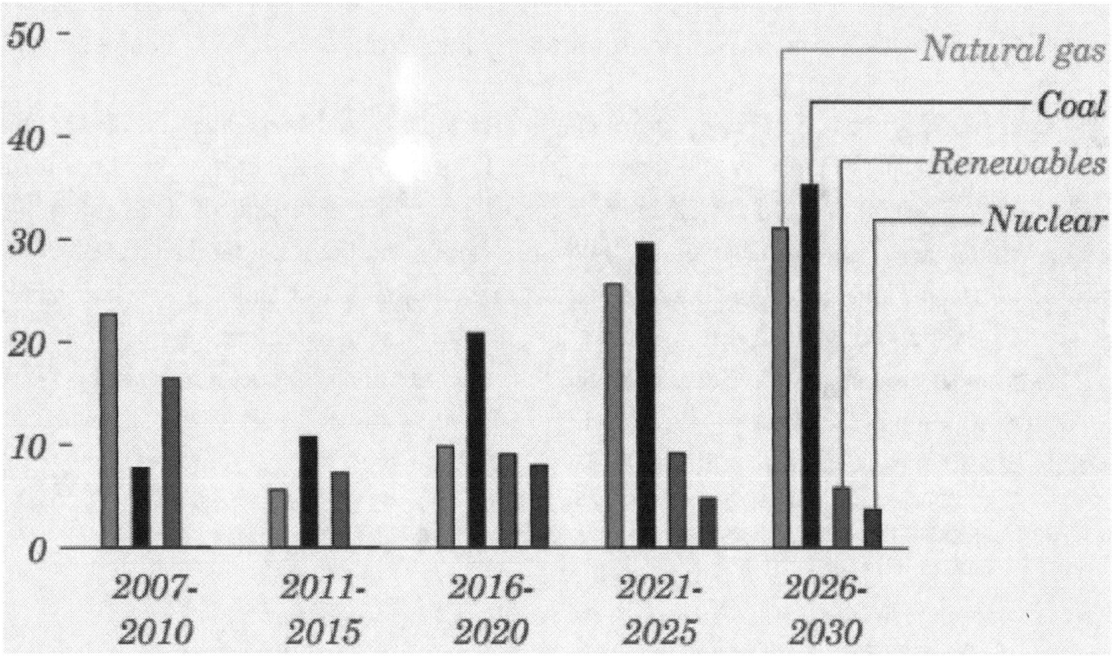

Figure 5. Electricity Generation Capacity Addition Projections by Fuel Type, 2008 (Gw) [30]

As Figure 5 shows, in 2008, it was projected that coal would begin to dominate electricity production additions starting in 2011. Much of this can be attributed to the increasing cost of natural gas, as shale fracking was still in its infancy in 2008. Additionally, renewables, which see a large spike in 2007-2010, were projected to fall by the wayside after 2010 as they begin to be dwarfed by both natural gas and coal.

However, as a result of the shale fracking boom, the reduced cost of renewables, and increased government pressure to reduce carbon emissions, these 2008 projections do not reflect what has actually happened in recent years. Figure 6 below shows the electricity generation capacity additions for 2013.

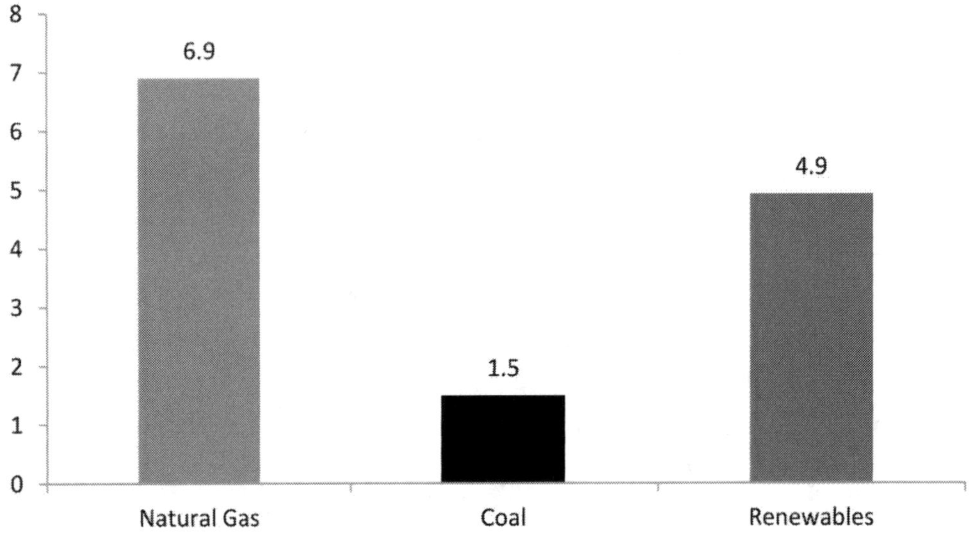

Figure 6. Electricity Generation Capacity Addition by Fuel Type 2013 (GW) [Adapted from 31]

As seen in Figure 6, natural gas dominated new capacity additions, accounting for more than half of the electricity capacity additions in 2013. Renewables made up roughly one-third of new additions, which is attributed to the completion of a number of large scale solar thermal plants in Arizona and California as well as "falling [photovoltaic] costs, aggressive state renewable portfolio standards, and continued federal investment tax credits" [31]. Coal makes up a meager 11% of new capacity additions, in stark contrast to the projections shown in Figure 5.

Long term projections have also changed to reflect the changing economic and political landscape surrounding electricity production. Figure 7 below shows the EIA projections for electricity sector capacity additions from 2014.

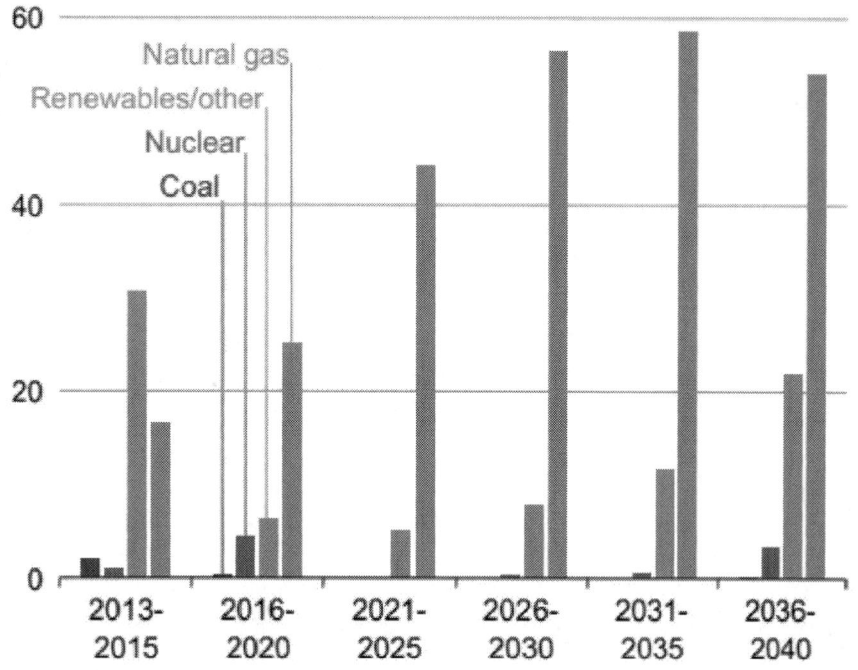

Figure 7. Electricity Generation Addition Projections by Fuel Type 2014 (GW) [32]

As Figure 7 shows, natural gas dominates new additions through 2040, mainly as a result of low gas prices. Coal additions fall virtually to 0 as a result of increased emissions regulations. Most alarmingly, the addition of renewables is actually lower than those projected in 2008 for the period between 2016 and 2035. This is mainly due to the extremely low cost of natural gas relative to any other fuel source, making it take up a larger share of the new additions in the 2014 projections than conventional sources did in the 2008 projections. It is important to note, however, that these projections only show a reference case where no additional government actions are taken.

5.1 Policy Action Effects: Overall Carbon Emissions

Despite the recent transition from coal to natural gas, as well as the increased onset of recently-cheap renewables, carbon emissions from the electricity production sector are projected to increase over the next several decades if current laws and regulations remain in effect. This is mainly attributed to the shift towards natural gas-based electricity production and the shift away from renewables and nuclear as a result of cheap natural gas for the foreseeable future [33]. However, if some form of carbon tax is implemented, carbon emissions could decrease instead increase, as seen in Figure 8 below.

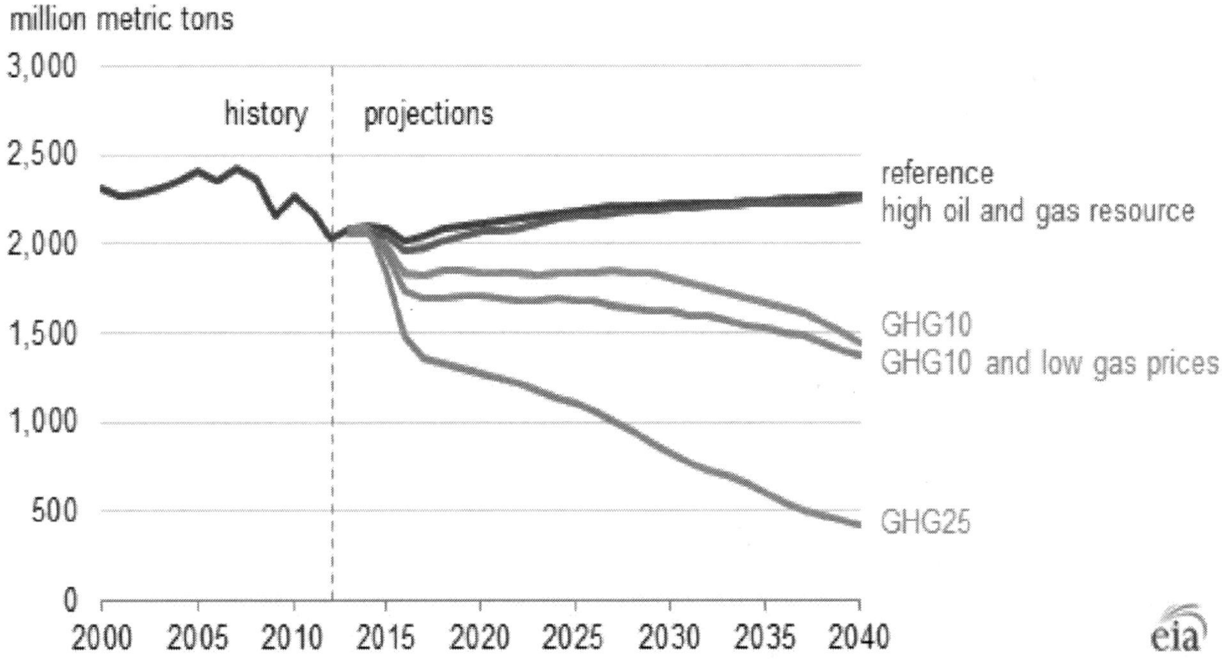

Figure 8. Electricity-Related CO_2 Emissions in Five Cases, 2000-40 [33]

Figure 8 takes into account five different potential cases. The reference case reflects no change in laws or regulations surrounding electricity production and assumes relatively constant economic factors relative to those in 2014. As seen in Figure 8, the reference case shows a 12% increase in emissions up through 2040 [33].

The GHG 10 and GHG 25 cases both refer to the implementation of a carbon tax (or equivalent cap-and-trade permit cost) in 2015 of $10 / metric ton of CO_2 or $25 / metric ton of CO_2, respectively. The GHG 10 case shows a 16% lower CO_2 emission rate in 2025 and a 36% lower CO_2 emission rate in 2040 relative to the reference case. The GHG 25 case shows an even greater decrease in CO_2 emission rates, with a 49% reduction in 2025 and an 82% reduction in 2040. In both cases, there is a large dip in emission rates before 2020, partially due to an accelerated shift away from coal and partially due to an increased implementation of renewables [33].

The High Oil and Gas Resource case assumes the same conditions as the reference case, except that oil and gas resources become more abundant than currently projected. In this case, CO_2 emissions decrease ever so slightly more than in the reference case as a result of increased shift towards natural gas, but by 2040 emission rates are roughly the same as the reference case as a result of producers expressing and increased favor towards natural gas relative to renewables [33].

Finally, the GHG 10 and Low Gas Prices case effectively combines the GHG 10 and High Oil and Gas Resources cases. In this case, reduction of CO_2 emissions before 2020 is greater than those seen in the GHG 10 case, mainly due to an even greater shift from coal to natural gas. However, because a greater share of this shift away from coal is given to natural gas and not renewables, CO_2 emissions do not decrease as drastically after 2030 compared to the

GHG 10 case; it can be extrapolated that CO_2 emissions would actually begin to be higher than the GHG 10 case sometime between 2040 and 2050 [33].

Overall, these cases suggest that some form of action needs to be taken in order to produce a marked reduction in carbon emissions from the electricity production sector. If nothing changes, CO_2 emissions will increase slightly over the next 40 years, in contrast to the significant emission reductions we"ve seen since 2005. Emissions get even worse with an increased gas resource and lower gas prices, which may result with the advancement of fracking and other extraction technologies. Based off of this projection, the only way to continue a trend of CO_2 emission reduction is to accelerate the shift away from coal and ensure that an increased share of that shift is made up of renewables or nuclear.

5.2 Policy Action Effects: Conventional Fuels

When assessing the viability of the cases listed in the previous section, it is important to understand the effects of each action on each specific electricity production method. As discussed previously, there is already a shift towards natural gas and away from coal in response to lower gas prices and increased carbon emission regulations. Figure 9 below shows how possible policy actions may augment that shift by showing the projected natural gas electricity generation in the five previous cases.

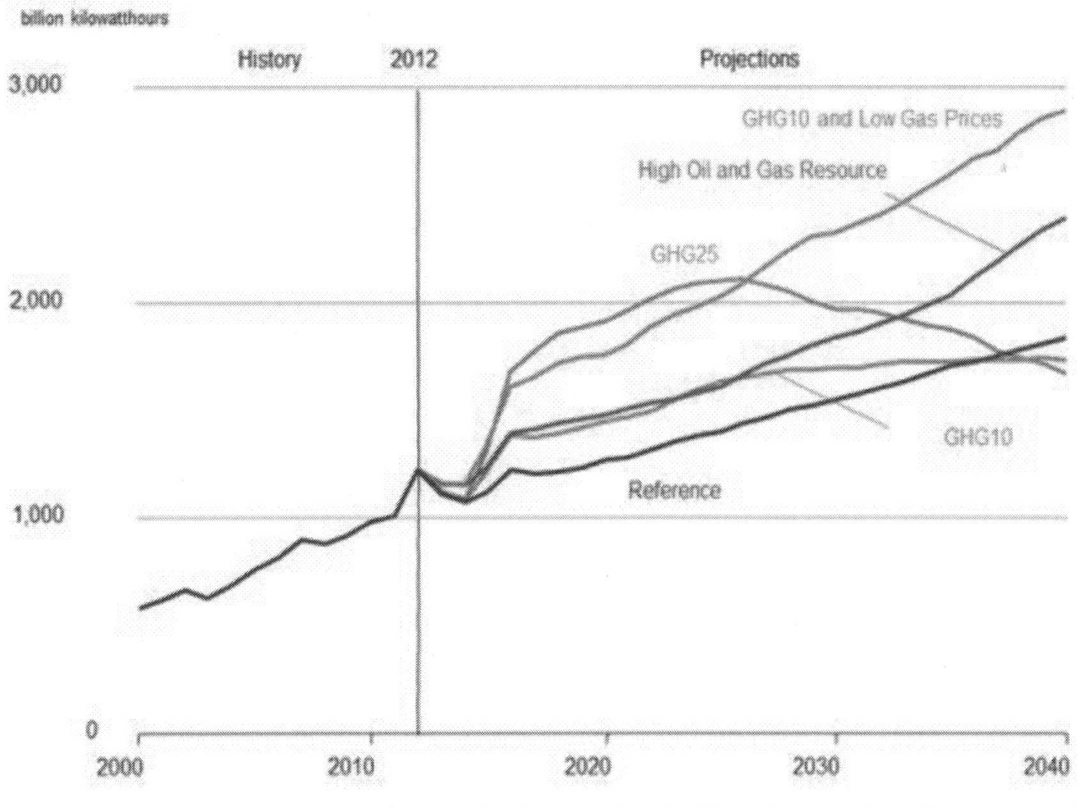

Figure 9. Natural Gas Electricity Generation in Five Cases, 2000-40 [34]

As seen in Figure 9 above, electricity production via natural gas is projected to increase over the next several decades in all cases. In the reference case, we see a relatively stable and constant increase of natural gas fired plants as they replace retiring coal plants; in 2012, 37% of electricity generation comes from coal, and in 2040 that share is 32%, marking a relatively slow shift [34]. With a high gas resource, this shift is more dramatic in the near future, and the long term rate of increase in natural gas plants is also slightly higher than the reference case.

The most interesting part of this projection, however, is the effects of a carbon tax. In all GHG cases, the shift towards natural gas in the near future is increased relative to the reference case. In the GHG 10 case, the share of electricity from coal is 28% in 2025 and 19% in 2040. For the GHG 25 case, the shift is even more drastic, with the coal share being 10% in 2025 and 1% in 2040 [34]. In both cases, the increase in natural gas production is curbed by the increased implementation of nuclear of renewables: in the GHG 10 case, natural gas electricity production levels off around 2030, while in the GHG 25 case, it actually decreases starting in 2025. However, if gas prices remain low, natural gas electricity production continues to increase through 2040, reaching the highest level of any of the five test cases.

5.3 Policy Action Effects: Renewables

Despite the fact that natural gas dominates new electricity production additions, renewables are also seeing an increase in implementation as the electricity production shifts away from coal. However, without some form of government action, the rate of renewable additions will be dwarfed by those of natural gas, leading to the increased carbon emissions seen in Figure 8. Figure 10 below shows the effects of some possible government actions by plotting the projected renewable electricity generation in four cases.

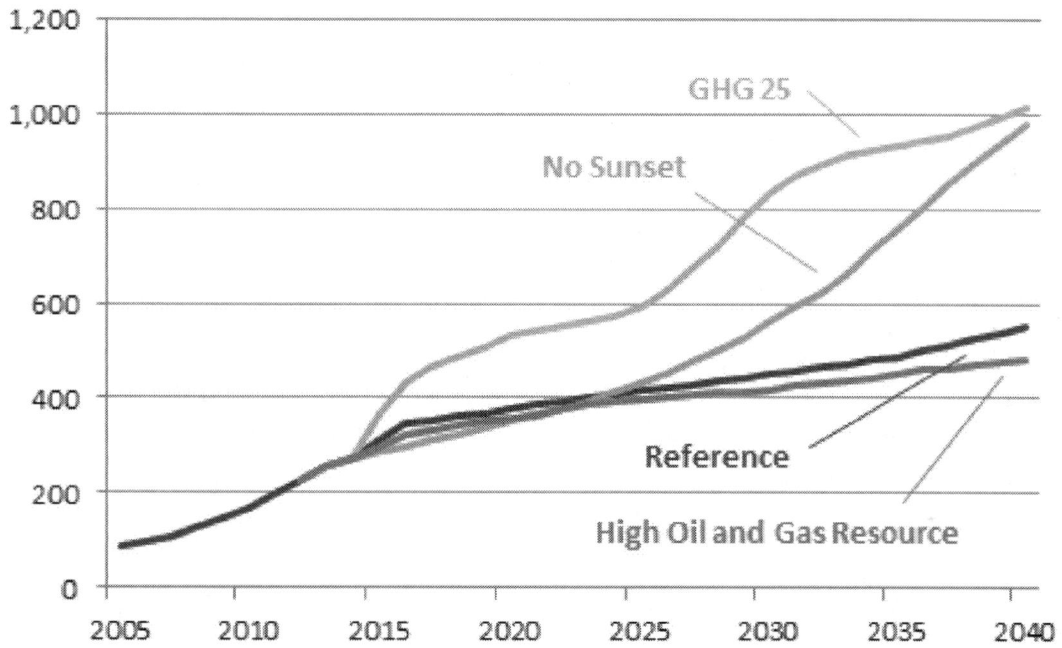

Figure 10. Renewable Electricity Generation in Four Cases, 2000-40 (billion kWh) [Adapted from 35]

In all cases, electricity generation from renewables increases over the next few decades. However, for the reference and high gas resource cases, the rate of increase in renewables is less than that seen between 2000 and 2014, mainly due to the increased implementation of natural gas power plants. The no sunset case shows the effects of extending renewable tax credits (i.e. the installment tax credit and the production tax credit) indefinitely in their current forms. In this case, the growth seen between 2000 and 2014 remains relatively constant through 2025, at which point the rate of implementation of renewables begins to increase steadily. In the GHG 25 case, a large increase is seen just after 2015 in response to the implementation of the carbon tax, after which the rate of implementation of renewables begins to level off. However, around 2025, electricity generation from renewables begins to increase dramatically once again, as more projects come into completion in response to the increased cost of natural gas power plants, whose production begins to decline as seen in Figure 9.

Compared to the reference case, both the no sunset and GHG 25 cases see disproportionate increases in solar and wind relative to other renewables like hydro or geothermal. When it comes to solar, electricity production in both cases is triple that of the reference case in 2040; for wind, it"s double. In both cases, geothermal actually sees slightly lower electricity production in 2040 compared to the reference case, and hydro sees relatively little growth between 2014 and 2040 in all cases [35].

6.0 Policy Recommendations

The electricity production sector in the US is rapidly changing as a result of the shale fracking boom. While the large shift from coal to natural gas will improve carbon emissions on a per kwh basis, the low price of natural gas is also reducing the amount of renewable sources entering the market. If no government action is taken, the growth of electricity demand will still mean an increase in carbon emissions from the electricity production sector over the next 30 years due to the low prices of natural gas. If the government truly wishes to use natural gas as a bridge fuel to a low-carbon electricity sector, a carbon tax needs to be implemented and tax credits for renewables need to be extended.

A carbon tax is necessary to appropriately reflect the societal cost of carbon-emissions-heavy technology. However, a large carbon tax is not necessary, and would cause unnecessary instability as the market acted swiftly to adjust. Instead, a low carbon tax that progressively increases is the most responsible way to implement such a change. Based off the projections in Figures 8 and 9, a carbon tax of $10 / metric ton of CO_2 would lower carbon emissions in the electricity production sector without drastically changing the economic viability of natural gas facilities. This tax would increase at a rate of 5% per year in order to incentivize investment in further carbon emission reduction technology. For a cap-and-trade program equivalent, this means the acceptable level of carbon emissions each year would be reduced, effectively increasing the cost of emissions a proportional amount.

On the other side of things, renewable technology still needs some assistance to be competitive with conventional fuels, as seen in Figure 3. However, the current split between the investment tax credit and the production tax credit makes the development of new and improved renewable technologies difficult, as it is more difficult to predict the economic environment these technologies will emerge into. A single, long-term tax strategy would offer economic stability and predictability that promotes private investment that has been yet unseen. In this way, a single, long-term tax strategy should be used to incentivize the use of all types of renewables in the electricity sector.

Of the two programs currently available, fully embracing the production tax credit makes the most sense, as it incentivizes the implementation of energy efficient and energy dense technologies[2]. It also incentivizes the development of improvements that reduce the upfront cost of these technologies, which help open them up to small scale residential and commercial consumers. The current rate of 2.3 cents per kwh is an appropriate start-point for such a tax program, but with the implementation of a carbon tax, indirect incentive towards renewables increases proportionately. The increasing nature of the carbon tax combined with the increased rate of improvement for renewable technologies actually means that the production tax credit can be slowly reduced. As such, a 5% reduction per year to mirror the carbon tax rate increase would be appropriate.

In both the carbon tax and the production tax credit, both programs need to be comprehensive and long term in order to allow the electricity producers to accurately model the economic environment that they will be entering with new projects. Due to the long-nature of construction of electricity production facilities, as well as the long operating life-time of these

[2] Again, this credit would not incentivize non-profit utility providers. Alternative programs, not covered in the scope of this paper, could be used to incentivize these utility providers.

facilities, stability in the tax structure from the government is paramount. The unexpected expiration of the production tax credit in 2013 mainly to political gridlock is unacceptable, and should be avoided in the future if at all possible.

All in all, if such a two pronged approach of a $10 / metric ton carbon tax combined with an extension of the production tax credit is used, the government will turn a net profit as a result. Based on the projections in Figures 9 and 10, there will be an average profit of $5.5 billion per year between 2015 and 2040 under such a program, with yearly profits ranging between $4.5 billion and $6.1 billion [adapted from 40 and 41]. Keep in mind that these projections only factor in the revenue from natural gas facilities; factoring in the revenue of coal should increase revenue further, despite the increase in cost from the increased use of renewables.

It is further recommended that the money made from this program be reinvested in the loan programs for electricity producers. While the reintroduction of the section 1705 program may be beneficial, it seems more beneficial to use this money to offset or eliminate the CSC cost burden on borrowers of section 1703 loans. With the accelerated retirement of coal, reduced or eliminated CSC costs would incentivize increased use of nuclear power, which acts as a more practical and carbon neutral base load power supply alternative compared to natural gas.

7.0 References

[1] Obama, B. (2014, January 28). *President Barack Obama's State of the Union Address.* Speech presented at the House Chamber, Capital Building, Washington, DC. Retrieved from http://www.whitehouse.gov

[2] EPA. (2014, June 2). *Learn About Carbon Pollution From Power Plants*. Retrieved from http:// www2.epa.gov

[3] Clement, S. (2014, June 2). A huge majority of American support regulating carbon from power plants. And they're even willing to pay for it. *The Washington Post*. Retrieved from http://www.washingtonpost.com

[4] Koch, W. (2014, June 3). EPA carbon emission limits may shift energy mix. *USA Today*. Retrieved from http://www.usatoday.com

[5] Union of Concerned Scientists. (2014, January 31). Production Tax Credit for Renewable Energy. Retrieved from http://www.ucsusa.org

[6] Davenport, Coral. (2014, May 12). Amid Pipeline and Climate Debate, Energy-Efficiency Bill Is Derailed. *New York Times*. Retrieved from http://www.nytimes.com

[7] EIA. (2014, May 7). *Annual Energy Outlook 2014*. "Market Trends: Electricity demand". Retrieved from http://www.eia.gov

[8] EIA. (2014, May 21). *Electric Power Monthly*. "Table 1.1. Net Generation by Energy Source: Total (All Sectors), 2004 – March 2014". Retrieved from http://www.eia.gov

[9] EIA. (2013, May). *State-Level Energy-Related Carbon Dioxide Emissions, 2000-2010.* "Table 2. 2010 state energy-related carbon dioxide emissions by fuel". Retrieved from http://www.eia.gov

[10] EIA. (2011, February). *Emissions of Greenhouse Gases in the United States 2009*. "Table 5. Greenhouse gases and 100-year net global warming potentials". Retrieved from http://www.eia.gov

[11] Meier, P.J. (2002, August). *Life-Cycle Assessment of Electricity Generation Systems and Applications for Climate Change Policy Analysis*. pg 70. Madison, Wisconsin: University of Wisconsin, Fusion Technology Institute. Retrieved from fti.neep.wisc.edu/pdf/fdm1181.pdf

[12] EPA. (2012, December 4). *What is Acid Rain?*. Retrieved from http://www.epa.gov

[13] EPA. (2014, March 10). *Basic Information | Mercury*. Retrieved from http://www.epa.gov

[14] EPA. (2013, September 25). *Natural Gas | Clean Energy*. Retrieved from http://www.epa.gov

[15] EIA. (2011, March 31). *Emissions of Greenhouse Gases in the US*. Retrieved from http://www.eia.gov

[16] World Nuclear Association. (2014, May). *World Energy Needs and Nuclear Power*. Retrieved from http://www.world-nuclear.org

[17] EPA. (2013, October 22). *Nuclear Energy | Clean Energy*. Retrieved from http://www.epa.gov

[18] Nuclear Energy Institute. (2014). *Life Cycle Emissions Analyses*. Retrieved from http://www.nei.org

[19] EPA. (2012, February 28). *Solar Energy | Energy and Global Climate Change in New England*. Retrieved from http://www.epa.gov

[20] EPA. (2014, May 1). *Wind Energy | Energy and Global Climate Change in New England*. Retrieved from http://www.epa.gov

[21] DOE. (2012, April 17). *Energy Department Report Finds Major Potential to Increase Clean Hydroelectric Power*. Retrieved from http://energy.gov

[22] EPA. (2014, May 1). *Geothermal Energy | Energy and Global Climate Change in New England*. Retrieved from http://www.epa.gov

[23] EIA. (2014, April). *Levelized Cost and Levelized Avoided Cost of New Generation Resources in the Annual Energy Outlook 2014*. Retrieved from http://www.eia.gov

[24] American Clean Energy and Security Act of 2009. H.R. 2454 of the 11th Congress. Retrieved from http://www.opencongress.org/bill/111-h2454/actions_votes

[25] Hulse, Carl. (2010, July 22). Democrats Call Off Climate Effort Bill. *New York Times.* Retrieved from http://www.nytimes.com

[26] 26 U.S.C. §§46, 48

[27] DOE. *Section 1703 Loan Program.* Retrieved from http://energy.gov

[28] DOE. *Section 1705 Loan Program.* Retrieved from http://energy.gov

[29] Wilson Sonsini Goodrich & Rosati. *Status Update – Department of Energy Title XVII Loan Guarantee Program.* Retrieved from http://www.wsgr.com

[30] EIA. (2008, June). *Annual Energy Outlook 2008.* "Early Capacity Additions Use Natural Gas, Coal plants Are Added Later". Retrieved from http://www.eia.gov

[31] EIA. (2014, April 8). *Half of power plant capacity additions in 2013 came from natural gas.* Retrieved from http://www.eia.gov

[32] EIA. (2014, May 7). *Annual Energy Outlook 2014.* "Most new capacity uses natural gas and renewables". Retrieved from http://www.eia.gov

[33] EIA. (2014, June 9). *Power sector CO_2 emissions sensitive to policy changes and natural gas supply.* Retrieved from http://www.eia.gov

[34] EIA. (2014, May 7). *Market Trends: Emissions.* "Energy-related carbon dioxide emissions are sensitive to potential policy changes". Retrieved from http://www.eia.gov

[35] EIA. (2014, April 29). *Renewable electricity projections show growth under alternative assumptions in AEO2014.* Retrieved from http:// www.eia.gov

392327 Life cycle assessment of vaccine supply chain in developing countries

Abstract:

People in low-and middle-income countries are most vulnerable to environmental, climate change and health impacts. It is known that strategies to reduce greenhouse gas (GHG) emissions indirectly result in wide ranging public health co-benefits. In particular, immunization supply systems in these countries are not optimized to reduce environmental impacts. For example, there are approximately 100,000 kerosene refrigerators that are in use these countries for vaccine cold chain storage resulting in annual GHG emissions of 82 million Kg CO_2 emissions. This study tries to capture the total emissions of entire vaccine supply chain from cradle to grave. This supply chain includes vaccine manufacturing and packaging, transportation, cold storage and disposal phase. For each unit process there are multiple alternative technologies with different life cycle emissions. This study derives these emissions using literature data, commercial databases such as ecoinvent and EIO methodology. The emissions are translated in climate, health and ecosystem impacts using impact assessment methods. These results could be important for immunization policy makers for expanding the green vaccine supply chain.

389663 The Study of Metal Oxide Modified Limestone As CO2 Sequestration Materals

Abstract:

Many scientists have devoted themselves to CO_2 sequestration research because of the global warming. Limestone can be used for CO_2 sequestration in cement industry and had been widely reported. But the surface "sintering" of limestone during the recycling will was dramatically decrease its efficiency. So here we use some metal oxides such as MgO or Fe_2O_3 to modify the morphology of the limestone particles thus to increase the CO_2 sequestration capability of limestone. The efficiency can be increased over 8% after adding 1% MgO particles. The mechanism and the morphology changes have been proved with SEM, optical microscope, XRD, etc.

363753 Simple Method of Deposition of CuO Nanoparticles on a Cellulose Paper and Its Antibacterial Activity

Abstract:

We report a simple in situ technique to deposit nanoparticles of copper oxide on the cellulose paper which forms a stable layer of CuO. Process of deposition has insignificant effect on the integrity of the paper. This method can be used to deposit copper oxide on any substrate provided that it is wettable and tolerates relatively high pH. Same method was applied to deposit CuO on commercial textile and polyethersulfone flat sheet. The resultant CuOpaper shows superior antibacterial activity against Gram-positive and Gram-negative bacteria. This antibacterial effect is due to release of copper ion from CuO nanoparticles on the paper.

386702 Use of a Commercial-Scale Reverse Osmosis Laboratory Module to Illustrate Interception Technology Applications in Water Pinch Network Design

Abstract:

Cost-effective water allocation within chemical processes continues to be a critical component for future sustainability of large chemical and polymer manufacturing plants. Thus, designing networks for allocating and recycling process water is a key competency for chemical engineering students. Furthermore, reverse osmosis is an example of an interception technology that can be used to purify water streams to some acceptable level prior to recycle or reuse. A commercial-scale laboratory module coupled with water pinch technology software are used in the senior chemical engineering lab at Vanderbilt University to reinforce these design concepts and to illustrate the use of laboratory generated data to scale-up to a larger industrial water interception and recycle system.

Influence of Stabilizer Size and Chelation Strength on Iron Nanoparticle Oxidation

Nikki S. Rentz

Lauren F. Greenlee

Today, an increasing number of water contaminants are being identified and measured in our water sources. In addition, as the demand on water sources grows world-wide, cities will have to increasingly rely on water recycling, water reuse, and impaired water sources. Traditional methods of water treatment are not always effective at removing smaller water contaminants, such as heavy metals and industrial chemicals, and the increasing number and concentrations of contaminants will be more and more challenging to eliminate. Iron, an abundant element found on Earth, can be used to synthesize different structured nanoparticles, including iron metal and iron oxide morphologies. In particular, iron core-iron oxide shell particles are known to be used for treatment of contaminated groundwater. Based on this interest in iron-based nanoparticles and cleaning water to allow for more reusable water, we are currently investigating the interactions between synthesized core-shell iron nanoparticles and the different organic stabilizers used during synthesis. After synthesis, these stabilizer compounds are subsequently adsorbed to the surface of the iron nanoparticles and can affect the reactivity and lifetime of the nanoparticles in an aqueous environment.

In this poster, we will present results on the water-mediated oxidation of iron nanoparticles stabilized with several types of organic compounds. We tested four different stabilizer molecules and used a quartz crystal microbalance (QCM) instrument to measure water adsorption and nanoparticle oxidation as a function of relative humidity. Relative humidity values of 10 %, 50 %, and 90 % were tested and compared with 100 % water. All measurements were performed in the presence of air, assuming equilibrium dissolved oxygen concentrations. With QCM, we investigated the mass change of each nanoparticle sample over time while the sample was exposed to changes in relative humidity. The results indicate little oxidation at 10 % and 50 % relative humidity, while at higher relative humidity values there appeared to be oxidation occurring. In addition, within the four different stabilizers tested, there was a significant relationship between nanoparticle oxidation and both the size and chelation strength of the stabilizer. However, the chelation strength of the stabilizer seems to be a more important parameter in controlling oxidation than stabilizer size.

Immobilization Carbonic Anhydrase on Magnetic Polymer Microsphere for Accelerating CO_2 Absorption into a Carbonate Solution

Zuoming Zhou, Fujun Pan, Guohua Jing*

Department of Environmental Science & Engineering, Huaqiao University, Xiamen 361021, China

Carbon dioxide (CO_2) is a greenhouse gas that has been shown to be a major contributor to global warming. The attention of many researchers has been attracted to develop efficient and low energy-consuming technologies for capturing the CO2 emitted from fossil fuel-fired power plants. Carbonic anhydrase (CA) is, currently, the most efficient catalyst for the reversible reaction of CO_2 hydration and HCO_3^- dehydration. Application of carbonic anhydrase in the field of CO_2 capture and storage technology will make absorption process with highly efficient and low energy consumption. However, enzymes generally are not stable at elevated temperatures and degrade over time. Immobilization of enzymes on solid carriers is perhaps the most commonly used strategy for improving the operational stability of biocatalysts and increasing the flexibility of reactor design.

In this paper, Magnetic polymer microspheres were used to immobilize carbonic anhydrase (CA) by covalent binding for the biomimetic carbonation reaction. The core magnetic material (Fe_3O_4) in the support induced an easy recovery of the catalyst by the use of a magnetic field. The physicochemical properties of the support were determined by Brunauer-Emmett-Teller (BET), X-ray diffraction spectroscopy (XRD), transmission electron microscopy (TEM), Fourier transform infrared (FTIR) spectroscopy, vibration sample magnetometry (VSM), and thermogravimetric analysis (TGA). The effect of various parameters, such as pH, temperature and enzyme dosage, on the properties of the immobilized CA was investigated using an uniform $U_9(9^6)$ tests. The catalytic activities for the free and immobilized CA were estimated using paranitrophenyl acetate (p-NPA) as the substrate. The optimum pH, temperature and kinetic parameters of immobilized and free CA (K_m and V_{max} values) were also evaluated. For immobilized CA, the K_m value was 6.091 mM and the V_{max} value was 0.091 μmol min^{-1} ml^{-1}, whereas for the free CA, the K_m value was 8.077 mM, and the V_{max} value was 0.027 μmol min^{-1} ml^{-1}. The thermal analysis and storage stability test results indicate that immobilized CA can retain higher activity compared with the free CA. Reusability studies suggest that immobilized

CA retained 47.6 % of its initial activity after 6 cycles of use. Besides, the hydration of CO_2 by the free and immobilized CA enzymes in a Carbonate Solution was experimentally quantified.

Optimal Design of Domestic Water-Heating Systems through Solar Collectors

Aurora de Fátima Sánchez-Bautista,[1] José Ezequiel Santibañez-Aguilar,[1] José María Ponce-Ortega,[1] Fabricio Nápoles-Rivera,[1] Medardo Serna-González[1], and Mahmoud M. El-Halwagi[2,3]

[1]*Chemical Engineering Department, Universidad Michoacana de San Nicolás de Hidalgo, Morelia, Michoacán, México, 58060*

[2]*Chemical Engineering Department, Texas A&M University, College Station, TX, USA, 77843.*

[3]*Adjunct Faculty at the Chemical and Materials Engineering Department, Faculty of Engineering, King Abdulaziz University, Jeddah, 21589. P. O. Box 80204, Saudi Arabia.*

A significant portion of the household usage of energy may be consumed to heat cold water that is used in the shower, washbasin, dishwasher, laundry, etc. To satisfy these energy demands, fossil fuels (i.e., natural gas or LP gas) are typically burned in a boiler to heat the fresh water that is stored into a cold water tank. This usage of fossil fuels in the heating application leads to greenhouse gas emissions (GHGE) that contribute to the global warming problem. The current environmental problems associated to the GHGE have enhanced the research of renewable forms of energy such as solar energy; this way, the use of solar collectors has been considered as an attractive solution to satisfy the energetic demands of hot domestic water. These devices use the available solar radiation to heat cold water and store this hot water in a recirculating tank to use it when it is required. Nonetheless, because the solar radiation is not equally available through the year, a backup boiler that burns fossil fuels is needed to compensate the energy requirements.

Moreover, there are several problems identified in its implementation, which are: the selection of the solar collector is based on standards taken from design codes developed for different countries accounting exclusively for the number of inhabitants in a home also, the solar collector is designed with a lower capacity than the one required to meet the hot water demand and the operation and design of the integrated system has not been optimized simultaneously in previous works. In this way, the optimal integrated system is an important goal to achieve, since it features the correct trade-off between the capacity of the storage tank of hot water and the solar collector that allows storing enough hot water in the periods when the solar radiation is available to use it in the periods when there is not solar radiation. In this context, this work presents a multi-objective optimization formulation for the optimal design of a water-heating system for homes. The proposed model accounts for the available solar radiation in the specific place where the solar collector is installed and the goal is to target economic and environmental objectives by optimizing the design and operating conditions including the optimal hot water storage and distribution.

The addressed problem considers a location for a house with known number of inhabitants; the problem consists of determining the optimal integrated solar collector-boiler-storage system to satisfy the hot water demands while accounting for economic and environmental aspects. The superstructure considers a set of human activities U. These human activities require specific amounts of hot water at different temperatures. Additionally, the hot water flow rate for each activity is different and it depends on the period of the day and the season of the year when the flow rate is required.

The economic objective function considers the minimization for the total annual cost including the annualized capital costs of the solar collector, the isolated tank to store hot water, the boiler and new pipes, in addition to the operating costs associated with fossil fuels and pumping, minus the tax credits obtained for the reduction of GHGE. Additionally, the environmental objective function accounts for the minimization of GHGE.

$$
\begin{aligned}
TAC = {} & K_F\left(CF^{SC} + CV^{SC}(A^{SC})^{\alpha^{SC}}\right) + \sum_{m \in MONTH} \sum_{t \in TIME} D_m Q_t^{Boil} CV^{Fuel} \\
& + K_F\left(CF^{Boil} + CV^{Boil}(CAP^{Boil})^{\alpha^{Boil}}\right) \\
& + K_F\left(CF^{Tank} + CV^{Tank}(V^{THW})^{\alpha^{Tank}}\right) - TAXCREDIT(TONGHGED)
\end{aligned}
$$

$$
TONGHGED = \sum_t \sum_m \frac{Q_{t,m}^{SC}}{PC^{fuel}} (FG) D_m
$$

The proposed model was applied to several case studies to demonstrate the proposed methodology. These case studies are for different cities of Mexico (i.e., Morelia, Chihuahua, Mexico City, Monterrey and Guadalajara) and include different number of inhabitants; however, the proposed model is general and this can be used for solving different cases. **Figure 1** shows the general representation of the case studies. Each case study considers the solar radiation for the corresponding city, for each hour during an average day per each month.

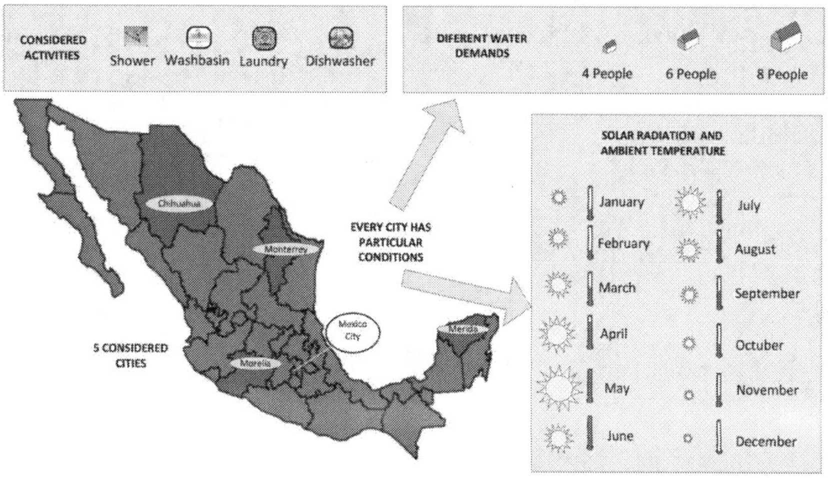

Figure 1. Locations of the different cities considered in the scenarios analyzed.

Figure 2 shows that the percentage of reduction of tons of GHGE is significant for all the scenarios considered using the proposed scheme; this is because the large amount of heat provided by the solar collector in each case analyzed. On the other hand, **Figure 3** shows the contribution to the total annual cost for the different elements associated to the heating water system; it is important to note that the major contribution to the cost is the one associated to fossil fuels; however, this proportion is not constant for the different number of inhabitants in the houses as well as in the different places analyzed. Furthermore, the annualized cost for the boiler is the smallest one, but the one associated to the solar collector is the second greatest contribution to the total cost. This way, the cost associated to the solar collector represents a significant initial investment (in addition to the cost associated to the capital cost for the hot water storage tank). However, the implementation of the solar collector is reflected in the reduction of fossil fuels as well as the decrement in the GHGE.

In addition, the solution of the optimization formulation identifies the optimal design and operation of the system. The optimization formulation was applied to several cases at various Mexican cities with different number of inhabitants. Finally, results show that the selected solar collector does not follow the standard heuristic rules because the location and the demand patterns of the inhabitants are important factors that need to be considered in the optimal design.

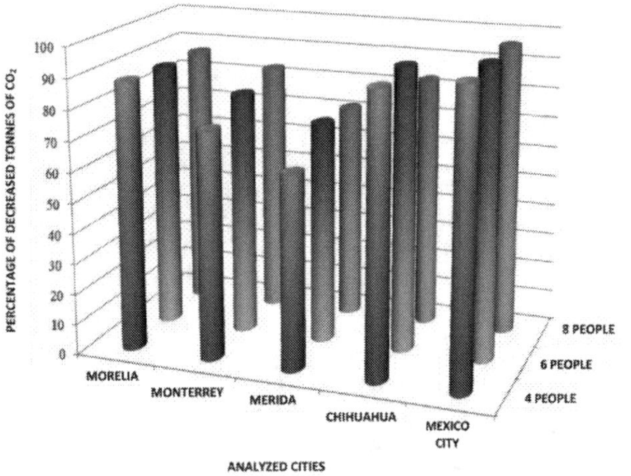

Figure 2. Percentage of GHGE decreased due to the use of the solar collector.

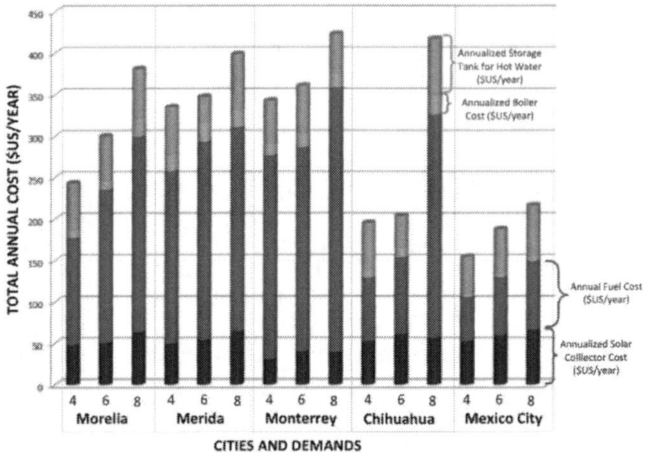

Figure 3. Individual contribution to the total annual cost for different locations and number of inhabitants.

372382 Simultaneous Design of Water Reusing and Rainwater Harvesting Systems in a Residential Complex

Abstract:

Recently, the available water for human consumption has decreased and at the same time the water demands have increased. This way, the problems associated to satisfy the water demands have increased. Therefore, this paper proposes an optimization model for satisfying the water demands in a housing complex involving rainwater harvesting, storage and distribution as well as the simultaneous design of water networks into the housing complex for recycling, reusing, regenerating and storing reused water. Several strategies such as mixing, segregating and storing reused water have been considered. The optimization model accounts for mass and component balances for the different units considered and the problem is considered as a multi-objective optimization, where one objective is the minimization for the fresh water consumption and the other objective is the minimization of the total annual cost accounting for the capital and operating costs for the units involved in the entire horizon. The proposed model accounts for the variability in the water demands through the different hours of the day and for the different seasons of the year. Furthermore, the seasonal dependence for the rainwater has also been considered in the optimization model. A case study of the city of Morelia in Mexico is presented; in this case study the residential complex of Villas del Pedregal was considered. Results show that significant reductions in the total fresh water consumption were obtained with the proposed methodology, and also that the total cost was decreased because the reduction in the cost associated for the fresh water. Also, the results show that the simultaneous implementation for the rainwater harvesting and the recycle and reuse network allows a better solution than the case when only one of these strategies is implemented.

Mineralization of CO_2 with industrial solid waste and natural ore

Liang Bin[1,2*], Zhu Jiahua[2], Li Chun[2], Yue Hairong[2], Lu Houfang[2], Ye longpo[2], Wang Chao[2], Wang Yufei[1], Xie Heping[1]

1 Multi-phases Mass Transfer and Reaction Engineering Laboratory, College of Chemical Engineering, Sichuan University, Chengdu 610065, China;
2 Center of CCUS and CO_2 Mineralization and Utilization, Sichuan University, Chengdu 610065, China

*Corresponding author: liangbin@scu.edu.cn

In past centuries, utilization of carbon-rich fossil fuels, coal, oil and natural gas, has led to the amount of CO_2 in the atmosphere rising from 22 billion tons before the industrial revolution to 31.6 billion tons in 2013 (IEA, 2013). The increase in CO_2 emissions arguably contributes to the warning climate due to ''greenhouse effect'', which resulted in the sea level rise of 38cm, and global temperature rise of 1.9°C.

CO_2 Capture Storage (CCS) was intensively investigated and locally practiced over world. The risks induced by CCS, like the geometric risk, underground water contamination as well as leakage from storage sites, were occasionally argued in scientists. Utilization is a promising alternative routine to cut CO_2 emmision while using CO_2 as a resources. Thermodynamically, CCU uses CO_2 in a routine along with the Gibbs Free Energy (GFE) reduction. CCU technologies explored and demostrated in this work are the mineralization of CO_2 with solid waste like phosphogypsum and $CaCl_2$, and natural ore K-feldspar. They are more atractive because they convert CO_2 into useful products.

Mineralization with Phosphogypsum (PG, 70%～80% $CaSO_4.2H_2O$) in ammonum mediam converts CO_2 to calcium carbonate and produces ammonium sulfate. Phosphogypsum is a by-product in the phosphouric acid production, it is deposed as solid waste in a rate of about 50 million tons per year in China. The process simultaneously captures CO_2 from flue gas, eliminate the PG waste and produce fertilizer. The calcium carbonate can be used as the cement feeding stock, which reduce the mining of lime stone. The reaction is writen as:

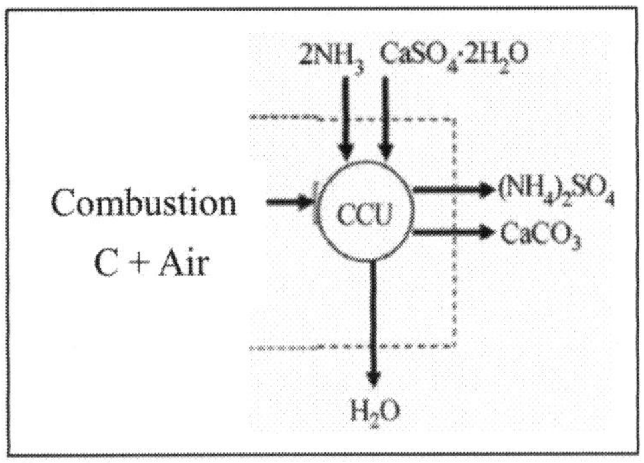

$$C + [O_2 + 3.761N_2]_{mix} + CaSO_4{\cdot}2H_2O + 2NH_3 \rightarrow CaCO_3{\downarrow}(s) + (NH_4)_2SO_4 + H_2O + 3.761N_2$$

This process has a GFE change of -146 kJ/mol C, it is a spontaneous process.There is a pilot device having been constructed in Sichuan, China. It converted 100 Nm^3/h flue gas and have been operated. The feeding-product balance for a commercial device is listed in following table:

Comsumption		Products	
CO_2, t/a	8×10^4	$(NH_4)_2SO_4$, t/a	2.4×10^5
Gypsum, t/a	3.127×10^5	$CaCO_3$, t/a	1.818×10^5
NH_3, t/a	6.18×10^4		

Another alternative CCUS (carbon capture, utilization and storage) method was proposed to fix CO_2 in the process of potassium extraction from insoluble K-feldspar. K-feldspar is an insoluble mineral of potassium and the potassium cannot be directly absorbed by plants. Many efforts have been made to produce potassium fertilizer from K-feldspar. In the conventional processes for the use of K-feldspar, the ore was roasted at an elevated temperature (e.g., >1300°C) to convert K components into soluble salts. The roasted ore was then extracted with hot water for the leach of potassium. However, these are energy consumption processes and not economically practicable.

A integrated technology by activating K-feldspar with calcium solid waste was investigated to leach potasium and mineralize CO_2. Activating K-feldspar with calcium chloride hexahydrate at 800-850°C, calcium ions exchanges with K^+ to release K^+ from alumina-silicate network, which make the potasium solvable. The calcium alumina-silicate was used to mineralize CO_2 after leaching potasium. The additive, calcium chloride, is a waste deposited in soda-ash production of Solvay method. It is a large environmental challenge for more than ten million tons of deposit per year. Therefore, this new process is also environmental benign due to the reduction of $CaCl_2$.

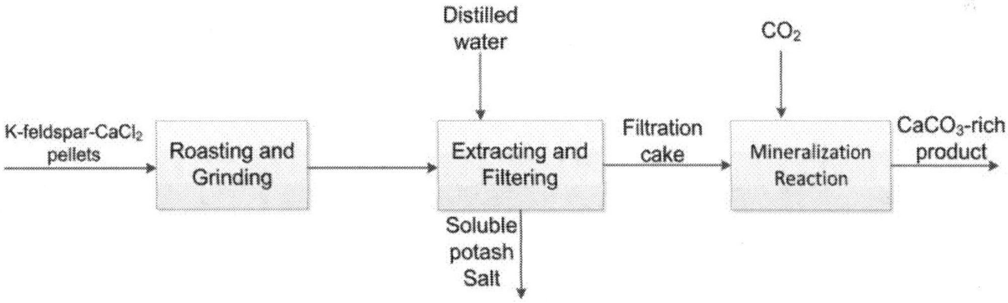

The technology was also applied by using PG as activation agent. A novel CO_2 mineralization approach using natural K-feldspar and phosphogypsum was developed to solidify CO_2, process phosphogypsum waste and produce K_2SO_4 simultaneously. K-feldspar was activated with $CaSO_4$ under 1200° and then mineralized with CO_2 to extact K in a hydrothermal condition. A K extraction ratio of ~87% and CO_2 mineralization ratio of ~7.7% (in K-feldspar weight) were obtained. In the high temperature reaction, the K^+ ions exchange with the Ca^{2+} ions for the formation of soluble salts, while the Ca^{2+} is converted to carbonates in the following hydrothermal mineralization process.

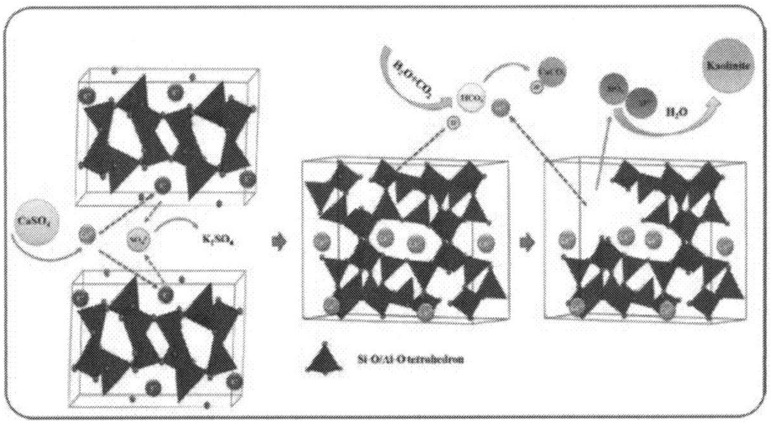

Summery

Chemical utilization of CO_2 and mineralization utilization was explored and practiced. It is more attractive because its thermodynamic favorite.

373363 Optimization of Electrode Design for Electrodialysis Reversal

Abstract:

The main advantages of electrodialysis reversal (EDR) technology are high water recovery and great resistance to scaling and fouling. These attractive characteristics make EDR particularly promising in a world of increasing water shortage where communities turn to treating impaired groundwater sources to supplement their supply. However, the relatively high cost of EDR today hinders widespread deployment of this technology, motivating efforts to improve its efficiency. This work targets the electrode, one of the principal components of an EDR device which provides the driving force for desalination.
The shape and functionality of electrodes impacts performance and stack life as well as the operation cost associated with the current utilization and efficiency. The geometric relationship between electrode and membrane also determines the distribution of current density throughout the surface of the electrode and membranes, which in turn dictates the effective limiting current for the membranes. The goal of this work is to develop methods of evaluating current distribution in the stack and compare the optimum electrode design.

375392 A Study on Effect of Temperature, COD and Influent Tds on Microbial Desalination Cells' Performance with an Approach to a Unique Predictive Model

Abstract:

For several years it has been demonstrated that microorganisms existing in the nature can break down organic substrates in order to generate electricity directly. However, controlling the generated electricity was just a theory for a long time. It has been barely two decades ago since scientists could find methods to overcome this challenge and academics became interested in Microbial Fuel Cells (MFC) which can convert the energy within a substrate to electrical energy by getting help from bacteria. Several ways of electron transfer have been discovered until now and there are many future applications available for MFCs. One of the newly revealed approaches is to desalinate water while treating wastewater and producing energy at the same time. This technology is called Microbial Desalination Cells (MDC). MDCs are basically MFCs with an additional chamber for feed water to be desalinated. MDCs are taking their very first steps into technological world and there are several obstacles on the way to bring them into industrial scale.

During this study a new approach for MDC technology is provided for desalination of brackish ground water. An artificial wastewater is used in the anode, and a solution containing potassium hexacyanoferrate is used as the catholyte in our reactors. Different principal parameters such as temperature, COD, and feed water TDS are studied thoroughly in order to achieve a general model for behavior of Microbial Desalination Cells.

377175 Controlled-Release Antimicrobial for Air Disinfection

Abstract:

Controlled-Release Antimicrobial for Air Disinfection

Yue Tak LAI[1], Hao Chen[1,3], Yan LI[2], Wei HAN[2] and King Lun YEUNG[1,2,*]

[1]Division of Environment, [2]Department of Chemical and Biomolecular Engineering,

[3]Nano Science and Technology Program, The Hong Kong University of Science and Technology, Clear Water Bay, Kowloon, Hong Kong SAR

*Corresponding author: kekyeung@ust.hk

Division: 09 Environmental Division

Subdivision: 09001

Poster Session

Bioaerosol droplet with pathogenic microorganisms is the main transmission route for the recent worldwide pandemic outbreaks, such as 2002 Severe Acute Respiratory Syndrome (SARS) and 2009 flu pandemic. In the form of bioaerosols, the microorganisms spread with the aid of air circulation by indoor air conditioning systems. Upon filtration, the bioaerosols are trapped by air filter media while remain viable. During microbial growth, the micoorganisms penetrate the filter and contribute to secondary contamination through the release of microbial VOCs and spores. Current technologies that address this problem such as electrostatic filtration, ultraviolet germicidal irradiation and ozone disinfection are often expensive and energy-intensive.

This work reports on a cost-effective antimicrobial coating which exhibits "release killing" by controlled-release of gaseous biocide (i.e., chlorine dioxide). The coating exhibits "anti-adhesion" property by using block copolymer and manipulating surface charge with various metal oxides. The encapsulating shell is also decorated with a catalytic-dyad to perform "contact-killing" activity. The antimicrobial coating was uniformly applied onto particulate air filters (HVAC and HEPA filters) by spray-deposition without altering the filtration properties.

The control release of volatile biocide remains long-term stable at elevated temperatures, and can maintain more than 2 log reduction in viable bacteria upon 10 minutes contact. Acceleration test shows the filter with antimicrobial coating has an operation lifetime for at least 1 year. Spectroscopic evidence and biological assays indicate the primary cause of cell inactivation is predominantly cell membrane peroxidation via ROS route. Catalytic process between the volatile biocide and the antimicrobial metal is believed to contribute to ROS generation.

381627 Impact of Fugitive Emissions on the Greenhouse Gas Emissions of Conventional Crudes

Abstract:

Fugitive emissions are unintended gases released to the atmosphere that can come from a variety of sources. These sources included equipment leaks, process venting, evaporation losses, venting, flaring, accidents and equipment failures. By nature they are variable because they depend on controlled and uncontrolled parameters that are specific to a particular process, which makes their estimation difficult. This study quantifies the fugitive emissions from ten different oil fields across the world to evaluate their impact on the well-to-wheel (WTW) greenhouse gas (GHG) emissions. Previous studies on these reservoirs either used an estimation of the associated gas loss at 0.5% or didn't consider them at all. To estimate fugitives, it was assumed a portion of the associated gas from the crude oil is lost to the atmosphere. The percentage gas loss was estimated from previous studies that determined the portion of natural gas released as fugitive emissions from oil and natural gas fields. The associated gas amount for each oil field varies as seen by the gas-to-oil ratio (GOR), which range from 366 scf/bbl-crude to 12,218 scf/bbl-crude for the Maya and Alaska North Slope oil fields respectively. Fugitive gas amounts were found by multiplying by the percentage of methane in the gas (specific to each location), which was converted to carbon dioxide equivalents to give the GHG emissions. These fugitive emissions were added to previous work by this group that calculated the WTW GHG emissions of these reservoirs. Results show that fugitive emissions increase with increasing GOR. For reservoirs with a high GOR, emissions are not negligible and should be considered in the life-cycle assessments of these conventional crudes. The sensitivity analysis indicates that the fugitive GHG emissions increase by 50% with just a one percent increase in the percentage of associated gas lost.

377180 Mini Pulsed Electric Field Device for Drinking Water Disinfection

Abstract:

Mini Pulsed Electric Field Device for Drinking Water Disinfection

Ka Wo LAM[1], CHEN Hao[2,4], Pik Shuen Karen HUNG[1,2], Oi Wa LEE[2], Siu Ming KWAN[2], Joseph Kai Cho KWAN[1,3], King Lun YEUNG[1,2,*]

[1]Division of Environment, [2]Department of Chemical and Biomolecular Engineering, [3]Health, Safety and Environment Office, [4]Nano Science and Technology Program, The Hong Kong University of Science and Technology, Clear Water Bay, Kowloon, Hong Kong, P.R. China.

Division: 09 Environmental Division

Subdivision: 09001

Poster Session

Abstract:

Clean drinking water supply is considered to be one of the most important milestones of public health. However, the microbial contamination of building water supply system is very common. The World Health Organization reported that diarrheal diseases from unsafe water kill 1.5 million children and causes two billion cases of illness each year worldwide.

Chlorination, ozonation, UV exposure and reverse osmosis are widely used in drinking water disinfection nowadays. Chlorination is also classified as the best drinking water disinfection technology. However, these technologies disinfect the drinking water accompanying with altering water quality. The discovery of Legionella in the water from the taps of Hong Kong's brand new government headquarters illustrates the gravity of the situation and its potential threat to public health.

Pulsed electric field disinfection is an effective, environmentally benign and safe technology for point-of-use disinfection of tap water based on the poration of microbial cell wall under electric field. The designed device is compact and portable. Low input voltage is applied to the system and it is able to attain 99.9 percent bacteria reduction in running tap water. As a purely physical disinfection method, pulsed electric field disinfection doesn't involve chemical addition, thus without altering the water quality. It is believed that this device is not only beneficial to the people in modern cities, but key to control the spread of water-borne diseases in most of the developing countries.

377203 Antimicrobial and Anti-Adhesion Coating for Water Filtration Membrane

Abstract:

Antimicrobial and Anti-adhesion Coating for Water Filtration Membrane

Hiu Pang Yu[1], Hao CHEN[1,3], King Lun Yeung[1,2,*]

[1]Department of Chemical and Biomolecular Engineering, [2]Division of Environment, [3]Nano Science and Technology Program, The Hong Kong University of Science and Technology, Clear Water Bay, Kowloon, Hong Kong, P.R. China

*Corresponding author: kekyeung@ust.hk

Division: 09 Environmental Division

Subdivision: 09001

Format: Poster

Abstract:

Water treatment by membrane filtration is one of the most reliable treatment process in water purification. However, microbial biofouling is one of the major obstacle for high permeability and high purity over a long-time operation. Microbial biofouling on the water filtration membrane is a main risk of public health because of the recontamination by pathogenic bacteria, viruses and fungi in the biofouling.

This works provide a solution to microbes growing by developing a anti-microbial and anti-adhesion coating which can reduce bacteria survive in filtration setup and prevent adhesion of microbes.

Quaternary ammonium cation is known to have a good bactericidal ability and zwitterionic compound is a good surfactant for anti-adhesion. Here 2-(diethylamino)ethylmethacrylate and [3-(methacryloylamino)propyl] dimethyl(3-sulfopropyl)ammonium hydroxide inner salt were used as antimicrobial and anti-adhesion agents, respectively. Poly (ethylene glycol) diacrylate (PEGDA) was chosen to form a strong cross-linked copolymer coating with the antimicrobial and anti-adhesion agents on the water filtration membrane. The membrane with antimicrobial and anti-adhesion coating is challenged with a high concentration of *E.coli* in water. High disinfection ability was obtained in liquid disinfection test. The coating was characterized by FTIR and XPS elementary analysis. SEM was applied for biofouling observation. The result shows that structure of the water filter membrane is beneficial to the biofilm formation, and the membrane with anti-adhesion coating can reduce the biofouling by the charged surface from zwitterionic polymer coating

389849 Well to Wheel Life Cycle Assessment of Greenhouse Gas Emissions of Transportation Fuels from Canadian Oil Sands

Abstract:

A significant portion of the U.S. imports of petroleum crude oil comes from Canada. Production from Canada's oil sands industry as well as the crude exports to the U.S. has been increasing substantially. Increasing environmental awareness to reduce greenhouse gas (GHG) emissions and policies such as such as low carbon fuel standard (LCFS) and European fuel quality directive demand appropriate quantification of GHG emissions in oil sands in comparison to conventional crudes.

This paper aims at an appropriate quantification of the life – cycle greenhouse gas (GHG) emissions for transportation fuels obtained from oil sands products- bitumen, dilbit and synthetic crude oil (SCO). This is achieved through development of theoretical engineering models to estimate the energy consumption in each unit operation – surface mining, steam assisted gravity drainage (SAGD), transportation, and upgrading in oil sands. This model has the potential to estimate energy consumption and GHG emissions for any specific project in oil sands based on the technical and operational parameters involved. The uniqueness of this model lies in its ability to predict the sensitivity of key project parameters and reservoir properties on energy consumption and GHG emissions. The results from this model have been validated with the results from published literature. This paper presents the estimated values of well to wheel life – cycle GHG emissions and contribution of each unit operation of transportation fuels, namely gasoline and diesel. The GHG emissions for transportation fuels range between 94.6 to 114.1 g/MJ of diesel and 102.2 to 116 g/MJ of gasoline, depending on the pathway of bitumen in oil sands. Vehicle emissions ranging from 69% to 79% are a major constituent of GHG emissions of transportation fuels. To address the needs of evolving oil sands industry in Canada, this project identifies GHG intensive areas and the opportunities for emissions reduction. This research adds to the knowledge base for comparing the GHG intensity of oil sands products to the conventional crudes imported in the U.S.

Title: Advanced Ozone Membrane Reactor for Treatment of Endocrine Disrupting Compounds in Water

Authors:
Tiphaine CORBET[1], Yakub FAM[1], Liping LI[2], Ying LI[3], Rafael SERRACUESTA[1], Wei HAN[1] and King Lun YEUNG[1,2,3,*]
[1]Department of Chemical and Biomolecular Engineering
[2]Division of Environment
[3]Environmental Engineering Program
The Hong Kong University of Science and Technology (HKUST)
Corresponding author: kekyeung@ust.hk

Session: 09 ENVIRONMENTAL DIVISION
Topic: 09001 Poster Session: Environmental Division

Abstract text:
There is growing evidence that a large number of chemical compounds found in common household products ranging from pharmaceuticals, cosmetics, personal care products and detergents, contaminate surface and ground water and can survive in common wastewater treatment process Many of these compounds are endocrine disruptors that interfere with the normal function(s) of the endocrine system that regulates growth and development, reproduction, the immune system and neurobehavior. An advanced ozone membrane (AOM) reactor that combined ozone oxidation, adsorption and membrane separation was designed for the treatment of endocrine disrupting compounds (EDCs) in water. The compact treatment unit uses membranes as ozone distributor, reaction contactor and water separator, and can achieve total conversion of EDC and up to 40% of TOC removal. This work investigates the development of an efficient catalyst, to be coated on the contactor membrane, for faster and more complete ozone treatment of EDCs and its degradation in water. It is expected to remove potentially toxic EDC, shorten treatment time, increase clean water production and allow the design of a compact treatment unit.

Carbamazepine, caffeine, diclofenac sodium, potassium hydrogen phthalate, bis(2-ethylhexyl) phthalate and diethyl phthalate were chosen as model EDCs. The catalysts, pure □-alumina, □-alumina supported metal oxide, pure activated carbon, activated carbon supported iron oxide and zero valent iron, were prepared, characterized and tested for EDC treatment performance in the following experiments: ozonation alone, adsorption alone, catalytic ozonation and ozonation followed by adsorption. The samples were collected during the experiment and analyzed by UV-Vis spectroscopy, high-performance liquid chromatography and TOC analyzer to determine EDC and total organic compound concentration. Metal leaching is measured by inductively coupled plasma.

Semi-batch reactions indicated that the pure □-alumina and □-alumina supported metals (Cu, Co, Mn) showed low catalytic activities and the removal process is mainly by adsorption and homogeneous reaction catalyzed by leached metals. EDCs were completely removed in a few minutes when activated carbon was used. The TOC removal by ozonation alone can only achieved 20% in 1 h while 70% of TOC can be removed in 5 minutes when activated carbon (0.2 g/L) was used. The TOC removal by catalytic ozonation was higher than adsorption after ozonation, indicating a catalytic effect. Activated carbon supporting zero valent iron was more active than activated carbon supporting iron oxide and raw activated carbon.

Based on the above-mentioned experimental results, the mechanism of the reaction is proposed and a mathematical model is developed to predict the performances of the membrane reactor. In addition, the effects of water matrix for instance, humic acid and mineral ions (Ca^{2+}, Na^{+}, SO_4^{2-}) are taken into consideration in this study. More importantly, bioassays such as ER-CALUX, YES are employed to monitor the (anti-) estrogenic and (anti-) androgenic effect of EDCs and their by-products during the treatment process.

Greenhouse Gas Calculator for Tracking Combustion and Process Upset Emissions: Methodology and Visual Representation for Simple Ethylene Process Base Case

Fahd M. Mohammed, Kazi Monzure Khoda, Fadwa T. Eljack*
Department of Chemical Engineering, College of Engineering,
Qatar University, P.O. Box-2713, Qatar
Email: Fadwa.Eljack@qu.edu.qa

Flaring is recognized as a common worldwide environmental issue that contributes to global warming. Flare emissions contribute to more than 50% of the annual Certified Emissions Reductions (624 million tons) according to Kyoto's Clean Development Mechanisms[1]. A simple Ethylene process base case is studied and simulated to account for GHG emissions through common abnormal situations within the process. The GHG calculator is developed as a part of a project that investigates and develops mitigation methods and approaches to combat emissions during process upsets. The focus of this work is to illustrate the a mass balance based methodology to count GHG emissions from fuel combustion and process upset flaring that follows international guidelines such as the EU commission for monitoring and reporting GHG emissions and Intergovernmental Panel on Climate Change (IPCC) emissions factors. In addition, it uses international standards such as Deutsches Institut für Normung to evaluate wasted energy content of process flared streams depending on mixture Wobbe Index. Finally, In this work, we will show the estimation of CO2 emissions cost according to carbon tax systems in various focal countries through different time periods.

[1] The world bank - Russia, Kazakhstan Lead Way to Reduce Gas Flaring and Lower Emissions http://web.worldbank.org/WBSITE/EXTERNAL/TOPICS/EXTSDNET/0,,contentMDK:22949562~menuPK:64885113~pagePK:7278667~piPK:64911824~theSitePK:5929282,00.html

380230 Tower Gardens and Solar Technology

Abstract:

Tower Gardens and Solar Technology

Description

The complete development and components of a solar system will be described in favor of a tower garden. Tower gardens are vertical aeroponic growing system designed to allow the cultivation of plants in restricted places. This system will be completely powered by a renewable source: sunlight.

Characteristics of a Tower Garden

- It requires very minimal space of 3 ft. diameter.
- It utilizes aeroponics which is more effective, cleaner and requires less work compared with traditional farming.
- It uses 90% less of water and land than are used in traditional farming.
- It recycles 100% of its nutrients and water.
- It has been proven to have 30% higher yields of product and faster maturation rates.

UAB's Research Involvement

- UAB's Community Gardens: comparison between 6 traditional raised beds and 6 tower garden systems (variables monitored: time to maturation, weight, size, yield, produce quality, water usage per lb of produce, weather impact, resistance to bacteria and insects).
- Aldridge Botanical Gardens: observation of the performance of tower garden systems in the natural ecosystem (tower garden systems placed near the beehives to observe their influence on bees activity)
- Gadsden State Community College: experimental research aiming to determine the key elements crucial for vertical farming

Examples of places where the invention can be used:

- Hospitals
- Power irrigation for crops
- Places where an electric installation is too expensive.
- Invention will increase the number of points that can be achieved under the LEED code: Renewable Energy.

Tower Garden and Solar Panels

Solar panels will be installed to provide electricity for the water pumps in the tower gardens. Our sustainable focus in the tower gardens will be reinforced with electricity provided by the sun. This solar panel system will provide enough electricity to keep the pump on and the water running through the tower garden. Using solar panels will also help place the tower garden in any place, without the need of an electrical outlet.

381871 Investigation of the Influence of Pulsed Corona Discharges on Benzotriazole Degradation

Abstract:

The main objective of this research is to investigate alternative means to eliminate from our environment the so-called "contaminants of emerging concerns," (CECS). The case study in this instance involves the degradation of Benzotriazole under the influence of pulsed corona discharge in a point-to-point electrode, batch mode and isothermal reactor. With a view to optimizing the process of degradation of the contaminant in our reactor, the ultimate goal is understanding the kinetics of the reactions that occur during this degradation process. Key preliminary results of several important factors affecting the degradation of the contaminant will be reported as well a detailed plan for future studies will be included.

MODELING OF PHYSICAL-CHEMICAL VARIABLES BEHAVIOR IN RIVERS USING THE STREERTER-PHELPS MODEL

A predictive mathematical model like Streeter-Phelps has two mechanisms that describe the dissolved oxygen behavior (organic matter degradation and aeration). This model was used in order to identify the distance between a food factory discharge point and the completely organic matter degradation point. The oxygen deficit was the critic variable of assimilation and the mass balances were calculated using Matlab. The distance between the discharge point and the completely degradation point rise with the contaminant concentration and the speed of the river.

386458 Dynamic Model of Stabilization Pond Systems

Abstract:

Wastewater generation is inevitable and its discharge into surface waters leads to environmental problems such as odor, eutrophication, depletion of dissolved oxygen, loss of biodiversity and also health risk. For these reasons, standards for wastewater discharge have been enforced and are expected to become stricter. Resulting in a growing interest area of study [1]. Wastewater treatment in stabilization ponds mainly results from settling and complex symbiosis of bacteria and algae where the oxidation of organic matter is accomplished by bacteria in presence of dissolved oxygen supplied by algal photosynthesis and surface re-aeration [2].

In this work different configurations of stabilization ponds are considered. Dynamic modeling of each pond was implemented within a dynamic optimization environment and the whole system was simulated during a time horizon of four months. A detailed mechanistic model is constructed, based on first principles of mass conservation, of different types of systems of anaerobic, aerobic and facultative ponds in series. The main objective of the work is to compare the performance of the different configuration pond system, by the amount of organic matter in the effluent of the treatment plant.

Model takes into account dynamic mass balances of biomass of algae, the main groups of bacteria: heterotrophic bacteria, autotrophic bacteria, fermenting bacteria, acetotrophic sulphate reducing bacteria and acetotrophic methanogenic bacteria. Also, mass balances for organic load are formulated, such as slowly biodegradable particulate COD, inert particulate COD, fermentation products, inert soluble COD, and fermentable readily biodegradable soluble COD. For nutrients, ammonium and ammonia nitrogen, nitrate and nitrite nitrogen, sulphate sulphur and dissolved oxygen. Finally, molecular nitrogen and methane emissions are considered in the model.

For the whole time horizon, we find that the conventional configuration is better than the actual configuration of the wastewater treatment plant under study [3]. Even though the differences are not as high as expected, this fact influences the total energy consumption by the aerators of the aerobic ponds. The results provide useful information on the complex relationship among micro-organisms, nutrients and organic matter concentration, as well as information about the impact of modification in the pond system that can be used to improve the control of the effluent composition.

References

1. Sah, L., Rousseau, D. P. L., Hooijmans, C.M., 2012. Numerical Modelling of Waste Stabilization Ponds: Where Do We Stand? Water Air Soil Pollut, 223, 3155-3171.
2. Beran, B., Kargi, F., 2005. A dynamic mathematical model for wastewater stabilization ponds. Ecological Modelling 181, 39-57.
3. Iturmendi, F., Estrada, V.G., Ochoa, M.P., Hoch, P.M, Diaz, M.S., 2012. Biological Wastewater Treatment: Dynamic Global Sensitivity Analysis and Parameter Estimation in a System of Waste Stabilization Ponds, Computer Aided Chemical Engineering, 30, 212-217.

387707 Design and Optimization of the Distillation and Dehydration Steps for Bio-Ethanol Production

Abstract:

Bio-ethanol production is a rapidly growing market due to the possible uses of the anhydrous product in energy generation. This production however, is weighed down by energetic demands and environmental effects.

This paper examines the distillation and dehydration steps of a bio-ethanol production plant complementing a sugar mill. The distillation step is optimized by performing rigorous energy analysis and heat integration on the process utilizing the resources made available by the sugar mill, while pervaporation is used in the dehydration step due to the formulation and use of a pervaporation module integrated into a process simulation engine, several schemes and membrane materials are assessed and compared in light of their energy demands and membrane area required.

The two steps are then combined and an energy analysis is conducted on the resulting process.

389129 The Morphology Modification of Limestone By Different Forms of Carbon Additives for CO2 Sequestration

Abstract:

The limestone was used to sequestrate the CO_2 in cement industry had been widely reported. The efficiency was dramatically reduced after several cycles due to the "sintering" of limestone. Here we found that we can improve the CO_2 sequestration capability of limestone by adding different carbon additives to modify the limestone morphology. We have compared the influence of different forms of carbon and found that we can improve the efficiency over 18% and 8% by adding carbon black and bamboo carbon respectively. However, it will decrease by adding layered graphite type materials such as, graphite and graphene. The experimental results were discussed and characterized by SEM, particle distribution, XRD, optical microscope, and its corresponding mechanism was proposed in the paper.

389217 The Enhancement of CO2 Sequestration of Limestone in Cement Industry By Carbon Black Modification

Abstract:

The CO_2 sequestration is becoming more and more important due the global warming. The investigation of CO_2 sequestration in cement industry is becoming even more important in China. Limestone, as the raw material in cement industry, was used to sequestrate the CO_2 had been reported. However, the efficiency was dramatically reduced due the surface "sintering" during the recycling. Here we report a promising method to increase the CO_2 sequestration ability of limestone by simply adding some carbon black. The efficiency was increased over 17% compare to the pure limestone. The mechanism of the improvement was discussed and analylized by SEM, XRD and optical microscope.

Primarily investigation of mixed adsorbents for the removal of copper and methylene blue from aqueous solutions

With most recent studies being focused on the development of advanced chemical adsorbents, this paper investigates the possibility of using two natural low-cost materials for selective adsorption. Heavy metals have many environmental issues such as animal and plant death, reduced rates of reproduction and the decrease of ecosystem diversity [1]. While dyes are difficult to treat as the colour tends to hold strong even after the conventional removal processes [2]. There are more than 100,000 commercially available dyes with over 700,000 tons produced annually. The presence of some of these dyes even in very small amounts i.e. < 1 $mg.dm^{-3}$ is undesirable [3]. Dyes-Heavy metal loaded wastewaters are highly coloured, have a fluctuating pH and regularly contain large amounts of suspended oils. Dyes and heavy metals find their way into the environment, primarily dissolved or suspended in water every day and create serious damage to the environment by changing the biological life balance in rivers and lakes [4]. Adsorption is a well-established and powerful practice for the removal of dyes and heavy metal pollutants from wastewaters. For these reasons, thousands of investigations have used uncountable adsorbents to remove heavy metals [5] or dyes [6]. Most of these studies focus on single-component aqueous solutions and recently, more research papers have been published dealing with the adsorption of dyes and metals in multi-component aqueous systems [7]. Also, composite adsorbents have been established to adsorb dyes and heavy metals from wastewater such as chitosan/montmorillonite composites [8] and fly ash/activated carbon [9]. In previous studies, composite materials were used to selectively adsorb a particular pollutant. Ho and co-workers proved that amino functionalized silica solids can selectively adsorb dye anthraquinone blue in the presence of methylene blue, while, from the same dye mixture, carboxy-functionalized materials adsorb methylene blue selectively [10]. Prediction of multicomponent adsorption is still one of the most challenging problems in the adsorption field [11]. There is little data available to provide more understanding about the interactions between adsorbents and developing design models for such practical systems is very important [12]. This is the case for many composite materials as well as adsorption processes with multi-adsorbate and multi-adsorbent systems. Dolomite, with a structure of alternative layers of calcite and magnesite, is a potential inexpensive and readily available adsorbent [13]. Many studies showed that dolomite and charred dolomite have the potential to act as adsorbents for reactive dyes and heavy metals [14]. Tea, processed leaves from *Camellia sinensis* species, is the second most common drink being consumed after water with an average daily drinking of 20 billion cups. The producers of making tea drinks face a problem in disposing of the spent tea leaves after the extraction [15]. The feasibility of using the richly available solid waste, spent tea leaves, for the removal of heavy metals and dyes has also been extensively investigated [16]. Hence, the aim of this research is to clarify the chemical and physical interferences between the two materials, namely tea waste and dolomite, when used for the removal of copper and methylene blue from aqueous solutions. In this paper, the term adsorption will refer to biosorption onto tea waste (TW) and adsorption onto dolomite (DO). Copper ions and methylene blue dye adsorption onto tea waste and dolomite using different factors, such as reaction contact time and pH of solution as well as initial metal and dye concentration on selective adsorption was examined. Recognizing the fundamental mechanisms and behaviour would be useful for practical multi-adsorbent packed columns.

Primarily Results

Methylene blue was efficiently removed by TW over the pH range (2–10) studied, which is attributable to the electrostatic attraction between the negatively charged TW surface and the cationic MB dye molecules [17]. These results demonstrated that pH had little effect on the capacity of the TW+DO system, specifying that the system has good potential for practical use. Cu(II) and MB were almost completely removed by dolomite and tea waste after 6 hrs at room temperature (20 °C). The adsorption of Cu(II) and MB dye was fast in the initial stage due to the large number of available binding sites, resulting in a concentration gradient. The removal rate gradually declined with time until equilibrium was achieved. The maximum removal efficiencies of copper ions and MB onto TW and DO were approx. 90% at pH of 5□7. The adsorption capacities of Cu(II) and MB were reduced when the two materials were mixed together; this reduction was more significant in the case of Cu(II). The adsorption capacity of the mixed adsorbent (TW+DO) was reduced to approx. 45% and 85% for Cu and MB, respectively. The Cu sorption capacities were effect more by the mixing of TW and DO materials. This reduction is attributed to the change in the sorbent and solution properties such as pH_{PZC} and $pH_{equilibrium}$ [18]. However, results indicate that these pollutants can still be efficiently removed with high selectivity onto the mixed adsorbents. EDX analysis coupled with FTIR investigation showed that surface complexation was the main route of removal for Cu and MB by DO. Using data from the pH study it can be concluded that physisorption was responsible for most of Cu and MB adsorption onto TW even if chemisorption was also contributing to the removal of MB by TW.

The Langmuir and Freundlich isotherms were able to represent adsorption of copper ions and methylene blue by both materials in different systems. The adsorption capacity of DO for Cu is significantly higher than TW and the trend is reversed for MB. It can also be observed that the combination of TW and DO adsorbents will reduce the adsorption capacity towards Cu, while increasing the MB adsorption capacity. In general, the sorption capacities were higher for MB than Cu, which may be due to the kinetic experiment being carried out without any pH adjustment.

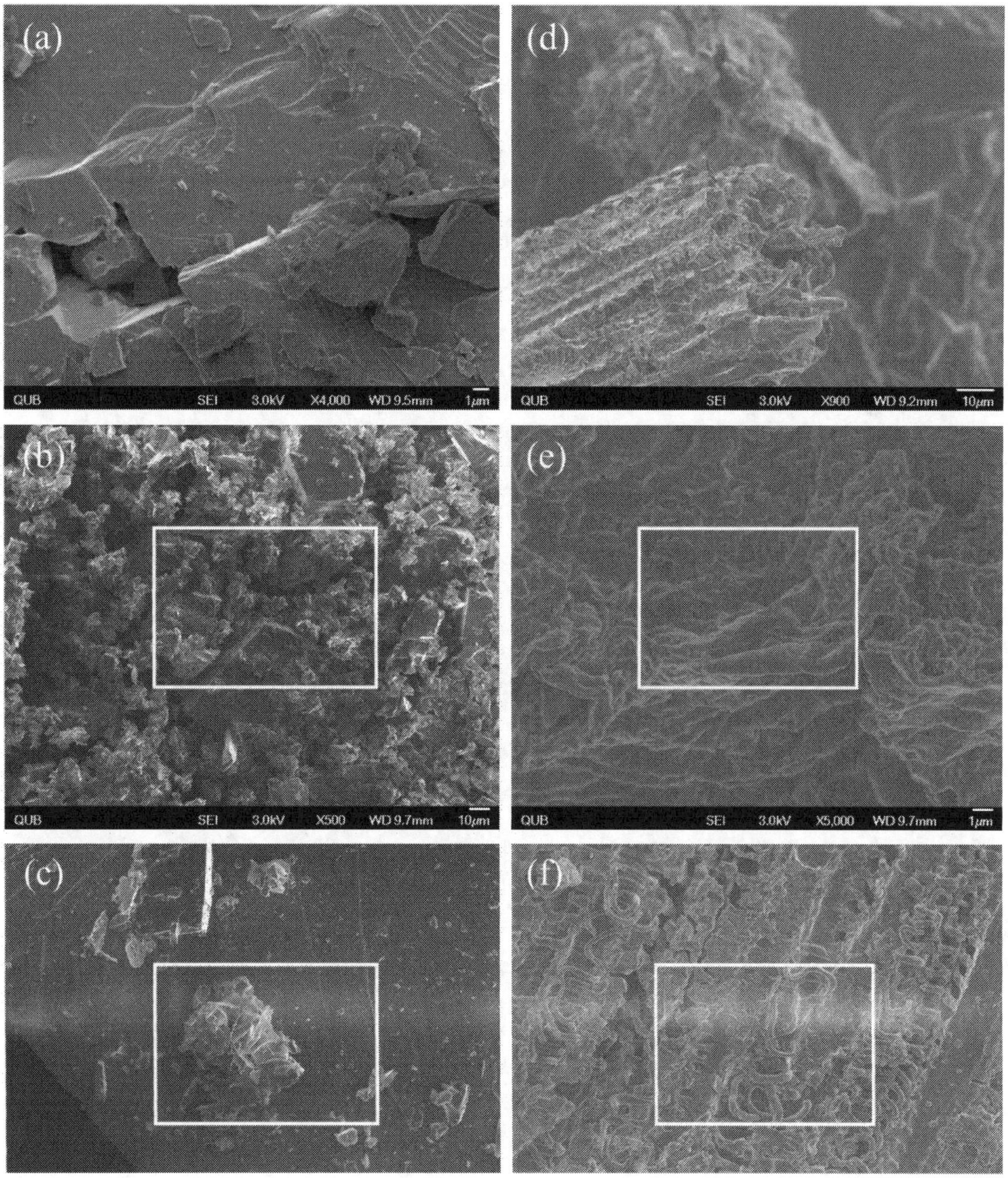

SEM images of Dolomite and Tea Waste before adsorption experiments (a) and (d), after Cu adsorption (b) and (e) and after MB adsorption (c) and (f) respectively.

References

1. Brown, P.A., J.M. Brown, and S.J. Allen, *The application of kudzu as a medium for the adsorption of heavy metals from dilute aqueous wastestreams.* Bioresource Technology, 2001. **78**(2): p. 195-201.
2. Maria Visa, Cristina Bogatu, and A. Duta, *Simultaneous adsorption of dyes and heavy metals from multicomponent solutions using fly ash.* Applied Surface Science, 2010. **256**: p. 5486-5491.
3. Al-Degs, Y.S., et al., *Solid-phase extraction and simultaneous determination of trace amounts of sulphonated and azo sulphonated dyes using microemulsion-modified-zeolite and multivariate calibration.* Talanta, 2008. **75**: p. 904-915.
4. Walker, G.M., et al., *Kinetics of a reactive dye adsorption onto dolomitic sorbents.* Water Research, 2003. **37**: p. 2081-2089.
5. Yuttasak Chammui, et al., *Removal of arsenic from aqueous solution by adsorption on Leonardite.* Chemical Engineering Journal, 2014. **240**: p. 202-210.
6. Yahya S. Al-Degs, et al., *Solid-phase extraction and simultaneous determination of trace amounts of sulphonated and azo sulphonated dyes using microemulsion-modified-zeolite and multivariate calibration.* Talanta, 2008. **75**: p. 904-915.
7. George Z. Kyzas, Nikolaos K. Lazaridis, and M. Kostoglou, *On the simultaneous adsorption of a reactive dye and hexavalent chromium from aqueous solutions onto grafted chitosan.* Journal of Colloid and Interface Science, 2013. **407**: p. 432-441.
8. Ely, A., et al., *Copper and nitrophenol removal by low cost alginate/Mauritanian clay composite beads.* Chemical Engineering Journal, 2011. **178**(0): p. 168-174.
9. Visa, M., L. Isac, and A. Duta, *Fly ash-activated carbon powder composites for dyes and heavy metals removal.* Advanced Materials Research, 2009. **79-82**: p. 243-246.
10. Ho, K.Y., G. McKay, and K.L. Yeung, *Selective adsorbents from ordered mesoporous silica.* Langmuir, 2003. **19**: p. 3019-3024.
11. Khouloud, A.A. and E.W. Dale, *Prediction of adsorption from multicomponent solutions by acti-vated carbon using single-solute parameters. Part II - proposed equation.* AAPS PharmSciTech, 2002. **3**: p. 1-8.
12. Choy, K.K.H., J.F. Porter, and G. McKay, *Intraparticle diffusion in single and multicomponent acid dye adsorption from wastewater onto carbon.* Chemical Engineering Journal, 2004. **103**(1–3): p. 133-145.
13. Brady, P.V., J.L. Krumhansl, and H.W. Papenguth, *Surface complexation clues to dolomite growth.* Geochimica et Cosmochimica Acta, 1996. **60**: p. 727-731.
14. Allen, S.J. and B. Koumanova, *Decolourisation of water/wastewaterusing adsorption: review.* Journal of the University of Chemical Technology and Metallurgy, 2005. **40**: p. 175-192.
15. Hameed, B.H., *Spent tea leaves: A new non-conventional and low-cost adsorbent for removal of basic dye from aqueous solutions.* Journal of Hazardous Materials, 2009. **161**(2–3): p. 753-759.
16. Mangwandi, C., et al., *Development of a value-added soil conditioner from high shear co-granulation of organic waste and limestone powder.* Powder Technology, 2014. **252**(0): p. 33-41.
17. El Qada, E.N., S.J. Allen, and G.M. Walker, *Adsorption of Methylene Blue onto activated carbon produced from steam activated bituminous coal: A study of equilibrium adsorption isotherm.* Chemical Engineering Journal, 2006. **124**(1–3): p. 103-110.

18. Moreno-Castilla, C., *Adsorption of organic molecules from aqueous solution on carbon materials.* Carbon, 2004. **42**: p. 83-94.

365266 Biodegradation of Contaminants in Karst Groundwater, a Dual Continuum Model

Abstract:

The advection dispersion equation (ADE) as applied to pipe flow often successfully models solute transport and biodegradation along major karst groundwater features. However, the residence time distribution (RTD) generated by the transient solution of the ADE as applied to pipe flow often under estimates biodegradation of karst groundwater contaminants due to its failure to predict the long upper tail of tracer response curves. The slower solute transport associated with flow in small fractures adjacent to karst conduits may explain the tendency of tracer response curves to have a long upper tail compared to the diminishing tail predicted by the ADE as applied to pipe flow. It is in this context that the case is made for a dual continuum transport model for karst. In the dual continuum model, the two adjacent continua share a common boundary. Due to the extreme differences in the flow regimes near the common boundary, the velocity vectors at the boundary are characterized by very steep gradients. A dual continuum model was developed using an adaptive finite element mesh system that addresses these steep velocity gradients using fuzzy type switching between the flow equations for the adjacent continua. This effectively interpolates the common boundary as a continuous polynomial and avoids the introduction of high frequency spatial components into the solution. The model was developed as a staged solution in that it solves for the steady state velocity and pressure fields and then uses these fields as inputs for the second stage solution of the advection dispersion equation describing contaminant transport. Initial testing of the model demonstrates that it is capable of modeling a wide variety of karst tracer response behavior ranging from near Gaussian or symmetric responses to bimodal responses.

367512 CFD and Response Surface Modeling of Flare Performance: DRE/CE Vs. Soot

Abstract:

As indicated in recent TCEQ and EPA reports, the flare destruction and removal efficiency (DRE) and combustion efficiency (CE) can drop below the 98% threshold under certain high air/steam-assisted conditions even in compliance with 40 CFR 60.18. Flare operators have a special concern about soot emission, i.e., to avoid visible smoke to comply with EPA's visibility regulations. In this work, the width of the operating window for flares to minimize soot emissions (i.e. be "smoke-free") while maintaining high destruction efficiencies (e.g., DRE>98%, CE>96.5%) will be discussed. A response surface model based on experimental flare efficiencies and soot emission data and correlating DRE/CE/soot to vent gas composition, heating value, steam/air injection, tip diameter, and jet velocity will be presented. Combustion mechanisms with soot precursor species developed using CHEMKIN were deployed in conjunction with soot models built in FLUENT. Computational fluid dynamics (CFD) software, FLUENT, were used to model flare cases in 2010 TCEQ flare study as well as in 1983/1984 EPA flare studies for which soot/DRE/CE data are available. These data will be used to select combustion mechanism, soot/turbulence/turbulence-chemistry interaction models, and model parameters. The CFD results will be compared with experimental industrial/laboratory scale flare soot/DRE/CE data.

This work presents a unique forum to discuss the potential trade-offs in emissions at different operating conditions, and ultimately to develop optimal control strategies to maximize environmental performance.

AIChE 2014--ENVIRONMENTAL DIVISION_09F04 Fundamentals of Environmental Kinetics & Reaction Engineering

Development of Detailed Kinetic Mechanisms for Olefins Pyrolysis

Kun Wang, Stephanie M. Villano, and Anthony M. Dean*

Chemical and Biological Engineering Dept., Colorado School of Mines, Golden, CO 80401

*corresponding author: amdean@mines.edu

Abstract: Hydrocarbons, whether derived from petroleum, natural gas, or alternative sources, will be the main energy source for the foreseeable future. One of the critical issues involving reactions of hydrocarbon fuels is the production of molecular weight growth (MWG) species (e.g., deposits in pyrolysis or soot in combustion). Deposit or soot formation can reduce the efficiency of the technical application, e.g., of the reforming catalyst in a solid oxide fuel cell. Molecular weight growth species contained in exhaust gas emissions also constitute an environmental and health problem. Predictions of the kinetics of molecular weight growth have proved to be difficult. Compared to alkane pyrolysis[i], larger amounts of heavier products are formed during olefin pyrolysis at the same level of fuel conversion. Thus, an analysis of olefin pyrolysis has the potential to provide more information on the kinetics of MWG generation.

The pyrolysis of propene, 1-butene, 2-butene and isobutene was studied in flow reactor experiments at an absolute pressure of ~0.83 atm to examine the effect of temperature, inert gas dilution, and residence time on the product distributions. In addition to measurements of conversion and major product formation, multiple molecular weight growth species ranging from C_5 to C_{10} were quantified. In all cases, the data exhibited consistently good carbon and hydrogen mass balances.

A kinetic model consisting of ~8000 reactions and ~520 species was developed. The reactions and rate constants in the model were based on an extensive series of high-level electronic calculations that served as the basis for assignment of rate rules for the various types of free radical reactions. This model was originally developed to describe the pyrolysis of ethane and was updated in this work by improving and expanding the kinetic descriptions regarding resonantly-stabilized radicals[ii]. These include: (1) Recombination reactions between two stabilized radicals, including the self- and cross-recombination, which then cyclize after H-abstraction forming a new stabilized radical; (2) Addition reactions of stabilized radicals to olefins and conjugated dienes, and alkynes that then cyclize; (3) CBS-QB3 calculation of H-abstraction reactions by resonantly-stabilized radicals leading to another resonantly-stabilized radical. Other critical updates regarding molecular weight growth include recombination reactions between resonantly-stabilized radicals and alkyl radicals, unimolecular molecular reactions (e.g., cyclization, intramolecular isomerization, and β-scission), disproportionation reactions between two resonantly-stabilized radicals, and ene/retro-ene reactions. Detailed discussions can be found elsewhere[ii, iii]. The model was able to capture the fuel conversion, and major/minor species production as well as describe the MWG kinetics up to C_{10} for all four hydrocarbons studied. The model predictions were also found to be in good agreement with literature data that were collected under different conditions. Rate and sensitivity analyses were used to identify the critical reactions.

Key words: Detailed kinetic mechanism; Pyrolysis of olefins; Molecular weight growth kinetics

---

[i] Al-Shoaibi, A., Ph.D. Thesis, Colorado School of Mines, 2008.

[ii] Wang K.; Villano M. S., and Dean A. M., "Reactions of Resonantly-Stabilized Free Radicals that impact Molecular Weight Growth Kinetics" (Submitted)

[iii] Wang K.; Villano M. S., and Dean A. M., "Experiment and Kinetic Modeling of Molecular Weight Growth Kinetics during Olefin Decomposition: 1-Butene Pyrolysis" (Submitted)

Development of a new low-GWP refrigerant composed of HFO-1123 (trifluoroethylene)

Toshiyuki Tanaka[1], Hidekazu Okamoto[1], Katsuya Ueno[1], Jun Irisawa[1], Tetsuo Otsuka[1], Tatsuhiro Nogami[1], Ritsu Dobashi[2]

[1] Research Center, Asahi Glass Co., Ltd., Yokohama-shi, Kanagawa, Japan

[2] Department of Chemical System Engineering, The University of Tokyo, Tokyo, Japan

Key Words : GWP, refrigerant design, safe, HFO-1123(trifluoroethylene)

Abstract

Refrigerants with lower GWP is required globally. HFO-1123 (trifluoroethylene), despite its fascinating properties as a refrigerant for air-conditioners, is known to disproportionate to cause sudden increase in temperature and pressure under specific conditions.

Through our intensive research on HFO-1123 reaction behavior and how to prevent the disproportionation reaction, we found that adoption of an additive had the ability to prevent the disproportionation and made it possible to design a refrigerant that won't cause disproportionation under arbitrary conditions. This enables us to successfully develop new refrigerants composed mainly of HFO-1123 that is safe as well as having quite a low GWP, by the compositional design of refrigerant gases.

1. Introduction

Hydrofluorocarbon (HFC) products are widely used as a refrigerant for air-conditioning systems. Commercially available HFC refrigerants, such as HFC-410A[1], has excellent properties as a refrigerant, but their high global-warming potential (hereinafter "GWP"[2]) is now at issue because when air-conditioners being disposed of, a large proportion of refrigerant gases stored in them is emitted to the atmosphere without destroyed chemically. This results in contributing to global warming. Therefore, refrigerants with lower GWP is strongly required globally.

HFO-1123 (trifluoroethylene, CF_2=CHF) is a strongly attractive candidate because its GWP is extremely low (GWP of HFC-410A = 2090, GWP of HFO-1123 = 0.3[4]) and the performance as a refrigerant composed mainly of HFO-1123 is equivalent to that of conventional refrigerants[3].

Despite HFO-1123 being so fascinating, it has a downside. Unlike chemical materials adopted in conventional refrigerants, HFO-1123 is known to disproportionate under specific conditions and in the presence of ignition sources. The formula of the disproportionation reaction is described as follows[5].

$$CF_2{=}CHF \rightarrow 0.5CF_4 + 1.5C + HF + 250\ \text{kJ/mol} \quad (1)$$

As shown in the formula, this reaction is highly exothermic, which means once this reaction is initiated, it might propagate and result in both temperature and pressure soaring.

Needless to say, air-conditioners have to be a safe device as well as being an environmentally-friendly one, so such an increase in pressure and temperature would not be allowed, not only under normal operating conditions but also under failure modes in air-conditioners.

Table 1. Normal operating conditions in air-conditioners

Pressure (cycle top) [MPa]	Temperature (cycle top) [°C]
6	130

Studies about failure modes (reachable pressure, temperature, and possible ignition energy, *etc.*) are now in progress.

This report focused on the study on HFO-1123 disproportionation behavior and the development of method to prevent the disproportionation reaction in general.

2. Experimental Section

2-1. Evaluation of auto ignition behavior

Fig. 1 showed a schematic view of the experiment apparatus. An ellipsoidal vessel (1.38 L or 10 L volume) was adopted. This vessel was equipped with a pressure sensor and a temperature sensor to monitor the pressure and temperature inside the vessel.

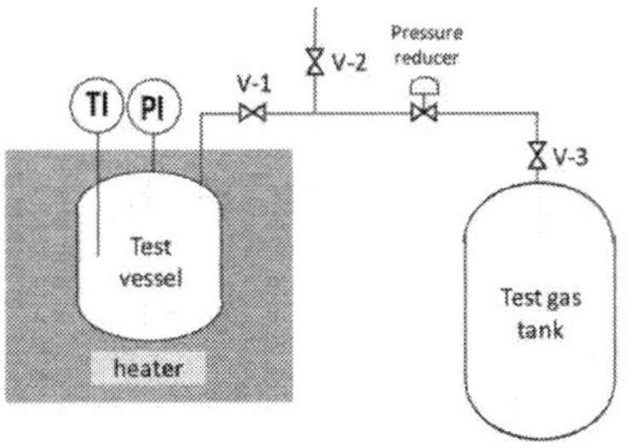

Fig. 1. Schematic view of the experimental apparatus
PI : pressure sensor TI : temperature sensor

Experimental procedures were as follows. First, the test vessel was set at a predefined temperature (= initial temperature), followed by vacuumization using a vacuum pump through V-1 and V-2. Then, the test gas was introduced into the test vessel through V-1 and V-3 at a predefined pressure (= initial pressure) which was controlled by the pressure reducer. When the pressure reached at a predefined pressure, V-1 was closed. By the constraints of the equipment, the initial pressure was limited below 1 MPa.

When the rapid increase in pressure and temperature was observed, it was judged that auto ignition occurred. Otherwise, it was judged that auto ignition didn't occur.

2-2. Evaluation of forced ignition behavior

Fig. 2 showed a schematic view of the experimental apparatus. An autoclave (0.65 L volume) was adopted. This autoclave was equipped with a pressure sensor, a temperature sensor and a rapture disk. Sensors were used to monitor the pressure and temperature inside the autoclave. For the safety reason, the burst pressure of the rapture disk was designed to be lower than the pressure capacity of the autoclave, so that the pressure inside the autoclave wouldn't exceed the pressure capacity of the autoclave.

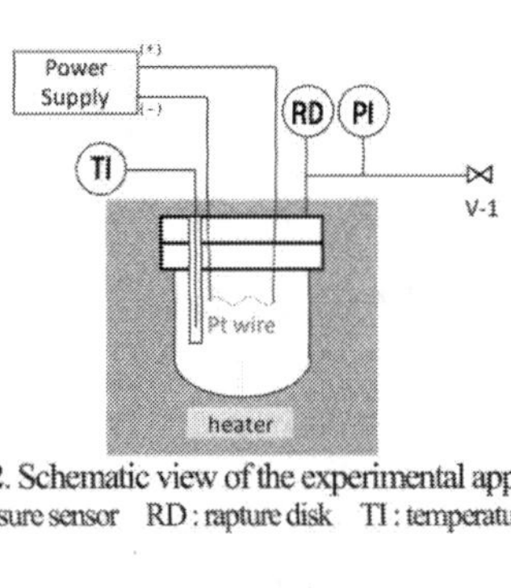

Fig. 2. Schematic view of the experimental apparatus
PI : pressure sensor RD : rapture disk TI : temperature sensor

Experimental procedures were as follows. First, the autoclave was vacuumized using a vacuum pump, then test gas was introduced into the autoclave through V-1. Then the autoclave was set at a predefined temperature (= initial temperature) and a predefined pressure (= initial pressure) using a heater. Unlike auto ignition test explained in 2-1, a pressure more than 10 MPa was tolerable in this autoclave, so the initial pressure up to 10 MPa was experimentally examined. Application of the current to the platinum (Pt) wire resulted in the evolution of Joule heat. This made the Pt wire melted at around Pt melting point (1768 ℃), and this was defined as an ignition. To vary the ignition energy, different diameter of Pt wire was adopted. Ignition energy was determined by the time integration of the voltage and the current applied to Pt wire.

Table 2. Ignition energy by Pt wire melting

Entry	Pt diameter [mm]	Pt length [mm]	Ignition energy [J]
1	0.1	36	2
2	0.25	36	9
3	0.5	36	30
4	1	36	150

After the ignition, when the rapid increase in pressure and temperature was observed, it was judged that the disproportionation propagated. On the other hand, when no change in pressure and temperature was observed, it was judged that the disproportionation didn't propagate.

2-3. Evaluation of the dimerization reaction rate of HFO-1123

The dimerization reaction rate of HFO-1123 was determined by using the same equipment described in 2-1. Here the vacuumization of the vessel and the introduction of the test gas were conducted at room temperature. Then the vessel was heated at the rate of ca. 5 ℃/min. The reaction rate was determined by measuring the temperature and pressure change.

3. Results and Discussion

3-1. Auto ignition of HFO-1123

It's important to investigate the auto ignition behavior of a refrigerant to know if the refrigerant can be operated safely in air-conditioners.

By using the experimental procedure shown in 2-1, HFO-1123 (CF_2=CHF) auto ignition behavior was investigated.

Auto ignition is a term commonly used in combustion phenomena that proceed in the presence of oxidizers, so note that auto ignition in this study is different in that no oxidizer is included.

3-1-1. Effect of temperature and pressure

Using a test vessel with 1.38 L volume, the effect of initial conditions (pressure, temperature) on auto ignition behavior was investigated. (Fig. 3)

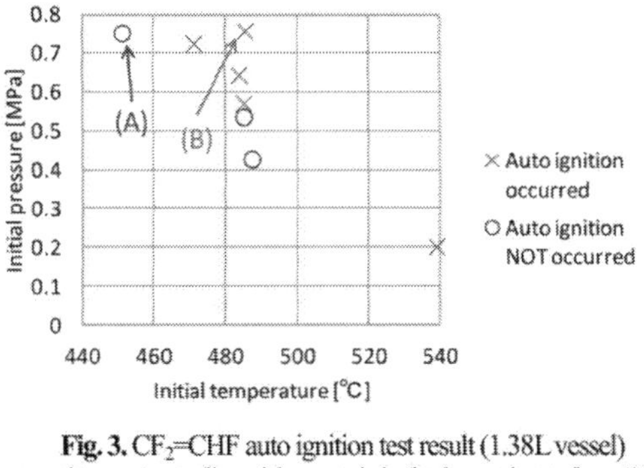

Fig. 3. CF_2=CHF auto ignition test result (1.38L vessel)
Pressure and temperature profiles and the gas analysis after the experiment of case (A) and (B) are shown in Fig. 4, 5 and 6.

Fig. 4 and Fig. 5 showed the examples of the pressure and temperature profiles. And Fig. 6 showed the gas analysis results of after the experiments.

In both cases, the drop in temperature was first observed due to the introduction of the test gas which was cool (around 20 ℃), then the increase in temperature and the drop in pressure were observed. In case (A) (Fig. 4), the pressure kept on decreasing to reach at nearly half that of the initial pressure. This was judged that auto ignition didn't occur. The gas analysis after the test (Fig. 6) showed the formation of various compounds such as a C336. C336 ($C_4H_2F_6$) was estimated to be formed by the dimerization of CF_2=CHF.

The dimerization reaction formula is described as follows.

$$CF_2{=}CHF \rightarrow 0.5C_4H_2F_6(c336) + 83.7\ kJ/mol \quad (2)$$

The reaction heat was calculated using Gaussian09 (CBS-QB3).

On the other hand, case (B) (Fig. 5) showed a rapid increase in pressure and temperature after 15 seconds of induction period. This was judged that auto ignition occurred. The gas analysis after the test (Fig. 6) gave CF_4 exclusively, so it was revealed that auto ignition of HFO-1123 corresponded to the transition to disproportionation (eq. (1)).

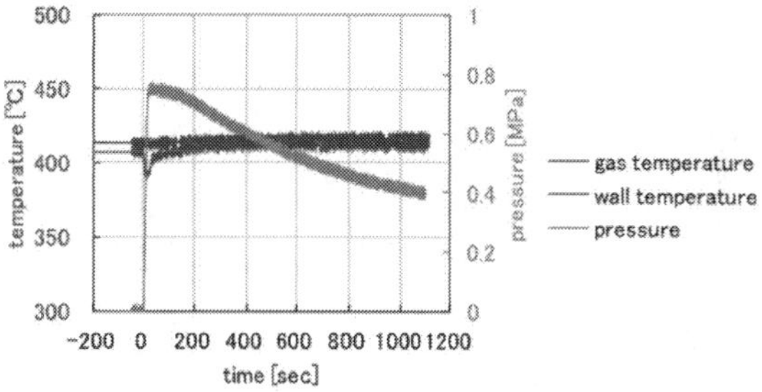

Fig. 4. Temperature and pressure curve (Case (A))
Initial pressure : 0.73 MPa, Initial temperature : 450 ℃
t = 0 was when the test gas began to be introduced into the test vessel.

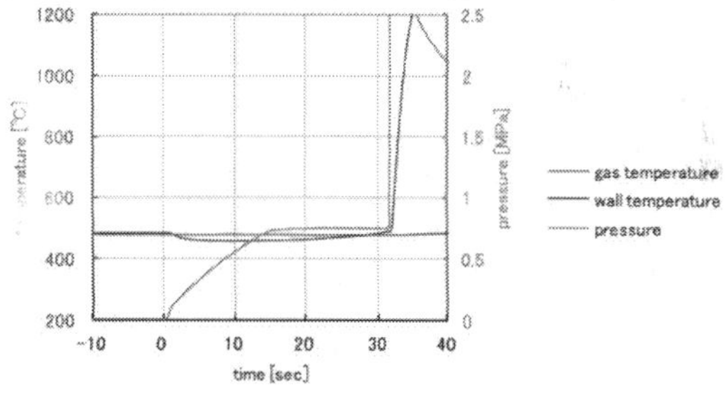

Fig. 5. Temperature and pressure curve (Case (B))
Initial pressure : 0.76 MPaG, Initial temperature : 485 ℃
t = 0 was when the test gas began to be introduced into the test vessel.

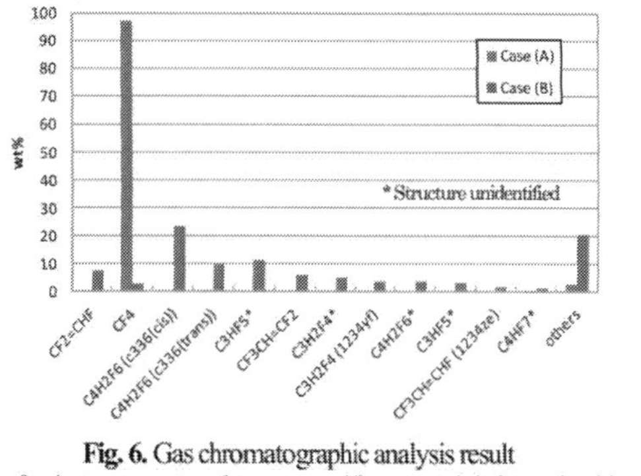

Fig. 6. Gas chromatographic analysis result
The gas after the test was pretreated to remove acidic compounds before analyzed by gas chromatography, so acids like HF (hydrogen fluoride) would not be detected here.

These phenomena were very similar to tetrafluoroethylene (CF_2=CF_2)

auto ignition behavior. As described in the literature[6], the dimerization reaction of tetrafluoroethylene (eq. (3)) precedes the auto ignition, or the transition to the disproportionation (eq. (4)), because the activation energy of the dimerization reaction is smaller than that of the disproportionation reaction. Whether the auto ignition occurs or not depends on the balance between the heat generation rate by the dimerization and the heat dissipation rate through the vessel wall.

$CF_2{=}CF_2 \rightarrow 0.5C_4F_8(\text{cyclic}) + 103\ \text{kJ/mol}$ (3)
$CF_2{=}CF_2 \rightarrow C + CF_4 + 257\ \text{kJ/mol}$ (4)

As shown in the literature[6], the influence of pressure on tetrafluoroethylene's MITD was formulated by applying the simplified theory on spontaneous thermal ignition from Semenov[7].

$\ln(p_{initial}) = a / MITD + b$ (5)
where
$p_{initial}$ = initial pressure
MITD = minimum ignition temperature of decomposition
a, b = constant obtained by fitting equation (5) to the experimental data

Minimum Temperature of Decomposition (MITD) is defined as the lowest temperature of a hot surface, e.g. the vessel walls, at which self ignition for the decomposition of a gas occurs without the presence of an oxidizer like air.

In the same manner, application of the test results shown in Fig. 3 to equation (5) gave Fig. 7 and equation (6).

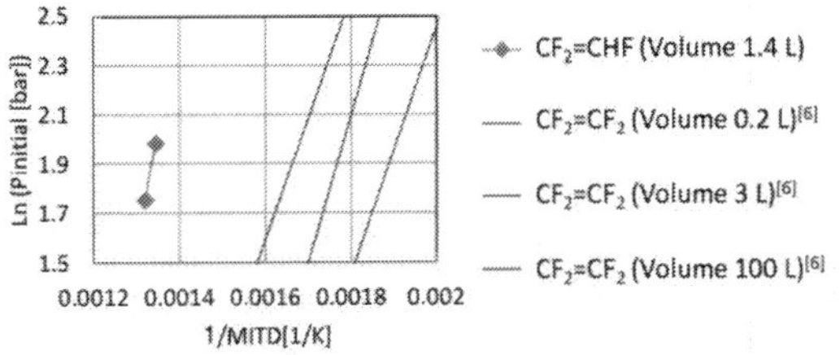

Fig. 7. Ln ($P_{initial}$) vs 1/MITD curve

$\ln(p_{initial}\ [\text{bar}]) = 5496/ MITD_{CF2=CHF} - 5.57$ (6)

Application of equation (6) gave Table 3. MITD at 6 MPa (cycle top) is calculated to be 296 ℃. That is, if $CF_2{=}CHF$ is adopted as a refrigerant, the possibility of auto ignition can be eliminated under normal operating conditions when the vessel volume is 1.38 L.

Table 3. Calculated MITD of $CF_2{=}CHF$ (1.38L)

pressure [MPa]	1.0	2.0	3.0	4.0	5.0	6.0
MITD [℃]	425	369	340	320	307	296

3-1-2. Effect of the vessel volume

In case of tetrafluoroethylene, that MITD depends on the vessel volume, as well as the pressure, is known (Fig. 7). The larger the vessel volume is, the lower MITD is at the same initial pressure. This is because larger vessel volume means smaller specific surface area, resulting in lower heat dissipation rate through the vessel wall.

To examine the $CF_2{=}CHF$'s MITD dependency on the vessel volume, auto ignition experiments were conducted using the same procedure shown in 2-1 and 10 L volume vessel.

As shown in Fig. 8, comparing at around the same pressure (0.7 – 0.8 MPa), MITD got lowered by tens of degrees C when the vessel volume was changed from 1.38 L to 10 L. Therefore it can be concluded that larger vessel volume means lower MITD at the same pressure.

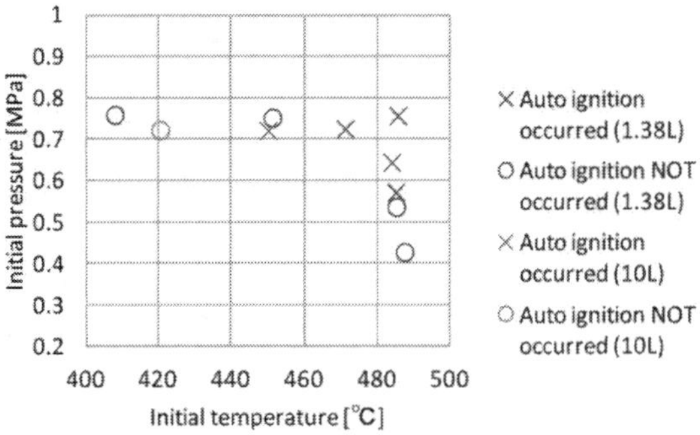

Fig. 8. $CF_2{=}CHF$ Auto ignition test result (1.38L and 10L vessel)

3-2. Forced ignition of HFO-1123

It's important to investigate the forced ignition behavior of a refrigerant because there is a possibility that ignition sources like a spark is generated in air-conditioners in their failure modes.

By using the experimental procedure shown in 2-2, HFO-1123 ($CF_2{=}CHF$) forced ignition behavior was investigated. Here, ignition energy was fixed at 30 Joules in all cases.

The relationship between initial conditions (pressure, temperature) and the results (whether the disproportionation propagation occurred or not) were shown in Fig. 9.

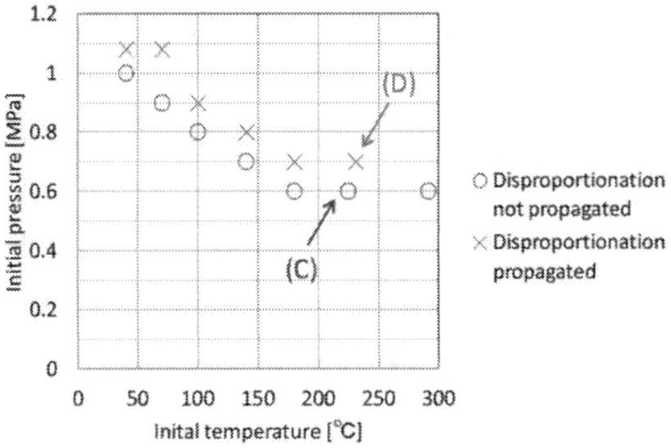

Fig. 9. Forced ignition test result
Ignition energy = 30 J (Pt diameter = 0.5 mm, length = 36 mm)
Pressure profiles of case (C) and case (D) were shown in Fig. 10.

Fig. 10 showed the examples of the pressure profile (Case (C) and (D)). Temperatures could not be measured with enough sensitivity because the sensor was set in a sheath. (Therefore, the temperature sensor was used mainly to measure the initial temperature.) Case (C) showed no change in pressure, and this was judged that the disproportionation didn't propagate, whereas case (D) showed a rapid increase in pressure, and this was judged that the disproportionation propagated.

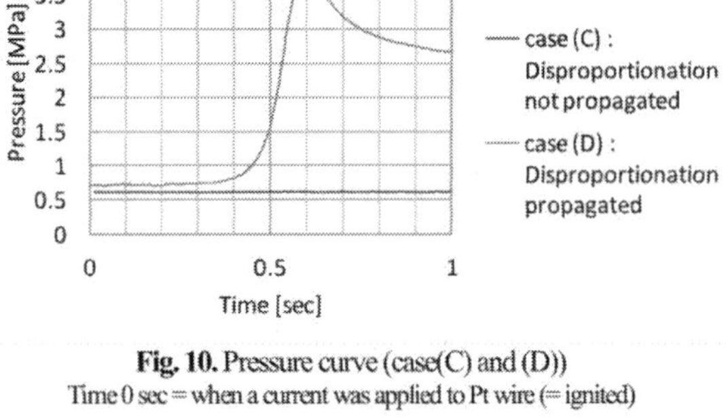

Fig. 10. Pressure curve (case(C) and (D))
Time 0 sec = when a current was applied to Pt wire (= ignited)

Gas chromatographic analysis of the gases after the experiments gave interesting results. The analysis of case (D) revealed the formation of more than 10wt% of CF_4, confirming that the disproportionation propagated. On the other hand, CF_4 was also detected in case (C), but the concentration was much lower (several hundreds ppm). This meant that even when ignition didn't cause the rise in pressure, the disproportionation reaction itself occurred near the hot surface of Pt wire, but quenched. This phenomenon could be explained as follows. The high temperature generated by the ignition (Pt melting point 1768 ℃) gave CF_2=CHF gas near Pt surface enough energy to exceed the activation energy of the disproportionation. However, when the heat dissipation rate around the ignited spot was faster than the heat generation rate by the disproportionation, the ignition didn't lead to the propagation of the disproportionation.

Therefore, from the safety perspective, the important thing in considering forced ignition behaviors is not to prevent the disproportionation reaction itself, but to prevent the propagation of the disproportionation.

Fig. 9 revealed that in the presence of an ignition source, CF_2=CHF can cause sudden increase in pressure and temperature under normal operating conditions of air-conditioners (~6 MPa, ~130℃). We assumed that the possibility of the evolution of an ignition source cannot be eliminated completely. Therefore, in the next stage we developed the method to make a refrigerant more difficult to cause the disproportionation propagation.

3-3. Auto ignition of HFO-1123/R32 mixture

We explored additives to make a refrigerant composed of CF_2=CHF more difficult to disproportionate. In our study, R32 (CH_2F_2) was adopted as an additive because the mixture of CF_2=CHF and CH_2F_2 forms pseudo azeotropic composition, which means high performance as a refrigerant can be achieved[3].

First, auto ignition behavior of HFO-1123/R32 mixture was investigated using the experimental procedure shown in 2-1.

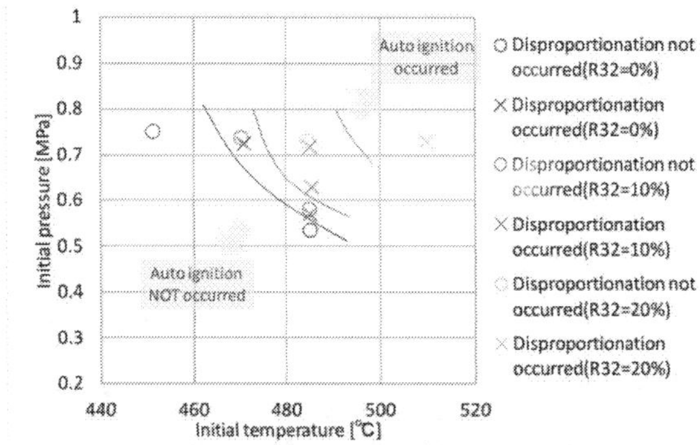

Fig. 11. Auto ignition test result (1.38 L vessel)
R32 = 0%, 10%, 20% corresponded to the composition of CF_2=CHF/R32 = 100/0, 90/10, 80/20 [wt/wt], respectively.

As shown in Fig.11, the higher R32 concentration was, the higher MITD of the CF_2=CHF/R32 mixture gas became. That is, addition of R32 had the effect to make a refrigerant safer in terms of auto ignition.

The quantitative effect of CH_2F_2 on preventing auto ignition will be discussed in 4-1.

3-4. Forced ignition of HFO-1123/R32 mixture

Using the experimental procedure shown in 2-2, the forced ignition experiments of the mixture gases were conducted.

3-4-1. Effect of pressure, temperature and composition

The effect of the initial condition (pressure, temperature) and the composition of the test gas on the forced ignition behavior was thoroughly investigated as shown in Fig. 12. (The ignition energy was fixed at 30 J.)

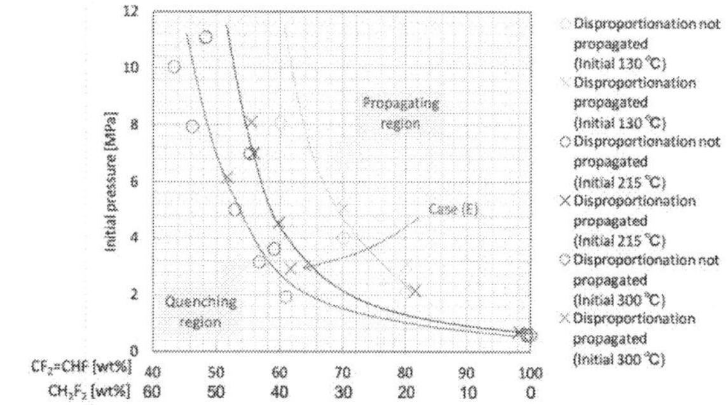

Fig. 12. Forced ignition test result (CF_2=CHF and CH_2F_2 mixture)
The analysis result of the gases before and after the ignition of case (E) was shown in Fig. 13.

When the initial temperature was fixed at a specific value, the higher CH_2F_2 ratio was, the higher the initial pressure in which the disproportionation propagated became. When the initial pressure was fixed at a specific value, the higher CH_2CF_2 ratio was, the higher the initial temperature in which the disproportionation propagated became. This meant that increasing the ratio of CH_2F_2 made a mixture gas more difficult to cause the disproportionation propagation, that is, safer as a refrigerant in terms of forced ignition.

The analysis result of the gases (case (E)) before and after the ignition was shown in Fig. 13.

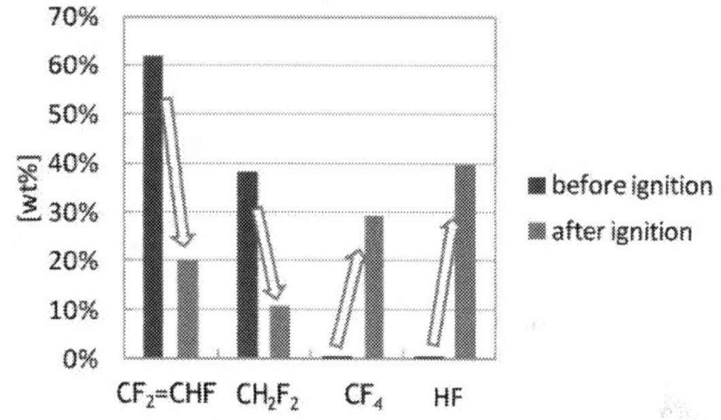

Fig. 13. Gas chromatographic analysis result (case (E))

Fig. 13 revealed that both CF_2=CHF and CH_2F_2 reacted proportionally to form CF_4 and HF. Although not shown in Fig. 13, the formation of carbon was also confirmed as the autoclave after the test was filled with carbon soot. These facts implied that not only the disproportionation reaction (1) but also the reaction (7) proceeded.

$CH_2F_2 \rightarrow C + 2HF + 95$ kJ/mol[8] (7)

The reaction (7) also is exothermic, but the reaction heat is much smaller than that of (1). We speculated on of the effects of CH_2F_2 on preventing the disproportionation propagation as follows. The disproportionation reaction first occurred only near the ignition source, where the temperature was high enough to exceed the activation energy of the disproportionation reaction, and whether this reaction propagated or not depended on the balance between the heat generation rate and the heat dissipation rate around the hot spot. Higher CH_2F_2 concentration in the gas mixture meant lower heat generation rate because the reaction heat of (7) is much lower than that of (1). On the other hand, the heat dissipation rate doesn't depend on the composition of the gas so much. This resulted in the higher CH_2F_2 concentration leading to the prevention of the disproportionation propagation.

3-4-2. Effect of ignition energy

Whether the disproportionation propagates or not was expected to depend also on the ignition energy.

The influence of ignition energy was evaluated by using the experimental procedure shown in 2-2 and by changing the diameter of Pt wire.

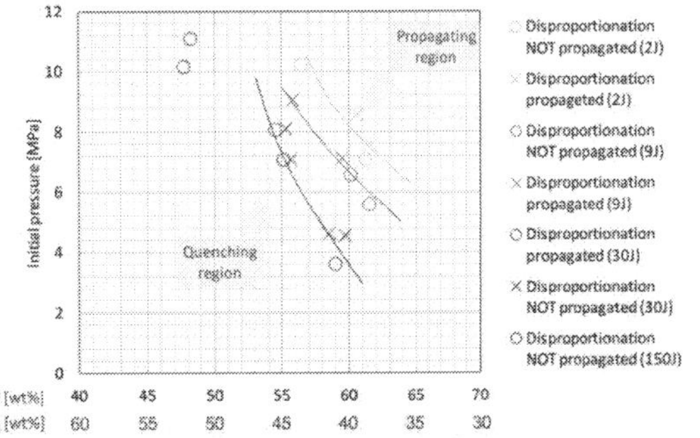

Fig. 14. Forced ignition test result of the CF_2=CHF and CH_2F_2 mixture.
Initial temperature was fixed at 215 ℃.
2J, 9J, 30J, 150J showed the ignition energy.

As shown in Fig. 14, when the initial pressure was fixed at a specific value, the lower the ignition energy was, the higher CF_2=CHF ratio was required for the disproportionation to propagate. This meant that the lower the ignition energy was, the more difficult for a mixture gas to cause disproportionation propagation.

3-5. Determination of CF_2=CHF dimerization reaction rate

As discussed in 3-1-1, the auto ignition of CF_2=CHF was supposed to be triggered by the reaction heat of the dimerization reaction. (eq. (2))

To clarify the reaction rate constant of eq. (2), we conducted a test described in 2-3. The experimental conditions were shown in Table 4.

Table 4. Experimental conditions

Run	test gas	Initial temperature [℃]	Initial pressure [MPa]
1	CF_2=CHF = 100%	20	0.4
2	CF_2=CHF/CH_2F_2 = 50/50 wt%	20	0.4

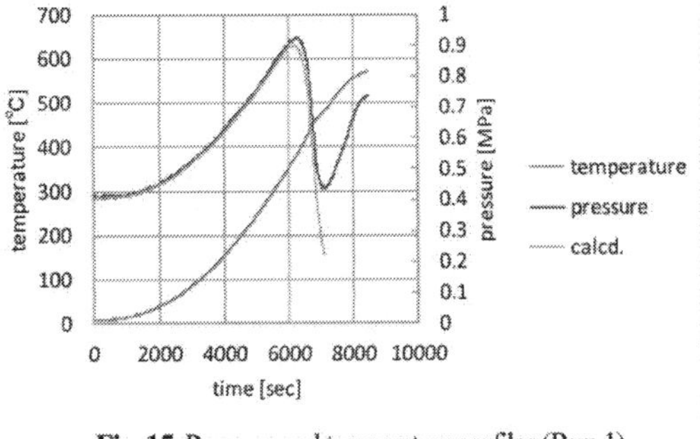

Fig. 15. Pressure and temperature profiles (Run 1)
Calcd. : calculated pressure profile using the rate constant in table 5.

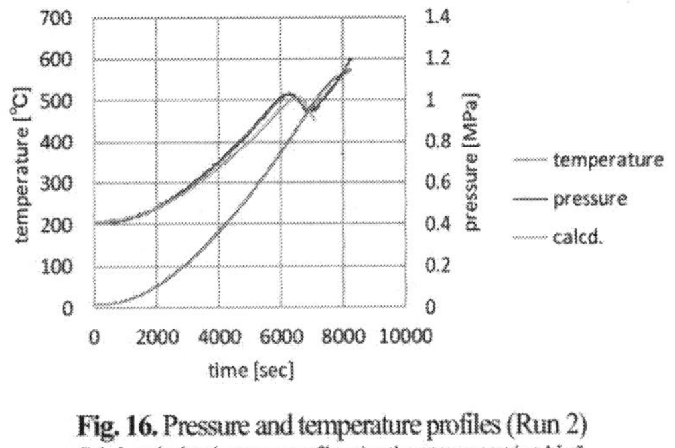

Fig. 16. Pressure and temperature profiles (Run 2)
Calcd. : calculated pressure profile using the rate constant in table 5.

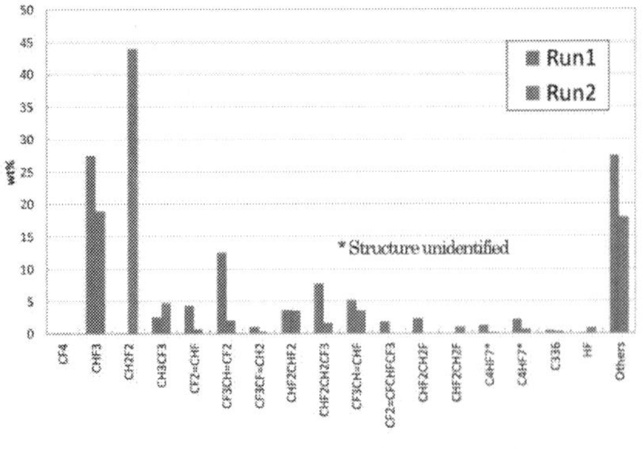

Fig. 17. Gas chromatographic analysis result

After reached at about 600 ℃, the test gas was analyzed by gas chromatography. As shown in Fig. 17, in both cases CF_4 was not detected, implying that the disproportionation reaction (eq. (1)) didn't occur by the gradual heating to 600 ℃. In case of Run 2, the concentration of CH_2F_2 remained near 50wt%, implying that CH_2F_2 didn't react at 600 ℃ and functions as an inert gas.

Reaction rate constant of CF_2=CHF dimerization (eq. (2)) was determined by fitting the pressure profile. We assumed that CH_2F_2 didn't react, so the drop in pressure was attributed only to the dimerization of CF_2=CHF. This approximation was supported by another experimental fact. When the heating of CF_2=CHF was stopped at 300 – 350 ℃, the CF_2=CHF dimer c336 exclusively was detected other than CF_2=CHF. This implied that the dimerization occurred first, followed by the decomposition of the dimers to cause pressure rise (after 7000 sec in Fig. 15-16) and to form various compounds (Fig. 17).

Table 5. Determined reaction rate constant of dimerization

frequency factor A [L/mol/s]	activation energy E [kJ/mol]	reaction order
1.0×10^{10}	1.65×10^{2}	2

As shown in Fig. 15 and 16, the pressure profile was well fitted by the reaction rate shown in Table 5 until the decomposition started after 7000 sec.

4. CFD analysis

To elucidate the mechanism of the disproportionation reaction behavior and further to make it possible to extrapolate the experimental results to other conditions like higher pressures, CFD (computational fluid dynamics) calculation was conducted by using ANSYS Fluent.

In this report, CFD analysis was applied only to auto ignition behavior.

To simulate the reaction behavior, Fluent requires values shown below.

- chemical reaction formula
- reaction rate constant $k = A\exp(-Ea/RT)$
- standard enthalpy change of formation ΔH_f
- specific constant at constant pressure Cp
- viscosity μ
- coefficient of thermal conductivity λ
- coefficient of diffusion D

Properties, all of which are the function of temperature except ΔH_f, were calculated in the following manner unless being listed in NIST Chemistry WebBook[8]. ΔH_f, Cp of each compound were calculated by using Gaussian09 (CBS-QB3). μ, λ and D of each compound were calculated by using Lennard-Jones parameters (σ, ε/k_B). [9]

$$\mu = 2.67 \times 10^{-6} \frac{(M_w T)^{1/2}}{\sigma^2 \Omega_\mu} \quad (8)$$

$$k = \frac{15\,R}{4\,M_w} \mu \left(\frac{4\,c_p\,M_w}{15\,R} + \frac{1}{3} \right) \quad (9)$$

$$D_{ij} = 0.00188 \frac{\left(T^3 \left(\frac{1}{M_{w,i}} + \frac{1}{M_{w,j}} \right)\right)^{1/2}}{p_{abs}\,\sigma_{ij}^2\,\Omega_D} \quad (10)$$

Table 6. properties adopted in CFD calculation

	Δ H_f [kJ/mol]	Cp [J/kg/K]	σ [Å]	ε /k_B [K]
CF_2=CHF	-474	$2.00\times10^2+2.85\times10^0\times T-2.67\times10^{-3}T^2+1.20\times10^{-8}T^3-2.09\times10^{-10}T^4$	4.90	279
c336	-1115.4	$4.06\times10^1+3.47\times10^0\times T-3.22\times10^{-3}T^2+1.20\times10^{-6}T^3-2.09\times10^{-10}T^4$	6.26	409
CH2F2	-450.7	$1.99\times10^2+2.48\times10^0\times T-1.39\times10^{-3}T^2+3.45\times10^{-7}T^3-2.82\times10^{-11}T^4$	4.37	304

Lennard-Jones parameters (σ, ε/k_B) were calculated as follows.

At First, each molecular structure was determined by the optimization using Gaussian09 software M062X/6-31+G** level of theory.

Non-bonding interaction energy was given by the sum of coulombic interaction energy and the van der Waals interaction energy. For coulombic interaction, each atomic charge in molecules was determined by Merz-Kollmann method by using M062X/6-311+G** level of theory. The van der Waals interaction parameter was taken from Dreiding II force field. [10]

Each molecule was assumed to be van der Waals sphere atoms.

step1: Two molecules of each compound were set randomly in a space in more than 50000 ways and made to interact with each other in a point contact at van der Waals radius sphere atom.

step2 : The average distance between the centers of two molecules and the interaction energy between the two molecules in Step 1 were calculated.

step3 : By applying step 1 and step 2 to fluorine compounds whose σ and ε/k_B are already known in the literature[11], correlation equations between values listed in the literature and the calculation results from step 1 and step 2 were obtained. (Table 7-8 and Fig. 18-19)

step4 : By applying step 1 and step 2 to compounds whose σ and ε/k_B are not listed in literatures, such as CF_2=CHF and its dimer c336, the average distance and the interaction energy were obtained. Applying the correlation equations obtained in step 3 to these values gave σ and ε/k_B (Table 6).

Table 7. Calculated average distance and σ[11]

	calculated value [Å]	σ [11] [Å]
CF_4	4.45	4.66
CF_2=CF_2	4.83	5.16
CF	3.46	3.64
CF_2	3.82	3.98
CF_3	4.16	4.32
CHF_3	4.35	4.33
C_2F_6	5.31	5.51
CH_2F_2	4.24	4.08
CF_2=CHF	4.74	unknown
CF_2=CHF dimer	6.01	unknown

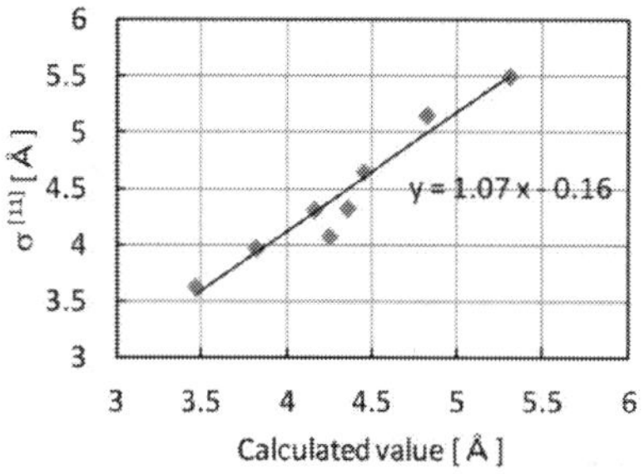

Fig. 18. Correlation between calculated average distance and σ[11]

Table 8. Caclulated interaction energy and ε/k_B [11]

	calculated value [kcal/mol]	ε/k_B [11] [K]
CF_4	-0.19	134
CF_2=CF_2	-0.82	203
CF	0.43	94
CF_2	0.15	108
CF_3	-0.23	121
CHF_3	-1.44	240
C2F6	-1.08	195
CH_2F_2	-2.10	318
CF_2=CHF	-1.82	unknown
CF_2=CHF dimer	-3.29	unknown

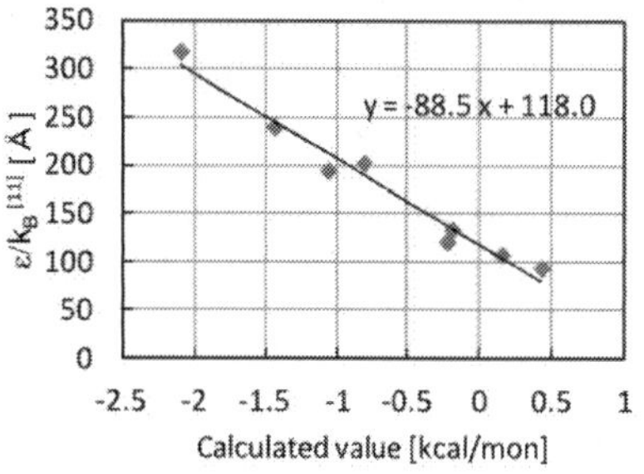

Fig. 19. Correlation between calculated interaction energy and ε/k_B [11]

A 2D geometry using the axial symmetry modus was adopted to reduce the computational load.

4-1. Auto ignition behavior

CFD calculation of CF_2=CHF auto ignition behavior was conducted by considering the dimerization reaction, in analogy with tetrafluoroethylene.[12] CH_2F_2 was treated as an inert gas.

The reaction rate of dimerization reaction rate constant was set at the values shown in Table 5. In all cases, the temperature of the vessel wall (= initial temperature) was fixed at the initial temperature during the calculation.

The initial conditions of CFD calculation were shown in Table 9, all of which except Entry 8 were examined experimentally. (Fig. 3, 8, 11)

Table 9. Conditions of CFD calculation

Entry	composition CF_2=CHF [wt%]	composition CH_2F_2 [wt%]	Vessel volume [L]	Initial temperature [°C]	Initial pressure [MPa]	Experimental results 1)
1	100	0	1.38	450	0.76	○
2	100	0	1.38	485	0.76	×
3	100	0	1.38	485	0.42	○
4	90	10	1.38	485	0.72	×
5	80	20	1.38	485	0.73	○
6	80	20	1.38	510	0.73	×
7	100	0	10	450	0.72	×
8	100	0	10	450	0.72	No data

1) ○ : auto ignition didn't occur

× : auto ignition occurred

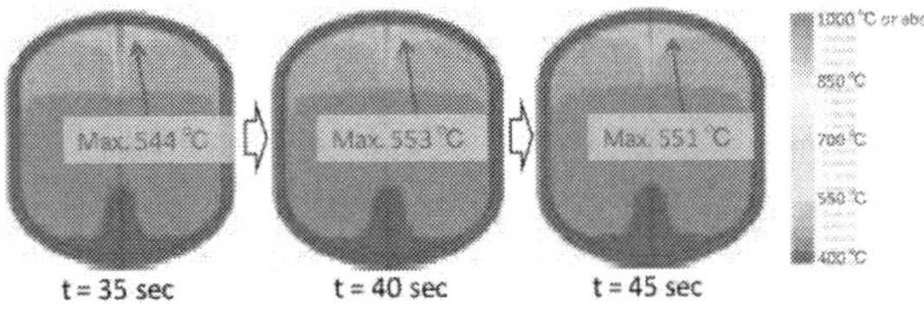

Fig. 20. CFD calculation result (Entry 1)

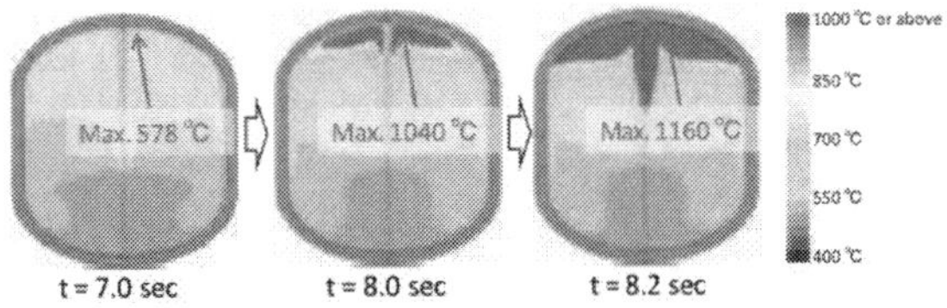

Fig. 21. CFD calculation result (Entry 2)

CFD calculation result of Entry 1, where auto ignition didn't occur experimentally, was shown in Fig. 20. First, hot region was generated in the superior spot of the vessel. This was because the gas heated by the dimerization reaction heat moves upward because of the buoyancy. The temperature kept on rising to reach the maximum temperature 553 ℃ (t = 40 sec). Then the temperature began to drop (t=45 sec).

The temperature rise can be explained by the dimerization reaction heat, whereas the temperature drop can be explained by the decrease in CF_2=CHF concentration as a result of CF_2=CHF consumption and the dimer formation. The decrease in CF_2=CHF concentration means lowered heat formation rate, resulting in the heat dissipation rate via the vessel wall exceeding the heat formation rate.

On the other hand, CFD calculation result of Entry 2, where auto ignition occurred experimentally, showed the different behavior. It reached so high as 1000 ℃ in 8 seconds. It can be interpreted that the higher initial temperature of case 2 than case 1 allowed the temperature runaway owing to faster heat formation rate. We speculated this temperature runaway by the dimerization reaction enabled the transition to the disproportionation.

Fig. 22-26 showed the CFD calculation results. In each case, the moment when it reached the maximum temperature durng the calculation, or when it showed temperature runaway was depicted.

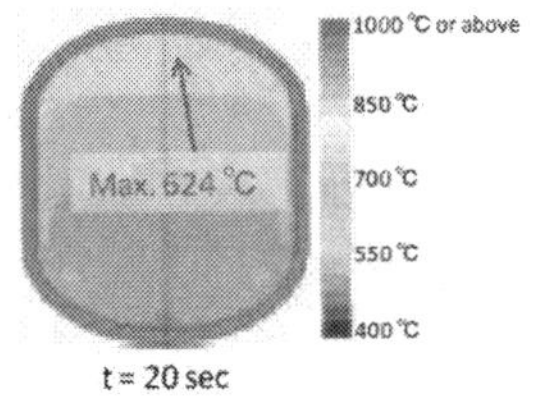

Fig. 22. CFD calculation (Entry 3)

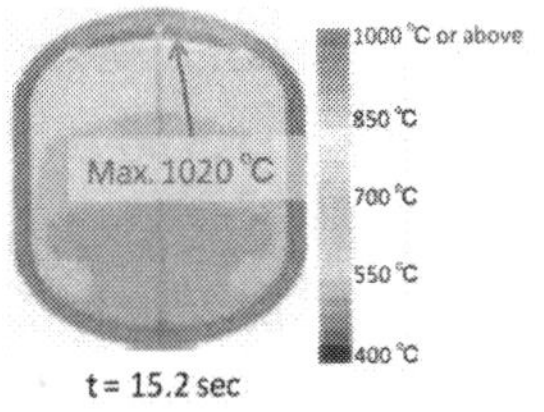

Fig. 23. CFD calculation (Entry 4)

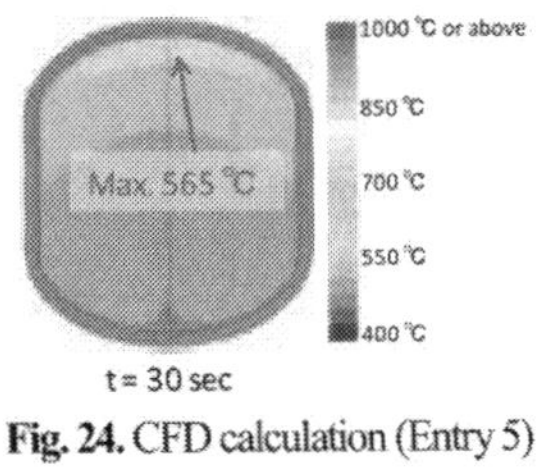

Fig. 24. CFD calculation (Entry 5)

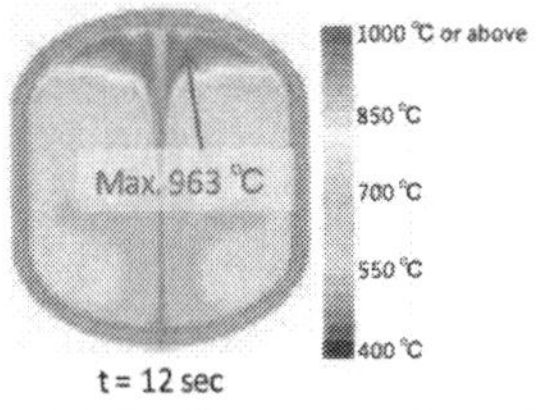

Fig. 25. CFD calculation (Entry 6)

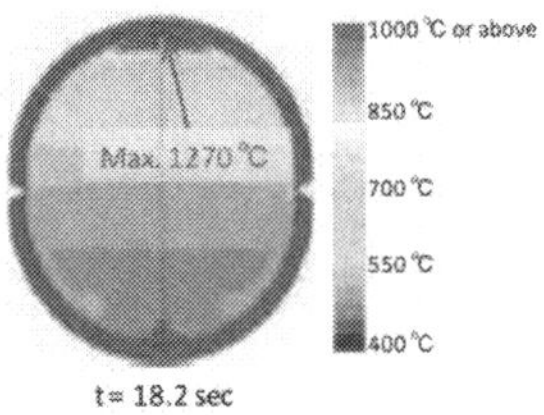

Fig. 26. CFD calculation (Entry 7)

As shown in Fig. 20-26, when auto ignition didn't occur experimentally (Entry 1, 3, 5), CFD calculation gave the temperature rise of around 100 ℃ or less. On the other hand, when auto ignition occurred experimentally (Entry 2, 4, 6, 7), CFD calculation gave temperature runaway to reach near 1000 ℃ or more. Entry 1-7 includes the cases whose initial pressure, initial temperature, the composition of the test gas, and the vessel volume, were different. From the fact that in all cases CFD calculation results (whether the temperature runaway occurred or not) coincided well with the experimental results (auto ignition occurred or not), it could be concluded that the simple modeling that considers only the dimerization reaction of CF_2=CHF explained well the nature of auto ignition behavior of CF_2=CHF.

Finally, CFD calculation was applied to assess CH_2F_2 effect on lowering MITD quantitatively.

Table 10. Comparison of Entry 1 and 8 (Table 9)

Entry	composition		Vessel volume [L]	Initial temperature [℃]	Initial pressure [MPa]	ΔT_{max} 1)
	CF_2=CHF [wt%]	CH_2F_2 [wt%]				
1	100	0	1.38	450	0.76	**103**
8	**80**	**20**	1.38	**485**	0.76	**97**

1) The maximum temperature difference between the vessel wall and the hot spot in the vessel observed during the calculation.

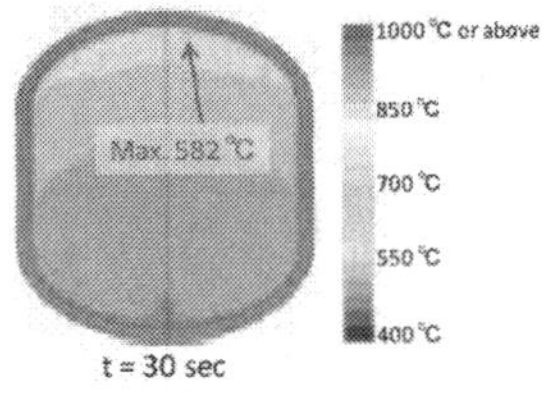

Fig. 27. CFD calculation (Entry 8)

Fig. 27 showed the CFD calculation result when the temperature reached the highest point. Entry 1 and Entry 8 had the same vessel volume and initial pressure, and the composition of the gas and the initial temperature were different.

In both cases, temperature rise were about 100 ℃, implying that the balance between the heat formation rate and the heat dissipation rate were almost the same. Therefore we speculated that addition of 20wt% of CH_2F_2 had the ability to lower MITD by 35 ℃ (= 485 ℃ - 450 ℃).

By applying this CFD calculation, whether auto ignition occurs or not at arbitrary conditions can be expected.

4-2. Forced ignition behavior

Forced ignition behavior was revealed experimentally up to high pressures (~10 MPa) or high CH_2F_2 concentration (~50 wt%). However it's meaningful to reveal forced ignition behavior theoretically to understand the disproportionation mechanism and how to prevent it.

Unlike auto ignition behavior, forced ignition causes the disproportionation reaction (eq. (1) and eq. (7)) near the ignition source regardless of the disproportionation propagating or not. Therefore the reaction rate of the disproportionation reaction is necessary to calculate, which is yet to be determined. Therefore CFD analysis for forced ignition behavior is not established yet.

At the present, we found the effect of CH_2F_2 in its smaller reaction energy compared to CF_2=CHF when disproportionated, as described in 3-4-1. But we assume CF_2=CHF can have the effect like lowering the disproportionation reaction rate. To test the hypothesis, we are now studying the elementary reactions by quantum-chemical approach. We are planning to publish a more elaborate and precise modeling in the following report.

5. Conclusions

The reaction behavior of HFO-1123 (CF_2=CHF) and R-32 (CH_2F_2) was intensively investigated. HFO-1123 alone was found to have the possibility to cause the disproportionation reaction if an ignition source breaks out in some failure mode.

Adoption of R32 (CH_2F_2) as a second component was elucidated to have the ability to prevent the disproportionation reaction both in auto ignition and forced ignition.

In auto ignition study, the effect of temperature, pressure, the composition of the gas, the vessel volume was revealed experimentally and further CFD analysis made it possible to anticipate the auto ignition behavior at arbitrary conditions.

In forced ignition study, whether the disproportionation propagates or not was revealed experimentally up to high pressure (~10MPa) or high CH_2F_2 concentration (~50wt%).

Further studies about the worst case scenario are now in progress, which include the possibility study about the failure mode in air-conditioners (reachable pressure, temperature, and possible ignition energy).

By combining the study described in this paper and the further studies about the worst case scenario, we will clarify the composition of refrigerants composed of HFO-1123 that are substantially safe, as well as having low GWP, and highly efficient.

Acknowledgements

Financial support of this research by New Energy and Industrial Technology Development Organization is gratefully acknowledged.

References

1. Mixture of HFC-32 (difluoromethane) and HFC-125 (pentafluoroethane) (50/50wt%). GWP = 2090.
2. A relative measure of how much heat a greenhouse gas traps in the atmosphere in a specific time interval (here 100 years is adopted as a time interval), expressed as a factor of carbon dioxide (whose GWP is standardized to 1)
3. M. Fukushima, M. Hashimoto, *In Proceedings of JRAIA International Symosium Kobe 2014.* Kobe, Japan, 2014.
4. Measured by AIST (National Institute of Advanced Industrial Science and Technology)
5. Ya. A. Lisochkin, et al., *Combustion, Explosion, and Shock Waves*, 42(2), 2006, 140-143.
6. Fabio Ferrero, et al., *J. Loss Prev. Process Indust.*, 25, 2012, 1010-1017.
7. Irvin Glassman (1987). *Combustion (2nd ed.)*, Academic Press.
8. Reaction heat was calculated using standard enthalpy change of formation in *NIST Chemistry WebBook* (http://webbook.nist.gov/chemistry/).
9. McGee, Henry A. (1991). *Molecular Engineering*, New York.
10. S.L.Mayo, et.al., *J.Phys.Chem.*, 94(26), 1990, 8897-9809.
11. Available from <http://www.eureka.im/1296.html>, (accessed 2014-11-01)
12. Fabio Ferrero, et al., *Chem. Eng. Trans.*, 31, 2013, 817-822.

The removal of nitric oxide (NO) from simulated flue gas in a bubble column reactor at atmospheric pressure is investigated using combined aqueous persulfate ($Na_2S_2O_8$) and ferrous ethylenediamine tetraacetate (Fe^{II}-EDTA) systems. 0.1 M persulfate solution with 0.01 M Fe^{2+} was used as the optimum amount of absorption-oxidation reagents and molar ratio of Fe^{2+} and EDTA was found to be 1:1 for maximum NO conversion. NO absorption experiments were carried out at 23-70 ^{0}C. Iron species' (Fe^{2+}, Fe^{3+} and Fe^{II}-EDTA) concentrations were measured spectrophotometrically to understand the simultaneous and synergistic relationship between persulfate and Fe^{II}-EDTA in NO removal. The results show significant improvement in NO removal compared with thermally and Fe^{2+} activated persulfate systems (at lower temperature 25-30% and at higher temperature 5-10%) and almost 100% NO conversion can be achieved at 70 ^{0}C. The detailed chemistry, mechanistic aspects and kinetics, and the effects of persulfate and EDTA concentrations, and pH will be discussed.

References:
Adewuyi, Y.G.; Khan, M.A.; Sakyi, N.Y. Ind. Eng. Chem. Res. 2014, 53, 828.
Adewuyi, Y.G.; Sakyi, N.Y. Ind. Eng. Chem. Res. 2013, 52, 11702.
Adewuyi, Y.G.; Sakyi, N.Y. Ind. Eng. Chem. Res. 2013, 52, 14687.
Khan, N.E.; Adewuyi, Y.G. Ind. Eng. Chem. Res. 2010, 49, 8749

Modeling and Analysis of Light-off Behavior of DOC

Richa Raj, Michael P Harold and Vemuri Balakotaiah
Department of Chemical and Biomolecular Engineering, University of Houston, Houston TX 77204

Recent advances in low temperature combustion (LTC) diesel engines lead to higher net efficiencies and lower NO_x and particulate matter (PM) emissions as compared to current diesel engine technologies. However, the LTC engine has lower exhaust temperature that results in higher levels of CO and hydrocarbons (HCs) than that of today's engine. The design of the diesel oxidation catalyst (DOC) for treatment of exhaust gases from LTC engines is more challenging. It is important to study the light-off behavior of exhaust gases at lower temperature and higher levels of CO and HCs (CO = 1000 - 4000 ppm and HCs = 2000 - 3000 ppm). In this study, we have developed global kinetic model for co-oxidation of CO and C_3H_6 over commercial diesel oxidation monolithic catalysts (Pt/Al_2O_3). The kinetic model coupled with (1+1) -dimensional (1D in fluid phase + 1D in washcoat phase) model of a monolith channel with washcoat diffusion is used to present a detailed analysis of light-off behavior. In particular, we have studied the effect of inlet feed composition, washcoat diffusion, non-uniform catalyst loading (both within the washcoat and along the length) and space velocity on light-off.

In order to develop global kinetic model we used transient light-off data for CO and C_3H_6 oxidation on Pt/Al_2O_3 diesel oxidation catalyst published in literature [1]. The model is also validated using new set of experimental data. Fig. 1 shows both the experimental data and model predictions for CO and C_3H_6 light-off. The model developed can predict independent CO and C_3H_6 oxidation light-off curves as well as light-off curves corresponding to co-oxidation of CO and C_3H_6. The model captures the inhibition of C_3H_6 on CO light-off and similarly, inhibition of CO on C_3H_6 light-off. The results show that on increasing inlet C_3H_6 concentration there is increase in both CO and C_3H_6 light-off temperature. Similarly, on increasing CO inlet concentration there is increase in both CO and C_3H_6 light-off temperature. The model is also able to capture the hysteresis observed experimentally for both CO and C_3H_6 oxidation. The simulation results show increase in light-off temperature for both CO and C_3H_6 on decreasing the PGM loading.

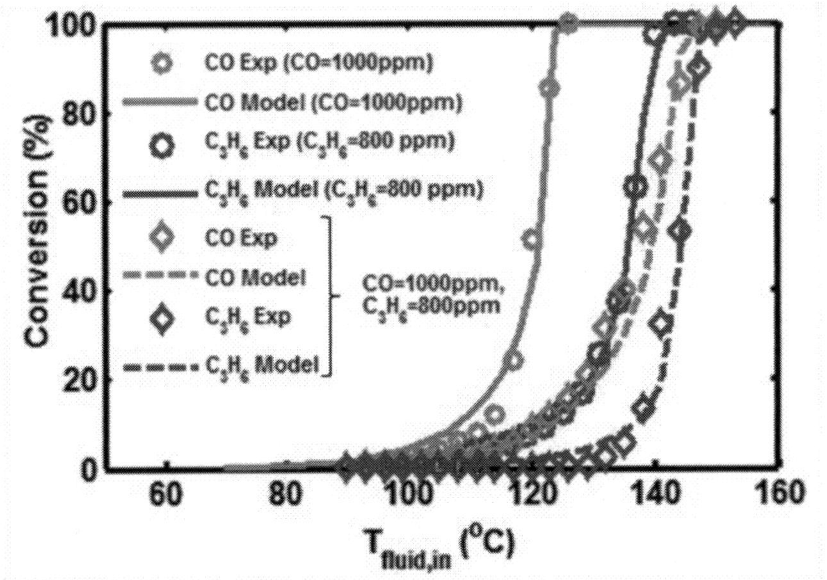

Figure 1 Fitted light- off curves of CO and C_3H_6 (CO = 1000 or 0 ppm, C_3H_6 = 800 or 0 ppm, O_2 = 10 %, H_2O = 10%, CO_2 = 10%, balance gas: N_2)

References

[1] A. Abedi, R. Hayes, M. Votsmeier, W. S. Epling, Inverse hysteresis phenomena during CO and C_3H_6 oxidation over a Pt/Al_2O_3 catalyst. Catal. Lett. 142 (2012) 930-935.

Kinetic Study of Anaerobic Digestion of Sewage Sludge/ High Organic Strength Industrial Waste

POOJA SHARMA, U K GHOSH, A K RAY
Department of Polymer & Process Engineering, Indian Institute of Technology, Roorkee, Saharanpur Campus, Saharanpur (247001 UP), India

email: sharma_poojas@rediffmail.com , ghoshuk_iitr@yahoo.com, amiyakumarray@gmail.com
Mobile No.09012205303, 09855406044

ABSTRACT

Sludge / High organic strength waste generated from municipal and industrial wastewater treatment plants are highly organic in nature, causes environmental pollution if disposed directly into environment and require treatment prior to disposal. Anaerobic digestion is favourable method for the stabilisation of the highly organic waste and production of useful product methane. This method produces less sludge as compare to aerobic process during degradation of organic compounds. Microorganisms responsible for anaerobic degradation of organic compounds are highly sensitive to external conditions as well as internal conditions inside the digester such as inhibitory agents, pH, Volatile fatty acids (VFA) etc. Various dynamic and steady state kinetic models have been developed by various researchers for the description of specific growth rate and also for predicting the behaviour of microorganisms for anaerobic degradation of organic compounds. In the present work kinetic models and kinetic parameters such as yield coefficient (Y), specific growth rate (μ), decay coefficient (k_d), half velocity coefficient (K_s) etc. have been studies.

Keywords: Anaerobic digestion, kinetic model, microbial kinetic parameters and organic waste

INTRODUCTION

The present work describes the anaerobic digestion process kinetic models developed by various researchers and also reviews on kinetic parameters for different kind of waste. Various kind of waste such as high strength industrial effluent generated from distillery spent wash[10] , ice cream wastewater[8] , olive mill wastewater[3] , two phase olive pomace [4] etc., municipal solid waste[1], industrial and municipal sludges require treatment prior to disposal. Anaerobic degradation of complex organic compound is favourable method for disposal of different kind of waste. Various kind of anaerobic digesters such as complete mix digester with recycling, complete mix digester without recycling, upflow anaerobic sludge blanket, upflow anaerobic packed bed reactor, anaerobic expanded bed reactor, anaerobic fluidized bed reactor, anaerobic sequencing batch reactor depending upon the nature of waste to be treated, process conditions to be maintained and inhibitory agents. The schematic process kinetics for the degradation of organic compounds inside the digester is shown in figure 1. Anaerobic digestion is most favourable process for the reduction of environment pollution load by treating high strength waste and production of methane. Anaerobic digestion process often suffers instability problem results in failure of digester. Various kinetic models have been developed for describing process kinetics such as Monod, Haldane, Andrews, Contois, Mosey etc describing process kinetics and also considering the factors that causes process failure. Rate limiting step in most of the models describing anaerobic digestion kinetics is methanogenesis. In some cases solubilisation of particulate substrate is considered as rate limiting step.

Anaerobic digestion process involves three steps: In the first step hydrolysis of complex organic compounds. After solubilisation, fermentation of soluble organic compounds and production of

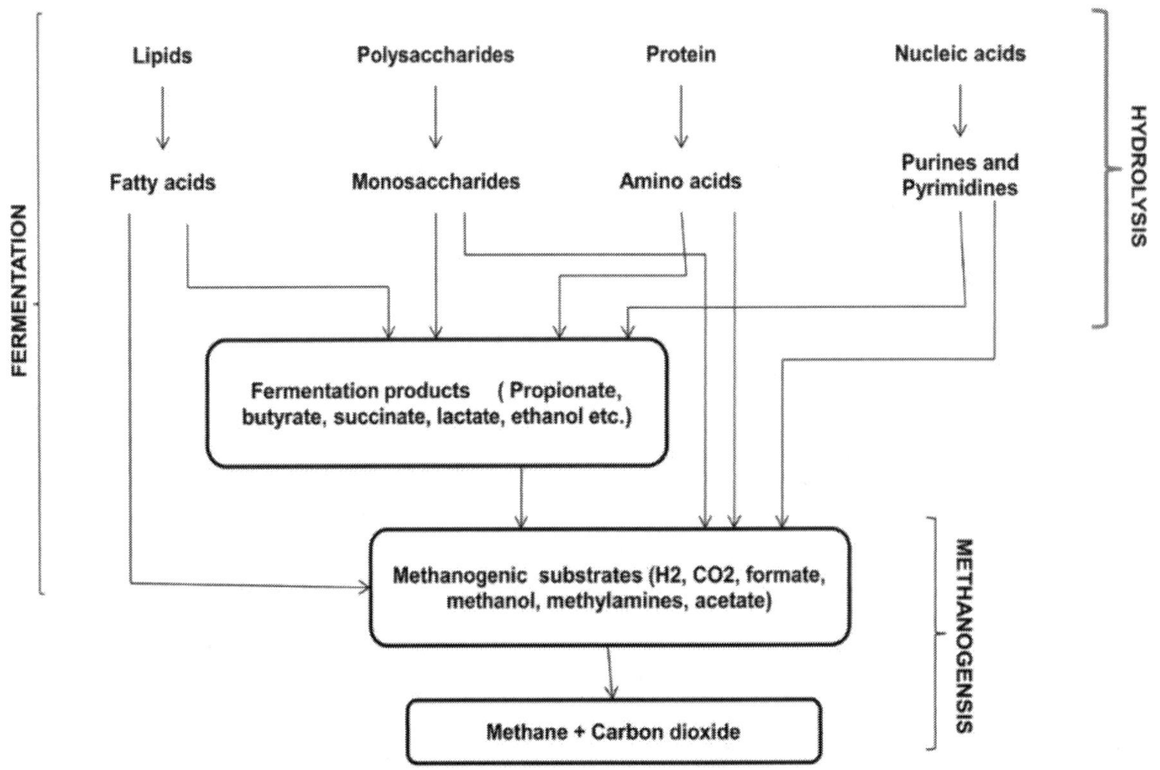

Figure 1: Schematic process description of anaerobic degradation of organic compounds [11]

acetate, hydrogen and carbon dioxide (precursor of methane formation). In the methanogensis, production of methane from acetate, and also from hydrogen and carbon dioxide.

REVIEW ON KINETIC MODEL FOR ANAEROBIC DIGESTION PROCESS

Various microbial growth kinetic models as shown in Table 1 have been developed for describing the performance of anaerobic digester and also able to predict behaviour under dynamic conditions by substituting rate expressions in mass balance equations.

Monod kinetic model which is based on Michaelis-Menten equation, relationship between specific growth rate and growth limiting substrate in dissolved form [11]. Contois kinetic model describes the specific growth rate of micro-organism dependent on growth limiting substrate in both influent and effluent from the digester [6]. Andrews developed model for anaerobic digestion process considering the effect of unionized volatile fatty acids on the growth kinetics of methanogens and the rate limiting step is methanogensis[2] . Chen and Hashimoto kinetic model is based on substrate utilization and methane production [5].

An excellent review paper describing the kinetics and modelling of anaerobic digestion process in which properly mention the models using unionized volatile fatty acids inhibition , models using total volatile fatty acids inhibition, models considering different composition of wastewater and also models using hydrogen as the key parameter[7] .

In 1977, an International Anaerobic Modelling Task Group was established in Japan as a common platform for the establishment of anaerobic digestion model for complex organic wastes.

Table 1: Microbial growth kinetic models

Sr. No.		Kinetic Model
1.	Monod (1942)	$\mu = \frac{\mu_m S}{K_S + S}$
2.	Tessier (1942)	$\mu = \mu_m(1 - \exp(-\frac{S}{K_S}))$
3.	Moser (1958)	$\mu = \frac{\mu_m S^{\lambda}}{K_S + S^{\lambda}}$ ($\lambda > 0$)
4.	Contois (1959)	$\mu = \frac{\mu_m S}{Bx + S}$
5.	Powell (1967)	$\mu = \frac{\mu_m}{2K_S} (K_S + S - \sqrt{(K_S + S^2 - 4K_S S}$
6.	Andrews (1971)	$\mu = \frac{\mu_m S}{1 + K_S / HS + HS / k_I}$
7.	Chen and Hashimoto (1978)	$\mu = \frac{\mu_m S}{S_o . K + (1 - K) S}$

MICROBIAL KINETIC PARAMETERS

Different kind of waste to be treated anaerobically having different kinetic parameters depending upon process conditions such as temperature, types of reactor, nature of waste and also any kind of inhibitory substances that effect microbial activity. Determination of kinetic parameters with the help of modelling helps in predicting the behaviour.

The kinetic coefficient values used for the treatment of sewage sludges in the incomplete anaerobic digestion process are K_{SA} (8.0 g/g/day), K_{VM} (6.2 g/g/day), Y (0.3), K_I (1 g/dm^3), K_d (0.1 day^{-1} for acidogens) and (0.015 day^{-1} for methanogens) [9].

Biokinetic coefficient values for distillery spent wash for K_d (0.05 day^{-1}), Y (0.5), μ_m (1.563 day^{-1}) and K_s (24 mg/L) [10].

Kinetic coefficient values for anaerobic digestion of ice cream wastewater by anaerobic contact process are Y (.2116), B (0.4818), K_s (0.4028 g/L) and μ_m (0.7844 day^{-1}) [8].

ACKNOWLEDGEMENT

This work has been supported by Ministry of Human Resources & Development (MHRD).

NOTATION

μ	Specific growth rate (day^{-1})
μ_m	Maximum specific growth rate (day^{-1})
S_o	Initial concentration of substrate (g/m^3)
S	Growth limiting substrate concentration (g/m^3)
K_s	Half velocity concentration (g/m^3)
B	Contois Kinetic constant
HS	Unionized substrate concentration (g/m^3)
K_I	Inhibition constant (g/m^3)
K_{SA}	Maximum specific substrate utilization rate for acidogens
K_{VM}	Maximum substrate utilization rate for methanogens
Y	Yield coefficient
K_d	Decay coefficient (day^{-1})

REFERENCES

1. Adak, A., Mazumder, D., Bandyopadhyay, P., Anaerobic digestion of municipal solid waste. International Journal of Civil and Environmental Engineering (2011), 3, 177-182.

2. Andrews, J.F., Graef, S.P., Dynamic modelling and simulation of the anaerobic digestion process. Advances in Chemistry, American Chemical Society (1971), 126-162.

3. Borja, R., Garrido, S.E., Martinez, L., Romas-Cormenzana, A., Martin, A., Kinetic study of anaerobic digestion of olive mill wastewater previously fermented with Aspergillus Terreus. Process Biochemistry (1993), 28, 397-404.

4. Borja, R., Martin, A., Sanchez, E., Rincon, B., Raposo, F., Kinetic modelling of the hydrolysis, acidogenic and methanogenic steps in anaerobic digestion of two-phase olive pomace. Process Biochemistry (2005), 40, 1841-1847.

5. Chen Y.R., Hashimoto A.G., Substrate utilization kinetic model for biological treatment processes. Biotechnology and Bioengineering (1980), 22, 2081-2095.

6. Contois, D.E., Kinetics of bacterial growth: relationship between population density and space growth rate of continuous culture. J Gen Micobial (1959), 21, 40-50.

7. Gavala H.N., Angelidaki I., Ahring B.K., Kinetics and modelling of Anaerobic digestion process. Advances in Biochemical Engineering/ Biotechnology (2003), 57-91

8. Hu, W.C., Thayanithy, K., Forster, C.F., A kinetic study of anaerobic digestion of Ice-cream wastewater. Process Biochemistry (2002), 37, 965-971.

9. Mendoza, R.B., Sharratt, P.N., Modelling the effects of imperfect mixing on the performance of anaerobic reactors for sewage sludge treatment. J Chemical Technol. Biotechnol.(1998), 71, 121-130.

10. Prakash, N.B., Sockan, V., Raju, V.S., Anaerobic digestion of distillery spent wash. ARPN Journal of Science and Technology (2014), 4, 134- 140.

11. Tchobanoglous, G., Burton, F.L., Stensel, H.D., Wastewater engineering: Treatment and reuse (2003), 630-631.Jana, A.K., Chemical Process Modelling and Computer Simulation (2008), 70-74.

Impact of hydrocarbon fuel structure on anaerobic biodegradation via the fumarate addition reaction: An *ab-inito* and kinetic modelling study.

Abstract

Bioremediation is considered to be one of the most effective and environmentally sound approaches to regenerating oil-contaminated soil and aqueous resources. Although the aerobic biodegradation of hydrocarbons has been relatively well researched, significant questions remain concerning the metabolic pathways and reaction mechanisms governing the anaerobic route to hydrocarbon biodegradation. One of the primary effective biochemical strategies employed by anaerobic bacterial cultures to digest hydrocarbon substrates is the fumarate addition mechanism. This mechanism involves the addition of a hydrocarbon radical to fumarate to obtain a succinate derivative. The succinate derivative is subsequently metabolized via decarboxylation and β-oxidation pathways. The fumarate addition mechanism has been implicated in the anaerobic biodegradation of a variety of hydrocarbon substrates including, toluene, ethyl-benzene, n-hexane, dodecane *etc.*, and is known to involve an enzymatic radical mechanism. The proposed mechanism involves the generation of a hydrocarbon radical by a sulfur centered radical (located in the enzyme) followed by its addition onto fumarate and the regeneration of the sulfur centered radical and the product succinate derivative. Presented here are the results of high level electronic structure calculations in the gas-phase on a model system, where we evaluate the energetics of the fumarate addition reaction mechanism with respect to alkanes (butane) and aromatics (toluene). The potential energy surfaces (PES) describing these reactions and their respective activation barriers and thermodynamics are used to determine rate constants, which have been utilized in a kinetic model. The presence of species like sulfur and oxygen in these systems is found to result in hydrogen bonding. As a result, the PES are more complicated due to the presence of a pre-reaction complexes that dramatically impact the energetics. These results and their connection to the enzymatic system shall be discussed.

Biodiesel production via transesterification in a mixed carbon dioxide-methanol system with a heterogeneous catalyst

Lindsay Soh, Chun Chi Chen, Thomas A. Kwan, Eric J. Beckman, Julie B. Zimmerman

Biodiesel offers a sustainable alternative to petroleum based transport fuels, but in order for biodiesel to be viable, the production process must be improved to allow for a net positive energy balance and greater economic efficiency. In this work, the use of carbon dioxide (CO_2) as a green solvent is shown to improve heterogeneously catalyzed biodiesel production via advantages in terms of system phase behavior, mass transport, and selectivity. A mixed CO_2 and methanol system was used for transesterification of triglycerides into fatty acid methyl esters (FAME, biodiesel), and a fundamental approach was taken to understand the effect of system phase behavior of system yields. The system was shown to successfully transesterify triolein into methyl oleate at moderate pressures and temperatures below 100°C in the presence of Nafion® NR50, a heterogeneous catalyst. The effect of pressure, temperature, methanol loading, and subsequent influence on system phase behavior was shown to have significant impact on product yield and composition. Results indicate that high FAME yields (>98% methyl oleate at 95°C) are achieved when the reaction is carried out in a three phase system (not including the solid catalyst as a separate phase) where it is assumed that the reaction mainly occurs in the now methanol rich phase now expanded with CO_2. The success of this CO_2-expanded liquid is likely attributed to increased solubility of triolein in methanol as well as increased mass transfer. Further work on system kinetics show that differences in substrate as well as catalyst active site availability can influence product fractionation. Optimization of catalyst utilization as well as differential reaction kinetics between system species can enhance system selectivity potentially producing a more effective and efficient process.

381811 Production of Biochar and Combustible Gas from Co-Pyrolysis of Agricultural Plastic Wastes and Animal Manures

Abstract:

Researchers report that manure-derived biochar, the solid product obtained from pyrolyzing animal manures, has considerable potential both to improve soil quality and reduce water pollution. One of obstacles in obtaining manure biochar is the high energy requirement of the pyrolyzing wet animal manures. The combustible gas produced from manure pyrolysis generally does not provide enough energy to sustain the pyrolysis process. Supplementing this process may be achieved with spent agricultural plastic films; these feedstocks have large amounts of available energy. Plastic films are often used in soil fumigation. They are usually disposed in landfills, which is wasteful, expensive, and environmentally unsustainable. The objective of this work was to investigate both the energetics of co-pyrolyzing swine solids with spent plastic mulch films and the characteristics of its gas, liquid, and solid byproducts. Based on laboratory-scale pyrolysis experiments, the heating value of the product gas from co-pyrolysis was found to be much higher than that of natural gas; furthermore, the gas had no detectable toxic fumigants. Energetically sustaining pyrolysis of the swine solids through the energy of the product gas could be achieved by co-pyrolyzing dewatered swine solids with just 10% spent plastic mulch. If more than 10% plastic mulch is used, the co-pyrolysis would generate surplus energy which could be used for power generation. Biochars produced from co-pyrolyzing plastic mulch and swine solid were similar to that from pyrolyzing swine solid alone based on their surface areas and ^1H NMR spectra. The results of this study demonstrated the potential of using pyrolysis technology to manage two prominent agricultural waste streams (plastic mulch wastes and swine solids) while producing value-added biochar and a power source that could be used for local farm operations. Pilot-scale experiments are being conducted currently and the highlights will be presented at the meeting.

375578 Hydrothermal Liquefaction of Wet Algal Biomass with/without Catalysts

Abstract:

A comprehensive study of hydrothermal liquefaction was conducted with and without catalysts. The strains of algal biomass used in this work were *Galdieria sulphueria*, *Coelastrella* and *Chlorella sorokiniana*. Experimental parameters studied in this work were reaction temperature (180-330°C), 10-20% biomass loading, molarity (~1.0M), and 30 min. reaction time. HCl and H

The yields of biocrude oil, bio-char, water soluble compounds and gaseous products were calculated based on the experimental results. The high heating values (HHV) of biocrude oil and bio-char were determined with a bomb calorimeter. The energy recovery calculations were performed based on the HHV values of biocrude oil and bio-char. from the biomass was calculated The biocrude oils and water soluble organics were characterized with High Throughput Time of Flight Mass Spectrometer (HT TOF-MS) and GC-MS. Based on the experimental results and biochemical compositions of the biomass used, a theoretical model has been developed to predict the biocrude oil yield.

Biorefinery on biocathodes of microbial desalination cells

Bahareh Kokabian and Veera Gnaneswar Gude
Civil & Environmental Engineering Department, Mississippi State University
Mississippi State, MS 39762, USA

Microbial desalination cells (MDCs), a recent technological discovery, allow for simultaneous wastewater treatment and desalination of saline water with concurrent electricity production. The premise for MDC performance is based on the principles that bioelectrochemical (BES) systems convert wastewaters into treated effluents accompanied by electricity production and the ionic species migration (i.e. protons) within the system facilitates desalination.

One major drawback with microbial desalination cells (MDCs) technology is its unsustainable cathode chamber where expensive catalysts and toxic chemicals are employed for electricity generation. Introducing biological cathodes may enhance the system performance in an environmentally-sustainable manner. This study describes the use of algae as sustainable biocatalyst/biocathode in MDCs. Since alga is a photosynthetic microorganism, the availability of light as well as the electron-donating anodic process may have a have significant effect on the biocathode performance. A series of experiments evaluating these effects prove that algae perform better under natural light/dark cycles and higher COD concentrations do not necessarily improve the power density. A maximum power density of 1.1 W/m^3 NCC (0.77 W/m^3 NAC) was observed for a COD concentration of 500 mg/L while the same for 1000 mg/L of COD was 0.987 W/m^3 NCC (0.69 W/m^3 NAC). This study confirms the beneficial use of algae as biocathode in MDCs. Further, the algal biomass can be utilized for recovery of valuable biocommodities such as renewable fuels (biodiesel) and food supplements via biorefinery processes.

380873 Life-Cycle Energy Use and Greenhouse Gas Emissions of Biofuels Production from Sweet Sorghum

Abstract:

Sweet sorghum is a promising feedstock to produce renewable and sustainable biofuels to alternative fossil fuels. Utilizing the juice, grain and bagasse to produce bioethanol has been studied extensively. In this study, the life-cycle energy use and green house gas (GHG) emissions of biofuels production from sweet sorghum based an advanced solid state fermentation technology (ASSF) is discussed. The bagasse obtained from ASSF fermentation was converted into diverse biofuels by various conversion pathways such as burning, anaerobic fermentation, enzymatic saccharification fermentation. The final productions were power, biogas and bioethanol. The distillation process of ethanol in ASSF stage was slightly adjusted according to different conversion pathways.

All the biofuels production from sweet sorghum based ASSF can achieve fossil energy saving, especially the biofuel-biogas pathway. Compared to GHG emissions from conventional gasoline, bioethanol-biogas production can reduce GHG emission above 70%. If the fertilizer for planting sweet sorghum is replaced partially by the biogas slurry and residues, the reduction increased further.

The results of this paper provide a comprehensive insight into life-cycle energy use and GHG emissions of multiple biofuels production pathways from sweet sorghum. It is benefit to evaluate the proper integrated process for biofuels production from sweet sorghum based the local planting condition and consideration of capital investment.

MANCHESTER 1824
The University of Manchester

AIChE

Title: Improving a process site sustainability through waste heat recovery
Paper Number: 392984
Authors: Gbemi Oluleye, Megan Jobson & Robin Smith

L287c – 1 2014 AIChE Annual Meeting Centre for Process Integration © 2014

The Challenge!

AIChE

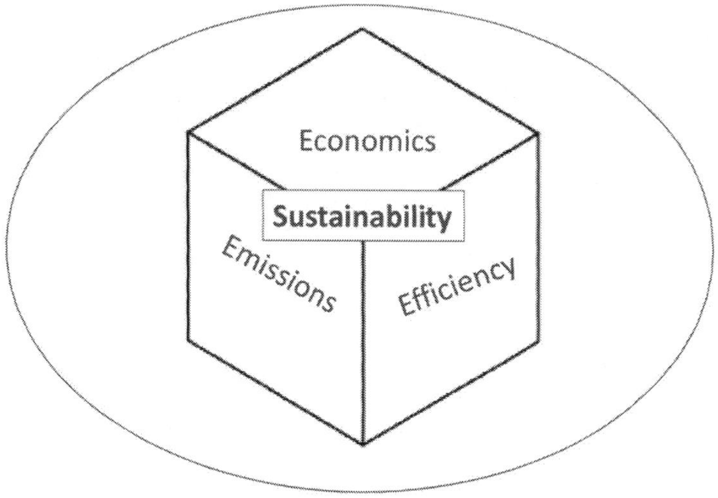

Can waste heat recovery improve process sites sustainability?

L324a – 2 2014 AIChE Annual Meeting Centre for Process Integration © 2014

AIChE

CONTENTS :

1. Introduction
2. Background
3. Objectives
4. Methodology development
5. Methodology
6. Case study
7. Conclusions and future work

L324a – 3 2014 AIChE Annual Meeting Centre for Process Integration © 2014

AIChE

1. Introduction

L324a – 4 2014 AIChE Annual Meeting Centre for Process Integration © 2014

Introduction

AIChE

- Waste heat recovery has received interest in recent years
  - ✓ Energy sources becoming less available
  - ✓ Energy sources becoming less affordable
  - ✓ Conventional energy sources becoming less acceptable
- Two-thirds of input energy for electricity generation in the USA is lost as heat during conversion processes[1]

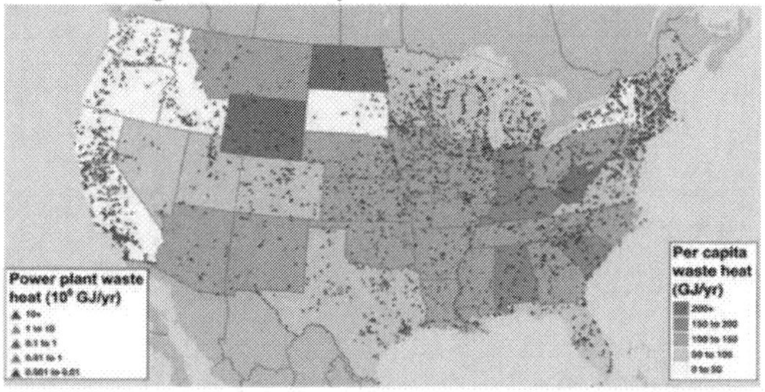

[1] Rattner AS, Garimella S. Energy harvesting, reuse and upgrade to reduce primary energy usage in the USA. Energy 2011; 36(10): 6172-83.

Introduction

AIChE

- The pinch concept has been introduced
  - ✓ Thermodynamically feasible energy targets (including heat and power targets) are set
  - ✓ Maximum energy recovery can be guaranteed

"What happens to the residual heat when a site is pinched?"

- Optimization strategies have also been introduced
  - ✓ An efficient and systematic decision-making approach to ensure capital energy trade-offs are captured is provided

"What happens when heat is available at low temperatures and there is no demand for the heat on-site?"

AIChE

2. Background

AIChE

Background

- Waste heat is available from site processes
- For a single process, the process grand composite curve can be used to show the potential for heat recovery

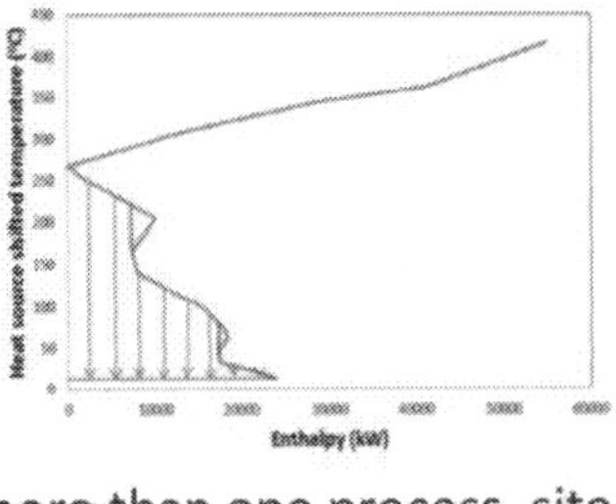

composite temperature-enthalpy curves provide a powerful tool showing the energy flow variation with temperature

- For more than one process, site profiles can be used

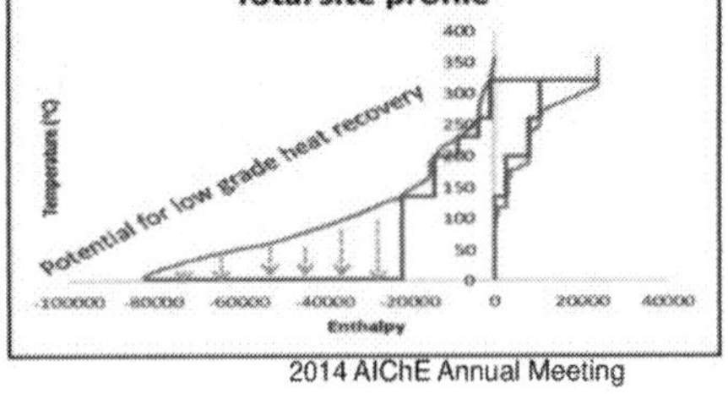

Heat from different units available over a wide temperature range

Background

- Waste heat is available from site cogeneration system and fired heaters designed to satisfy site heat and power demand

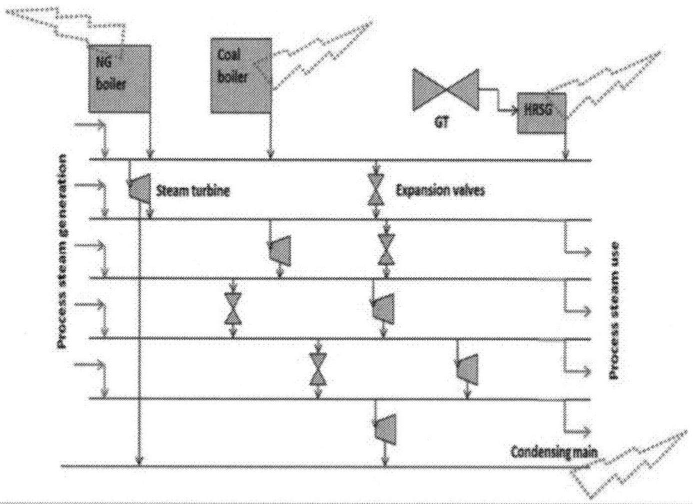

Heat available from different sources over a wide temperature range

Background

- Useful energy can be recovered from waste heat

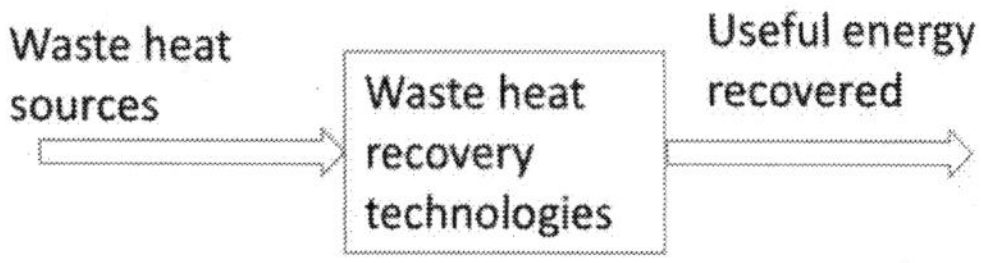

- Diverse mature and commercialized technologies exist to recover useful energy from waste heat

Organic Rankine cycle: Power generation
Working fluid: benzene and cyclopentane

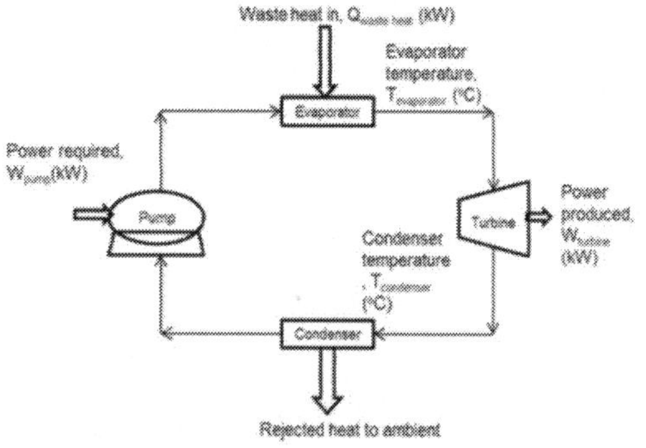

Background

- Diverse mature and commercialized technologies exist to recover useful energy from waste heat

LiBr/$H_2$0 Absorption chiller:
Chilling provision

LiBr/$H_2$0 Absorption heat pump:
Thermally upgraded heat

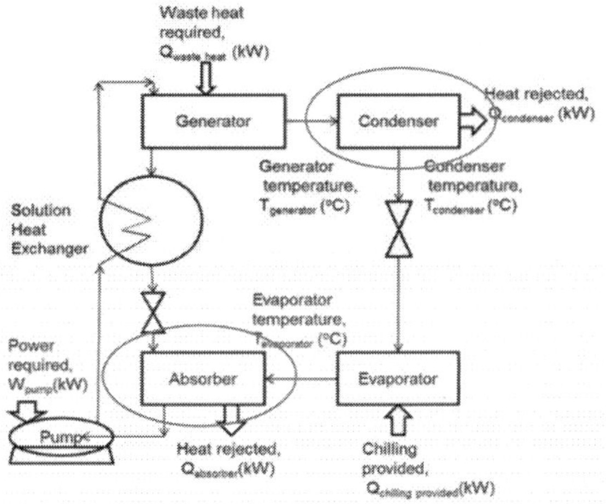

Heat is recovered from the condenser and absorber of an absorption heat pump

AIChE

Background

- Diverse mature and commercialized technologies exist to recover useful energy from waste heat

Mechanical heat pumps:
Mechanically upgraded heat
Working fluid: Ammonia

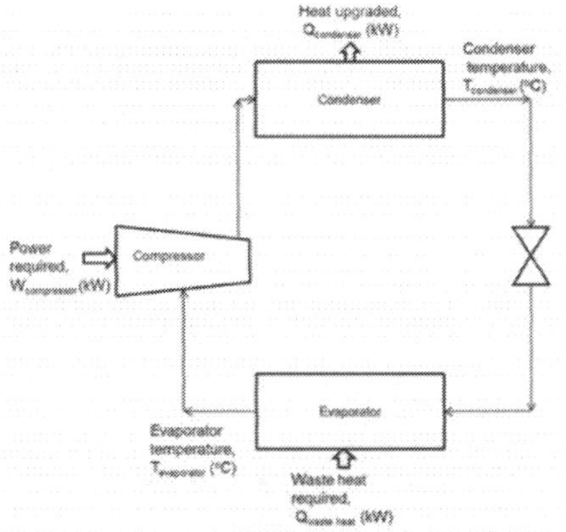

Heat exchangers:
steam and hot water

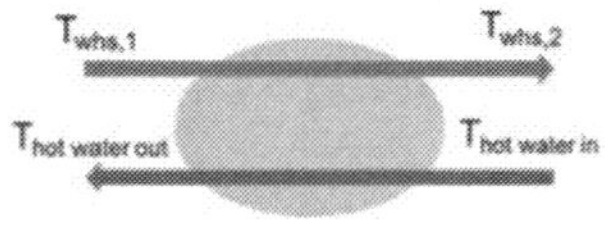

Background

AIChE

- How can the useful energy recovered be utilized?

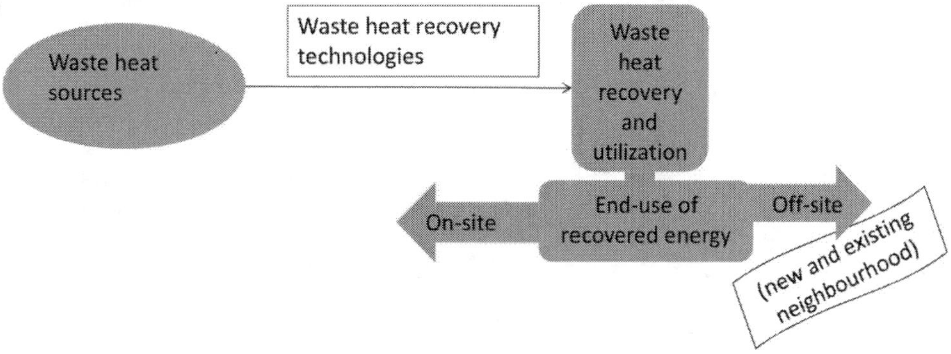

Background

AIChE

- Diverse opportunities exist to use the recovered energy

Form of Recovered Energy	On-site opportunities	Off-site opportunities
Electricity	• Reduce import on-site • Satisfy site electricity demand	• Electricity export • Charging port for electric vehicles
Chilling	• Space cooling • Gas turbine compressor inlet air chilling • Process chilling	• Space cooling
Heat	• Boiler feed water preheating • Space heating • Steam generation	• Heat export to new buildings • Heat export to existing buildings • Heat for agriculture

AIChE

Summary: Background

- Waste heat available over a wide temperature range from different sources
- Different forms of energy can be recovered from industrial waste heat
- Various opportunities exist to use the recovered energy from industrial waste heat
- The operational limits of the waste heat recovery technologies and physical limitations of a site dictate what waste heat it is physically possible to recover

AIChE

3. Objectives

Objectives

- Identify and evaluate options for recovering useful energy from industrial waste heat
- Evaluation of options done based on
  - ✓ Economics: Change in site operating costs*
  - ✓ Emissions: Change in site CO_2 emissions
  - ✓ Efficiency: Change in site energy efficiency
- Investigate how waste heat recovery can improve a process site sustainability

**including capital cost of technologies*

4. Methodology development

Methodology Development

AIChE

- To support the proposed methodology, there is need to

1. Represent the waste heat sources in a systematic way
2. Evaluate how much useful energy can be recovered
3. Determine what form of energy to recovery and temperature
4. Determine how to use the energy recovered i.e. utilization opportunities

L324a – 19 2014 AIChE Annual Meeting Centre for Process Integration © 2014

Methodology Development (1)

<Waste Heat Source Profile>

AIChE

- Representing energy profiles
  - ✓ For a process using composite curves and grand composite curves
  - ✓ Total site profile also introduced for several processes on a site
- These concepts can be extended to represent the waste heat sources in a process site[3]

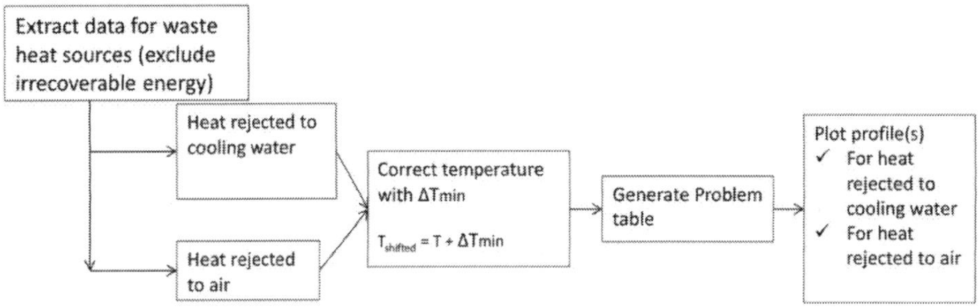

3 Oluleye G., Jobson M., Smith R., Perry S.J., 2014, Evaluating the potential of a process site for waste heat recovery, Chemical Engineering Transactions, 39, 1069-1074 DOI:10.3303/CET1439179

L324a – 20 2014 AIChE Annual Meeting Centre for Process Integration © 2014

Methodology Development (2)

<Useful Energy Recovered>

AIChE

- Work generation

$$\eta_{ideal} = 1 - \frac{T_{condenser}}{T_{evaporator}}$$

$$\eta_{real} = \frac{W_{turbine} - W_{pump}}{Q_{waste\ heat}}$$

$$\eta_{real} = Factor_{ORC} \cdot \eta_{ideal}$$

$$Factor_{ORC} = \alpha . \eta_{ideal} + \beta$$

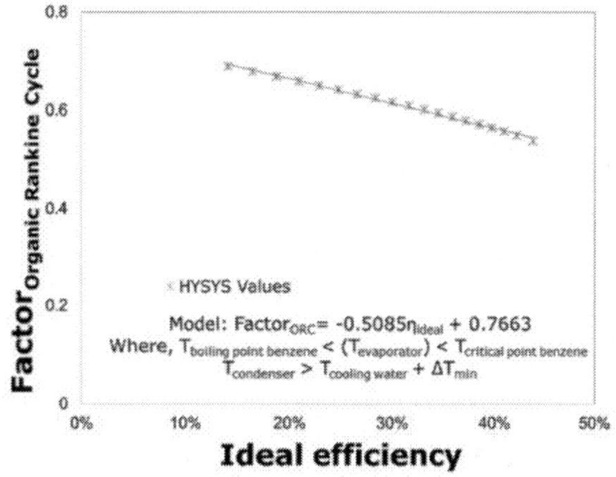

Oluleye G., Jobson M., Smith R., Perry S.J., 2014, Evaluating the potential of a process site for waste heat recovery, Chemical Engineering Transactions, 39, 1069-1074 DOI:10.3303/CET1439179

L324a – 21 2014 AIChE Annual Meeting Centre for Process Integration © 2014

Methodology Development (2)

< Useful Energy Recovered >

AIChE

- Chilling provision

$$COP_{AbC,ideal} = \left(1 - \frac{T_{condenser}}{T_{generator}}\right)\left(\frac{T_{evaporator}}{T_{condenser} - T_{evaporator}}\right)$$

$$COP_{AbC,real} = \frac{Q_{evaporator}}{Q_{waste\ heat} + W_{pump}}$$

$$COP_{AbC,real} = a.(factor_{AbC}) + b$$

$$COP_{AbC,real} = factor_{AbC} . COP_{ideal}$$

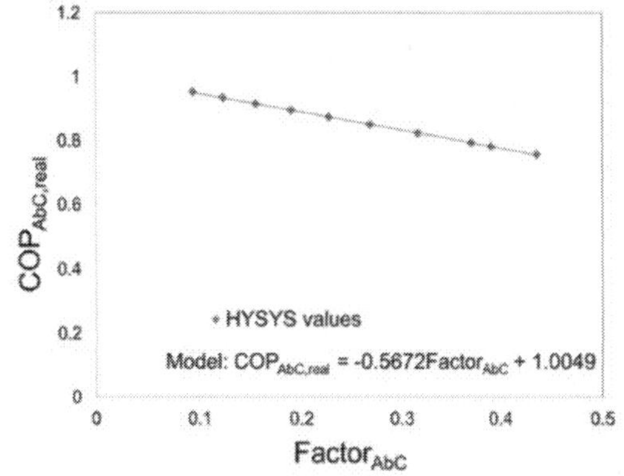

Oluleye G., Jobson M., Smith R., Perry S.J., 2014, Evaluating the potential of a process site for waste heat recovery, Chemical Engineering Transactions, 39, 1069-1074 DOI:10.3303/CET1439179

L324a – 22 2014 AIChE Annual Meeting Centre for Process Integration © 2014

Methodology Development (2)

AIChE

< Useful Energy Recovered >

- Thermally upgraded heat

$$COP_{AHP,ideal} = 1 + \left(1 - \frac{T_{condenser}}{T_{generator}}\right)\left(\frac{T_{evaporator}}{T_{condenser} - T_{evaporator}}\right)$$

$$COP_{AHP,real} = \frac{Q_{condenser} + Q_{absorber}}{Q_{waste\ heat} + W_{pump}}$$

$$COP_{AHP,real} = factor_{AHP} . COP_{ideal}$$

$$COP_{AHP,real} = a.(factor_{AHP}) + b$$

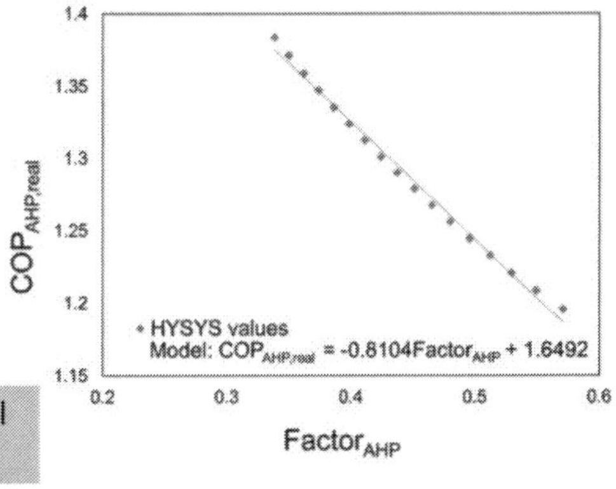

Similar models developed for mechanical heat pumps

Oluleye G., Jobson M., Smith R., Perry S.J., 2014, Evaluating the potential of a process site for waste heat recovery, Chemical Engineering Transactions, 39, 1069-1074 DOI:10.3303/CET1439179

Methodology Development (3)

AIChE

- Different forms of energy can be recovered from industrial waste heat
- Determine the best quality of heat to use to generate work, chilling, direct heat, mechanically upgraded heat and thermally upgraded heat
  - ➢ Heat recovery temperatures
- Using the exergy degradation associated with heat recovery technologies
  - ➢ use energy according to its quality
  - ➢ reduce the exergy destruction accompanying the energy utilization/conversion process

The objective is to minimize the exergy degradation accompanying the energy conversion process

Methodology Development (3)

AIChE

- Exergy degradation measures deviation from ideal performance of technologies

$$Exergy\ degradation = \frac{Performance_{ideal} - Performance_{real}}{Performance_{ideal}}$$

  - ➢ Ideal performance is the maximum possible thermodynamically
  - ➢ Real performance is corrected for by the exergy output factor
- Exergy output factor
  - ➢ Includes 'usefulness' of the heat source (waste heat)

$$Exergy\ Output\ Factor = \frac{Output\ exergy\ tranfer\ by\ heat}{Input\ exergy\ transfer\ by\ heat}$$

$$Exergy\ transfer\ by\ heat = 1 - \frac{T_{ambient}}{T_i}$$

$$Performance_{real}{}' = (Factor_{Technology} * Performance_{ideal}) * Exergy\ Output\ Factor$$

The ideal performance is synonymous with the ideal efficiency for an Organic Rankine cycle and ideal COP for an absorption system and mechanical heat pump, while the Factor accounts for irreversibility's in the technology components and working fluid non-ideality

Illustrative Example: Consider a Heat Exchanger

AIChE

- Consider two forms of recovered heat
  - ✓ Hot water at 80°C
  - ✓ LP steam generation at 130°C

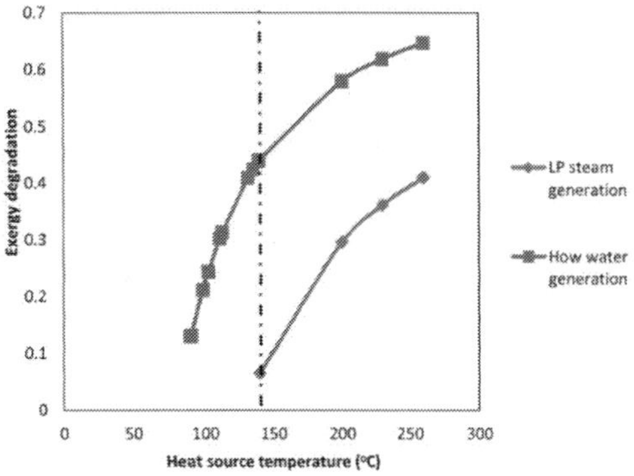

- Exergy destruction associated with steam generation is lower
- The heat source temperature for HW and steam generation should be close to the sink temperature
- In both cases, temperatures only separated by a ΔT_{min}

Methodology Development (3)

AIChE

- Similar models developed to evaluate exergy degradation associated with other technologies
  - Organic Rankine cycles
  - Absorption heat pumps
  - Mechanical heat pumps
  - Absorption chillers

Exergy degradation associated with technologies can be used to determine the form of energy to recover and its temperature, it is also necessary to determine how to use the recovered energy i.e. utilization opportunities

Methodology Development (4)

AIChE

- Ranking criterion for utilisation opportunities captures trade off between
  - ✓ CO_2 emissions reduction potential
  - ✓ Economics (costs and benefits)
- Criterion measures the economic potential associated with reduced CO_2 emissions resulting from waste heat recovery
- Estimation of the criterion depends on end-user (On- or off-site)

$$RC_{\text{on-site opportunities}}\ (£/kg) = \frac{\text{Financial Benefits } (£/y) - TAC_j(£/y)}{\text{reduced } CO_2 \text{ emissions (kg/y)}}$$

$$RC_{\text{off-site opportunities}}\ (£/kg) = \frac{\text{Profit}(£/y)}{\text{reduced } CO_2 \text{ emissions (kg/y)}}$$

TAC: Total annualised cost, j: technology

Oluleye G., Jobson M., Smith R., 2014, A hierarchical approach for evaluation of waste heat utilization opportunities, Chemical Engineering Transactions, 39, 1093-1098 DOI:10.3303/CET1439183

Example: Application of the Ranking Criterion

AIChE

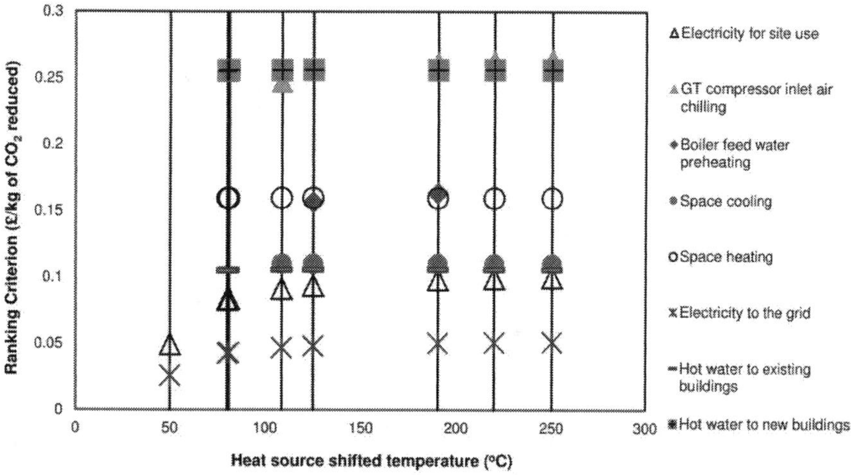

Hierarchy of opportunities depends on the waste heat source temperature

* Heat distribution cost to buildings not included in this analysis
* Temperature shifted by ΔT_{min}

Summary: Methodology development

AIChE

- A systematic way of representing the heat sources is presented i.e. the waste heat source profile
- Simple models for evaluating the recovered energy
- Exergy analysis introduced to determine the form of energy to recover and temperature
- An environ-economic criterion presented to rank and select waste heat utilization opportunities
- The operational limits of the waste heat recovery technologies and physical limitations of a site dictate what waste heat it is physically possible to recover

❑ How can a system be designed to utilize the available waste heat in a process site?
❑ Can waste heat recovery improve a process site sustainability?

AIChE

5. Methodology

L324a – 31 | 2014 AIChE Annual Meeting | Centre for Process Integration © 2014

Methodology

AIChE

- Using thermodynamic insights to determine
  - ➢ The available waste heat
  - ➢ Preliminary heat recovery temperatures
  - ➢ Working fluids for technologies
  - ➢ Recovered form of energy
  - ➢ Final heat recovery temperature
- Using the ranking criterion to determine
  - ➢ How to use the energy recovered i.e. the utilization opportunities
- Design evaluation
  - ➢ Useful energy recovered
  - ➢ Change in site energy efficiency
  - ➢ Change in site operating costs*
  - ➢ Change in site CO_2 emissions

$$\text{* Change in site operating cost} = \frac{\text{Financial Benefits (£/y)} - TAC_j(£/y)}{\text{Site operating cost before heat recovery (£/y)}}$$

L324a – 32 | 2014 AIChE Annual Meeting | Centre for Process Integration © 2014

Methodology

AIChE

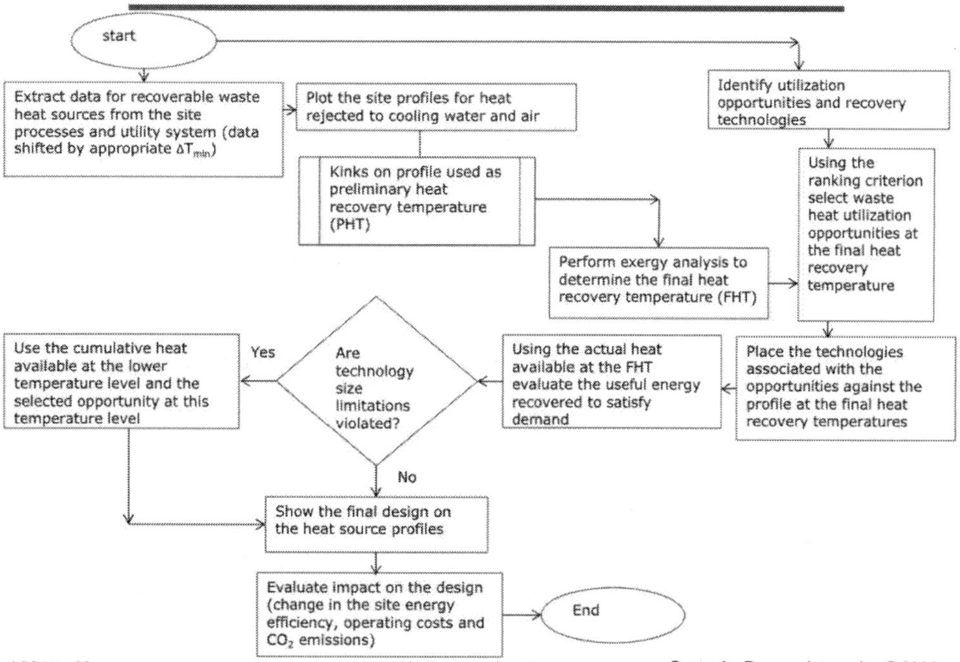

AIChE

6. Case Study

Case Study

AIChE

- Petroleum refinery[5]
  - 7 processing units: hydrotreaters (naphtha, kerosene and diesel), vacuum distillation unit, platformer, visbreaker and fluidised catalytic cracking unit
  - Cogeneration system (gas turbine, boilers, steam turbines etc.)
  - Refrigeration system
  - Cooling water system
  - Fired heaters
  - Site energy efficiency is 50%
  - Neighbourhood demand for heat: 2.4 MW
  - Power export allowed

Energy Prices (DECC, IAG 2012)

- Industrial electricity price: 13.4 p/kWh
- Industrial natural gas price: 3.2 p/kWh
- Domestic electricity price: 16.7 p/kWh
- Domestic natural gas price: 5.0 p/kWh
- Grid emission factor: 0.485 kg/kWh
- Natural gas emission factor: 0.193 kg/kWh

Capital costs

- Organic Rankine cycle [6]: 3510 £/$kW_{electricity}$
- Absorption chiller[6]: 374 £/$kW_{chilling}$
- Discount rate: 15%
- Operating hours: 8760

5. Fraser and Gillespie (1992) The application of pinch technology to retrofit energy integration of an entire oil refinery. Trans IChemE, vol. 70, Part A
6. Oluleye G., Jobson M., Smith R., 2014, A hierarchical approach for evaluation of waste heat utilization opportunities, Chemical Engineering Transactions, 39, 1093-1098 DOI:10.3303/CET1439183

Case Study

AIChE

- Design a waste heat recovery system for this refinery site
- Summary of methodology

1. Extract data for recoverable waste heat sources and plot the site profiles*

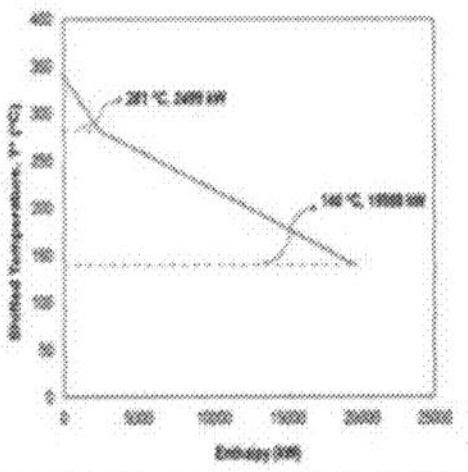

- Temperature shifted by appropriate ΔT_{min}
- Use kinks on profile as preliminary heat recovery temperatures (PHT)
- Select working fluids

2. Perform exergy analysis to determine final heat recovery temperatures and form of energy to recover

3. Using the Ranking Criterion select utilization opportunities for the recovered form of energy

4. Show selected opportunities on the heat source profiles* taking into account demand for recovered energy

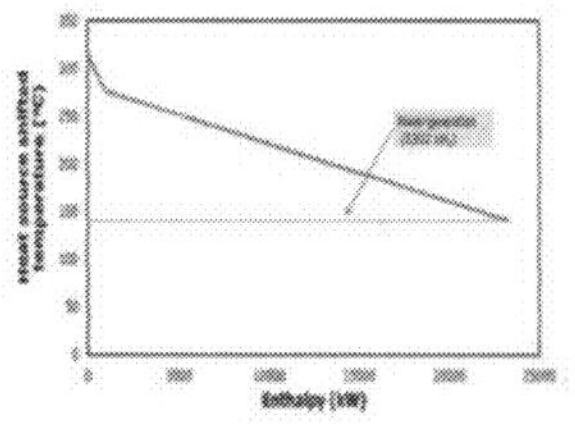

5. Evaluate the design/Results
   - ✓ Change in the site energy efficiency
   - ✓ Change in site operating costs
   - ✓ Change in site CO_2 emissions

Case Study

AIChE

❖ Heat rejected to cooling water

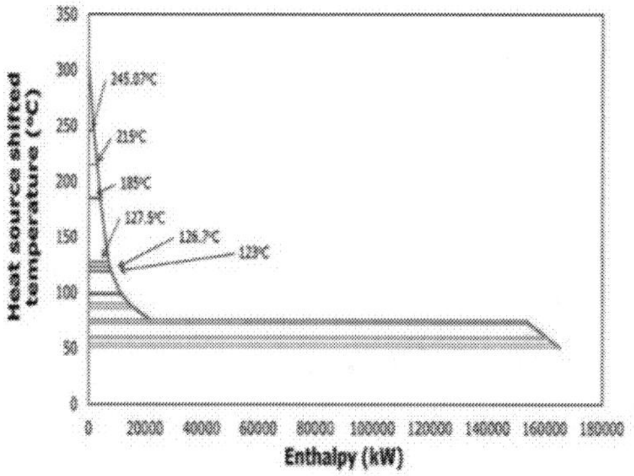

Kinks on profile used as preliminary heat recovery temperatures

❖ Heat rejected to air

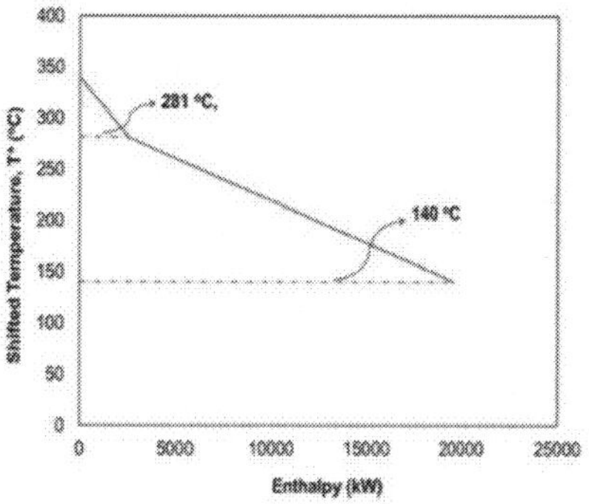

Organic Rankine cycle
- Above 80°C: benzene
- Below 80°C: cyclopentane

Absorption chiller
- Lithium Bromide/water

Mechanical Heat pump
- Ammonia

Case Study

AIChE

❖ Exergy analysis

➢ Select final heat recovery temperature and recovered form of energy to minimize exergy degradation associated with technologies

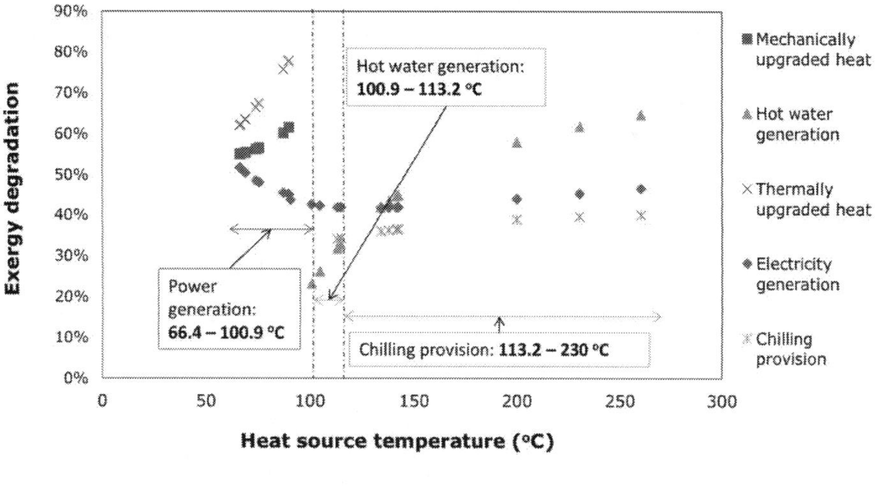

Case Study

AIChE

❖ Selected heat utilization opportunities

➢ Used the ranking criterion at the final heat recovery temperature

Final heat recovery temperature, °C	$Q_{available}$ (actual), kW	$Q_{available}$ (cumulative), kW	Utilization opportunity	Technology
113.2	7,526	11,731	• **GT inlet air chilling** • **Process chilling** • **Power generation**	• Lithium bromide/water absorption chiller • ORC using benzene
100.9	4,493	16,223	• **Hot water to new buildings** • **Power generation**	• Heat exchanger • ORC using benzene
66.4 – 100.9	148,893	165,116	• **Power generation**	• ORC using cyclopentane and benzene as working fluid

Case Study

AIChE

❖ Waste Heat Recovery System Design

✓ Design for heat that was rejected to cooling water

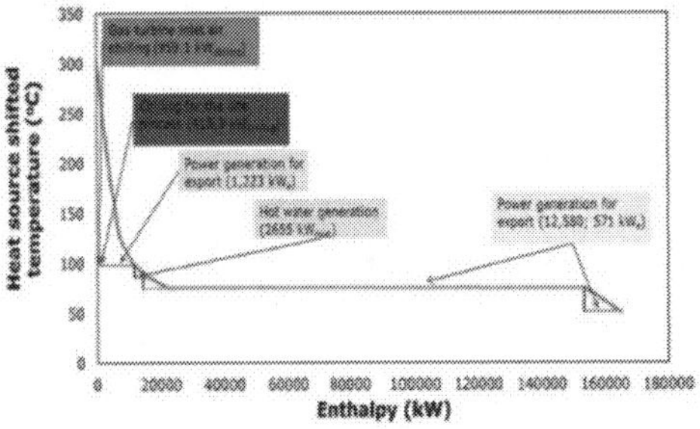

✓ Design for heat that was rejected to air

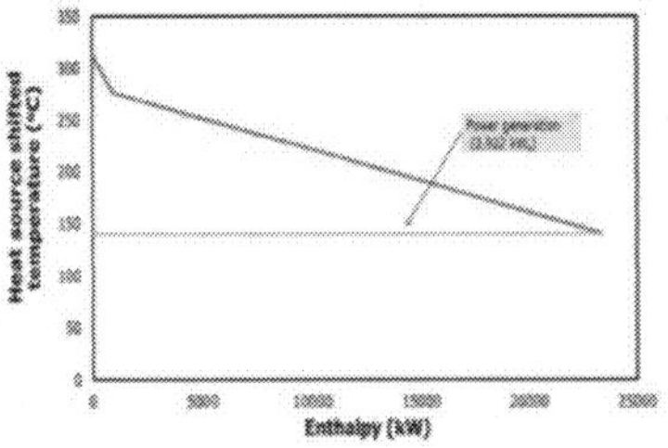

❖ Design Evaluation

Technologies required
- 4 Organic Rankine cycles
- 1 heat exchanger
- 2 absorption chillers

❖ Total useful energy recovered
- ➢ 8% increase in site energy efficiency
- ➢ 12% of available waste heat recovered

❖ Potential to reduce emissions
- ➢ Reduces site emissions by 10.73%

❖ Economics*
- ➢ Reduces the site operating cost by 24.19%

*heat distribution cost to buildings not included in this analysis

Case Study

AIChE

❖ Design Evaluation

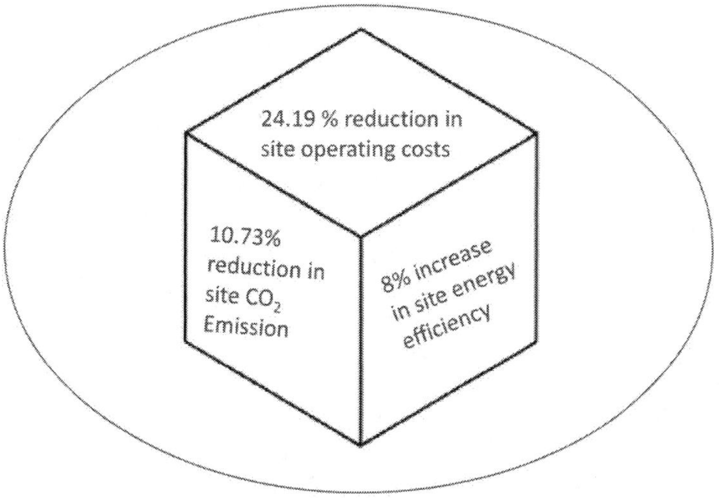

Improved site Sustainability

*heat distribution cost to buildings not included in this analysis

Summary: Case Study

AIChE

- There is potential in industrial waste heat to improve a process site sustainability
- A win-win situation for the site and it's neighbourhood since emissions and costs are reduced
- More sinks for the recovered energy can increase the benefits

AIChE

7. Conclusions and future work

L324a – 43 | 2014 AIChE Annual Meeting | Centre for Process Integration © 2014

Conclusions & Future work

AIChE

- ❑ Waste heat recovery can indeed improve the sustainability of a process site
- ❑ Waste heat from both the site process and utility system is taken into account
- ❑ Mathematical models of waste heat recovery technologies presented
- ❑ Improving performance of technologies is required to maximize the useful energy recovered
- ❑ More heat sinks are required to maximize the useful energy recovered

Future work

- Include costs associated with heat distribution for off-site use of recovered energy
- Include complex configurations of waste heat recovery technologies
- Develop an optimization framework to account for integrating waste heat recovery with a process site utility system

L324a – 44 | 2014 AIChE Annual Meeting | Centre for Process Integration © 2014

AIChE

Thank You ☺

The financial support from Process Integration Research Consortium (PIRC) is sincerely acknowledged

Multiscale connectivity for chemical mixture toxicity assessment

D.A. Sarigiannis

Environmental Engineering Laboratory, Department of Chemical Engineering, Aristotle University of Thessaloniki, 54124 Greece

Introduction

Toxicity of chemical mixtures is only partially addressed by the current state of environmental health sciences. The so-called 'cocktail' effect becomes all the more complex due to the large number of possible combinations of chemicals and other (physical or biological) stressors in the environment. This has hampered the development of rigorous methodologies for tackling the issue of environmental mixture safety. Currently, our scientific understanding and policy for environmental mixtures are based largely on extrapolating from, and combining, data in the observable range of single chemical toxicity to lower environmental concentrations and composition - i.e. using higher dose data to extrapolate and predict lower dose toxicity. More precise approaches to characterize toxicity of mixtures are needed. A major obstacle to the development of effective mixture risk assessment methodologies is the possibly infinite ways of combining chemicals into actual environmental mixtures [1]. It should be noted, however, that although in theory the number of combinations of chemicals or stressors is infinite, the number of biological processes is finite. Therefore, in considering an integrated approach for risk assessment, it makes more sense to work on the finite biological processes that may be affected by human exposure to these mixtures rather than the infinite combinations of chemicals and stressors [2].

Integrated health impact assessment of environmental stressor mixtures would need to follow a 'full chain' approach to take into account all relevant health stressors and their interaction. Application of the full chain approach entails considering all possible exposure pathways via the environment and lifestyle choices. It also encompasses considering the effects of co-exposure to relevant stressors and how risk modifying factors such as age, diet, gender, and time window of exposure affects the final physiological response. It is clear that successful application of this approach poses demanding data requirements both in terms of environmental monitoring and in terms of biological and clinical data interpretation. What is most important is the need for comprehensive data interpretation of the molecular, biochemical and physiological processes that couple exposure to health outcome.

This requires forging a new paradigm for interdisciplinary scientific work in the area of environment and health. We shall call this the connectivity paradigm for chemical risk assessment, denoting an approach that builds on the exploration of the interconnections between the co-existence of multiple stressors and the different scales of biological organisation that together produce the final adverse health effect [3]. Connectivity marks a clear departure from the conventional paradigm, which seeks to shed light on the identification of singular cause-effect relationships between stressors and health outcomes. It entails creating a new way of combining health-relevant information coming from different disciplines, including (but not limited to) environmental science, epidemiology, toxicology, physiology, molecular biology, biochemistry, mathematics and computer science [4]. In this paper we discuss the integration of these different information classes into a unique framework to better inform and support public health impact assessment of chemical mixtures in the environment.

Methodology

In the context of applying systems biology approaches to chemical mixtures risk assessment, we move forward towards the definition of a co-exposure biology using optimally physiology and systems biology based modeling and data mining and assimilation algorithms. The aim is to couple the systems biology approach to the gene-environment interactions with the corresponding "physiome" approach. The variable levels of biological organization involved in this holistic view of mixture toxicology and cumulative exposure and risk assessment suggest that different technologies need to be brought to bear in order to obtain a comprehensive view of how co-exposure to multiple chemicals affects the overall phenotypic response of individuals. Technological variability introduces the need for better data integration and assimilation and for the development of novel data analysis and hypothesis generation and testing procedures in order to best elucidate the biological mechanisms underlying mixture toxicity. On the basis of the data currently available, the best available techniques, and the need for data integration, a tiered approach has been developed and outlined herein to support the connectivity approach to the assessment of health risk associated to combined exposure to multiple chemical stressors [5]. This exposure biology approach to mechanism-based risk assessment of environmental chemical mixtures can be tackled with an integrated, multi-layer computational methodology, ideally comprising the following steps:

a) Characterization of exposure factors quantifying the parameters that affect human exposure to environmental chemicals, such as time-activity relationships, seasonal and climatic variation, and consumer choice. These exposure factors can be used to derive aggregate and cumulative exposure models, leading in probabilistic exposure assessments. Aggregation can be done across exposure pathways and routes and even across different exposure scenarios, if the relevant exposure metric or the imputable biological or physiological effect can be related to these scenarios. For instance, exposure to volatile organic compounds (VOCs) such as benzene or toluene and mixtures thereof may occur both from environmental media and in specific occupational settings. A cumulative exposure scenario for these substances would have to take stock of the actual variability of exposure across these different settings throughout typical days for the same period in an individual's lifespan.
b) Current toxicological state of the art combines estimations of biologically effective dose with early biological events to derive dose-effect models, which can be used in combination with the probabilistic exposure estimates to derive biomarkers of exposure and/or effect. Combined use of epidemiological, clinical and genetic analysis data may shed light on the effect of risk modifying factors such as lifestyle choices and DNA polymorphisms. Observation of real clinical data and /or results of biomonitoring, if coupled with the exposure/effect biomarker discovery systems, can produce biomarkers of individual susceptibility and thus allow estimations of individual response to toxic insults. Toxicogenomics, comprising transcriptomics, proteomics and metabolomics, and adductomics (considering adducts of xenobiotics not only to DNA but also to proteins such as albumin) are key technologies to this kind of analytical and data interpretation process.
c) The analysis of the biomarker data (including results on biomarkers of exposure, effects and individual susceptibility) results in the integrated assessment of risk factors. Use of information on risk factors with molecular dosimetry data (i.e. estimation of the actual internal and biologically effective dose of xenobiotic substance found in the target organ and, indeed, perturbing cellular response) enables population risk studies to be done, by converting generic exposure profiles into population risk metrics having taken into account inter-individual variability of response and exposure uncertainty.

The connectivity approach was applied on data from a Europe-wide campaign on environmental and biological monitoring of a virtually ubiquitous mixture of volatile organic compounds, i.e. benzene, ethylbenzene, toluene and o-, m- and p-xylene, aldehydes such as formaldehyde and acetaldehyde and a complex mixture of polyaromatic hydrocarbons, which is typical of combustion products [6]. The full array of –omics technologies outlined above were applied to samples of indoor air and dry blood spots and urine of exposed subjects from almost all European Union capitals. Exposure assessment was completed with detailed time activity diaries and questionnaires regarding smoking and dietary habits (especially with regard to alcohol consumption) [7,8]. Gene expression results were processed using clustering algorithms to derive heat maps demonstrating clearly the differences in the biological perturbations caused by parts of the indoor and ambient airborne chemical mixtures in the sampled sites.

Results

Dry blood spot samples from a subset (n=50) of the population which participated in the personal exposure study were analysed with the –omics technologies and state-of-the-art bioinformatics software. The Agilent MicroArray Express was used for analysis of the blood samples and comparison with controls and GeneSpring, the Agilent data analysis system, and in-house bioinformatics analysis software developed on R and Stata was used for –omics data analysis.

The sample analysis followed a hybrid, tiered approach starting from an agnostic search in whole genome mRNA extracted from the biological samples after appropriate sampling and storage at -80°C. The population data were clustered by group of dominant source of exposure to airborne chemicals including ambient air in cities, typical airborne mixtures in dwellings, schools, kindergartens and other indoor professional environments (public buildings) and samples from people occupationally exposed to fly ash from coal-fired thermal power plants. $p < 0.005$ was used as test of statistical significance revealing the level of gene expression modulation against the controls – in this case only the genes that were up- or down-regulated by more than two-fold compared to the controls were selected. This agnostic search indicated the presence of exposure-specific signatures using gene expression data, especially when considering not just the number but also the loci of the genes that showed the most important differences in expression levels after exposure to the respective mixtures of xenobiotics. Both common parts and clearly distinct regions of the genes with significant modulation in expression were identified when comparing biological samples of subjects exposed to ambient air chemicals against the ones of subjects exposed non-professionally to indoor chemicals and professionally to fly ash (figure 1). Comparative analysis showed that ambient air chemicals in urban settings across Europe affect a wide spectrum of genes (n=376), most of which are up-regulated (n' = 209); a significant number of genes in the same samples was down-regulated (n''=167). Fly ash had an almost as high an effect on the human genome. Here there were fewer genes that showed modulation in expression (n=214). Unlike ambient urban air chemicals, fly ash from power plants resulted in limited up-regulation (n'=66) and more important down-regulation of gene expression (n''=148). Indoor air chemicals had the smallest effect on gene expression modulation. In this case the total number of genes that modulated their expression levels was contained (n=145). Up-regulated genes were dominant (n'=92) and down-regulated genes were relatively limited (n''=53). A number of genes showed very distinct expression patterns when cross-comparing samples from subjects exposed to different types of airborne chemicals. These could be isolated to serve as exposure signatures at the molecular level. Clearly, such genomic biomarkers of exposure do not imply causal associations with adverse health outcomes. However, they may serve as seed information to the formation of biologically plausible mechanistic hypotheses on adverse outcome pathways (AOPs). These hypotheses help in limiting the investigation space for AOPs without compromising the agnostic nature of the initial analyses.

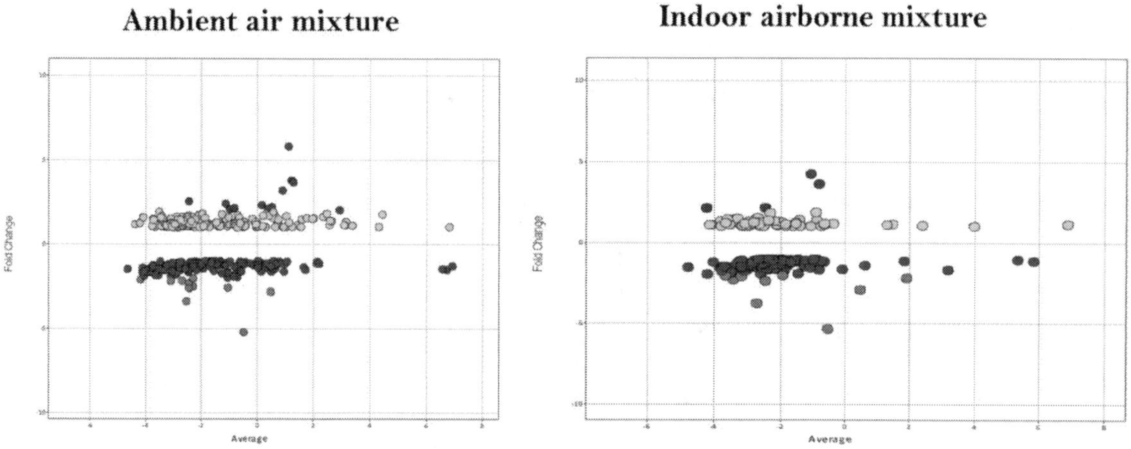

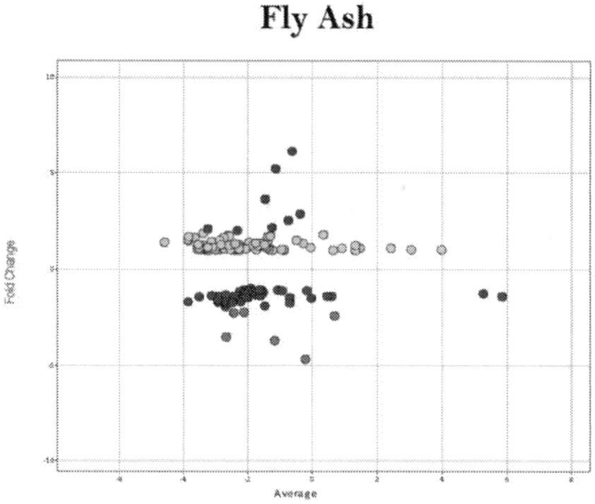

	# genes	Up	Down
Ambient air	376	209	167
Indoor air	145	92	53
Fly ash	214	66	148

Figure 1. Comparison of gene expression modulation from biosamples of population exposed to different types of airborne chemical mixtures: (a) ambient air chemicals (top left); (b) characteristic airborne chemicals found in indoor settings (top right); (c) chemicals found in fly ash from thermal power plants using fossil fuel (bottom left). The table at the bottom right of the figure shows the total number of gene expressed differentially in subjects exposed to the different airborne chemical mixtures and the number of genes up- and down-modulated respectively.

The results of the agnostic transcriptome search on biological samples with regard to the identification of potential genomic signatures that can serve as reliable exposure biomarkers were corroborated by means of in vitro testing. For this reason, samples of the airborne chemicals taken in the various environments considered in this study were extracted and applied on cell lines covering both lung epithelial cells (A549) and skin keratinocytes (Ha-CaT). Cluster analysis of the gene expression results are depicted in the heat map shown in figure 2. It is very clear that indoor chemicals show distinctly different patterns of gene expression modulation in both types of matrices, especially when compared to the mixture of indoor air chemicals and PAHs from the ambient air in urban settings. It should be noted that the latter corresponds to realistic total exposure patterns of the human population in European metropolitan areas. Exposure to ambient air chemicals only shows intermediate results, indicating that co-exposure to ambient and indoor air chemical mixtures in cities may have more than additive effects on gene expression modulation.

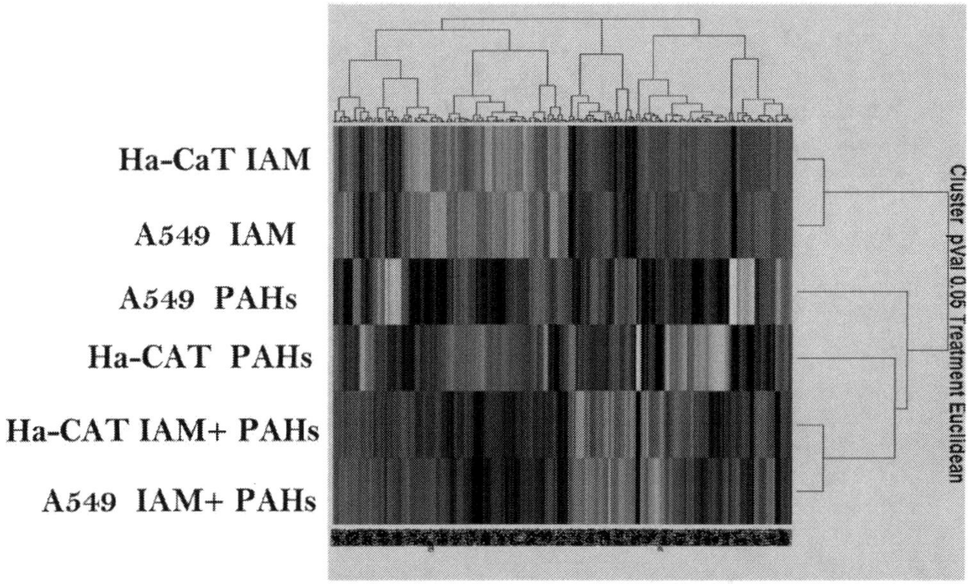

Figure 2. Heat map representing comparative cluster analysis of airborne chemical mixtures on dermal (Ha-CAT) and lung epithelial (A549) cells. IAM: indoor air mixture; PAHs: the mixture of 19 polyaromatic hydrocarbons commonly found in the ambient air in European metropolitan areas

After the agnostic tier it is possible to identify not only single genes that have shown significant modulation in expression levels, but also determine the biological pathways that are regulated by gene networks that were significantly modulated with regard to their induction levels from exposure to xenobiotics. Pathway analysis using Agilent GeneSpring and the on-line PANTHER – Protein ANalysis THrough Evolutionary Relationships) Classification System (http://www.pantherdb.org). showed that two key pathways, the p53 and oxidative stress were differentially modulated from specific chemical families such as aldehydes, while specific genes or gene sequences could be characterized as molecular markers of exposure. In figure 3 comparative analysis of the gene regulation induction after exposure to different chemical classes of xenobiotics, which comprise a large part of the indoor air mixture found in European dwellings is shown, focusing on the genetic network regulating oxidative stress response at the cellular level. More specifically, the samples of indoor air chemical mixtures taken from different European cities were decomposed and sequentially extracted to analyze separately the effect on gene expression of major chemical families, including: (a) carbonyl compounds such as aldehydes, (b) phenols and other aromatics, and (c) terpens.

Aromatics had the highest impact in terms of inducing oxidative stress responses. Terpens had the second highest impact in terms of activation of parts of the oxidative stress regulatory network. Aldehydes, and in particular formaldehyde, induced among others a specific gene, p38, which is central to the regulation of oxidative stress response in human cells. This gene (p38) could serve as a reliable biomarker of oxidative stress (intermediate effect) associated to exposure to formaldehyde, since no other treatment seemed to modulate its expression levels.

Similar results were found when exploring the modulation of the gene network that regulates the function of the p53 pathway, controlling thus cell cycle and death and thus the onset of carcinogenesis in case of faulty operation and down-regulation. In this case the indoor air chemical mixture as a whole has the highest impact. It seems to induce the expression of the overall regulatory network thus leading to significant up-regulation compared to the controls. Aromatics and terpens have similar size effects on gene network regulation. Terpens are more involved in processes regulating apoptosis, as well as inhibition of angiogenesis and metastasis. Aldehydes are involved in processes leading to inhibition of angiogenesis and metastasis too. Thus, different chemical families show different toxicity profiles, not only when associating them with phenotypes of disease such as cancer, but also, and possibly more importantly, when elucidating their role in biological pathways of toxicity that may (or may not) lead to adverse health outcomes.

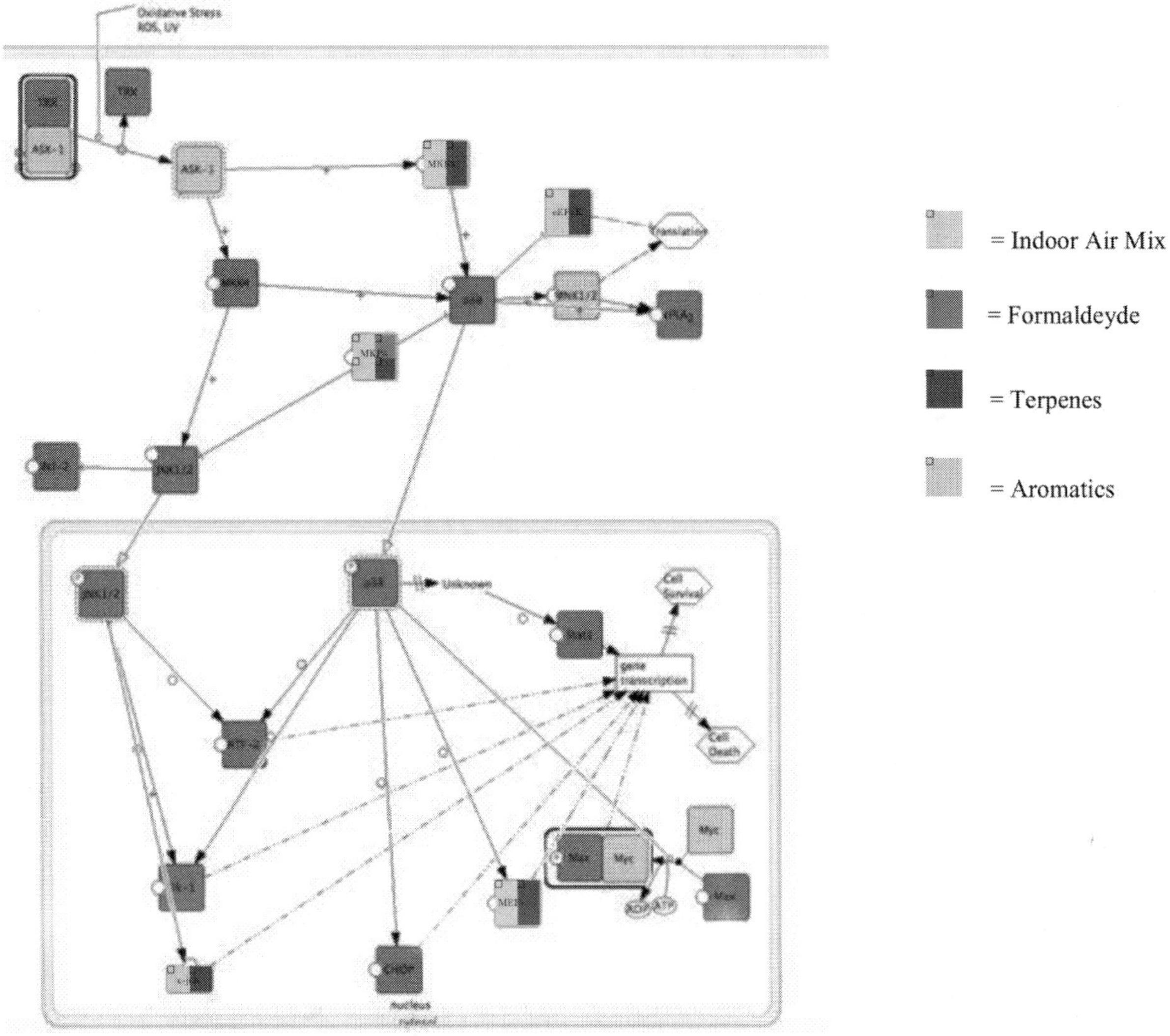

Figure 3. Modulation of the oxidative stress pathway in blood samples of subjects exposed to different airborne chemical families reveals molecular signatures of co-exposure

From the chemical speciation of the air samples taken in the different European urban settings in this study, typical airborne chemical mixtures were identified for indoor and ambient air as a function of latitude. For example, with regard to volatile organic chemicals, compounds that are essentially ubiquitous in Europe, air mixtures were richer in benzene in the south compared to the north of Europe, where toluene seemed to be more abundant. Following through to the biological processes that were perturbed by exposure to these airborne chemical mixtures we found that during acute (short-term) exposure signal transduction and mRNA transcription were modulated the most following an inverse dose-response function. When chronic (longer-term) exposure results were analyzed, protein metabolism, mRNA transcription regulation and cell proliferation and differentiation were the main mechanisms that were modulated following a normal dose-response behavior. The salient results of this tier of the connectivity analysis are given in figure 4 for mixture A (the benzene-richer mixture, typical of the European south) and for mixture B (the one that was relatively richer in toluene and other chemically similar solvents, typical of central and northern Europe).

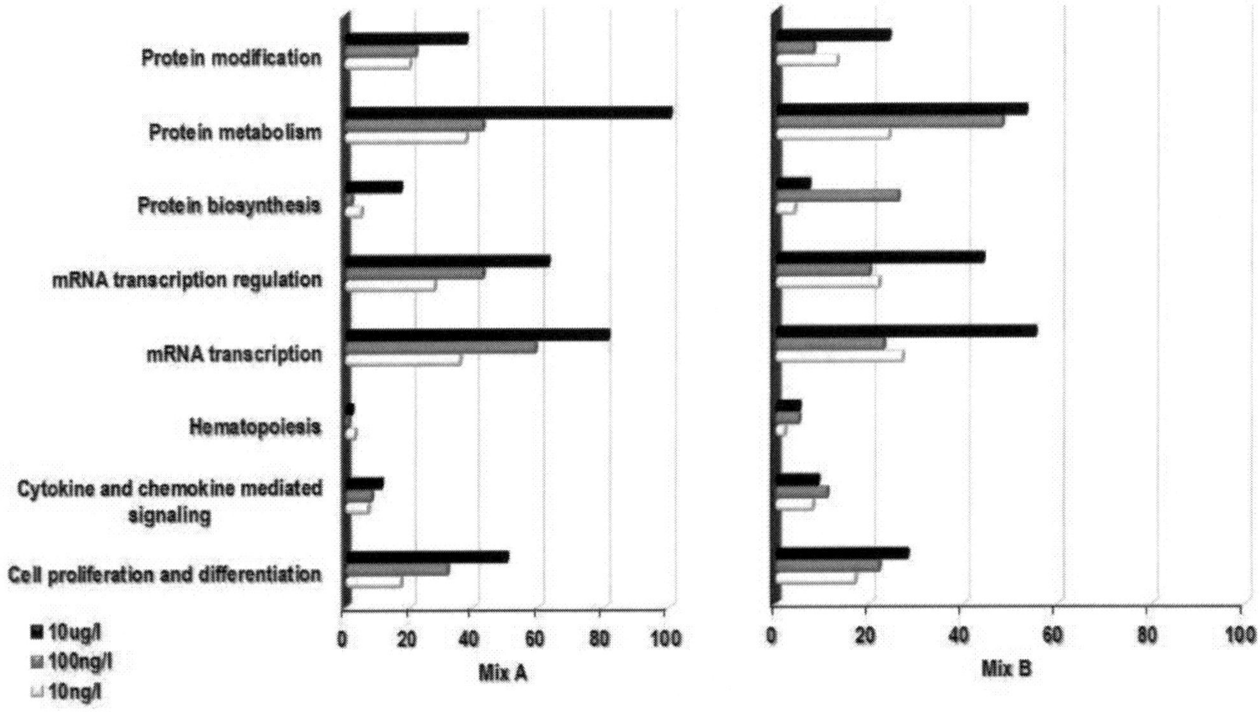

Figure 4. Biological processes induced by the two characteristic types of indoor air mixtures in Europe

In addition, the mixtures that were richer with non-carcinogens were the ones that had the highest impact on the biological processes induced during short-term exposure; on the contrary, mixtures richer with carcinogens such as benzene induced a higher response in terms of biological process induction to the individuals who were exposed longer. This behavior was confirmed by untargeted metabolomics profiling, which showed a relative increase in benzene metabolites such as s-mercapturic acid as well as free, non-metabolized benzene. Phenotypic observations confirm that chronic exposure to carcinogenic VOCs such as benzene could increase the risk of leukemia. Taking into account the metabolic processes and interactions (e.g. competitive inhibition) that regulate the effective metabolism of the VOCs to which the European population was exposed we estimated the actual risk of cancer from the combined exposure to such airborne chemical mixtures released from fuel and consumer products. Focusing even further on the molecular underpinning of the induced modulation of the human capacity to metabolize VOCs under co-exposure conditions, targeted analysis of the CyP450 was performed using Taqman Fluidigm microarrays. Results showed a significant variation in the expression of genes involved in the coding of the enzymes involved in benzene and toluene metabolic chains. This variation results in reduction of the amount of available enzymes and thus enhances the effect of competitive inhibition for the limited amount of receptor sites at the metabolically active tissues. The results show suppression of CyP450 metabolic capacity after co-exposure to the four VOCs. These findings were translated into quantitative change in the metabolic rates captured in the biology-based dose-response (BBDR) model developed in-house to estimate cancer risk by applying dynamic flux balance analysis and coupling the gene regulatory network with the ADME model.

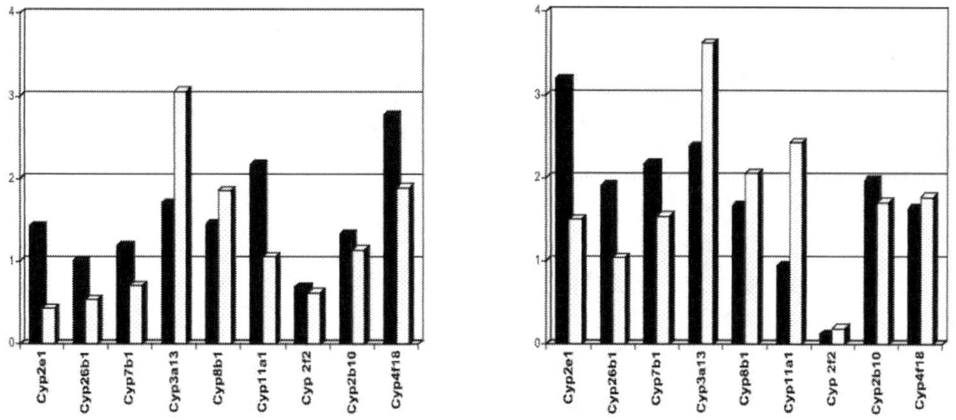

Figure 5. Comparison of CyP450 expression modulation for benzene-rich (black bar) and toluene-rich (white bar) quaternary BTEX mixtures after 4 (left figure) and 24 h (right figure) of exposure.

Concluding Remarks

Following the integrated multiscale connectivity methodology coined herein it is possible to identify the molecular pathways and biological processes that underlie the onset or exacerbation of disease phenotypes associated to co-exposure to chemical mixtures. This approach sheds light to the respective pathways of toxicity and assigns causal associations between environmental stressors and human health.

References

1. Mason AM, Borgert CJ, Bus JS, Mumtaz MM, Simmons JE, Sipes IG (2007) Improving the scientific foundation for mixtures joint toxicity and risk assessment: contributions from the SOT mixtures project – introduction, *Toxicology and Applied Pharmacology* 223:99–103.
2. Liao KH, Dobrev ID, Dennison JE Jr, Andersen ME, Reisfeld B, Reardon KF, Campain JA, Wei W, Klein MT, Quann RJ, Yang RSH (2002) Application of biologically based computer modeling to simple or complex mixtures, *Environmental Health Perspectives* 110(S6):957-63.
3. Workman CT, Mak HC, McCuine S, Tagne JB, Agarwal M, Ozier O, Begley TJ, Samson LD, Ideker T. (2006) A systems approach to mapping DNA damage response pathways. *Science* 312(5776):1054-9
4. Kitano H (2002) Computational systems biology, *Nature* 420:206-10.
5. Sarigiannis, D., Gotti, A., Cimino Reale, G. and Marafante, E. Reflections on new directions for risk assessment of environmental chemical mixtures, *Int. J. Risk Assessment and Management,* (2009) 13(3/4): 216–241.
6. Kotzias, D., Geiss. O., Tirendi, S., Barrero-Moreno, J., Reina, V., Gotti, A., Cimino-Reale, G., Casati, B., Marafante, E., Sarigiannis, D. Exposure to multiple air contaminants in public buildings, schools and kindergartens – the European indoor air monitoring and exposure assessment (AIRMEX) study, *Fresenius Environmental Bulletin* 18 (5A): 670-681 (2009).
7. Sarigiannis D.A., Karakitsios S.P., Gotti A., Liakos I.L., Katsoyiannis A. Exposure to major volatile organic compounds and carbonyls in European indoor environments and associated health risk. *Environment International* (2011) 37: 743-65.
8. Karakitsios S., Asikainen A., Garden C., Semple S., De Brouwere K., Galea K.S., Sánchez-Jiménez A., Gotti A., Jantunen M., Sarigiannis D. Integrated exposure for risk assessment in indoor environments based on a review of concentration data on airborne chemical pollutants in domestic environments in Europe. *Indoor and Built Environment* (2014) doi: 10.1177/1420326X14534865.

AIChE The Global Home of Chemical Engineers

Double Containment Piping Systems: A Fail-Safe, Leak-Free Solution

Darin Johnson, P.Eng.
Applications Engineer
IPEX USA LLC

2014 AIChE Annual Meeting
EHS and Sustainability

Tuesday, November 18, 2014

AIChE The Global Home of Chemical Engineers

Agenda

- Why Double Containment Piping Systems?
- System Design Considerations
  - Material Selection
  - Thermal Expansion and Contraction
  - Layout
- Leak Detection
- System Installation
- Applications

2

Why Double Containment Piping Systems?

- December 1988 – A revision to the ***Resource, Conservation, and Recovery Act*** in the United States regarding underground piping.
  - Required a majority of underground tanks and their associated pipes intended to transport/store petroleum fluids and/or chemical waste be provided with special means to prevent leaks against the surrounding environment.
- U.S. Environmental Protection Agency (EPA) requirement:
  - Code of Federal Regulations (CFR), Title 40, Part 280, Regulations Pertaining to Underground Storage Tanks.
  - Code of Federal Regulations (CFR), Title 40, Part 302.4, Designation of Hazardous Substances.

3

Why Double Containment Piping Systems?

- **Code of Federal Regulations (CFR 40 part 280) requirements:**
  - **Underground Tanks and associated Piping storing:**
    - Petroleum Products
    - Hazardous Substances (over 3200, CFR Title 40 Part 302.4)
    - Hazardous Wastes
  - **Contain and divert all releases to a proper collection system until they are detected and removed**
  - **Containment may be via trench or double wall piping**
  - **Containment must host leaking product for 30 days**
  - **Must be inspected (monthly or yearly) via leak detection**

4

Why Double Containment Piping Systems?

- **Above Ground Storage Tanks**
  - If the associated buried piping consists of 10% or more of the tank volume, it also must be contained.
  - If the piping is less than 10% of the volume of the aboveground tank, and you choose not to double contain the buried piping, ***responsibility for any clean-up or remediation resulting from a release will still fall on the Client.***

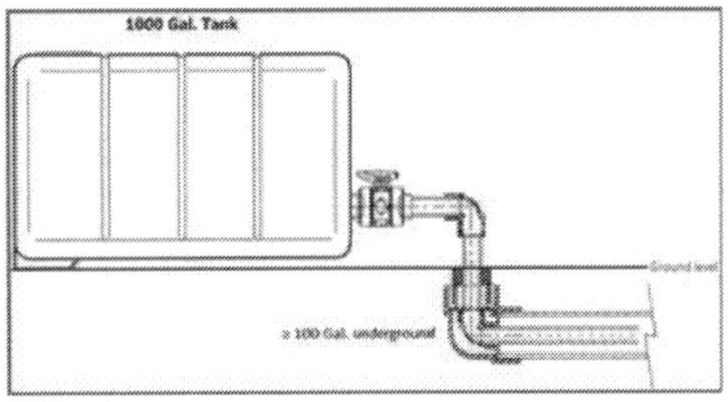

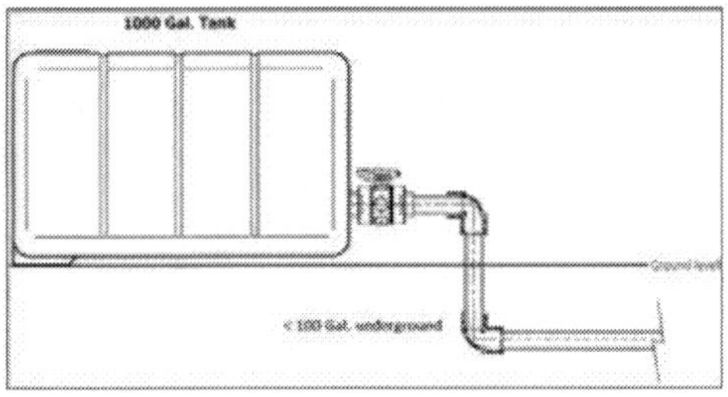

5

System Design - Nomenclature

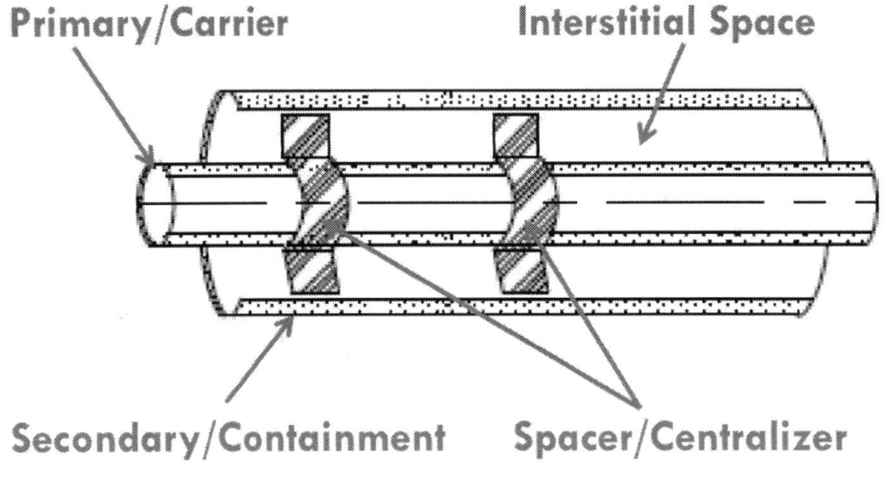

6

System Design - Material Selection

- **Material Selection**
  - Primary Pipe
    - Chemical Compatibility
    - Pressure Capabilities
    - Thermal Properties
  - Secondary Pipe
    - Installation Environment
    - Loading (Static & Live) for buried applications
    - Leak detection requirements
    - UV requirements

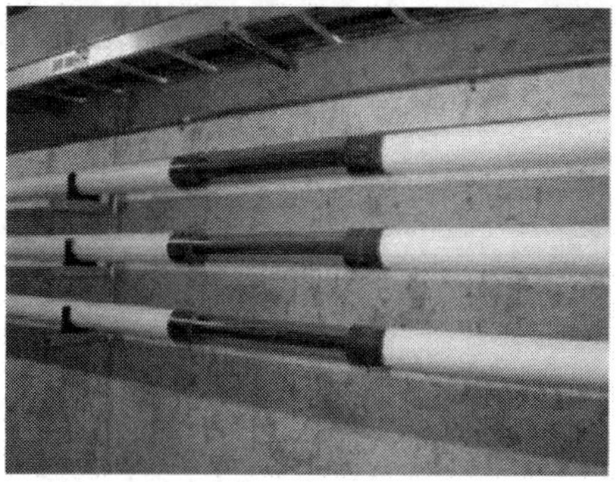

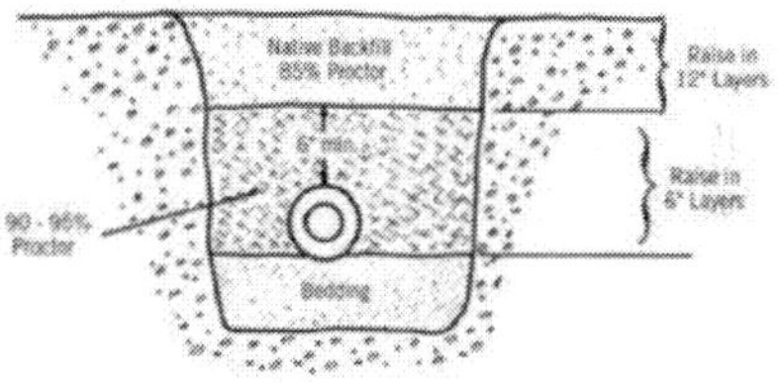

System Design – Thermal Expansion and Contraction

- **Thermal Expansion & Contraction**
  - Two basic thermal relationships may be present:
    - When only one of the two systems experiences thermal expansion/contraction.
    - When both systems experience different magnitudes of thermal expansion/contraction.
  - The Floating System
  - The Restrained System
  - Partially Restrained System

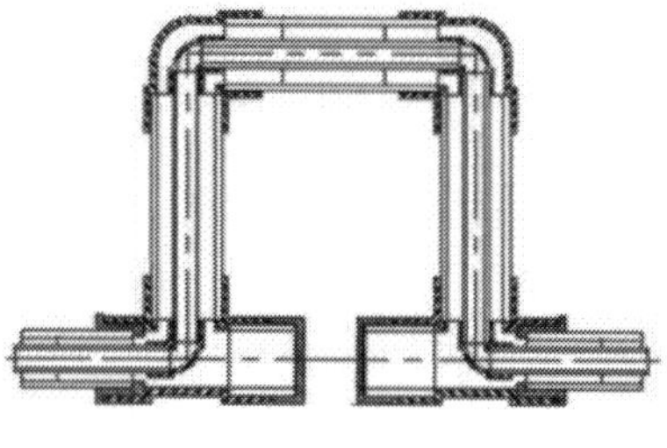

AIChE
The Global Home of Chemical Engineers

System Design – Layout

- There are three major issues that must be considered when designing the piping layout
  - OVERALL size of the double containment components
  - Proper venting and draining of the system (primary and secondary)
  - Termination of the secondary system at different points in the design:
    - Transition points (double wall to single wall)
    - Floor Drains
    - Clean-outs

9

AIChE
The Global Home of Chemical Engineers

Leak Detection

- The following forms of leak detection are acceptable as per 40 CFR 280 Regulations:
  - Drainage Applications
    - Visual
    - Inspected every 30 days
  - Pressure Applications
    - Electronic Low point sensors
    - Cable
    - Visual should be used in conjunction with electronic monitoring
    - Inspected once per Year

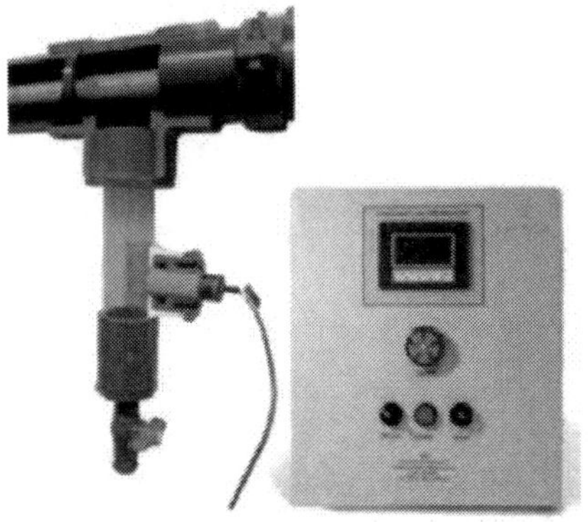

System Installation

- Designers should be familiar with the various methods of installing DC systems
  - Chemical Welding (Solvent Cementing)
  - Gasketing
  - Butt Fusion
  - Socket Fusion
  - Electrofusion
  - Bonding
  - Hot Air Welding
- Each manufacturer should provide installation instructions
- All installers should be trained and certified

11

Applications

- Underground Applications
- WTP & WWTP
- Oil & Gas
- Hospitals
- Museums
- Food and Beverage
- Pharmaceuticals
- Laboratories
- Data Centers

Conclusion

- Secondary material does not need to be the same as the primary material
- Secondary connections do not need to be the same as the primary connections
- Layout according to secondary pipe sizing
- The containment system is a failsafe, temporary system that is not required to operate within the same parameters as the primary piping
- Systems should only be installed by trained personnel
- Engineered systems that are designed to compensate for the interaction between primary and secondary pipe should be specified
  - Never allow for jobsite assembled systems

380428 Humic Substances in Treatment of Water Contaminated with Hydrocarbons

Abstract:

Humic substances facilitate the degradation of hydrocarbon contaminants in water, making them bioavailable and stimulating microbial metabolic activity. In this paper the effect of adding humic substances to water samples contaminated with diesel and oil, using a microbial consortium with bacteria of the genus Pseudomonas was studied. The variation is analyzed in the concentration of pollutants by determination of organic carbon, inorganic carbon, total carbon, chemical oxygen demand and biomass production in periods of 12 hours. A 80 to 95% was obtained without the presence of humic substances and of 35 to 70% when a concentration of 1000 mg / L in the samples was added. The values of organic carbon, and chemical oxygen demand are higher in the samples with added humic substances, so that the removal of hydrocarbon contaminants could be higher. However, the tests are not selective for hydrocarbon contaminants or humic substances.

389498 Sustainability Assessment and Performance Improvement of Electroplating Systems

Abstract:

The surface finishing industry in general, the electroplating industry in particular, is one of the most polluted industries in the U.S. In electroplating systems, various types of chemical solvents and plating solutions are always unnecessarily consumed; process wastewater frequently contains high concentrations of chemicals and metals that are dragged out from the cleaning and plating steps; different types of toxic or hazardous chemicals may enter the working environment through emission; and energy consumption for process heating and ventilation are excessive. While the economic, environmental, and societal performance of the industry has been significantly improved over the past two decades, the industry is still significantly challenged and is seeking sustainable development pathways.

Different from the chemical, petrochemical, automotive and many other industries, the surface finishing industry does not have its own development roadmap. As the first step, the sustainability performance of an industry should be assessed. However, there exists no sustainability metrics system in general. In this study, we propose three sets of sustainability indicators to assess electroplating system's economic, environmental, and social sustainability. We then introduce a systematic sustainability assessment method to quantify sustainability performance, and a method to prioritize improvement measures. The applicability of the metrics system and the efficacy of the sustainability assessment and improvement methods will be demonstrated through two industrial case studies.

389644 Uncertainties of Ozone Increments Caused By Industrial Startup Flaring

Abstract:

Ground-level ozone is one of six common pollutants defined by US EPA (Environmental Protection Agency). It can cause health effects, including lung diseases. HRVOC (Highly Reactive Volatile Organic Compounds) emissions from industries are highly responsible for ground-level ozone pollution, especially around industrial district. Flare emissions from MSS (maintenance, startup and shutdown) & events are major sources of HRVOC emissions.

Most of air quality modeling treats industrial flares as constant emissions, which means the emissions keep constant for the whole scenario days, and the total amount of emissions equal to the total emissions reported to environmental authorities. However, real flare emissions from MSS and events are time-various. At certain time, flare emissions could be very huge, and at other time, flare emissions could be very small. During emission peaks, the ozone formation could be much higher than a flat emission curve.

Our previous works showed that time-various flare emissions could increase the 8-hour average ozone concentration by 15 ppb, which is 20% of current ozone standard (75 ppb). However, different locations may cause different results.

In this paper, different locations of time-various startup flare emissions are considered for air quality modeling. Emission data are obtained from dynamic simulation of a typical ethylene plant. The dynamic simulation results were validated with real plant data and emission data of startup events. The time-various flare emission data were added into Houston-Galveston-Brazoria (HGB) 2006 episode base case modeling as a point emission source. The flare emission point was placed at different locations which were obtained from center of industrial areas.

Simulation results gave a cluster of maximum ozone and ozone increment curves. The mean value curve, standard deviation curve, 90% and 100% uncertainty regions were generated from ozone curves. These curves showed the average and region of maximum ozone, ozone increment of the whole area and at monitoring stations.

391685 Experimental Investigations of a Salinity Gradient Solar Pond Under the Northern Cyprus Climate Conditions

Abstract:

Today, renewable energy sources gain importance day by day. It is crucial to develop devices and processes to supply energy from non-polluting and renewable energy sources for sustainable development of the world. Solar pond is an example of such devices that basically collects solar energy and stores it as thermal energy for a long period.

Temperature in solar ponds can reach up to 100°C indicating that thermal energy from solar ponds can be useful to various applications with low grade energy demand. It was reported that the solar ponds have the annual collection efficiency in the range of 15–25 %. The heat obtained from solar pond can be converted into electric power even at low temperatures. Solar pond plants depending on their size can generate rated power up to 5 MW itself or up to 80 MW as a hybrid solar electric generation system. One of the most important advantages of solar pond compared to the other energy sources is lower investment cost. Another important property is that solar pond is environmentally friendly in particular when it is used for electricity generation by driving a thermo-electric device or an organic Rankine cycle engine. The CO_2 emission for such processes is zero when a single solar system is used for electricity generation. Also, it was reported that the CO_2 emission can be reduced up to 65% when a hybrid system is operated for power generation. Solar ponds normally consist of three different salinity layers. The first layer, known as the upper convective zone (UCZ), is located at the top of the pond, and contains the least salinity level. The second layer, whose salinity level increases with depth, is called non-convective zone (NCZ). This layer is responsible to act as an insulator to prevent heat from escaping to the UCZ, maintaining higher temperature at deeper zones. The last layer made of a saturated salt solution, is responsible for energy storage, and is known as the lower convective zone (LCZ). The performance of the solar pond may decrease with the increasing evaporation rate, and decreasing the salinity gradient. Thus, it is critical to prevent these phenomena. Northen Cyprus is enriched in solar energy in particular during summer. The performance of a solar pond in this region has not been evaluated as yet. To take advantage of solar energy in Northern Cyprus, a salinity gradient solar pond is constructed and operated at Middle East Technical University Northern Cyprus Campus (METU NCC) located at Guzelyurt, Northern Cyprus. The pond is well insulated and made of stainless steel with 1m height, 0.64 m diameter and 1mm thickness. Several thermometers are installed in the pond at each salinity layer to monitor the temperature over time. Three sampling valves are installed to the pond to monitor the salt (NaCl) concentration at each layer over time. To adjust the salt concentration to its original level, two input ports were installed to wash the UCZ with fresh water or add salt solution to LCZ as required. The solar pond is planned for operating for a long term enabling a realistic comparative study with the existing energy storage system, flat-plate solar collectors, installed at the METU NCC. Results concerning the operation of the experimental pond are presented and discussed.

374895 Using Soybean Derived Crude Glycerol As Co-Digestate in Sewage Sludge Anaerobic Digester to Increase Biogas Production

Abstract:

In an effort to convert waste streams to energy in a green process, glycerol from biodiesel manufacturing has been used to increase the gas production and methane content of biogas within a mesophilic anaerobic co-digestion process using primary sewage sludge. Glycerol was systematically added to the primary digester from 0 to 60 percent of the organic loading rate (OLR). The optimum glycerol loading range was from 25% to 60% OLR. This resulted in an 82% to 280% improvement in specific gas production. Following the feeding schedule described, the digesters remained balanced and healthy until inhibition was achieved at 70% glycerol OLR. This suggests that high glycerol loadings are possible if slow additions are upheld in order to allow the bacterial community to adjust properly. Waste water treatment plant operators with anaerobic digesters can use the data to increase loadings and boost biogas production to enhance energy conversion. This process provides a safe, environmentally friendly method to convert a typical waste stream to an energy stream of biogas.

375022 Kinetic Modeling of Hydrothermal Liquefaction of Algal Biomass

Abstract:

A kinetic model for hydrothermal liquefaction (HTL) of algal biomass is proposed to predict the product distribution. This model is developed based on the experimental results obtained from a batch reactor. The experiments were conducted to study the effects of variations in biochemical composition of algal biomass produced at different time points of a heterotrophically grown culture under photosynthesis adaptation. The developed model is validated with the HTL experimental results available in literature. Along with the model development, the product distribution behavior is also studied at different experimental conditions. A micro bomb calorimeter is used to determine the high heating value (HHV) of the biocrude oil and bio-char. The biocrude oils and water soluble organic compounds were characterized with the High-Throughput Time of Flight Mass Spectrometer (HT TOF-MS) and GC-MS. Various analytical methods are used to study the behavior of lipids, proteins, amino acids, carbohydrates and other water soluble compounds.

378720 Utilization of Solid Residual Wastes Arising from Woody Biomass Gasification

Abstract:

Biomass Gasification usually generates significant amount of solid residues i.e. char and ashes as by-products, which usually contains high concentration of harmful compounds and heavy metals. As a result, disposal of these solid residual wastes is creating grave risks to human health. Majority of them is dumped or used in low-valued methods such as using as a land-fill material. However, the cost of landfilling is now dramatically increasing due to the presence of harmful compounds, strict environmental regulations and limited availability of landfill space. From those points of view, it is essential to develop beneficial uses of these solid residual wastes to solve the concerns associated with their disposal.

In this work, **ashes**arising from woody biomass gasification have been successfully refined into the active CaO catalyst for biodiesel production. Calcium carbonate was found to be the main component in woody biomass gasification bottom ash, which can be completely transformed into the active CaO catalyst after calcination under air at 800 °C or above. The obtained CaO catalyst exhibits high biodiesel production activity, over 90% yield of biodiesel can be achieved at the optimum reaction condition (methanol-to-oil molar ratio, 20:1; catalyst, 5 wt. %; temperature, 65 °C).

Moreover, it was found that ashes arising from woody biomass gasification also have high contents of potassium (1.92 g/kg), phosphate (0.47 g/kg) and other elements which are in the desired range for planting. Therefore, in this work, **ashes** have also been used as agricultural fertilizer by mixing it with soil in different ratios and used for planting of water spinach (*Ipomea aquatica*) which can be fully grown within 6 weeks. Performance of the ashes-derived fertilizer was investigated by a growth rate (i.e. height and weight) and heavy metal accumulation of the water spinach.

On the other hand, **char** (another by-product from woody biomass gasification) has been employed as a precursor of activated carbon (one of the most important microporous adsorbents), due to its highly porous structure, its higher surface area and larger pore volume. In this work, activated carbon has been developed from char by chemical activation using potassium hydroxide as an activating agent. The effects of operating conditions of activation process, such as temperature, activating agent to char mass ratio, and nitrogen flow rate, on the textural and chemical properties of the activated carbon product were investigated. From BET analysis, the surface area of char was less than 10 m^2/g and increased to more than 500 m^2/g after the activation process.

The utilization of solid residues arising from woody biomass gasification as a source of above materials not only provides a cost-effective and environmental friendly way of recycling those solid residual wastes, reducing its negative environmental effects, but also produces highly-valuable products, offering the economic benefits for biomass gasification process.

378766 Toxicity Assessment of Bottom Ash from Biomass and Sewage Sludge Co-Gasification

Abstract:

The harmful substances in sewage sludge, such as heavy metals, can pollute the environment and pose severe threat to human health. Gasification, a process that converts organic or fossil based carbonaceous materials into carbon monoxide, hydrogen and carbon dioxide, can be used to treat sewage sludge and turn it into solid residues which consist mainly of inert mineral ash. The bottom ash can be 8-15% of the original volume and is usually for landfill. However, toxins such as most heavy metals still persist in the ash and can be possibly leached out into environment by rain or surface water. Therefore, it is important to examine the possible toxic effects of the bottom ash on human health.

In this study, toxicity analysis was conducted for sewage sludge and bottom ash – feedstock and products respectively, from the co-gasification of sewage sludge and woody biomass process in a fixed-bed downdraft gasifier. Extractions of both sewage sludge and bottom ash exhibited toxic effects on human cells of HepG2 and MRC-5 when they were present in high concentrations in the cell culture medium, represented by the decreased cell viability and abnormal cell morphology. The toxic effect of these materials was found to be more significant with longer duration of exposure, and the toxicity could be greatly alleviated upon further dilution. The effects of PH value, heavy metals and PAHs were investigated to evaluate their roles in the toxicity observed.

Single-Step Synthesis of Biodiesel from Crude *Jatropha curcas* Oil using Chlorosulfonic acid catalyst

Hanif A. Choudhury[a], Pulkit Srivastava[b], Vijayanand S. Moholkar[C], Sai Gu[a]

[a]. School of Engineering, Cranfield University,Cranfield, Bedfordshire MK43 0AL,England

[b] Department of Chemical Engineering, National Institute of Technology (NIT), Tiruchirapalli – 620 015, Tamil Nadu, INDIA

[c]. Dept. of Chemical Engineering, Indian Institute of Technology Guwahati, Guwahati, Assam 781 039, India

Abstract: High free fatty acid (FFA) containing vegetable oils are cheap feedstock for biodiesel production. Biodiesel production from cheap feedstocks offers an alternate option for nations with limited edible oil resource. However, these oils offer difficulties in transesterification due to higher FFA content. A novel single-step process with chlorosulfonic acid catalyst for ultrasonic biodiesel synthesis using feedstock with high free fatty acid content is investigated. Jatropha curcas oil has been used as the model feedstock with methanol as alcohol. The distinct merit of chlorosulfonic acid is that it catalyzes both esterification and transesterification reactions. Moreover, chlorosulfonic acid also counteracts inhibition caused by water formed during esterification, which is the cause for very slow kinetics of acid catalyzed transesterification. In addition, sonication of the reaction mixture also causes strong micro-mixing and emulsification that enhances the transesterification kinetics. Statistical optimization of the process shows 93% yield for 8.5 wt % catalyst, 20:1 alcohol to oil molar ratio and temperature of 333 K. Peculiar feature of this process is that high yield is seen at moderate temperature and molar ratio, which are much smaller than that for conventional sulfuric acid catalyzed processes. The activation energy for the process (57 kJ/mol) is at least 3X lower than the energy for sulfuric acid catalyzed transesterification.

384889 The Chemistry of the "Herycnite Cycle" Solarthermal Water Splitting Reactions

Abstract:

Solar thermal water splitting (STWS) is a promising technology for producing hydrogen gas economically, efficiently and cleanly. Two-step, metal oxide based STWS cycles generate H_2 by a metal oxide undergoing sequential high temperature reduction and water re-oxidation. The "hercynite cycle", which is based on iron reduction in a spinel structure, is attracting increased attention due to its ability to undergo isothermal water splitting. However, the hercynite STWS mechanism was not well understood. Through a combination of computational and experimental studies using periodic boundary condition density functional theory, high temperature XRD and EDS we have determined that the hercynite cycle operates via an O-vacancy mechanism rather than the displacement reaction mechanism as previously suggested. The O-vacancy mechanism operates by the formation and filling of O-vacancies in the spinel structure during the reduction and oxidation steps respectively, rather than via the phase changes which characterize the displacement reaction mechanism.

386194 Hydrogen Production from Biomass Via Microbial Electrolysis

Abstract:

Production of gasoline and diesel from plant biomass requires significant deoxygenation of the biomass. Biooil produced via pyrolysis contains over 30 wt.% oxygen requiring significant amount of hydrogen for upgrading to biofuels. We report on use of microbial electrolysis for extraction of hydrogen from water-soluble carbon compounds associated with biooil, improving the hydrogen as well as carbon conversion efficiency of the process.

The conversion of key components of the biooil aqueous phase to hydrogen will be reported. The primary parameters for development of the microbial electrolysis cell (MEC) process include current density, coulombic efficiency, applied voltage and organic loading. The relationship between these parameters will be discussed. The microbial consortium active in the anode chamber represents a complex community with exoelectrogens and fermenting organisms. The substrate specificity and stability of the consortium for treatment of the pyrolysis-derived biooil aqueous phase will be presented. A preliminary current density of 0.56 mA/cm2 has been achieved thus far, along with the complete removal of acetic acid, furfural, and hyxdroxymethylfurfural within 48 hours (undetectable by HPLC) with an organic loading rate of 2.5 g/Lr/day. These results demostrate the ability of MEC systems to remove racalcitrant compounds found in biooil and shows the potential of these systems for increasing the overall energy efficiency in related bioprocesses through removal of inhibitory compounds, water recycling, and hydrogen production.

390294 The Power of Electricity & Chemicals Co-Production

Abstract:

We present a methane-based process for the production chemicals and power via the utilization of steam methane reformer. Computer aided simulation via UNISIM are used to provide a sufficiently accurate application of the reactions that take place. Economical analysis provides interesting findings in support of the new design in which the chemicals produced as co product is utilized as a new source of revenue. Alternative designs are still being tested.

393601 Low temperature H_2 generation from thermochemical water-splitting reaction using complex redox materials

Abstract:

H_2 can be efficiently produced via thermochemical water-splitting process using redox materials. Among the redox materials investigated, $NiFe_2O_4$ exhibited higher H_2 volume generation at water-splitting and regeneration temperatures of 900°-1100°C. High energy is typically needed to achieve higher temperatures, therefore, H_2 generation via thermochemical water-splitting process is economically less attractive. To make this process cost effective, we have synthesized novel ferrite materials by using sol-gel technique and further investigated their H_2 generation ability at lower temperatures. Although there are thermodynamic limitations associated with materials' regeneration abilities at lower temperatures, we believe that the materials synthesized have some merits as they exhibited H_2 generation via water-splitting at lower temperatures. The H_2 generation ability of these materials was investigated by performing five thermochemical cycles under isothermal condition. Solid-state diffusional parameters and charge transport properties of these redox materials were measured using EIS measurements, which were performed using GAMRY EIS 300 system by applying an A.C. voltage of 200 mV at 0.01 Hz - 300 kHz. The conductivity of the redox materials was determined from the Nyquist plots generated at the reaction temperatures. The activation energy and diffusivity of redox materials was calculated using Warburg time constant. Specific surface area (SSA) and porosity of these materials were analyzed using BET surface area analyzer. Additionally scanning electron microscopy (SEM) and transmission electron microscopy (TEM) were used to characterize the grain growth.

Uncertainty and Sensitivity Analysis in Sustainable Process Design – Environmental Indicators

Carina L. Gargalo[1], Gurkan Sin *[1]
Computer Aided Process Engineering Center, Department of Chemical and Biochemical Engineering, Technical University of Denmark, DK-2800 Lyngby, Denmark
*gsi@kt.dtu.dk

The synthesis and design of sustainable chemical processes is a complex decision-making problem that imposes a high impact on the profitability of process industries. It starts with the selection of raw material, product portfolio and definition of the process technologies followed by a full sustainability analysis (environmental, economic and social), finally arriving to a multi-criteria/objective decision making procedure. As these decisions lead to large capital investments, any uncertainties in data, models or assumptions that may impact decisions need to be systematically identified and managed, which sets the objective for this contribution.

Due to climate changes and government policies, the environmental sustainability is one of the biggest concerns to achieve sustainable chemical processes, thus being one of the most important criteria in the decision making procedure. The Life Cycle Assessment (LCA) methodology, defined by ISO 14040 standards (Vanegas, 2003), is therefore commonly applied to estimate environmental performance of process or product. It comprises four parts: (i) definition of goal and scope, (ii) inventory analysis, (iii) impact assessment and, (iv) interpretation of results. The life cycle inventory analysis involves the data collection and calculation procedures that quantify significant inputs and outputs of a system. These data also compose the input to part (iii) - Life cycle impact assessment (LCIA). LCIA is aimed to evaluate the importance of potential environmental impacts. It involves elements such as, among others, to relate inventory data to impact categories and to model the inventory data to impact categories. Thus being largely dependent on estimations (e.g. potency factors), data collection (inventory) and forecasts of future data whose uncertainty has to be taken into account and properly managed.

Even though several solution approaches have been proposed in the literature for this class of problems, environmental sustainability metrics under uncertainty remains a big challenge that has not been yet addressed comprehensively to the best of our knowledge. In this study, we extend the LCA methodology, models and database with uncertainty and sensitivity analysis to provide a systematic methodology to identify inaccuracies in data and models and to propagate these to the estimation of environmental indicators which will be used in the context of sustainability analysis. To this end, engineering standard Monte Carlo technique will be used for uncertainty analysis which consists of the following steps (Sin et al 2009): (1) identification of uncertain data and characterization of uncertainty using statistical distribution, (2) Latin Hypercube Sampling (LHS) technique will be used for sampling from uncertainty domain (multi-dimensional) and (3) Evaluation of environmental indicators for each LHS sample and (4) interpretation within the context of sustainability analysis of process design The step 4 is input to the computer-aided sustainable process design framework proposed by Gargalo et al. (2014), where multi-criteria decision making under uncertainty will be used for analyzing the impact of uncertainties in the environmental indicators for selecting the best process design.

In order to highlight the main features of the extended methodology for LCA, namely impact of uncertainties in environmental indicators on sustainable process design, a complex biorefinery case study from the edible oil industry is formulated and solved. Glycerol, side-product from biodiesel plants, is used as raw material to produce a platform of value-added chemicals that are usually delivered by the petro-chemical industry. The optimal processing design under uncertainty is obtained and a procedure to manage and act against uncertainty are proposed and analyzed.

References:

Vanegas, J. a. (2003). Road map and principles for built environment sustainability. *Environmental Science & Technology*, *37*(23), 5363–72. Retrieved from http://www.ncbi.nlm.nih.gov/pubmed/14700321 (downloaded on December 2013)

Gargalo, C. Duque, J. P. & Sin, G. (2014) Computer-Aided framework for Sustainable Process Design - targeting conceptual and detailed engineering phases. AIChE proceedings 2014

Sin G, Eliasson Lantz A and Gernaey KV (2009) Good modelling practice (GMoP) for PAT applications: Propagation of input uncertainty and sensitivity analysis. Biotechnol. Prog., 25,1043-1053.

383793 A Techno-Economic, Life-Cycle Modeling Framework for Emerging Technology Assessments in the U.S. Chemical Industry

Abstract:

Driven by rapid growth of global economy and population, the increasing demand for energy and more severe environmental problems is stimulating the development and adoption of emerging technologies in chemicals industry for reduction of energy consumption and adverse environmental impacts. The assessment of the energy/environmental/economic impacts of these emerging technologies is critical because these evaluations provide companies and policy makers with useful references for decision making in investment of research and promotion of emerging technologies. But current assessment models are not able to offer generic and flexible approaches for evaluating technologies across different feedstocks and products, nor do they deliver vigorous methods to assess emerging technologies in the absence of process data and over the temporal and spatial scales relating to investment decisions[1-3].

In this research, a novel and credible modeling framework is developed to evaluate the net impacts, including energy, emissions, and economic implications, of pathway/technology changes for chemicals systems effects across the entire U.S. economy from a life cycle perspective over spatial and temporal scales. The model integrates techno-economic analyses, LCA (life-cycle assessment) and chemical process design to provide guidance on the reduction potential of different emerging technologies, the technologies and operational changes required, and the investment cost for such technologies. The results given by the model can shed the light on the understanding of the important system drivers for energy, emissions, resource use, and costs and at the same time figure out certain technologies critical for "step change" reductions in the economy-wide impacts of chemicals production.

The modeling approach and several case studies will be presented in oral presentation. A concept model for the case of ammonia is shown here for demonstration. The results indicate that 170 million GJ of life-cycle energy consumption of U.S. ammonia production can be saved by adopting a novel auto-thermal reforming technology by 2040, as a result, 9.5 million tons of CO_2-equiv emissions can be avoided. With respect to other novel technologies, the model evaluates their net impacts in different scenarios of economic and resources projections. Expanding on this modeling framework for other bulk chemical production, this model allows for the assessment of economy-wide potential reductions of emerging technologies for a variety of chemical productions. This modeling framework can also serve as a useful tool for reasonable projections and robust comparisons over temporal and spatial scale to answer the question of how do different feedstocks and/or technologies compare for a given target product over time and space, which will be useful for long-term national energy efficiency and GHG emissions mitigation planning.

References

[1] EIA. (01/02). *The National Energy Modeling System: An Overview* Available: http://www.eia.gov/oiaf/aeo/overview/

[2] ETSAP. (12/11). *MARKAL*. Available: tp://www.iea-etsap.org/web/Markal.asp

[3] M. D. Tabone, J. J. Cregg, E. J. Beckman, and A. E. Landis, "Sustainability Metrics: Life Cycle Assessment and Green Design in Polymers," *Environmental Science & Technology*, vol. 44, pp. 8264-8269, 2010/11/01 2010.

Rare Earth Oxide Production: Quantification of Life Cycle Environmental Impacts

George G. Zaimes[1], Berlyn J. Hubler[2], & Vikas Khanna[1,]*

[1]**Department:** Civil and Environmental Engineering
[1]**Institution:** University of Pittsburgh

[2]**Department:** Golisano Institute for Sustainability
[2]**Institution:** Rochester Institute of Technology

*Corresponding Email: khannav@pitt.edu

Abstract: Rare Earth Elements (REE) are a collection of 17 chemical elements comprised of the 15 lanthanides as well as scandium and yttrium, and are critical to the functionality of multiple modern commercial technologies including electric vehicles (batteries and magnets), wind turbines (magnets), fluorescent lighting (phosphors), photovoltaic films (coatings), catalytic converters, medical devices, and various national defense applications. REEs are a relatively abundant resource, however they are often widely dispersed and found in low concentrations, resulting in energy intensive and environmentally taxing mining, extraction, and refining processes. As such, REE are typically more resource intensive and costly to recover as compared to traditional ores such as iron or coal. In the past, the U.S. domestically produced enough REEs to meet internal demands, but now relies primarily on imports from China due to lower-cost labor and regulations. Currently, 95% of global rare earth oxides (REO) are produced in China, and the largest REE mine in China is located in Bayan Obo, Inner Mongolia. Resultant global REE supply and demand dynamics and monopolistic market conditions have raised significant US economic and national security concerns, and have motivated efforts for rare earth mining and exploration in North America. Accordingly, REEs are of significant U.S. national interest, as these chemical elements are integral to the design of emerging clean energy and energy efficient technologies and are vital to US national security and economic well-being. Despite their key importance in multiple industries, to-date there has been little emphasis on environmental systems analysis of REE production. Rapid growth in industrial sectors such as wind power and electric vehicles could result in heightened global demand for REEs. As such, assessing the broader ramifications of REE production on human health and the environment is crucial for guiding the sustainable development of these industries.

In this study, a cradle-to-gate LCA was conducted to evaluate the environmental impacts and resource intensity of producing REO from the Bayan Obo mine located in Inner Mongolia, China. Analysis indicates that the production of heavy REO is resource intensive and consumes over 12 times more primary energy as compared to traditional materials such as cast iron. The roasting of rare earth sulfates had the greatest contribution to overall life cycle environmental impacts, comprising 55% of overall life cycle GHG emissions and 58% of total primary energy consumption. These findings are significant, as the material demands for REOs are projected to increase in the near future due to their use in alternative energy technologies.

Life Cycle Analysis of Oil Shale Production Using GREET

Shyam Paudel, Daniel Steiner, and Joseph D. Smith, PhD.
Missouri University of Science and Technology, Rolla, Missouri, USA

ABSTRACT

Limited petroleum resources are a source of volatility in the price of liquid transportation fuels. In addition, they contribute to trade deficits for the United States and present national security challenges. To resolve these concerns, exploitation of rich oil shale deposits in the western United States has been suggested as a source of increasing domestic petroleum production. Life cycle analysis to determine the sustainability of shale oil production is the main focus of this work.

Life cycle analysis of shale oil has been used to investigate the sustainability of this unconventional hydrocarbon resource. The Greenhouse Gases, Regulated Emissions, and Energy Use in Transportation model (GREET) developed by Argonne National Laboratory provides a framework for life cycle analysis of various energy sources, but does not currently include modelling of shale oil production or processing. This work focuses on developing a comprehensive model of shale oil production from recovery of ore through the creation of crude oil. The extraction of shale oil is an energy-intensive process through which large volumes of greenhouse gases are released. By analyzing different methods of mining and transportation of oil shale ore, the energy return on investment and pollutant emissions can be estimated.

After considering various pathways for shale oil production, its sustainability has been compared to conventional petroleum resources. As well, areas for further research in improving the sustainability shale oil production has been identified.

377802 Dimethyl Ether Chemical Storage Cycle for Uninterrupted Renewable Power

Abstract:

With increasing global demand for limited fossil fuel resources and escalating atmospheric greenhouse gas (GHG) emissions, there is growing interest to develop technologies for harnessing intermittently available renewable energy sources such as solar energy. In terms of end use of primary energy, electrical power generation is among the fastest growing sectors and also represents a major stationary source (~40%) of global CO_2emissions.

To date, the use of non-hydro renewable energy for electrical power generation has thus far been limited to peak shaving rather than base-load applications. In the absence of efficient large-scale energy storage options, the intermittent nature of renewable energy sources remains the grand challenge for baseload power generation. For example, to provide uninterrupted power supply of 100 MW from solar energy in the US (with solar energy being available for 4.8 h of a 24 hour-day), approximately 2 GWh of electrical energy storage is needed.

Carbon storage cycles (CSC) involving cyclic transformation of carbon atoms between carbon dioxide and carbon fuel have the potential to achieve high storage efficiency (~54–59%) for GWh-level energy storage with much reduced storage volumes compared to other options. During the period of renewable energy availability, the cycle utilizes stored liquid carbon dioxide to synthesize a carbon fuel and then store it in liquid state. When renewable energy is not available, the carbon fuel is oxidized to deliver electricity. The produced carbon dioxide is captured, liquefied and stored for reuse during energy availability [1].

Here, we review the novel features of the CSC, and three exergy based metrics that allow systematic screening of carbon fuel to be utilized in the cycle and present the process simulation details and results of dimethyl ether chemical storage cycle [2]. Dimethyl ether is identified as one of the attractive carbon fuel candidates by the exergy based metrics and the resulting ~57% storage efficiency is in accordance with the prediction of the metrics.

[1] Al-musleh EI, Mallapragada DS, Agrawal R. Continuous power supply from a baseload renewable power plant. Applied Energy 2014;122:83.

[2] Gençer E, Al-musleh E, Mallapragada D, Agrawal R. Uninterrupted Renewable Power through Chemical Storage Cycles. Current Opinion in Chemical Engineering 2014.

383568 Plasma-Based Advanced Oxidation: Effects of Reactor Design on the Degradation of Organic Contaminants

Abstract:

To improve the feasibility of plasma-based water treatment technology and asses the possibility of the technology scale up, a study was conducted to identify and characterize the design parameters that influence treatment effectiveness and the physical phenomena that determine treatment efficiency. The effect of several reactor design parameters, broadly grouped into three categories, on the degradation of 5 mg/L Bisphenol A (BPA) and 5 mg/L Rhodamine B (RhB) was investigated. These parameters include: (1) reactor geometry (e.g., electrode material and shape, electrode spacing, reactor diameter, and ground electrode plate area), (2) electrode configuration, and (3) contact between plasma and the treated solution.

Results indicate that changing the reactor geometry, particularly minimizing the distance between the high voltage electrode and the grounded electrode, increasing the reactor radius, and increasing the area of the grounded electrode plate increase the peak current (i.e., input power) and thus the removal rates of BPA and RhB. Discharges in a gas contacting water were found to be about four times as effective as discharges directly in water. Prior to maximizing the contact between plasma and the treated solution, the described reactor optimization increased the extent of BPA removal from 10% to 90% in less than 1 h and RhB from 5% to 95% in less than 30 minutes.

The contact between the plasma and the treated solution was found to be strongly influential. The effect of contact was investigated by testing several new reactor configurations featuring a discharge between a high voltage and a grounded electrode placed above water surface and within the liquid feed stream as it fell from above the surface of the solution. These high-contact configurations were twice as effective as the optimized gas phase discharge. A final alteration involved generating bubbles on the liquid surface by introducing the gas feed through a ceramic diffuser beneath the liquid surface, to present a more favorable interface for the plasma to travel along and to increase mass transfer of plasma radicals into the treated solution. This strategy again doubled the efficiency, yielding a G_{50} value about 2 orders of magnitude higher than that of the original reference configuration with both electrodes immersed in water.

Synthetic organic Azo dyes are among the most dangerous contaminants in industrial wastewater due to presence of aromatic amines in their structure. Conventional treatment systems are not adequate for treating azo dyes because they are designed to resist biodegradation and they have high chemical stability. Pulsed corona Discharge is within the Advanced Oxidation Techniques (AOT's) and it is a promising technology for treating this kind of wastewater. One key advantage of AOT`s is mineralization of organic contaminants with no secondary pollution problem for numerous contaminants and under optimal conditions of degradation. The pulsed corona discharge (PCD) causes the formation of hydroxyl radicals, hydrogen peroxide, and aqueous electrons among other powerful oxidants and these are useful to achieve the effective degradation of many contaminants which is, usually, not possible by other degradation techniques.

In this study, a Pulsed Corona Streamer Discharge batch reactor, with point to point electrodes, and with a batch mode is tested for the decomposition of Amido black 10B in wastewater. The concentration of reactive species such as H_2O_2 and O_3 and their role on decomposition of azo dyes in different frequencies (60Hz ad 120 Hz) are determined. The effect of addition Zeolite on the decomposition of Amido black 10B in comparison with nano- TiO_2 is also investigated. Results will show the different degradation extents reached under the different conditions and suggestions for further work.

371052 Novel Spatially and Temporally Confined Microplasmas for Advanced Oxidation

Abstract:

A continued discussion of a study on dc discharges in aqueous environments is offered. The focus remains on non-thermal plasma confined to micrometer spatial scales on sub-nanosecond to nanosecond temporal scales. Specific target compounds are used in order to demonstrate the previously observed efficacy of plasma technology. Each regime such as pulsed, spark, corona, glow, and partial discharges of these charge densities, in terms of their respective mediums, are explored. The time scale is used to control the development of microplasma discharge events within specific current density regimes as an efficiency promoter. Nanostructured and microstructured emitter electrodes are still employed under the high electric fields at the micrometer scale. Discharges in liquid at 365V on the nanosecond scale with nanoemitter electrodes is observed as reported. Significant oxidative degradation of chemical colorimetrics is observed for ultra-short residence times. Several reactor configurations have been explored including analogs to macroscopic systems. The reaction engineering kinetics of these confined microplasmas will be discussed for these massively paralleled systems. Improvements in discharge efficiency are related to reductions in mass transfer in microfalling films and continuous microdrip configurations. Optical imaging of the interaction of the discharge's electric field with the fluid will be offered if time permits as an exploratory tool as well as spectral grating of frequency for unobserved metastables based on photoionization intensity and energy.

389778 Photocatalytic Degradation of Methamphetamine Using UV/TiO2

Abstract:

Methamphetamine (MAT) is a prescription drug and a substance of abuse; it cannot be effectively removed in wastewater treatment plants (WWTPs). As a result, MAT has been found at 61-405 ng/L in hospital effluents, WWTP influents and effluents, as well as surface waters in Taiwan. Even at trace level, MAT can pose risks of long-term exposure, bioaccumulation, and biomagnification on aquatic organisms. This project evaluates the effectiveness of eliminating MAT from waters using existent UV disinfection process in WWTPs in the presence of TiO_2 photocatalyst. We tested the use of TiO_2 (Degussa P25) under UV illumination (9 W at 365 nm) for the photocatalytic degradation of MAT to define the removal kinetics. Removal was investigated under varied experimental parameters including TiO_2 loading, MAT concentration, and pH; MAT and its intermediates throughout treatment were monitored by HPLC-MS/MS, along with TOC and IC analyses to determine the extent of mineralization. The results showed elimination of 100 ppb of MAT within 30 min of irradiation of TiO_2 at 0.03 g/L at pH 7. The degradation of MAT followed pseudo-first order kinetics conforming to the Langmuir-Hinshelwood model. Complete mineralization of 10 ppm of MAT was achieved in 180 min with 0.1 g/L of TiO_2 at pH 5, whereas the organic nitrogen was converted predominantly to NH_4^+. Photocatalytic process with UV/TiO_2 shows promise in arresting the release of MAT and its intermediates into the water environment.

Visible Light Mediated Heterogeneous Photo-Fenton Oxidation of Endocrine Disrupting Compounds in Wastewater and Water Using CdS/Carbon nanotube-Iron Oxide Composite Photocatalyst

Jihyun R. Kim and Eunsung Kan. Department of Molecular Biosciences and Bioengineering, University of Hawaii at Manoa, Honolulu, HI 96822, USA

Significant amounts of bisphenol A (BPA), an endocrine disrupting chemical, is released from manufacturing plants of plastics, epoxy resins, dental material and flame retardants. Due to its wide usage, BPA has been detected in urban and industrial wastewater, drinking water and landfill leachates. Recent studies reported that estrogenic activity of BPA at very low concentrations affects various aquatic organisms (i.e., 40-1000 μg/L) and humans (i.e., $2x10^{-4}$ μg/L) and can cause hormonal disruption inducing infertility and breast cancer.

Heterogeneous photo-Fenton oxidation of Bisphenol A (BPA), a model endocrine disrupting compound found in wastewater and water, was carried out by the CdS and iron oxide-coated multi-walled carbon nanotube (CdS/MWCNT-Iron oxide) under visible light irradiation. The main objective of this study was to investigate the removal and mineralization of BPA in a short time during the photo-Fenton reaction. The photo-Fenton reaction with the CdS/MWCNT-iron oxide relied on combination of the photocatalytic oxidation by CdS and Fenton oxidation using the iron oxide while the iron oxide also acted as an electron acceptor. The SEM and TEM image analysis confirmed the uniform distribution of the nano-scale CdS and iron oxide (i.e., 10-20 nm) over the MWCNT. The photo-Fenton reaction resulted in fast and high removal and mineralization of BPA (BPA removal of 100 %, TOC removal of 89 % in 1 h) which were much higher than those by the dark Fenton oxidation and the photocatalytic degradation alone. The optimum molar ratio of [BPA]:[H_2O_2] (1:15) in the photo-Fenton oxidation showed that the process would be a cost-effective solution for fast and effective removal of BPA in water. The photocatalytic activity of the CdS/MWCNT-iron oxide was maintained effectively even after several cyclic experiments. In addition, the photocatalyst (CdS/MWCNT-Iron oxide) was easily separated after the Fenton reaction under the magnetic field. Based on the literature and these results, the possible mechanisms associated with the visible light-driven heterogeneous photo-Fenton of BPA were discussed. Further works will include the evaluation of cytotoxicity/estrogenic activity of BPA oxidation products and optimization of the process by process modeling of the photoreactor.

ID# 369173

TECHNICAL, ECONOMIC AND ENVIRONMENTAL VIABILITY OF OFFSHORE CO_2 REUSE FROM NATURAL GAS BY DRY REFORMING

Bruna C. de S. Lima, Federal University of Rio de Janeiro, Rio de Janeiro, RJ, Brazil
Ofélia Q. F. de Araújo, Federal University of Rio de Janeiro, Rio de Janeiro, RJ, Brazil
José L. de Medeiros, Federal University of Rio de Janeiro, Rio de Janeiro, RJ, Brazil
Cláudia R. V. Morgado, Federal University of Rio de Janeiro, Rio de Janeiro, RJ, Brazil

ABSTRACT

The recent discoveries of Pre-Salt layers in Brazil require process developments for enhanced sustainability as these reservoirs have oil with associated natural gas exhibiting an expressive amount of CO_2 (FORMIGLI, 2007; GAFFNEY, 2010). The challenge is thus to expand the supply of this fossil raw material to produce energy and chemicals in a carbon-constrained economy. In this scenario, CO_2 capture and reuse techniques are important routes for a moderate transition from the present fossil-based economy to a long-term sustainable future. This work evaluates two process alternatives for a floating unit consisting of physical absorption of CO_2 from natural gas, with propylene carbonate, producing a natural gas stream poor in CO_2 and a stream consisting of a mixture of CO_2 and hydrocarbons. Two alternatives are evaluated for processing the CO_2 rich stream to yield methanol. Technical, economic and environmental analyses of the process flowsheets are performed for comparison purposes. Alternative 1 combines dry and steam reforming in one reactor (one pot reactor, Bi-Reforming), while Alternative 2 segregates the two reactions: dry reforming (Dry-Reforming) occurs in one reactor and water gas-shift reaction takes place in a subsequent reactor. The two Alternatives are evaluated with ASPEN HYSYS® (ASPENTECH Inc) simulator for calculation of energy and mass balances necessary to assess economic and environment performances. The economic evaluation employed the software Capital Cost Estimation (CAPCOST) for calculations of CAPEX and OPEX. The software Waste Reduction Algorithm (WAR, EPA) was used to evaluate the potential environmental impacts, by PEI (Potential Environmental Impact). Results indicate that the performance of Alternative 1 was better than Alternative 2: (a) methanol production (17905 kg/h) 4 times higher, (b) CAPEX (U$$75.497.127), 21,7% lower, (c) Sales Revenue (U$$144,890,686) 181% greater, and (d) the index for environment impact (868 PEI/hr) 47,7% lower.

INTRODUCTION

The recent discoveries of Pre-Salt layers in Brazil require process developments for enhanced sustainability as these reservoirs have oil with associated natural gas exhibiting an expressive amount of CO_2 (FORMIGLI, 2007; GAFFNEY, 2010). Carbon dioxide is one of the greenhouse gases and its emission should hence be avoided or reduced. The challenge is thus to expand energy supply in a carbon-constrained economy. In this scenario, CO_2 capture and reuse techniques are important routes for a moderate transition from the present fossil-based economy to a long-term sustainable future (ARAUJO et al., 2014).

In this sense, sustainable principles for designing natural gas conditioning and utilization processes must be inserted in the synthesis and operation phases. The sustainability concept states that a sustainable progress is only possible when considered, simultaneously, economics, environmental and social aspects (Johnston et al., 2007). With this vision, this work presents a study of natural gas production, concomitantly with CO_2 capture and reutilization in methanol synthesis as well as in enhanced oil recovery (CO_2-EOR).

Two offshore process alternatives are herein evaluated where CO_2 is captured from natural gas, by physical absorption with propylene carbonate (PC), producing a natural gas stream poor in CO_2 and a stream consisting of a mixture of CO_2 and hydrocarbons, which is processed by Bi-Reforming (Alternative 1) and Dry-Reforming (Alternative 2), to yield methanol. This solvent has a high affinity for CO_2 and its use has advantages for the environment, since PC is a nontoxic and biodegradable chemical (GADELHA, 2010), and its production route involves the utilization of CO_2 as a reactant along with propylene oxide (DAI et al., 2010), a relevant aspect in a life cycle approach.

The selection of PC as solvent has an important advantage: its regeneration does not require heat input, and involves an expansion valve and a flash vessel. Also relevant in the scope of the downstream processing of the CO_2 rich stream is that PC has also affinity for hydrocarbons, yet lower than its affinity for CO_2. This aspect is often viewed as a disadvantage. However, this will be a benefit for the proposed alternative processes since the hydrocarbon which "contaminates" CO_2 will react with CH_4 from natural gas in dry reforming reaction.

Dry reforming to yield synthesis gas (syngas) is one alternative to reduce CO_2 emissions, adding value to it (ARAUJO et al., 2014). The interest in this reforming route is due to the production of a syngas with a H_2/CO molar ratio suitable for a variety of products, and because the reaction consumes two types of greenhouse gases, CO_2 and CH_4 (ZHANG et al., 2003, GANGADHARAN et al., 2012). This reaction can be represented as

$$CH_4 + CO_2 \rightarrow 2CO + 2H_2 \qquad \Delta H = +247.3 \text{ kJ/mol} \tag{1}$$

The biggest limitation to the dry reforming process is the availability of a suitable catalyst. This reaction requires high temperature, what contributes to carbon deposition. In CO_2 reforming, coke deposition on the catalyst is reported to be very fast. However, when CO_2 reforming occurs simultaneously with steam reforming, coke deposition is drastically reduced. (GANGADHARAN et al., 2012). This combination, named bi-reforming (Eqs. 2, 3 and 4), can generates syngas with a H_2/CO ratio of 2, ideal for the synthesis of methanol (OLAH et al., 2009).

Steam reforming: $CH_4 + H_2O \rightarrow CO + 3H_2$ $\qquad \Delta H = +206.3$ kJ/molCO$_2$ (2)

Dry reforming: $CH_4 + CO_2 \rightarrow 2CO + 2H_2$ $\qquad \Delta H = +247.3$ kJ/molCO$_2$ (3)

Bi-reforming: $3CH_4 + 2H_2O + CO_2 \rightarrow 4CO + 8H_2$ (4)

Although dry reforming generates syngas H_2/CO ratio suboptimal of for methanol synthesis, the water gas-shift (WGS, Eq. 5) reaction is responsible for adjusting the syngas ratio (GANGADHARAN et al., 2012). The reversibility of Eq 5 (WGS and RWGS) is important, once that H_2/CO ratio can be adjust according process needs just changing the reaction conditions.

$$H_2 + CO_2 \leftrightarrow CO + H_2O \qquad \Delta H = +51 \text{ kJ/mol} \tag{5}$$

On the other hand, syngas from bi-reforming is appropriate for methanol production (Eq. 6).

$$CO + 2H_2 \rightarrow CH_3OH \qquad \Delta H = -90.6 \text{ kJ/mol} \tag{6}$$

This work presents the synthesis of two processes for processing the CO_2 rich stream, Alternative 1 and Alternative 2, to yield methanol. Technical, economic and environmental analyses of the process flowsheets are performed for comparison purposes. Alternative 1 combines dry and steam reforming in one reactor (*one pot reactor*), while Alternative 2 segregates the two reactions: dry reforming occurs in one reactor and water gas-shift reaction takes place in a subsequent reactor.

The two Alternatives are evaluated with Aspen HYSYS® simulator for calculation of energy and mass balances necessary to assess economic and environment performances, employing Peng-Robinson EOS. The economic evaluation employed Capital Cost Estimation Software (CAPCOST) for calculations of CAPEX and OPEX, and the Waste Reduction Algorithm Software (WAR) was used to evaluate the potential environmental impacts. It is worth noting that besides the environmental motivation, the alternatives were driven by the fact that methanol is one of the most important commodity chemicals, used as raw material in several processes, what would bring economic advantages, and it is a route of gas to liquid(s).

This work was conducted at CE-GN – Center of Excellence in Natural Gas, a joint venture between UFRJ and PETROBRAS/CENPES.

PROCESS SIMULATION

The two processes - Bi-Reforming (Alternative 1) and Dry Reforming(Alternative 2) - were evaluated with Aspen HYSYS® simulator for calculation of energy and mass balances necessary to assess economic and environment performances. The first process steps relates to CO_2 capture and are common to both alternatives, which is fed with raw a natural gas stream with composition shown in Table 1.

Table 1. Natural Gas Composition

Component	% mol
CH_4	60
CO_2	20
C_2H_6	12
C_3H_8	06

The study assumes the following process premises:

- The natural gas, previously dehydrated, enters at 70 bar and 40 °C with a volumetric flow rate of 1MMNm³/d.;
- Treated gas meets selling specifications (ANP, 2008), i.e., CO_2 poor gas leaves PC absorption columns exhibiting less than 3% in CO_2.
- Propylene Carbonate mass flow was set up according to the molar ratio of CH_4 to CO_2 necessary for reforming reactions, resulting in a value of 435ton/h.

Figure 1 shows the Aspen HYSYS® process flow diagram (PFD) of the CO_2 PC absorption process, common for Dry Reforming and Bi Reforming processes.

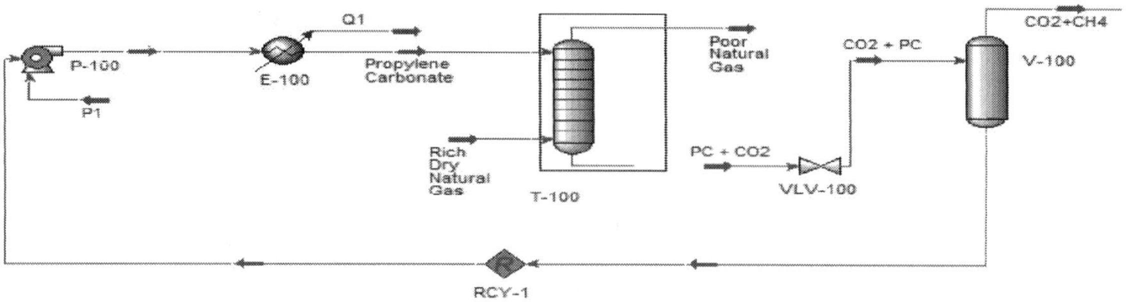

Figure 1. Process flow diagram for CO_2 absorption with PC - Dry Reforming and Bi Reforming processes.

The natural gas stream enters at the bottom of absorber column (T-100), countercurrent to Propylene Carbonate (absorbent). The stream with PC and CO_2 is recovered at the bottom of T-100 and goes to a flash vessel (V-100) for PC regeneration. Natural Gas correctly specified is the top product of absorber. The PC rich stream that leaves the bottom of T-100 is depressurized (at VLV-100) and enters in V-100 for PC regeneration. Then, the gas that leaves V-100, predominant in CO_2 and CH_4, but with residual hydrocarbons components, goes to reaction and the liquid product (PC) is recycled and returns to T-100.

Bi-Reforming Process

The stream 'CO_2 + CH_4' is divided into two parts: one goes to a compression train, for CO_2 reinjection, and the other goes to reaction. In the downstream section, this stream goes to a reactor where both dry and steam reforming reactions occur, shown in Figure 2.

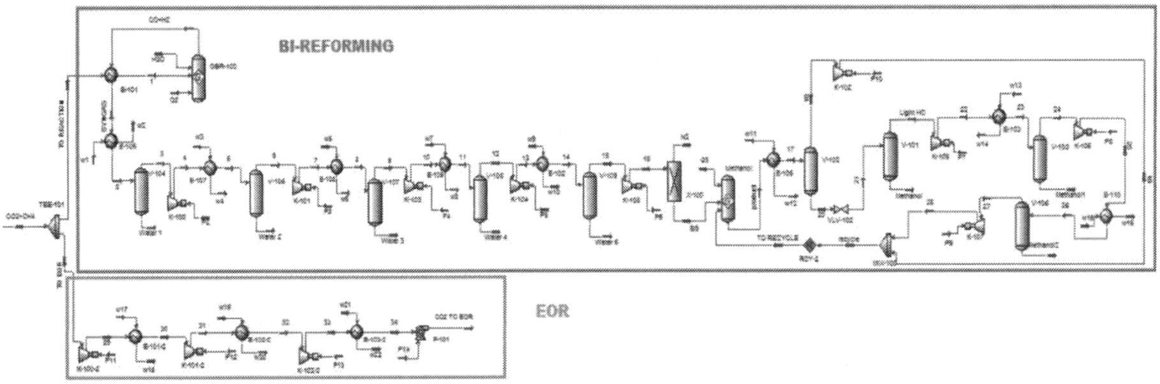

Figure 2. Process flow diagram for EOR and Bi-Reforming process – Alternative 1.

The stream 'TO EOR' is pressurized to 400 bar to meet requirements of injection for the CO_2-EOR. Table 2 shows the operational conditions for compressors: compression ratio, suction pressure, discharge pressures and discharge temperatures. After the compression train, pump P-101 finally pressurizes the stream that goes to EOR. The criteria utilized for compression was that temperature discharge is limited to 170 °C.

Table 2. Operational conditions of compressors for EOR

Compressor	Suction Pressure (bar)	Discharge Pressure (bar)	Discharge Temperature (°C)	Compression ratio
K-100-2	2.0	9.5	167.2	4.75
K-101-2	9.0	47	166.6	5.20
K-102-2	46.5	240	169.3	5.16

On the other hand, the stream 'To Reaction' proceeds to reactor GBR-100, passing through E-101 for energy integration. The reactor model used for simulation was the Gibbs Reactor, which works by finding the equilibrium state with the lowest Gibbs free energy. At the feed ratio of CO_2/CH_4 of 1:1, the optimum temperature for dry reforming reaction is between 870 and 1040 °C to prevent carbon deposition at pressure of 1 atm (WANG, 1996). However, this process combines dry and steam reforming. For steam reforming, the reaction temperature is between 700 °C and 980 °C, and pressure around 20 bar (Pacífico, 2004). In this case, GBR-100 operates at 900°C and 1 bar, to favor dry reforming. Water vapor is added in reactor at 10.81 ton/h. Table 3 shows the gas composition (% mol) before and after reaction.

Table 3. Gas composition (% mol) before and after Bi-Reforming

Before Reaction		After Reaction	
CH_4	35.10	CH_4	0.040
CO_2	35.10	CO_2	3.890
C_2H_6	19.20	C_2H_6	-
C_3H_8	10.40	C_3H_8	-
N_2	0.200	N_2	0.040
CO	-	CO	27.71
H_2	-	H_2	57.86

After reaction, the stream 'SYNGAS' passes through a sequence of compressors, heat exchangers and vessels, in order to eliminate water and pressurize gas to 70 bar for methanol synthesis. These heat exchangers utilize seawater for refrigeration. It is important to add that before reaction, nitrogen is removed with a splitter component, to simulate molecular sieve adsorption process (CAVENATI et al, 2006).

For methanol production from synthesis gas, the operational conditions for reaction are: temperature around 260 °C and pressure between 50 and 100 bar (AQUINO, 2008, KIM et al., 1998). In this work, Methanol reactor operates at 260 °C and 70 bar. The stream leaving the Methanol reactor goes to V-102, a flash vessel at high pressure. The stream that leaves V-102 at top returns to the Methanol reactor, while the liquid stream (bottom product of V-102) is depressurized to 5 bar for liquid methanol recovery at V-101. Light hydrocarbons and residual methanol are pressurized at K-105, refrigerated at E-103, and methanol is recovered at V-103. The same sequence occurs with K-106, E-100 and V-106. Finally, compressor K-107 pressurizes the remaining hydrocarbons that return to Methanol reactor.

Dry Reforming Process

Similarly to the Bi-Reforming process, the stream 'CO_2 + CH_4' is divided into two parts and the stream "TO EOR" has the same destiny on both processes. On the other hand, the stream 'To Reaction' proceeds to the GBR-100 reactor, where occurs exclusively the dry reforming. This stream passes through E-101 for energy integration. In this case, GBR-100 operates at 870°C and 1 bar. After reaction, stream 'SYNGAS' passes through a compression train, with four compressors and five intercoolers, which utilize seawater for refrigeration. This stream is pressurized to 70.5 bar before entering the Adjust reactor. Water vapor is added at a rate of 2 ton/h, and the water shift reaction occurs at 90 °C. The product stream for this reactor goes to a split component operation, used to simulate N_2 removal by molecular sieve adsorption. In the sequence, stream "feed" enters in the Methanol reactor, which operates at 260 °C and 70 bar. The stream that leaves the Methanol reactor has a high quantity of unreacted methane, carbon monoxide and hydrogen. Hence, after methanol recovery, it will be necessary to recycle this stream to GBR-100. For recovering methanol, stream "product" is refrigerated by E-105, E-112 and E-104 before enter in V-101. The top product from this vessel goes to recycle, and the bottom product is depressurized and enters in V-102 for light hydrocarbon removal from methanol. Methanol is recovered at V-102, and those hydrocarbons are recycled. Stream "Recycle 1" works as a cooler for E-105 and is heated before entering a turbine (K-105) to reach GBR-100 condition. After K-105, this stream passes through E-112 also working as a cooler to reach a higher temperature to enter in GBR-100. This stream is divided and part of it goes to CRV-100 and is combusted. "Recycle 2" stream goes directly to GBR-100, since it already has the same pressure of this reactor. Figure 3 shows the PFD for this process.

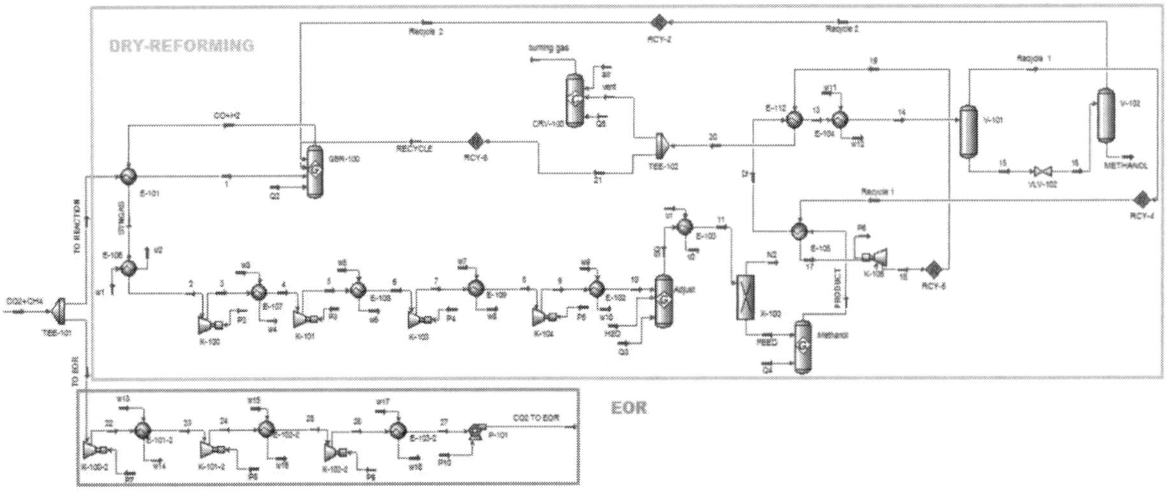

Figure 3. Process flow diagram for EOR and Dry-Reforming process – Alternative 2

ECONOMIC ANALYSIS

The two process alternatives are evaluated economically, through CAPEX (Capital Expenditure) and OPEX (Operational Expenditure) calculations. A prior step to the CAPEX calculation is sizing of process main equipment necessary for cost estimation through correlations (TURTON et al., 2009). Raw materials costs, costs of utilities, operating labor (direct costs) and others general expenses were equally estimated based on Turton et al. (2009). Equipment sizing was performed according to procedures described in the literature (e.g., CAMPBELL, 2004; PERRY, 1997; TOURTON et al., 2009). Economic premises used in the evaluation of the economic performance of the proposed alternatives are presented in Table 4.

Table 4. Economic Premises Adopted

Economic Variable	Value Adopted
CEPCI (Chemical Engineering Plant Cost Index)	572.6
Operating labor cost	U$$136.363/year
Conditioned natural gas price	U$$ 16/GJ
Raw natural gas price	U$$ 8/GJ
Propylene carbonate price	U$$ 1.19/kg
Methanol price	U$$ 0.81/kg
CO_2 for EOR	1ton – 2barrel: U$$118.94
Offshore electricity price	U$$ 0.154
13X Molecular Sieve	U$$ 1.60/kg
Catalyst	U$$ 10/kg
Desalinated water	U$$ 0.25/m^3
Water for refrigeration	none

For the economic analysis, the MS Excel spreadsheet CAPCOST, developed by Turton et al (2009), was employed, allowing the estimation of Capital Expenditure and Operational Expenditure.

Table 5 presents the main results from Bi-Reforming (Alternative 1) and Dry-Reforming (Alternative 2) Processes, indicating the Total Module Cost per piece of equipment and the resulting CAPEX for each one. Compressors, turbines and pumps are responsible for the major share of CAPEX. The CAPEX of Dry-Reforming process is higher when compared to Bi-Reforming, mainly because Dry-Reforming requires more compression steps. On the other hand, the OPEX of Bi-Reforming is higher than the OPEX of the Dry-Reforming. Raw material and operating labor costs are similar. The utilities costs are responsible for the major share of OPEX for both processes. However, Bi-Reforming requires a large amount of electric energy due to the GBR-100 reactor. Table 6 shows these results.

Table 5. Equipment Cost and CAPEX

Total Module Cost (U$$)	Bi-Reforming	Dry-Reforming
Compressors, turbines and pumps	16,941,000	22,208,000
Drives	3,519,800	3,485,000
Heat Exchanger	3,000,465	3,275,000
Vessels	665,400	486,900
Reactors	280,100	450,061
Absorber column	317,000	317,000
Adsorbers	441,944	1,923,945
CAPEX (U$$)	**25,165,709**	**32,145,906**

Table 6. Operational Cost (OPEX)

Cost (U$$/year)	Bi-Reforming	Dry-Reforming
Raw Material	15,933,520	15,917,834
Utilities	79,130,000	60,200,000
Operating Labor	409,089	681,815
OPEX (U$$/year)	**122,575,216**	**101,189,415**

The revenue considers the sum of methanol and natural gas sales, and also takes in account that for each ton of CO_2 reinjected, 2 barrels of oil is produced and sold. The revenue is calculated based on the price market for each product and their rates of production. The results are presented at Table 7, which shows that Bi-Reforming process (Alternative 1) has the highest revenue. Methanol production for this process (17960 kg/h) was 4 times higher than for the Dry-Reforming Process (Alternative 2).

Table 7.Revenue Sales

Revenue (U$$/year)	Bi-Reforming	Dry-Reforming
Natural Gas	7,906,748	7,925,725
CO_2 to EOR	14,107,236	13,995,744
Methanol	122,876,702	29,542,288
TOTAL (U$$/year)	**144,890,686**	**51,463,757**

ENVIRONMENTAL ANALYSIS

The WAR Algorithm (CABEZZAS et al.,1999) was used to quantify the environmental performance of the alternative processes, based on eight categories of environmental impacts, equally weighted:

- Human Toxicity Potential by Ingestion (HTPI)
- Human Toxicity Potential by Exposure (HTPE)

- Aquatic Toxicity Potential (ATP)
- Terrestrial Toxicity Potential (TTP)
- Global Warming Potential (GWP)
- Ozone Depletion Potential (ODP)
- Smog Formation Potential (PCOP)
- Acidification Potential (AP)

WAR measures performance as output rate of PEI (Potential of Environmental Impact). PEI can be described through rates of inlet (Îin), outlet (Îout) and generation (Îgen) of impact in a given system. At steady state, PEI input and output fluxes are constant, and their difference equals the PEI generation (Eq.7). Then, if Îgen is negative, it can be conclude that the process reduces the environmental impact.

$$0 = \hat{I}in - \hat{I}out + \hat{I}gen \tag{7}$$

The mass flow and composition obtained from ASPEN HYSYS simulations were used in the WAR algorithm, with calculations presented in Figure 4. For both processes, PCOP is the category that most contributes to PEI, followed by HTPI and TTP. That is because methanol has high scores for these categories. Comparing the total impact for these processes, Bi-Reforming has a lower impact (868 PEI/hr) than Dry-Reforming (1661 PEI/hr). Another conclusion that arises is that Îgen is negative for Bi-Reforming and positive for Dry-Reforming. Consequently, the Bi-Reforming process shows superior environmental performance when compared to the Dry-Reforming alternative.

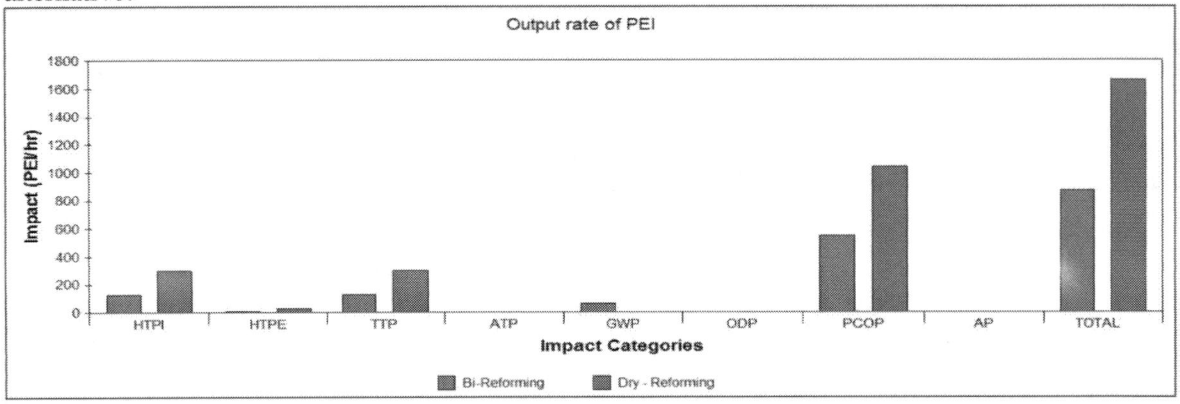

Figure 4. Potential Environmental Impact for Bi and Dry-Reforming Processes

DISCUSSION AND CONCLUSIONS

This work combines CO_2 capture to CO_2 utilization, producing methanol, natural gas and CO_2 for EOR, in an offshore installation. Two alternatives for converting CO_2 and CH_4 into methanol were evaluated - Bi-Reforming and Dry-Reforming, on economic and environmental grounds. It is worth noting that the work evaluates a hybrid alternative encompassing gas-to-liquid technology (bi-reforming and dry-reforming), that allows natural gas deployment to remote locations, gas conditioning to pipeline transportation, which would service main large-scale consumption points, and CO_2-EOR. Furthermore, both proposed processes can be implemented in a methanol floating production system, an MFPSO, where the regeneration of PC would present a competitive advantage over chemical absorption in aqueous solutions of ethanolamines or membranes, since weight and space are the key considerations for MFPSO plants. The methodology adopted employed ASPEN HYSYS®, for calculation of mass and energy balances, necessary for equipment sizing needed in economic evaluation (using CAPCOST) and measuring environmental performance (with WAR Algorithm). The main results for Bi-Reforming and Dry-Reforming Processes are displayed in Table 8 and Table 9, respectively.

Table 8. Main Results for Bi-Reforming Process

	Poor Natural Gas	CO_2 TO EOR	Methanol	Methanol 1	Methanol 2
Volumetric flow (m^3/h)	248.7	27.38	24.05	0.1624	0.1208
Mass flow (kg/h)	13190	14230	19130	133.3	107.7
Composition (% mol)					
CH_4	90.19	35.08	-	0.01	0.10
CO_2	1.67	35.12	3.45	13.63	-
C_2H_6	3.31	19.19	-	-	-
C_3H_8	0.63	10.43	-	-	71.84
N_2	4.20	0.18	-	-	-
CO	-	-	-	0.02	0.27
H_2	-	-	-	-	0.05
H_2O	-	-	4.13	0.73	0.05
CH_3OH	-	-	92.42	85.61	27.69

Table 9. Main Results for Dry-Reforming Process

	Poor Natural Gas	CO_2 TO EOR	Methanol
Volumetric flow (m^3/h)	249.0	27.05	5.52
Mass flow (kg/h)	13210	14230	4341
Composition (% mol)			
CH_4	90.14	35.04	0.05
CO_2	1.69	35.16	0.28
C_2H_6	3.33	19.18	-
C_3H_8	0.64	10.44	-
N_2	4.20	0.18	-
CO	-	-	-
H_2	-	-	-
H_2O	-	-	0.6
CH_3OH	-	-	99.07

The main difference of the proposed processes is the fact that at Bi-Reforming just one reactor is utilized for both dry and steam reforming. On the other hand, for Dry-Reforming two reactors are conceived: one for dry reforming and other to shift reaction. This difference poses a large discrepancy of methanol productivity, as methanol production by Bi-Reforming was 17905 kg/h, 4 times higher compared to Dry-Reforming. It was also observed that dry reforming occurs not only for methane, but also for ethane and propane, and that these last reactions are more favored than the first one. In consequence, methane is not consumed, but produced in the Dry-Reforming alternative, and requires recycle to the reformer reactor. This conclusion explains the low production of methanol for this process.

The economic analysis shows that the Bi-Reforming process has advantages due to the larger methanol production. Although its OPEX is higher than the OPEX for Dry-Reforming, methanol production capacity is sufficiently high to compensate the increased operational expenditure. For By-Reforming, the CAPEX was U$$25,165,709, considering an Onshore Plant, at United States. It was assumed that the offshore cost is the double of onshore cost, and that the "Brazilian Cost Factor" (cost for a plant installed at Brazil) is about 1.5. Then, the real CAPEX for this process would be U$$75,497,127. For OPEX, this process has a cost of U$$122,575,216, and the Revenue Sales U$$144,890,686, what turns Bi-Reforming technically and economically feasible.

The potential environmental impact of both processes were calculated (in PEI/hour) as 868 PEI/h and 1661 PEI/h, respectively for Bi-Reforming and Dry-Reforming. This result shows that Bi-Reforming is environmentally more favorable than Dry-Reforming. Another important point is that the potential impact generation rate is negative for Bi-Reforming, what strengthens the conclusion of environmental feasibility.

Therefore, the preliminary calculations indicate that Bi-Reforming is technically, economically and environmentally attractive. Future studies are required such as (among others): (i) sensitivity analysis to design and economic premises (e.g., cost of offshore desalinated water, price of raw natural gas), (ii) flowsheet optimization, (iii) study impact of production scale production, (iv) analysis of EOR process with pure CO_2; (v) detailed analysis of N_2 removal.

ACKNOWLEDGMENT

The authors gratefully acknowledge the financial support from the National Petroleum Agency (ANP) – PRH41, for scholarship of Bruna C.S. Lima.

REFERENCES

Aquino, A. S. **Analysis of alternative synthesis routes for chemical sequestration of co2: methanol, syn gas and acetic acid production**. 2008. 87f. Monografia - Curso de Engenharia Química, Departamento da Escola de Química, Universidade Federal do Rio de Janeiro, Rio de Janeiro, 2008.

ARAUJO, O. Q. F.; MEDEIROS, J. L. de; ALVES, R. M. B. CO_2 utilization: a process systems engineering vision. In: MORGADO, Claudia do Rosário Vaz ; ESTEVES, Victor Paulo Peçanha (Ed.). **CO_2 sequestration and valorization.** InTech, 2014. cap.2. ISBN 978-953-51-1225-9.

CABEZAS, H.; BARE, J. C.; MALLICK, S. K. Pollution prevention with chemical process simulators: the generalized waste reduction (WAR) algorithm - full version. **Computers and Chemical Engineering**, v. 23, n. 4–5, p. 623–634, 1999.

CAMPBELL, J. M. **Gas conditioning and processing: the equipment modules**. 8. ed. Norman: John M Campbell & Company, 2004. 2 v.

CAVENATI, S.; GRANDE, C. A.; RODRIGUES, A. E. Separation of CH4/CO2/N2 mixtures by layered pressure swing adsorption for upgrade of natural gas. **Chemical Engineering Science**, Elsevier, v. 61, p. 3893-3906, 2006.

DAI, W. L., et al. Synthesis of propylene carbonate from carbon dioxide and propylene oxide using Zn-Mg-Al composite oxide as high-efficiency catalyst. **Catalysis Letters**, v. 136, n. 1-2, p. 35-44, 2010.

FORMIGLI, J. **Pre-salt reservoirs offshore Brazil: perspectives and challenges**. Miami. nov. 2007.

GAFFNEY, CLINE & ASSOCIATES. **Exame e avaliação de dez descobertas e prospectos selecionadas no play do pré-sal em águas profundas na Bacia de Santos, Brasil.** CG/JW/RLG/C1820.00/gcaba.1914. ANP, 2010.

GANGADHARAN, P.; KANCHI, K.C.; LOU, H. H. Evaluation of the economic and environmental impact of combining dry reforming with steam reforming of methane. **Chem. Eng. Res. Des.**, v. 90, n. 11, p. 1956–1968, 2012.

JOHNSTON, P. et al. Reclaiming the definition of sustainability. **Env. Sci. Pollut. Res.** v. 14, n. 1, p. 60–66, 2007.

KIM, W., et al. Methanol synthesis from syngas over supported palladium catalysts prepared using water-oil-microemulsion. Applied Catalysis A: General, v. 169, p. 157-164, 1998.

OLAH, G.A. et al. Bi-reforming of methane from any source with steam and carbon dioxide exclusively to metgas (CO–2H2) for methanol and hydrocarbon synthesis. **J. Am. Chem. Soc.**, v. 135, n. 2, p. 648–650, jan. 2013.

PACÍFICO, J. A. **Cinética do processo de reforma catalítica do metano com dióxido de carbono. Aplicação à modelagem e simulação da operação em reator de leito fluidizado.** 2004. 159f. Dissertação (Mestrado em Engenharia Química) – Departamento de Engenharia Química, Universidade Federal de Pernambuco, Recife, 2014.

PERRY, R. H.; GREEN, D. W.. **Perry's Chemical Engineers's Handbook**. 7th ed.. McGraw-Hill, 1997.

TURTON, R. et al. **Analysis, Synthesis, and Design of Chemical Processes**. 3. ed. New Jersey: Prentice Hall, 2009.

WANG, S. B.; LU, G. Q. M.; Millar, G. J. Carbon dioxide reforming of methane to produce synthesis gas over metal-supported catalysts: state of the art. **Energy Fuels**, v. 10, n. 4, p. 896–904, 1996.

ZHANG, X., et al. Carbon dioxide reforming of methane with Pt catalysts using microwave dielectric heating. Catalysis Letters, v. 88, n. 3–4, p. 129-139, 2003.

371846 Improved CO2 Capture Process : Rich Vapor Recompression with Split Flow

Abstract:

CO_2 capture process using MEA scrubbing is the most mature technology for reducing the CO_2 emission into the atmosphere. However the MEA scrubbing process requires a lot of heat energy for solvent regeneration in the stripper reboiler. The reboiler heat requirement is a function of preheating duty and stripper reflux ratio. The reboiler heat requirement decreases with preheating duty increases in the lean/rich heat exchanger. When the stripper reflux ratio decreases, the reboiler heat duty is reduced also. In order to reduce the reboiler heat requirement, large number of advanced configurations have been suggested already. The split flow configuration, well known advanced configuration, reduces the stripper reflux ratio by lowering the stripper top temperature. As a part of cold rich solvent is fed to the stripper top without passing through the lean/rich heat exchanger, the cold rich solvent is directly cooling off the stripper top. Although this split flow configuration reduces the stripper reflux ratio, the heat exchanger preheating duty is decreased due to the cold side flow rate reduction. It is hard to reduce the stripper reflux ratio while maintaining the preheating duty. In order to solve this problem, this paper suggests an innovative configuration of MEA scrubbing process; rich vapor recompression with split flow. The rich vapor recompression is a new configuration which is vaporizing the cold rich solvent at the lean/rich heat exchanger. It is similar concept with lean vapor recompression which is vaporizing the hot lean solvent at the flash drum. However the energy reduction effect is different when split flow is combined. In the suggested configuration, rich vapor recompression with split flow process, a part of cold rich solvent is fed to the stripper top without passing through the lean/rich heat exchanger. The other part of cold rich solvent is vaporized at the lean/rich heat exchanger. The generated vapor is fed to the stripper bottom as steam energy source. The remaining liquid is fed to the stripper middle. Although the heat exchanger inlet feed flow rate is reduced, the cold side heat capacity increases through the solvent vaporization. Consequently, the stripper reflux ratio is decreased with maintaining the preheating duty in the lean/rich heat exchanger. As a result, the reboiler heat requirement was 2.91 GJ/tonCO_2, 3.07 GJ/tonCO_2, and 3.38 GJ/tonCO_2 in the rich vapor recompression with split flow, lean vapor recompression with split flow and conventional MEA process, respectively.

378844 Development of CO2 Capture and Utilization Technologies for Alkaline Wastes for Building Waste-to-Resource Supply Chain: Theoretical Consideration

Abstract:

In this study, accelerated carbonation of alkaline solid wastes using a rotating packed bed (RPB) for "CO_2 capture" and their utilization as "green construction materials" have been developed for economic feasibility and environmental sustainability. The CO_2 capture capacity and overall energy consumption of the developed carbonation processes were critically assessed from the engineering and environmental aspects. The challenges of CO_2 capture by carbonation of alkaline waste were to accelerate the reaction and exploit the heat of reaction to minimize energy and material losses. Therefore, the objectives of this research work were to evaluate the performance of RPB process by introducing life cycle assessment (LCA) and response surface methodology (RSM) for minimizing the environmental impacts and maximizing the overall CO_2 capture capacity.

In addition, a prediction model for carbonation within an RPB was developed by combining the process chemistry, reaction kinetics, and mass transfer. The enhancement factors for mass transfer coefficient in case of carbonation of different types of wastes within an RPB were determined accordingly. The important factors include particle size of solid waste, reaction temperature, rotation speed, and liquid-to-solid (L/S) ratio. According to the observations of TGA, XRD and SEM-EDX, the alkaline wastes were found to be successfully carbonated with CO_2 in an RPB, where calcite ($CaCO_3$) was identified as the main product.

The LCA was utilized to determine the net performance of carbon offsetting by considering (1) the energy consumption and carbon emission associated with RPB operation; (2) benefits of offering alternative materials for cement or concrete; (3) the CO_2 caused by raw material pre-processing & conditioning of carbonate product for final products. The functional unit is set to be capture of 1 kg CO_2 by carbonation processes within the system boundary while carrying out the LCA. Energy consumption of the overall processes including grinding, sieving, pressuring, heating, pumping, and rotating processes was measured. A balance for material and energy flows is drawn up for calculating and analyzing the investigated system, which consists of all input (e.g., land, energy, raw materials, auxiliary agents, etc.) and output (e.g., products, emissions, waste, etc.)

Furthermore, the developed RPB process were designed and analyzed by RSM using Design Expert statistical software to visualize the effects of operating parameters on the carbonation conversion of solid wastes for overall maximization. Two types of response parameters including the carbonation conversion of alkaline waste and environmental impact were systematically evaluated in this study. The factors affecting the above two types of responses were coded with low and high levels using central composite design (CCD) method. The carbonation conversion and environmental impact could be maximized and minimized, respectively, by observation of the location of the saddle point on the 3D-surface imaging from different sets of response surfaces. With the results of the prediction model, LCA and RSM, the best achievable technology (BAT) for carbonation using an RPB was proposed for system optimization. It was thus concluded that accelerated carbonation of alkaline wastes using an RPB is an effective and efficient method for CO_2 capture and utilization due to its higher mass transfer rate and carbonation conversion.

380371 CO2 Fixation through Carbonation of Waste Cement and Concrete

Abstract:

CO_2 capturing and fixation processes with waste concrete via mineral carbonation reactions have been developed. The waste concrete is either waste concrete generated from demolished concrete buildings (waste concrete, hereafter) or concrete sludge. Waste concrete is composed of aggregates and waste (hydrated) cement, and waste cement powder of which the diameter is under about 100 μm would be separated in the recycling process of aggregates. The concrete sludge is waste fresh concrete in the form of slurry generated at the concrete using sites. Both concrete wastes are alkaline containing calcium that could be used as a capturing and fixation material for CO_2 via carbonation reaction. Several options are considered as the CO_2 fixation process with these waste concrete, and process feasibility has been evaluated based on the laboratory experimental studies. The followings are the options examined in this study.

(1) Direct fixation of CO_2 with concrete sludge

This option is a direct fixation process of CO_2 with waste concrete sludge. Flue gas is introduced to the slurry of concrete sludge, and CO_2 will react with hydrating cement components, mainly calcium hydroxide, to form calcium carbonate. The effect of hydration reaction time, gas flow rate, water content, CO_2 concentrations on the CO_2 fixation performances were examined in a small-scale laboratory experiments.

(2) Indirect multistep fixation of CO_2 with waste cement powder

This option is four-step process composed of (i) dissolution of calcium ions into an acid, (ii) CO_2 capture with an alkaline solution, (iii) formation of calcium carbonate by mixing the above two solutions, and (iv) regeneration of the acid and the alkaline solutions via electrodialysis. Parameters affecting the process performances were examined for each step in laboratory experimental studies.

Based on the experimental studies, the feasibilities of these processes have been evaluated in views of the process power consumption as well as the cost.

14AIChE
Annual Meeting, Atlanta, GA

Technological Trends in CO_2 Capture, Transport and Utilization

Rita M. Brito Alves
Fernanda M. Fontes
José L. Medeiros
Ofélia Q. F. Araújo

USP
H₂CIN

*Motivation

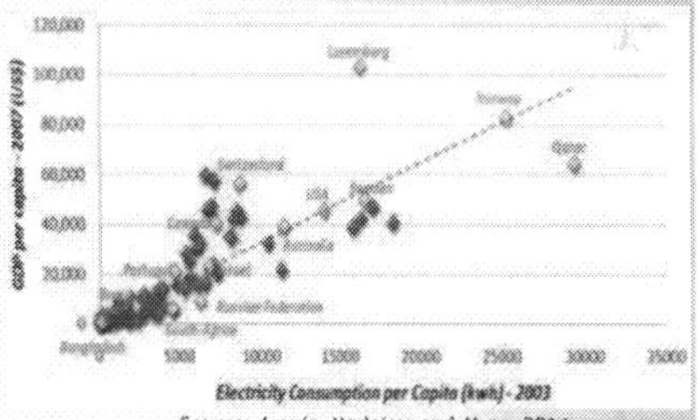

Source: Araujo, Medeiros and Alves, 2014

*The development of economies results in increased energy consumptions.

*In the coming decades, the supply of such expanding demand will still remain based on fossil fuels technologies.

*Fossil fuel based energy is recognized as carbon emitter with serious consequences for the environment.

*Measures have been taken not only to reduce but also to try to reverse the current situation of intensification of global warming due to the greenhouse effect.

*These measures include reduction of the dependence on fossil fuels by modern society, and methods to control the emission of greenhouse gases, especially CO_2.

*The challenge is thus to expand energy supply in a carbon-constrained economy.

*Motivation

* According to the World Economic Forum (King, 2010), no truly low-carbon technology will be able to penetrate the mass market in the short term.
* In the context of energy and climate policy, governments place much emphasis on carbon capture and geological storage (CCS) as an option that can reduce substantial amounts of greenhouse gas emissions.
* However, there is also the possibility of CO_2 utilization in building material production, for fuels or in the chemical industry, which could contribute to a green economy.
* On the other hand, it is reasonable to expect that CO_2 conversion and biomass-based processing alternatives will coexist for a while with fossil-based technologies.
    * So, the use of installed fossil processing infrastructure, with co-processing of biomass and fossil feedstock and capturing and utilization of emitted CO_2 is the "*escape route*" for a moderate transition from the present to a long-term sustainable future.

*Objectives

* The main objective of this work is to to build technological maps in order to support decision-makers aligned to initiatives for reduction of CO_2 emission.
* In this framework, this work presents the CO_2 Capture Cycle as well as promising alternatives of its reutilization.
    * The work is oriented to cover CO_2 as feedstock, i.e., use of CO_2 as raw material to chemicals and fuels with economic application.

*CCS

* Carbon capture and storage (CCS) is most commonly defined as the capture of CO2 from an industrial or power-sector point source combined with its transport and its storage in geological formations (IEA, 2009).
* CCS in some sectors provides cost-effective emission reductions, but has significant shortcomings:
  * Costs of CCS are considered to be highest for the capture component, but storage and especially transport costs can also be significant.
  * the potential storage capacity has uncertainties
  * public resistance to CCS has been increasing
  * and it costs energy.

CCS

* According to the IPCC (2005) the components of CCS are classified as being in different stages of technological maturity.
* Some options, such as capture from a limited number of industrial processes, are considered mature as they are implemented routinely, for instance in hydrogen production and gas processing plants.
* Other components are seen as in or just beyond the demonstration phase – significant-scale demonstrations are operative but the technology is not yet proven at full scale or in commercial conditions in the power sector.

CCS - CO_2 Capture

* Current state of the art post-combustion CCS uses a variety of capture agents; however these are mainly based on amines.
  * Corrosivity/volatility/hazardous emission
  * Capex
* There is therefore considerable effort required to produce new capture agents with better environmental credentials.
  * Recent studies have shown that solid capture agents such as functionalized silicas, poly(ionic liquid)s, activated carbons and calcium oxide can be used to not only enhance capture but also reduce the capital expenditure required on plant construction.

CCS - CO_2 Storage

* Geological storage

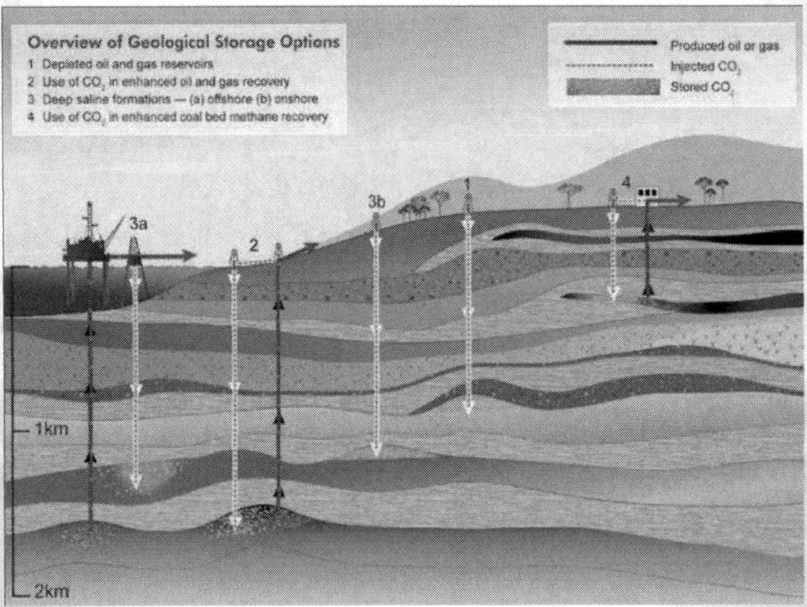

Methods for storing CO_2 in deep underground geological formations. Two methods may be combined with the recovery of hydrocarbons: EOR (2) and ECBM (4)

*CCS - CO_2 Storage

*Geological Storage

*Geological storage of CO_2 is accomplished by injecting it in dense form into a rock formation below the earth's surface.

*The option of storing CO_2 in coal beds and enhancing methane production is still in the demonstration phase

*Geological storage of CO_2 is ongoing in three industrial-scale projects (~1 $MtCO_2$/yr): the Sleipner project in the North Sea, the Weyburn project in Canada and the In Salah project in Algeria.

*About 3-4 $MtCO_2$ that would otherwise be released to the atmosphere is captured and stored annually in geological formations.

*CCS - CO_2 Storage

*CO_2-EOR

The US has been commercially operating EOR programs for over 40 years.

It currently provides nearly 300,000 barrels of oil per day in North America from 125 distinct EOR projects

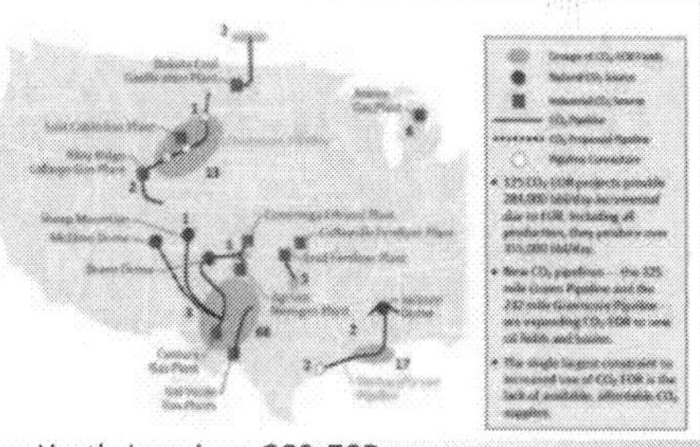

North American CO2-EOR sources

Source: Oil&Gas Journal, 2012

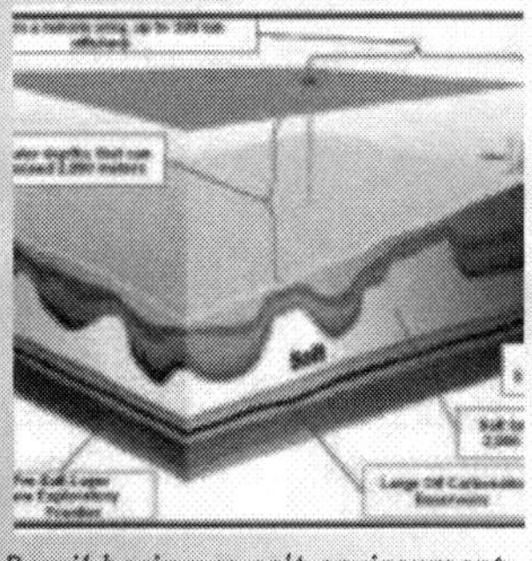

Brazil basin pre-salt environment.

Outside North America, only a few CO_2-EOR projects are underway in Brazil, Turkey, and Trinidad.

However, the development of universally agreeable predictive models and comprehensive monitoring techniques must still be established.

*CO_2 Storage

* EOR and CGS constitute, under present conditions, the only CO_2 processing chain that has some steps ready to be put into operation at high scales.
* These encompass some separation technologies for CO_2 post-combustion capture, CO_2 compression and CO_2 transportation via long pipelines.
* EOR and CGS can be reinforced if oxy-fuel technologies finally attain industrial maturity, which may occur within the next 5 decades.
* Pre-combustion technologies are also potential powerful contributors to EOR-CGS systems, which should be fully developed within the next 5 decades.
* Pre-combustion relies on some separation technologies also present in post- combustion alternatives.

*Transport of CO_2

* Pipelines today operate as a mature market technology and are the most common method for transporting CO2.
* In some situations or locations, transport of CO_2, as a liquid, by ship may be economically more attractive, particularly when the CO_2 has to be moved over large distances or overseas.
* Road and rail tankers also are technically feasible options.

* Alternative routes for CO_2 valorization

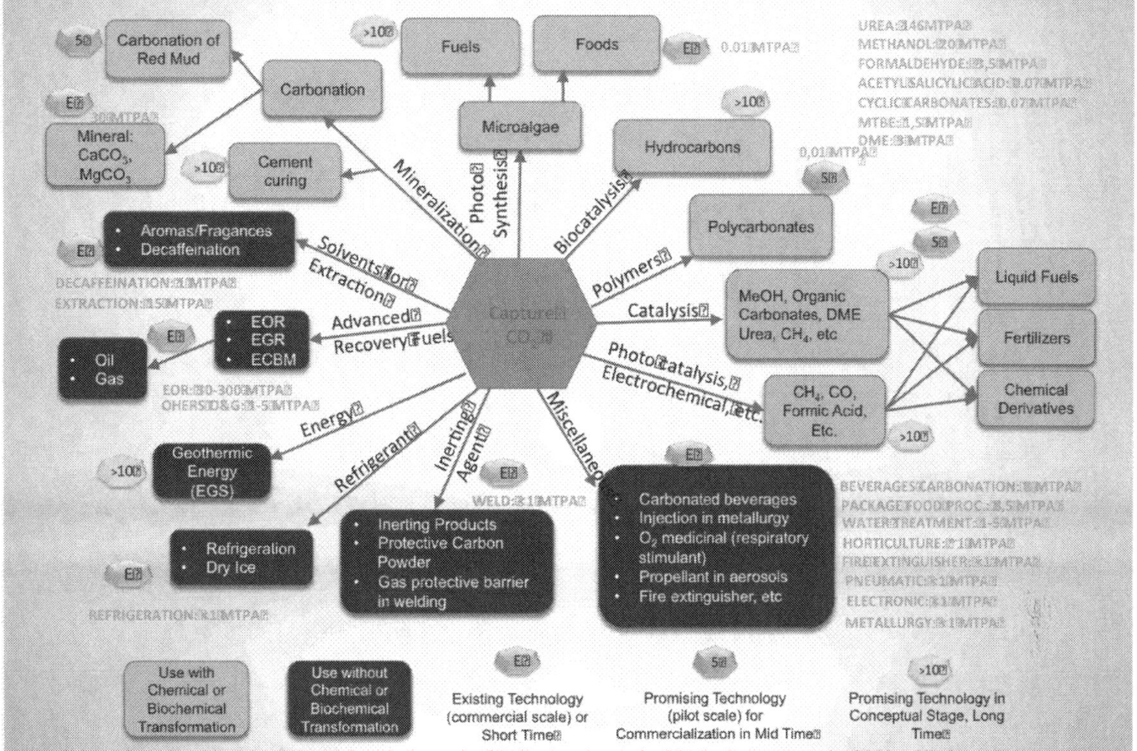

Built based on Gresser (2011), Bucin-Dumitriu et al. (2013), Araujo et al. (2014), Gouveia et al. (2008), Gürtler (2012).

* CCU - direct use

* GEOTHERMAL ENERGY GENERATION

Rather than using water to extract heat from deep underground to produce geothermal energy, it uses carbon dioxide.

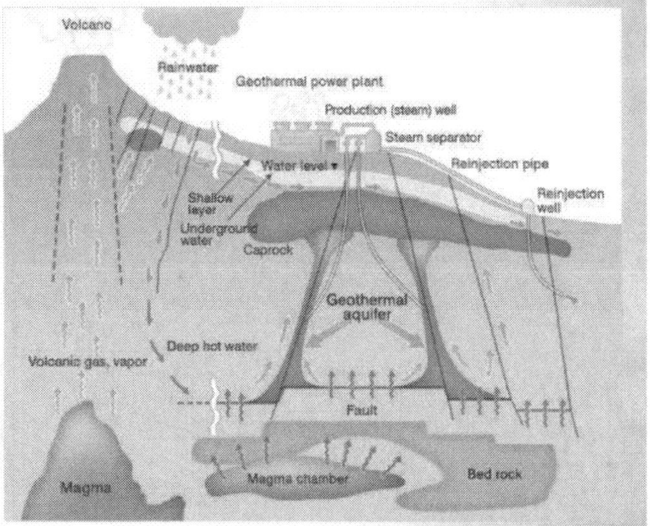

Source: Geothermal Energy, Serial No. 87 (July 1999)

* CCU- CO_2 conversion to chemicals

* The strategic objectives include:
  * rather than treating CO_2 as waste, it can be regarded as a chemical feedstock for the synthesis of other chemicals;
  * use CO_2 to produce added-value chemicals useful to the industry and able to reduce the penalty asssociated to the costs of capture process;
  * use CO_2 as benign fluid for processing or a way to recover energy and reduce emissions;
  * use CO_2 recycle as energy renewable source to preserve the carbon sources, aiming sustainable development.

CCU should not be regarded as an alternative technology to CCS but as a complementary technology

* CCU- CO_2 conversion to chemicals

* The main barriers to CO_2 utilization are:
  * the costs associated to CO_2 capture, purification and transport.
  * the unfavorable thermodynamics of many conversions
  * the requirement of energy for chemical conversion of CO_2
    * that means that there will be an energy cost associated with utilization
  * limited size market
  * small investments/incentives and lack of industrial commitment to increase CO_2 based chemicals; and
  * lack of social and economic drivers.

*CCU- CO_2 conversion to chemicals

*Chemicals from CO_2

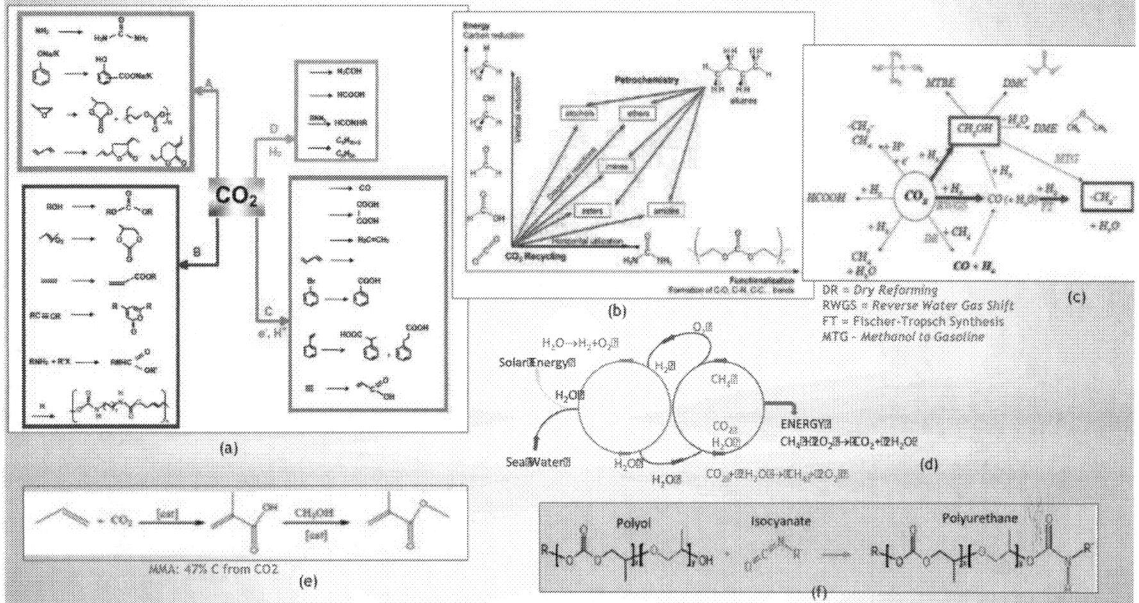

(a) Chemicals from CO_2, Aresta (2010); (b) "Diagonal Conversion", Gomes et al. (2012); (c) Synthesis of Liquid Fuels/Energy Storage, Hashimoto (2011); (d) Energy from H_2O, CO_2 and solar energy, adapted from Carrera (2011); (e) MMA (Wang, 2011); (f) Polyurethane, Gürtler (2012).

*CCU- CO_2 conversion to chemicals

*Chemicals from CO_2

Reagents and products generated by CO_2 reaction.

CO-REAGENT	CO_2 - products	CO-REAGENT	CO_2 - products
Alkane	*Syngas*, acids, esters, lactones	Substituted hydrocarbon [(c)]	Acids, esters, lactones, polycarbonates
Cycloalkane	Acids, esters, lactones	Alkyne	Lactones, organic carbonates, polycarbonates
Active H compounds	Acids, esters, lactones	Epoxy	Carbonates, (co)polymers (polycarbonates)
MonoAlkene	Acids, esters, lactones	NH_3 and Amine	Simetrical Ureias, amino acids, (co)polymers
Diene[(a)]	Acids, esters, lactones [(b)]	Diamine	Ureias, carbamates, (co)polyemers, (polyureias)
CycloAlkene	Acids, esters, lactones, (co)polymers	Imines	Carbamates, (co)polymers (uretane)

Source: Kargari e Ravanchi (2012)

*CCU- CO_2 conversion to chemicals

*CO_2 Converion to Synthetic Fuels

*Technological routes for CO_2 conversion to synthetic fuels (synfuel)

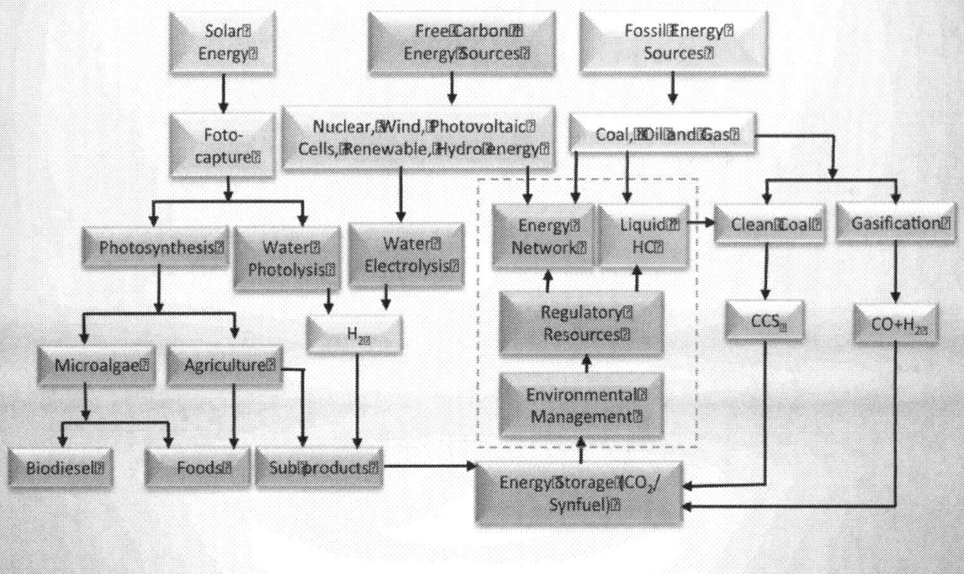

Source: Amoureux, Siffert e Massue (2011)

*CCU- CO_2 conversion to chemicals

*CO_2 Converion to Synthetic Fuels

*Solar route for Synthesis Gas from CO2

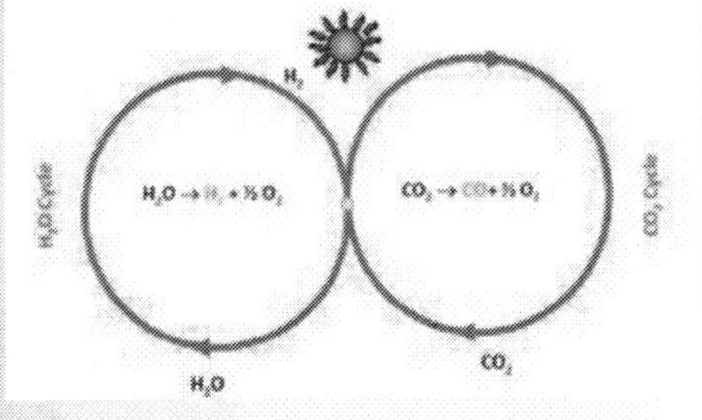

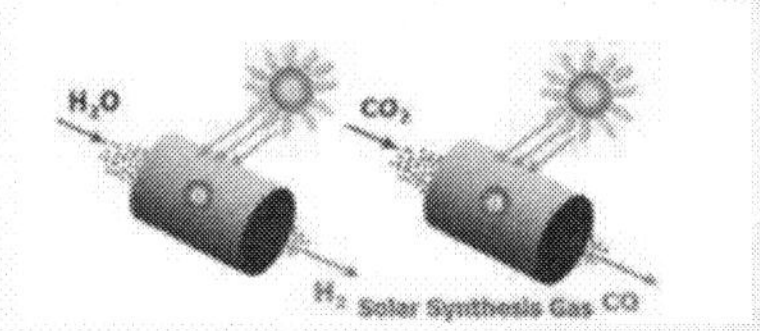

*Sustainable storage of H_2 e Carbon

$H_2 + CO \longrightarrow C_XH_Y$		Liquid Fuels - Fischer-Tropsch Process
$4H_2 + CO_2 \longrightarrow CH_4 + 2H_2O$		Gaseous Fuels - Methane/Sabatier Process
$H_2 + CO \longrightarrow C_XH_Y$		Plastics

Source: Konstandopoulos (2011)

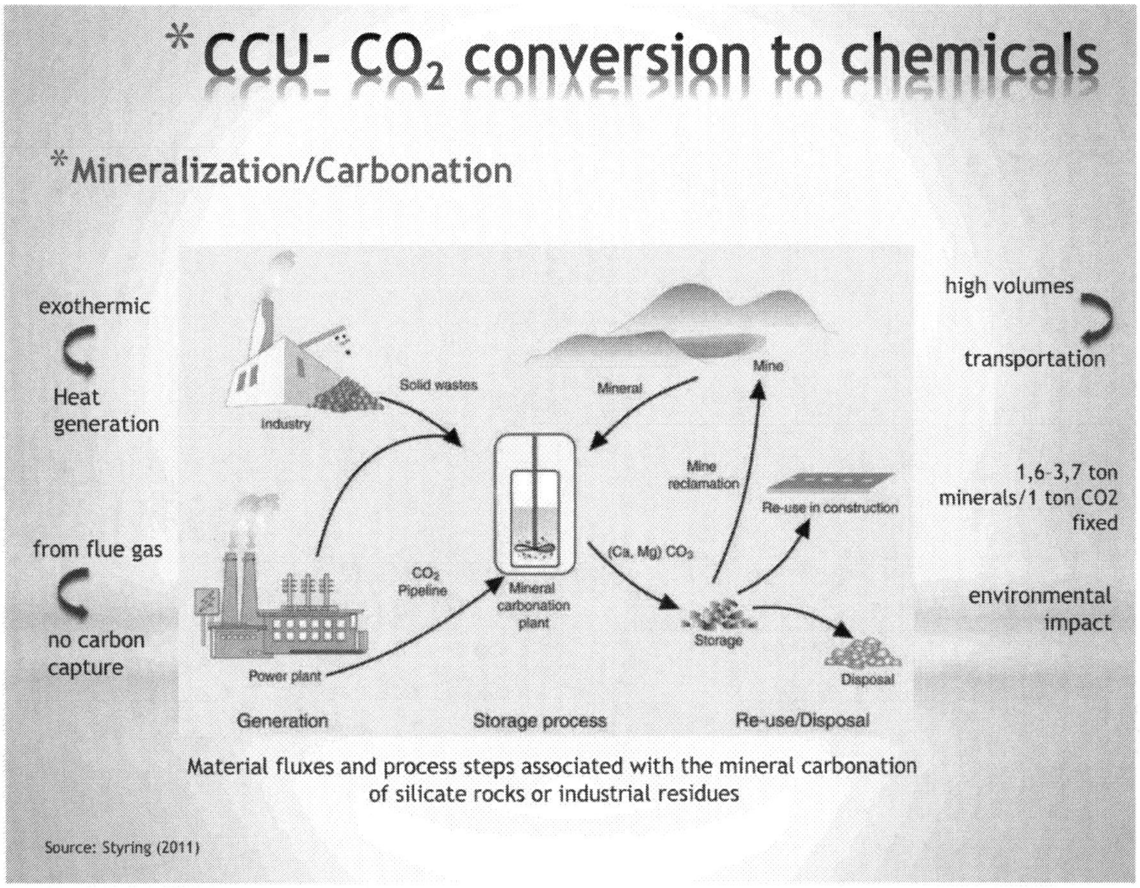
CCU- CO2 conversion to chemicals
Mineralization/Carbonation
exothermic
Heat generation
from flue gas
no carbon capture
Industry
Solid wastes
Mineral
Mine
Mine reclamation
Re-use in construction
(Ca, Mg) CO3
CO2 Pipeline
Mineral carbonation plant
Storage
Disposal
Power plant
Generation
Storage process
Re-use/Disposal
high volumes
transportation
1,6-3,7 ton minerals/1 ton CO2 fixed
environmental impact
Material fluxes and process steps associated with the mineral carbonation of silicate rocks or industrial residues
Source: Styring (2011)

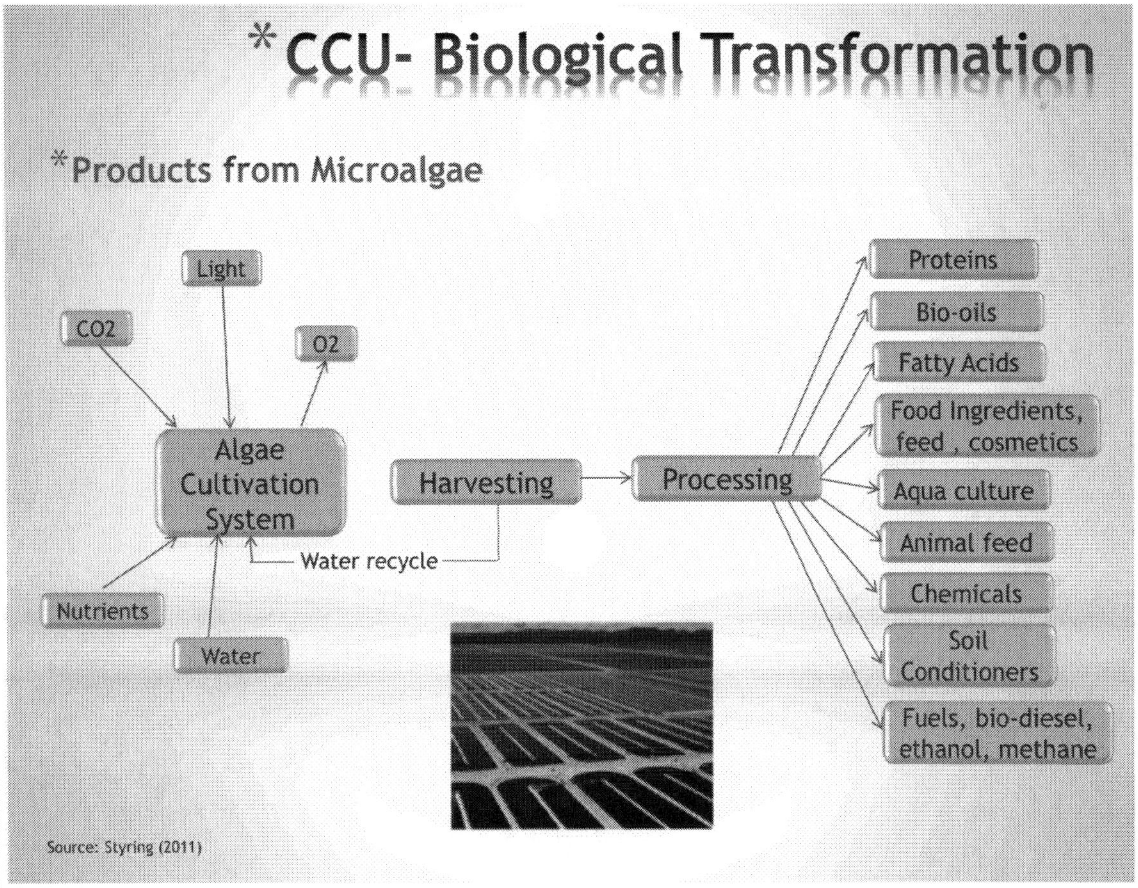
CCU- Biological Transformation
Products from Microalgae
Light
CO2
O2
Algae Cultivation System
Harvesting
Processing
Water recycle
Nutrients
Water
Proteins
Bio-oils
Fatty Acids
Food Ingredients, feed , cosmetics
Aqua culture
Animal feed
Chemicals
Soil Conditioners
Fuels, bio-diesel, ethanol, methane
Source: Styring (2011)

*CCU

*Technological maturity in CO_2 Utilization

CATEGORY	TECHNOLOGY	RESEARCH	DEMONSTRATION	ECONOMICALLY VIABLE	MATURE MARKET
CO2 TO FUELS	Hydrogen (Renewable Methanol)		■		
	Hydrogen (Formic Acid)	■			
	Microalgae (Biofuels)	■	■		
	Photocatalysis	■			
	Catalysts (nanomaterials)	■			
INCREASE PRODUCTIVITY	Power cycles with supercritical CO_2	■			
	Urea Boosting				■
ADVANCED RECOVERY	EOR				■
	EGR		■		
	ECBM		■		
MINERALIZATION	Cement Production		■		
	Concrete Healing			■	
	Carbonation Red Mud			■	
	Mineralization with Carbonate	■	■		
CHEMICALS	Sodium Carbonate				■
	Polymers	■			
	Other Chemicals (e.g., acetic acid)	■			
	Microalgae (for Chemicals)		■		

Fonte: Bocin-Dumitriu et al. (2013)

*Technological Roadmaps

*This work highlighted the state of art in technologies of *Carbon Capture and Utilization (CCU)* in order to

- *Identify the main drivers that will promote the development of technologies of capture, transport and utilization of CO_2
- *Identify the main technologies and classify them according to their technology readiness level (TRL)
- *Identify the Initiaves and technolgical perspectives
- *Build a technological roadmap able to associate short, mid and long term milestones to the technological solutions and organize them in a timeline.
- *Allow a broad vision about the most promising technologies in the short, mid and long term.

WHY, WHAT, WHEN & WHO

*Technological Roadmaps

*Roadmap Structure:

*Timeline

* *Current (or Today) represents technologies already in use
* *1 to 5 years includes the technologies in advanced stage of development or scale almost commercial
* *5 to 10 years are projects in demonstration or pilot plant scale and which efficiency has been demonstrated.
* *10 to 15 years are projects still in test or lab scale, which efficency hasn't been yet proved. As the case of researches recently published in literature.

*Technological Roadmaps

*Roadmap Structure:

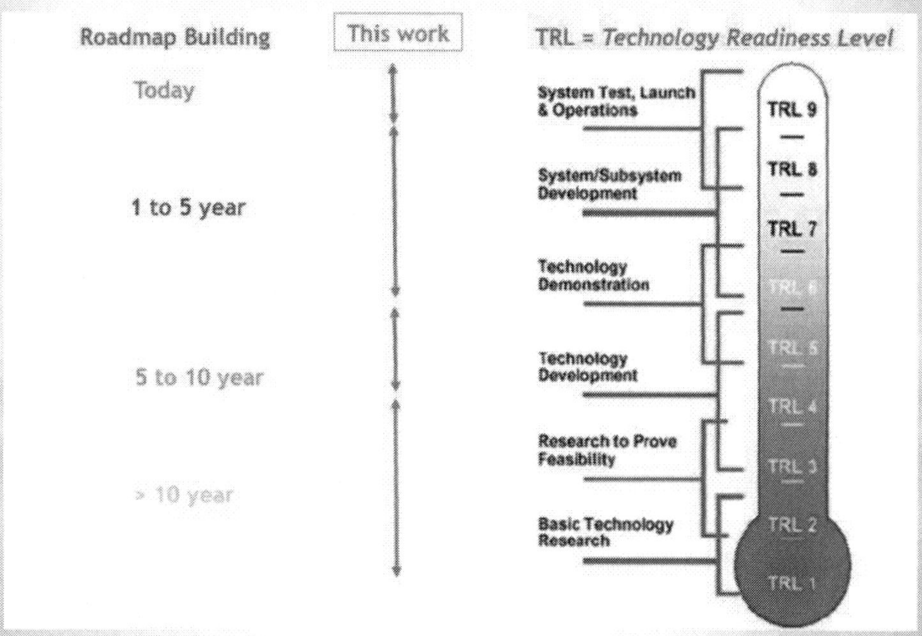

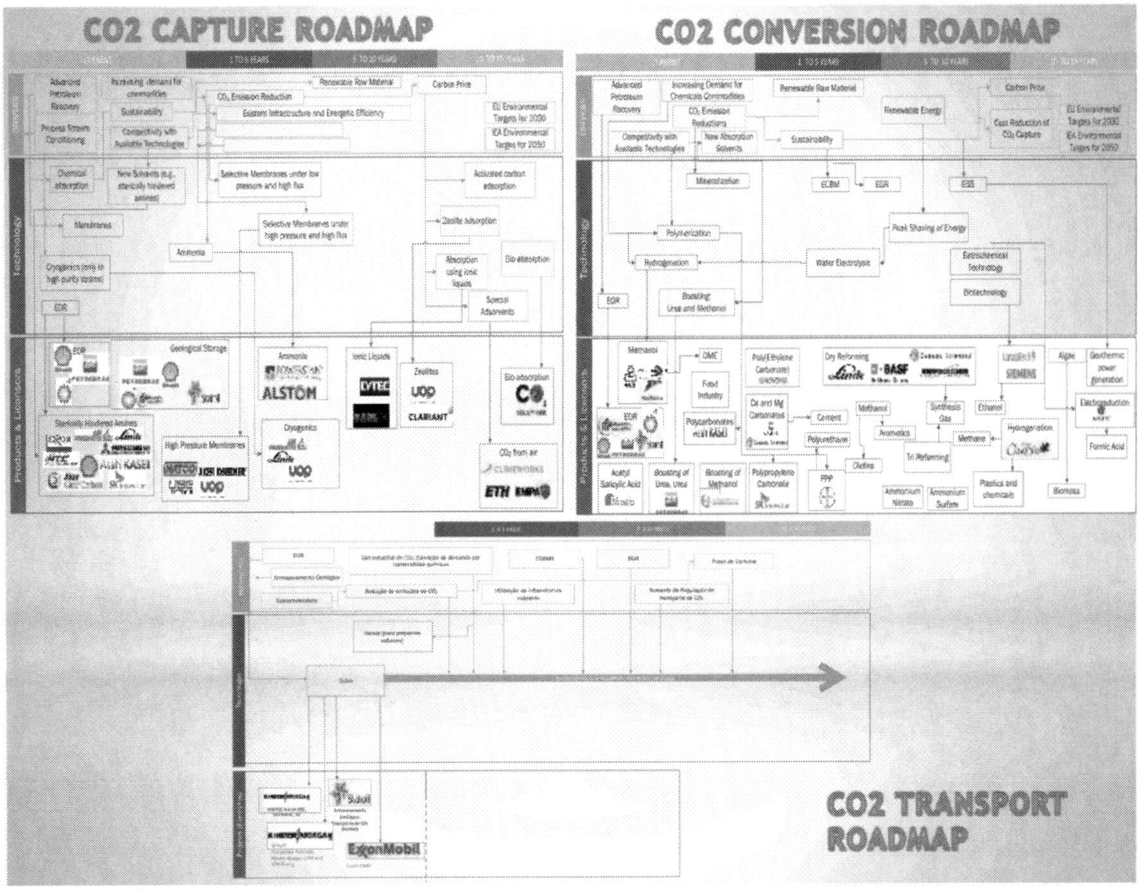

*Technological Roadmaps -Conclusions

*The cost of CO_2 capture, storage and utilization technologies should be analyzed to evaluate the project feasibility.

*The most promising technology for CO_2 capture is chemical absorption with amines.

*Regarding transportation technologies there is no doubt that the ducts are the present and the future. They are a safe, and they have already significant maturity in behavior and structure (due to similarities with pipelines for gas) and they are cheaper when compared to other forms of transportation.

*Researches on technologies for CO_2 utilization indicate that the EOR and geological storage destinations are the methods for which we have more knowledge and they are / will be able to absorb large amounts of CO_2, having a significant effect on reducing CO_2 emissions.

*Technological Roadmaps -Conclusions

- *Concerning CO_2 conversion, the roadmap shows scenarios with opportunities for next decades.
- *An important conclusion is that the driver for CO_2 utilization technologies is not the mitigation of emissions but the increasing in price of fossil raw materials, which would promote technologies based on CO_2 as a raw material - cheap and abundant due to emissions regulation.
- *On the other hand, CCU alone cannot realistically remediate all emissions because of the volumes involved and the potential markets for the individual products.

Acknowledges

Braskem

CAPES

Contacts

rmbalves@usp.br

ofelia@eq.ufrj.br

*Thank you!

390952 Effects of Contaminants to Physical Solvents for Pre-Combustion CO2 Capture

Abstract:

The pre-combustion capture process involves converting fossil fuel into syngas, primarily a mixture of carbon monoxide and hydrogen, usually by gasification. During pre-combustion, coal is first transformed into hydrogen and CO_2. In a CO_2 separation unit, the carbon dioxide is separated from the hydrogen using a solvent, such as ionic liquids (ILs). This process has the potential of capturing and removing 90 percent of CO_2 emissions from power plants - major contributors to global warming.

ILs are salts with many advantages over conventional pre-combustion CO_2 capture materials, including low melting points, longer lifetimes and higher stability over a range of process conditions. While ILs for CO_2 capture are a popular research topic, less attention has been focused on the impact of contaminants, i.e., water and coal derived *contaminants* (primarily H_2S), on ILs for pre-combustion carbon capture process. We explore the use of physical solvents for pre-combustion carbon capture. Four solvents have been under investigation, including PDMS, hybrid-PDMS, and two types of ILs, i.e., P888 allyl Tf_2N and allyl pyridinum Tf_2N, with different initial viscosity. A 1-liter Continuous Stirred Tank Reactor (CSTR) equipped with a 2-liter gas buffer tank has been used to measure CO_2 isotherms and CO_2 breakthrough curves at different temperatures. For CO_2 solubility and kinetics measurement without contaminants, a dynamic pressure-step method was applied, while a continuous flow operation was used for CO_2 breakthrough measurement in the presence of H_2O and H_2S. Effects of H_2O and H_2S on these solvents physiochemical properties will be discussed. This study has established an understanding of the influence of precombustion contaminants, i.e., H_2O and H_2S, on solvents physiochemical properties, including CO_2 solubility, CO_2 absorption kinetics, viscosity, and so on.

364033 Self-Emulsification of Alkaline-Dissolved Clove Bud Oil By Whey Protein, Gum Arabic, Lecithin, and Their Combinations

Abstract:

Essential oils (EOs) are well-known for their antioxidant and antimicrobial activities and have been widely used as functional ingredients in food, pharmaceutical and cosmetic products. However, poor water solubility and high volatility during processing are two major obstacles for utilizing EOs as preservatives in food matrices or as sanitizing agents. In the present study, a novel self-emulsifying process was studied to prepare clove bud oil (CBO) emulsion without any specialized equipment or organic solvents. In this method, the major CBO component (eugenol) was deprotonated at alkaline conditions, which was accelerated by heating. Heating 10% v/v CBO in 3 M NaOH at 110 °C for 10 min was observed to be optimal conditions to obtain a transparent CBO solution. The alkaline-dissolved CBO was mixed at an overall CBO concentration of 1% v/v in aqueous neutral solutions containing 1% w/v naturally-occurring emulsifiers - whey protein concentrate (WPC), gum arabic, lecithin, and their combinations. Emulsions were obtained after neutralization to pH 7, during which eugenol became re-protonated and encapsulated in situ by emulsifiers. CBO extracted from emulsions had identical UV-visible absorption spectrum as native CBO, and the peak in reversed-phase HPLC had a same elution time and peak area as the native eugenol. These results verified the maintained structure of eugenol after self-emulsification. The freshly prepared emulsions had hydrodynamic diameters in the range of 160–540 nm, but their dimension and stability were dependent on the emulsifiers. The sample self-emulsified with WPC only had the smallest particles, followed by the combination of all three emulsifiers and the WPC-lecithin mixture, while that with gum arabic alone had the largest particles. The emulsifier blend generally resulted in better stability of the emulsion particle size than individual compounds. Emulsions were evaluated for their antimicrobial activities against *Escherichia coli* O157:H7 ATCC 43895, *Listeria monocytogenes* Scott A and *Salmonella* Enteritidis. Emulsions generally exhibited better activity than same amounts of ethanol-dissolved CBO, showing lower minimum inhibitory and bactericidal concentrations and stronger inhibition of bacterial growth in tryptic soy broth. Our findings indicated that this novel self-emulsifying process may find promising applications in preparing EO emulsions using low cost food ingredients for use as antimicrobial food preservatives.

OIL EXTRACTION FROM SPENT COFFEE GROUNDS USING ADVANCED TECHNIQUES

Javier A. Dávila, Moshe Rosenberg, Gonzalo Taborda, Carlos A. Cardona
Universidad Nacional de Colombia, Manizales. University of California, Davis, Universidad de Caldas

2014 AIChe Annual Meeting

1. INTRODUCTION

SPENT COFFEE GROUNDS (SCG)

Coffee is one of the most important agroindustrial products in Colombia generating appreciable amounts of residues

654,000 T/Year

- Accounts the 10% of the total weight of the grain
- Contains bioactive compounds (Chlorogenic acids)
- Oil content (4% - 25%)
- 44% of fiber (Cellulose, hemicellulose)

2

1. INTRODUCTION

SPENT COFFEE GROUNDS (SCG)

Products

- Cellulose
- Hemicellulose

- Alcohols
- Biofuels
- Bioproducts
- Furfural
- PHB
- Xylitol
- Acids

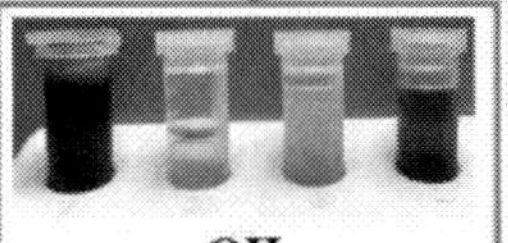

OIL

- Food
- Pharmaceutical
- Biofuels

(Rodriguez, Cenicafe, 2011)

3

1. INTRODUCTION

CHEMICAL COMPOSITION OF SCG

Component	(%)
Nitrogen	1.64
Protein	10.30
Ash	0.59
Fiber (Hollocelulose)	39.72
Lipids	25.83
Extractives	21.92

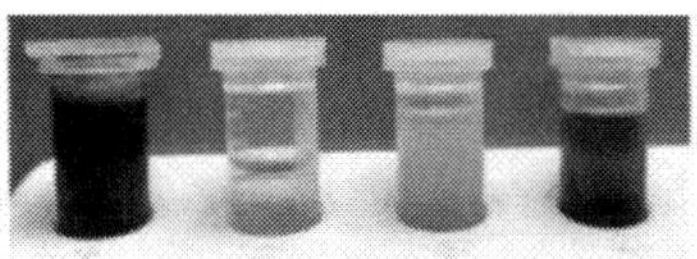

Value Added Product

4

1. INTRODUCTION

EXTRACTION TECHNIQUES

Most common techniques used:

- Mechanical extraction (High pressure)
- Solvent extraction (Hexane)
- Supercritical extraction (CO2 at supercritical conditions)

6

1. INTRODUCTION

HOW TO EXTRACT OIL FROM SCG

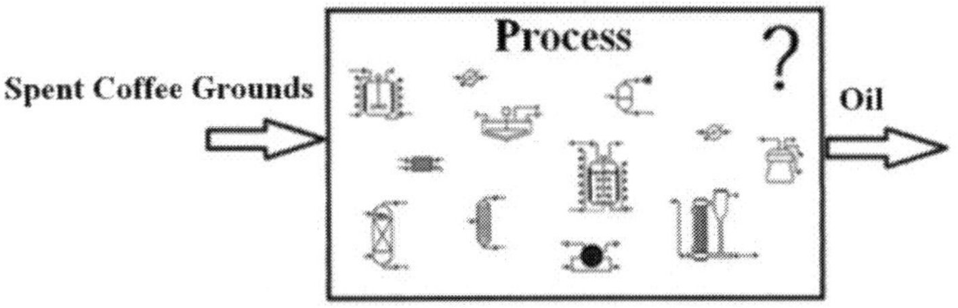

Yields?
Cost of production?

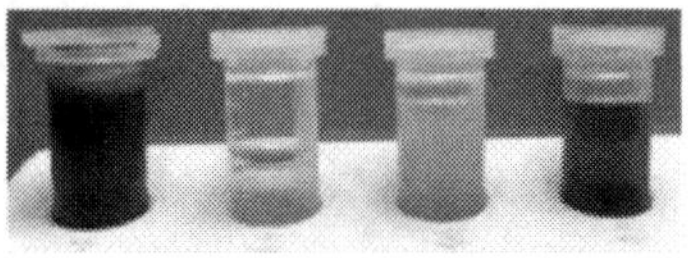

7

1. INTRODUCTION

INTEREST OF THIS WORK

Oil from Spent Coffee Grounds

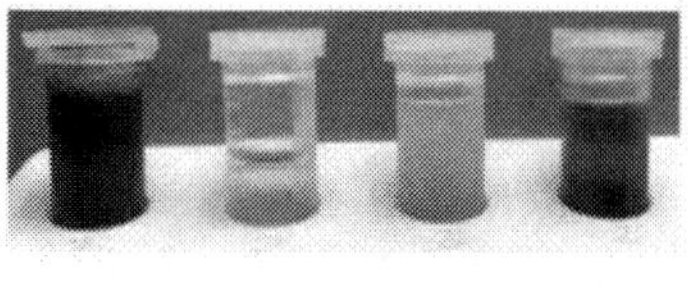

Motivation:

Extract oil from SCG using three different techniques and obtain a preliminary techno-economic analysis of the extractions

2. METODOLOGY

METHODOLOGY

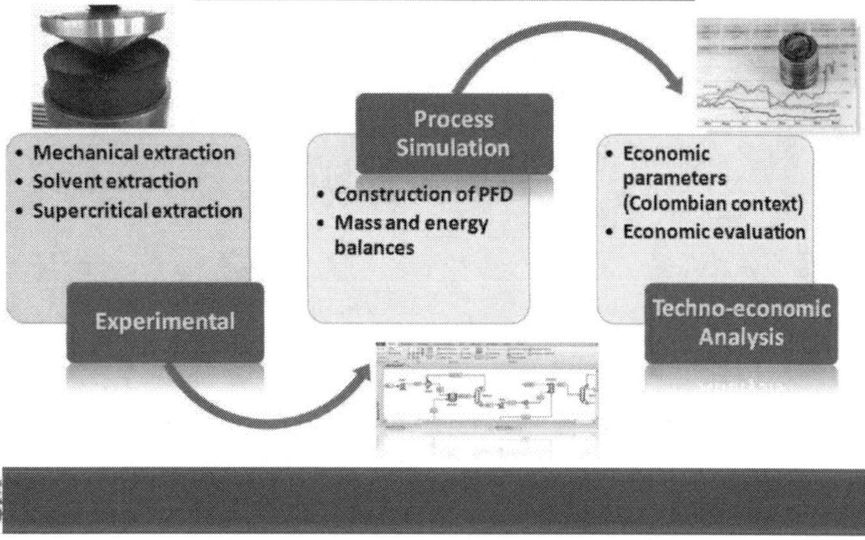

2. METODOLOGY

MECHANICAL EXTRACTION

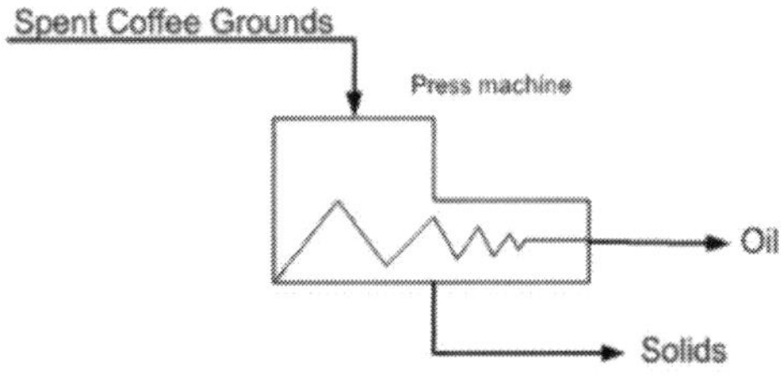

MECHANICAL EXTRACTION

- Pressure: 60 bar
- Temperature: 50, 70, 90 °C

2. METODOLOGY

SOLVENT EXTRACTION

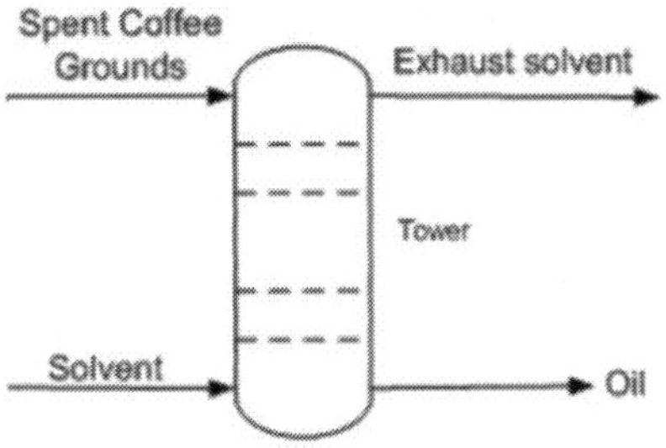

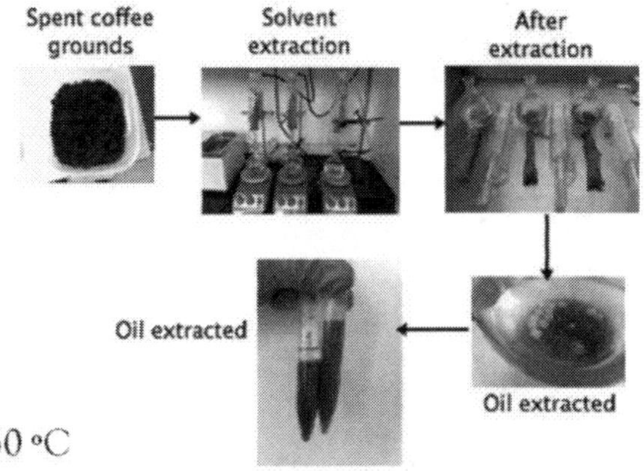

SOLVENT EXTRACTION

- Solvent: Hexane
- Temperature: 90, 120, 160 °C
- Time: 24 h

2. METODOLOGY

SUPERCRITICAL EXTRACTION

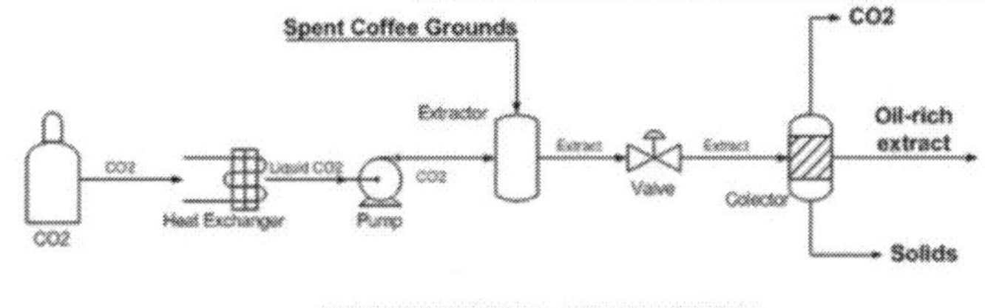

SUPERCRITICAL EXTRACTION

Supercritical Extraction

- Solvent: CO_2
- Temperature: 45 °C, pressure 300 bar
- Time: 5 h

2. METODOLOGY

PROCESS SIMULATION

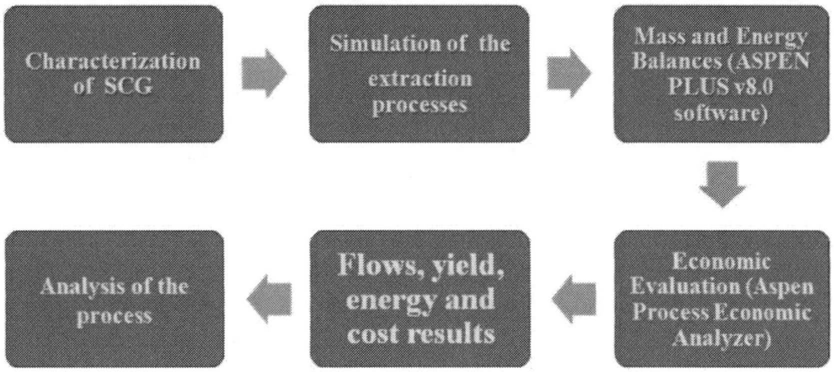

3. RESULTS

EXPERIMENTAL RESULTS

Yields

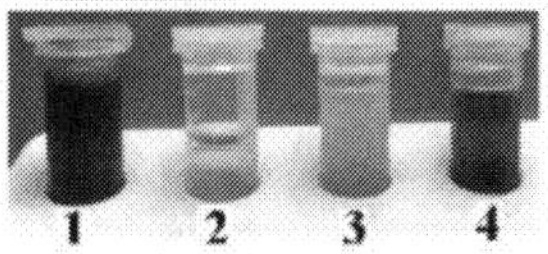

Oil 1: 60 bar, 50, 70, 90 °C
Oil 2: Hexane, 90 °C
Oil 3: Hexane, 120 or 160 °C
Oil 4: Hexane, 120 or 160 °C evaporating inmediately.
Supercritical extraction

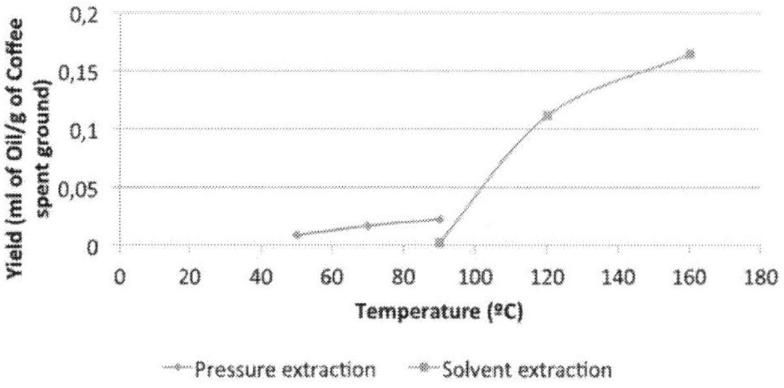

Melanoidins
Liposoluble dyes

Technology	Oil (g/g)	Yield (%)
Solvent extraction	0.165	63.88
Mechanical extraction	0.02	8.96
Supercritical extraction	0.2066	80

14

3. RESULTS

ECONOMIC ANALYSIS

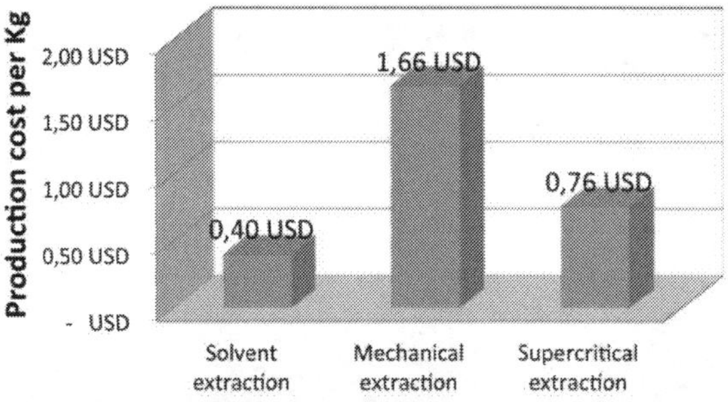

Market price of Palm Oil 0.64 USD/kg

(Index Mundi, 2014)

15

3. RESULTS

ECONOMIC ANALYSIS

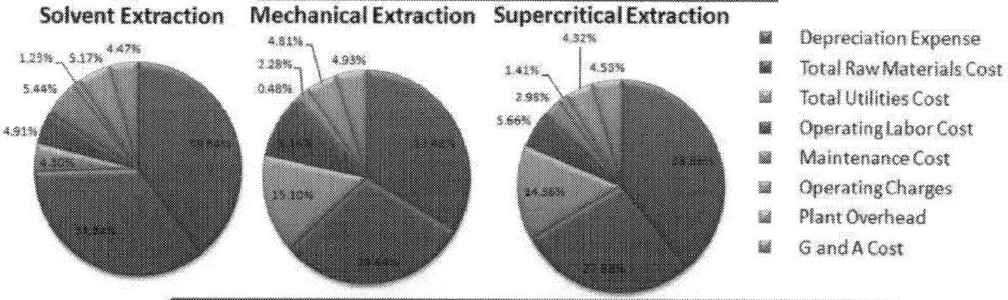

PRODUCTION COST PER YEAR

Solvent Extraction (1.048 million USD/Year)

Mechanical Extraction (0.617 million USD/Year)

Supercritical Extraction (1.211 million USD/Year)

3. RESULTS

OIL EXTRACTED

Oil extracted from SCG

Free Fatty Acids (FFA) were measured according to *AOCS Ca* 5*a* – 40 official method (Rukunudin et al., 1998)

Technology	Acid Index (% FFA)
Solvent extraction	3.26
Mechanical extraction	3.08
Supercritical extraction	2.95
Palm Oil	2.4 (Rincón et al., 2009)
Coffee Oil from SCG	2.99 (Ramírez, 2008)

Acid Index values ranges from 0.1 % to 4.7 % for biodiesel production.

4. CONCLUSIONS

CONCLUSIONS

1. Oil extracted from SCG is a potential raw material to be for other purposes.
2. Solvent extraction is the most interest technique to obtain oil from SCG.
3. Acid Index indicates that oil from SCG can to be used for biodiesel production.

18

REFERENCES

REFERENCES

1. Rodriguez Valencia Nelson. 2011. Recent experiences on the use of byproducts from coffee. CENICAFE 2011.
2. Price Indexes. Palm oil price market. 2014. Available in www:indexmundi:com
3. Rincon M. Sandra Milena, Martinez C. Daniel Mauricio. Analysis of the properties of palm oil in the industries development. 2009. Palmas Vol 30 No. 2.
4. Ramirez Ovalle Lourdes Mara. Assessment of the yield extraction and characterization of oil from coffee. 2008. Bachelor thesis, Guatemala Universidad de San Carlos de Guatemala.

19

Microcapsulation of Blackberry antioxidants. Modeling and simulation

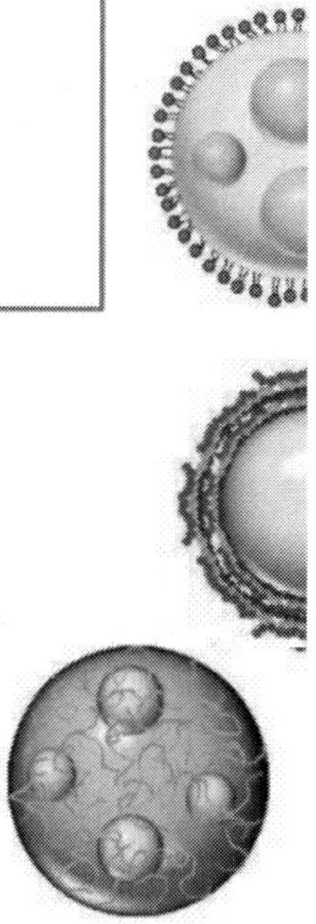

Miguel A. Rojas[1], Javier A. Davila[1] Moshe Rosenberg[2], Carlos A. Cardona[1*]

[1] **Biotechnology and Agribusiness Institute, Department of Chemical Engineering Universidad Nacional de Colombia Sede Manizales, Manizales -Colombia**

[2] **University of California Davis, Food Science & Technology Department**

marojasre@unal.edu.co; jadavilar@unal.edu.co; mrosenberg@ucdavis.edu; *ccardonaal@unal.edu.co;

14AIChE
Annual Meeting, Atlanta, GA

1. INTRODUCTION

The Interest in Designing functional foods

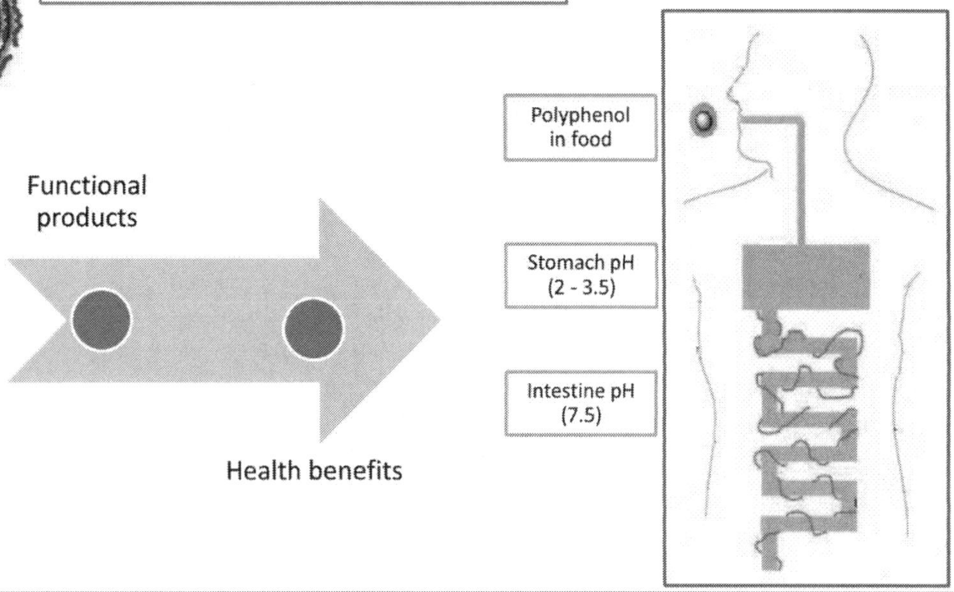

1. INTRODUCTION

Foods rich in polyphenols

- Red Wine
- Green tea
- Coffee
- Pomegranate
- Blackberry
- Nuts
- Prune
- Cholocate

Development of natural products with high contents of benefical compounds

Significant market for functional foods

1. INTRODUCTION

Foods rich in polyphenols

- Red Wine
- Green tea
- Coffee
- Pomegranate
- Blackberry
- Nuts
- Prune
- Cholocate

Blackberry (*Rubus glaucus benth*) is an exotic tropical fruit that have high content of natural antioxidants, phenolic compounds and anthocyanins.

In 2013, Colombia produced 105,285 tonnes of blackberry from 11,986 ha.

1. INTRODUCTION

After the extraction of active ingredients they need to be immediately protected against biological activity

Food manufacturers

- Polyphenols
- Antimicrobial
- Flavors
- Preservatives

Problems

Chemically and physically unstable

They are incompatible with the product matrix

Lack the appropriate functional attributes

1. INTRODUCTION

After the extraction of active ingredients they need to be immediately protected against biological activity

Food manufacturers

- Polyphenols
- Antimicrobial
- Flavors
- Preservatives

Challenges

Overcome by incorporating the active ingredients into some kind of delivery system before it is introduced into the final product

1. INTRODUCTION

Microencapsulation is a potential solution to the loss of biological activity and maintain the stability of polyphenols during food processing and store

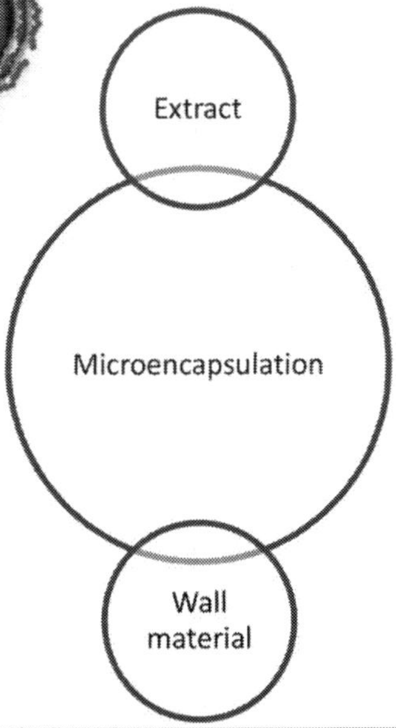

Shell Materials used for microencapsulation purposes

- Polysaccharides: Maltodextrin, Dextrin, Cyclodextrin, Cellulose, gum Arabic, Alginate, Pectin
- Protein: Soy, Wheat, Corn, Gelatin, Casein, Whey protein Isolate, Caseinate
- Polymers: Chitosan, Ethyl cellulose, Polyethylene glycol
- Lipids: Hard fat, Glycerides, Phospholipids
- Waxes: Beesswax, carnuba wax

1. INTRODUCTION

Microencapsulation is a potential solution to the loss of biological activity and maintain the stability of polyphenols during food processing and store

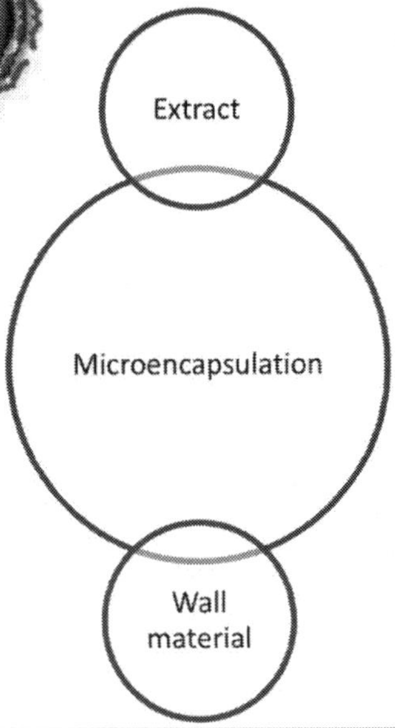

Microencapsulation methods used within the food industry.

Physical	Chemical
Spray drying	Phase separation
Spray cooling	Solvent evaporation
Extrusion	Interfacial polarization
Coextrusion	Liposomes

2. OBJECTIVE

- The objective of this study was to evaluate the influence of microencapsulation by spray drying of blackberry extract on the antioxidant activity and phenolic content and also to obtain the economic analysis of the process

3. METHODOLOGY

Extraction Process → Sample Preparation → Spray Drying → Chemical Analysis → Modeling and Simulation → Economic Evaluation

Supercritical fluid extraction

This Blackberry extract need be protected

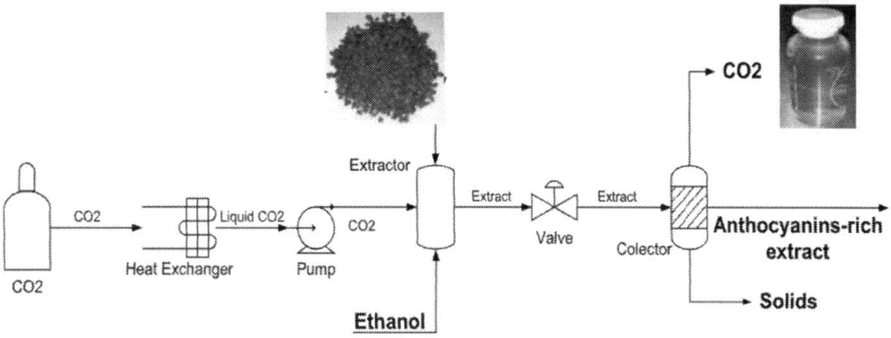

- Presion: 150 - 300 bar
- Temperature: 45 °C

3. METHODOLOGY

- Extraction Process
- **Sample Preparation**
- Spray Drying
- Chemical Analysis
- Modeling and Simulation
- Economic Evaluation

Delivery System — Designed to encapsulate, protect and release the active ingredients

Blackberry extract

Maltodextrin 20 DE (Solution 6% w/w)

Core in wall solution

3. METHODOLOGY

- Extraction Process
- Sample Preparation
- **Spray Drying**
- Chemical Analysis
- Modeling and Simulation
- Economic Evaluation

Microencapsulation Method.

Conditions:

Flow rate 0.49 kg/h.

Inlet air temperature:160 and Outlet air temperature: 80ºC

Nozzle atomizer: 0.7mm diameter

Drying flow rate: 0.36 m3/h

Aspirator flow rate: 35 m3/h.

Total Solid content: 30%

Initial phenolic compounds [mg GAE/ 100g]: 267.21

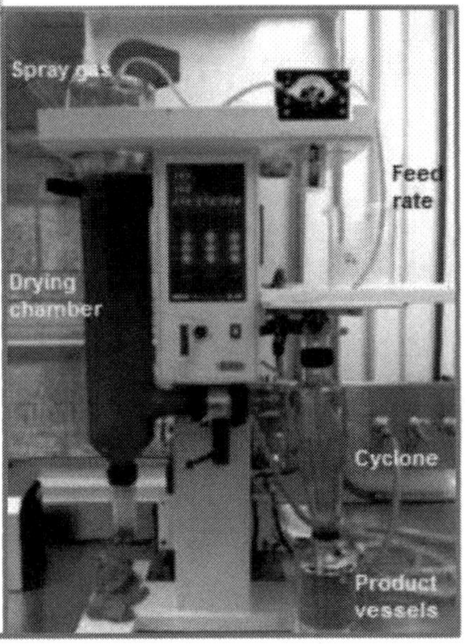

3. METHODOLOGY

Extraction Process → Sample Preparation → Spray Drying → Chemical Analysis → Modeling and Simulation → Economic Evaluation

Characterization of the blackberry extract microencapsulated:

Moisture content: (AOAC)

Water activity:(Hygrometer)

Total phenolic compounds (TPC): Folin Ciocalteu method

Total phenolic retention (TPR):

$$TPR = \frac{TPC\ in\ BEM\ (mg/100g)}{TPC\ in\ feed\ (mg/100g)}\ x\ 100$$

Antioxidant capacity: (DPPH method)

3. METHODOLOGY

Extraction Process → Sample Preparation → Spray Drying → Chemical Analysis → Modeling and Simulation → Economic Analysis

Blackberry Extract → Core in Wall solution

Biopolymer → Hydration → Core in Wall solution

Core in Wall solution → Final emulsification → Spray drying → Encapsulated powder

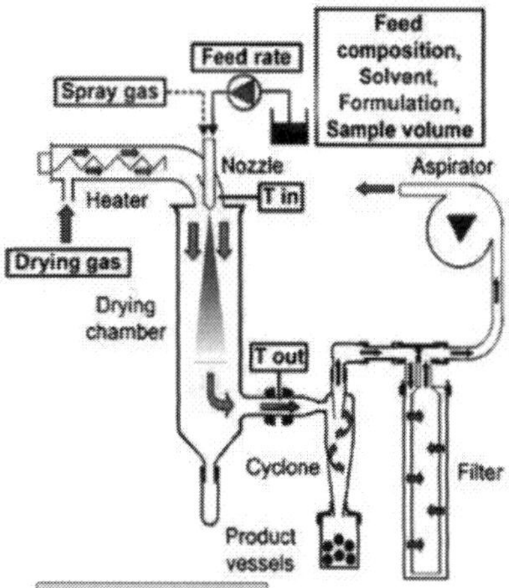

3. METHODOLOGY

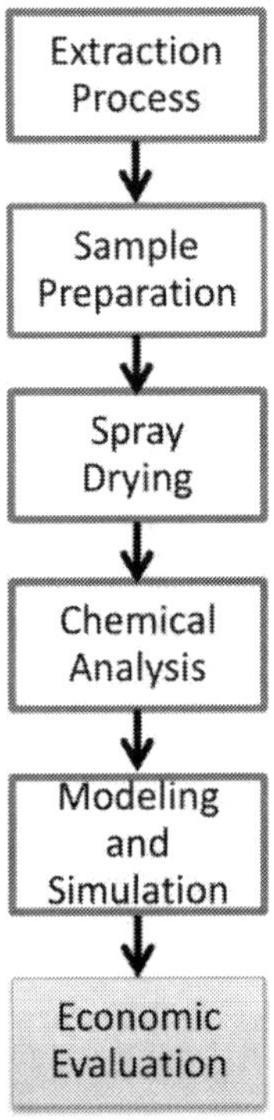

The economic analysis was performed by calculating the production cost per unit of product (e.g. kg). The total production cost considers the total of raw material, utilities, operating labor, maintenance costs, operating charges, plant overhead and general administrative costs. The costs of personnel labor, water and electricity were obtained from typical prices in Colombia.

4. RESULTS AND DISCUSSION

4.1 Chemical Analysis of Blackberry Extract (BE) and Blackberry Extract Microencapsulate [BEM]

Chemical analysis	[BE]	[BEM]
Moisture content [%]	92.47	3.42
Water activity [aw]	0.97	0.26
Phenolic compounds [mg GAE/ 100g]	267.21	253.30
Total phenolic Retention [%]	N.A	94.79
Antioxidant activity DPPH [EC_{50} mg sample /mg DPPH]	21.33	30.23

N.A= Not applicable

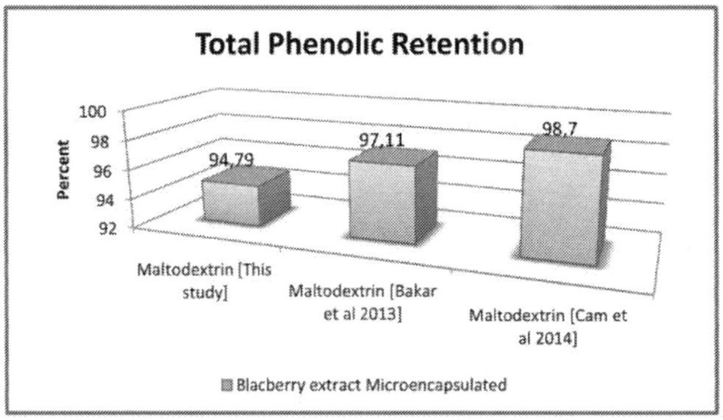

A delivery system should be capable of incorporating an appropriate amount of the active ingredient into the final product

It is important to obtain a high retention of the phenolic compounds within the deliver system until they are released at the required site of action.

4. RESULTS AND DISCUSSION

Chemical Analysis of Blackberry Extract (BE) and Blackberry Extract Microencapsulate [BEM]

Chemical analysis	[BE]	[BEM]
Moisture content [%]	92.47	3.42
Water activity [aw]	0.97	0.26
Phenolic compounds [mg GAE/ 100g]	267.21	253.30
Total phenolic Retention [%]	N.A	94.79
Antioxidant activity DPPH [EC_{50} mg sample /mg DPPH]	21.33	30.23

N.A= Not applicable

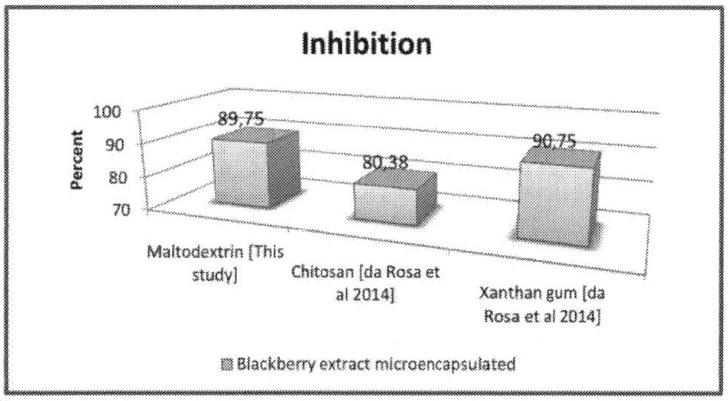

The results from the antioxidant activity studies indicated that the encapsulation procedure had no untoward effect on phenolic compounds.

The phenolic compounds present in the blackberry extract microencapsulated showed a greater antioxidant activity with a high inhibition of 89.75%

4. RESULTS AND DISCUSSION

Modeling and Simulation

Distribution Cost in the production of BEM

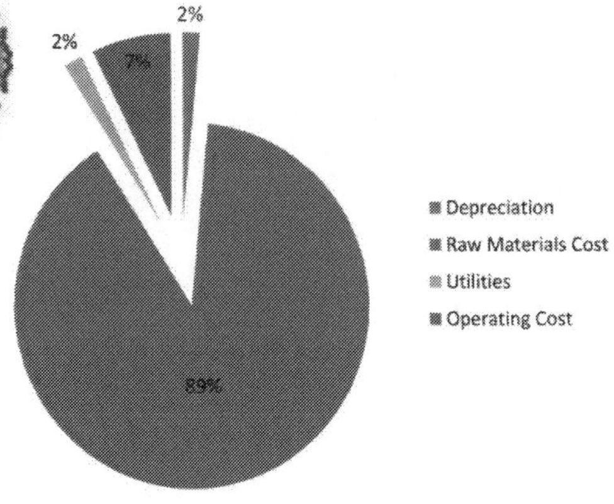

In this context, it can be observed that the price of the raw material has a great influence on the distribution cost in the production of BEM.
The utilities had a low percentage of the total cost

Economic parameters used for BEM estimation cost		
Raw material		
Blackberry Extract[a] (Solvent system CO_2 + EtOH)	USD /kg	7.63
Maltodextrin[c]	USD/ kg	0.38

[a]Previous work
[b]http://www.icis.com/

4. RESULTS AND DISCUSSION

Economic evaluation

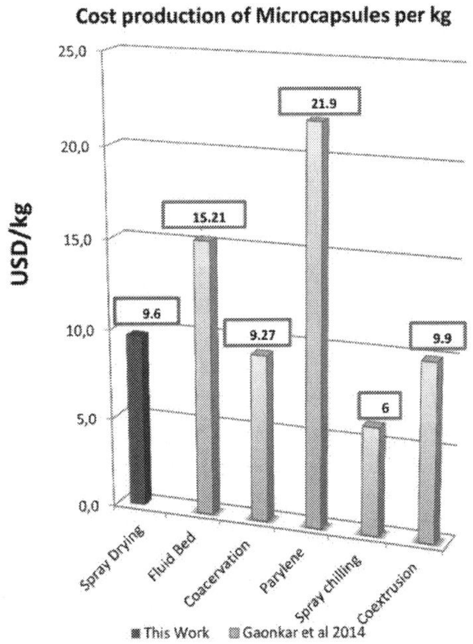

The process economic evaluation largely depends on some factors that must be taken into account before claiming that the BEM process is not economically viable.

Factors:

- Cost and content of the active ingredient to entrap
- Operating conditions
- Water content in the initial feed mixture
- Small scale and high scale

5. CONCLUSIONS

- The BEM obtained by spray drying process had an high retention of 95% of the total phenolic compounds and high inhibition of 89% of free radicals
- The use of polyphenols can not be considered without the protection provided by microencapsulation
- The benefits gained with microencapsulation of blackberry extract should outweigh the additional cost associated with the process of encapsulation

6. ACKNOWLEDGEMENTS

- The authors express their acknowledgments to the Universidad Nacional de Colombia at Manizales
- Research group of Chemical, Catalytic and Biotechnological Processes.

7. REFERENCES

- T. Egan, J.-C. Jacquier, Y. Rosenberg, and M. Rosenberg, "Cold-set whey protein microgels containing immobilised lipid phases to modulate matrix digestion and release of a water-soluble bioactive.," *J. Microencapsul.*, vol. 31, no. 2, pp. 184–92, Jan. 2014.
- T. Egan, J.-C. Jacquier, Y. Rosenberg, and M. Rosenberg, "Cold-set whey protein microgels for the stable immobilization of lipids," *Food Hydrocoll.*, vol. 31, no. 2, pp. 317–324, Jun. 2013.
- C. C. Ferrari, S. P. M. Germer, and J. M. de Aguirre, "Effects of Spray-Drying Conditions on the Physicochemical Properties of Blackberry Powder," *Dry. Technol.*, vol. 30, no. 2, pp. 154–163, Feb. 2012.
- Da Rosa, C. G., Borges, C. D., Zambiazi, R. C., Rutz, J. K., da Luz, S. R., Krumreich, F. D., ... Nunes, M. R. (2014). Encapsulation of the phenolic compounds of the blackberry (Rubus fruticosus). *LWT - Food Science and Technology, 58*(2), 527–533. doi:10.1016/j.lwt.2014.03.042
- Veggi, P. C., Cavalcanti, R. N., & Meireles, M. A. a. (2014). Production of phenolic-rich extracts from Brazilian plants using supercritical and subcritical fluid extraction: Experimental data and economic evaluation. *Journal of Food Engineering, 131*, 96–109. doi:10.1016/j.jfoodeng.2014.01.027
- Santos, D. T., Veggi, P. C., & Meireles, M. A. a. (2010). Extraction of antioxidant compounds from Jabuticaba (Myrciaria cauliflora) skins: Yield, composition and economical evaluation. *Journal of Food Engineering, 101*(1), 23–31. doi:10.1016/j.jfoodeng.2010.06.005
- Anilkumar Gaonkar, Niraj Vasisht, Atul Khare and RS. Microencapsulation in the food industry. A practical Implementation Guide. (Elsevier AP in an imprint of, ed.). San Diego, CA; 2014.

THANK YOU FOR YOUR KIND ATTENTION

Corresponding Author: Carlos Ariel Cardona Alzate.
Km 9 Vía al Aeropuerto La Nubia, Manizales, Caldas
tel. +57 68879400 ext 55354
e-mail: ccardonaal@unal.edu.co
Universidad Nacional de Colombia at Manizales

383760 Liquid-Phase Electrical Discharge Plasmas Rapidly Inactivate Pathogenic and Spoilage Microorganisms in Water

Abstract:

This work seeks to reduce food wastage caused by microbial spoilage through the development of an energy efficient and effective food sterilization technology applicable to the liquid foods manufacturing sector.

We have developed an energy efficient plasma technology that can rapidly achieve 5-log or greater reduction of food spoilage organisms in water. The technology, which creates plasma directly in a liquid, sterilizes microorganisms by combinations of chemical (hydrogen peroxide and oxidative oxygenated species), physical (shockwaves and UV light), and electrical effects.

Through the manipulation of plasma chemistry, we were able to reduce the treatment time for selected microorganisms from 30 min to ~ 50 sec while achieving greater than 5-log reduction, and have noted differences in optimal plasma chemistries for different organisms of interest. The extent of sterilization appears to be determined by the plasma reactor design, input power, plasma electrode material, and electrode polarity. The work also discusses the mechanism of sterilization and investigates the possibility of using the technology to treat fruit juices.

Modeling the Drying Kinetics of Carrot Particles in a Closed Chamber Following Forced Convection and Electric Field Treatment

A.B. Jemai[a], L. Khezami[b], M. HadjKali[a], and E. Vorobiev[c]

a. King Saud University, College of Engineering, Chemical Engineering Dept., P.O. Box 800 Riyadh 11421- KSA.
b. Al Imam Mohammed Ibn Saud University, Faculty of Sciences-Department of Chemistry. P.O.Box: 90950 Riyadh: 11623 Kingdom of Saudi Arabia.
c. University of Technology of Compiegne, Chemical & Industrial Process Engineering Department

The impact of forced convection and Pulsed Electric Field pretreatment on the drying kinetics of grating type carrot particles, placed in a closed chamber, have been studied based on continuous relative humidity measurements ($\Delta\Phi$) of circulating air. The particles were placed in a laboratory-scale closed chamber (30g particles capacity) which can be used for the subsequent pressing operation and drying air circulation. The kinetics of moisture content loss was indirectly deduced from a normalized form of the relative humidity difference between the entering and leaving air streams. Using two thermo-hygrometers, the continuous measurements of relative humidity difference between entering and exiting air were used in a normalized form to study the kinetics of drying for various air speed. The effect of an electric field pre-treatment of the particles was also illustrated using this experimental setup. About 20% reduction in drying time could be achieved following PEF pretreatment.

Keywords: Agro-food particles processing, Relative humidity, Modeling Drying kinetics, Forced convection, Electric fields.

383942 Antimicrobial Peptide Segments from Soy Protein for Use in Food Safety

Abstract:

Antimicrobial peptides (AMPs) kill microbial cells through insertion and damage/permeabilization of the cytoplasmic cell membranes and has applications in food safety. Soy protein may be such an attractive, cost-saving candidate for commercial consideration because the protein subunits have amino acid sequences that contain several α-helix or 3-10 helix domains which possess characteristics of AMPs. A methodology for identification of AMPs from soy protein is proposed. They were identified from soy β-conglycinin (7S) and glycinin (11S) based on (i) number of amino acids, (ii) positive charge, (iii) hydrophobicity and (iv) hydrophobic moment. Explicit solvent molecular Dynamics (MD) simulation was employed to assess the secondary conformation of these peptides in POPC/POPG bilayers to mimic their permeation action on the cell membrane of microorganism. The effects of number of peptides, their orientation and hydrophobic moment on the deformation of membrane and formation of water channel were investigated. A mathematical model for the prediction of minimum inhibitory concentration for deactivation (antimicrobial activity) was proposed accounting for free energy of formation of a pore by an aggregate of peptides. The antimicrobial activity was tested by using synthetic selected peptides against *Listeria monocytogenes* and *E. Coli* and compared with predictions.

James E. Van Wyk P.E.

Process Development for Sustainable Food and Biochemical Production

AIChE Annual Meeting Atlanta GA

19 June 2015

Making Food Safe in Light of New Demands

1. Introduction

Food products are non-Newtonian. This means that there are few if any imperial formulas that can be used to predict heat transfer properties. The most reliable data is from laboratory data and field results. This is compounded by the fact that if a new ingredient is added to an existing product, the properties can change dramatically.

The concern in the design of a food heat exchanger system is to insure that all potentially harmful bacteria are destroyed. When a new ingredient is added it can change the heat properties to an extent that a previous safe product becomes unsafe.

There are numerous examples that can be shown. In this discussion we will focus on two products which have become popular in the current market

1. Orange juice which has now other juices added such as tropical fruits
2. Tomato soup which has products added such a bits of potatoes, carrots and other vegetables

While admittedly there could be many more these two were chosen because of their time in the market as a stand-alone product and how changes drastically changed the way they are processed.

2. Product Particulars

   a. Orange Juice- Orange juice has been on the market for a number of years. The trend since 1990 has been for NFC or not from concentrate. This actually represents the same product as you would get if you squeezed it yourself. Since the orange juice is blended form various types, the consumer gets a consistent product that always tastes the same

      In the early 2000's the demand for single strength orange juice had peaked. In fact the per person consumption of orange juice has dropped. Added is the fact that major producers were attempting to move orange juice from an only breakfast drink to a drink consumed any time of the day.

      To achieve these objectives marketing people added tropical juices to orange juice to add variety of flavor. Many different juices were tried. The one that appeared to be the best taste wise was cranberry.

      From a thermal point cranberry changed the characteristics tremendously. The juice no longer behaved as orange juice with an additive. It behaved as an entirely different product.

      The way we learned was due to spoilage. Orange which is pasteurized has a shelf life of 60-80 days. This product had a shelf life of 10 days.

      It should be noted that no human became ill from the product. It simply spoiled very quickly

   b. Tomato Soup

      Tomato soup has been available in cans for a long time. The can are retorted (Cooked in can). This actually causes overcooking and can give a burnt flavor to the product. Since 1980 there has been a market demand for a fresher flavor. In order to do this the soup has to be pasteurized and then aseptically filled in a container.

Plain tomato soup sales were either flat or slightly declining. In order to increase volume it was decided to add ingredients. Small diced potatoes and carrots were added.

The issue became how can be we be sure that the potatoes and carrots are adequately heat treated. These products constitute an added mass that not only affects the heat transfer properties but also insurance that they have been heat penetrated to insure safety.

3. Laboratory Results
   a. Orange Juice. We first did a side by side calculation to see what properties might have changed.
      i. Viscosity- No change
      ii. Thermal Conductivity No change
      iii. Heat Capacity-Significant change A decrease of almost 20 %
   b. Tomato Soup We first did a side by side calculation to see what properties might have changed.
      i. Viscosity- A decrease in dynamic viscosity of 29%
      ii. Thermal Conductivity No change
      iii. Heat Capacity-Significant change A decrease of almost 20 %
   c. Laboratory Conclusions-
      i. The design of the heat exchanger had to be modified to meet the change in heat capacity. We either had to add surface or lower the processing rate

ii. With the tomato soup we found that the particles were not heated sufficiently. This meant that the soup had to be exposed to the heating for a longer period. The resonance time went from 10 min to almost 18 minutes. This meant that either we add surface or lower the processing rate,

4. Field Results

   a. Orange Juice-The juice processing rate is closely coupled to the rate of filling. Specifically the processing rate is 100 gpm to coincide with a ½ gallon filling rate of 80 gpm. This meant the only option for us was to increase the heat transfer surface. Since we were using a plate heat exchanger this was accomplished with almost no difficulty. We increase the area to what the laboratory test would indicate. This however proved to be not enough. We still had to increase by an additional 5% to achieve the pasteurization temperature.

   b. Tomato Soup- This proved to be a more difficult fix. We could not increase the surface area due to space limitations and cost. This was being processed in a tubular heat exchanger. We therefore had to slow the flow rate. The actual processing rate was about 30% less. From an engineering point of view this was acceptable. The economics at running at a lower rate proved not the case. Once the product was established in the market an new line as built for the soup

5. Conclusions

   a. Assume you know nothing- You cannot simply add an ingredient and assume that it will behave as before. Since the majority if not all food products are non-Newtonian, laboratory testing must be done each time.

b. Look at economics-Can you add surface area with a minimal amount of cost? Adding this cost will the product still be profitable? In some cases this may kill a project.

c. Never put a product out for public consumption until you are 110 % sure

386751 Low Temperature Batch Conversion of Cellulosic Biomass

Abstract:

The project aims at developing small, affordable batch reactors that communities in developing countries can use to produce clean cooking gas from biomass. Our approach extracts clean combustible gases such as methane and hydrogen using low cost locally available catalytic ingredients such as sodium chloride and sodium carbonate. Compared to uncatalyzed biomass conversion, the onset of pyrolysis/gasification shifts to significantly lower temperatures in the presence of catalysts. One possible explanation is that the ions in solution migrate in between cellulosic strands/fibers and disrupt the interstrand hydrogen bonding. Upon drying, the salt nucleates and crystallizes in between the fibers, leading to mechanical strain and loss of structural integrity of the fibers. The recovered biochar can be used as a supplemental source of fuel, while the gas evolved can be collected inside the pressure vessel and connected to a gas burner. This allows the use of a wide range of biomass sources without forcing a community to radically change their way of living, thereby increasing the probability of acceptance of this new method.

385688 Influence of the Synthesis Method on Cr Leaching during the Liquid Phase Photocatalysis of Cr-TiO2

Abstract:

There has been an increasing interest in developing catalysts for visible light photocatalytic degradation of volatile organic compounds in the past decade. The Liquid Flame Spray Pyrolysis (LFSP) is a favorable technique for controlling crystal structure, particle size and its distribution. Cr modified TiO_2 have shown a higher efficiency in utilizing the visible light of the solar spectrum. In our present studies, we are analyzing the influence of synthesis method has been investigated on structure, photocatalytic degradation of VOCs and the effect of leaching of the catalyst for pH range of 3-11. For this purpose, Ti-Cr of ratio 40 nanomaterials are synthesized by using co-precipitation, sol-gel, and compared with our previously reported Cr-TiO_2 synthesized by LFSP method. ICP-MS has employed to discover the concentrations of the metals leached at the different pHs. Among the various catalysts, the catalyst synthesized by LFSP Cr-TiO_2 exhibit excellent photocatalytic activity compared to the nanomaterials prepared by other synthesis methods. Temperature programmed reduction (TPR) and UV–vis spectra profiles clearly indicated that only Cr^{6+} phase is present in the LFSP made material sol-gel and co-precipitation synthesized catalyst have Cr^{3+} species along with Cr^{6+} species. The photodegradation ability of TiO_2/Cr catalyst with Ti/Cr atomic ratio of 40 was highly related to the existence of Cr^{6+}species which strongly interacted with TiO_2.The reduction peaks in Cr-doped TiO_2 shifted to much lower temperatures, due to the increase in the reduction potential of titania and chromium species. The developed catalyst also demonstrated an advantage of low metal leaching during the catalytic reaction, avoiding secondary metal contamination to the treated wastewater. The photocatalytic performance, characterizations and the effect of the leaching of Cr incorporated TiO_2 material synthesized by the three different methods will be presented in the presentation.

Electrochemical wastewater treatment; industrial applications

D.Woisetschläger[1)], M. Siebenhofer[2)]

[1)] VTU Engineering GmbH
Parkring 18
8074 Grambach/Graz, Austria

[2)] Institute of Chemical Engineering and Environmental Technology
Graz University of Technology
Inffeldgasse 25c/II
8010 Graz, Austria

Agenda

- Introduction
- Tasks and Targets
- Process development: Electrochemical oxidation
- Full scale electrolysis
- Conclusion and outlook

Advanced Oxidation Processes

- Photolysis
- UV-Peroxide
- PEROXON
- Fenton-Reaction
- Photo-Fenton process
- Photo catalysis
- Ozonisation
- Electrochemical Oxidation (EAOP)

Boron-doped diamond electrodes

- Non-active electrode
- Inert surface, low absorption property
- High oxygen evolution overvoltage

$$M + H_2O \rightarrow M(HO^\bullet) + H^+ + e^-$$

- High redox potential
  E°(OH•/H2O = 2.80 vs NHE)
- Non-selective reaction with organic pollutants

VTU engineering Electrochemical wastewater treatment for industrial applications TU Graz

Tasks and Targets

Task

- Implementation of EAOP for the treatment of persistent pollutants
- Treatment of industrial/municipal wastewaters
- Full scale electrochemical oxidation plants

Approach

Characterisation of wastewater specification → Selection and Combination of treatment technologies → Bench/Pilot scale tests → Full scale Electrolysis

Address Economics (cost estimation of investment / operation cost)

VTU engineering — Electrochemical wastewater treatment for industrial applications — TU Graz

Oxidation of organics

- Theoretical charge [Ah/g]
- Main parameters
  - Electrode material
  - COD
  - Exchange current density
  - Fluid dynamic conditions
  - Applied current density
- Operation boundaries
  - Charge transfer controlled region
  - Mass transfer controlled region
- Evaluation of charge-, mass- and heat transfer

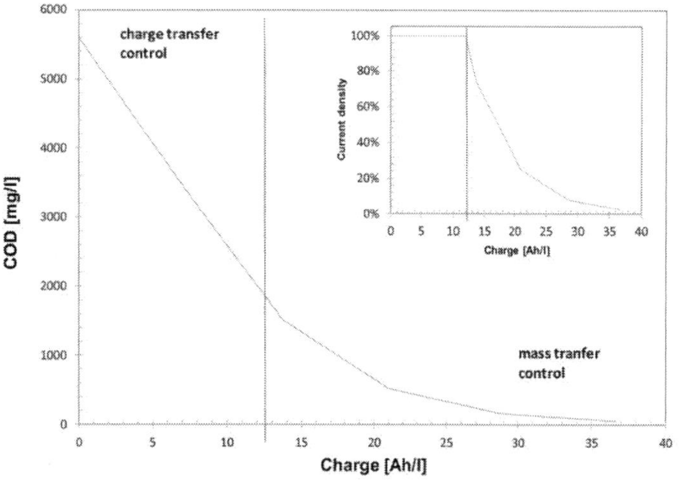

CEET 5 TU Graz I **Institute of Chemical Engineering and Environmental Technology** CEET

VTU engineering — Electrochemical wastewater treatment for industrial applications — TU Graz

Electrochemical wastewater treatment

- Anode materials for the oxidation of organics
- Degradation trends of organic model substances
- Effect of side reactions
  - Formation of halogenated organic compounds (AOX)
  - Fouling, solid deposit
  - Foam formation
- Treatment of industrial wastewaters
  - Chemical industry
  - Dye industry
  - Pharmaceutical industry
  - Pulp and Paper industry
  - Landfill leachate
- Scale-up
- Cost estimation
  - Operation cost
  - Investment cost

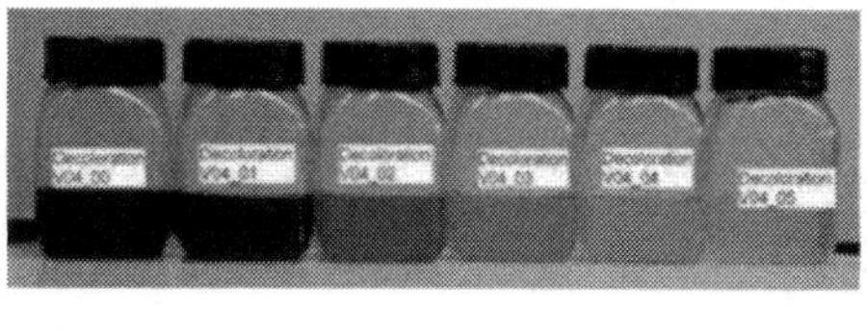

CEET 6 TU Graz I **Institute of Chemical Engineering and Environmental Technology** CEET

Laboratory Setup

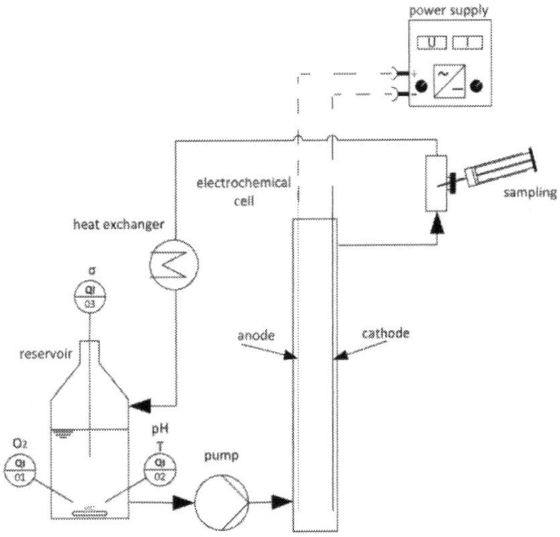

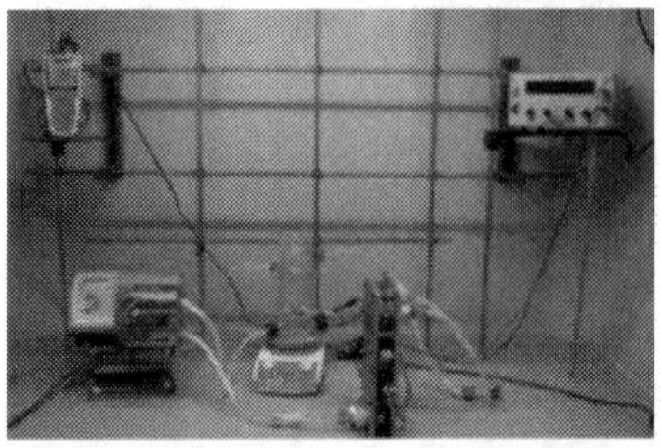

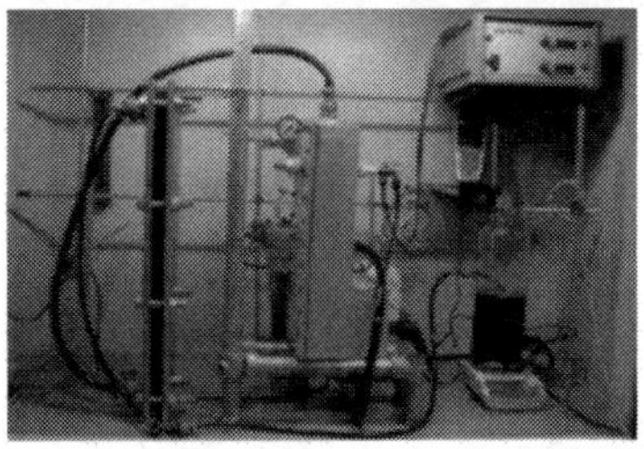

Laboratory Setup

Electrolysis I		
BDD anode area	0.00175	[m²]
Batch volume	0.2 – 0.4	[l]
Current density	0 – 2000	[A/m²]
Inter-electrode gap	4.5	[mm]
Membrane assembly possible	Yes	
Cell	Self construction	

Electrolysis II		
BDD anode area	0.02	[m²]
Batch volume	1 – 3	[l]
Current density	0 – 2000	[A/m²]
Inter-electrode gap	3 – 5	[mm]
Membrane assembly possible	yes	
cell	Electro MP Cell	

Elektrolysis III		
BDD anode area	0.075	[m²]
Batch volume	5 - 10	[l]
Current density	0 – 1000	[A/m²]
Inter-electrode gap	2	[mm]
Membrane assembly possible	no	
cell	DiaCCon „Bärbel“	

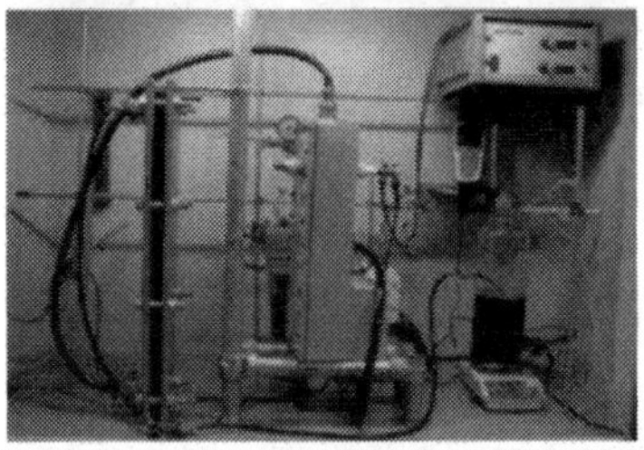

Anode material: active/non-active electrodes

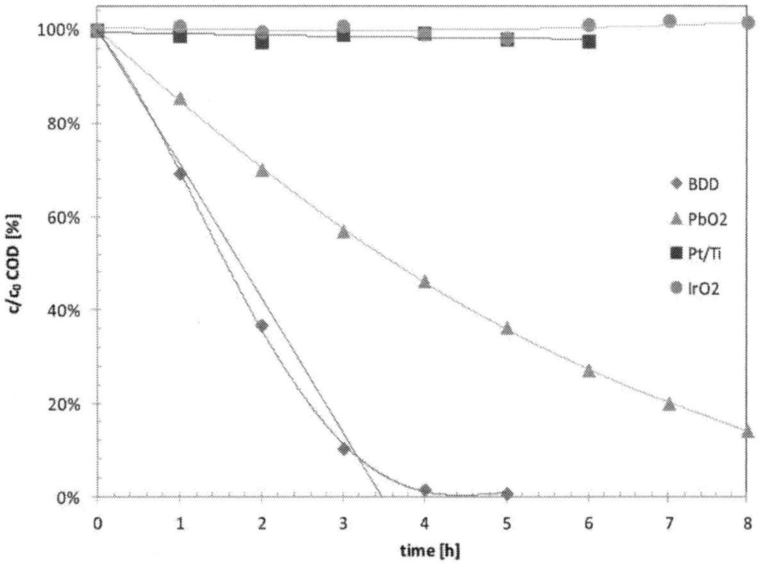

Oxidation trends of glucose with different electrode materials: $COC_{0,\,BDD}$ = 10.000 [mg/l], $COC_{0,\,PbO2}$ = 5.000 [mg/l], $COD_{0,\,Pt/Ti}$ = 5.000 [mg/l], $COD_{0,\,IrO2}$ = 10.000 [mg/l]; T = 298 [K]

Electrochemical wastewater treatment for industrial applications

Oxidation trends of organics on BDD-electrodes

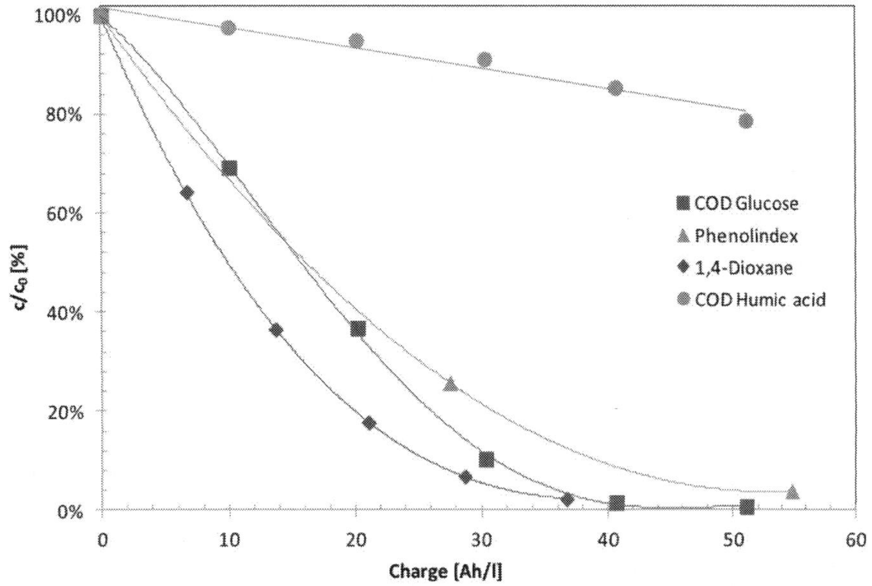

Degradation trends of water constituents with BDD-electrodes: $c_{0,COD}$ = 10.000 [mg/l], $c_{0,Phenolindex}$ = 70 [mg/l], $c_{0\ 1,4\text{-}Dioxane}$ = 14.000 [mg/l]; T = 298 [K]

Side reactions: Formation of halogenated organic compounds (AOX)

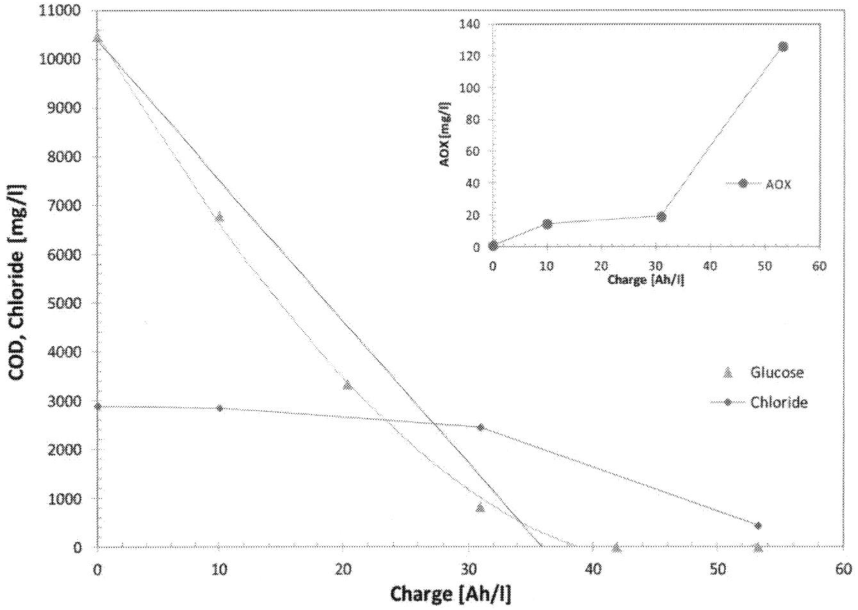

Formation of halogenated organic compounds: COD_0 = 10,000 [mg/l], $Chloride_0$ = 3,000 [mg/l], i = 500 [A/m²]; T = 298 [K]

Electrochemical wastewater treatment for industrial applications

Limiting current density / mass transport limitation

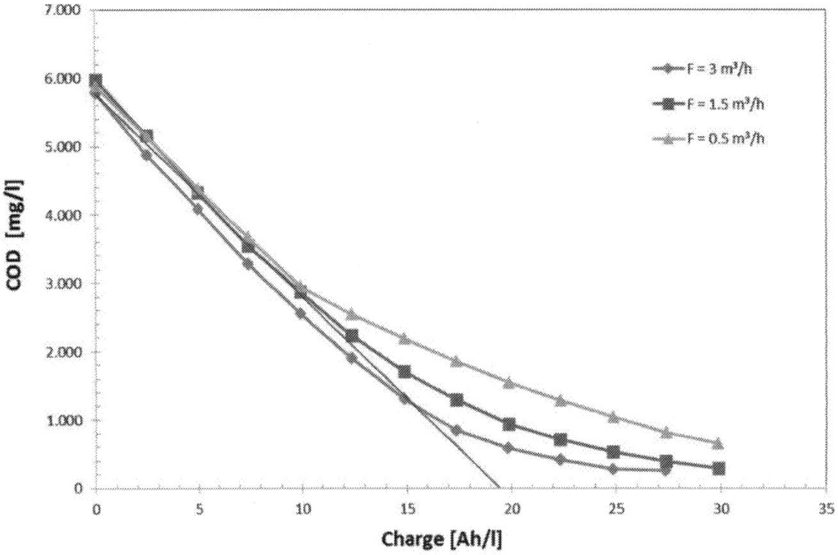

Oxidation trends of glucose at different flow rates: $c_{0,COD}$ = 6.000 [mg/l], i = 1000 [A/m²], T = 298 [K]; Setup: coolox Pilot cell: A_{BDD}=0.075 m²

Treatment of industrial wastewaters

Wastewater	Target	Charge consumption	Comment
Dye industry	colour, COD removal	10 [Ah/l]	Laboratory scale, i = 1000 [A/m²]
Pulp and Paper industry	Phenol	15 [Ah/l]	Laboratory scale, i = 500 [A/m²]
Pharmaceutical industry	1,4-Dioxane	21 [Ah/l]	Laboratory scale, i = 500 [A/m²]
Chemical industry	COD removal	30 [Ah/l]	Laboratory scale, i = 1000 [A/m²]
Chemical industry	AOX removal	80 [Ah/l]	Laboratory scale, i = 1000 [A/m²]
Landfill leachate	COD removal, colour	90 [Ah/l]	Laboratory scale, i = 500 [A/m²]
Production process	Phenol, COD removal	110 [Ah/l]	Laboratory scale, i = 1000 [A/m²]

Mobile Full scale plant

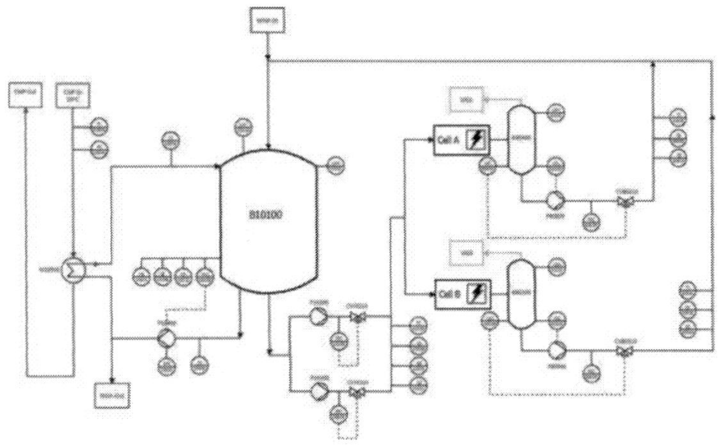

Elektrolysis Plant – Full Scale		
BDD anode area	2 x 1.05	[m²]
Batch volume	200 – 800	[l]
Current density	0 – 1500	[A/m²]
Inter-electrode gap	2	[mm]
cell	VTU coolox®	

Technical Data	
COD Performance	max. 0.5 [kg_{COD}/h] @ 1000 [A/m²]
Specific energy consumption	min. 20 [kWh/kg_{COD}]

Mobile Full scale plant

Elektrolysis Plant – Full Scale		
BDD anode area	2 x 1.05	[m²]
Batch volume	200 – 800	[l]
Current density	0 – 1500	[A/m²]
Inter-electrode gap	2	[mm]
cell	VTU coolox®	

Technical Data	
COD Performance	max. 0.5 [kg_{COD}/h] @ 1000 [A/m²]
Specific energy consumption	min. 20 [kWh/kg_{COD}]

Electrolysis cell - Design

Active electrode area max.	3.6 m²
Max. current	3,600 A_{DC}
Current density max.	1,000 A/m²
Voltage max.	3 - 30 V_{DC}
Operating Temperature min. /max.	+ 5 / + 60 °C
Operating Pressure min./max.	0 / 1.0 bar_g
pH-value	1.0 – 13.0
Materials	PP (housing), Viton (sealings)
Anode material	Boron-doped-diamond (BDD) on niobium backbone
Cathode material	Stainless steel
Sight glass	2 x per cell (borosilicate glass)
Inter-electrode-gap	2.0 mm
Dimensions (L x W x H)	540 x 400 x 1220 mm
Weight	95 - 135 kg

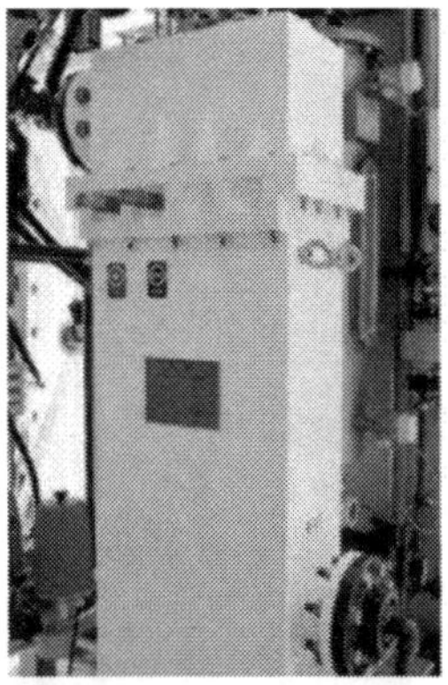

Number of anodes/stacks installed	Electrode area installed [m²]	Required current [A]	max. COD degradation [mg/h]
8 / 1 stack	1.2	1,200	357,600
10 / 1 stack	1.5	1,500	447,000
12 / 1 stack	1.8	1,800	536,400
16 / 2 stacks	2.4	2,400	715,200
20 / 2 stacks	3.0	3,000	894,000
24 / 2 stacks	3.6	3,600	1,072,800

Performance full scale plant

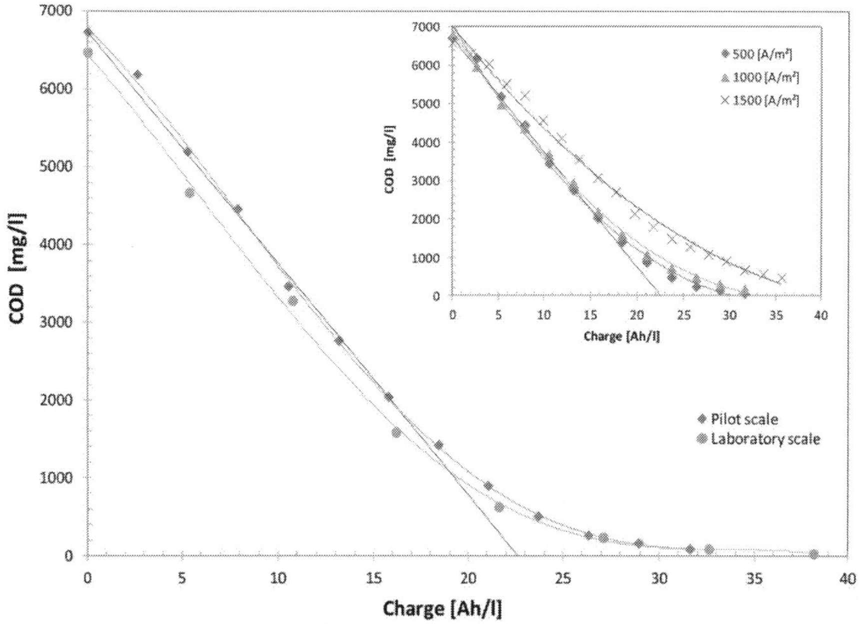

Comparison COD degradation laboratory (A = 0.075 m²) and pilot plant (A = 1.05 m²); Inset: Variation of the current density, COD degradation pilot plant (A = 1.05 m²); T = 298 [K]

Case Study: Phenol wastewater

Industrial wastewater

COD	6600 - 6800 mg L^{-1}
TOC	2480 - 2600 mg L^{-1}
Phenolindex	60 - 65 mg L^{-1}
Chloride	52 – 72 mg L^{-1}
Ammonium NH_4-N	1100 - 1200 mg L^{-1}
Calcium	80 – 120 mg L^{-1}
Magnesium	30 - 35 mg L^{-1}
Sulfate	2500 - 3000 mg L^{-1}
pH	6.8 – 7.2
Conductivity	9.6 -10 mS cm^{-1}

Target

- Electrochemical treatment of phenol wastewater in pilot scale
- Phenol degradation, COD removal

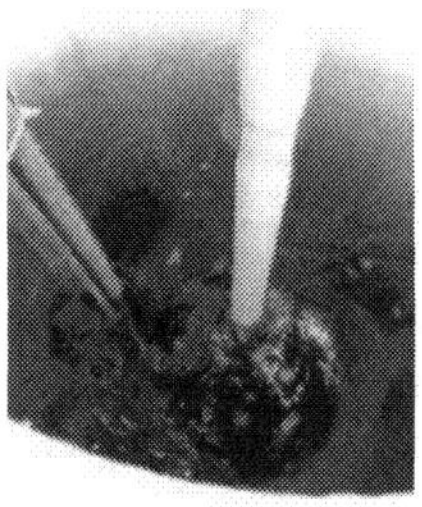

Electrochemical wastewater treatment for industrial applications

Case study: Phenol wastewater

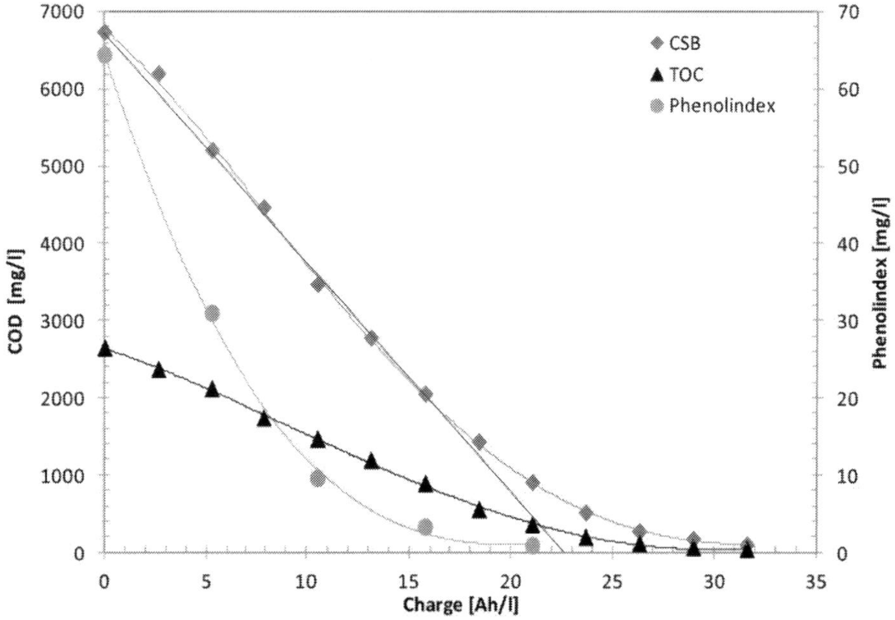

Degradation trend of organic substances in industrial wastewater: i = 500 [A/m²], V = 410 [l], A = 1.05 [m²]; T = 298 [K]

Electrochemical wastewater treatment for industrial applications

Conclusion and Outlook

- Process development in laboratory / pilot scale for individual requirements
  - Results of manifold wastewaters/effluents from different industry sectors
  - Investigation oxidation trends, side reactions, maintenance procedures
- Development of Full Scale electrolysis cell: Coolox 2.0
  - Modular cell housing
  - Simple and efficient cell design
  - Max. 3.6 m² BDD electrode area (2 stacks per cell)
- Research on BDD-electrodes
  - Cost reduction
  - Improved stability

Electrochemical wastewater treatment for industrial applications

Thank you for your attention!

Dr. Dieter Woisetschläger

dieter.woisetschlaeger@vtu.com

VTU Engineering GmbH
Parkring 18
8074 Grambach/Graz, Austria

Prof. Dr. Matthäus Siebenhofer

m.siebenhofer@tugraz.at

Institute of Chemical Engineering and Environmental Technology
Graz University of Technology
Inffeldgasse 25c/II
8010 Graz, Austria

Mathematical Model of a Thin Film Slurry Reactor for the Degradation of an Azo-Reactive Dye by Solar Photo-Fenton Process.

Abdón Parra López[1], Lidia Yokoyama[1], Fabiana Valéria da Fonseca[1].

1. Departamento de Processos Inorgânicos, Escola de Química, Universidade Federal de Rio de Janeiro, Brasil.

ABSTRACT

Advanced oxidation processes (AOPs) have been considered as an attractive alternative to the treatment of wastewaters containing organic pollutants of difficult degradation. Among the possible advanced oxidation process combinations, solar assisted Photo-Fenton process it's an interesting technology in places with a high annual insolation.

In the present work, solar assisted Photo-Fenton is tested in the degradation of a reactive azo dye Vermelho Drimaren X-6BN. The experiments were developed in a thin film slurry solar reactor (TFR) conformed by a stirred tank reactor in series with a flat plate reactor, and consisted in a series of experiments conducted to determine the best reaction conditions, taking as an interest parameters the effect of pH, the initial concentration of hydrogen peroxide ($[H_2O_2]$), and the relation between the concentration of hydrogen peroxide and catalyst ($[H_2O_2]/[Fe^{+2}]$). These experiments were made according to a statistical factorial design (2^3), after the statistical analysis all the effects of the independent parameters were found significant, and the best reaction conditions found were: pH 3,0, $[H_2O_2]_0$ 500 mg/L and $[H_2O_2]/[Fe^{+2}]$ 20:1 for a 100 mg/L solution of azo dye Vermelho Drimaren X-6BN. Under these reaction conditions a 90% dye removal was achieved after 10 minutes of reaction.

Additional a series of experiments were developed under these conditions, to determine the kinetics constants for different operation flow rates. From these results dependence between the observed kinetic constant and the mean residence time of the reactors was identified.

This relationship between the observed kinetic constant and the residence time could be explained by a mathematical expression obtain from the Nusselt´s falling film analysis coupled with the mass balance and the reaction kinetics. The final expression describes in a good agreement the operation of the reactor.

Key words: Advance oxidation process, AOP´s, Solar assisted Photo-Fenton, Thin film Reactor, Model, Dye degradation.

PERFORMANCE OF ACTIVATED CARBON SUPPORTED CATALYST DURING WET OXIDATION OF PULPING EFFLUENT

Mr. Bholu Ram Yadav, Centre for Environmental Science and Engineering, Indian Institute of Technology Bombay, Powai, Mumbai, INDIA
Anurag Garg, Centre for Environmental Science and Engineering, Indian Institute of Technology Bombay, Powai, Mumbai, INDIA

Abstract

The performance of activated carbon supported non-noble metal catalysts (Cu/Ce/AC and Cu/Mn/AC) was studied for catalytic wet oxidation (CWO) reaction performed at moderate operating conditions (120 – 170°C temperature and 0.5 MPa of oxygen partial pressures) for pulping effluent. After CWO with Cu/Ce/AC and Cu/Mn/AC catalysts at 170°C temperature, the COD reduction from the wastewater was 67% and 63%, respectively while only 44% COD reduction was obtained in the absence of catalyst. After the catalytic reaction, pH of the treated wastewater using Cu/Ce/AC catalyst was reduced to 6.0 from an initial pH of 8.5. For WO and CWO (with Cu/Ce/AC catalyst reactions), the TOC time based data was fit in second order rate equation. The CWO rate constants at different temperatures were 2- 4 times to the non-catalytic reaction. The activation energies of non-catalytic WO and CWO were found to be 37 kJ mol^{-1} and 21.55 kJ mol^{-1}, respectively. Further investigation on the catalytic stability and its reusability are suggested for future studies.

Keywords: Wet oxidation; Pulping liquor; Heterogeneous catalysts

Introduction

In a pulp and paper mill, the pulping process generates a high strength colored liquid stream primarily containing not easily biodegradable compounds. The dark brown color of the effluent is primarily due to the presence of lignin. Wet oxidation (WO) is one of the potential pre-treatment alternatives for the high strength waste streams. Using this process, the removal of intense color and biodegradability enhancement can be obtained. However, the oxidative degradation requires high temperature (125-350°C) and pressures (0.5-20 MPa) to achieve desired oxidation in presence of air or molecular oxygen [1]. The reaction follows free radical mechanism and occurs in three steps: initiation, propagation and termination. To facilitate the occurrence of WO, the formation of high oxidation potential specie (e.g. HO·) is desired. The generation of such radicals can be accelerated in the presence of a suitable catalyst even at less severe conditions. In the past, several activated carbon supported non-noble based metal catalysts have been used for the removal of organic compounds from various industrial effluents [2-5].

For the present study, the performance of two bimetallic catalysts supported on activated carbon (Cu/Ce/AC and Cu/Mn/AC) was investigated for the oxidative degradation of the diluted pulping liquor.

Materials and Methods

The major characteristics of diluted wastewater (obtained from a wood based kraft pulp and paper mill) were as follows: pH = 11.23, chemical oxygen demand (COD) = 16000 mg/l, biochemical oxygen demand (BOD_5) = 2600 mg/l and total organic carbon (TOC) = 4500 mg/l. The wastewater is not amenable to the biological treatment process due to low biodegradability (*i.e.*, 0.16).

Two heterogeneous catalysts, Cu/Ce/AC and Cu/Mn/AC (mass ratio = 3:3:4) were prepared by impregnation method. The detailed of preparation procedure and characterization (actual composition of catalyst, X-ray diffraction (XRD), BET surface area, average particle diameter and the particle size distribution) of the synthesized catalysts have been discussed in our earlier publication [5].

WO/CWO reactions were carried out in a 0.7 L capacity high pressure stainless steel (SS) reactor (material of construction: SS-316) equipped with all necessary provisions for temperature control, periodic sampling etc. For a typical WO/CWO runs, 0.25 L wastewater was charged into the reactor without or with 3 g/l fresh catalyst (for CWO runs). During the reaction, moderate temperature (T = 120-170°C) was maintained while the initial oxygen pressure (P_{o2}) was 0.5 MPa. The reactor contents were mixed with continuous at a stirrer speed of 1000 rpm. The liquid samples were collected periodically and analyzed for COD and TOC. BOD_5 was measured for the samples collected only after 3 h reaction.

pH and TOC of the liquid samples was measured with a pH meter (Model and Make: Polmon, LP-139S, India) and TOC analyzer (Model and Make: Shimadzu, TOC-VCSH, Japan), respectively. The standard closed reflux method was used to determine COD [6]. BOD_5 of pulping effluent (treated and untreated) was determined by the modified Winkler's method [6]. Color was detected at 465 nm using a UV spectrophotometer (UV-1800, Shimadzu, Japan).

Results and Discussion

Effect of Temperature

The effect of temperature during WO/CWO reaction of wastewater (adjusted pH = 8.5), was studied in a range of 120-170°C and the final results are shown in Figures 1 and 2. The figures indicate that increase in temperature favored overall TOC reductions for both WO and CWO reactions. It was observed that TOC reduction during WO at 120°C temperature was only 9% which was increased to 42% by increasing the temperature to 170°C temperature. The highest COD and color reductions at 170°C temperature were found to be 44% after the reaction.

For CWO reaction with Cu/Ce/AC catalyst, the highest TOC and color removals of 65% and 71%, respectively were achieved. With Cu/Mn/AC catalyst the results were inferior (TOC removal = 58% and color removal = 61%). After CWO with Cu/Ce/AC and Cu/Mn/AC catalysts at 170°C temperature, the COD reductions from the wastewater were 67% and 63%, respectively. The pH of treated wastewater (in the presence of Cu/Ce/AC catalyst at 170°C temperature) was reduced to 6.04 from an initial pH of 8.5. The pH drop may possibly due to the formation of low molecular weight acidic species. The BOD_5/COD ratio was doubled (*i.e.* 0.32) after the Cu/Ce/AC catalyzed reaction at 170°C temperature.

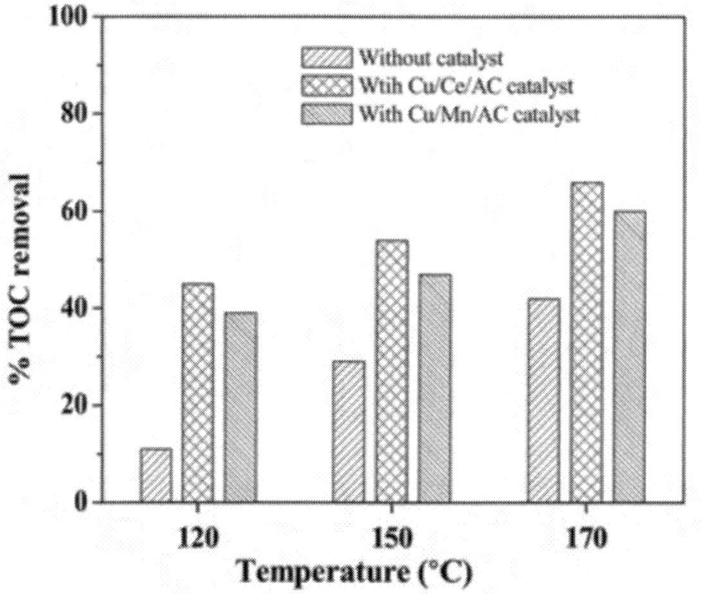

Figure 1. TOC reduction during WO/CWO (Po_2 = 0.5 MPa, Catalytic concentration = 3g/l, Reaction time = 3 h)

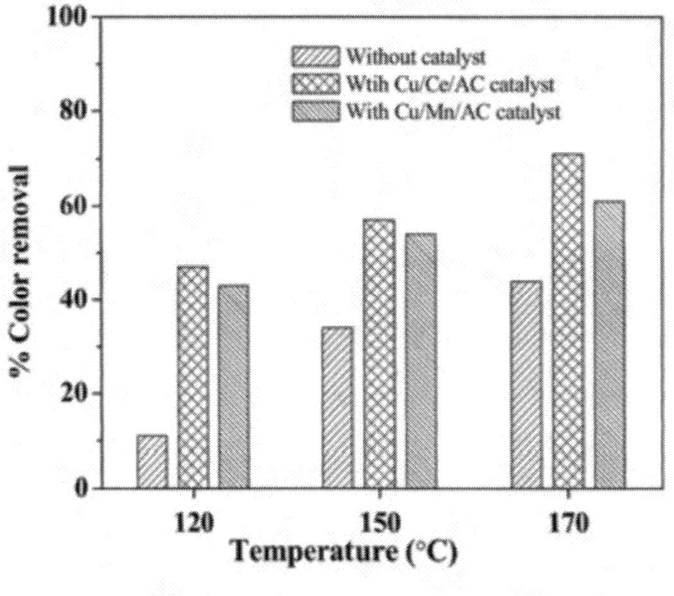

Figure 2. Color removal during WO/CWO (Po_2 = 0.5 MPa, Catalytic concentration = 3g/l, Reaction time = 3 h)

Kinetic Study of WO and CWO Reactions

To determine the kinetic parameters, (i.e., specific rate constant and activation energy), TOC degradation (obtained from non-catalytic WO and CWO) data was fit to the second order rate equation. The rate expression for a second order reaction can be written as

$$-\frac{dC}{dt} = kC^2 \quad \text{...........1}$$

On integration and applying the limits, the following expression can be obtained:

$$\frac{1}{C} - \frac{1}{C_0} = kt \quad \text{..........2}$$

where C = TOC of the wastewater at time t, C_o = initial TOC and k = rate of reaction. The data at all temperatures could be fitted well in equation 2. The results from the kinetic study are presented in Table 1. The reaction rates at different temperatures during CWO were 3.5 to 5.0 times higher than non-catalytic WO reaction.

The activation energy of the two reactions was calculated using the Arrhenius expression:

$$k = k_o e^{\frac{-E_a}{RT}} \quad \text{.........3}$$

where k_o = specific rate constant, E_a = activation energy of the reaction, R = universal gas constant. The Arrhenius plot is shown in Figure 3.

Table 1. Rate constants obtained for non-catalytic and catalytic (with Cu/Ce/AC catalyst) WO reactions

S. No.	Temperature (°C)	Non-catalytic WO		CWO (with Cu/Ce/AC catalyst)	
		k (L mg^{-1} min^{-1})	Regression coefficient (r^2)	k (L mg^{-1} min^{-1})	Regression coefficient (r^2)
1.	120	2×10^{-7}	0.931	8×10^{-7}	0.913
2.	150	5×10^{-7}	0.984	1×10^{-6}	0.983
3.	170	7×10^{-7}	0.991	2×10^{-6}	0.928

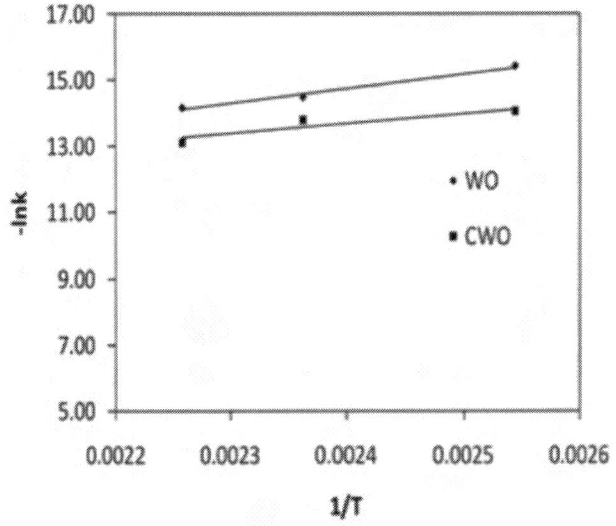

Figure 3. Arrhenius plot

The activation energy for non-catalytic WO and CWO were found to be 37.00 kJ mol^{-1} and 21.55 kJ mol^{-1}, respectively, whereas the specific rate constants for the two reactions were 0.0.017 min^{-1} and 0.0015 min^{-1}, respectively.

Conclusions

The major conclusions drawn from the study are listed below:

- Cu/Ce/AC catalyst showed the best performance with an overall TOC and color reduction of 65% and 71% compared to Cu/Mn/AC catalyst (TOC reduction = 58%). In the absence of catalyst, TOC reduction was only 44% at 170°C temperature.
- WO/CWO followed second order reaction model. The values of rate constants suggest the faster degradation of organics in the presence of catalysts.
- The activation energies of non-catalytic WO and CWO, 37 kJ mol^{-1} and 21.55 kJ mol^{-1}, respectively, suggested that the reaction in the absence of catalyst was more temperature sensitive than CWO.

Acknowledgments

The first author acknowledges the financial support of Ministry of Human Resources and Development (MHRD), New Delhi, India for pursuing doctorate degree. We also recognize the support of IIT Bombay to provide the research facility.

References

1. Mishra, V. S., Mahajani, V. V. and Joshi, J. B. (1995), "Wet air oxidation," *Industrial and Engineering Chemistry Research*, 34, pp. 2-48.
2. Trawczyński, J. (2003), "Noble metals supported on carbon black composites as catalysts for the wet-air oxidation of phenol," *Carbon*, 41, pp. 1515–1523.

3. Suarez-Ojeda, M. E., Stüber, F., Fortuny, A., Fabregata, A., Carrera, J. and Font, J. (2005), "Catalytic wet air oxidation of substituted phenols using activated carbon as catalyst," *Applied Catalysis B: Environmental*, 58, pp. 105–114.
4. Yadav, B. R. and Garg, A. (2013), "Ferulic Acid Degradation by Wet Oxidation Process," *International Journal of ChemTech Research*, 5(2), pp. 654-658.
5. Yadav, B. R and Garg, A. (2014), "Catalytic wet oxidation of ferulic acid (a lignin model compound) in the presence of non-noble metal based catalysts at mild conditions," *Chemical Engineering Journal*, 252, pp. 185–193.
6. American Public Health Association (APHA), (2005), "Standard methods for the examination of water and wastewater," 21st ed., APHA, Washington, DC.

CATALYTIC WET OXIDATION OF PHENOLIC COMPOUNDS AT MODERATE TEMPERATURE AND PRESSURE

Rajendra Mohite, Centre for Environmental Science and Engineering, Indian Institute of Technology Bombay, Mumbai, INDIA
Anurag Garg, Centre for Environmental Science and Engineering, Indian Institute of Technology Bombay, Mumbai, INDIA

Abstract

In the present work, a mixture of phenolic compounds (phenol, o-cresol and 2,5-dimethylphenol) was treated by catalytic wet oxidation (CWO) process at moderate temperature (130-160°C). The total phenolics concentration was 10 g/L and the initial pH was raised to ~13.0 by adding NaOH. Such highly alkaline phenolic streams are emerged from petroleum refinery. The wet oxidation (WO) reaction was performed in a high pressure reactor using oxygen as oxidant (initial oxygen pressure = 0.8 MPa). For the catalytic reaction, 5% (by wt) Cu/AC catalyst was synthesized in the laboratory which was characterized using different techniques such BET surface area, XRD analysis and FTIR. The WO and CWO experimental results showed that performance of WO process was improved with the addition of the catalyst. The initial reaction pH of the wastewater also exhibited significant impact on the oxidation reaction.

Keywords: Phenolic compounds; Catalytic wet oxidation, Cu/AC heterogeneous catalyst

Introduction

Rapid technological developments are causing an increase in the environmental pollution. The natural water bodies become contaminated by the pollutants present in untreated or partially treated wastewater. Phenols are one of the mostly used chemical in different process industries. The wastewater from petroleum refineries contains high concentration of phenolic compounds which are not removed by conventional biological treatments. Such pollutants should be removed by some chemical or physico–chemical process during the pre-treatment step.

Wet oxidation (WO) is a promising hydrothermal technology for non-biodegradable waste streams which are neither too dilute nor very much concentrated. It involves the oxidation of organic and inorganic substances present in liquid phase by using air or pure oxygen at elevated temperatures (125 – 350°C) and pressures (0.5 – 20 MPa) [1]. In order to reduce the severity of the process, suitable catalysts (preferably heterogeneous) can be used. In earlier studies, homogeneous as well as heterogeneous copper has been used as an effective WO catalyst [2-4]. For heterogeneous catalysts, the active metal specie is supported by a high surface area material like activated carbon (AC). The majority of studies were focused on the degradation of phenol but limited studies are available on the oxidation of a mixture of phenolic compounds.

Hence the present work was aimed to investigate the oxidative degradation of phenolic compounds in the presence of 5 wt% Cu/AC catalyst at mild temperature and pressure conditions. The characteristics of the heterogeneous catalyst are also reported.

Experimental

Synthetic Wastewater and Catalyst

The synthetic wastewater was prepared by dissolving equal mass concentrations of phenol, *o*-cresol and 2,5-dimethylphenol in tap water (total phenolics concentration = 10 g/L). In the wastewater, NaOH was also added to raise its pH to around ~13.0 which is generally found in the refinery effluent. Among other characteristics, chemical oxygen demand (COD) and total organic carbon (TOC) of the wastewater were ~ 24000 mg/L and 7800 mg/L, respectively.

The analytical grade powdered AC and copper nitrate trihydrate were used to prepare catalyst. To prepare the catalyst, copper (5 wt%) was impregnated on acid treated AC and calcined at 500°C for 4 h. The detailed procedure of catalyst preparation is described elsewhere [5]. Pure oxygen was used as oxidant for the catalytic hydrothermal reaction.

Experimental Procedure and Analytical Methods

WO or CWO experiments were carried out in a 0.7 L a stainless-steel (SS 316) high pressure reactor. The reactor has an impeller mounted on a vertical shaft, electrical heating device with temperature controller, gas introduction port, treated wastewater sampling port and pressure relief valve. The oxidation reaction was performed with 0.3 L wastewater at a temperature of 130–160°C and oxygen partial pressures of 0.8 MPa. For catalytic runs, the COD to catalyst dose ratio of 8.0 was used. To start the oxidation reaction, pure oxygen was purged into the reactor after reaching the desired reaction temperature with rigorous agitation of the reactor contents (stirrer speed = 1000 rpm). During 4 h reaction, the samples were withdrawn periodically and analyzed for TOC and COD.

Analytical Methods

The Brunauer–Emmett–Teller (BET) surface area of the catalyst was determined using N_2 adsorption with a Micromeritics ASAP-2000 machine. The synthesized heterogeneous catalyst was alone analyzed by X-Ray Diffraction (XRD) technique with Cu–Kα as the X-ray source. The phases present in the catalyst samples were identified using the Joint Committee on Powder Diffraction Standard (JCPDS) data files.

Dynamic Laser Scattering (DLS) analyzer (LS 13320, Beckman Coulter, U.S.A.) was used to obtain particle size distribution in the catalyst. The solid catalyst sample was also subjected to FTIR analysis (Bruker Cooperation, Germany) for the identification of surface groups on the catalyst surface. TOC of the liquid samples was determined using a TOC analyzer (Shimadzu, TOC-VCSH, Japan). COD was determined by standard closed reflux method [6].

Results and Discussion

Catalyst Characterization

The results for the catalyst characterization are summarized in Table 1. The BET surface area of the mesoporous catalyst was higher (~ 924 m^2/g) compared to the untreated AC. The catalyst was pore size was ranged from 2-50 nm.

Table 1. Properties of AC and Cu/AC catalyst

Parameter	AC	Cu/AC
	XRD analysis	
CuO peak position (2θ)	-	35.6°, 38.7°, 48.8° and 61.6°
C peak position (2θ)	26.6° and 81.4°	26.6° and 81.4°
	BET analysis	
A_{BET} (m^2/g)	874.81	924.66
	FTIR	
Alcohol, carboxylic acids, esters absorption (nm)	1556, 1385, 1117	1701, 1555, 1157
	DLS analysis	
Average particle size (μm)	-	~ 45

In the diffraction pattern of Cu/AC catalyst, CuO phase was observed at 35.6°, 38.7°, 48.8° and 61.6° positions. The location of peaks for CuO was confirmed by comparing with XRD spectra reported in one of previous study [1]. According to the DLS analysis results, the particle size distribution of the catalyst particles was similar to the untreated AC and the average particle size catalyst was ~45 μm. The surface chemistry of the AC and Cu/AC was registered with FTIR.

WO experimental results

The results from the WO/CWO experimentation are illustrated in Figure 1. At lower temperatures (i.e., 130 and 140°C), the TOC reductions during CWO were higher compared to the non-catalytic WO. For instance, the TOC removals during CWO at these temperatures were 44 and 50% respectively. The corresponding TOC removals in the absence of catalyst were 35 and 36%, respectively. Further increase in temperature exhibited no impact of catalyst and the TOC reduction was ~ 50% for WO as well as CWO at 160°C temperature. A TOC reduction of 2% and 9% was also observed during WO and CWO reactions, respectively. The COD removal was ranged from 44-50% for non-catalytic WO and 48-63% for CWO process at different temperatures.

By adjusting the initial pH of the synthetic wastewater to 9.0 from the 12.8 before the start of reaction showed significant increase in the TOC removals. After CWO, TOC and COD of the wastewater were reduced by 70 and 78%, respectively. This result indicates the occurrence of free radical chain reaction during the oxidative degradation of organics. Free radicals are generally scavenged by alkalinity species (e.g., OH^-). In highly alkaline environment, the destruction of free radicals can be attributed to the presence of OH^- ions resulting in inferior organic removal compared to the reaction carried out at an initial pH of 9.0.

The average oxidation state of the carbon (AOSC) was determined using COD and TOC reduction to get an idea of the nature of compounds may have produced during the process. Higher values of the AOSC at lower temperatures (i.e. 130°C) found increase in the formation of short chain organic compounds such as carboxylic acids.

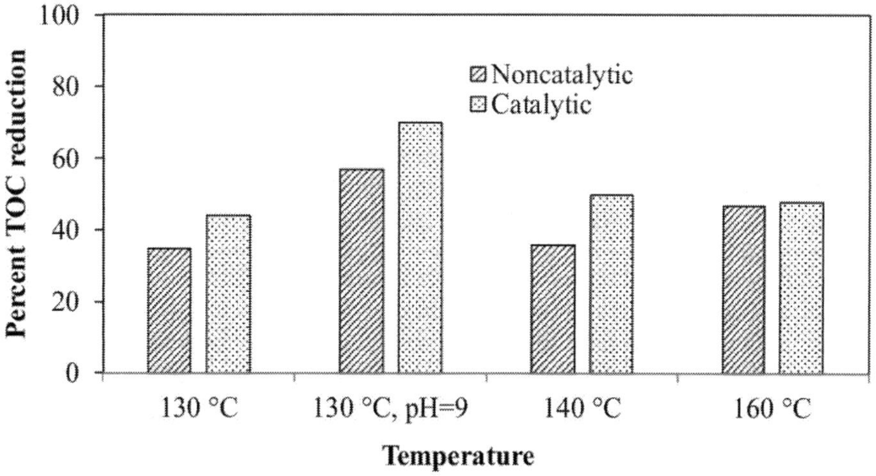

Figure 1. Effect of temperature and pH on TOC reduction (Initial P_{O2} = 0.8 MPa, Catalyst concentration = 3 g/L, Reaction time = 4 h, Initial pH = 12.6)

Pseudo first order kinetic model was used to determine the reaction rate during oxidation reaction. The rate constant was in the range of 0.0021–0.0055 min^{-1}, lowest for noncatalytic WO performed at 130°C and highest for catalytic WO performed at pH=9 and T=130 °C.

Conclusions

The diffraction pattern of the catalyst powder verified the presence of copper oxide on the activated carbon support. The mesoporosity of the catalyst could have facilitated the diffusion of the reactant molecules into the pores and enhanced the degradation of the pollutant compared to non-catalytic WO. The mild CWO process showed the encouraging results for the degradation of phenolic mixture presented in synthetic wastewater. A higher TOC removal (upto 78%) could be achieved during CWO process. The TOC reduction data with time was fitted well in the pseudo first–order reaction model. The rate constant for CWO was much higher compared to WO reaction.

Acknowledgements

The authors are thankful to Council for Scientific and Industrial Research (CSIR), New Delhi for providing the financial support to carry out the presented study. We are also thankful to Sophisticated Analytical Instrumentation Facility (SAIF) and Chemical Engineering Department, IIT Bombay for their timely help in the characterizing the catalyst.

References

1. Mishra, V. S., Mahajani, V. V. and Joshi, J. B. (1995), "Wet air oxidation," *Industrial and Engineering Chemistry Research*, 34, pp. 2-48.
2. Garg, A., Mishra, I. M. and Chand, S. (2010), "Oxidative phenol degradation using non-noble metal based catalysts," *Clean,* 38(1), pp. 27-34.

3. Garg, A. and Mishra, A. (2013), "Degradation of organic pollutants by wet air oxidation using non-noble metal based catalysts," *Journal of Hazardous, Toxic, and Radioactive Waste, ASCE,* 17, pp. 89-96.
4. Lal, K. and Garg, A. (2014), "Catalytic Wet Oxidation of Phenol under Mild Operating Conditions: Development of Reaction Pathway and Sludge Characterization," *Clean Technologies and Environmental Policy*. (In Press, Available online)
5. Abdullah, A. H., Yuan, W. W., and Yaziz, M. I. (2010), "Decolourisation of reactive orange 16 by activated carbon and copper oxide catalysts supported by activated carbon," *Journal of Physical Science*, 21(2), pp. 29-40.
6. American Public Health Association (APHA). (2005), "Standard methods for the examination of water and wastewater," 21st Ed., American Public Health Association, Washington, DC.

384103 Establishing Analytical Adsorption Methods for Pore Characteristic Evaluations in Gas Shales

Abstract:

The advent of hydraulic fracturing, when combined with vertical and horizontal drilling, has allowed for increased production of natural gas and liquids from tight shale formations. The rapid ascent of shale sourced gas has created a need for analytical evaluation of shale, in order to determine permeability and gas-in-place. The majority of the work has focused on permeability and flow properties, as this relates directly to hydraulic fracturing effectiveness and production rates. Researchers have also focused on the adsorption properties of shale, in order to evaluate potential gas-in-place and understand how gas transports from micropores to macropores to fractures. The literature, to this point, has been unfocused, and pointed towards evaluating as many shale samples from the broadest perspective as possible. However, analytical adsorption approaches appropriate for investigating homogenous samples, such as activated carbon or zeolitic material, has limitations when applied to highly heterogeneous samples. Shale, which has high variability at all scales, from the formation level to the microscopic level, is a particularly difficult challenge.

This work focuses on creating a standard analytical method for the evaluation of gas shale, focused on low-pressure adsorption techniques. The outgas temperature will be evaluated to determine effects on adsorption isotherms, with emphasis on potential changes in pore volume. The samples will be evaluated with 3 probe gases (N_2 at 77 K, Ar at 87 K and CO_2 at 273 K). Isotherms will be evaluated with DFT techniques to determine pore size distribution, while using information from Fourier transform infrared spectroscopy (FTIR), scanning electron microscopy (SEM), elemental analysis and total and inorganic carbon contents to support the gas adsorption observations. The thorough analytical approach presented in this work will present a method that researchers can use to properly evaluate pore characteristics of shale gas, which can be used to support research in flow models and gas-in-place estimates.

A Generalized Model for Gas-Solid Adsorption Equilibria

Austin Ladshaw (aladshaw3@gatech.edu), Sotira Yiacoumi (sotira.yiacoumi@ce.gatech.edu), Costas Tsouris (tsourisc@ornl.gov), David DePaoli (depaolidw@ornl.gov), Ronghong Lin (rlin01@syr.edu), Lawrence Tavlarides (lltavlar@syr.edu)

Modeling of single-component gas-solid adsorption at equilibrium is typically accomplished through the use of isotherms such as the Freundlich, Langmuir, Tóth, and other similar models. These models are relatively simple to adapt for describing experimental data, but their simplicity limits their ability to be used with different sets of data. To more generically describe the equilibrium requires a more generic isotherm model, such as the Generalized Statistical Thermodynamic Adsorption (GSTA) model. This model has the potential to describe a wide variety of adsorption equilibrium data because it has a variable number of adjustable equilibrium parameters. In conjunction with collected equilibrium data, this generic isotherm can be used to accurately describe single-component systems. However, in real adsorption processes, the systems of interest never involve just a single adsorbing species. Therefore, this isotherm model alone is insufficient for adsorption equilibrium modeling, but could be combined with other multicomponent adsorption theories to yield a more complete description of the system. One such theory is the Ideal Adsorbed Solution Theory (IAST), which considers that all adsorbed species on the surface of an adsorbent behave as an ideal solution. This assumption, however, is often problematic as ideality rarely occurs in real systems. In order to address this issue, we must consider the non-idealities between the different adsorbing molecules at the surface, typically through the use of an activity model. This approach causes additional complications as many activity models involve binary interaction parameters that must be determined through experiments and, as the number of components in a system increases, the number of experiments that must be performed increases exponentially. However, we can predict these binary interaction parameters by considering the limiting behavior of each species' individual isotherm. We have developed a complete and comprehensive set of adsorption-equilibrium modeling tools, from single-component optimization to multi-component adsorption predictions, which are fully dependent on temperature, pressure, and gas composition. The developed software includes all the necessary components to provide the equilibrium parameters for each single species, using the GSTA isotherm, and use that information to make a multi-component equilibrium prediction for any number of adsorbing species. The model has been successfully examined using single- and multi-component data taken from the open literature, and data obtained in our laboratories.

382404 Binding of SO3 to Fly Ash Components: CaO, MgO, Na2O and K2O

Abstract:

Oxy-combustion proposes to incorporate carbon capture and storage technologies into existing coal-fired power plants to lower carbon emissions. It does so by separating out the N_2 prior to combustion, resulting in a concentrated CO_2 as a final product that can be easily captured and stored. Sulfur oxide (SO_X) emissions can be lowered due to high concentrations of SO_3in the oxy-combustion flue gas resulting in sulfur retention on fly ash and ash deposits in the furnace. It has been shown that the extent of sulfur retention is heavily dependent on the alkali and alkaline earth metal (AAEM) species in coal such as Na, K, Mg and Ca and the sulfur retention increases as the AAEM:S molar ratio increases. While higher sulfur retention on the ash particles reduces the emission rates, it creates problems utilizing the fly ash for cement and concrete production.

Although there have been previous experimental studies, the mechanism of sulfur retention on fly ash is not well understood. In this study, a variety of computational and experimental methods were employed to examine the binding of SO_X species to the AAEM species. Firstly, density functional theory (DFT) calculations were performed using Vienna *ab initio* Simulation Package (VASP) to investigate the binding mechanism for SO_3 on several metal oxides that compose fly ash, namely CaO, MgO, Na_2O, and K_2O. Calculations were also run where SO_3 was co-adsorbed with H_2O to examine if its presence would facilitate further reaction of the SO_3 on the oxide surface. Secondly, FTIR spectroscopy will be employed to investigate the AAEM oxides after exposure to a SO_2/SO_3 mixture plus a variety of other flue gas components. FTIR results along with DFT calculations will be used to create a more complete picture of the SO_X binding mechanism on the metal oxides.

Metal Organic Frameworks for Selective Adsorption of t-Butyl Mercaptan from Natural Gas

Grace Chen, Christopher W. Jones, William J. Koros

AICHE Annual Meeting
Abstract ID# 386269
November 19, 2014

Georgia Institute of Technology

Introduction – Scope & Motivation

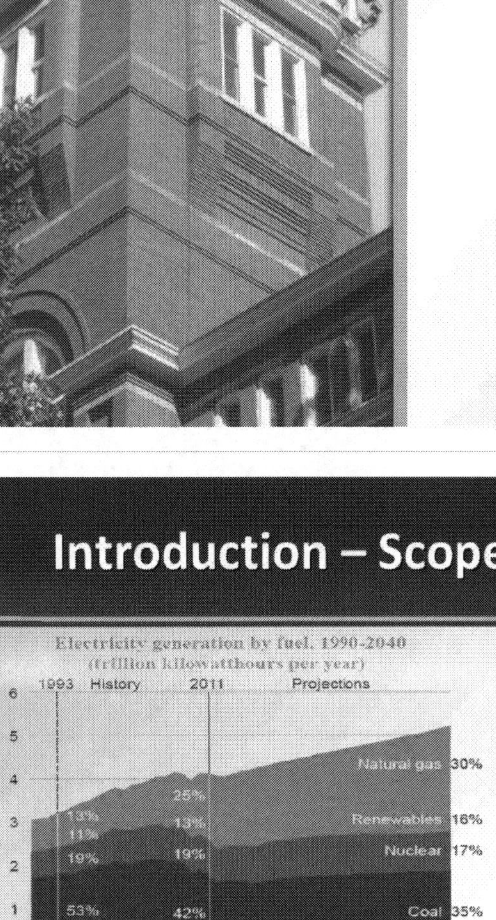

Source: U.S. EIA, Annual Energy Outlook, 2011.

Natural Gas to Replace Coal

- Produces less CO_2, CO, and NO_x than coal
- Negligible amounts of SO_2, mercury, and ash
- Easier to transport
- Growing domestic supply
- > 300,000 miles of transmission pipelines

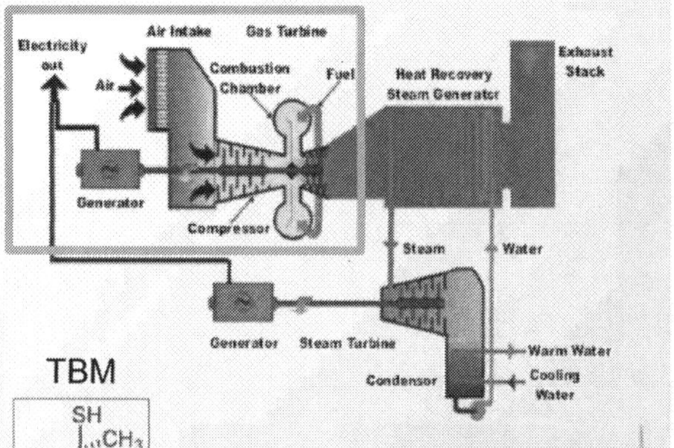

Corrosion in Gas Turbines

- Sulfur present in fuel (odorants) reacts with alkali material present in air
- Combustion product is a corrosive, reactive alkaline sulfate
- Corrosion reduces turbine efficiency and engine lifetime
- Motivation to prevent/delay corrosion

TBM

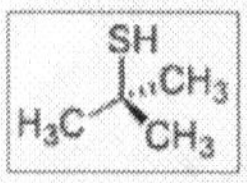

Georgia Tech

2

Background – Corrosion Prevention

Corrosion Prevention Strategies:

- Remove alkali compounds from air and fuel
- Use corrosion-resistant alloys in turbine
- Recondition affected areas
- Remove sulfur compounds from fuel

Sulfur Removal Technologies:

- Catalytic Hydrodesulfurization: Add H_2 to sulfur compound to produce H/C and H_2S
- Selective Catalytic Oxidation: Add O_2 to sulfur compound to produce sulfoxides, and sulfones
- Selective Adsorption on Solids: Sulfur compounds attracted to and bound to surface sites of sorbent

Sorbent Materials:

- Activated carbons
- Silica supported metals
- Metal oxides
- Zeolites
- Metal organic frameworks

System Configuration:

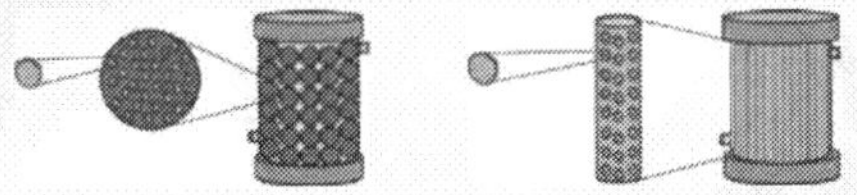

- Pellet packed bed
- Fiber module

Georgia Tech

3

Background – MOFs and Zeolites

Zeolites

- Zeolites are well studied aluminosilicate materials
- Tunable by modifying cation
- Commercially available, robust
- Have shown high sulfur adsorption capacities with gaseous fuels, but hindered by moisture

Metal Organic Frameworks

- MOFs are new hybrid materials - metal clusters linked by organic ligands
- Tunable by modifying metal or linker
- Water stability issue and many not commercially available
- Have shown very high sulfur adsorption capacities with liquid fuels

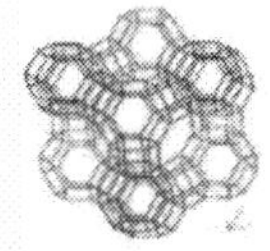

Faujasite

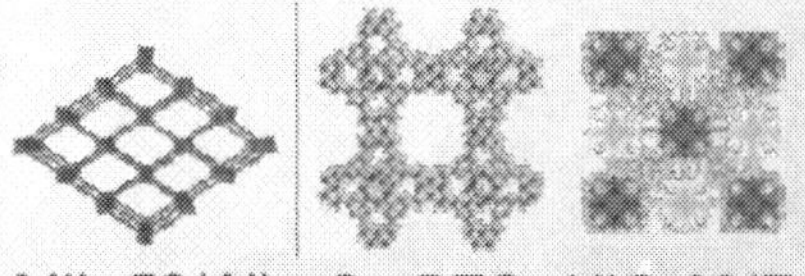

MIL-53(Al) Cu-BTC UiO-66(Zr)

Georgia Tech

Materials Selection - MOFs

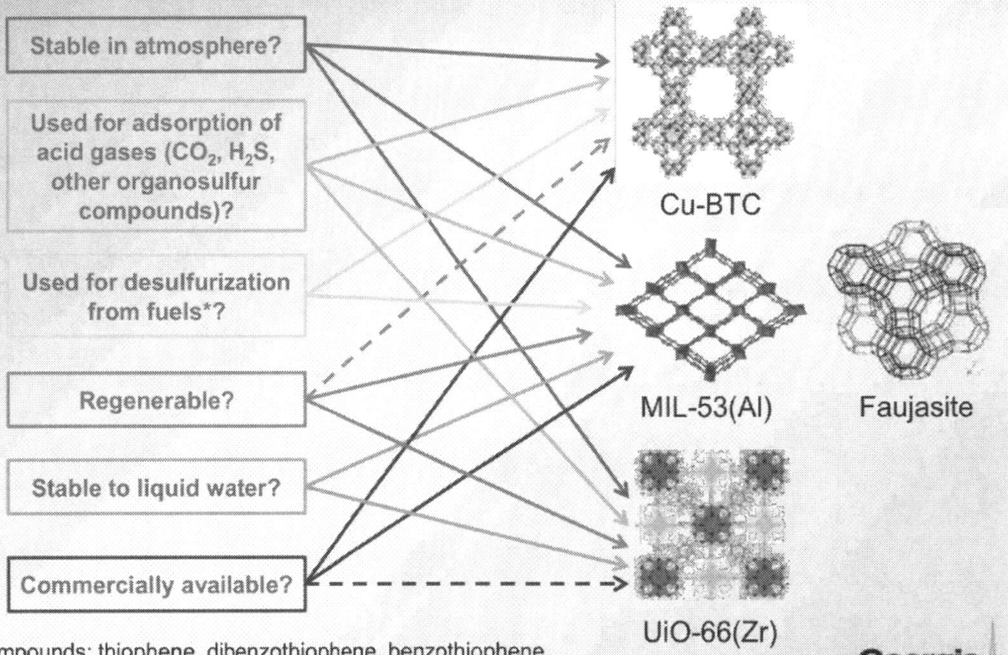

*Sulfur compounds: thiophene, dibenzothiophene, benzothiophene, tetrahydrothiophene
*Model fuel: trimethylpentane, octane, dodecane, hexane

Georgia Tech

Experimental Procedure

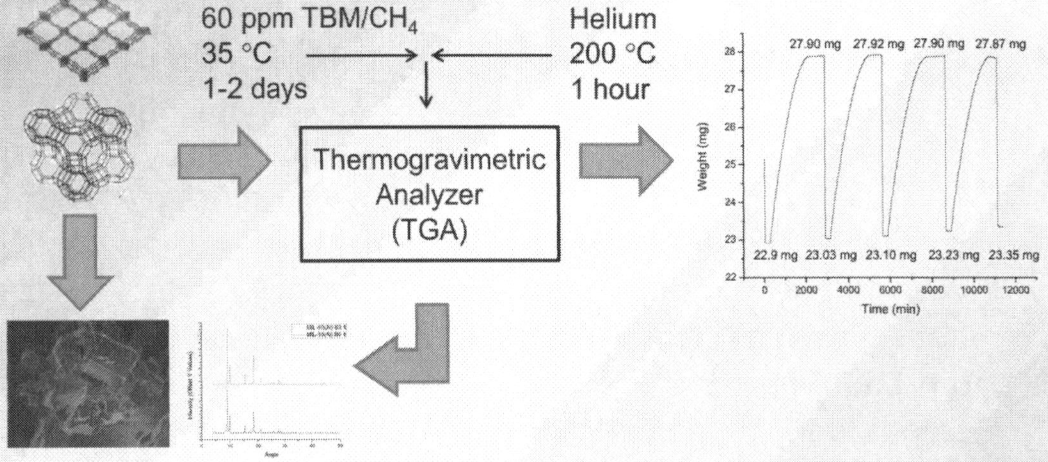

1. Synthesize MOFs and purchase zeolites
2. Characterize materials: BET, XRD, XPS, SEM, TGA, Raman, UV-Vis
3. Adsorption with TGA
4. Characterize used sample for structural changes

6

Georgia Tech

Results - Sorbent Characterization

	BET Surface area (m^2/g)	t-plot micropore volume (cm^3/g)	Particle size range (µm)
MIL-53(Al)	1329	0.54	1-4
Cu-BTC	1504	0.55	3-7
UiO-66(Zr)	1002	0.40	1-4
NaY	868	0.35	1-3

Si/Al ratio: 5.1 for NaY

- Both MOF and zeolite materials displayed type I isotherms without hysteresis, which is typical for microporous materials.
- All particle sizes within range for fiber spinning (future work)

Georgia Tech 7

Results – Uptake & Estimated Selectivity

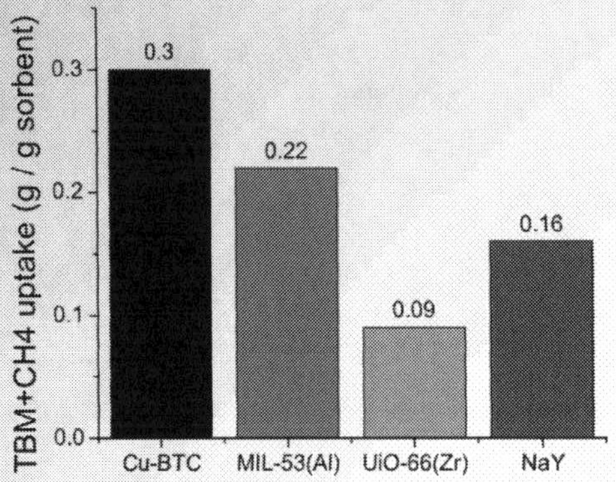

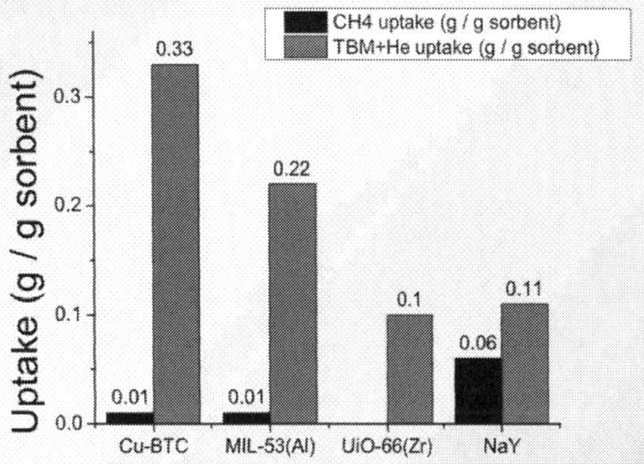

	Minimum Selectivity
Cu-BTC	228,000
MIL-53(Al)	263,000
UiO-66(Zr)	198,000
NaY	14,000

$$S_{TBM,CH_4} = \frac{x_{TBM,adsorbed} \;/\; x_{CH_4,adsorbed}}{y_{TBM,bulk\ gas} \;/\; y_{CH_4,bulk\ gas}}$$

Georgia Tech 8

Results - Cyclic TBM Adsorption

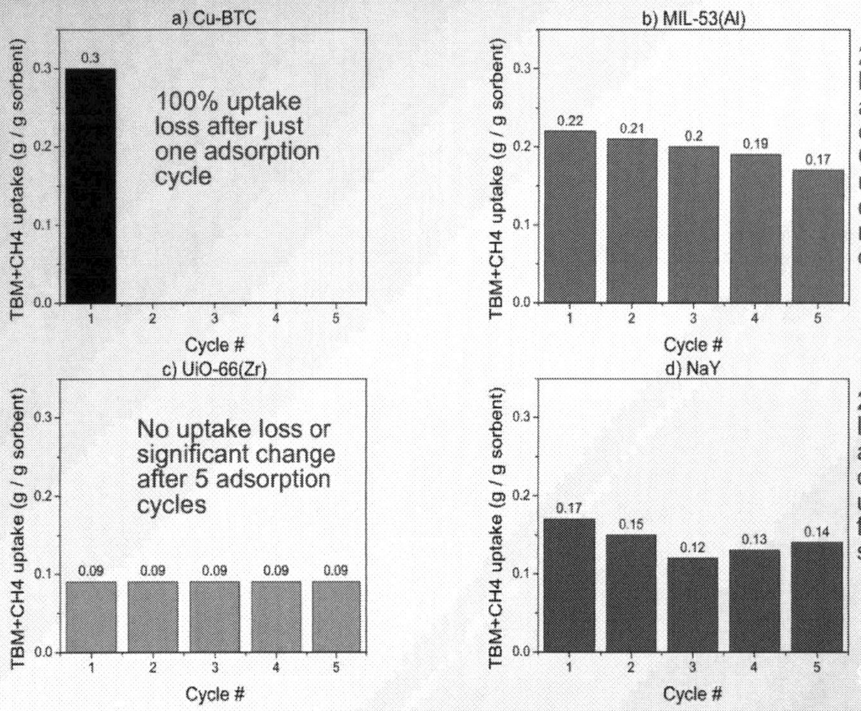

20% uptake loss after 5 adsorption cycles - Continual uptake loss in each cycle, no tapering off

24% uptake loss after 5 adsorption cycles - Initial uptake loss followed by stabilization

Georgia Tech

9

Results - X-Ray Diffraction Patterns

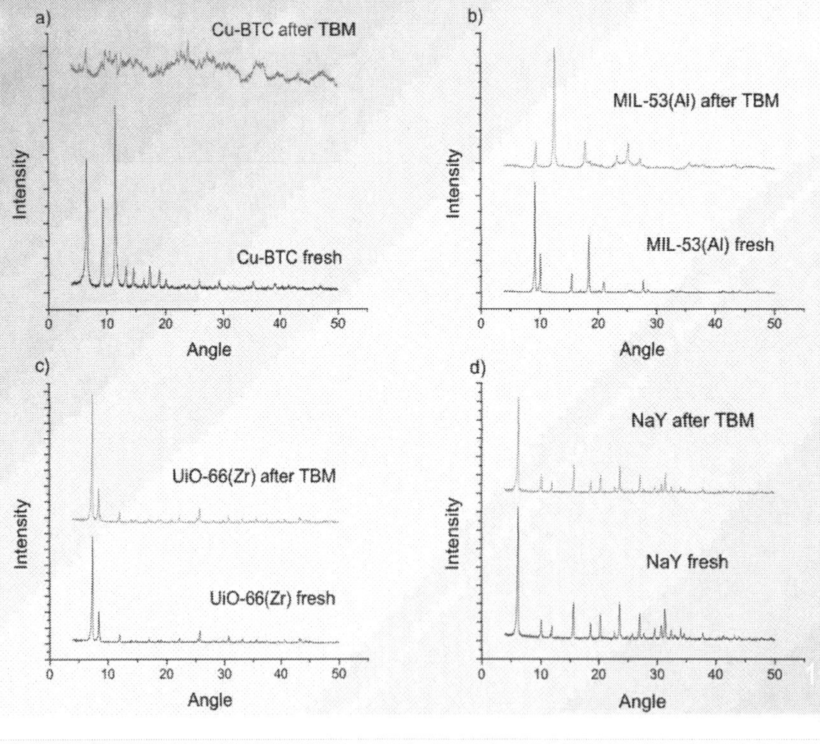

Georgia Tech

10

Results - X-ray Photoelectron Spectroscopy

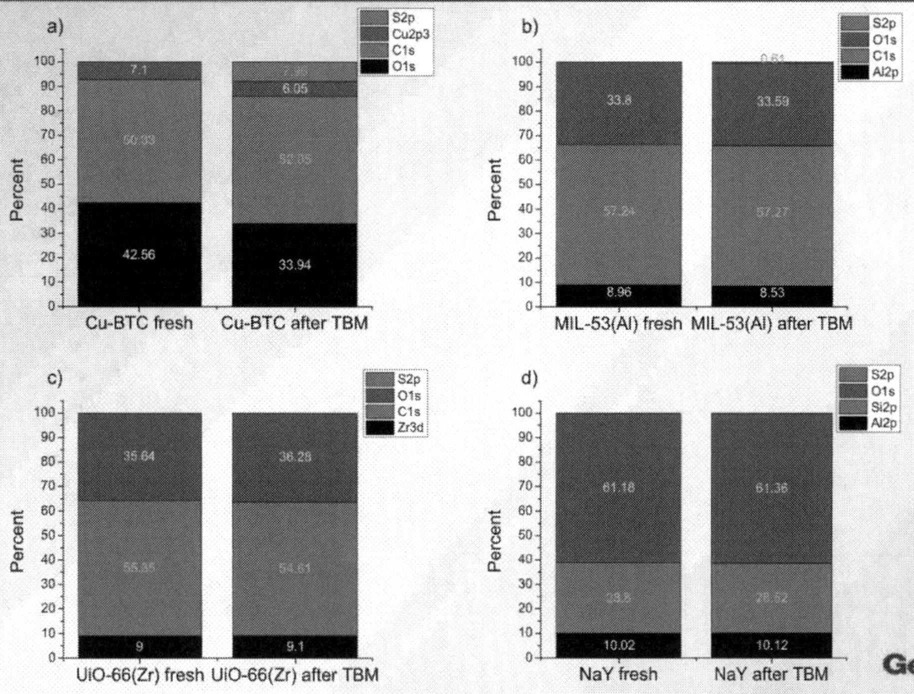

Georgia Tech

Conclusions and Future Work

- MOF UiO-66(Zr) is promising as a new regenerable, selective, and stable sorbent for odorant removal from natural gas
  - For similar uptake values, UiO-66(Zr) has better selectivity towards TBM over methane / impurities and regenerability than benchmark NaY zeolite
- Cu-BTC and MIL-53(Al) uptakes are very high but adsorption is highly irreversible with TSA and inert purge and / or destructive towards the sorbent

- Fiber spinning for UiO-66(Zr) and zeolite NaY
- Implementation in fiber module adsorption system

Georgia Tech

12

Thank you

- Dr. Chris Jones
- Dr. William Koros
- Jones and Koros group
- General Electric
- DOE

Georgia Tech

13

Results - Raman Spectroscopy

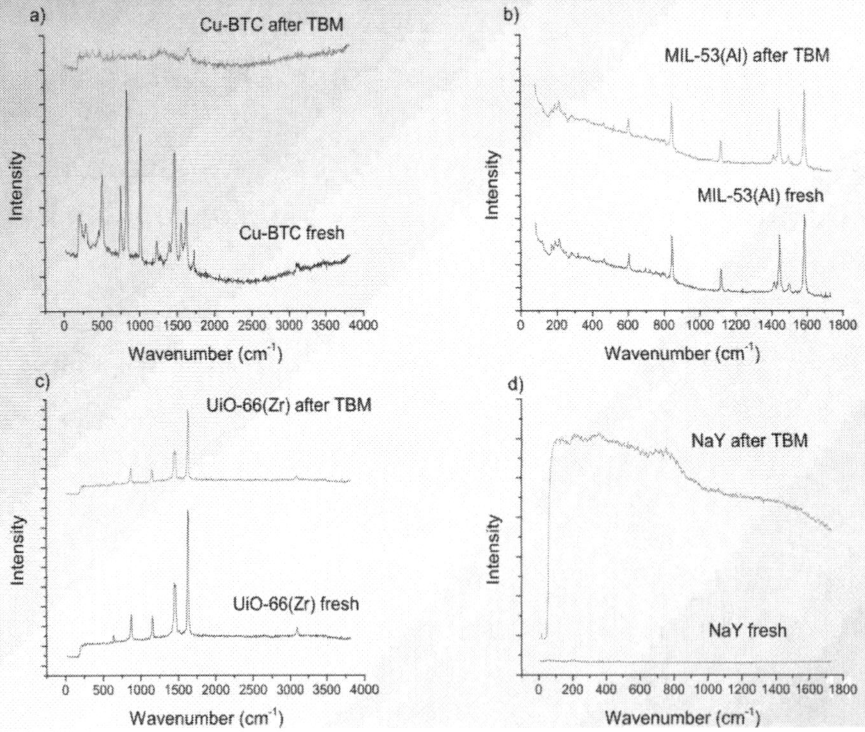

Georgia Tech

MIL-53(Al) Cyclic Regenerability

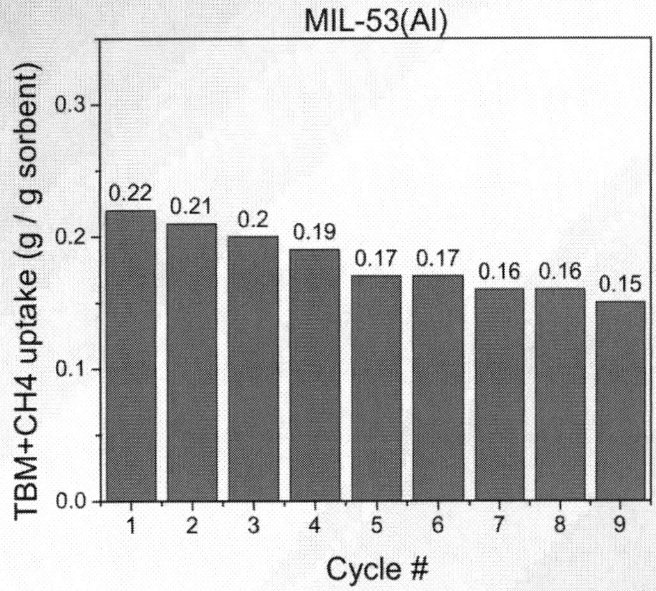

Georgia Tech

15

UV-Vis Spectroscopy

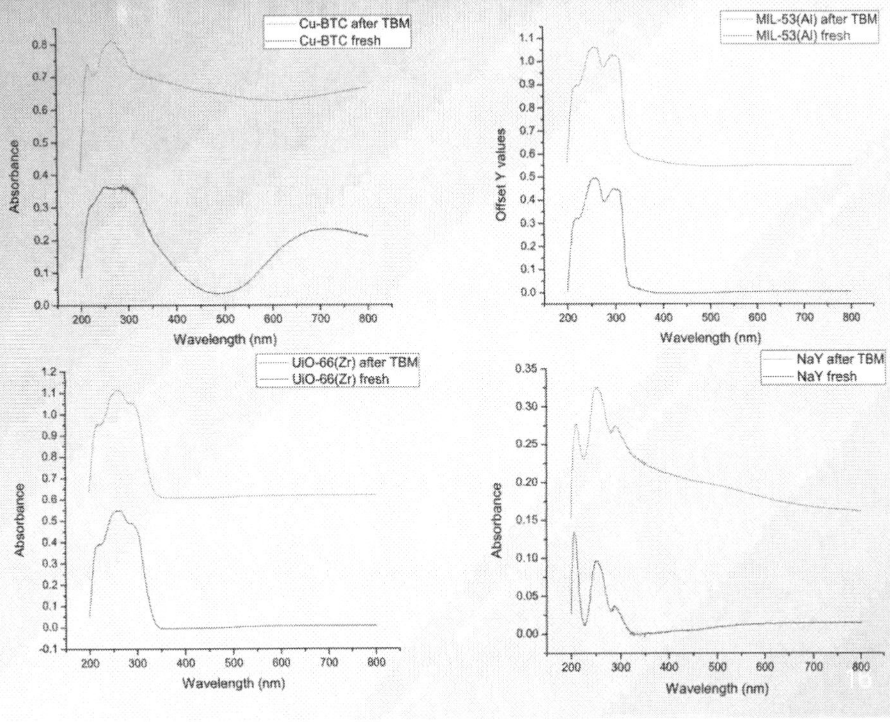

Georgia Tech

16

Sorbent Characterization

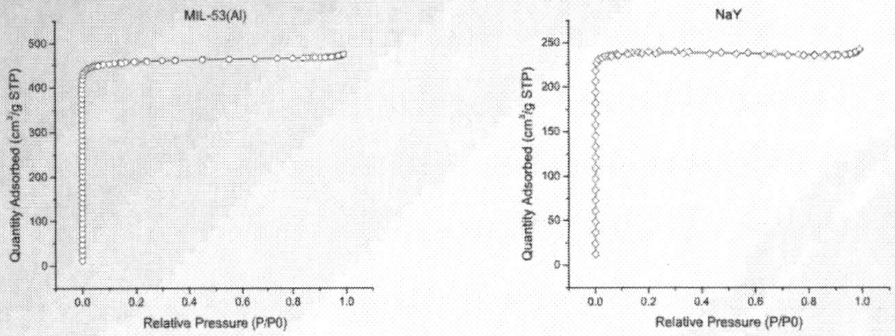

- Both MOF and zeolite display type I isotherms without hysteresis, which is typical for microporous materials.
- As expected, MOFs exhibit a very high surface area and pore volume as compared to zeolite materials.

Georgia Tech

17

Atomic Layer Deposition Assisted Layer-by-Layer Growth of Metal Organic Frameworks on Polymer Fiber Mats for Hazardous Gas Adsorption

Junjie Zhao,[†] *Bo Gong,*[†] *Paul C. Lemaire,*[†] *William T. Nunn,*[†] *Eric C. Stevens,*[†] *Fahim I. Sidi,*[†] *Matthew A. Browe,*[§] *Gregory W. Peterson,*[§] *Mark D. Losego,*[†] *and Gregory N. Parsons*[†]*

[†]Department of Chemical and Biomolecular Engineering, North Carolina State University, 911 Partners Way, Raleigh, NC 27695, United States.
[§]Edgewood Chemical Biological Center, 5183 Blackhawk Road, Aberdeen Proving Ground, MD 21010, United States.

Metal organic framework (MOF) thin films deposited on flat substrates can show relatively smooth surface texture, whereas most reports of MOF thin films grown on polymer fibers show low conformality with many individual crystal domains. The adsorption capacities of the MOF thin films on fibers are often not quantified. Here we report a new method to achieve uniform layer-by-layer (LbL) MOF growth on fibers using an ALD Al_2O_3 nucleation layer. Cross-sectional TEM images show a well-defined core@shell structure of the MOF-functionalized fiber. We analyze the thickness and mass increase data of LbL HKUST-1MOF thin film on ALD-coated polypropylene fibers, and find the growth rate to be 288~290 $ngcm^{-2}$ for each LbL cycle. ALD-coated polypropylene fiber mat with 40 LbL cycles exhibits a BET surface area of up to 93.6 $m/g_{MOF+fiber}$, and breakthrough test results also show its high dynamic loadings for NH_3 (1.37 $mol_{NH3}/kg_{MOF+fiber}$) and H_2S (1.49 $mol_{H2S}/kg_{MOF+fiber}$). These results all demonstrate the high adsorption capacities of our MOF-coated fiber mats. In addition, this synthesis route is generally applicable to various polymer fibers, and the fiber@ALD@MOF structure is also promising for membrane separation, catalysis, chemical sensing and many other applications beyond gas filtration. This presentation will highlight the MOF synthesis approach, including discussion of the differences observed for LbL vs batch solvothermal synthesis on fibers with and without ALD surface preparation.

Key Words: Metal organic frameworks, HKUST-1, atomic layer deposition, adsorption, polymer fiber

Amino-Pillared Nanosheet (APN) for High Performance Carbon Dioxide Capture

Christopher Cogswell
Northeastern University Chemical Engineering Department

Carbon dioxide capture is considered one of the primary challenges currently facing the engineering and scientific community. One approach to capture which has been suggested is the use of solid porous adsorbents which have been functionalized with amine moieties, due to the high capture capacity and surface areas present in these materials as well as their low heats of adsorption[1]. However, using current platforms such as clays or zeolites the addition of amine molecules to the solid support has been seen to lead to a large decrease in the kinetics of carbon dioxide adsorption, due to the porous channels being blocked by the impregnated amine molecules[2]. This has limited the capture capacities possible for these adsorbents, and greatly hindered their applicability in industry. In order to face these challenges, we are proposing the use of a porous material with multiple diffusion channels of different sizes, which can be preferentially impregnated with amine molecules while leaving intact pore channels for the diffusion of carbon dioxide into the total volume of the solid[3].

Here we present the amino-pillared nanosheet (APN) hybrid as a promising platform that will answer the current challenges faced by solid amine adsorbents for CO_2 capture. APN synthesis is carried out through a general method involving the swelling of the interlayer space through intercalation with quarternary ammonia cations, followed by the creation of silica pillars between the layers via the introduction of silica complexes followed by calcination[3]. Lamellar structured materials such as MCM-22-P[4] and AMH-3[5] are employed as the precursor structures used in the APN synthesis. Finally, amine molecules are functionalized onto the porous structure preferentially in the large mesopore space generated via pillaring due to their large molecular size which cannot fit into the micropores present in the nanosheet layers. The resulting complex offers unique CO_2 capture functionality due to its dual pore systems, providing separate channels for the diffusion of carbon dioxide through the material and capture onto the impregnated amine species. Owing to the presence of the micropores dedicated for CO_2 diffusion toward the internal adsorption sites, the APN will achieve high amine loadings as well as high CO_2 adsorption capacity without deteriorating the CO_2 adsorption kinetics, which is usually seen due to pore blocking at high amine loadings. Further control of the site of amine binding is possible by introducing amino-siloxane groups or quarternary amine molecules of various sizes to the crystal, which can only covalently bond to the silanol sites present in the interlayer space[7]. The CO_2 capture performance of the APN hybrids is evaluated in terms of their CO_2 capture capacity and adsorption kinetics, and the crystals are characterized by various techniques such as physisorption, XRD, and FTIR to elucidate the relationship between the APN structure and the CO_2 capture properties.

References

1) Choi S. et al; Chemsuschem, *2* (2009), 796-854
2) Drese, J.H. et al; *Micro and Meso Mater. 151* (2012), 231-240
3) Maheshwari S. et al; J. of Catalysis, *272* (2010), 298-308
4) Choi S. et al; Enviro. Sci. and Tech., *45* (2011), 2420-2427
5) Corma A. et al; Nature, *396* (1998), 353
6) Jeong H.K. et al; Nature Mater., *2* (2003), 53-58
7) Kim W. et al; Langmuir, *27* (2011), 7892-7901

Carbon Dioxide Capture Using Elastic Layered Metal-Organic Frameworks: Experimental and Computational Evaluation

Elastic layered metal-organic framework (ELM) adsorbents are a group of latent porous crystalline materials that undergo abrupt reversible transitions from a nonporous collapsed structure to an expanded porous structure through cooperative adsorption of guest molecules between layer planes. These materials have attracted interest due to their unique gating adsorption and high selectivity for simple gas molecules like CO_2 and show promise as a new generation of gas separators.

We evaluate the suitability of ELM-11 $[Cu(BF_4)_2(bpy)_2]$ (bpy = 4,4′-bipyridine) and ELM-12 $[Cu(OTf)_2(bpy)_2]$ (OTf = trifluoromethanesulfonate) for use in CO_2 capture using a combination of experimental and computational methods. Hydrothermal cycling is performed and the effects of water vapor adsorption on the stability of the ELM-11 and ELM-12 frameworks are reported. Breakthrough curves of CO_2/CH_4 mixtures on the ELM-11 structure are also reported and discussed. Results from these experiments are compared with the results of grand canonical monte carlo (GCMC) simulations and density functional theory (DFT) calculations. Differences in adsorption characteristics between ELM-11 and ELM-12, as well as the general implications for CO_2 capture on flexible MOFs will also be discussed.

378700 Effects of Flue Gas Contaminants, Regeneration Conditions, and Amine Modification on the Adsorption of Carbon Dioxide By MIL-101(Cr)

Abstract:

MIL-101(Cr) has drawn much attention in recent years due to its high stability compared with other metal-organic frameworks. In this study, three trace flue gas contaminants (H_2O, NO, SO_2) were each added to a 10 vol% CO_2/N_2 feed flow and found to have a minimal impact on the adsorption capacity of CO_2. In dynamic CO_2 regeneration experiments, complete regeneration occurred in 10 min at 328 K for temperature swing adsorption-N_2-stripping under a 50 cm^3/min N_2 flow and at 348 K for vacuum-temperature swing adsorption at 20 KPa. Almost 99% of the pre-regeneration adsorption capacity was preserved after 5 cycles of adsorption/desorption under a gas flow of 10 vol% CO_2, 100 ppm SO_2, 100 ppm NO, and 10% RH, respectively. Strong resistance to flue gas contaminants, mild recovery conditions, and excellent recycling efficiency make MIL-101(Cr) an attractive adsorbent support for CO_2 capture. MIL-101(Cr) was functionalized with various amine (tetraethylenepentamine, pentaethylenehexamine and polyethyleneimine) to improve the ability of CO_2 adsorption. The amine-functionalized MIL-101(Cr) was characterized with various experimental methods including N_2 adsorption/desorption isotherms, X-ray diffraction, Fourier transform infrared and thermogravimetric analysis. Both the adsorption isotherms and the isosteric heats of different adsorption capacities were obtained from experiments. Avrami's fractional order kinetic model was selected to describe the CO_2 adsorption behavior. The adsorption capacity of amine-functionalized MIL-101(Cr) increased significantly.

387725 Effect of C/N Ratio on Microbial Lipid Production with the Oleaginous Yeast Lipomyces Starkeyi

Abstract:

Neutral lipids have been used as raw materials for the production of biodiesel. Since agricultural and plant-derived neutral lipids are often utilized in food preparations, alternative sources of lipids are being investigated. A number of microorganisms possess the ability to produce significant fractions of their dry weight as neutral lipids. The oleaginous yeast, *Lipomyces starkeyi*, is a promising candidate for industrial production of neutral lipids which will ultimately be converted into biofuels. The yeast grows rapidly on various carbon sources and can produce lipids upto 75% of its dry weight. Lipid accumulation most commonly occurs when the nutrient supply is exhausted in the presence of excess carbon at the stationary phase. Under nutrient limiting conditions, yeast cells accumulate lipids as a means of storage when energy sources are abundantly available and when the cellular mechanisms for the microbial synthesis are active. In this research work, experiments are conducted by varying the C:N ratio in the growth medium, and its effect on cell growth and the accumulation of lipids. The carbon source utilized in this work is laboratory grade sweet potato starch as well as starch from raw sweet potato. All the experiments were conducted in the shake flasks as well as in the fermenters with controlled pH, and temperature conditions. Different feeding strategies will also be studied and discussed with this presentation. Cell mass, starch content and lipid content were analyzed periodically. The cell mass was recorded by weighing the sample before and after drying. Lipid content was measured by Nile red florescence. The supernatant from centrifugation or filtrate was used to measure starch concentration by iodine method.

392434 Recovery of Polyphenol-Enriched Sugarcane Molasses Solution By Ultrafiltration

Abstract:

Blackstrap molasses was diluted with deionized water and purified using ultrafiltration hollow fiber membrane. Evaluation of permeate flux and fouling were compared using polysulfone membrane with MWCO of 500 kDa and 100 kDa. Permeate flux was observed every 5-minute interval and steady time was recorded. The transmembrane pressure was varied from 10 to 15 psi, permeate flux changes, which is dependent on the pressure, was observed and recorded. At the end of ultrafiltration, data on final permeate flux were collected for each transmembrane pressure and for each membrane used. Fouling study was performed at total recycling mode by observing the permeate flux decline with time at transmembrane pressures of 15, 20, and 25 psi. The performance of the two hollow fiber membranes used were compared in terms of ultrafiltration permeate flux and fouling behavior. The membrane rejection with respect to polyphenol content of the initial feed and final permeate were calculated to complete the comparison for the two membranes. The results showed that both the membranes have clarified the permeate solution but the rejection for the 100 kDa membrane was the greatest, around 0.7. For both membranes, fouling data were comparable, but final flux in 100 kDa was much lower by 20% as compared to 500 kDa. Removal efficiency of the unwanted molecules such as potassium and calcium crystals were more significant in 100 kDa membrane. Thus, by evaluating used membranes on the enrichment of polyphenols from molasses, it was concluded that 100 kDa membrane has been more appropriate.

388825 β-1,3-Glucan Production By Metabolically Engineered Agrobacterium Sp. from Cellulosic Sugar

Abstract:

β-Glucans are polysaccharides of D-glucose monomers linked by β-1,3-glycosidic bonds and have been used for immune and heart health functional ingredients in food industry due to their functions as "biological response modifiers" including activating the immune system. In addition, they can be utilized as growth promoting agent of domestic animals without use of antibiotcs in feed industry and lots of companies currently use β-glucan as a functional ingredient in cosmetics. Production of β-1,3-glucans from cellulosic sugar, such as cellobiose, is important as it establishes a more sustainable technology for the synthesis of this commercial product, and allows a cheaper renewable substrate, cellulose, to be used. While cellobiose could be potentially used through a hydrolysis mechanism, in which the glucose dimer is hydrolyzed into two glucose molecules, we have decided to pursue a more energetically favorable mechanism, which relies on cellobiose phosphorylase. Cellobiose phosphorylase is an enzyme specific to cellobiose and catalyzes the phosphorolytic cleavage of cellobiose, generating glucose and glucose-1-phospahte which can be directly used for biosynthesis of sugar nucleotides and polysaccharides including β-1,3-glucan. In order to construct a strain producing β-1,3-glucan from cellobiose, the gene encocding cellobiose phosphorylase (CBP), , was introduced to *Agrobacterium* sp. for improved cellobiose utilization in of this bacterium. Five promoters, including three *Agrobacterium* sp. promoters and two commercial promoters, were tested for expression of CBP. Two engineered strains, with promoters P_{HCE} and P_{ExoC}, produced β-glucan 11.6 g/L and 10.3 g/L, respectively, in shake flask cultivation after 5 days. Compared to that of control strain, the two engineered strains produced 2.5~3.0 folds higher β-glucan. The production level in engineered strains are also higher than that from sucrose (7~8 g/L), the best carbon source for β-glucan production. In 5L jar fermenter cultivation, both engineered strains produced 24.5~25.8 g/L of β-glucan from 50 g/L cellobiose, which is as much as twice of the parent strain. Therefore, engineering of *Agrobacterium* sp. for production of β-glucan via cellobiose phosphorolytic reaction is an effective strategy for high-yield production of β-glucan from cellobiose. The availability of metabolically engineered strain capable of producing β-glucan from cellobiose paves the way for its production from cellulose.

Chemical Imaging of Cheese Microstructure using ATR Microimaging Spectroscopy

Ramazan KIZIL* kizilr@itu.edu.tr
Istanbul Technical University, Chemical Engineering Department. Maslak 34469 Istanbul.

The new avenue for vibrational spectroscopy appears to be chemical imaging of complex materials from their own molecular fingerprint at micrometer scale under a spectrometer-coupled optical microscope through consecutive scanning with an array detector of a large number of small size classical semiconducting MCT detectors. The findings through chemoinformatic images provide two-dimensional spatial distribution of species which has ever revealed before at the molecular level.

Determination of food structure at the microscopic level is expected to significantly contribute to the modern food technology whose prime interest lies in development of controlled processes and operations to preserve, transform and create or destroy food structure. ATR imaging system combines a high refractive index (n=4.01) hemispherical Germanium internal reflection element with a Cassegrain-style reflective objective is capable of producing images at a spatial resolution which is four times greater than transreflectance and transmission techniques.

This talk covers developments in the imaging IR spectroscopy technology and its application on microstructure analysis of a heterogeneous food product, cheese. The spatial distribution of proteins and fats in the cheese matrix was determined from their unique structure related absorption on contrast images at spatial resolution of 1.56 micrometer. It was shown that cheddar cheese, feta cheese and cream cheese all has all different patterns of fat distribution in the texture. In addition the size of fat globules differs greatly from type to type. A sample image, which is obtained from a feta cheese sample, is shown below to illustrate how spatial distribution of chemical constituents appears in the IR chemical imaging maps.

A B C

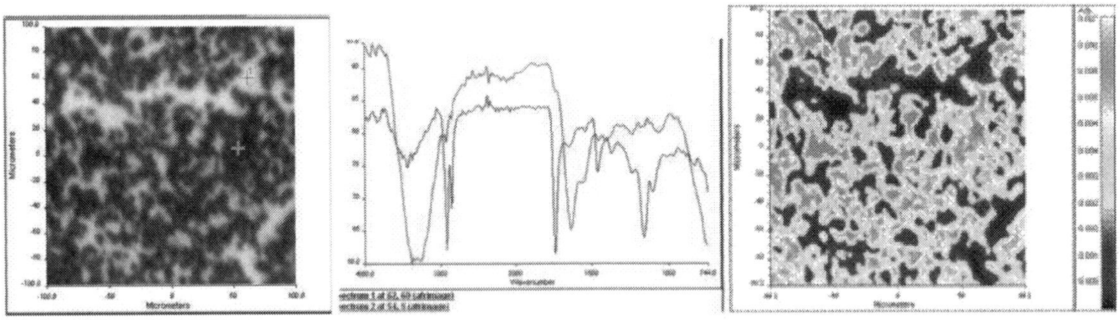

Figure 1. ATR imaging of Feta Cheese A) Raw image based on pattern recognition analysis, B) Each pixel marked on A has full spectral information C) Chemical imaging of cheese based on the amide I vibration absorption to map out spatial distribution of protein in the texture, the red regions depict protein rich sections.

385352 Study on Swine Manure Foaming to Understand Its Major Reason and Its Relations to Industrial Products

Abstract:

In the past few years, a sharp increase of pit-foaming is discovered in swine production facilities. The foams on the swine manure hold significant amount of methane gas, limited the space for manure storage and creates a potential safety hazard. Cases of flash fires and explosions were reported recently in Midwestern states, causing significant assets loss and worker injuries. This research combined with the field sampling and lab simulation to study the foaming mechanism, including manure compositional analysis, microbial community analysis and surfactant identification. Microbial community analysis results show that unlike many foaming cases in wastewater treatment facilities, the manure foaming was not induced by the massive growth of filamentous bacteria. Instead, it is indicated from the lab simulations that the increased concentration of long chain fatty acids together with solid particles in manure solution could initiate the generation of stable foam bubbles. Analysis on solid particles suggests that protein contributes as the major surfactant in manure solution, and further proteomic study will specify which group of protein is mainly involved in this foaming. This research also explored the possibility of some feeding factors that may be related to this foaming phenomenon, like dried distiller grains solubles (DDGS) addition in pig diets and lower digestibility of feed in the pork production. Since DDGS is a major by-product of corn ethanol industry, this will help to improve the industrial process design and provide more applicable products.

Economic analysis of different stabilization methods of antioxidants

Miguel A. Rojas[1], Moshe Rosenberg[2], Carlos A. Cardona[1*]
[1] Biotechnology and Agribusiness Institute, Department of Chemical Engineering
Universidad Nacional de Colombia Sede Manizales, Manizales -Colombia
[2] University of California Davis, Food Science & Technology Department
marojasre@unal.edu.co; mrosenberg@ucdavis.edu; *ccardonaal@unal.edu.co;

1

14AIChE
Annual Meeting, Atlanta, GA

1. INTRODUCTION

The growing interest on personal health

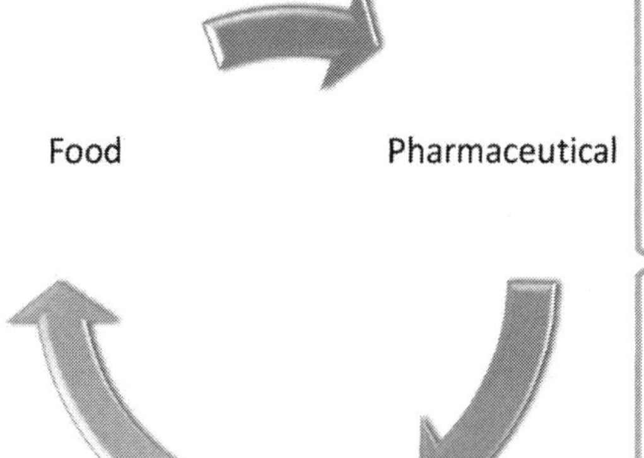

- Flavors
- Antioxidants
- Colors
- Enzymes
- Antimicrobials
- Vitamins
- Minerals
- nutraceuticals

2

1. INTRODUCTION

The growing interest on personal health

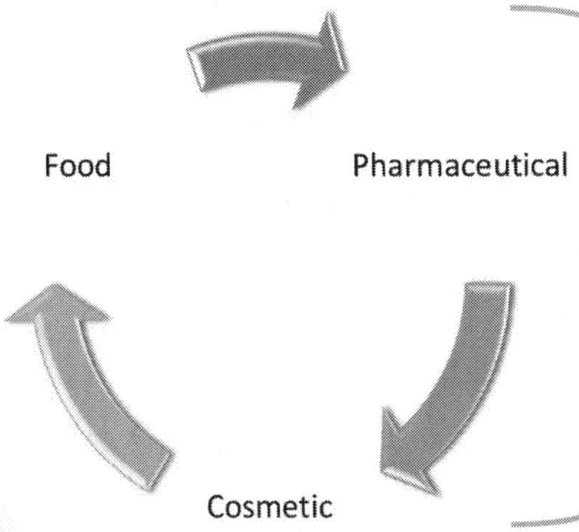

Antioxidants

- Unstable
- Unplease taste

3

1. INTRODUCTION

After the extraction of active ingredients they need to be immediately protected against biological activity

Food manufacturers

- Antioxidants
- Antimicrobial
- Flavors
- preservatives

Challenges

Overcome by incorporating the active ingredients into some kind of delivery system before it is introduced into the final product

4

1. INTRODUCTION

- A wide variety of microencapsulation techniques are available for utilization in the food industry to encapsulate, protect, and release these active ingredients

Choose the most appropiate system

- Physicochemical properties
- Labelling and legal requirements
- Economic factors

5

Research Group in Chemical, Catalytic and Biotechnological Processes
Universidad Nacional de Colombia - Sede Manizales

1. INTRODUCTION

The delivery system should be capable of being economically manufactured from inexpensive ingredients

Economic Factors

- Active ingredients microencapsulated
- Microencapsulation methods
- Encapsulating materials

Prebiotics and probiotics	Herbs and spices
Antioxidants	Carotenoids

6

Research Group in Chemical, Catalytic and Biotechnological Processes
Universidad Nacional de Colombia - Sede Manizales

1. INTRODUCTION

The delivery system should be capable of being economically manufactured from inexpensive ingredients

Economic Factors

- Active ingredients microencapsulated
- Microencapsulation methods
- Encapsulating materials

Mechanical	Spray Drying
	Spray Freezing
Chemical	Gelification
	Coacervation
Physical-Chemical	Interfacial polymerization

7

Research Group in Chemical, Catalytic and Biotechnological Processes
Universidad Nacional de Colombia - Sede Manizales

1. INTRODUCTION

The delivery system should be capable of being economically manufactured from inexpensive ingredients

Economic Factors

- Active ingredients microencapsulated
- Microencapsulation methods
- Shell materials

- Proteins
- Simple sugars
- Carbohydrates/gums
- Lipids
- Emulsifiers
- Cellulose

8

Research Group in Chemical, Catalytic and Biotechnological Processes
Universidad Nacional de Colombia - Sede Manizales

2. OBJECTIVE

- The objective of this work perform an economic analysis of the most commonly microencapsulation methods in antioxidants and analyze their economic viability.

9

Research Group in Chemical, Catalytic and Biotechnological Processes
Universidad Nacional de Colombia - Sede Manizales

3. METHODOLOGY

Extract Selection → Stabilization method → Modeling and simulation → Economic analysis

Chemical analysis	Total phenolic Retention [%]	Microencapsulation method
Blackberry Extract Microencapsulated [BEM][a]	94.79	Spray Drying
Bilberry extract microencapsulated [BBA][b]	95	Gelation
Blackberry Extract microencapsuled [BPM][c]	80.52	Complex Coacervation

[a]Previous work
[a]Betz et al. 2012
[C]Zheng et al 2011.

10

Research Group in Chemical, Catalytic and Biotechnological Processes
Universidad Nacional de Colombia - Sede Manizales

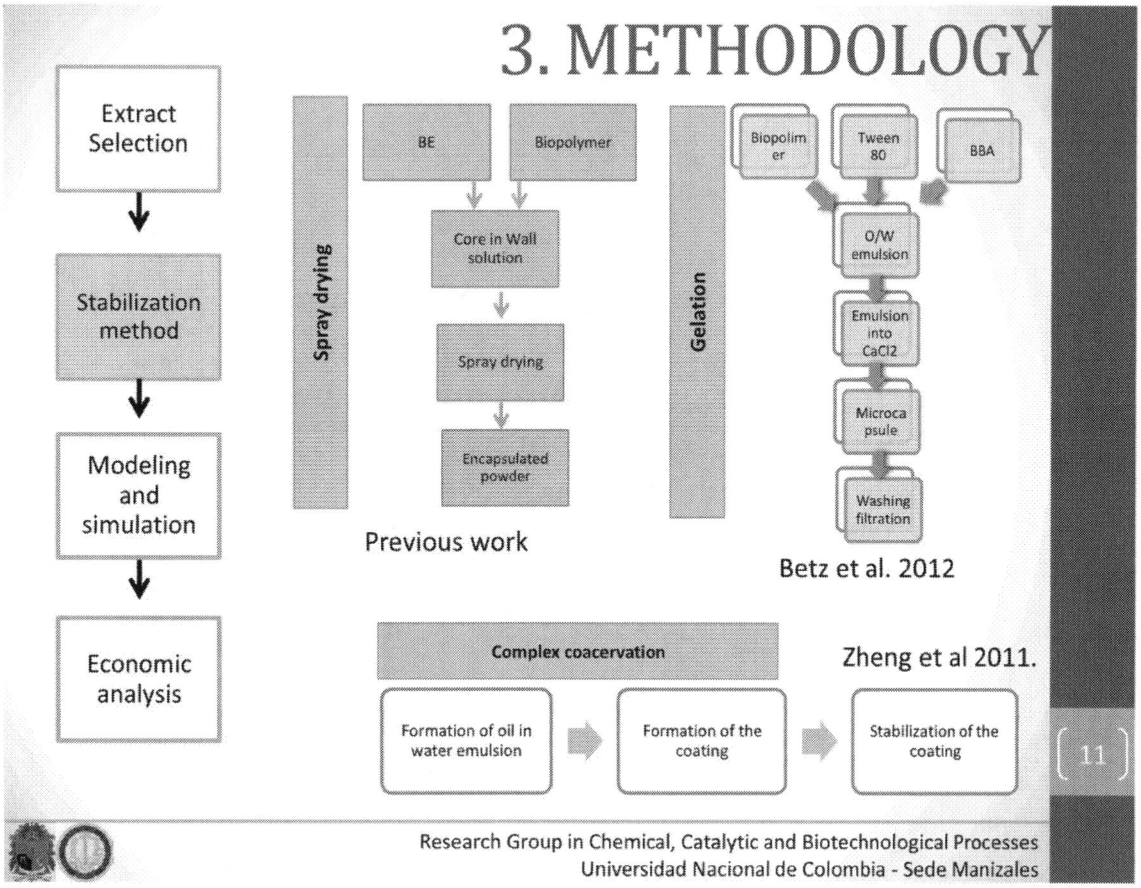
3. METHODOLOGY
Extract Selection
Stabilization method
Modeling and simulation
Economic analysis
Spray drying
BE
Biopolymer
Core in Wall solution
Spray drying
Encapsulated powder
Previous work
Gelation
Biopolim er
Tween 80
BBA
O/W emulsion
Emulsion into CaCl2
Microca psule
Washing filtration
Betz et al. 2012
Complex coacervation
Zheng et al 2011.
Formation of oil in water emulsion
Formation of the coating
Stabilization of the coating
11
Research Group in Chemical, Catalytic and Biotechnological Processes
Universidad Nacional de Colombia - Sede Manizales

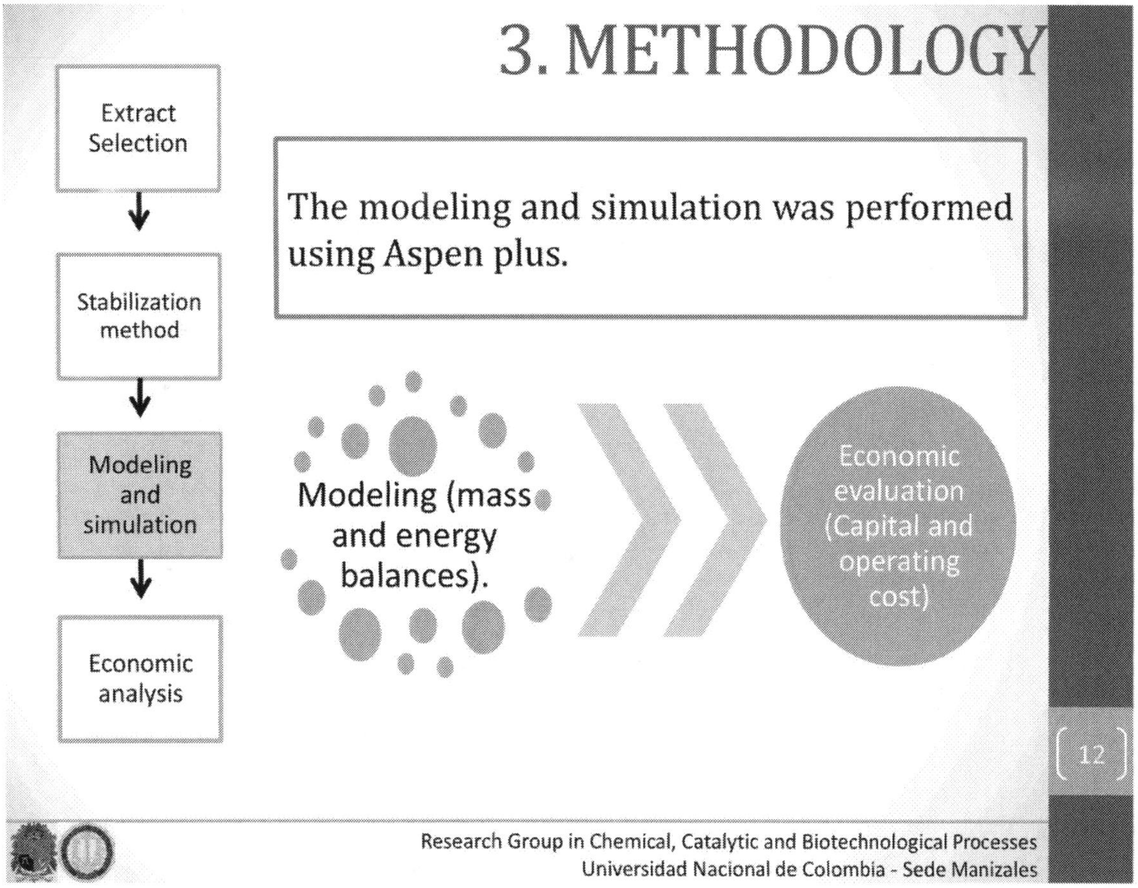
3. METHODOLOGY
Extract Selection
Stabilization method
Modeling and simulation
Economic analysis
The modeling and simulation was performed using Aspen plus.
Modeling (mass and energy balances).
Economic evaluation (Capital and operating cost)
12
Research Group in Chemical, Catalytic and Biotechnological Processes
Universidad Nacional de Colombia - Sede Manizales

3. METHODOLOGY

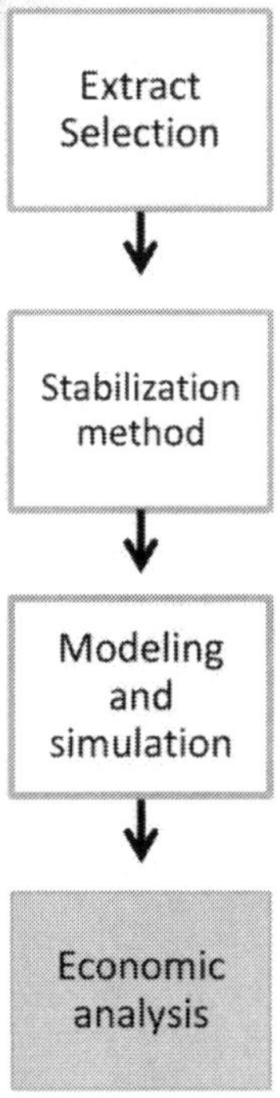

Economic analysis

The economic analysis was performed by calculating the production cost per unit of product (e.g. kg). The total production cost considers the total of raw material, utilities, operating labor, maintenance costs, operating charges, plant overhead and general administrative costs. The costs of personnel labor, water and electricity were obtained from typical prices in Colombia.

4. RESULTS

Economic parameters used for antioxidant extract microencapsulation [AEM] estimation

Raw material		
Extract[a] (Solvent system CO_2 + EtOH)	USD /kg	7.63
Soybean oil[b]	USD /kg	0.6
Soy lecithin	USD /kg	1.4
Arabic gum[b]	USD/ kg	3.8
Sunflower Oil[b]	USD/ kg	1.3

[a]Previous work
[b]http://www.icis.com/

Modeling and Simulation

In this context, it can be observed that the price of the raw material has a great influence on the distribution cost in the production of microcapsules.
The utilities had a low percentage of the total cost

Distribution Cost in Gelation, Spray drying and coacervation Methods.

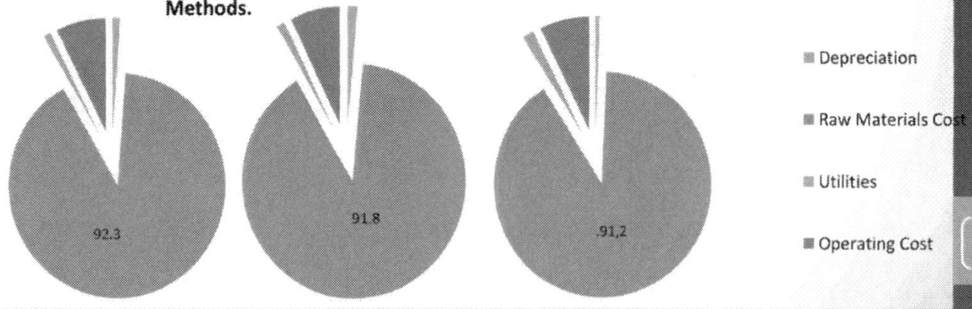

4. RESULTS

Economic evaluation

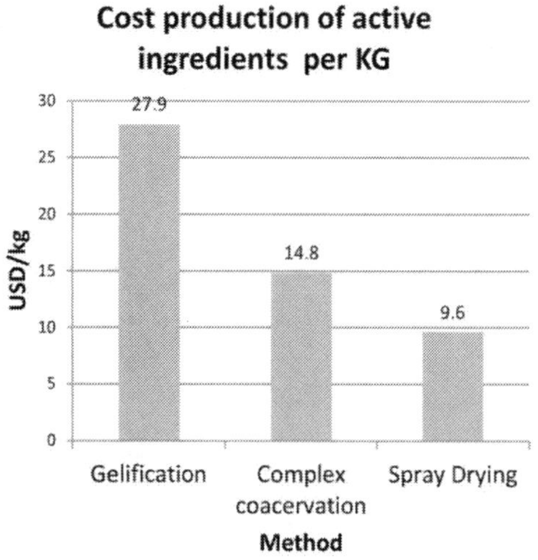

The process economic evaluation largely depends on some factors that must be taken into account before claiming that the microencapsulation method are not economically viable.

Factors:

- Cost and content of the active ingredient to entrap
- Operating conditions
- Water content in the initial feed mixture
- Small scale and high scale

15

5. CONCLUSIONS

- From an economic point of view, the type of microencapsulation method used determine the overall cost of the process, which is an important consideration for the food industry where profit margins are often tight.
- The benefits gained with microencapsulation of active ingredients should outweigh the additional cost associated with the process of encapsulation

16

6. AKNOWLEDGEMENTS

- The authors express their acknowledgments to the Universidad Nacional de Colombia sede Manizales
- Research group of Chemical, Catalytic and Biotechnological Processes.

17

Research Group in Chemical, Catalytic and Biotechnological Processes
Universidad Nacional de Colombia - Sede Manizales

7. REFERENCES

- T. Egan, J.-C. Jacquier, Y. Rosenberg, and M. Rosenberg, "Cold-set whey protein microgels containing immobilised lipid phases to modulate matrix digestion and release of a water-soluble bioactive.," *J. Microencapsul.*, vol. 31, no. 2, pp. 184–92, Jan. 2014.
- T. Egan, J.-C. Jacquier, Y. Rosenberg, and M. Rosenberg, "Cold-set whey protein microgels for the stable immobilization of lipids," *Food Hydrocoll.*, vol. 31, no. 2, pp. 317–324, Jun. 2013.
- C. C. Ferrari, S. P. M. Germer, and J. M. de Aguirre, "Effects of Spray-Drying Conditions on the Physicochemical Properties of Blackberry Powder," *Dry. Technol.*, vol. 30, no. 2, pp. 154–163, Feb. 2012.
- Da Rosa, C. G., Borges, C. D., Zambiazi, R. C., Rutz, J. K., da Luz, S. R., Krumreich, F. D., ... Nunes, M. R. (2014). Encapsulation of the phenolic compounds of the blackberry (Rubus fruticosus). *LWT - Food Science and Technology*, *58*(2), 527–533. doi:10.1016/j.lwt.2014.03.042
- Veggi, P. C., Cavalcanti, R. N., & Meireles, M. A. a. (2014). Production of phenolic-rich extracts from Brazilian plants using supercritical and subcritical fluid extraction: Experimental data and economic evaluation. *Journal of Food Engineering*, *131*, 96–109. doi:10.1016/j.jfoodeng.2014.01.027
- Santos, D. T., Veggi, P. C., & Meireles, M. A. a. (2010). Extraction of antioxidant compounds from Jabuticaba (Myrciaria cauliflora) skins: Yield, composition and economical evaluation. *Journal of Food Engineering*, *101*(1), 23–31. doi:10.1016/j.jfoodeng.2010.06.005
- M. Betz, B. Steiner, M. Schantz, J. Oidtmann, K. Mäder, E. Richling, and U. Kulozik, "Antioxidant capacity of bilberry extract microencapsulated in whey protein hydrogels," Food Res. Int., vol. 47, no. 1, pp. 51–57, Jun. 2012.
- L. Zheng, Z. Ding, M. Zhang, and J. Sun, "Microencapsulation of bayberry polyphenols by ethyl cellulose: Preparation and characterization," J. Food Eng., vol. 104, no. 1, pp. 89–95, May 2011.

18

Research Group in Chemical, Catalytic and Biotechnological Processes
Universidad Nacional de Colombia - Sede Manizales

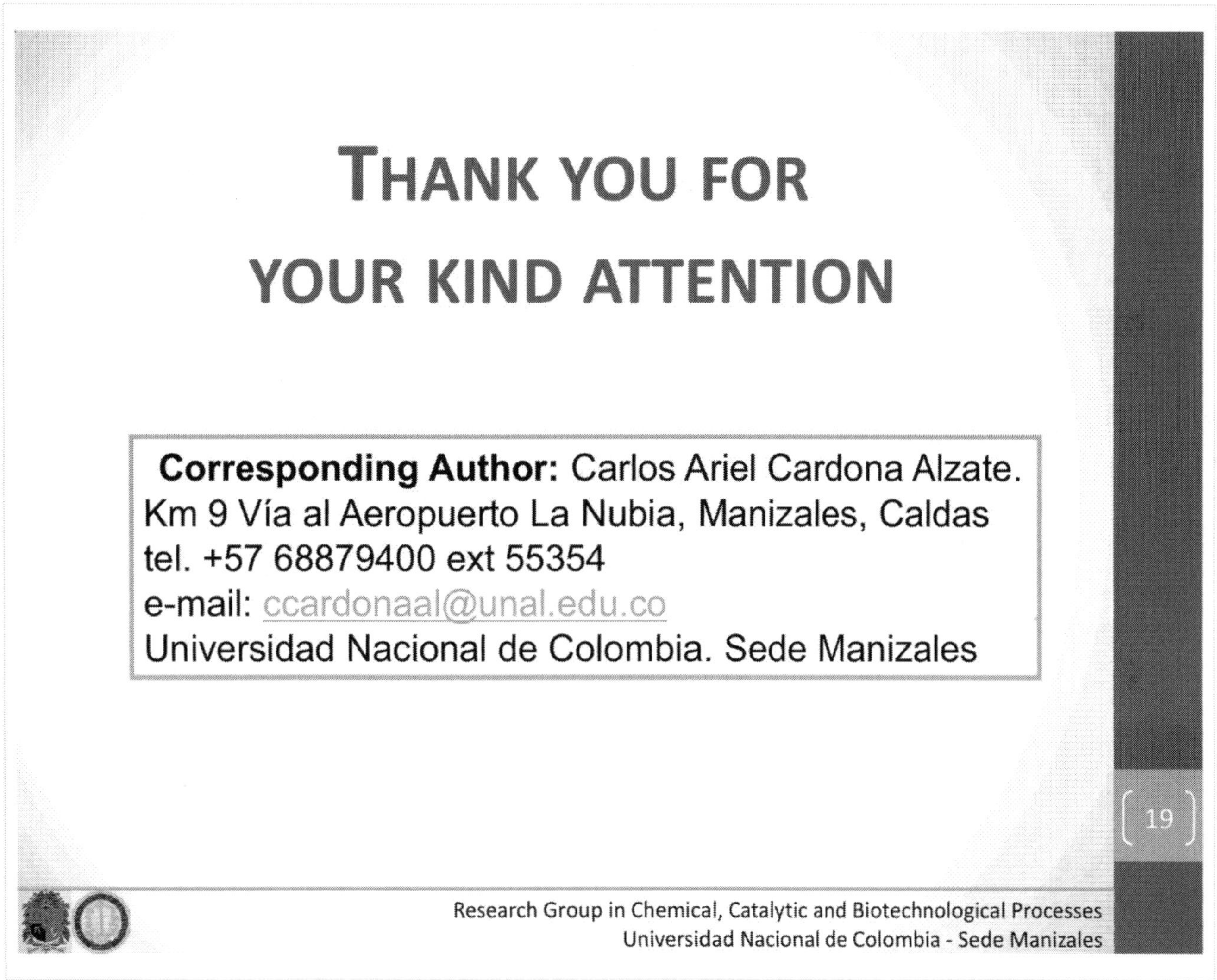
THANK YOU FOR
YOUR KIND ATTENTION
Corresponding Author: Carlos Ariel Cardona Alzate.
Km 9 Vía al Aeropuerto La Nubia, Manizales, Caldas
tel. +57 68879400 ext 55354
e-mail: ccardonaal@unal.edu.co
Universidad Nacional de Colombia. Sede Manizales
19
Research Group in Chemical, Catalytic and Biotechnological Processes
Universidad Nacional de Colombia - Sede Manizales

Preveting Paper Towels from Bacteria Growth Using Selenium Nanoparticles

Qi Wang[1], Thomas J. Webster[2]
[1]Department of Bioengineering Department, [2]Department of Chemical Engineering, College of Engineering, Northeastern University, Boston, MA 02115, U.S.A.

Introduction

There are wide spread bacterial contaminations on various paper products, such as paper towels hanging in sink splash zones or those used to clean surfaces, filter papers used in water and air purifying systems, and wrapping used in the food industry, which may lead to the potential spread of bacteria and consequent health concerns. In this study, selenium nanoparticles were coated on normal paper towel surfaces through a quick precipitation method, introducing antibacterial properties to the paper towels. Their effectiveness at preventing biofilm formation was tested in bacterial assays involving *Staphylococcus aureus*, *Pseudomonas aeruginosa*, *Escherichia coli* and *Staphylococcus epidermidis*. The results showed that there were significant and continuous bacteria inhibition with about a 90% reduction from 24 to 72 hours for gram-positive bacteria including *S. aureus* and *S. epidermidis*. The selenium coated paper towels also showed significant inhibition of gram-negative bacteria like *P. aeruginosa* and *E.coli* growth at about 57% and 84%, respectively, after 72 hours of treatment. This study established a promising selenium-based antibacterial strategy to prevent bacterial growth on paper products, which may lead to the avoidance of severe health concerns.

Materials and Methods

Selenium nanoparticles were synthesized and coated on paper towels (MB550A Hand Towel, Tork Advanced, cut into round chips, 7.01mm in diameter) through a simple and quick precipitation reaction, which involves glutathione (reduced form, GSH) (97%, TCI America, Portland, OR) and sodium selenite (99%, Alfa Aesar, Ward Hill, MA) mixed at a 4:1 molar ratio. Selenium nanoparticles were formed and precipitated on the surface of paper towels immediately following the addition of NaOH as visualized by a color change of the reactant solution from colorless to clear red. The paper towel samples were coated for 30s under a 200 rpm (round per minute) agitation to ensure a uniform coating. The coated substrates were rinsed in deionized water five times and soaked in water for 24 hours to remove the free, non-adherent, selenium nanoparticles and remaining reactants. Then, the selenium coated paper towels were left in petri dishes for several days to completely dry before being used in any further experiments or tests. The uncoated samples that were used as a control group in the following experiments were prepared by treating paper towels with the same procedure for selenium coated paper towel samples described above except that sodium hydroxide was not added so that the reaction did not occur.

Four bacterial cell lines were involved in the tests to evaluate the growth of various types of bacteria on selenium coated and uncoated paper towels. Bacteria were propagated in 30 mg/mL tryptic soy broth (TSB) for 20 hours. Each sample was treated separately with 1 mL of the prepared bacterial solutions (10^6 bacteria/ml) and were cultured for either 24, 48 or 72 hours in an incubator (37°C, humidified, 5% CO_2). For those samples that were cultured for 48 and 72 hours, the media was changed with 1 mL of sterile and fresh TSB every 24 hours. After the treatment, the samples were rinsed with a PBS (phosphate buffered saline) solution twice and placed into 1.5ml microfuge tubes with 1ml of PBS. These microfuge tubes were shaken at 3000 rpm for 15 minutes on a vortex mixer to completely release the bacteria attached on the surface into the solution. Solutions with bacteria from each sample were spread on agar plates and bacteria colonies were counted after 20 hours of incubation. All bacterial inhibitory tests were conducted in triplicate and repeated three times. Data were collected and the significant differences were assessed with the probability associated with a one-tailed Student's t-test. Statistical analyses were performed using Microsoft Excel.

Results and Discussion

The results of bacteria assays involving *S. aureus*, *P. aeruginosa*, *E. coli* and *S. epidermidis* showed a high effectiveness for the selenium coated paper towels at inhibiting bacteria growth. As shown in Figure 1, the selenium coatings significantly inhibited the growth of *S. aureus* by about 89% after 24, 48 or 72 hours on the surface of paper towels. Similarly, the selenium coated paper towels had 90.66%, 92.50% and 92.16% less *S. epidermidis* attached (Figure 2) compared with the uncoated paper towels after 24, 48 and 72 hours, respectively. From 24 hours to 48 hours, there was a large increase in the numbers of bacteria, which means the bacteria propagated quickly during this period on paper towels without a selenium coating. However, on the surface of selenium coated paper towels, there was not an obvious increase in the bacteria numbers from 24 hours to 48 hours. In addition, after incubation for 48 hours and 72 hours, the uncoated paper towel was saturated by bacteria, indicating the formation of a matured biofilm, while the selenium coated paper towels still had a significantly reduced number of bacteria after 72 hours. Overall, the numbers of *S. aureus* and *S. epidermidis* on the selenium coated paper towels remained at a low level not increasing from 24 to 48 to 72 hours, indicating successful inhibition to their growth and biofilm formation.

In the bacterial assays with *P. aeruginosa*, the results in Figure 3 also showed a successful inhibition to the growth of *P. aeruginosa* by 55% and 84% after 48 or 72 hours, respectively, on the surface of selenium coated paper towels. The total amount of *P. aeruginosa* increased from 24 hours to 48 hours to 72 hours on paper towels without selenium coatings, while on the surface of selenium coated paper towels, the amount of *P. aeruginosa* decreased from 24 hours to 48 hours and further decreased dramatically from 48 hours to 72 hours. Therefore, the activity of *P. aeruginosa* was successfully inhibited on selenium coated paper towels and the inhibition increased as the time of treatment increased. Moreover, the results of the bacterial inhibitory assays with *E. coli* are shown in Figure 4. The selenium coated paper towels inhibited *E. coli* growth by about 50%~60% compared with the uncoated paper towels after treatment for 24 hours, 48 hours and 72 hours. Overall, compared with the uncoated paper towels, the selenium coated paper towels showed a significant inhibition to the growth of *E. coli* continuously from 24 hours to 72 hours.

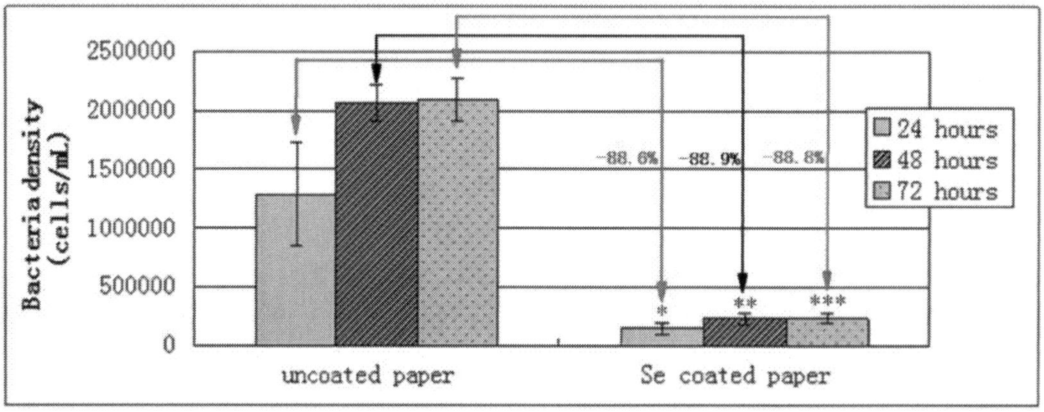

Figure 1. The growth of ***Staphylococcus aureus*** on the surface of paper towels. There was about 90% inhibition on bacteria growth compared with control groups. Data=Mean ± standard deviation, n=3; *, **, ***$p<0.05$ compared with the control group (uncoated paper) after either 24, 48 or 72 hours.

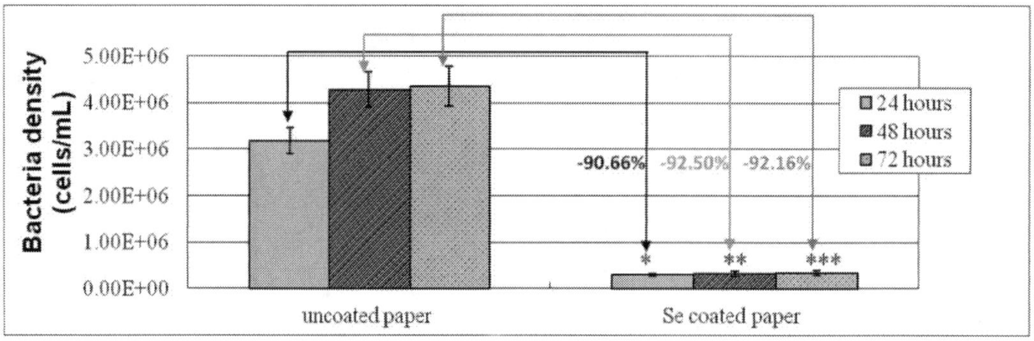

Figure 2. The growth of ***Staphylococcus epidermidis*** on the surface of paper towels. There was more than a 90% inhibition of bacteria growth compared with control groups. Data=mean ± standard deviation, n=3; *, **, ***$p<0.05$ compared with the control group (uncoated paper) after either 24, 48 or 72 hours.

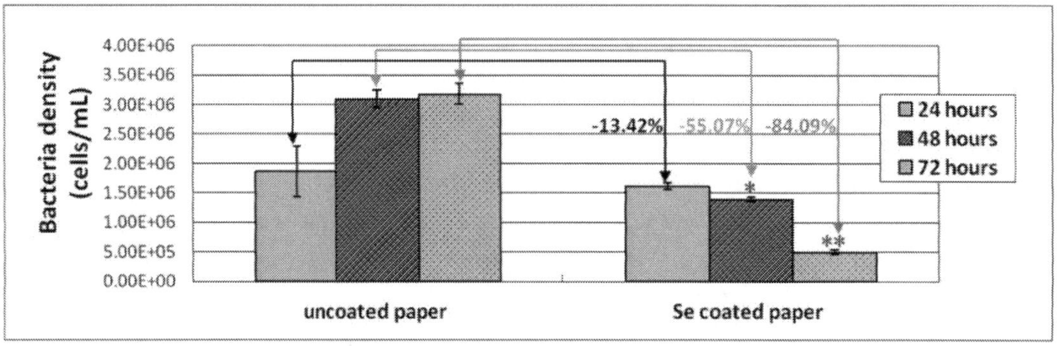

Figure 3.The growth of ***Pseudomonas aeruginosa*** on the surface of paper towels. Data=mean ± standard deviation, n=3; *, **$p<0.05$ compared with the control group (uncoated paper) after either 48 or 72 hours.

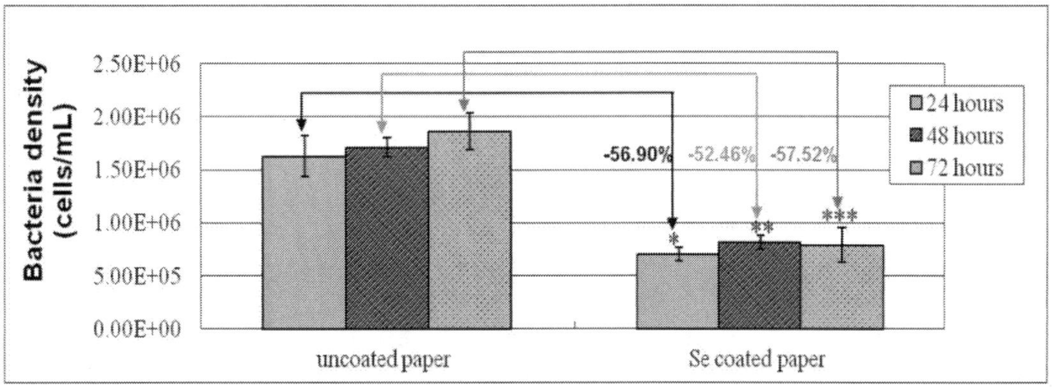

Figure 4. The growth of ***E. coli*** on the surface of paper towels. Data=mean ± standard deviation, n=3; *, **, ***$p<0.05$ compared with the control group (uncoated paper) after either 24, 48 or 72 hours.

Conclusions

Selenium nanoparticles were synthesized and coated on the surface of paper towels through a simple precipitation process which happened within one minute. Impressively, compared with the paper towels without selenium coatings, the selenium coated paper towels significantly inhibited the growth of *S. aureus*, *S. epidermidis*, *P. aeruginosa* and *E.coli* after 24, 48 or 72 hours. There were significant and continuous bacteria inhibition with about a 90% reduction from 24 to 72 hours to the growth of gram-positive bacteria including *S. aureus* and *S. epidermidis*. The selenium coated paper towels showed a relatively lower effectiveness at inhibiting gram-negative bacteria like *P. aeruginosa* and *E.coli* with about a 57% and 84% reduction, respectively, after 72 hours of treatment. Therefore, this study suggested that coating paper products with selenium nanoparticles may be an effective way to decrease various gram-positive and gram-negative bacteria growth on paper products, which might be used for potentially important applications for antimicrobial purposes in the food packaging industry and in clinical environments.

Acknowledgements

The authors thank Northeastern University for funding.

385940 Novel Tools in Genome-Scale Metabolic Flux Modeling to Identify Metabolic Engineering Targets and Predict Microbial Phenotypes

Abstract:

Several new tools have been developed that further enable genome-scale metabolic flux modeling to (i) design metabolic engineering strategies, (ii) select microbial hosts and formulate culture media that favor desired products, and (iii) provide metabolic flux predictions that more closely resemble those determined by ^{13}C-isotopomer metabolic flux analysis (^{13}C-MFA). The Flux Balance Analysis using Flux Ratios (FBrAtio) approach enables the incorporation of specific flux distributions at critical metabolite directly into the stoichiometric matrix of a genome-scale model. Here, we provide detailed instruction on how this is implemented and provide an account on how this method was used to engineer record level cellulose accumulation in the model plant *Arabidopsis thaliana*. Next, the Total Membrane Influx constrained Flux Balance Analysis (ToMI-FBA) tool will be described and used to show that (i) L-valine addition can increase isobutanol production by *Bacillus subtilis*, (ii) cellobiose addition increases ethanol selectivity in *Clostridium acetobutylicum* ATCC 824, and (iii) *B. subtilis* may be an optimal host for artimisinate production. Finally, a novel tool involving minimization of the Total Unconstrained eXchange flux by a Genetic Algorithm (TUX GA) will be described and used to show that biomass equation optimization improves metabolic flux predictions. Here, significantly improved agreement between predicted and ^{13}C-MFA determined fluxes was obtained for autotrophic and heterotrophic growth of *Synechocystis* PCC 6803 using the TUX GA approach. These results were further improved by the incorporation of FBrAtio.

393623 Sustainable Winemaking – Challenges and Opportunities

Abstract:

In the making of wine one can expect to generate 2 gallons to as high as 10 gallons of water for every gallon of wine produced. This water is purely for cleaning/sanitation since only the water contained in the grape can be added to wine. The variation is largely driven by the scarcity/cost of available water, which suggests conservation is intimately tied to economics and the economic conditions are changing at an accelerated rate due to the severity and suddenness of the drought in the western US. A more proactive approach is to consider the problem of water use reduction, treatment (including energy recovery) and synergistic use of the solids streams that exit a typical winery which will be the topic of this talk.

393625 The Challenges of Water Use Reductions and Food Safety in Dairy Processing

Abstract:

The UC Davis Viticulture & Enology Department is developing technologies necessary to realize a self-sustainable, zero carbon winery. Through the use of rainwater capture, filtration and recycling, green cleaning chemistries, Clean-In-Place technology, carbon capture, as well as use of non-carbon energy sources, UCD proposes to develop the world's first energy positive, water positive, and zero carbon winery. Advanced data systems will enable analysis and optimization of processing technologies. Our goal is to develop systems and processes that will enable the food and beverage industry to significantly reduce their water and energy requirements as well as show how process wastewater can be reduced to nearly zero.

THE CHALLENGES OF FOOD SAFETY AND WATER USE REDUCTION IN DAIRY PROCESSING

Montgomery Bohanan and Joe Herrud, Leprino Foods, Denver, CO

Overview

We will discuss the challenge of limiting water usage in conflict with food safety requirements in the dairy industry. As the food industry on a whole strives to deliver the safest food products possible, cleaning requirements and demands continue to increase every year - which is compounded by the fact that the majority of new equipment installations have greater volume demands for cleaning water/fluid. The ability to re-use water is dependent upon and limited by the Federal Pasteurized Milk Ordinance (PMO), and by the infrastructure in place throughout the industrial facilities. With improved equipment designs and the possibility of using enzymes or other cleaning methodologies that are much less water intensive, a significant reduction in water consumption could be made across the dairy manufacturing segment. Chemical engineers working on water reduction around sanitation practices will be vital to simultaneously achieving food safety and water consumption goals within the dairy industry.

Food safety is becoming a greater and greater concern each year as negative health events drive food growers, processors, and manufacturers to work harder to eliminate food-borne risks. The efforts usually result in more cleaning/sanitizing and those steps typically generate more water use. And at the same time, companies are also working hard on reducing the amount of water consumed in their operations to meet sustainability goals or to reduce costs for usage and/or waste water treatment. The availability of potable water in many arid regions (which is more typical in many of the higher dairy production areas of the country) coupled with the ability to treat the additional wastewater generated each day in the sanitation process becomes a balancing act for every organization regarding cleaning and sanitizing and food safety.

Historically, the variables used to achieve the desired level of cleanliness for a clean in place (CIP) system in any unit operation has consisted of:

- Time,
- Temperature,
- Concentration, and
- Mechanical action.

Each variable is dependent upon the others and also upon the soil type and soil load. Let's review the criticality of each of these variables and how they relate to sanitation/food safety.

Time

With enough time, a just about any cleaning strategy may be able to attain the desired soil removal and bacteriological kill. Increased cleaning time could lead to lower temperature requirements or lower cleaning chemical concentration needed. From a food safety standpoint, time is critical in microbial growth and the development of biofilms - so that should also be taken into account. And to maximize capital utilization, the amount of time the equipment is down has to be limited - because time is money. Uptime is vital. Additionally, in most dairy manufacturing processes, sanitation must begin after no longer than every 24 hours of continuous operations (as required in the PMO). Therefore, the ability to sanitize in an efficient amount time will dictate the daily throughput of each manufacturing facility. Typically cleaning time is minimized rather than maximized.

Temperature

Temperature ranges to inhibit bacteriological growth when storing food are typically below 40 degrees F or higher than 140 degrees F. Also, to achieve the necessary kill level as required during the pasteurization step in dairy operations, temperature is time dependent. As an example, raw milk must be held to 161 degrees F for a minimum of 15 seconds for proper pasteurization according to the PMO. So the higher the temperature, the shorter the amount of time needed to achieve desired bacteriological kills. Still, in the thermal process the food flow must pass through the food safety danger zone (between 40 - 140 F), typically twice. These thermal processes for storing, pasteurizing, and otherwise handling food products effect the soil load on equipment surfaces and therefore effect the cleaning and sanitizing process. Soil residues in the food safety danger zone are especially worrisome as they can become microbiological harborages. Typically, cleaning a process requires a temperature minimally 10 F higher than the processing temperature for hot processes. Additionally, the reactivity of cleaning solutions is dependent upon the energy supplied by the temperature of the solution. Temperature can also be dictated by the nature of the soil to be removed; for example in the dairy industry butterfat typically melts at 128 F so wash solutions would exceed that temperature to ease removal of the soil, and usually exceed 140 F. And on the other hand excessively high temperatures can denature or burn on soils, thus making them even harder to remove.

As stated above in the section discussing time needed for the CIP process, there is a balance between the amount of energy utilized to attain desired temperatures and the amount of time the system can be down for sanitation. Typically, cleaning temperatures are limited by energy requirements and equipment tolerance.

Concentration

Water may be the "universal" solvent, but not all soils are water soluble. To improve the efficacy of cleaning and shorten the time required to remove soils, chemicals are used in cleaning solutions dependent upon the soil to be removed and the equipment or surface to be cleaned. Alkaline and/or chlorine solutions have typically been used to remove organic soils, while acid solutions have been used to remove mineral soils by helping to dissolve or emulsify the soil into the cleaning solution.

In the dairy industry, the organic soil load typically consists of butterfat, proteins, and carbohydrates; while the inorganic soils are minerals found in milk such as calcium or minerals deposited from the water source or alkaline cleaners. The amount of cleaning time needed can be shortened with higher concentrations of caustics or acids, but there is a point of diminishing return - when the costs of the higher concentrations outweigh the time saved, and when the higher chemical concentrations begin to damage equipment or prove too costly in the waste water treatment plant.

Sanitizers used in providing bacteriological kill after cleaning are EPA registered pesticides and must be used following label instructions. The efficacy of sanitizing solutions is dependent upon time, temperature and concentration. In a CIP application, the sanitizing solution needs to make contact with the surfaces with enough exposure time to achieve bacteriological kills, and enough volume to ensure complete coverage. For no-rinse sanitizing solutions, a specific concentration range must be maintained to ensure enough chemical to effectively kill microorganisms but still not exceed the maximum level allowed for safe human consumption/exposure. In some instances, this can require significant amounts of potable water to ensure coverage and exposure time on large pieces of equipment or long or complicated line circuits with multiple valve clusters. While critical to ensuring food safety, most sanitizer chemicals prove detrimental to the waste water treatment flora, which can add significant cost to the overall operation.

Typically, chemical concentrations are limited due to usage costs and to waste water treatment and handling costs (COD).

Mechanical Action

The sanitation fluid used in the CIP process must have enough energy to make it to all the surface areas of the equipment, the ability to penetrate biomass or "soil" to be able to chemically dislodge, and be able to scour/remove all residues. The minimum velocity desired in piping systems is typically 5 to 7 feet per second to attain the turbulence needed. Laminar flow will not ensure all soil films are removed from the surfaces. As far as face velocities on other unit operations, the numbers can vary significantly. The flow of the cleaning solution in these applications typically delivers mechanical action through impingement provided by spray devices. Traditionally, supplying improved mechanical action through increased flow rates and thus increased water usage has been the accepted practice. Additionally, as companies have tried to optimize production, the size of equipment and especially piping has increased significantly. The large diameter piping requires large flow rates to fill the entire pipe to ensure coverage of all surfaces plus turbulent flow - when the needed action and contact is only on the inner surface area of the pipe. So, now with cost optimization and water conservation efforts becoming more prevalent, new and improved methods for delivering mechanical action need to be explored.

Additionally, CIP cleaning requires at least two and sometimes several water rinse steps. The initial rinse step in a CIP program is designed to remove as much of the gross soils (and easily removed soils) from the system as possible (targeting 90 - 95% of the soil load). This allows the chemicals in the cleaning solution to work off the hard the remove soils - the films and hardened soils, and can require a large amount of water if a large amount of product is left in the system. Also, there are rinses after each wash step to assist in remove the dissolved soils and to ensure there are no chemical residues left behind. Sometimes these rinses are necessary to ensure incompatible chemical solutions do not inadvertently mix and potentially damage surfaces or equipment. Also, the rinse prior to sanitizing must not only ensure removal of all soil and chemical residues so that microorganisms are not protected from exposure to the sanitizer, but the rinse also is necessary at times to ensure the temperature of the surfaces are appropriate for the sanitizer application.

Even if these variables were always accepted as key to achieving proper sanitation, they are typically incorporated into the design as an afterthought following equipment efficiency, throughput, and capital costs. Cleaning of equipment and processes is often an afterthought to operations, and are therefore difficult to cost or performance optimize. So to achieve sanitation targets in such designed systems and processes, it has been necessary to design equipment to include:

- Higher flow rates and volumes (to increase mechanical action),
- More CIP nozzles and segments (to achieve the necessary coverage),
- Higher temperatures (to maximize solution reactivity), and/or
- Higher concentrations or more effective chemicals (to increase solubility using wetting agents, surfactants, and detergency).

All of which can have a negative impact on water usage, energy usage, and waste water treatment – ultimately the cost of production.

Design for Success

If we were truly optimizing the sanitation process during the design phase of equipment or processes, the following would be considered:

- Minimize residue/biofilm create or left on equipment during operations
  - Keep material deposits to a minimum
    - Eliminate hot spots favorable to biological growth or soil degradation
    - Self cleaning velocities during processing
    - Correct surface finishes or improved surface finished to enhance soil removal
- Make equipment easier to clean
    - Don't have places for bacteria to hide
    - Surface finishes

      No dead spots
    - No ledges
    - Easily and completely drains
    - Accessibility for cleaning
- Ability to remove soils without water (Potential yield improvement)

- Dry clean
    - Line "pigs" or other air purging devices
    - New methods for purging the system or equipment
- Easy teardown to allow manual removal instead of using water
- Correct style, number, and location of CIP nozzles
    - Design equipment around a spray pattern to minimize wash solution needs instead of adding spray balls to create coverage in an existing piece of equipment
    - Utilize nozzles that maximize coverage and impingement for the volume of water used.
    - Develop new CIP impingement devises.
    - Treat CIP systems and solutions as a limited utility resource and design equipment around those limitations and availability.

Sanitation with Minimal Water or Chemicals

Just pumping more water and additional sanitation chemicals can't always be the answer for proper cleaning. What other options should be investigated to make step changes in the amount of water used?

- Alternative dry cleaning, such as:
    - Dry ice/CO2 residue removal/pretreatment
    - Air pulse or turbulent air blow
- Enzymatic cleaners to break down soils into more readily soluble forms
    - Develop/utilize enzymes that are recoverable and reusable
    - Potential for high concentrations that will not damage equipment while minimizing water usage
- Cleaning chemical recovery and re-use, such as
    - Membrane recovery of caustic/acid
    - Mineral precipitation from acid or of caustic
    - Phase separation from acid or caustic
    - Removal of sanitizing agents from water prior to traditional waste water treatment
    - Associated water recovery from the process listed above
- Different chemical combinations, such as
    - solvent based cleaners
    - acid override washes
    - nanotechology or encapsulation
- Different methods for delivering/applying cleaning chemicals
    - Automated foam application, followed by minimal water impingement
    - Automated and/or robotic brushes and/or spray devices
    - Nanotechnology or encapsulation

The majority of the cleaning techniques/strategies we've discussed are pertaining specifically to processing equipment and CIP systems. There are many other examples of cleaning requirements that each facility must tackle that deal directly with environmental conditions and water consumption at the facility. What are the other practices that can be put into place to address:

- COP (Clean-Out-of-Place) tanks
- Manually cleaning (Foam, Boosted Pressure) and rinsing
- Proper hand cleanliness
- Foot traffic exposure
- Forklift contamination
- Tool contact and cleanliness

Conclusion

Balancing food safety with water reduction is a challenging task within the dairy industry. There are existing best practices we can put in place today to help achieve reduction goals while not sacrificing food safety requirements. But to make a step change for the future, new designs and "out of the box" thinking will be needed. Who will discover the disruptive technology that will allow a reduction in water consumption while maintaining - or improving - plant hygiene and food safety? Who are the creators of "what could be"? You!

393627 Maximizing Water and Energy Efficiency in Food Processing While Minimizing Negative Environmental Impact

Abstract:

The long term health of the food processing industry depends on development of technologies and practices that maximize efficiency of water and energy use while minimizing negative environmental impact. Realizing sustainable food processing hinges on understanding how water and energy use are intertwined throughout food processing facilities down to the level of individual unit operations. The water-energy nexus (WEN) assessment is a framework for auditing all supply and demand points for water and energy within a food processing facility. The WEN analysis provides processors with a complete breakdown of how water and energy use are distributed throughout their processing pipeline and identifies targets for improving efficiency (e.g., by identifying inefficient equipment or opportunities for water and heat recovery). To demonstrate, the results of a WEN analysis from a tomato processing facility are provided. Additionally, opportunities for improving management strategies for food processing solid and liquid wastes are discussed. These include advancing technologies for biofuel production and emerging technologies for agricultural application of food processing organic residues and microbial desalination of wastewater streams.

391474 Carbon Footprint of Agrarian Systems

Abstract:

Carbon footprint is an important sustainability metric that quantifies the climate impacts (interms of carbon dioxide equivalents) of human activity. This approach is well researched, understood and established for systems at national scale (countries), and city scale (cities and towns). Several methodologies including IPCC and World Resource Institute (WRI) have developed standardized protocols for estimating carbon footprint. These protocols have been used by local governments to develop sustainability strategies with a focus on climate mitigation. However, these protocols and methodologies fail to capture the entire picture when applied to regional agrarian systems due to their unique characteristics. Agrarian systems are primarily production oriented (e.g. grow food) as against urban systems which are demand centers (e.g. Consume food and energy). In this paper we present a carbon footprint methodology that is specifically tailored for rural agrarian systems. This methodology captures unique characteristics of agrarian systems such as land use emissions. We present a case study related to the application of the developed carbon footprint methodology to San Luis Basin, Colorado. Further, the use of carbon footprint along with other metrics to assess sustainability of the region is discussed. Finally, future sustainability of the region is assessed through scenario modeling and analysis.

391194 Process Sustainability Evaluation for the Recovery of Metal from Spent Batteries

Abstract:

Sustainable development has become a challenge with the importance of environmental, social and economic dimensions requiring the integration of multidisciplinary efforts to achieve quality of life enhancement without affecting the availability and function of ecological goods and services. Contributions to these efforts should be holistic, for example through the advancement of resource (material and energy) efficiency, improved material and product recycling, life cycle considerations, etc. An important system that should be studied and improved with the goal of sustainable development is batteries. Because of their environmental and efficiency benefits, lithium batteries have been chosen as the most standard portable source energy for many electronic devices. With the expansion of the lithium ion (Li-ion) battery market, especially transportation (automotive) applications require the minimization of production costs, material use, and energy demand. In addition, because of the high abundance of spent batteries and the serious environmental, health, and safety impacts associated with their disposal, the recovery of valuable metal from used batteries has gained high relevance. Therefore, developing and proposing sustainable solutions for battery recycling and metal recovery should be prioritized.

In other words, the development of sustainable processes needs quantitative evaluation approaches to describe and demonstrate when a particular process alternative is more sustainable. Hence, the GREENSCOPE tool provides this needed quantitative assessment of sustainability by employing a set of indicators capable of transmitting and translating process performance, feedstock, utility, equipment, and output information on sustainability measurement scales. These process indicators are designed to provide assessments in the areas of environment, efficiency (material), energy, and economics.

GREENSCOPE is presented in conjunction with a sustainability performance assessment for the recovery of metals from spent lithium ion batteries. A representative processes for metal recycling was designed and modeled. A process sustainability assessment was conducted to evaluate its technical benefits and to describe how well the process under study make use of mass and energy goods and services to recover a valuable material, achieve social and environmental needs, while maximizing their economic outcomes. Therefore, GREENSCOPE is proposed as a proficient tool for evaluating and designing recycling technologies for the recovery of valuable metals and identifying sustainable opportunities for improvements.

384530 Sustainability Control: Theoretical Aspects and Engineering Methods

Abstract:

Industrial practice on sustainability has been dominantly experience or case/scenario based, which has severely limited application success when tackling complex sustainability problems. This should be imputed to the fact that sustainability science is far from exact, and there is a serious lack of science-driven, generic approaches for practicing engineering sustainability.

It is known that engineering sustainability science has three main branches: sustainability measurement science, sustainability analytics, and sustainability decision-making science. There have been a variety of studies on methodological development with case studies in each branch. It seems a more urgent research need is to establish a theoretical framework that can integrate sustainability assessment, analysis and decision making as a system and perform these tasks synergistically. As sustainability is all about development and pathways, this has determined that sustainability development (SD) is a class of non-conventional dynamic control problem, and SD problems should and can be stage-wisely solved in a controllable way. This type of research should be critical in advancement of engineering sustainability study and be profound for science-driven methodological development in the whole spectrum of sustainability studies.

In this presentation, we introduce a concept called "sustainability control". We will discuss the issues of observability, predictability, and controllability of sustainability under uncertainty. We will then propose a holistic methodology for realizing strategic-tactical control of industrial sustainability by resorting to sustainability fundamentals, system control science, uncertainty theory, and engineering science. The development should help provide scientific answers to three questions: (i) why and what kind of industrial problems should be tackled as a sustainability control problem, (ii) what kind of sustainability control system should be synthesized and how to do so, and (iii) how to derive decisions for SD of industrial systems under uncertainty? In specific, we will introduce a predictive input-output modeling method, an interval-parameter-based system and technology assessment method, a multistage, hierarchical decision-making method, and a method for designing a hierarchical sustainability control system. The efficacy of the sustainability control methodology will be demonstrated by tackling a complex chemical manufacturing sustainability problem.

386193 Statistical Analysis of Global Environmental Impact Patterns Using a World Multi-Regional Input Output Database

Abstract:

In the recent past, the study of the international channels through which impacts are traded at a global scale has gained wider interest. In this context, environmentally extended multi-regional input-output tables (EEMRIO) are widely used to assess the impact of economic activities on the environment.These models are a valuable tool to attribute pollution or resources depletion to the final demand of a product or service following a consistent approach, which makes them very useful in the development of environmental policies.

A key point in their use and, more generally, in the area of environmental engineering, concerns how the environmental performance is assessed. A plethora of environmental indicators are available for quantifying the anthropogenic damage in different categories. Among them, those based on Life Cycle Assessment (LCA) principles have recently become the prevalent approach. Many methods for impact assessment have been presented so far (in general), and some of them have been incorporated into IO models (in a more specific field). In contrast, the relationships between impacts in different damage categories are poorly understood at both, the local and global scales. Particularly, at a global scale, little effort has been devoted to the study of the environmental impact patterns of nations. The analysis of these global impact fingerprints, however, could assist in the development of more effective environmental policies in several ways.

In this work, we apply a multivariate statistical analysis to data retrieved from a multi-regional input-output (IO) table covering 69 environmental indicators classified into 5 main categories: energy, emissions, material, water and land; and 41 countries.

We find that damages in different categories and also within the same one are highly correlated. In addition, our statistical analysis reveals that countries show a very similar environmental impact pattern. While the environmental fingerprint of nations is similar, the intensity is different depending on the level of development. Our findings might help to develop more effective environmental regulations.

MAXIMIZING SUSTAINABILITY OF ECOSYSTEM MODEL THROUGH SOCIO-ECONOMIC POLICIES DERIVED FROM MULTIVARIABLE OPTIMAL CONTROL THEORY

Rohan Doshi[1,3], *Urmila Diwekar*[2,3], *Pahola T. Benavides*[2,3], *Kirti M. Yenkie*[2,3], *Heriberto Cabezas*[4,5]

[1]Princeton University, Princeton, NJ 08544
[2]Department of Bioengineering, University of Illinois, Chicago, IL 60607
[3]Center for Uncertain Systems: Tools for Optimization & Management (CUSTOM), Vishwamitra Research Institute, Clarendon Hills, IL 60514.
[4]National Risk Management Research Laboratory, Office of Research and Development, U.S. Environmental Protection Agency, Cincinnati, OH 45268
[5]Department of Computer Science and Systems Technology, University of Pannonia, H-8200 Veszprem, Hungary

Abstract

Current practices in natural resources consumption are unsustainable and may eventually lead to ecosystem extinction. Sustainable ecosystem management is necessary to ensure that human and ecological needs of the present are satisfied without compromising the ability of future generations to meet their own. This paper uses a simple mathematical model of an integrated ecological and economic system representing our planet's sectors, including, but not limited to, natural, industrial, housing, and energy production sectors. The aim of the project is to maximize the sustainability of this system, using Fisher Information as a statistical measure as a measure of sustainability, and derive socio-economic policies using optimal control techniques. By controlling six policy parameters, we were able to sustain all the ecological mass compartments (which were not sustainable in the consumption increase scenario of the future), thus significantly increasing the lifespan of all the species in the model.

Keywords: Sustainability, Ecosystem Management, Optimal Control, Policy, Fisher Information, Multivariable

1. Introduction

Due to the recent globalization, the average standard of living of the human population has improved considerably. Apart from the conventional primary needs of food, energy, clothing, and shelter, many more commodities have become essential for a stable existence. Consequently, the average per capita consumption of natural resources is rising to unprecedented levels which the environment cannot provide for indefinitely. In fact, researchers have predicted that consumption of many resources will increase by about 50% in the next 50 years (Meadows et al. 1992). The intensification of human activities and its continuous evolution has encouraged different sectors of society to explore alternatives that counteract the negative impact of human development. Therefore, sustainability has emerged as a new discipline for analysis and evaluation that has different dimensions corresponding to political, social and economic sectors in relation to human activity. When the time dimension is added to sustainability, it becomes more similar to sustainable development, which has been defined as "the development that meets the needs of the present without compromising the ability of future generations to meet their own needs" (Brundtland 1987).

Mathematical models featuring key components of an ecosystem help in the study of sustainable development. These models yield qualitative and quantitative conclusions that aid policy makers evaluate technologies and ecosystem management strategies. Often times, these models focus on specific aspects of the environment. For example, Menshutkin et al. (2014) analyze the variety of mathematical models used to understand freshwater lake ecosystems, ranging from those formulated with systems of differential equations to those implemented using fuzzy logic. Chang et al. (2014) mathematically modeled greenhouse gas emissions from Fortune 500 companies to study their impact on the environment and derive policy conclusions. And, Llibre and Xiao (2014) study the global dynamics of a 3-dimensional Lotka-Volterra system, a mathematical representation of predator-prey population dynamics. Of particular interest, the literature features many attempts to incorporate optimal control – the development of time-dependent management strategies – with different ecological models; examples include maximizing forest carbon sequestration, sustaining shallow lake environments, and optimally harvesting fish populations (Sohngen and Mendelsohn 2003; Mäler et al. 2003; Chakraborty et al. 2011). Although all these studies can help us understand and manage specific aspects of our environment, a more comprehensive model of the planet's environment – including both human and natural aspects - is required to evaluate government policy effects across the many aspects of sustainable development. To fill this need, many research projects in the field of sustainability have aimed to incorporate economic features into ecological models to reflect human interactions with nature. Brock and Xepapadeas (2010) incorporated a basic economic-ecological system in their mathematical study of the management and regulation of semi-arid systems. Expanding the scope even more, Whitmore et al. (2006) incorporated a rudimentary economy into a twelve-cell ecological model. Kotecha et al. (2012) further enhanced Whitmore's model by incorporating biofuels to depict how energy production impacts sustainability. The Whitmore model has shown that in the current state of our planet, if per capita consumption increases as predicted, our planet will not be sustainable and species will start disappearing. Building upon Whitmore's and Kotecha's framework, we propose a novel application of multivariable optimal control for an integrated model of the planet to derive socio-economic policies in order to sustain the system. Since the government needs to anticipate how the ecosystem will react to not one but an entire array of policy options, we aim to look at multiple policy options which will be controlled simultaneously over time to guide the sustainable development of the system.

Sustainable development reaches across multiple disciplinary considerations; therefore, an equally interdisciplinary theory is required for measuring model sustainability. Previous literature has suggested Fisher Information (FI), a concept from information theory, be used to construct a basic theory of sustainability for ecological applications (Cabezas and Fath 2002). According to Frieden (1998), FI is a measure of the state of order or organization of a system. In a dynamically sustainable regime, the FI remains constant over time or oscillating about a constant value. If the FI is increasing over time, the system is migrating towards a state of increasing self-organization and is in a potentially sustainable state. But if FI is decreasing, the system's loss of organization indicates that it is unsustainable or transitioning to a different regime (Cabezas and Fath 2002). Shastri and Diwekar (2006) presented a mathematical definition of the sustainability hypothesis proposed by Cabezas and Fath (2002) based on FI. They assumed a system with *n* species, and calculated the time average FI_t using Equation 1.

$$FI_t = \frac{1}{T_c}\int_0^{T_c} \frac{a(t)^2}{v(t)^4}dt \tag{1}$$

$$v(t) = \sqrt{\sum_{i=1}^{n}\left(\frac{dx_i}{dt}\right)^2} \qquad (i=1, 2\ldots n) \tag{2}$$

$$a(t) = \frac{1}{v(t)}\left[\sum_{i=1}^{n}\frac{dx_i}{dt}\frac{d^2x_i}{dt^2}\right] \qquad (i=1, 2\ldots n) \tag{3}$$

In Equation 1, T_c is the time cycle, v(t) and a(t) are the velocity and acceleration terms, defined by Equation 2 and Equation 3 respectively. Here, the terms $\frac{dx_i}{dt}$ and $\frac{d^2x_i}{dt^2}$ represent changes found in the n state variables. In this work, we use an optimal control strategy to regulate the system's FI in order to achieve sustainability. The results presented here yield quantitative and qualitative conclusions about how different combinations of controlled economic, regulatory, and technological policy options can influence individual aspects of the environment as well as the overall ecosystem.

2. Materials and Methods

2.1 Model Description and Sensitivity Analysis

The model used for this work is a simplified natural ecosystem compounded with human interactions. The model is broken down into mass compartments, which are allowed for the tracking of mass transfers. Figure 1 gives an overall sense of all interactions occurring between the mass compartments. Refer to Online Resource Section 1 for more details on the model.

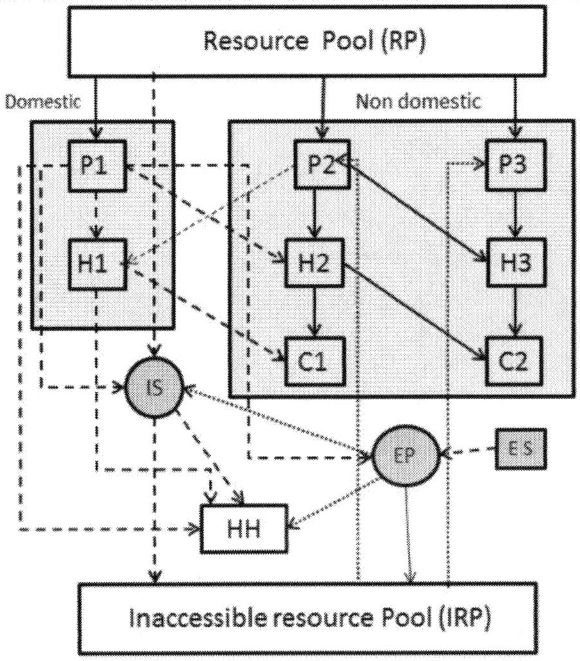

Fig. 1 Compartment of the economic-ecological model. This figure depicts the compartment model, which is comprised of : plants (P_i) which are the primary producers, herbivores (H_i), carnivores (C_i), humans households (HH), a resource pool and inaccessible resource pool (RP, IRP). The arrows represent the mass flows from one compartment (origination) to another compartment (termination), and all living compartments have an implied flow back to the resource pool that represents death. (IS) is the industrial sector, whereas (EP) and (ES) are the energy producers and the energy source compartment, respectively (Kotecha et al. 2012).

The dynamic interactions were formulated based on the Lotka–Volterra (LV) model. Usually these models are represented as differential equations, but the previous work, which formed the basis for this study, altered the LV model to be implemented through difference equations. Since the model presented here is an integrated economic-ecological model, the demand for mass from each compartment depends on a series of economic price equations. Both economic equations and ecological equations are governed by key coefficients. These coefficients address the efficiency of mass transfer and can be manipulated to decelerate or accelerate trends in the model over time. Therefore, altering parameters will impact the model's FI, and, thus, the sustainability. The model also relies on two key assumptions:

- This model is closed to mass but open to energy.
- Biofuels utilize mass from primary producers for meeting 30% of the human and industrial energy demand. The biofuel share of energy production is set to 30% to model the upper bound of biofuel energy production estimates for the future. Remaining energy needs are satisfied by conventional, finite fuel sources such as fossil fuels (Kotecha et al. 2012).

Additionally, the model is simulated for a time-span of two hundred years and can optionally be run with either the following scenarios:

- Base Case (BC): this represents a sustainable system that avoids extinction. The parameter values remain constant and there is no increase in consumption. The base case can serve as a standard for comparison since it can yield information about the dynamics of an ideal system.
- Consumption Increase (CI): certain coefficients related to economic demands for natural resources are increased as expected in the future. This causes many mass compartments to deviate from the BC and accelerate towards extinction, thus resulting in an unsustainable system.

In terms of implementation, the current model has been solved in Matlab. Kotecha et al. (2012) contains more details about this model. We included additional functions for performing optimal control in order to answer the guiding questions behind this work.

The model presented in this paper was also studied by Benavides et al. (2014). In this work, the model was represented as a network whose properties can be used to determine the controllability of the system. The controllability was determined as the minimum number of nodes that can guide the dynamics of the complete system. In that sense, they found that at minimum twelve nodes were needed to completely control the system; they argued that if the system is controllable, then, it is sustainable. Therefore, twelve parameters, all of which are able to be regulated by government policies, were explored in this work as potential candidates for the optimal control problem. Specifically, the sensitivity analysis is used to analyze each parameter's relative impact on the system's FI. For the sensitivity analysis, the values of these parameters were increased and decreased by 5%; then, its FI deviation from the base case was evaluated. The sensitivity analysis narrowed the list of candidate parameters from twelve to six. Half of the parameters were observed to have negligible effect on the system's FI, which indicates minimal impact on the system's sustainability. Despite not having twelve selected parameters to completely control the system, as the research of Benavides et al. (2014) would recommend, it was empirically found that these six parameters were sufficient enough to control the system for a finite amount of time (in this case, the model's simulated time duration of two hundred years). Table 1 in the Results and Discussion section summarizes the selected and non-selected parameters. Online Resource Section 2 also depicts the specifics of the methodology used for conducting the sensitivity analysis.

2.2 Optimal Control

Optimal control theory offers the mathematical tools necessary for steering models of engineered systems along desired pathways. It allows for the development of time-dependent strategies to optimize a system. The control problem consists of a set of mathematical equations describing a system involving time-dependent variables, also known as control variables, which can aid in controlling the performance of the system. For a given system, optimal control problems attempt to satisfy optimality criterion by determining a control law (Diwekar 2003). The optimality criterion is expressed as an objective function, which involves a maximization or minimization problem in terms of certain process variables.

In sustainable development, the challenge is to determine a mathematical objective that promotes the preservation of all aspects of the natural environment and human society. The objective can be formulated in two different ways: maximize the FI or minimize the variance of FI between the system under consideration and an ideal sustainable system. By maximizing FI, the system is pushed to a state that appears mathematically more sustainable. Alternatively, by minimizing FI variance, the system will be maintained in a sustainable state since the system's level of organization will strive to remain constant (Shastri and Diwekar 2006). We chose the latter approach.

As seen in Equation 4, the objective function minimizes the deviation of the system's FI from the FI of a sustainable base case (Shastri et al. 2008). The minimization function was chosen in order to follow the sustainable regime hypothesis, which states, "the system in a stable dynamic regime has a FI which is constant with time" (Shastri and Diwekar 2006). Furthermore, the maximization of FI may result in a regime change; although the system may be stable, the alternative system's characteristics may not be favorable to human existence. For this reason, we chose to minimize FI variance for the objective function.

$$J = \text{Min}_{DV} \sum_{i=1}^{10} (I(i) - I_c(i))^2 \quad (4)$$

Subject to: Model equations and bounds on decision variable DV (Kotecha et al. 2012).

Here, J is the objective function, $I_c(i)$ is the current FI, and I(i) is the ideal FI for a sustainable system. The objective function can be rewritten in terms of the state and decision variables. For more information please refer to Online Resource Section 3.

We use nonlinear programming (NLP) optimization methods to solve this problem. For NLP, we need to discretize the parameter space. Therefore, the total time is divided into ten equispaced time intervals. In the six parameter optimal control problems, six decision variables represent the six policy parameters selected earlier from the sensitivity analysis as they vary over time. Each decision variable has 10 dimensions to hold the optimized values of the policy parameter for each of the 10 time intervals. Since the model runs for 200 years and there are 10 time intervals, each time interval, therefore, symbolizes 20 years. It is worth noting that the decision variable's value during each time interval remains constant for the entire course of the 20 years.

The NLP optimization problem is solved in Matlab using the constrained optimization algorithm 'fmincon'. The varying values of the policy parameters in each of the time intervals are obtained and are then plugged into the model to see the effects on the ecosystem model. The results obtained from optimization are compared with the ideal case and the projected consumption increase scenario.

To put it all together, at each successive time step, a controlled parameter's value will change every 20 years to an optimized value, and, consequently, influence the system's FI. The parameter value is solved to satisfy the objective function to the optimal extent with the help of NLP. By satisfying the objective function during CI, the FI profile of the controlled parameter scenario will try to mimic the FI profile of the base case. Therefore, when examining CI, the controlled parameter scenario will follow a more sustainable trajectory than the uncontrolled parameter scenario.

3. Results and Discussion

3.1 Sensitivity Analysis

Below in Table 1, you will find a summary of the selected and non-selected parameter from the sensitivity analysis. Six of the twelve parameters were ultimately selected for optimal control. It is useful to be familiar the real world meanings, and hence the policy implications, associated with the selected parameters. The selected parameters come from all three categories of variables being explored in the model: regulatory, economic, and technological.

Table 1 Candidate policy options for the sensitivity analysis

Category	Name	Parameter	Mass-Flow	Meaning	Selected?
Regulatory Variables	Discharge Fee	d_{fee}	IS→IRP	Discharge fee imposed by government to the Industry sector for waste disposal.	Yes
	Energy Production Fee	d_{EE}^{fee}	EP→IRP	Discharge fee imposed by government to the energy producers.	No
Economic Variables	Plant Consumption	k_{P1}	P1→HH	Parameter relating the price of P1 to the demand of P1 by human households	Yes
	Animal Consumption	l_{H1}	H1→HH	Parameter relating the price of H1 to the demand of H1 by human households	Yes
	Agricultural Fee	F_{P1H1}	P1→H1	Parameter relating the price of P1 and the demand of P1 by H1	No
	Plant Material Fee	S_{P1IS}	P1→IS	Parameter relating the price of P1 to the demand of IS	No
	Industrial Energy Consumption	S_{EEIS}	EP→IS	Parameter relating the price of energy to the demand of IS	No
	Household Energy Consumption	O_{EE}	EP→HH	Parameter relating the price of energy to the demand of energy by Human household	No
	Household Biofuel Consumption	K_{EE}	P1→EP	Parameter relating the price of P1 to the demand of energy by Human household	No
Technologic variables	Theta	θ_{P1}	P1→IS	Amount of P1 required to produce a unit of industrial product IS	Yes
	Khat	$\hat{k}$	P2→H1	Constant value specified by the government. Represents the grazing of P2 by H1.	Yes
	Biomass to Biofuel Conversion Rate	λ_{biom}	P1→EP	Amount of biomass (P1) needed to produce a unit of biofuel	Yes

The sensitivity analysis of each parameters can give a basic intuition of how various compartments in the system react to changes in certain parameter values. For example, Figure 2 shows how the sensitivity analysis of the plant consumption parameter, k_{P1}, affects certain compartments (e.g. P1, H1, and C1). As seen in Table 1, the parameter k_{P1} relates the price of P1 (primary producer) to the demand of P1 by HH (human households). Therefore, increasing k_{P1} makes P1 more expensive for HH, which reduces HH demand for P1 and decelerates the depletion of P1. This is reflected in Figure 2A since the green profile, signifying a 5% increase in k_{P1}, has a slowed depletion relative to the blue profile, the base case. The green profiles in 2B and 2C green profiles indicate that herbivores and carnivores species (i.e. H1 and C1, respectively) also deplete at a slower rate relative to the base case; the slightly increased lifespan for these species can be attributed to a larger food source of P1 since P1 is not being as rapidly depleted through human consumption. However, Figure 2D's green profile shows how human per capita mass drops, indicating less P1 was available as a food source for human beings due to price increases.

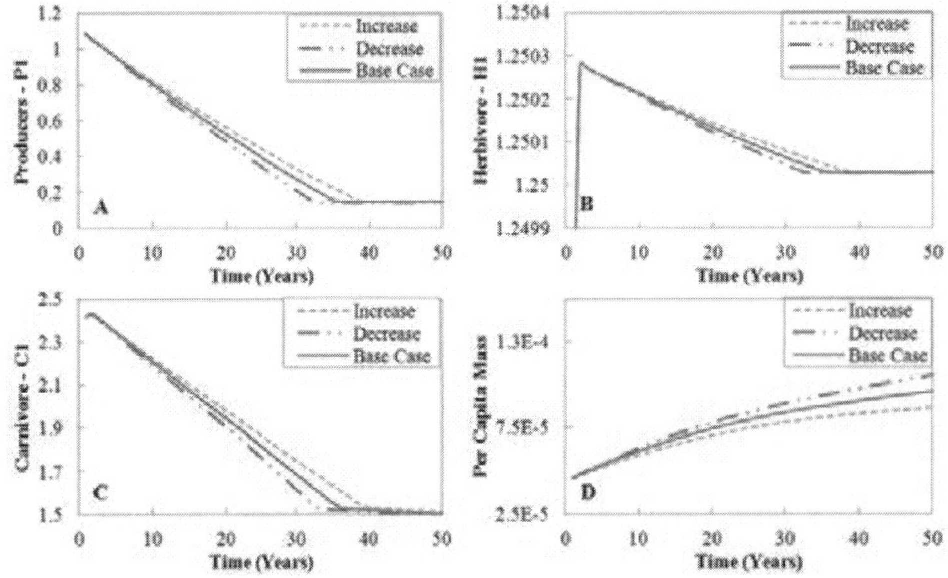

Fig. 2 Sensitivity analysis results for plant consumption (k_{P1}) without consumption increase. k_{P1} is increased and decreased by 5% to observe changes in P1, H1, C1, and per capita mass. The base case k_{P1} value is used as a standard of comparison

Looking to the real world, the conclusions derived from the sensitivity analysis provide keen insights about the relationship between human appropriation of net primary productivity (HANPP) and biodiversity. Our planet's primary producers generate a finite amount of biomass annually. This biomass provides humans with food, fiber, and fuel, but, the same biomass is a vital dietary resource for other animals and microorganisms. HANPP tells us how much of the planet's capacity to produce biomass is occupied by humans, serving as a useful metric on human alteration of the biosphere. During the last century, HANPP has doubled from around 13% to 25%. Furthermore, estimates predict that HANPP will raise to 44% globally by 2050 due to the increased usage of biofuels (Krausmann et al. 2013). The literature also indicates evidence that increases in HANPP correlate to decreases in biodiversity (Harbel et al. 2005; Rodríguez-Labajos and Martínez-Alier 2013). The reasoning matches the conclusions from the sensitivity analysis of the plant consumption parameter: if humans consume more biomass, then there is less biomass available to other species (e.g. herbivores and carnivores). To support this, by looking to the red profile in Figure 2 and examining the decrease in k_{P1}, one can see the increase in per capita mass of humans is correlated to the depletion P1, H1, and C1.

While the sensitivity analysis for other parameters can be found in Online Resource 2, there are still a few key findings worth mentioning. First, the sensitivity analysis for the animal consumption parameter (l_{H1}) suggests that increased consumption of herbivores will adversely affect carnivore populations while leaving primary producer populations unaffected. Furthermore, the sensitivity analysis for the biomass to biofuel conversion rate parameter (λ_{biom}) reveals that, when it comes to humans deciding how to utilize biomass, there is an inherent tradeoff between meeting human food and energy needs; basically, since the amount of biomass generated by primary producers in limited, allocating biomass for biofuel comes at the expense of biomass available for food, and vice versa. Applying this trend in the model to the real world provides insight on how U.S. subsidies for the production of corn ethanol may be responsible for cost increases in international food prices (Waage 2008). These discussion above reveals the type of qualitative conclusions about ecological systems a sensitivity analysis can yield.

3.2 Single Variable Optimal Control

The purpose for conducting single variable optimal control is to confirm the effectiveness of each parameter from the sensitivity analysis as a decision variable in control problems. Each of the six selected parameters from Table 1 were independently controlled and evaluated. To document an example of this process, this paper includes the results from controlling the plant consumption parameter (i.e. k_{P1}). Also, single variable optimal control is meant to serve as a basis for comparison when evaluating the impact of multivariable optimal control on sustainability.

Figure 3E compares the controlled and uncontrolled values of k_{P1}. The optimal control results suggest increasing k_{P1} for the first eighty years. Increasing k_{P1} would increase the cost for P1 by HH; therefore, humans will purchase less biomass. In the real world, this could be implemented through a tax on the purchasing on biomass. The model predicts that this policy strategy would decelerates the depletion of P1, H1, and C1, as predicted in Figure 3A, Figure 3B, and Figure 3C, respectively. To explain this, since P1 is more expensive for the first eighty years, humans will consume fewer primary producers during this time; the slightly larger plant population in the controlled scenario relative to the base case will act as an increased food source capable of supporting more herbivores and carnivores relative to the base case. Therefore, the slowed rate of depletion of P1 is responsible for the slowed rate of depletion of H1 and C1. After 80 years, the k_{P1} parameter reverts back to its default levels; the real-world policy analog would be the expiration of the biomass tax mentioned earlier. Figure 3D shows the FI for the 3 cases. The controlled parameter scenario has less deviance from the base case compared to the uncontrolled parameter scenario during consumption increase. Figure 3D, however, indicates that k_{P1} alone does not significantly increase the sustainability of the system since the FI of the controlled scenario still begins its decreases to zero after the 50th year, just like the uncontrolled scenario. Regardless of the improvements due to the controlling of the k_{P1} parameter, the system still goes extinct while running a consumption increase scenario. This motivates us to explore multivariable control.

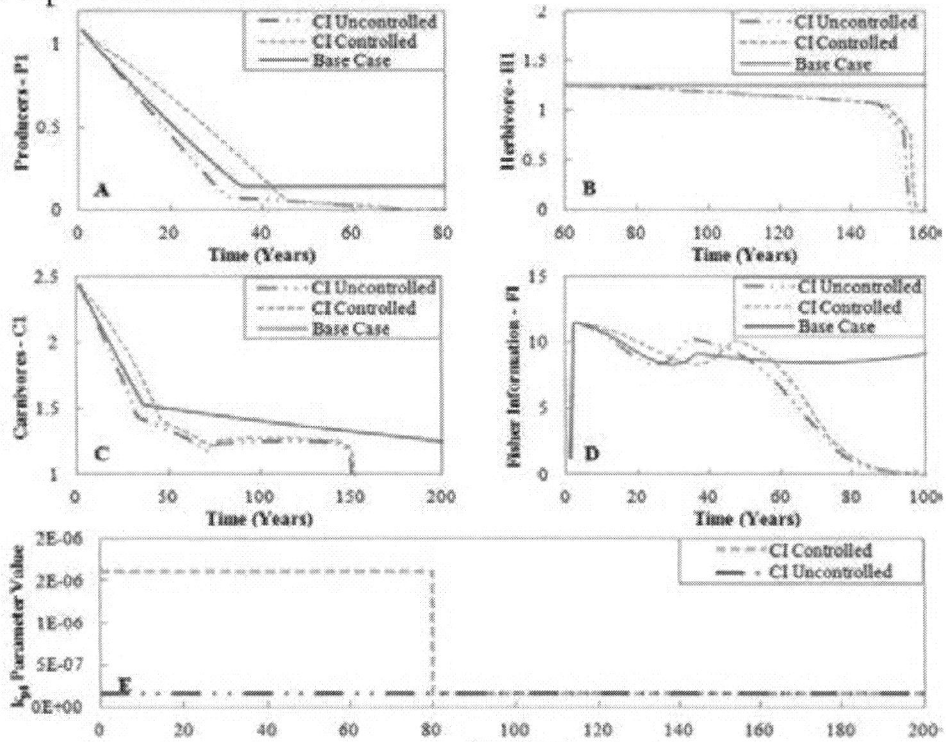

Fig. 3 Effects of controlled and uncontrolled plant consumption (k_{P1}) on P1, H1, C1, and FI. CI stands for consumption increase. The controlled parameter scenario (green profile) is aiming to emulate the sustainable base case (blue profile) and avoid extinction, a path taken by the uncontrolled parameter scenario (red profile). Also, the control profile for k_{P1} is included to show how its value varies with time relative to the constant uncontrolled parameter values.

3.3 Multivariable Optimal Control

In this section, we first present the results of the simultaneous control of three parameters: k_{P1}, l_{H1}, and λ_{biom}. As seen in Figures 4A, 4B, and 4C, there is a slight improvement from single variable optimal control since the lifespan of P1, H1, and C1 all increase by approximately 5 years. Figures 5A to 5C depict potential policy values for twenty year intervals if the government were to simultaneously implement and coordinate the three regulatory options represented by the three parameters. For example, looking to Figure 5 to determine a general policy strategy for the first twenty years, one may conclude to implement a tax increase for biomass (correlates to the increase in k_{P1}), slight tax increase for the purchasing of livestock (correlates to the minor increase in l_{H1}), and a relaxing of the standards which dictate the conversion rate of biomass to biofuel (correlates to the decrease in λ_{biom}). Still, as seen in Figure 4D, controlling three parameters is not enough to ensure sustainability during the consumption increase scenario; the green controlled profile still hits zero within a hundred years, which is only half of the model's two hundred year duration. This realization urges us to explore whether the optimization of six parameters would give enough control over the system to guide it towards sustainability for the model's full duration.

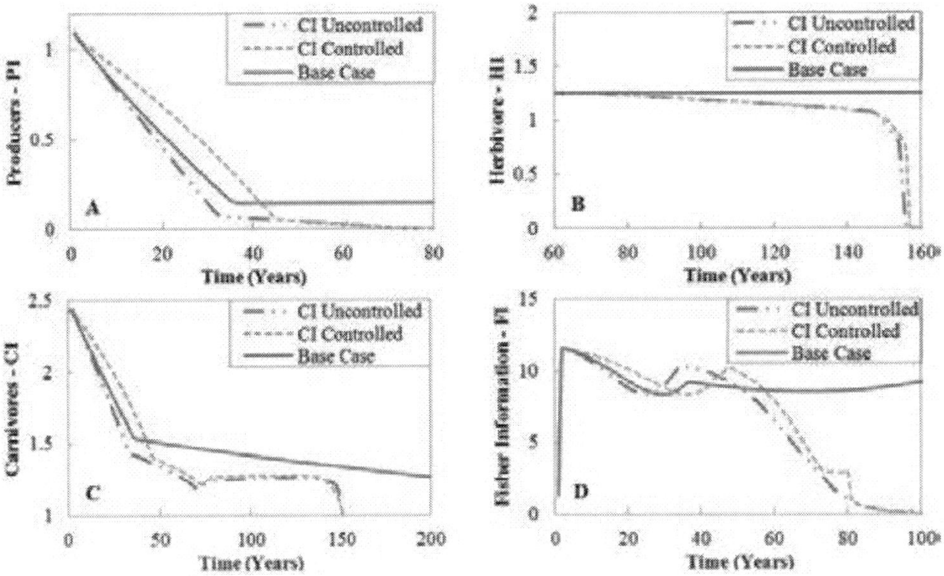

Fig. 4 Three parameter multivariable optimal control. Graphs A to D depict the effects of three parameter optimal control on P1, H1, C1, and FI. In the legend, CI stands for Consumption Increase. The controlled parameter scenario (green profile) is aiming to emulate the sustainable base case (blue profile) and avoid extinction, a path taken in the uncontrolled parameter scenario (red profile)

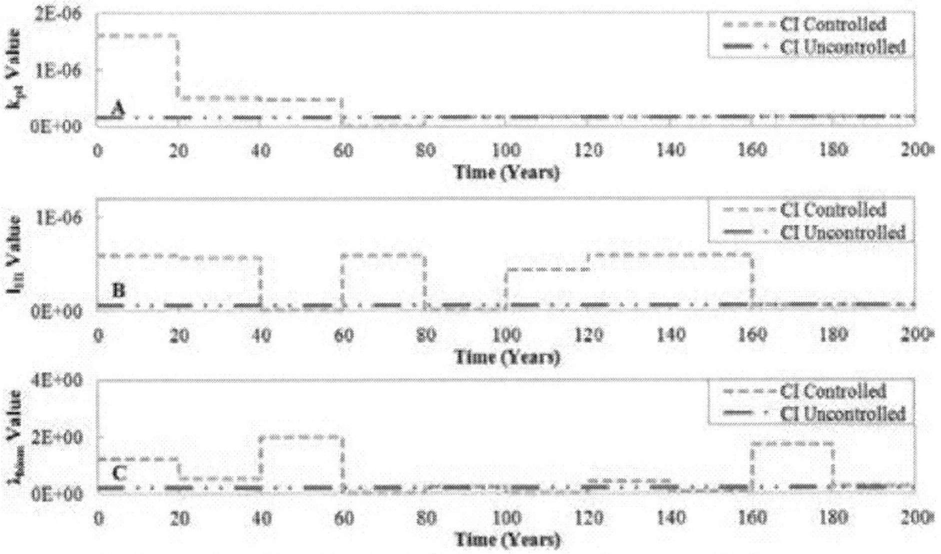

Fig. 5 Three parameter optimal control profiles. Graphs A-C show how the three controlled parameters vary with time relative to the constant uncontrolled parameter values.

Figures 6A to 6D present the results of dynamic optimization using all the six control variables. Only the most influencing compartments are presented whereas others are reported in Online Resource Section 4. Considering Figure 6D, it can be observed how the green profile is deviating from the red profile and better matching the blue profile, reflecting the extended lifespan of each compartment in the system (see Figure 6A to 6C). This indicates that optimal control helps the model to better match the behavior of the base case. Figure 7 graphs the value of the objective function, the accumulated FI variation from the base case, for the uncontrolled and controlled scenarios. By comparing the green profile to the red profile, one can see that the accumulation of FI variance was successfully minimized, proving the objective function was satisfied. Interestingly, in Figure 7, around year 70, the objective function value for the uncontrolled case dramatically began increasing since, according to Figure 6D, the model's FI approached zero, resulting in maximum deviation from the base case. Putting together all the conclusions from Figure 6 and Figure 7, one can see that minimizing FI variance by controlling six parameters sustained all the model compartments for at least 200 years despite increased resource consumption by humans – a key breakthrough. While one parameter and three parameter optimal control may not have been able to guide sustainable development for 200 years, six parameters optimal control finally managed to achieve to do so, a crucial extension of previous research on this model.

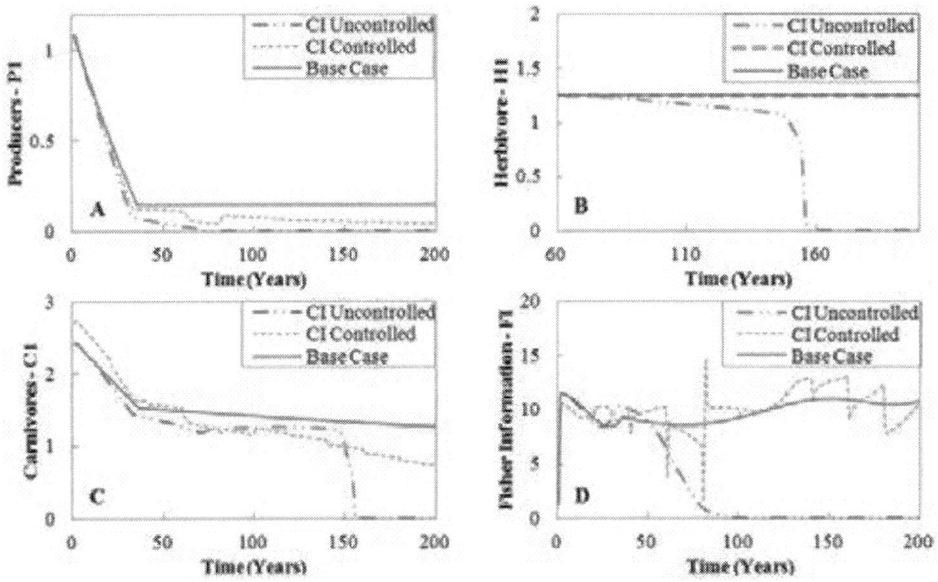

Fig. 6 Six parameter multivariable optimal control profiles. Graphs A to D depict the effects of multivariable optimal control on P1, H1, C1, and FI. In the legend, CI stands for Consumption Increase. The controlled parameter scenario (green profile) aims to emulate the sustainable base case (blue profile) and avoid extinction, a path taken by the uncontrolled parameter scenario (red profile)

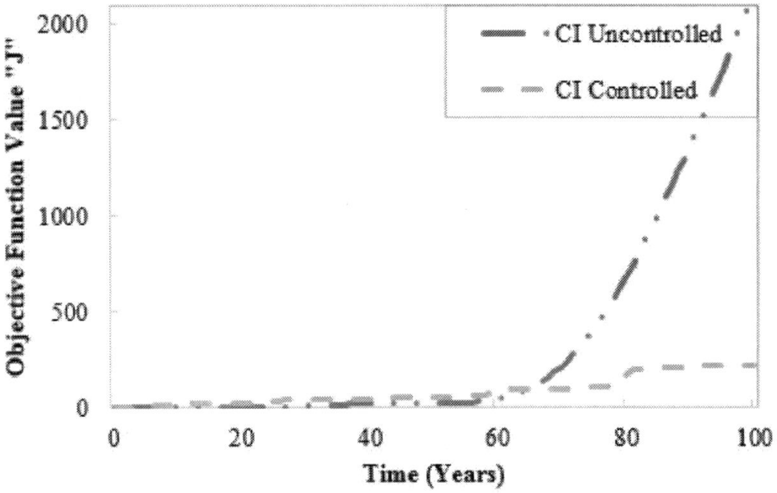

Fig. 7 Objective function profile for six parameter multivariable optimal control. The objective function increases as deviation from the base case FI value is accumulated. By comparing the green profile to the red profile, it is seen that controlling the model's six parameters significantly slows the accumulation of FI deviation from the base case.

The implementation of the control strategy of the six parameters is illustrated in Figure 8. In this figure, we can observe the fluctuation of each parameter every 20 years, which can represent a 20 year plan that the government can use. It has not escaped our notice that the parameter values change drastically over short periods of time between the time steps, which may be seen as unrealistic in a real-world policy context. Yet, the initial choice to do so helped prove the feasibility of controlling a complex system. To reconcile these two ideas, as a goal for future work, the model can be altered to facilitate a smooth change in parameter values over greater amounts of time. This will signifying the gradual adoption and implementation of government standards and legislation.

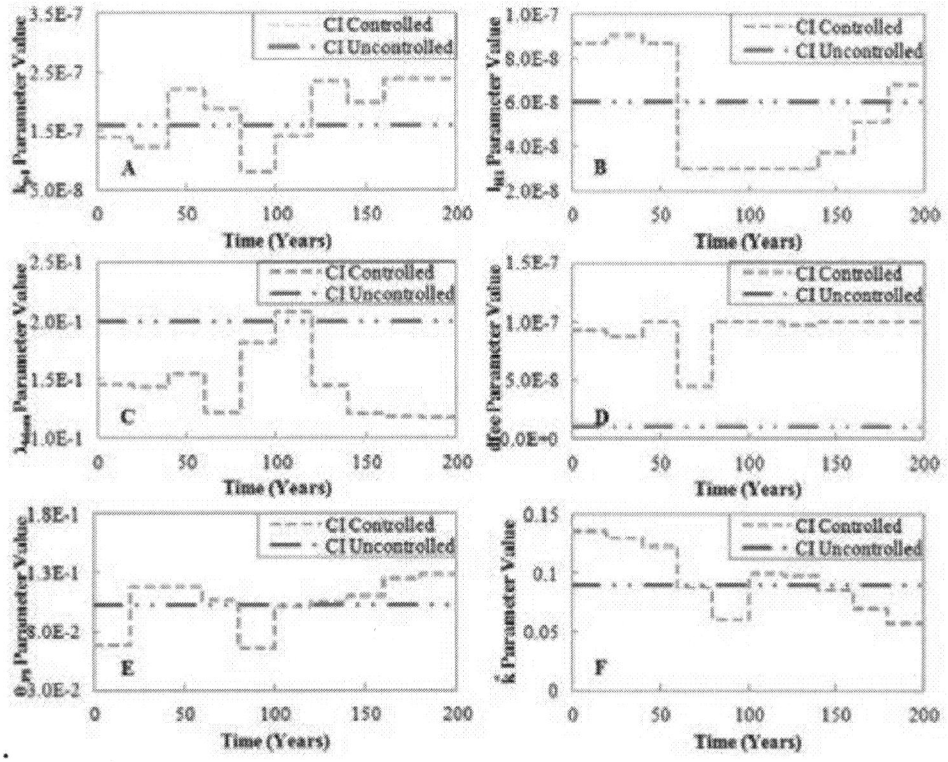

Fig. 8 Six parameter optimal control profiles. Graphs A-F show a comparison of the control profiles with the uncontrolled parameters and how their values vary over time

Comparing Figure 3D, 4D, and 6D, it is evident that having control over a greater number of decision variables (policy options) allows us to better control the FI, thus, the sustainability of the system. Additionally, we can conclude that multivariable optimal control is preferable over single variable optimal control. This reinforces the idea that there is not a single magical solution to the world's sustainability needs. Rather, a combination of regulatory strategies implemented to various degrees would be most effective.

4. Conclusion

In this work, we used optimal control techniques to guide a mathematical model of the planet away from extinction and towards a path of sustainable development for at least 200 years. We met original goals; none of the eight ecological compartments (P1, P2, P3, H1, H2, H3, C1, and C2) went extinct and growing resource demands from human society were met. This was a vital extension to previous work on this model since this is the first time multivariable optimal control has been successfully applied to achieve such long term sustainable development. We achieved this by controlling real-world policy analogs in the model: industrial waste discharge fees, biomass fees, animal fees, industrial efficiency regulation, grazing fees, and biomass to biofuel conversion rate standards. By grounding this work in real-world policy options within regulatory reach, we demonstrated how the strategic implementation of policies may have a significant impact on sustainability in practicality.

Also, we examined how the model reacted to manipulating different policy options and yielded both qualitative and quantitative conclusions relevant to sustainable ecosystem management. For example, the sensitivity analysis yielded insights into real-world policy challenges such as the trade-off between using biomass for food and fuel and the inverse relationship between HANPP and biodiversity. This work also showed how there is a critical level of control – a "tipping point" per se - necessary for guiding a complex system, such as the environment, along a path of sustainable development. While efforts to control one and three parameters failed to guide the system towards sustainable development, simultaneously controlling and optimizing six policy parameters managed to ensure system sustainability for at least 200 years. Then, models like the only developed in this work can be used by policy makers to figure out the necessary amount of system control needed for sustainable development. Ultimately, integrating sustainable ecosystem management with multivariable optimal control techniques and mathematical modeling can be useful for defining policy approaches that maintain the integrity of the environment and satisfy consumption demands.

Still, even this model is not perfect, which is why there is room for future work. As mentioned earlier in section 3.3, the model should be refined to transition between controlled parameter values gradually to reflect the time lag associated with the adoption of new government regulations. Another goal for the future is to incorporate new fuel sources such as shale gas and algae in order to further understand the role of energy production in determining decisions regarding sustainability.

5. Acknowledgements

This work was made possible and supported by VRI-CUSTOM (Vishwamitra Research Institute - Center for Uncertain Systems: Tools for Optimization & Management).

6. References

Benavides P, Diwekar U, Cabezas H (2014) Controllability of Complex Networks for Sustainable System Dynamics. Submitted to J of Complex Netw

Brock W, Xepapadeas A (2010) Pattern formation, spatial externalities and regulation in coupled economic – ecological systems. J Environ Econ Manag 59:149-164. doi: 10.1016/j.jeem.2009.07.003

Brundtland GH (1987) Our Common Future: Report of the World Commission on Environment and Development. Oxford University Press. http://www.un-documents.net/our-common-future.pdf

Cabezas H, Fath BD (2002) Towards a theory of sustainable systems. Fluid Phase Equilibria 194:3-14. doi: 10.1016/S0378-3812(01)00677-X

Chakraborty K, Das S, Kar TK (2011) Optimal Control of Effort of a Stage Structured Prey-predator Fishery Model with Harvesting. Nonlinear Anal Real World Appl 12:3452-3467. doi: 10.1016/j.nonrwa.2011.06.007

Diwekar, U (2003) Introduction to Applied Optimization. Norwell, MA

Chang DS, Yeh LT, Liu W (2014) Incorporating the carbon footprint to measure industry context and energy consumption effect on environmental performance of business operations. Clean Technol Environ Policy. doi: 10.1007/s10098-014-0785-9.

Haberl H, Plutzar, C, Erb KH, Gaube V, Pollheimer M, Schulz NB (2005) Human appropriation of net primary production as determinant of avifauna diversity in Austria. Agric Ecosyst Environ 100:119-131. doi: 10.1016/j.agee.2005.03.009

Frieden BR (1998) Physics from Fisher information: A Unification. Cambridge, UK

Kotecha P, Diwekar U, Cabezas H (2012) Model-based Approach to Study the Impact of Biofuels on the Sustainability of an Ecological System. Clean Technol Environ Policy 15:21-33. doi: DOI 10.1007/s10098-012-0491-4

Krausmann F, Erba KH, Gingricha S, Haberla H, Bondeaub A, Gaubea V, Lauka C, Plutzara C, Searchingerd TD (2012) Global human appropriation of net primary production doubled in the 20th century. Proc Natl Acad Sci. 110:10324-10329. doi: 10.1073/pnas.1211349110

Mäler K, Xepapadeas A, Zeeuw AD (2003) The Economics of Shallow Lakes. Environ Recour Econ 26:603-624. doi: 10.1023/B:EARE.000000735

Meadows DH, Meadows DL, Randers J (1992) Beyond the Limits: Confronting Global Collapse, Envisioning a Sustainable Future. Post Mills, VT

Menshutkin VV, Rukhovets LA, Filatov NN (2014) Ecosystem Modeling of Freshwater Lakes (Review): 2. Models of Freshwater Lake's Ecosystem. Water Resourc 41: 32-45. doi: 10.1134/S0097807814010084

Rodríguez-Labajos B, Martínez-Alier J (2013) The Economics of Ecosystems and Biodiversity: Recent Instances for Debate. Conserv Society 11:326-342. doi: 10.4103/0972-4923.125744

Shastri Y, Diwekar U (2006) Sustainable ecosystem management using optimal control theory: Part 1 (deterministic systems). J of Theor Biol 24:506-521. doi: 10.1016/j.jtbi.2005.12.014

Shastri Y, Diwekar U, Cabezas H (2008) Williamson, J. Is Sustainability Achievable? Exploring the Limits of Sustainability with Model Systems. Environ Sci Technol 42:6710-6716. doi: 10.1021/es800661x

Sohngen B, Mendelsohn R (2003) An Optimal Control Model of Forest Carbon Sequestration. Am J Agr Econ 85:448-457. doi: 10.1111/1467-8276.00133

Waage E (2008) U.S. Energy Policy Response to International Food Prices: The Costs and Effects of Corn Ethanol Production. Politics and Policy 36:792-804. doi: 10.1111/j.1747-1346.2008.00135.x

Whitmore HW, Pawlowski CW, Cabezas H (2006) Integration of an Economy under Imperfect Competition with a Twelve-Cell Ecological Model. United States Environmental Protection Agency. http://nepis.epa.gov/Adobe/PDF/P100778R.pdf

Understanding Resilience of Metro Systems in Polycentric Megacities: A Case Study of Delhi Metro Rail System

Metro systems have greatly benefitted metropolitan areas, both economically and environmentally. With the emergence of megacities (cities with population of over 10 million) around the world, metro systems begin to play an even more important role. By 2050, it is estimated that two-thirds of the world's population will live in urban areas. Rapid urbanization has boosted the demand for transportation, and further strained existing fragile transportation infrastructure in megacities, especially in developing countries. Additionally, studies have demonstrated that spatial organization of megacities impact the performance of urban transportation systems. Megacities with monocentric spatial organization have majority of employment opportunities, public services and transportation infrastructure located in the city center. On the contrary, polycentric organization is characterized by gradual decentralization of people and jobs from the central business district to less densely developed suburbs. While megacities in developed countries are relatively more monocentric (such as Tokyo or Paris), emerging megacities in most developing Asian and African countries grow in a polycentric fashion. In such polycentric megacities, random commuting pattern during peak hours, as well as poor metro connectivity in suburbs, pose major challenges for building resilience in metro systems.

In order to understand resilience of metro systems in polycentric megacities, we analyze the metro network structure and model the dynamics of urban spatial structure, by integrating graph theory approach with geospatial analysis. Here, resilience, an emergent property of a system, is defined as the metro system's ability to maintain its structure (overall connectivity of stations) and function (timely transportation of passengers) in the face of disruptions. In this study, we specifically look at the Delhi Metro Rail System (DMRS) in India, since it is a good example of an expanding complex metro network in a polycentric city of a developing country. Construction of DMRS is divided into four phases, with the third phase in progress. The total length of the network on completion is estimated to be 413 km (256 mi) of which 190 km (118 mi) is already in use.

We model DMRS as both, an unweighted-undirected network (stations are nodes and rail connection between them are edges), and a weighted-directed network (stations are nodes and passenger footfall represents the weight for each node). To begin with, we want to identify topological properties of the DMRS network at each construction stage. To do so, we use statistical methods such as frequency degree distributions, and perform robustness analysis by observing the impact of systematic node removal on the overall network structure. Subsequently, to identify specific sources of structural and functional vulnerability in DMRS network, we apply disruptive scenarios. We apply *station* and *line* disruptions to determine nodes that are extremely critical for the metro system, and whose removal has the largest impact on the overall network efficiency. Lastly, we develop a DMRS-geospatial model that includes information on ridership between stations, the physical distances between the stations and the neighborhood data on distribution of employment opportunities across the region. Using this geospatial model we identify vulnerabilities of specific stations in the context of polycentric urban structure of the Delhi region.

Resilience is a function of node-level and network-level vulnerabilities. Insights from our results are used to understand resilience of metro systems. Both statistical assessment of network topology and robustness analysis suggests that the DMRS network increasingly exhibits scale-free-like property with every construction stage. Results for analysis of the disruptive scenarios and the geospatial model indicate that metro stations that exhibit greater structural flexibility, in terms of the number of alternate routes between any two metro stations, and functional flexibility, in terms of the ability of passengers to be flexible with their routes, are less vulnerable to disruptions. The analysis also describes the implications of these results for the design of resilient metro networks, and highlights specific network features essential to build resilience of metro systems in polycentric cities.

Several industrial processes have rendered soil and groundwater contaminated in thousands of sites due to detrimental storage and disposal practices. These practices have resulted in the combined presence or co-existence of organic and inorganic contaminants at most sites. In this paper, the synergistic and antagonistic effects of trichloroethylene (TCE) and hexavalent chromium, Cr (VI) on the degradation of each contaminant were explored by simultaneous removal via redox reactions in the same system.

The extent and kinetics of trichloroethylene degradation using $KMnO_4$ in artificially spiked soil and aqueous systems were investigated. At molar ratios of 1:3 (stoichiometric), 1:5, 1:10 of TCE to $KMnO_4$, the 62.5%, 100%, and 100% of TCE were oxidized at neutral pH, respectively in aqueous systems. Under extreme pH conditions of 2.8 and 12 for stoichiometric molar ratio, 63.75% and 82.25% yield was achieved in a 3-hour time to equilibrium. In soil systems, the duration required for equilibrium degradation was longer with 62.8%, 96% and 100% conversions in 3 days. Also, the reductive transformation of toxic hexavalent chromium, Cr (VI) to non-toxic trivalent, Cr (III) using ferrous ion was performed. In a 3-hour period, 43.97% and 84.21% degradation was obtained for a ratio of 1:3 (Cr^{+6} to $FeSO_4$) and 1: 6 respectively in aqueous systems. This is fast reaction along with the precipitations of ferric ion and Cr (III) colloid in the system which ultimately quench or slow the reaction significantly. Under strong alkaline conditions, the degradation was more effective. However, pH variation does not largely impact the overall reaction at equilibrium. In soil, the conversions of 73% and 91.9% were achieved at molar ratios of 1:10 and 1:15 respectively in 3 hours. The result indicated that TCE oxidation by $K_2Cr_2O_7$ is not impacted by the presence of Cr (VI), but the reduction of Cr (VI) by ferrous ion was influenced greatly by the presence of TCE. Significance of these findings will be discussed in the presentation.

384040 Investigation of the Extraction and Recycle Potential of CCA Treated Wood Waste

Abstract:

Since the early 1970's, the most widely used preservative in the wood preservation industry has been chromated copper arsenate (CCA) treatment. Current estimations suggest that there may be as much as 240 million tons of CCA wood waste available for Disposal by the year 2020.

Until recently, landfilling the out of service materials was a generally accepted method of disposal. Recently, problems with soil and groundwater contamination have arisen, and the role of CCA impregnated wood waste in the matter has been confirmed. As a result, the need for an efficient and effective method of heavy metal separation form wood waste has become eminent.

The focus of this research is to recover the energy value from the wood while offering a recycling opportunity for the CCA components of the wood. In this research, the removal of the copper, chromium and arsenic from the CCA treat wood was examined using pyrolysis. Mass balances were conducted to determine if the metals were found in the pyrolysis char or in the pyrolysis liquid fraction. In addition, the metal contaminated char was subjected to electrokinetic treatment (EK) and the post treated char was examined. Result show that pyrolysis is an effective method for removing the Arsenic and some of the chrome from the char. By treating the char with EK the removal of almost all of the metals contaminates could be extracted from the char. The ultimate goal of this research was to evaluate pyrolysis and EK as a combine treatment process for recycle. The full scale potential of this process will also be discussed.

Development of a new paradigm in biochemical engineering: predicting the genetic regulation of aromatic pollutants degradation

Argyro Tsipa[1,2], Michalis Koutinas[3], Athanasios Mantalaris[1], Efstratios Pistikopoulos[2]

1. Biological Systems Engineering Laboratory, Imperial College London South Kensington Campus, London SW7 2AZ, UK
2. Centre of process Systems Engineering, Imperial College London South Kensington Campus, London SW7 2AZ, UK
3. Department of environmental science & technology, Cyprus University of Technology, Cyprus

Keywords: toluene degradation, genetic regulation, chromosome, TOL plasmid

In this work, we study *Pseudomonas Putida* which is a metabolically versatile soil bacterium. Its significance lies not only on the fact that it can metabolise numerous industrially important aromatic compounds (Pieper *et al.*, 2004) but on its biotechnological potential by producing a series of fine and bulk chemicals, resulting in a growing interest of studying specific metabolic pathways at the genetic expression and regulation levels (Ballerstedt *et al.*, 2007). This bacterium has several strains, but we focus on mt-2 because it degrades efficiently **toluene**. This strain can harbour TOL plasmid which specifies a metabolic pathway for toluene and m-xylene (Timmis, 2002). Toluene and m-xylene are aromatic pollutants belonging to the BTEX (Benzene-Toluene-Ethylbenzene-Xylene) group and their **degradation** leads to Krebs cycle intermediates which are essential for the biomass growth. In *P. Putida* mt-2, toluene oxidative catabolism occurs **through the chromosomal and TOL plasmid (pWW0) metabolic pathways**. TOL plasmid is a paradigm of global and specific genetic regulation due to the interactions between DNA-bending proteins, a set of sigma factors and the regulators encoded in it (Aranda-Olmero *et al.*,2005). Our objective is to investigate the gene expression and regulation of toluene in *P. Putida* mt-2 though the above metabolic pathways.

In TOL plasmid, the entrance of toluene activates xylR and xylS genes which encode the regulatory proteins XylR and XylS respectively. The activation of these proteins is the precursor of the expression of upper and meta operon of TOL plasmid. When expressed, these operons encode the necessary catalytic enzymes for the further degradation of toluene to Krebs cycle intermediates. The four transcriptional units mentioned, are controlled by four promoters, Pr, Ps, Pu, and Pm respectively (Ramos *et al.*, 1997).

In particular, Pr promoter boosts the expression of xylR gene which encodes XylR protein. The induction of toluene to the bacterial cell culture system provokes a series of regulatory events. XylR becomes a down-regulator for Pr promoter and up-regulator for the activation of Pu and Ps promoters. These promoters boost the expression of upper operon and xylS gene respectively. Afterwards, the enzymes encoded by the upper operon, catalyse the oxidative reaction of toluene to benzoate. In addition, XylS protein is encoded by xylS gene. Both XylS protein and benzoate activate Pm promoter which boosts the expression of meta operon. The enzymes encoded by the meta operon enable the further catabolism of benzoate to Krebs cycle intermediates.

Besides that, benzoate is the compound which activates the chromosomal pathway. In this pathway the transcriptional unit is benR gene which is driven by PbenR promoter. More specifically, PbenR promoter boosts the expression of benR gene which encodes BenR protein. With the appearance of benzoate in the biosystem, BenR becomes one of the up-regulators of Pm promoter of TOL plasmid's meta operon. Therefore **a cross-talk between the two pathways is observed** (Cowles *et al.*, 2000).

A more detailed genetic representation can be found in figure 1.In this figure the genetic information are presented in a genetic circuit which is like an electric circuit based on biological information (Weiss *et al.*, 2003).

A.

B.

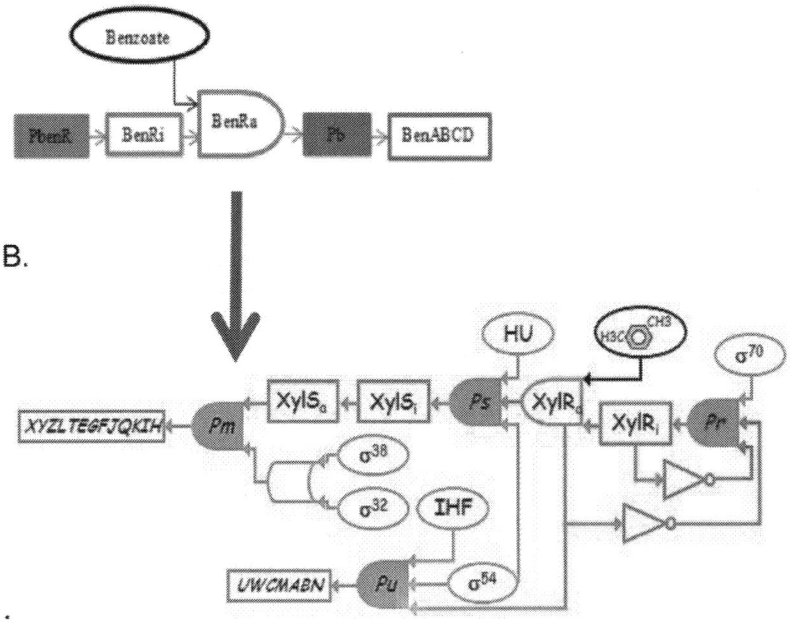

Figure 1: Genetic circuit of: A. the specific chromosomal pathway, B. the TOL plasmid metabolic pathway (Koutinas et al., 2011) and the up-regulation of Pm promoter of TOL plasmid by BenR protein of chromosome, i: inactivated form of protein, a:activated form of protein. :input, :ouput, :AND, :NOT

From experimental point of view, we measured the behaviour over time of the five promoters, since they are the driven power of the key transcriptional units' of our biosystem. The activity was measured via RT-PCR. Different concentrations of toluene were used, ranging between 0.4-1.3mM. In addition to that, toluene consumption and biomass growth were measured over time. The macro- and genetic- scale results in 1mM of toluene induction are presented in figure 2 below.

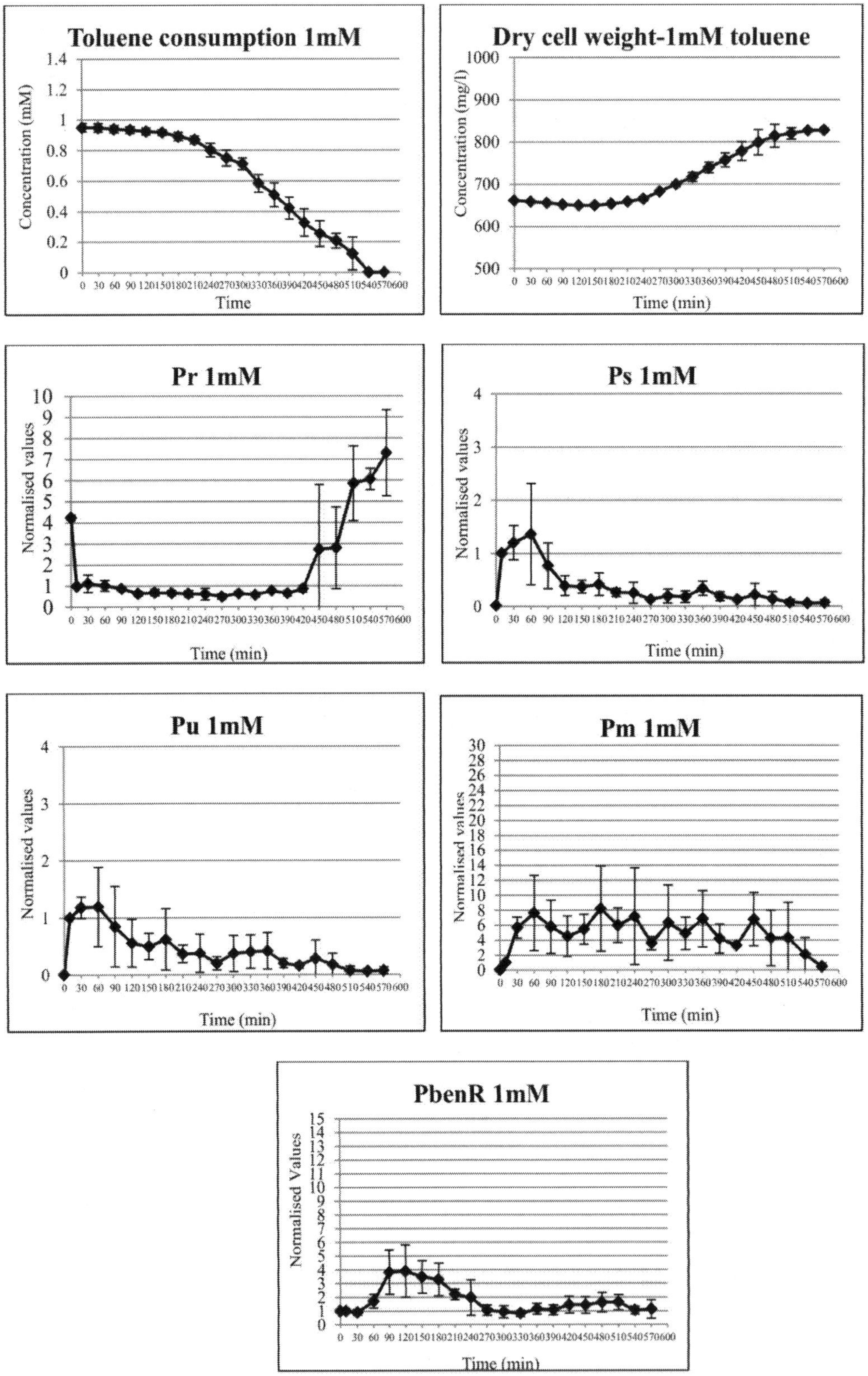

Figure 2: 1mM toluene consumption, dry cell weight (biomass growth), and the normalised levels of mRNA of Pr, Ps, Pu, Pm and PbenR promoters over the time until the depletion of toluene in the cell culture.

The results confirm the theoretical background. Pr's activity is kept in a basal level since is not directly influenced by the substrate entry, while Ps and Pu activity starts to increase when toluene is induced to the biosystem. Pr's activity increase towards the end of the process, can be attributed to the end of XylR down-regulation. After toluene's catabolism to benzoate, Ps and Pu activity decreases significantly. Pm's activation occurs with a delay as toluene has to be catabolised to benzoate in order Pm to be activated. Moreover, PbenR's activity is in a basal level until benzoate appears in the cell culture. Then, its activity starts to increase reaching a peak and decrease thereafter reaching the basal level. Pm's activity although reaching a peak and decreasing thereafter is kept in high levels compared to the other promoters' activity until the end of the process since the enzymes encoded by the meta operon are responsible for the further catabolism of benzoate to Krebs cycle intermediates. In addition, the differentiation of **Pm** among the promoters' behaviour in terms of the level of activation can be attributed to the fact that it is **activated by XylS and BenR protein**, pointing out the effect of the interaction between TOL plasmid and chromosome.
The experiments indicate that there is a close relationship between the key genes and the catabolism of toluene. Therefore, it would be extremely interesting to study the link between the intracellular information and macroscale observations. Towards this direction, a hybrid growth kinetic model has already been developed (Koutinas et al., 2011) using the same microorganism and studying m-xylene degradation in TOL plasmid. Our intention is to upgrade the existing model by integrating the chromosomal genetic information obtained through toluene entry in the biosystem. If we achieve capture of the experimental data by the upgraded model then its predictive capability will be tested revealing the connection of accounting genetic regulation to aromatic pollutants degradation.

References

Aranda-Olmedo, I., Marin, P., Ramos, J.L., and Marques, S., 2006. Role of the ptsN gene product in catabolite repression of the Pseudomonas putida TOL toluene degradation pathway in chemostat cultures. Appl. Environ. Microbiol. 72, 7418–7421.

Ballerstedt, H., Volkers, R.J.M., Mars, A.E., Hallsworth, J.E., dosSantos, V.A.M., Puchalka, J., vanDuuren, J., Eggink, G., Timmis, K.N., de Bont, J.A.M., Wery,J., 2007. Genomotyping of Pseudomonas putida strains using P. putida KT2440-based high-density DNA microarrays: implications for transcriptomics studies. Appl. Microbiol. Biotechnol. 75, 1133–1142.

Cowles C.E., Nichols N.N. and Harwood C.S., 2000. BenR, a XylS Homologue, Regulates Three Different Pathways of Aromatic Acid Degradation in Pseudomonas putida. Journal of Bacteriology, 182 (22), 6339-6346.

Koutinas, M., Kiparissides, A., Silva-Rocha, R., Lam, M., Martins Dos Santos, V., de Lorenzo, V., Pistikopoulos, E., Mantalaris, A., 2011. Linking genes to microbial growth kinetics – an integrated biochemical systems engineering approach. Metabolic Engineering, 13 (4), 401-413.

Pieper, D.H., Santos, V.A.P., and Golyshin, P.N., 2004. Genomic and mechanistic insights into the biodegradation of organic pollutants. Curr Opin Biotechnol 15: 215–224.

Ramos, J.L., Marques, S., Timmis, K.N., 1997. Transcriptional control of the Pseudomonas TOL plasmid catabolic operons is achieved through an interplay of host factors and plasmid-encoded regulators .Annu. Rev. Microbiol. 51, 341–372.
Timmis, K.N., 2002. Pseudomonas putida: a cosmopolitan opportunist par excellence. Environ.Microbiol.4,779–781.
Weiss, R., Basu, S., Hooshangi, S., Kalmbach, A., Karig, D., Mehreja, R., Netravali, I., 2003. Genetic circuit building blocks for cellular computation, communications, and signal processing. Nat. Comput. 2, 47–84.

357336 Studies on Evaluation of Acute Toxicity Factor for Selected Industrial Effluents Using Zebra Fish(Danio Rerio) As a Test Organism

Abstract:

Man in his attempt to lead a comfortable life, extracts all the resources from mother earth. In this process of utilization of useful materials he lets into the environment many hazardous substances, unaware of their effect on his own life. The rate of discharge of these materials into the ecosystem is more than that of their degradation by natural process, resulting in the accumulation of such substance in the environment. Accordingly the environment in the 21st century is becoming increasingly polluted; mainly water pollution has become a major threat to the existence of living organisms in aquatic environment. A huge quantity of pollutants in the form of domestic, agricultural run offs and industrial effluents are discharged directly or indirectly into the water bodies, which has severe impact on its biotic and abiotic environment. Before discharge of waste waters/effluents into the recipient, it is necessary to purify them with adequate treatment in a way and to an extent that does not threaten natural process and does not reduce the possibilities of their multi-functional use.

The aim of this study is to ascertain performance evaluation of Effluent Treatment Plants (ETP) of three different industries namely Steel sector, Pharma sector and Power sector with special reference to toxicity reduction using Zebra Fish as test organism. The 48-hrs acute toxicity factor T_F was introduced by the statutory authorities in India, for evaluation of industrial effluents and wastewaters is a number on geometrical scale by which the samples need to be diluted to ensure hundred percent survivals of fish under specified conditions. This bio-assay test can be considered as useful analytical tool for screening of chemical analysis and early warning system to monitor the different operational units of wastewater treatment plants. To evaluate the 48-hrs acute toxicity factor T_F, samples were taken from different operation units of ETP which includes physico-chemical units & biological unit, etc. This evaluation showed that improvements are still needed to increase their efficiencies for treating effluents. The results clearly indicate that, bio assay tests provide additional information on the toxicity potential of industrial effluent discharges and effectiveness of existing effluent treatment techniques in reduction of toxicity. So results of the present study clearly showed that physico-chemical parameters alone were not sufficient in obtaining reliable information on treated wastewater toxicity and that toxicity tests must be performed in order to guarantee the safety of aquatic organisms along with the pubic expectations. The combined use of toxicity and chemical parameters appears to be need in the quality control of wastewaters.

Key Words: Effluent Treatment Plant (ETP), Zebra Fish, Acute Toxicity Factor (T_F), Treatment Efficiency, etc.

Integrated Solar Photocatalytic Oxidation and Biodegradation for Degradation of Antibiotics and Endocrine Disrupting Compounds in Wastewater and Water

Jihyun R. Kim and Eunsung Kan. Department of Molecular Biosciences and Bioengineering, University of Hawaii at Manoa, Honolulu, HI 96822, USA

Combination of solar light-driven photocatalytic oxidation with subsequent biodegradation demonstrates high potential for treating wastewater and water containing pharmaceutical and endocrine disrupting compounds. The long-term and low-dose exposures of aquatic organisms and humans to pharmaceutical and endocrine disrupting compounds in the environment can lead to adverse impacts such as endocrine disruption, toxicity, and antibiotic resistance.

Although various advanced oxidation processes have been applied to degradation various pharmaceutical and endocrine disrupting compounds in wastewater and water, these processes are usually expensive and may lead to more toxic intermediates and byproducts. Thus, combination of advanced oxidation with biological treatment is proposed as a cost-effective treatment option to achieve high removal and mineralization of these emerging contaminants. In this study the photocatalytic degradation of sulfamethoxazole (antibiotics, SMX) and bisphenol A (endocrine disruptor, BPA) were carried out in an aqueous suspension using a biochar-supported TiO_2 catalyst under solar light irradiation. The photocatalytic oxidation of SMX and BPA transformed these biocalcitrant contaminants into much more biodegradable products while eliminating toxic functional groups in these contaminants. Subsequently, the oxidative products of BPA and SMX were completely degraded by the bacteria isolated from different soils at University of Hawaii. Recent results showed that the solar photocatalytic oxidation of SMX and BPA showed the high removal of SMX and BPA (i.e., 93-98%) and their oxidative products (i.e., 40-80%) with 6-8 h irradiation of solar light. Besides, the following biodegradation by the isolated bacteria resulted in further degradation of the oxidative products of SMX and BPA to achieve complete mineralization of SMX and BPA in water. The biotoxicity tests also confirmed that the oxidative products of SMX and BPA had negligible toxicity and consisted of mostly nontoxic and biodegradable products.

Based on these results, combination of solar photocatalytic oxidation with biodegradation could be a sustainable and cost-effective technology for the remediation of biocalcitrant and emerging contaminants in wastewater and water.

ENHANCED COMBUSTION OF RICE HUSK FOR POLLUTION ABATEMENT AND REGIONAL POTENTIAL OF SILICA FROM RICE HUSK ASH

Muhammad Suleman Tahir, Chemical Engineering, University of Gujrat, Gujrat, Pakistan.

ABSTRACT

To meet the energy requirements of industry, non-conventional sources of energy may be utilized for low cost energy solution. Rice husk feed boilers produces ash which contains 80% silica and 10-20% unburned rice husk. The ash is usually disposed-off as land filling and causing environmental issues. The disposal and transportation of rice husk ash is a challenge for regional industries in Pakistan. This research paper studies the reduction of ash formation from state of the art locally fabricated boilers and the potential of silica separation in Gujranwala division of Pakistan. The present research also investigates the effect of rice husk particle size on unburned rice husk and ash formation. The result shows that the combustion of crushed (50%) and pure rice husk produces less unburned rice husk with low ash formation and heating yield is also increased. A decreasing trend of unburned carbon is also observed from furnace with specific size of crushed rice husk and silica separation from rice husk ash is more feasible with this size.

INTRODUCTION

Rice is one of the main sources of food for the world population, especially in East and South East Asia and its support half population of the humankind. Each year the production of rice is approximately 600 million tonnes [1].

Pakistan is an agricultural developed country and average production of rice is 5.94 million tonnes each year since 2005-2011 as shown in Table 1. During the milling process, the outer cover of rice grain is removed which is known as rice husk. Rice husk is a light material having packing density of 122 kg/m^3 [2]. It has uniform size and irregular boat-like shape. It consists of high volatile and ash contents because of organic materials [3]. Rice husk is stored in an open atmosphere and therefore moisture contents may vary depending on the ambient conditions. Rice husk production data in Pakistan is also shown in Table 1.

Table 1. Yearly production of rice husk in Pakistan

Year	Production (1000 tonnes) paddy	Rice Husk (1000 tonnes)
2006-2007	5438.4	1631.52
2007-2008	5563.4	1669.02
2008-2009	6952.0	2085.6
2009-2010	6882.7	2064.8
2010-2011	4823.3	1446.99
5 year average	5931.96	1779.58

Depending upon the variety of the rice, the average yield of rice husk from rice may vary from 14% to 35% by weight [4], which could produce about 120 million tonnes rice husk globally (based on 20% by weight). Pakistan alone generated about 1.1 million tonnes of rice husk in year 2001-2002. Because of this high yield in quantity, it becomes very important to focus on its utilization but still it is treated as an agricultural waste. In developing countries, rice husk is disposed-off at landfill sites, used in bricks kiln or burnt which causes air pollution and environmental issues.

Pakistan is facing energy crisis since last few years, and one of the reason is the investment has not been made in the abundantly available alternative resources to make power generation. Some portion of rice husk is converting into useful products such as fertilizer, insulator, cement and concrete, adsorbent and catalyst [1]. Since the calorific value of rice husk is about 16720 kJ/kg, the trend of rice husk use as alternative fuel source for power plant and boilers is increased in developing countries including Pakistan. A study was carried out by All Pakistan Textile Mills Association (APTMA) to measure the potential of rice husk for power generation in the province of Punjab and the study revealed rice husk has the potential to produce 600 MW to 1,000 MW per annum. Rice husk is an alternative and renewable energy source, therefore this biomass has fascinated the researchers to develop method for efficient conversion of rice husk into useful energy such as electric power and thermal energy.

Fang et al. [3] discussed in his research that combustion efficiency of rice husk is 97% when rice husk was fluidized with the silicon sand and coal in the circulation fluidized bed. Madhiyanon et al. [5] investigations claimed that maximum combustion efficiency of rice husk was 99.8% by using short combustion chamber integrated with fluidized-bed combustor. In another study by Natarajan et al. [6] claimed that usually 80% combustion efficiency of rice husk is obtained in atmospheric bubbling fluidized bed reactor.

A waste rice husk ash is produced when rice husk is used in combustion process as a fuel for energy production and its composition depends on the combustion temperature. Pakistan is also among one of the main rice producing countries, therefore, the practice of using rice husk waste as a fuel is increasing, especially as a boiler fuel. The rice husk ash contains a little amount of metals and mainly 87-98 % silica [3]. When incineration is carried out of rice husk, white rice husk ash and carbonized rice husk ash are formed depending on complete and incomplete combustion respectively as shown in Fig. 1 and both the types have their own applications.

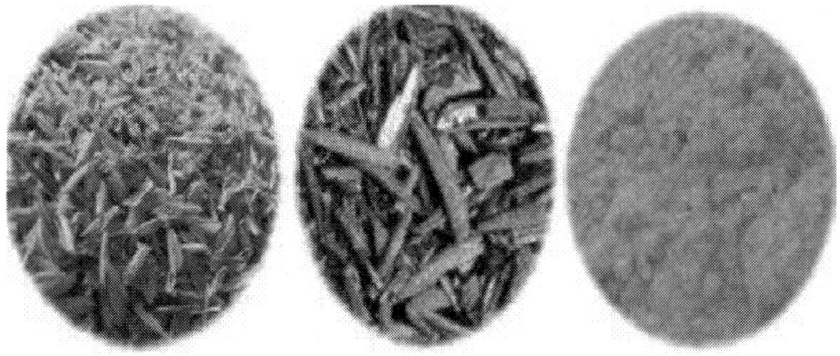

Figure1. (a) Rice husk (b) carbonized rice husk (c) Incinerated rice husk

Silica (SiO_2) is used in electronics and ceramic industries as a main raw material and in polymer industry as a filler/reinforcement due to its small-diameter. As rice husk ash contains up to 98% of silica, therefore, it is a viable and economical source for extraction of silica. Many processes have been developed by different researchers for the production of silica from rice husk and rice husk ash and every process have their own advantages and disadvantages [7]. The incineration of rice husk results in rice husk from which silica can be extracted or it can be directly obtain from rice husk in the forms of silicates but the most favorable method is to extract silica from the rice husk ash. The combustion conditions are very important, particularly, temperature for the structural transformation of the silica. The amorphous silica is obtained between 550-800°C and above of this temperature crystalline silica is formed and both forms have distinguished properties.

MATERIAL AND METHOD

The rice husk is collected from the local industry (Servis Industries Ltd, Gujrat, Pakistan). For the size reduction of rice husk, the Cushing/grinding is carried out in the ball mill to increase the surface area. The residence time in the ball mill is 1 hr. The ground rice husk from the ball mill is separated into different fraction by using sieve shaker. In sieve analysis, series of eight sieves are used having mesh # 8, 10, 15, 30, 40, 50, 60, 80 and 100 of US standard and the operational time is 30 min. The rice husk of eight different particle sizes is obtained which passed from the mesh# 8, 10, 15 so on and retained on mesh# 100. The heating of rice husk is carried out in the electric furnace (RHF 1600, Carbolite, USA) by taking four samples of 2gm from each sieve. These four samples from each sieve are heated in the furnace at temperatures of 600, 650,700 and 750°C respectively and %age rice husk ash is calculated. The %age rice husk ash of all 32 samples is calculated.

RESULTS AND DISCUSSION

Table 2 depicts the effect of mesh size and temperature on unburned (residue) of rice husk. Samples of rice husk with various mesh sizes used for investigation and rice husk combustion observed for 3hrs continuously at 600, 650, 700, 750°C.

Table 2. Effect of mesh size and temperature on unburned

Sample Size	Temperature (°C)	Rice husk ash (Wt.%age)
8/10	600 650 700 750	21 21 21 21
10/15	600 650 700 750	12 12 8 8
15/30	600 650 700 750	11 10 9 9
30/40	600 650 700 750	16 16 16 16
40/50	600 650 700 750	15 15 12 12
50/60	600 650 700 750	37 37 37 37
60/80	600 650 700 750	57 57 56 55

80/100	600 650 700 750	78 72 71 69.5

Composition of rice husk is described in literature [3] and given in Table 3 as well; rice husk composition indicates the combustible percentage. After combustion the unburned is considered as cellulose, lignin pentose or one of them along with silica.

Table 3. Composition of rice husk

Ingredients	Ash	Cellulose	Lignin	Pentose	Organic Components
%age	20	38	22	18	2

Various sizes (mesh) of rice husk like 8/10, 10/15, 15/30, 30/40, 40/50, 50/60, 60/80 and 80/100 mesh are used to investigate the combustion effect and the amount of unburned is calculated. It is observed that unburned rice husk (cellulose, lignin and pentose) is decreased for 8/10 and 10/30 mesh size while further increase in mesh size has increasing effect on unburned rice husk formation. The combustion system is optimized for 15/30 mesh at 650°C - 700°C, and 10/15 mesh at 600°C.

Potential of energy and silica from rice husk is calculated at regional level (especially Gujranwala division). Data collected from agricultural survey of Punjab Pakistan and production trend shown in Table 1, calculation shows a potential of 20,000 ton of silica per year and if we assume 75% of rice husk is consumed for power generation and ash is treated for silica separation 15,000 tons can be produced.

CONCLUSIONS

Effect of temperature and mesh size is investigated to optimize the process condition for combustion of rice husk to produce the energy.

The minimum (%age) rice husk ash obtained from the rice husk combustion at 700°C and 750°C is by mesh # 30.

By enhancing the temperature the rice husk ash (unburned) decreases and the optimum condition for the rice husk size mesh # 30 is about 700°C.

With the increase in particle size i.e., mesh size, an increase in the rice husk ash is observed. For validation more experimental data is required.

Regional positional for silica separation from rice husk is available.

REFERENCES

1. Adama F., Appaturi J.N. and Iqbal A. (2012), "The utilization of rice husk silica as a catalyst: review and recent progress," Catalysis Today, 190, 2-14.
2. Della V.P., Kuhn I. and Hotza D. (2002), "Rice husk ash as an alternate source for active silica production," Material Letters, 57, 818-821.
3. Fang M., Yang L., Chen G., Shi Z., Luo Z. and Cen K. (2004), "Experimental study on rice husk combustion in a circulating fluidized bed," Fuel Processing Technology, 85, 1273-1282.
4. Huang S., Jing S., Wang J., Wang Z. and Jin Y. (2001), "Silica white obtained from rice husk in a fluidized bed," Powder Technology, 117, 232-238.
5. Madhiyanon T., Sathitruangsak P. and Soponronnarit S. (2010), " Combustion characteristics of rice-husk in a short-combustion-chamber fluidized-bed combustor (SFBC)," Applied Thermal Engineering, 30, 347–353.
6. Natarajan E., Nordin A. and Rao N. (1998), "Overview of combustion and gasification of rice husk in fluidized bed reactors," Biomass and Bioenergy, 14, 533-546.
7. Witoon T., Chareonpanich M. and Limtrakul J. (2008), "Synthesis of bimodal porous silica from rice husk ash via sol–gel process using chitosan as template," Materials Letters, 62, 1476–1479.

370615 Innovative Design of an Integrated Biotrickling Filter System for Removal of Simulated Waste Gas Containing a Mixture of p-Xylene and Ethyl Mercaptan

Abstract:

Volatile organic compounds (VOCs) and odorous compounds discharged into the environment create ecological and health hazards. Among VOCs, a considerable number of BETXs (benzene, ethylbenzene, toluene and xylenes) into the environment originates from industrial facilities. Organic odorous pollutants are usually related to mercaptans (R-SH), organic sulfides (R-S-R), and disulfides (R-S-S-R). As far as mercaptans concerned, ethyl mercaptan is always used as an intermediate or starting material in the industrial manufacture of plastics, insecticides, and other organic synthesis process. Due to the high volatility, potential toxicity and the extensive use of these organic pollutants, most environmental releases partition to the atmosphere. As a consequence, it is inevitable to adapt effective technologies for the treatment of industrial VOC emissions, along with the more strict emission regulations in China.

Waste gas treatment processes using bioreactors have gained popularity in control of VOCs and odour, since they offer a cost effective and environment friendly alternative to conventional air pollution control technologies. However, industrial VOC emissions usually contained several recalcitrant pollutants (e.g., xylenes, toluene and mercaptans realased from the pharmaceutical industry in China), making it difficult for a conventional bioreactor to efficiently remove all of the target pollutants and strictly meet discharge limits in force. Therefore, one additional pretreatment step is required for removing these recalcitrant pollutants emitted from the industrial processes. Due to the high oxidation potential and reactivity of ozone and other reactive species, non-thermal plasmas technology has been widely used for removing recalcitrant pollutants from industrial VOC emissions. Nowadays, combination of a bioreactor with some physical-chemical approaches has attracted increasing attention in order to overcome the weaknesses of biodegradation-alone systems.

In this study, an innovative design of biotreatment system for the removal of the recalcitrant compounds by integrating pretreatment steps of non-thermal plasmas oxidation and water absorption with a biotrickling filter was studied. In this integrated process, a dielectric barrier discharge (DBD) reactor was adopted as a pretreatment step in order to transfer these recalcitrant compounds to easily biodegradable compounds. Furthermore, water absorption followed with the DBD reactor was designed to keep the stability of a biotrickling filter from the side-effect of overdose ozone, and simultaneously enhance the absorption of target pollutants into liquid medium. As the key process for the removal of VOCs and odorous compounds, a biotrickling filter inoculated with *Pandoraea* sp. WL1 and *Pseudomonas* sp. WL2 for removing a mixture of p-xylene and ethyl mercaptan, was firstly acclimated to steady-state period. Additionally, batch experiments were conducted to explicit the substrate interaction of p-xylene and ethyl mercaptan during the microbial degradation by strain WL1 and strain WL2, respectively. In these batch experiments, key interactions identified include inhibition of p-xylene degradation and enhancement of ethyl mercaptan degradation for a paired mixture relative to single substrate degradation by *Pandoraea* sp. WL1 and *Pseudomonas* sp. WL2, respectively. In this process, the integrated BTF system with a dielectric barrier discharge (DBD) reactor as the main pretreatment step showed a better performance (removal efficiency of 70%~100%) than that of the BTF-alone system (removal efficiency of 58%~77%) with total VOC concentrations of 300~1500 mg/m^3 (1:1 ratio for ethyl mercaptan paired with p-xylene). These indicate that this novel bioreactor could effectively overcome these operational limitations for poorly soluble and recalcitrant pollutants in the industrial application.

391639 Land Use Emissions and Environmental Impacts of Agricultural Systems

Abstract:

This poster presents results related to direct land use emissions and fluxes associated with agricultural activities. These emissions are further translated into environmental and climate impacts. The presented results relate to the case study of San Luis Valley in south central Colorado. The emissions (e.g. CO_2, N_2O etc.) are estimated using DNDC model and impact assessment methods are used to translate them into environmental impacts. We present results related different scenarios such as varying crops, cropping patterns, agricultural practices etc. We also discuss the relevance of these results in the context of regional sustainability.

377146 Development and Performance of a Novel Cu2+-Imprinted Membrane By Semi-Interpenetrating Polymer Network Technique for Selective Cu2+ Removal

Abstract:

Ion-imprinted membrane has received a great attention due to its high selectivity toward the target ions. In this study, a novel Cu^{2+}-imprinted polyvinyl alcohol (PVA)/ polyacrylic acid (PAA) membrane (Cu-IM) was prepared by semi-interpenetrating polymer network technique for selective removal of Cu^{2+} ions from aqueous solution. Chains of the PAA polymer complexed with the template Cu^{2+} ions were entrapped in the crosslinked PVA membrane matrix. The chemical structure and morphology of the Cu-IM were analysed by FT-IR and SEM. The batch adsorption studies showed that the adsorption of Cu^{2+} on Cu-IM was highly pH-dependent and the maximum uptake was reached at pH > 4.5. Based on Langmuir equation, the Cu-IM had a maximum adsorption capacity of 82.18 mg/g for Cu^{2+}, with a much higher affinity toward Cu^{2+} ions than that of non-imprinted membrane (NIM). Selective adsorption indicated that the Cu-IM exhibited a high selectivity for target Cu^{2+} ions in comparison with that of NIM, and showed a high selectivity for Cu^{2+} with respect to Zn^{2+}. Besides, the Cu-IM possessed a high reusability for Cu^{2+} uptake and could maintain 96 % of the adsorption capacity for Cu^{2+} after 8 adsorption-desorption cycles. In addition, the filtration studies indicated the Cu-IM could remove Cu^{2+} ions efficiently, with an effective treatment volume of 1710 bed volumes for initial Cu^{2+} concentration of 11 mg/L. Finally, the FTIR and XPS analysis demonstrated that carboxyl groups in PAA and hydroxyl groups in PVA play a key role in the adsorption of Cu^{2+} ions.

368169 Surfactant-Free Synthesized Mesoporous Carbon Nanocomposites Towards Efficient Pollutants Removal from Water

Abstract:

Water pollution is an increasingly serious issue for human beings and ecosystems. Materials with mesoporous structure are designed to remove the pollutants from wastewater with higher efficiency. Current techniques to control the mesoporous structure still rely on the template method, which involves the usage of expensive templates to regulate the mesoporous structure of desired materials and corrosive chemicals to etch out the template afterwards. Without template, mesoporous structure can be rarely achieved, which remains a challenge so far. Here, we developed a template-free method to synthesize mesoporous carbon nanocomposites without using templates and corrosive chemicals. More importantly, widely available natural resources, such as wood, cotton and *etc*, were used as carbon precursor and processed into mesoporous structure by a catalytic graphitization process. These nanocomposite materials show unique mesoporous structure and excellent pollutants removal property from wastewater. Highly efficient removal of both organic (organic dye) and inorganic (heavy metal ions) pollutants is demonstrated by kinetic study. This new technology provides a facile and economic approach to manufacture highly efficient adsorbents with multi-contaminants removal capacity.

AIChE Meeting

November 2014

Evaluation of Usability of Various Real Wastewaters in Microbial Fuel Cells

Anna I. Casasús
Luciana Bava
Jaime Lee
Kemira

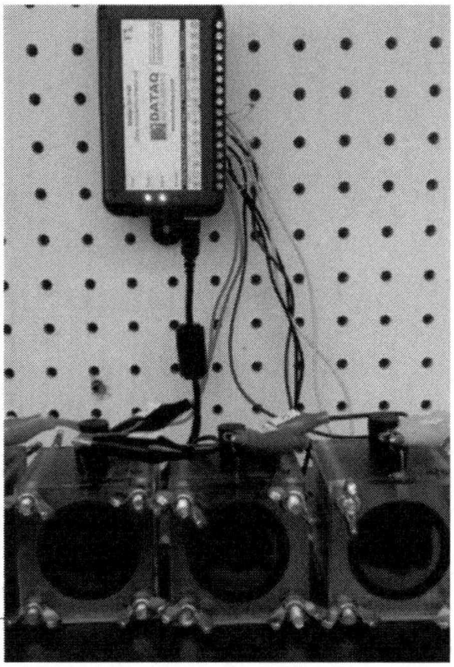

How does it work?

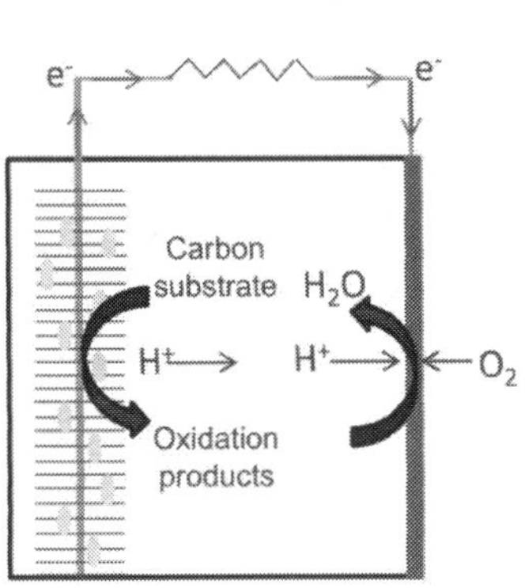

- Cell contains anode and cathode and is fed a liquid substrate (organic)
- Exoelectrogenic bacteria grow on the anode
- In the anode, bacteria oxidize a substrate, generating electrons and protons
  - Protons diffuse through water to the cathode
  - Electrons are transferred to the cathode through an external circuit
- In the air cathode, electrons combine with protons and oxygen to form water

What is the Benefit of MFCs?

- Water treatment generally requires energy input
- MFCs have the potential of water treatment with zero net energy or positive net energy

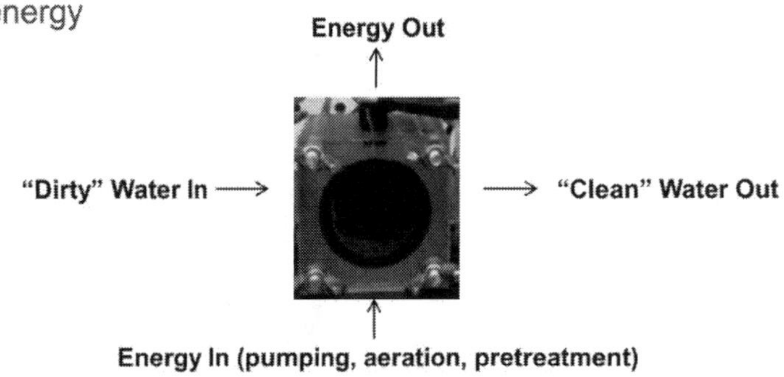

Electrode Designs

CATHODE (+)

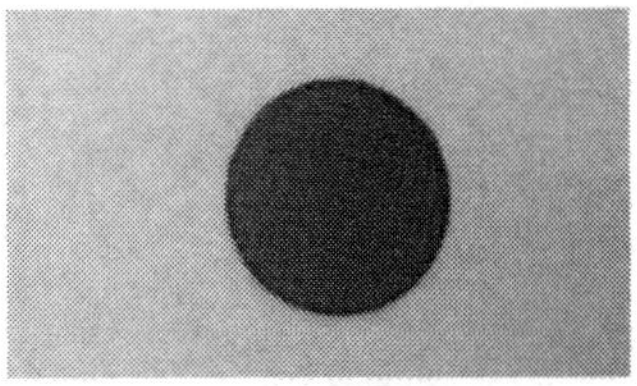

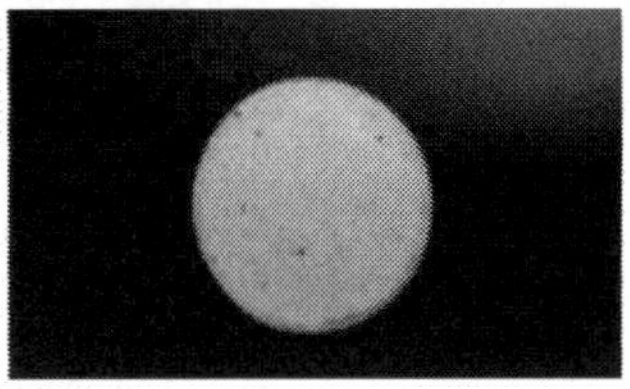

ANODE (-)

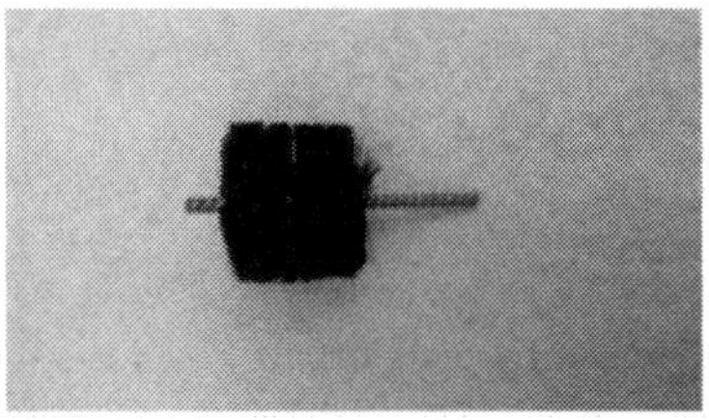

Experimental Methods

- Inoculation: Mix of bacteria, nutrients, and wastewater are fed to cell
- Start-up: Wastewater/nutrient alone is fed to cell
- Run: Includes acclimation to various resistances
- Performance measures
  - Water quality
  - Electrochemical measurements
  - Coulombic efficiency

Substrates

- Food and beverage
- Municipal
- Pulp and paper
- Oil and Gas

- With and without pretreatment

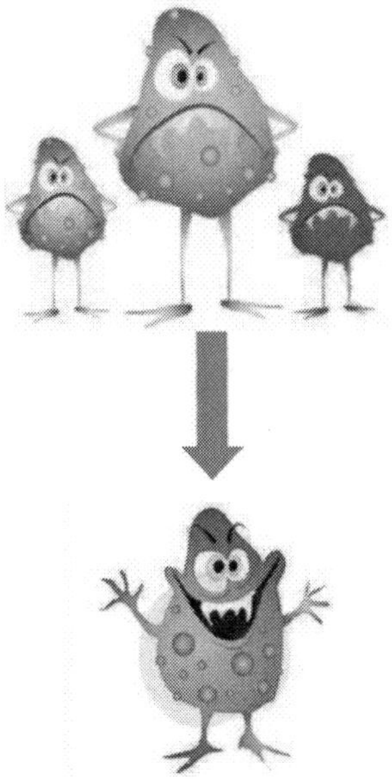

Results – Food and Beverage

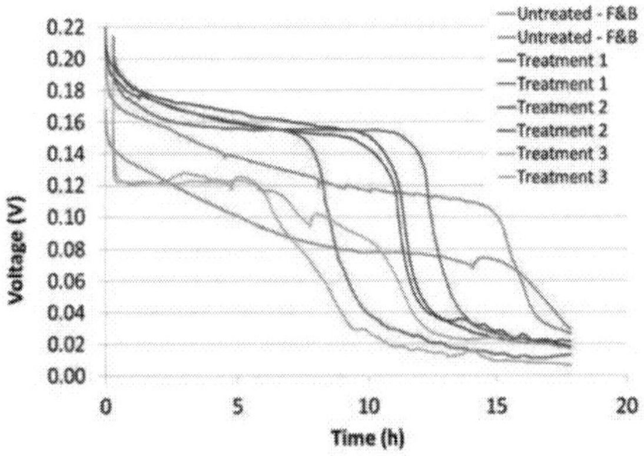

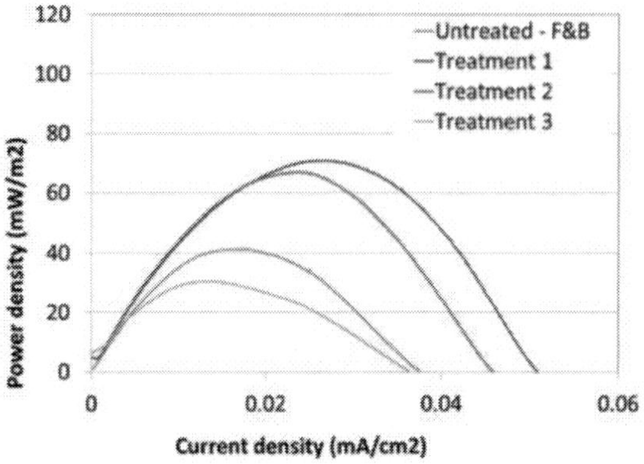

Results – Food and Beverage

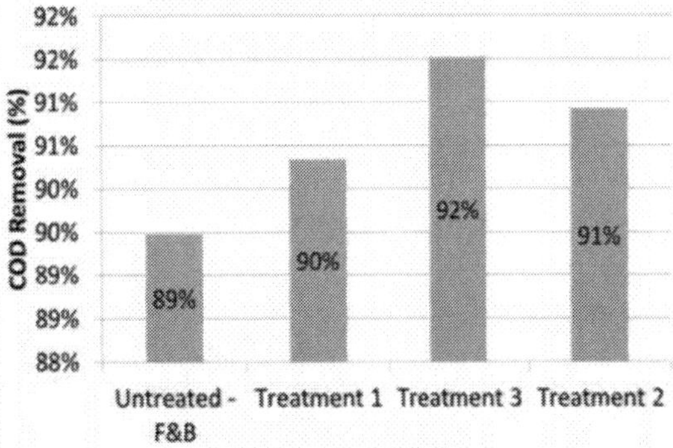

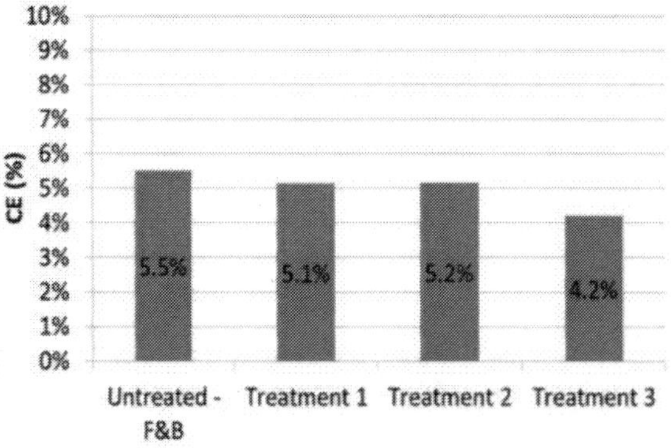

$$C_E = \frac{8 \int_0^t I\,dt}{Fv\Delta COD}$$

t = run time, I = current, F = Faraday's constant (electric charge/mol e^-), *v* = volume, and ΔCOD = change in COD during the run time.

Results – Municipal

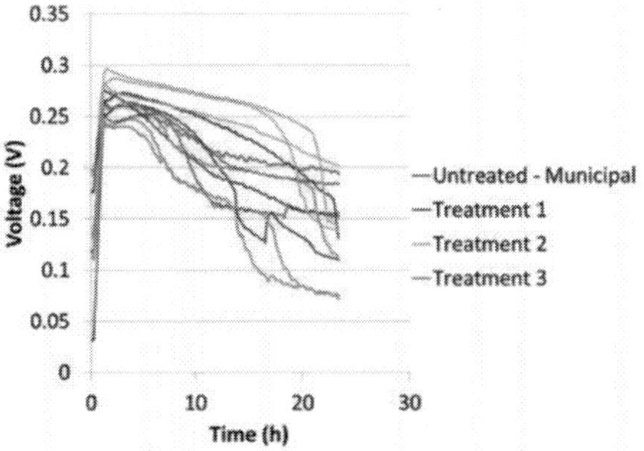

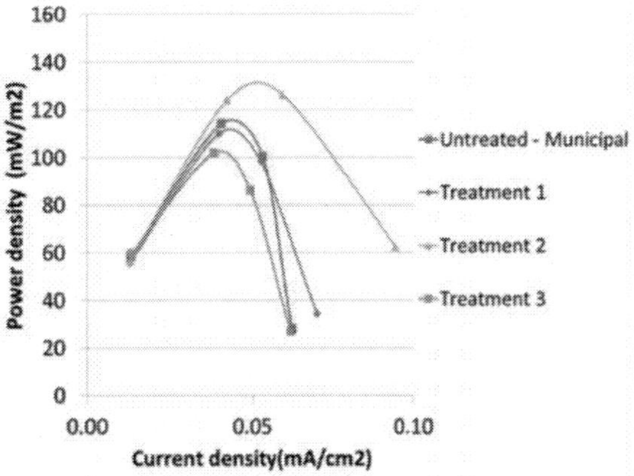

Results – Municipal

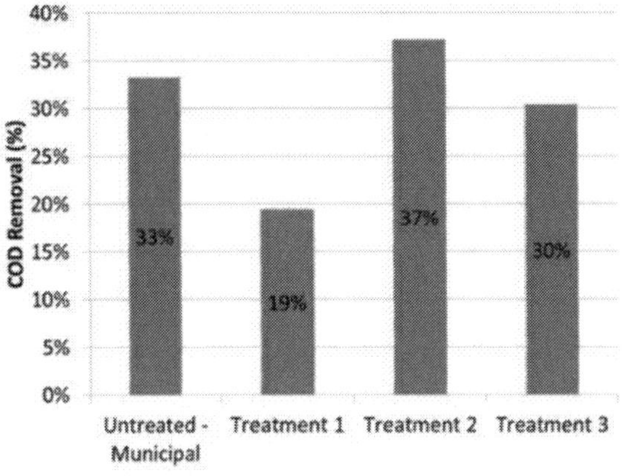

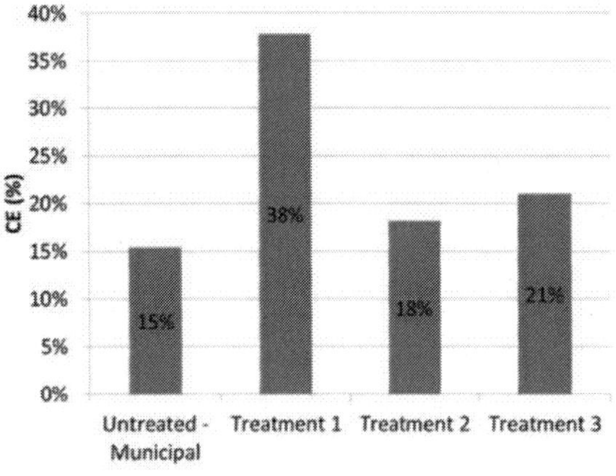

Results – Paper

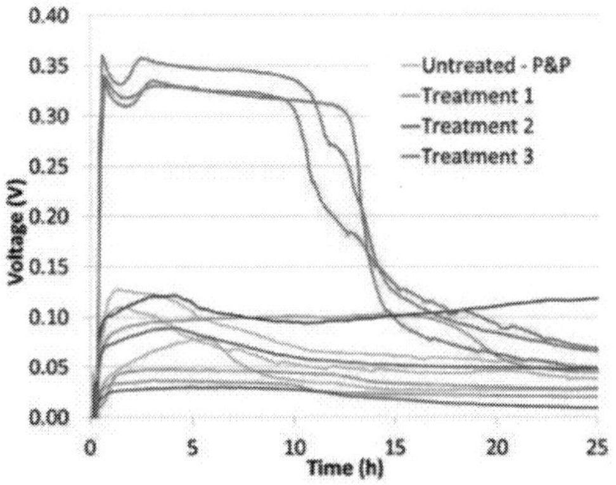

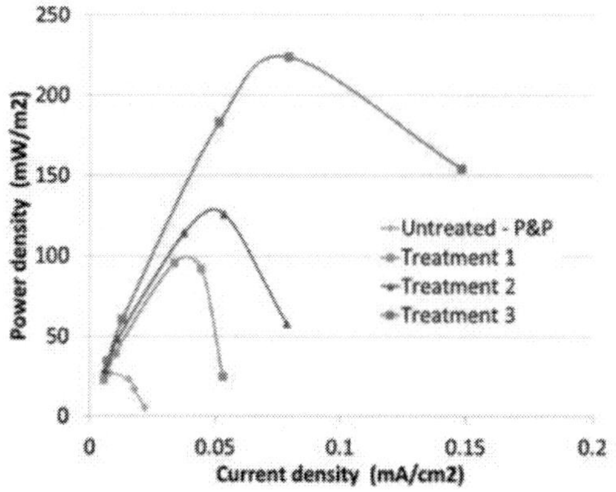

Results – Pulp and Paper

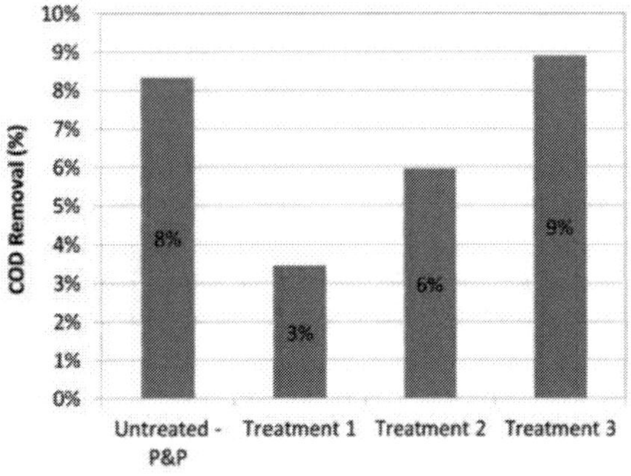

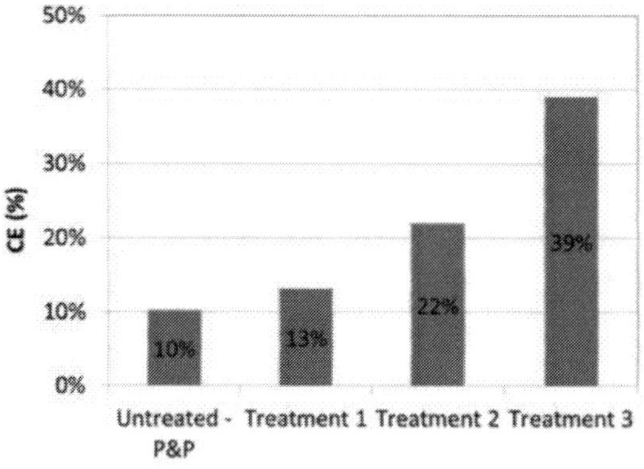

Results – Oil and Gas

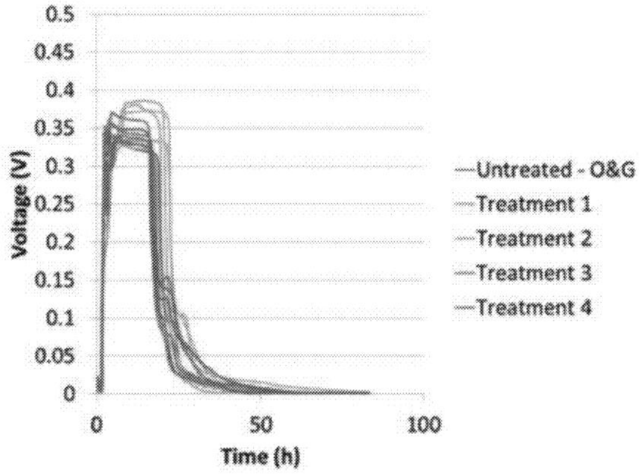

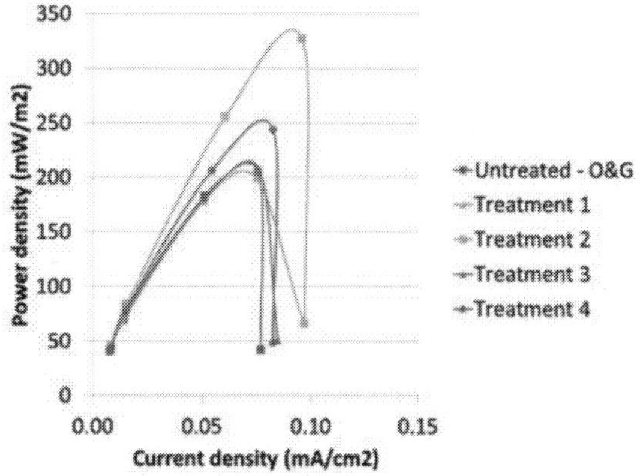

Results – Oil and Gas

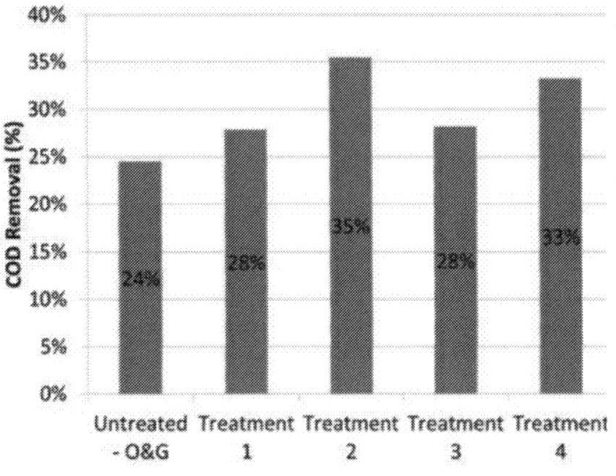

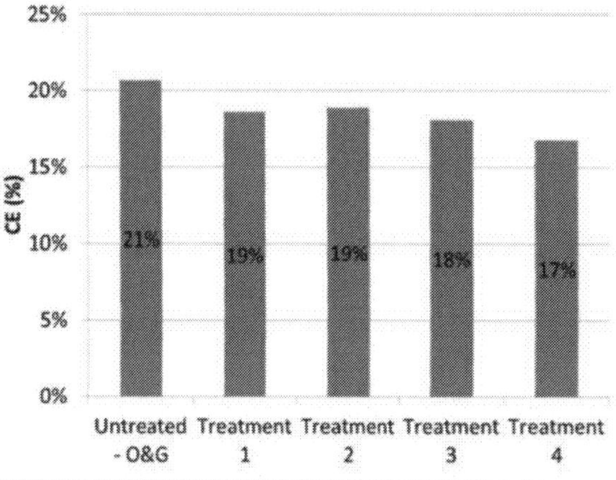

Conclusions

- Microbial Fuel Cells show potential for use in a number of industries
- Power densities greater than 300 mW/m^2 are possible with certain wastewaters
- Pretreatment of wastewaters may increase power density (over 7 times greater than without treatment, in some cases)
- In some cases, pretreatments resulted in increased COD removal and/or improved Coulombic Efficiency

Some Aspects of Hydrodynamics of Continuous Countercurrent Liquid-Solid System

Krishnan Nagarajan, T. Renganathan and K. Krishnaiah

Department of Chemical Engineering, Indian Institute of Technology Madras, Chennai, India

Abstract

Adsorption is an effective technique for treatment of dye containing effluents from textile and paper industries. Adsorption can be carried out in batch system, packed bed or fluidized bed. Among these contacting modes, fluidized bed offers the advantages like high degree of mixing and high mass transfer coefficient. However since the solids are in batch mode, the effluent dye concentration varies as a function of time. A steady effluent concentration can be achieved by using a continuous feed of solids. Higher concentration gradient can be obtained by continuously feeding the solids downwards countercurrent to the upflowing liquid. This is referred to as continuous countercurrent liquid–solid system (CCLSS) in the present work.

The continuous countercurrent liquid solid system can be operated as free system or as bottom restricted system (Lapidus and Elgin, 1957). In the present work hydrodynamic characteristics (pressure drop and holdup) of free system are studied using liquid under batch condition which is one of the limiting conditions of continuous flow system. This study will help in understanding the hydrodynamics of free system with continuous liquid flow. There are no studies reported on hydrodynamics of counter current liquid solid system with liquid in batch condition.

The experiments are carried out in a pilot scale column by distributing water from the bottom and feeding sand from the top. Pressure drop is measured using multi-limb U tube manometer. The solid holdup is measured by using the conductivity technique and the quick closing valve technique. It is observed that pressure gradient and solid holdup are almost uniform along the length of the column. The pressure drop and holdup increased with increase in solid flowrate.

The pressure drop measured in the present study for CCLSS with liquid in batch mode, is predicted using equation proposed by Foscolo et al.,(1983) for high voidage (low solid holdup) systems. The average solid holdup is predicted by adopting the Richardson-Zaki equation in terms of slip velocity (Bisschops et al., 2001).

References:

1. Bisschops, M.A.T. Luyben, K.Ch.A.M. and Van der Wielen L.A.M., "Hydrodynamics of Countercurrent Two-Phase Flow in a Centrifugal field", AIChE Journal, 47 (6), 1263-1276, (2001).

2. Foscolo, P.U. Gibilaro, L.G. and Waldram, S.P., "A Unified model for Particulate Expansion of Fluidized Beds and Flow in Fixed Porous Media", Chem. Eng. Sci., 38(8), 1251-1260, (1983).

3. Lapidus, L. and Elgin, J.C., "Mechanics of Vertical Moving Fluidized Bed Systems", AIChE Journal, 3, 63-68, (1956).

367672 Removal of Cadmium Ions from Aqueous Solution Using a New Low-Cost Biosorbent

Abstract:

In recent years, heavy metal pollution has increased as a result of industrial growth, especially in industries such as electroplating, mining, metal finishing, battery production, and welding. Among the heavy metals, one of the most hazardous is cadmium, which accumulates in the human body and causes cancer along with cardiovascular, neurological, and respiratory problems. Hence, it is crucial to sequester cadmium ions from industrial effluent before the contaminated effluent is discharged into the sewage system. There are several traditional treatment methods for removing heavy metals from aqueous effluents, including chemical precipitation, ion exchange, and membrane filtration. These techniques have high operational costs and require the continuous input of chemicals, so they are neither economical nor ecologically sustainable. Therefore, biosorption has gained significant attention as an effective alternative for sequestering toxic heavy metals. In this work, *Sorghum X drummondii* (Sudangrass), an inexpensive agricultural product, is used as an adsorbent for the removal of cadmium ions under a range of conditions, including variations in initial pH, temperature, contact time, adsorbent dose and initial cadmium ion concentration. Fourier Transform Infrared (FTIR) is used to characterize the surface of the biosorbent and also to investigate the functional groups involved in the process. Scanning Electron Microscopy (SEM) is used to observe the morphology of the biosorbent before and after the adsorption process. Well-known isotherm models are used to describe the equilibrium data, from which the parameters are calculated and reported. Two kinetic stages of adsorption are identified: a fast period, followed by a relatively slow period leading to equilibrium. Thermodynamic parameters, such as the change of free energy, enthalpy, and entropy, were also evaluated for the adsorption of cadmium ions onto Sudangrass.

386521 Environmental Remediation of Dense Non-Aqueous-Phase Liquids Using Multifunctional Iron-Carbon Nanocomposites

Abstract:

Dense non-aqueous phase liquids, such as trichloroethylene (TCE) and carbon tetrachloride (CCl_4), are widespread groundwater and soil contaminants which cause long-term environmental pollution. Extensive efforts have been carried out to develop the optimal remediation strategy. Here, we will introduce a composite with carbon submicrospheres serving as the supports and embedded nanoscale zero valent iron (NZVI) being the reactive agent to treat TCE and dechlorinate it to non-toxic products. This Fe-C composite unites the current prominent method of using NZVI with the hydrophobic nature of carbon, which enables the composites to target organic contaminants and facilitates the dehalogenation reaction by increasing the local concentration of TCE in the vicinity of iron particles through rapid TCE adsorption. Besides, hydrophobically modified chitosan (HMC) can be easily coated around the carbon supports to enhance the suspension stability of these submicrospheres in groundwater.

This composite has been proven to be effective and has several conspicuous advantages: (a) the synthesize process is a simple one-step process with inexpensive precursors and can be scaled up to manufacturing level (b) the NZVI particles are well dispersed on the carbon support, preventing them from aggregation, which eliminates the major challenge of using NZVI (c) composite particles are of the optimal size for transport through sediments (d) toxic intermediates such as dichloroethane ($C_2H_2Cl_2$) and vinyl chloride (C_2H_3Cl) can be sequestered within the carbon supports until they reacted to final light gas products.

The presentation will focus on the detailed materials synthesizing and characterization, and the significant aspects of adsorption, transport and reaction of this composite in remediation of groundwater contaminants will be discussed.

Adsorption and Oxidative Degradation of Bisphenol A on Surface Modified Iron-Amended Activated Carbon: Effects of temperature on adsorption and Fenton oxidation

Jihyun R. Kim and Eunsung Kan. Department of Molecular Bioscience and Bioengineering, University of Hawaii at Manoa, Honolulu, HI 96822, USA

For the past decade, BPA as an endocrine disrupting chemical has gained great attention, and become a public concern as they can cause adverse effects in aquatic organisms and human. The present study focuses on effective adsorption and oxidation of bisphenol A (BPA) in wastewater by using newly synthesized nano iron catalyst-deposited activated carbon.

The novel synthesis of nano iron/iron oxide-deposited granular activated carbon (Fe-GAC) was developed for adsorption and oxidation of BPA by amending ferrous iron onto the acid-pretreated GAC at the pH of iron solution higher than the pH at point of zero charge (pH, pzc) of the Fe-GAC. Batch experiments for the adsorption isotherm of BPA onto the Fe-GAC at 20 °C and pH 6 followed Langmuir isotherm-type adsorption, yielding maximum monolayer adsorption capacity (q_m) of 588.2 mg BPA/g GAC which was higher than the virgin GAC with q_m of 471.2 mg BPA/g GAC. The adsorption rate of BPA was increased by factor of 3.4 when temperature increased from 20°C to 80°C. It was exactly proportional to enhancement of diffusivity of BPA at increasing temperature indicating diffusion of BPA into the pores of the GAC would be the major limiting step in the adsorption process of BPA.

Since the heterogeneous Fenton oxidation of BPA in the Fe-GAC clearly showed strong dependence on H_2O_2 concentration, 9 to 108 mol H_2O_2/mol BPA led to enhance the BPA oxidation efficiency in the GAC from 62% to 99%. The Fenton oxidation of BPA at increasing temperature (20-80°C) also resulted in enhancing the oxidation efficiency and rate of BPA while increasing the diffusivity, desorption rate and reaction rate of BPA in the GAC. The pre-optimized conditions for the heterogeneous Fenton oxidation of BPA exhibited negligible accumulation of the major intermediates and oxidative products of BPA in the aqueous and GAC phases. The mathematical model based on diffusion, adsorption and oxidative reaction and Thiele-Modulus analysis indicated that the intraparticle diffusion limitation was the major limiting step among the Fenton oxidation of BPA in the GAC. Further works will include detailed analysis of diffusion, adsorption and oxidation in the GAC using the mathematical model and engineering analysis as well as the optimization of the operating conditions over the repeated cycles of adsorption and oxidation.

365690 Removal of Metronidazole and Dimetridazole from Aqueous Solution By Adsorption on Multiwalled and Singlewalled Carbon Nanotubes

Abstract:

Antibiotics are widely prescribed in human and veterinary medicine to prevent and cure bacterial infections. Antibiotics may not be biodegraded by conventional treatments, and they are being discharged in environmental media in their parent form or as transformation products. It is very well documented that the presence of antibiotics in surface water poses serious risks to human health. Nitroimidazoles are among the most widely produced and prescribed antibiotics in many countries. Metronidazole (MNZ) and dimetridazole (DTZ) are very commonly used nitroimidazoles.

The adsorption of the antibiotics MNZ) and DTZ on multiwalled (MWCNT) and single walled (SWCNT) carbon nanotubes was investigated in this work. The effect of the adsorbent-adsorbate interactions as well as the operating conditions (ionic strength, solution pH, temperature, chemical modification of the adsorbents, presence of other antibiotic) on the adsorption capacity were analyzed to substantiate the adsorption mechanism. The MWCNT were modified using a HNO_3 solution and the modified adsorbent was denoted as MWCNT-Ox. The adsorption capacity towards MNZ and DTZ markedly varied as function of the carbon material, decreasing in the following order: SWCNT > MWCNT > MWCNT-Ox, and depended not only on their surface area and pore size distribution, but also on their chemical nature. The adsorption of MNZ and DTZ was strongly influenced by the solution pH, but was not significantly affected by the ionic strength and temperature. Desorption equilibrium data of MNZ and DTZ on all carbon materials demonstrated that the adsorption was reversible corroborating the weakness of the adsorbent-adsorbate interactions.

The binary adsorption of MNZ and DTZ on SWCNT was studied in this work. The experimental binary adsorption data were interpreted using the following multicomponent adsorption isotherms: non-modified, extended and modified Langmuir, non-modified and modified Redlich-Peterson, extended Freundlich, and Sheindorf-Rebuhn-Sheintuch. The latter model best fitted the adsorption data of MNZ-DTZ on SWCNT. The competitive adsorption of MNZ-DTZ on SWCNT revealed that both antibiotics presented very similar antagonism toward the adsorption of the other antibiotic.

Synthesis of zirconium nanoparticles doped activated carbon fiber for optimized adsorption for arsenic removal

Arsenic contamination in ground water and industrial effluent has been a great concern globally for years. Arsenic is one of the highly harmful inorganic contaminants originating from either anthropogenic activities as well as natural erosion of arsenic containing mineral substances. Arsenic exposure may cause many diseases. It may be transported to different organs of body through blood circulation and causes a variety of adverse health effects. The maximum contaminant level (MCL) of arsenic in drinking water set by USEPA had been 50 $\mu g/L$. However, a new standard for arsenic in drinking water has been set at 10 $\mu g/L$ since early 2006, which creates a great challenge to the water industry.

To enhance the removal of arsenate and prevent the secondary contamination of nanoparticles, an activated carbon fiber (ACF) was synthesized by doping of zirconium nanoparticles onto the surface of ACF with chitosan as the coating support. In recent years, nano-sized materials become more attractive for water treatment. Compared to the traditional sorbents that have relatively larger particle sizes, nano-sized particles have such advantages as higher surface to volume ratio and better physicochemical properties for adsorption. Zirconium based nanoparticle sorbent has a significantly high adsorption affinity towards the inorganic arsenic species. However, the nanoparticles may cause secondary contamination because of its dispersion in the water. It is rather difficult to remove the nanoparticles after the adsorption. A better media such as activated carbons can be used to support the nanoparticles in order that the nanoparticle can still stay on the media instead of dispersing in treated water.

Activated carbon fiber (ACF) is extensively used in the field of air and water purification because it provides an extremely high absorption capacities compared with powdered and granular activated carbons. It has great mechanical and physical properties, and is rather stable in different chemical and physical conditions (e.g. high temperature. As a result, ACF can serve as a good supporting matrix for zirconium particles to remove arsenic from water. Chitosan (CTS) is a liner polysaccharide and made by treating shrimp and other crustacean shells. CTS manifest a good film-forming property, and it was made as a membrane which has both filtration and adsorption ability. Polyethylene glycol (PEG) is a highly water-soluble polymer and it can generate pores on the chitosan film to minimize the mass transfer and diffusion resistance in adsorption. Thus, the PEG can be used as a pore generator on the CTS film. The preparation for Zr-doped ACF was optimized by an orthogonal array design matrix. Mixtures of chitosan solution, PEG solution, and zirconium nanoparticle suspension of various component ratios were prepared according to a $L_9\ (3)^4$ orthogonal array design matrix.

Through the orthogonal experiments, these four factors were proved to have determined the determined the adsorption efficiencies to some extent. The optimal composition for ACF modification was 0.3% chitosan with low molecular weight, 0.2% PEG and the mass of zirconium nanoparticles was 0.6 g. It should be noted that the particles were not dispersed uniformly once the amount of zirconium nanoparticles exceeded 0.6 g in the impregnation solution. As a result, the maximum mass of zirconium nanoparticles was selected to be 0.6 g in 50-mL impregnation solution.

The optimized sorbent was characterized by scanning electron microscope. Bath adsorption experiments were conducted for arsenic removal by this material, including pH effect experiments, adsorption kinetics and isotherms experiments; natural organic matters and coexisting anions affect experiments. From the results, we can get the highest adsorption was achieved at the optimized pH of 3.0. The kinetics study showed that the adsorption equilibrium was achieved at around 30 h. The Langmuir equation well described the adsorption isotherm, and the maximum adsorption capacity of the sorbent was 20.5 mg/g at the optimized pH. The presence of coexisting fluoride and carbonate ions did not significantly affect the adsorption. In contrast, the co-existence of silicate, phosphate ions and humic acid retarded the uptake. The X-ray photoelectron spectroscopy analysis indicated that zirconium nanoparticles were immobilized on the modified ACF. $HAsO_4^{2-}$ as the dominant species of As (V) was adsorbed onto the sorbent. This Zr-doped ACF has good adsorption ability for arsenic removal.

388137 Integrated Membrane Bioreactor Process for Water Reclamation and Groundwater Recharge Applications

Abstract:

Among various technologies, integrated systems such as membrane bioreactor (MBR) processes have shown excellent potential for water reclamation, water reuse, groundwater recharge, and similar applications. Superior membranes with better aqueous transport and anti-fouling characteristics can make the technology more efficient and economical. In fact, they can significantly reduce energy costs that constitute a substantial fraction of total operation costs. The continuous flow hybrid MBR system offers several technical advantages over conventional biological processes in environmental applications: small size or footprint requirements for reactor systems, long solids retention times, and efficient retention of particulates, colloids, contaminants and microorganisms. The application of an adsorbent such as powder activated carbon (PAC) will defoul the membranes and maintain high permeate fluxes. An added advantage will be the ability of the PAC adsorbent to remove most trace-level or residual micro-pollutants during water reclamation. The micro-pollutants often include many endocrine disrupting chemicals (EDCs) such as pharmaceutical and personal care products (PPCPs), pesticides, and toxic heavy metals.

The treated effluent from such hybrid MBR processes must be generally low in suspended solids, biochemical oxygen demand (BOD), chemical oxygen demand (COD), total organic carbon (TOC) and most pathogens, and must be so as to meet groundwater recharge requirements. Water reclamation for reuse and groundwater recharge requires that tertiary treatment standards for various components be met, that can be briefly summarized as follows: (i) virus removal or inactivation must exceed 5 logs; (ii) total coliform levels must be below 2.2 coliforms/100 mL; (iii) turbidity must be below 2 NTU; (iv) TOC concentrations must not exceed 0.5 mg/L (of wastewater origin); (v) organic and inorganic contaminants must meet drinking water maximum contaminant levels (MCLs) (specified by the United States Environmental Protection Agency and/or the State Water Quality Control Board); (vi) lead and copper concentrations must conform to actions levels, and (vii) nitrate limits are based on the treatment technology. The general belief is that these standards could be achieved with high efficiency and favorable economics by the MBR technology using hollow fiber ultrafiltration (UF) membranes instead of employing more energy intensive processes such as nanofiltration and reverse osmosis.

The purpose of this study is to evaluate the performance of the MBR process in the tertiary treatment of reclaimed wastewater obtained from a wastewater treatment plant after secondary treatment. The study investigates the evaluation for characteristics of polyether sulfone (PES) UF hollow fiber membrane by measuring permeate flux patterns and organic removal measured in terms of TOC and ultraviolet absorbance at 254 nm (UV_{254}). The work also includes membrane fouling studies to evaluate the effect of surface fouling and internal pore fouling attributable to organic fouling. The influence of advanced oxidation processes (AOPs) such as ozonation and ozone/hydrogen peroxide oxidation on membrane fouling is also evaluated. A major concern is the influence of AOPs on the TOC removal by the membrane as well as on the permeate flux under different oxidation conditions. Under the influence of AOPs organic molecules are broken down into smaller fragments, due to free-radical reactions mostly pertaining to the hydroxyl radical. The fragmentation of organic molecules generally leads to the formation of more hydrophilic products. It is possible that pretreatment the application of AOPs may lower the surface fouling due to sorption and consequently lead to lower organic rejection. However, the smaller fragments have a larger propensity to pass through the membrane pores, causing marginally greater internal pore fouling and slightly reducing the permeate flux. These mechanistic aspects attributed to the breakdown of organic molecules by AOPs will be discussed. The study also investigates the effect of PAC at different concentration levels (ranging from 40 to 100 mg/L) on the organic removal and the control of permeate flux decline due to organic and biological fouling. The work further evaluates the effect of biofilm formation on the membrane at different concentration of *Escherichia coli* (*E.coli*) with regard to TOC removal and permeate flux decline pattern. Additionally the study also evaluates the use of membrane cleaning strategies using caustic cleaning for permeate flux recovery. Most importantly, the study also provides a comprehensive understanding of flux decline and TOC removal patterns under the influence of PAC with *E. coli*as well as those of AOPs.

Note: Corresponding author is Professor Massoud Pirbazari, pirbazar@usc.edu, 213-740-0592.

UNIVERSITY OF SOUTH FLORIDA

Controlling ammonia levels with natural materials to preserve bait and stored fish in freshwater and seawater

Wen Zhao[1], Fei Guo[1], Sarina Ergas[2] and Norma Alcantar[1]
1 Department of Chemical and Biomedical Engineering
2 Department of Civil and Environmental Engineering
University of South Florida, Tampa, FL

2014 AIChE Annual Meeting
Nov 20, 2014
Atlanta, GA

14AIChE
Annual Meeting, Atlanta, GA

1

Introduction

- Fresh and seawater fishing are the most popular outdoor recreational activities in Florida and throughout the U.S
- Fish are maintained to be alive by saturated aeration
- High concentrations of ammonia produced by fish waste can be lethal
- Traditionally, toxic ammonia is controlled by slightly lower pH or keeping changing new water

Ammonia Chemistry

- Free ammonia- NH_3
- Ionized ammonium – NH_4^+

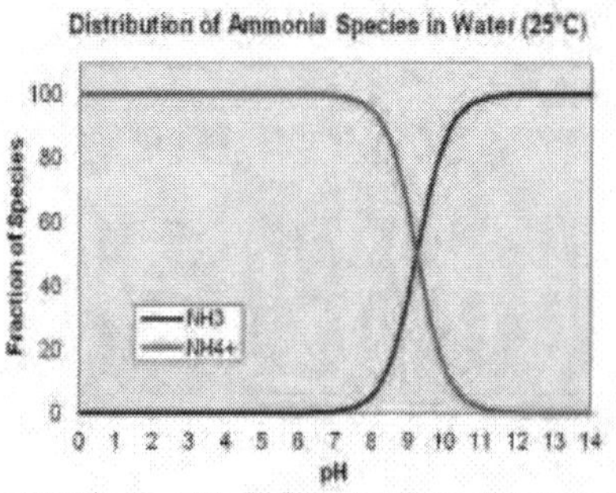

http://www.caslab.com/News/ammonia-all-youve-ever-wanted-to-know.html

$$NH_3 + H^+ \leftrightarrow NH_4^+ + H_2O$$

- Higher pH, more NH_3
- Less pH, more NH_4^+

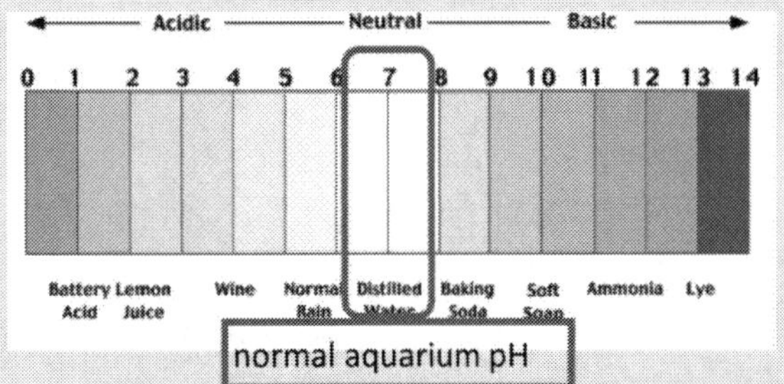

http://answers.seneye.com/en/water_chemistry/what_is_pH

Aim: pH=7

3

Objective

To develop a commercial product :

1. User-friendly
2. Sustainable
3. Affordable
4. Capable of extending the life expectancy of stored fish

by removing ammonia safely

4

Rationale

- Chabazite zeolite
  - Purchase from St. Cloud Mining Inc. (NM)
  - Form: Granules
  - Grain size: 1~2 mm
  - Color: Yellow/Tan
  - Si/Al ratio: 3.5 ± 0.06
  - Storage: DI water washed
    Dried at 100 °C oven
    Stored at room temperature

Table 1 Textural and physicochemical properties of chabazite*

Property	Value
Surface area	460 m^2/g
Ion exchange capacity	2.5 meq/g
Density	1.73 g/cm^3
Pore volume	0.468 cm^3/g

* The data were obtained from manufacture

Table 2 Chemical ingredients of chabazite (SEM-EDS)

Chemical	Percentage (%)
Si	31.65 ± 1.33
Al	9.04 ±0.35
Ca	0.83 ± 0.05
Mg	0.54 ± 0.05
Na	7.19 ± 0.61
K	1.02 ± 0.13

Outline

1. Why chabazite?
2. Concentration of ammonia over time
3. Maximum ammonium adsorption capacity
4. Real fish environment studies

6

Why chabazite?

- Ion selectivity series:
  $Ti^+ > K^+ > Ag^+ > Rb^+ > NH_4^+ > Pb^{2+} > Na^+ = Ba^{2+} > Sr^{2+} > Ca^{2+} > Li^+$
- Inexpensive
- Renewable
- Widely around the world
- High ammonium exchange capacity compare with clinoptilolite in this project

7

Concentration of ammonia over time

http://blogs.basspro.com/blog/bass-pro-shops-hampton-va/fish-feeding-and-tank-demos
http://harboarts.com/blog/tag/funny/page/2/
http://www.yoand.biz/pin-fish-coloring-pages-pinterest-hawaii-dermatology-picture-monster-truck-cartoon-images/

8

Ammonia accumulation rate determination

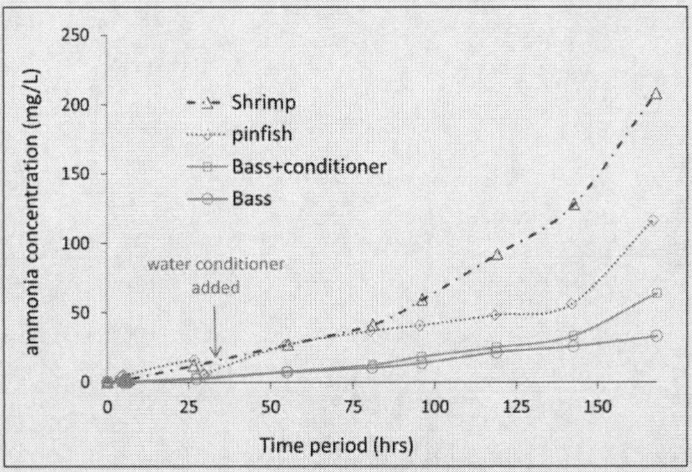

Results:

- Freshwater fish: 1.13 mg NH_3-N/fish/hr
- Seawater fish: 1.6 mg NH_3-N/fish/hr

9

Maximum ammonium adsorption capacity

- Pretreatment

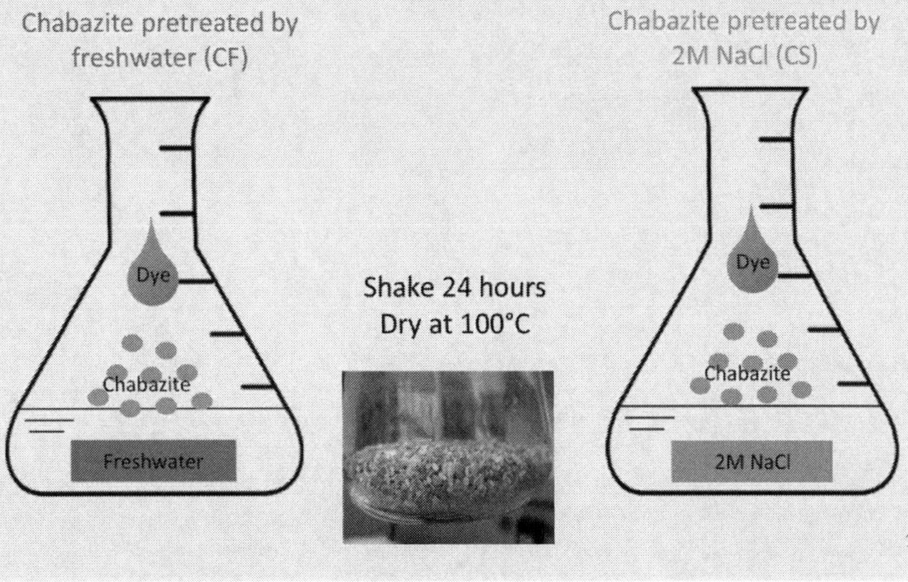

10

Rationale 2

- Freshwater
  - Obtained from USF Simmons park
  - pH: ~8
  - Alkalinity: 59.4 mg/L
- Blue dye (Triphenylmethane family)

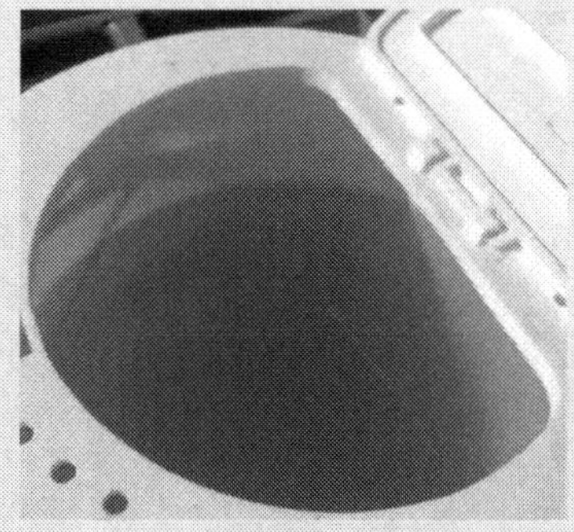

11

Characterization of pretreated chabazite

Chemical composition before and after pretreatment

- Na^+: CF has dramatic release
- K^+: no change
- Ca^{2+}: CF release half, CS release total
- Mg^{2+}: no change

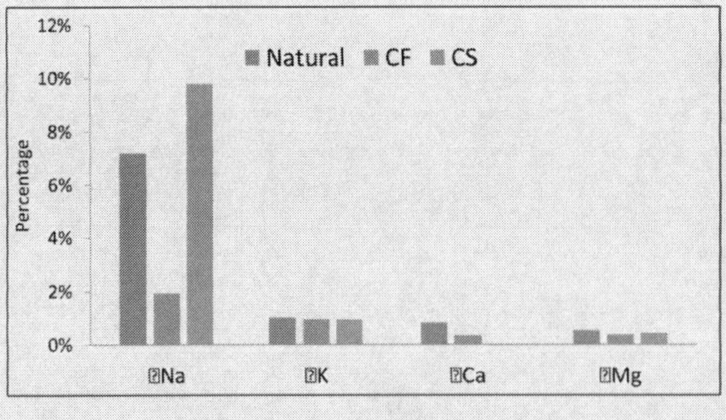

- Natural: natural chabazite
- CF: chabazite pretreated by freshwater
- CS: chabazite pretreated by 2M NaCl

12

Characterization of pretreated chabazite

Crystalline structure before and after pretreatment (XRD)

- CF: Same structure
- CS: Three peaks represent the existing of dye

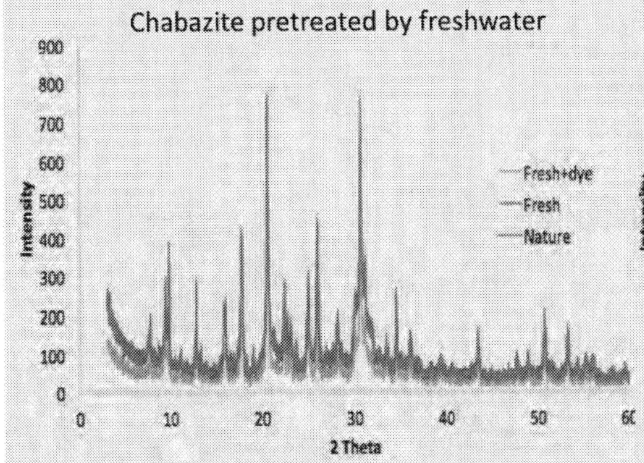

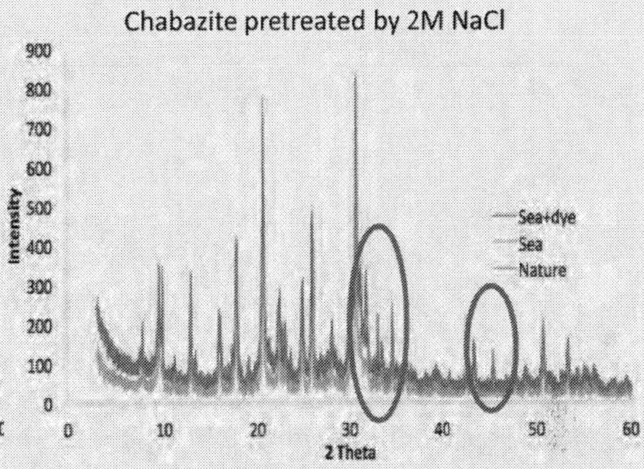

13

Ammonium adsorption isotherm

- Langmuir and Freundlich isotherm model
- The differences between traditional isotherm experiment and this experiment is mixing provided by aeration
- Synthetic water

Table 3 Chemical ingredients of synthetic water

Ion	Freshwater (g/L)	Seawater (g/L)
Na^+	0.075	10.780
K^+	0.00312	0.42
Mg^{2+}	0.024	1.32
Cl^-	0.193	19.290
Ca^{2+}	0.043	0.400
HCO_3^-	0.0048	0.200
SO_4^{2-}	0.096	2.66
Alkalinity	0.310	0.241

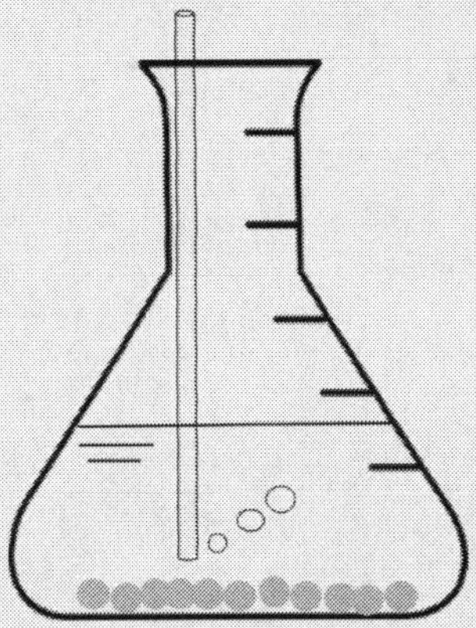

14

Ammonium adsorption capacity

Linear isotherm

Non-Linear isotherm

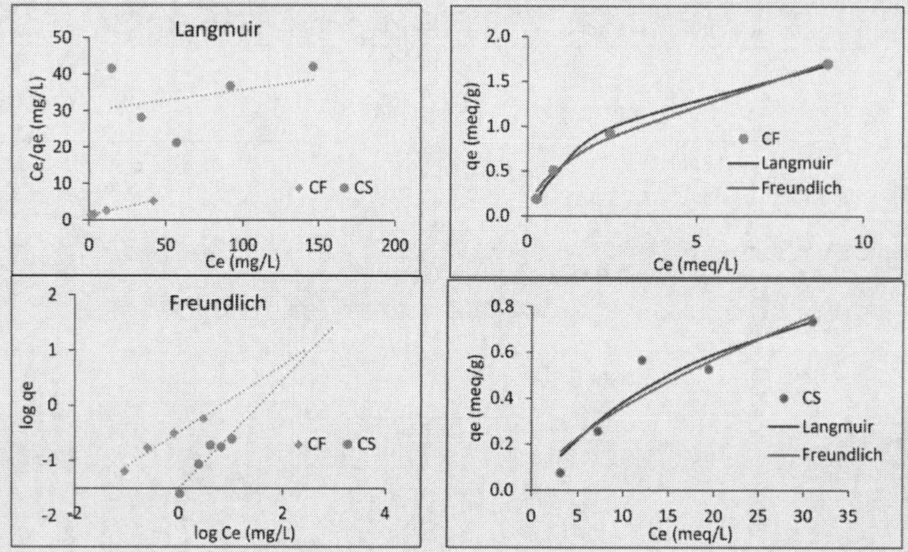

- CF: Fits both Langmuir and Freundlich model
  - Langmuir: q_0 = 0.75-0.77 meq/g
- CS: Fits Freundlich model

Ion exchange capacity	2.5 meq/g

15

Real fish environment studies

In-vitro experiment

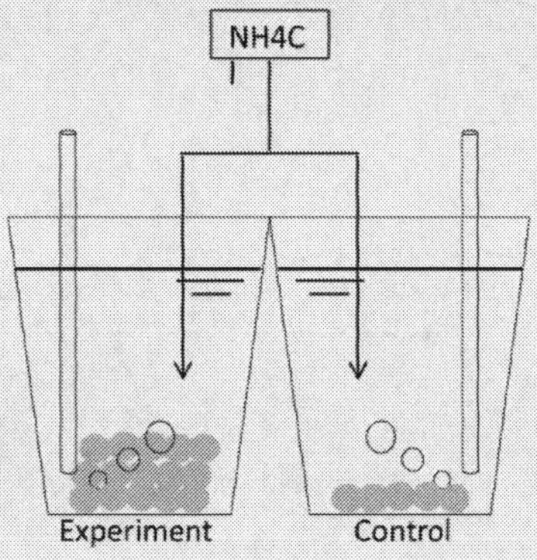

In-vivo experiment

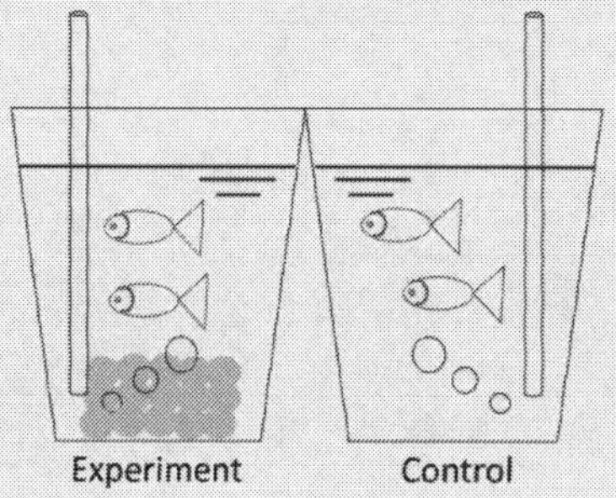

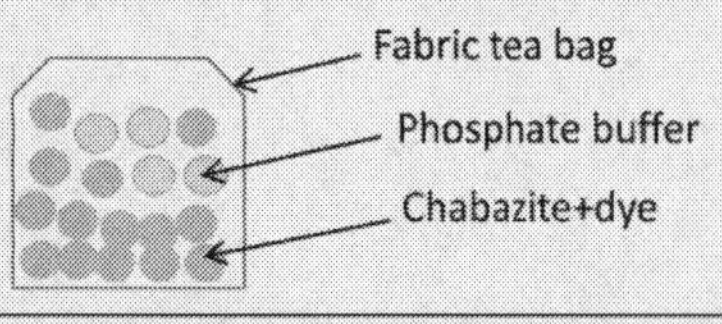

"Chabazite+ dye+ buffer" 3 in 1 hybrid

- The amount of chabazite determined from:
  1. Portability
  2. Ammonium adsorption isotherm

  So, 30g chabazite per bag was decided

16

In-vitro experiment (freshwater)

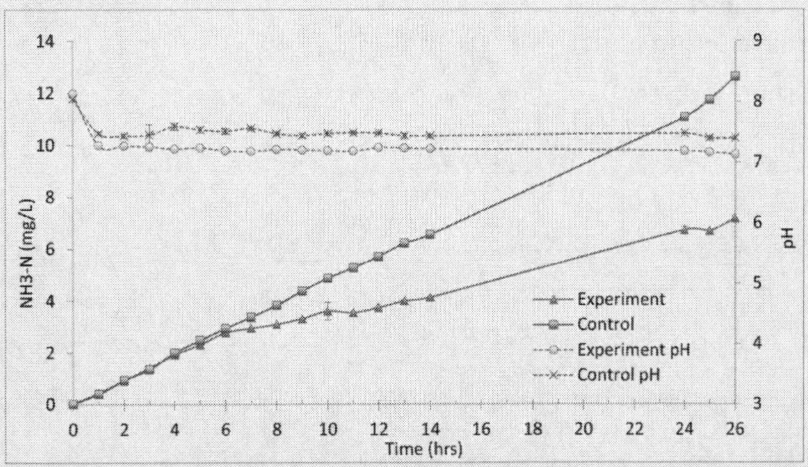

$k_{experiment}$= 2.5mg NH_3-N/hr
$k_{control}$= 4.71 mg NH_3-N/hr

17

In-vivo experiment (seawater)

Experiment

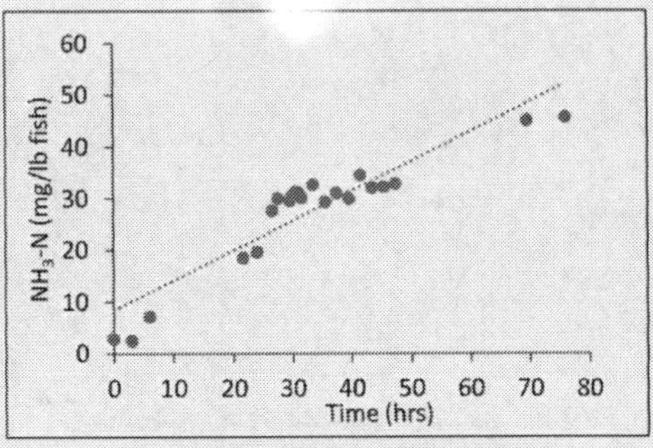

K= 0.57 mg NH_3-N/lb fish/ hr

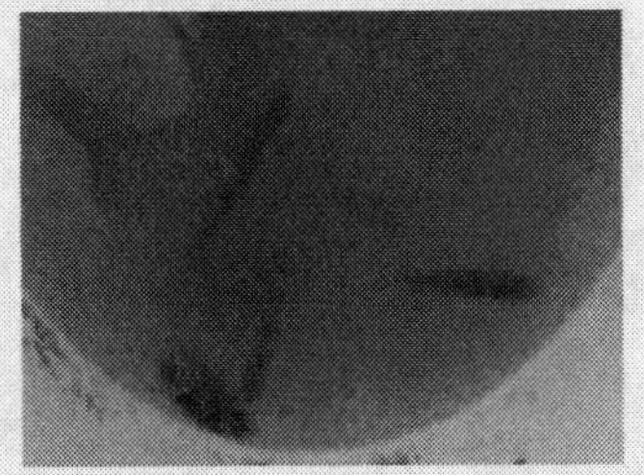

Control

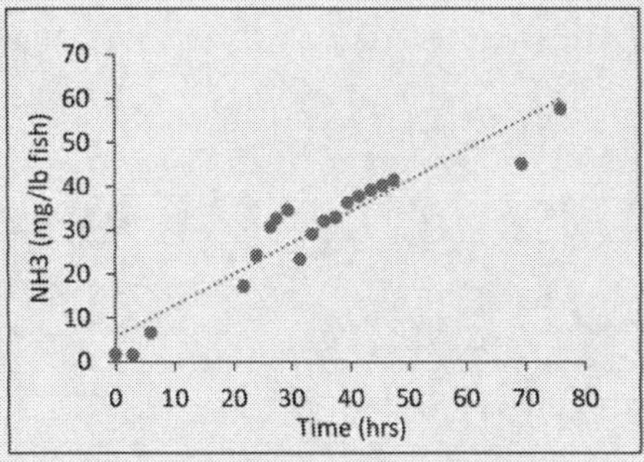

K= 0.76 mg NH_3-N/lb fish/ hr

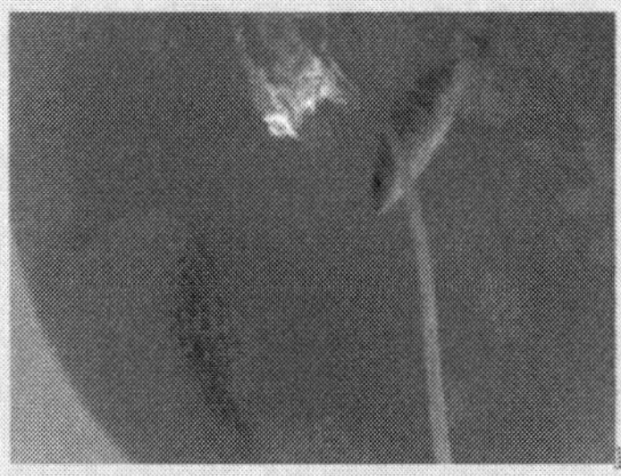

18

Economic analysis

- Commercial water conditioner price $4-$10
- “Chabazite+ dye+ buffer” 3 in 1 hybrid

Component	Price/day
Chabazite	$0.84
Buffer	$3.52
Bag	$0.56
Others	$1
Total	**$4.92**

- Chabazite zeolite is a renewable material
- Sustainable

19

Conclusions

Chabazite based ammonia removal product is

1. User-friendly
2. Sustainable Chabazite zeolit
3. Affordable “ rid
4. Capable of extending th tored fish

Component	Price/day
Chabazite	$0.84
Buffer	$3.52
Bag	$0.56
Others	$1
Total	**$4.92**

by removing ammonia safely

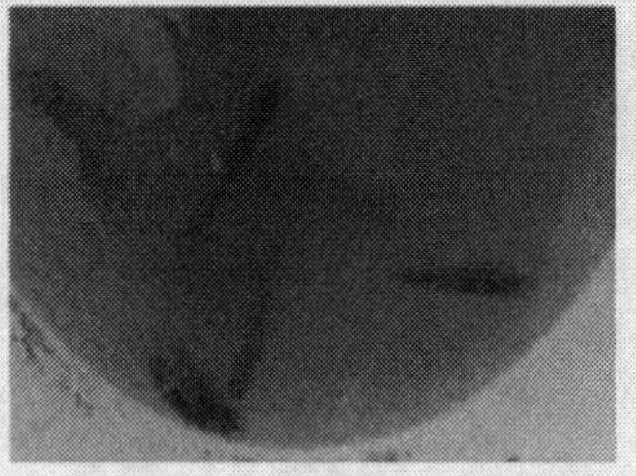

20

- Thank you for the following institutions and organizations for their financial support for this project:
  - Marine Metal Products Inc.
  - I4 FHTC Contract No. 13-22- Seed Grant
  - Gulf of Mexico Research Initiative via C-MEDS
- Thank you for Dr. Norma Alcantar and Dr. Sarina Ergas for their instruction
- Thank you for my group members help

- Thank you every audience
- <u>Questions?</u>

21

388636 Production Water Recovery Processing and Reuse

Abstract:

Production water is a product of natural gas and natural gas liquids recovery operations. It is primarily derived of flow back water from the well operations. The recovery, processing and reuse of production water is becoming increasingly the key to unlocking the potential of natural gas resources. Without the ability to process the volumes of production water, the use of water for the natural gas recovery operation becomes limited. The natural gas and the natural gas liquids that can be brought to market using the emerging drilling and recovery methods represents a significant long term opportunity.

Presented is a production water processing system. The integrated system identifies major constituent disposition. The final products include water meeting both re-injection and surface water reuse quality specifications. The removed constituents are produced in a form made available for feedstock to other downstream chemical operations. The non-recoverable wastewater and constituent are minimized. The economics of the system design are presented in generally accepted financial metrics for chemical operations investment.

388655 Polyaniline Coated Ethyl Cellulose with Improved Hexavalent Chromium Removal

Abstract:

The adsorption method has been widely used for the removal of toxic hexavalent Chromium (Cr(VI)) from wastewater.[1, 2] Cellulose is considered as a good adsorbent for Cr(VI) removal due to its sustainability and abound active sites.[3] But the lower adsorption capacity and slow removal rate limit its application. Cellulose has been wildly used as the adsorbents for Cr(VI) removal due to its cost effectiveness. Ethyl cellulose (EC) modified with 5.0, 10.0 and 20.0 wt% polyaniline (PANI) (PANI/ECs) were prepared by homogenous mixing the EC and PANI formic acid solutions. The synthesized PANI/ECs demonstrated a superior hexavalent chromium (Cr(VI)) removal performance than pure EC. The Cr(VI) removal percentage increased with increasing the PANI loading. PANI/ECs with 20.0% PANI loading were noticed to remove 2.0 mg/L Cr(VI) completely within 5 min, which is faster than the ethyl cellulose (>1 h). Faster Cr(VI) removal was observed in acidic solutions than in basic solutions. The Cr(VI) removal behavior by PANI/ECs followed the pseudo-second-order kinetic. The adsorption isotherm study demonstrated that the adsorbents follow the Langmuir model with calculated maximum absorption capacities of 19.49, 26.11 and 38.76 mg/g for the 5.0, 10.0, and 20.0 wt% PANI/ECs, respectively, which are much higher than that of EC (12.2 mg/g). The Cr(VI) removal mechanisms explored by FT-IR and XPS involved the reduction of Cr(VI) to Cr(III) by oxidation of PANI and EC.

[1] Gu, H., Rapole, S. B., Sharma, J., *et al.,* Magnetic polyaniline nanocomposites toward toxic hexavalent chromium removal. RSC Adv. **2012**, 2 (29), 11007-11018.

[2] Zhu, J., Wei, S., Gu, H., *et al.,* One-Pot Synthesis of Magnetic Graphene Nanocomposites Decorated with Core@Double-shell Nanoparticles for Fast Chromium Removal. Environ. Sci. Technol. **2011**, 46 (2), 977-985.

[3] Wang, S.L., Lee, J.F., Reaction mechanism of hexavalent chromium with cellulose. Chem. Eng. J. **2011**, 174 (1), 289-295.

389785 Study of Fluoride Adsorption Onto Nanostructured Zirconium-Manganese Based Particle

Abstract:

Fluoride is of a great concern as it is one of the major inorganic pollutants found in groundwater. In this study, a nanostructured zirconium-manganese binary hydrous oxide was prepared for the removal of fluoride from aqueous solution. Kinetics study revealed that the most uptake of fluoride took place within the first 6 h. The adsorption isotherm was well described by Langmuir equation, and the maximum adsorption of 71 mg/g was found at the pH between 2.5 to 3.5. The presence of humic acid and sulphate ions had limited influences on the adsorption. However, the presence of phosphate and silicate ions reduced the uptake of fluoride. The ionic strength had less significant effect on the adsorption. The adsorption of fluoride onto the sorbent surface was further confirmed by EDX, FTIR and XPS analysis. The presence of clear F 1s peak in the XPS analysis confirmed the binding of fluoride on the surface of sorbent. Finally, the cytotoxicity study by the MCF7 cell line demonstrated that the toxicity of nanoparticle was not an issue of concern. Our study illustrate that the nanoparticle may serve as a better material for the removal of fluoride from contaminated water.

390378 Removal of Chromium (VI) and Divalent Cations from Flowback Water By Graphene Oxide-Modified Natural Zeolite

Abstract:

Hydraulic fracturing in shale gas plays across the United States uses fluids and chemicals with high concentrations of scaling forming ions such as strontium and barium, and heavy metals such as hexavalent Chromium. Removal of these chemical from flowback water may increase the potential for recycling and reuse of flowback water. The Hexavalent Chromium may be a pollutant of concern associated with gas and oil-related wastewaters. Divalent strontium and barium may combine with sulfate and carbonate and build scaling precipitates.

Natural zeolites have demonstrated to be good ion-exchange and adsorbent materials. They can exchange and retain cations inside of their porous surface. Increasing the surface area of the natural material becomes a challenge to improve the adsorption removal efficiency of the natural zeolite. In this paper results on the adsorption of hexavalent chromium and divalent Strontium and Barium from flowback water using graphene oxide-modified zeolite are presented.

Adsorptive removal of humic acids and microbes by polyacrylonitrile-chitosan composite membrane

*Swapna Rekha panda, Munmun Mukherjee, Sirshendu De**

Abstract

The present study describes fabrication of PAN UF membrane using chitosan (CS) as an antibacterial coating material. The efficacy of cast membrane to remove natural organic matter in terms of humic acid (HA) content and microbes was evaluated using pond water as feed. The presence of amino groups in CS provides an appropriate adsorption site to the anionic HA as well as microbes. Chitosan coating lowers the MWCO and makes the cast membrane dense. Chitosan coating also increases the surface hydrophlicity of the composite UF membranes. SEM images confirm the formation of CS layer on the top membrane surface. The cast membrane exhibited high antimicrobial property and good rejection of humic acid. The flux decline was marginal due to the nominal hydraulic resistance offered by the humic acid deposits on the membrane surface. Reversible membrane fouling was identified. Whatever flux declination occurred due to irreversible fouling was, because of the adsorption of HA on the pores of PAN UF membranes. The adsorption of humic acid and microbes on composite membrane was confirmed by SEM and FTIR respectively.

1.0 Introduction

The use of membrane filtration is increasing day by day at an alarming rate to meet the stringent drinking water quality regulations [1]. Microfiltration (MF) membranes are capable to remove the suspended particles, turbidity, micro-organisms like bacteria, algae, and protozoan; while, ultrafiltration (UF) membranes can remove water born viruses and dissolved organic matters. [2]. Major issue in membrane separation is membrane fouling [3]. Recent literature highlights that, humic acids (HA) as a major natural occurring foulant in surface as well as stilled water [4, 5]. In addition to that, the presence of HA beyond an acceptable limit gives bad odor, color problems to water [6]. Additionally, HA also form complexes with heavy metals, synthetic organic chemicals [7], and chlorine present in water bodies during water treatment process forming carcinogenic disinfectant by-products [8, 9]. These major issues dragged into a motivated extensive research focusing on the use of a novel separation technique i.e. "membrane filtration", an easy, economic and lower waste generating method. Still flux declination and membrane fouling are the major inherent problems associated with membrane filtration. Coating, blending and grafting (by hydrophilic polymers) are the different means of membrane modification methods that could be adopted to enhance the antifouling properties of the membranes [10]. Chitosan is a hydrophilic, biocompatible, non toxic, antibacterial and biodegradable biopolymer of our interest [11, 12], used in our present study to coat the PAN hydrophobic membrane. Chitosan a natural adsorbent, having amino and hydroxyl groups in their active sites, used mostly to remove heavy metals, fluorides and dyes from polluted water [13].

2.0 Materials and Methods

PAN Co-polymer (average molecular weight 250 kDa) procured from M/s, Technorbital (Kanpur India), was used as a base polymer. Solvents, N, N- dimethylformamide (DMF) was purchased from M/s,

Merck (India) Mumbai Ltd. Polyethylene glycol of (PEG) average molecular weight 400 Da, 4 kDa, 6 kDa, 10 kDa, 20 kDa and 35 kDa was supplied by M/s, S R Ltd, Mumbai, India and dextran (average molecular weight: 70 kDa) was procured from M/s, Sigma Chemicals, USA. These neutral solutes were used to during the solute rejection study of the membranes. Distilled water was used as the non-solvent in the coagulation bath. All chemicals used were of analytical grade without further purification.

2.1 Preparation of casting solution

In our present study, polyacrylonitrile (PAN) UF membrane at a polymer concentration of 15 wt% in N,N-dimethylformamide (DMF) solvent was cast using phase inversion method. In the first step of casting of membranes, non-woven polyester fabric of thickness 118±22.8 μm (product number TNW006013, supplied by M/s, Hollytex Inc., New York, USA) was attached to a clean glass plate using adhesive tape. The PAN solution was drawn down the fabric using a Doctor's blade with an adjustable thickness fixed at 200 μm. Membranes was cast at room temperature (25 ± 2^0 C). The laboratory cast membrane was then subjected to coating with a hydrophilic biopolymer chitosan (CS), to enhance the hydrophlicity, filtering efficiency, permeability and antimicrobial activity of the membranes. A layer by layer method was adopted for the preparation of the final PAN-CS coated composite UF membrane using glutalaldehyde as a cross linking agent. The coating layer of chitosan was prepared by dissolving 0.6 gm of CS in 200 ml of 0.4 % (v/v) acetic acid solution with continuous stirring. Once the coating process was completed, the membrane was immediately dipped into a gelation bath containing 1N NaOH solution for 30 min to ensure the completion of neutralization process. Then the membranes were cleaned thoroughly to remove the excess NaOH adhered to the membrane surface. Finally, the membranes are cut into appropriate size with an effective area of 34 cm^2 and set for batch UF operation.

2.2 Ultrafiltration experiments

Ultrafiltration experiments were performed using a batch cell and pure water permeation flux was evaluated. Effective area of the membrane in the module was 34 cm^2. First, the cell was filled with 500 ml of distilled water and membranes were compacted at 690 kPa for 3 to 4 hours. The permeate flux was calculated by

$$J_w = Q / A.\Delta T \tag{1}$$

Where, Q is the volumetric flow rate of permeating water, A is effective membrane area, ΔT is the sampling time. Next, the value of steady state permeate flux was noted at five transmembrane pressure drop. A plot of J_w with transmembrane pressure drop resulted into a straight line through the origin. From the slope of this curve membrane hydraulic permeability was obtained. Whereas, cross flow filtration set up was used to do the other application studies like surface water treatment for microbes and humic acids removal.

2.3 Porosity (ε) and contact angle (CA)

Membrane porosity plays an important role in its performance. Membrane porosity was measured by the mass loss of wet membrane after its drying. The membrane, soaked with distilled water was weighed after mopping superficial water with filter paper. Then, the wet membrane was placed in an air-circulating oven at 60 °C for 24 h and then further dried in a vacuum oven before measuring the dry weight until a constant mass was obtained. From the two weights (wet sample weight, w_0 and dry sample weight, w_1), the porosity of the membranes was calculated using the following equation [14],

$$\varepsilon = \frac{w_0 - w_1}{\rho_w A l} \times 100 \tag{2}$$

Where, ε is the membrane porosity, A is the membrane surface area, l is the membrane thickness and ρ_w is the water density. In order to minimize the experimental errors, the membrane porosity of each sample was measured three times and the average values were reported. The contact angle was measured by a Goniometer, supplied by M/s, SEO, South Korea (Model number, Phonix 300), and using sessile drop method. The contact angle was measured at six different locations of the membrane and the average value was reported.

2.4 Molecular weight cut off (MWCO) of the membrane

Solute rejection measurements were carried out in a stirred batch cell. Solutions were prepared of 20 kg/m^3 using different neutral solutes. The neutral solutes in the permeate stream were measured using a refractometer of model Abbe type, supplied by M/s, Excel International Ltd., Kolkata (accuracy 0.0001). Rejection values were plotted against the molecular weight of solutes in a semi-logarithmic curve. Molecular weight corresponding to 90 % rejection was estimated as molecular weight cut off. The rejection was computed as,

$$R = 1 - C_p/C_0 \tag{3}$$

Where, C_p and C_0 are the concentration of solute in feed and permeate, respectively.

2.5 Scanning Electron Microscopy (SEM) of Ultrafiltration membranes

The skin of the membrane was carefully removed from the substrate. The membrane was fractured dipping it in liquid nitrogen for some time then sputter coated with gold. The coated membranes were viewed by JEOL, Japan of model ESM-5800 with an operating voltage of 20kv.

2.6 Antibacterial study of the UF membrane

The antibacterial activity of cast membrane was explored by means of colony forming unit (CFU) count and most probable number (MPN) method using tap water. All experiments were carried out in batch cell. Adsorption of *Escherichia coli* on MMM was also quantified.

2.7 Colony forming unit count

Diluted samples of o.1 ml was taken and incubated on agar plate using a spreader at 37°C for 24 hrs. This process was repeated for three plates. Colonies were counted visually. The average values of CFU/ml were reported. The following equation was used to calculate the number of colony forming units (CFU/ml)

$$\text{CFU/ml} = \frac{\text{CFU/plate} \times \text{dilution factor}}{\text{volume of sample taken (ml)}} \tag{4}$$

2.8 Most probable number (MPN)

MPN test was also carried out in feed and permeate samples to investigate the effectiveness of cast membranes on removal of coliforms. Separate analyses were conducted on five parts of each of three serial dilutions of feed and permeate. The individual portion was used to inoculate tubes of lauryl tryptose broth culture medium and incubated at 35°C for 48 hours. The presence of coliforms was indicated by turbidity or presence of gas in the culture medium. The MPN index was determined by comparing the pattern of positive results (the number of tubes showing growth at each dilution) with statistical tables. The obtained value was reported as MPN per 100 ml of the sample.

2.9 Membrane adsorption study

The adsorption experiments with the surface feed was performed in batch mode. Different concentrations feed solutions were prepared in a conical flask. A piece of membrane size 2×2 cm was dipped in to the solution. Sample bottles were kept in a continuous shaker at 35°C for 24 hours. Adsorption of humic acid on the membrane surface was measured using U-V spectrophotometer. For bacterial adsorption 100 ml *E. coli bacterial* culture solutions of different concentrations (1.32×10^6, 1.57×10^6, 1.77×10^6, 2.01×10^6 and 2.41×10^6 CFU/ml) were taken in six conical flasks. 0.1 g of membrane was dipped into each flask and the conical flasks were gently shaken at 37°C for 24 hours at 150 rpm. The adsorption efficacy of membrane was calculated by the difference between initial and final concentration of bacteria and adsorption capacity was expressed as number per gram of the membrane. Equilibrium adsorption of the membrane was carried out. The adsorption isotherm followed Langmuir isotherm.

$$q_e = \frac{kv_m C_e}{1+kC_e} \qquad (5)$$

3. 0 Results and Discussion

Pure water permeability of the membrane was determined and shown in Fig. 1, using Eq. 1 (discussed in section 2.3). As expected, higher TMP gives rise to higher flux. The results indicated the use of CS affected the pore size distribution and the membrane morphology significantly. The permeability was found to be 5.44×10^{-11} m/Pa.s. The corresponding porosity and contact angle data for the said membrane was found to be 58 and 65 ± 2^0, respectively. With the coating of CS, the membrane becomes porous but the membrane thickness was increased compared to the control membrane. The MWCO of the membrane is presented in Fig. 2. It was found that 90% of solute rejection was obtained for the solute molecular weight 85 kDa. Hence, the resulting membrane has a MWCO value found to be 85±2.2 kDa. The cross section and top views of the membrane was observed for the freshly prepared membrane and shown in Fig. 3(a) and 3(b), respectively. As seen from Fig. 3(a), the membrane has a porous structure with some finger like macrovoids that extends over the entire sub layer part. Whereas, the top layer i.e., membrane skin layer has consists of small tiny pores. The findings (Fig. 3a) indicate that, chitosan coating makes the membrane more porous. The top view of the membrane (Fig. 3b), is found to have some open pores visible at a magnification 3000X. However, the SEM images confirm the formation of CS layer on the top membrane surface.

The antibacterial activity of cast membrane was explored by means of colony forming unit (CFU) count and most probable number (MPN) method using pond water as a feed. In feed the concentration of total microorganisms (CFU/ml) and total coli forms (MPN/100ml) is 9.2×10^7 CFU/ml and 300/100ml respectively. However, the composite membrane completely removed total microorganisms (CFU/ml) and total coli forms (MPN/100ml). The adsorption study of HA on the membrane was evaluated spectrophotometrically at 254 nm. The amount of HA (solutes q_e) adsorbed was calculated from mass balance and plotted against equilibrium concentration to get the desired Langmuir isotherm shown in Fig. 4. The maximum adsorption capacity of the composite membrane was found to be 53 mg/g (Fig. 4a). The extent of flux decline with pond water as a feed was marginal for the cast membrane due to the nominal hydraulic resistance offered by the humic acid deposits on the membrane surface. Reversible membrane fouling was identified. Whatever scant flux declination occurred due to irreversible fouling was, because of the adsorption of HA on the pores of PAN UF membranes. Adsorption of *Escherichia coli* on PAN-CS composite membrane was also quantified. The bacterial adsorption using PAN-CS coated composite membranes followed Langmuir isotherm (Fig. 4b). The maximum adsorption capacity (V_m) and adsorption constant (k) were found to be 1.2×10^8 CFU/g and 1.1×10^{-6} ml/CFU, respectively. SEM image after bacterial adsorption of the composite membrane was reported at 5000X magnification. SEM image confirms adsorption of bacteria (*E. coli*) on the membrane surface as shown in Fig. 5.

4.0 Conclusion

Asymmetric PAN UF membrane was prepared by phase inversion method. The effect of CS coating on the fundamental characteristics features like pure water flux, contact angle, porosity, membrane morphology, MWCO etc., were investigated. The antibacterial study of the membrane complete removal of total microorganisms and total coli forms. However, rejection of HA by this membrane was achieved up to 80% only.

5.0 References

[1] AWWA Membrane Technology Research Committee. *J. AWWA* **1992**, *84*, 59.

[2] W. Yuan, A.L. Zydney, Humic acid fouling during ultrafiltration, *Environ. Sci. Technol.* **2000**, 34, 5043.

[3] Lahoussine-Turcaud, V. Wiesner, M. R. Bottero, J. Y. *J. Membrane Sci.* **1990**, *52*, 173.

[4] Y. Kaiya, Y. Itoh, K. Fujita, S. Takizawa, *Desalination* **1996**, *106*, 71.

[5] J.L. Bersillon, In *Future Industrial Prospects of Membrane Processes*; Cecille, T. L., Ed.; Elsevier: Oxford, **1988** 234.

[6] J. Lin, Y. Zhan, Adsorption of humic acid from aqueous solution onto un modified and surfactant-modified chitosan/zeolite composites, *Chem. Eng. J.* **2012** 200.

[7] C.J. Li, Y. Dong, D.Y. Wu, L.C. Peng, H.N. Kong, Surfactant modified zeolite as a dsorbent for removal of humic acid from water, *Appl. Clay Sci.* **2011** 52 353.

[8] A. Imyim, E. Prapalimrungsi, Humic acids removal from water by amino propyl functionalized rice husk ash, *J. Hazard. Mater.* **2010** 184 775.

[9] Q. Tao, Z.Y. Xu, J.H. Wang, F.L. Liu, H.Q. Wan, S.R. Zheng, Adsorption of humic acid to amino propyl functionalized SBA-15, *Micropor. Mesopor. Mater.* **2010** 131 177.

[10] J.F.Hester, P.Benerjee, A.M. Mayes, Preparation of protein-resistant surfaces on poly(vinylidene floride) membranes via surface segregation, *Macromolecules* **1999** 32 1643.

[11] D.A. Musale, A. Kumar, G. Pleizier, Formation and characterization of poly(acrylonitrile)/chitosan composite ultrafiltration membranes, *J. Membr. Sci.* **1999** 154 163.

[12] U.K. Aravind, J. Mathew, C.T. Aravindakumar, Transport studies of BSA, lysozyme and ovalbumin through chitosan/polystyrene sulfonate multilayer membrane, *J. Membr. Sci.* **2007** 299 146.

[13] W.S.W. Ngah, L.C. Teong, M.A.K.M. Hanafiah, Adsorption of dyes and heavy metal ions by chitosan composites: a review, *Carbohydr. Polym.* **2011** 83 1446.

[14] J.F. Li, Z.L. Xu, H. Yang, L.Y. Yu, M. Liu, Effect of TiO_2 nanoparticles on the surface morphology and performance of micro porous PES membrane, Appl. Surf. Sci. 255 (2009) 4725.

6.0 List of Figures

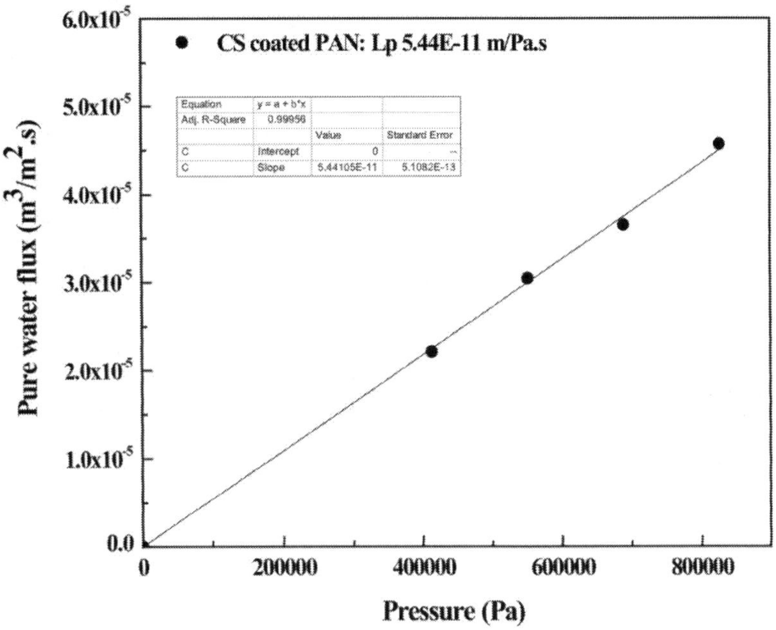

Fig. 1 The pure water flux of CS coated PAN membrane as a function of transmembrane press (TMP).

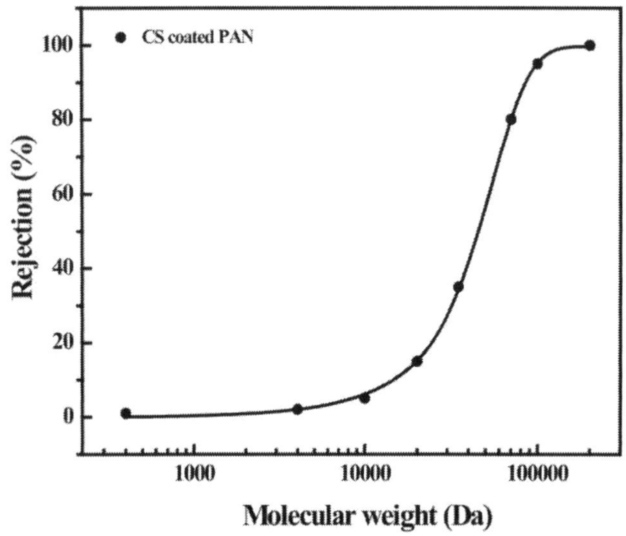

Fig. 2 The MWCO of CS coated PAN membrane.

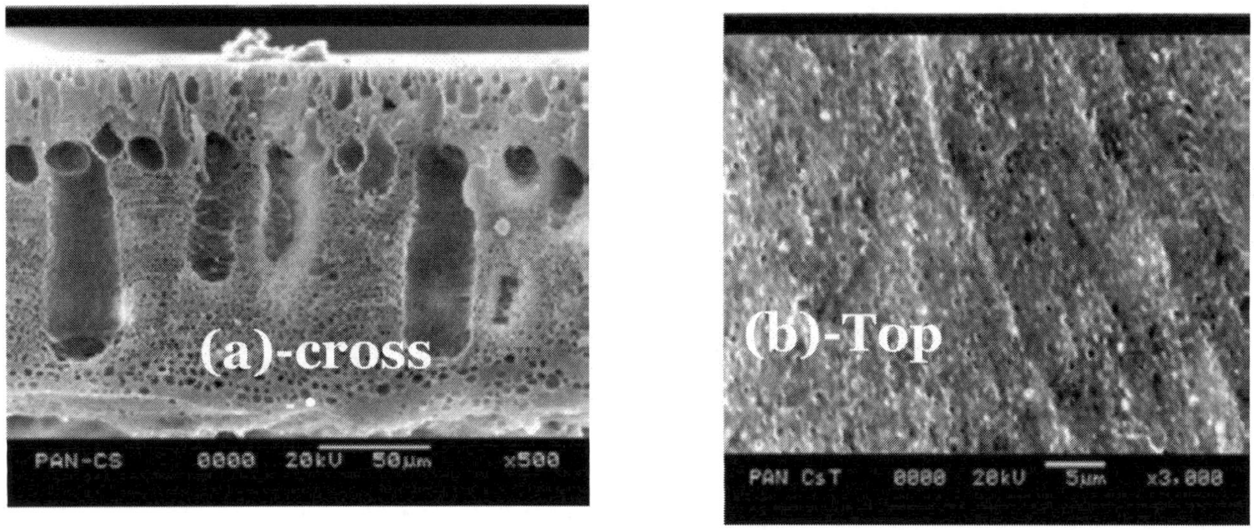

Fig. 3 Scanning electron microscopy of CS coated PAN membrane (a) cross section (b) top view.

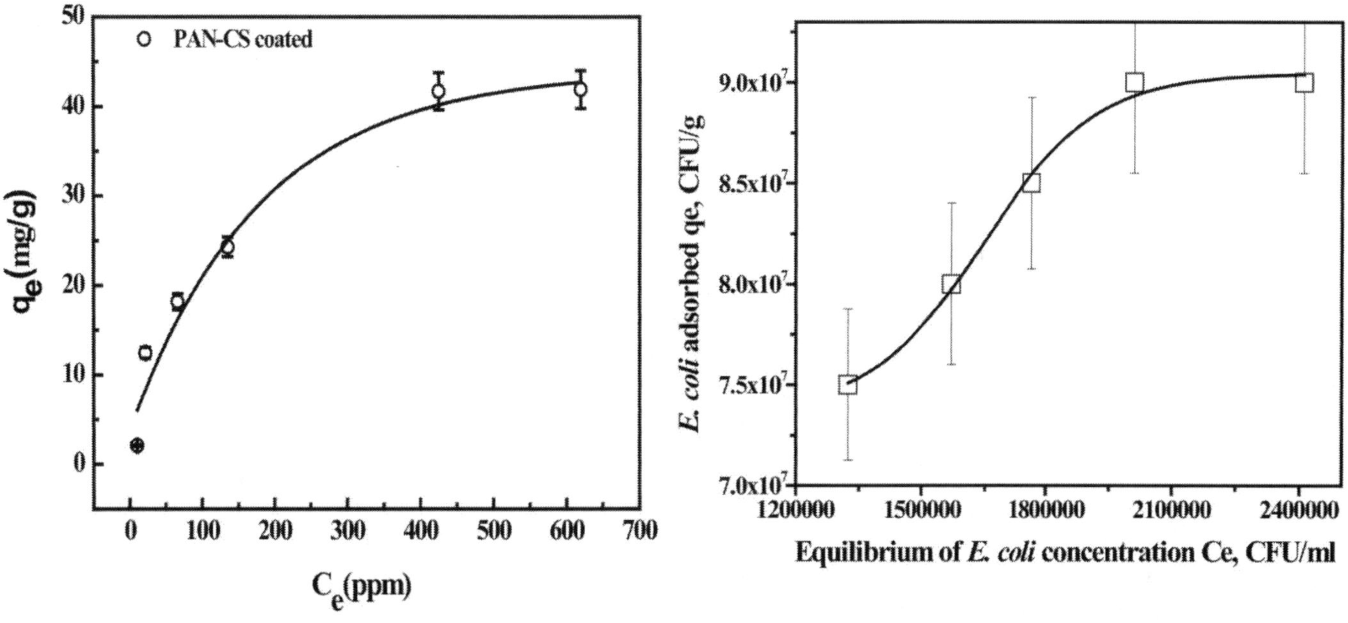

Fig. 4 Langmuir isotherm of CS coated PAN membrane (a) Adsorption of humic acid onto PAN-CS membrane (c) Adsorption of *E. coli* onto PAN-CS membrane.

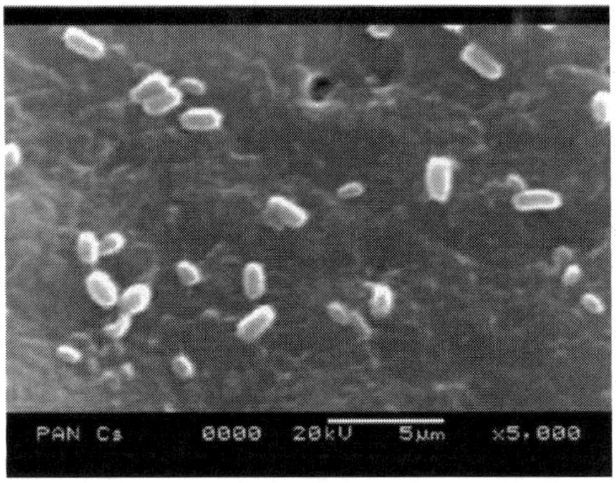

Fig. 5 SEM image showing adsorption of *E. coli* on the membrane top surface.

377360 Removal of Gaseous O-Xylene in the Two-Liquid Phase Biotrickling Filters and Airlift Bioreactors

Abstract:

O-xylene is the most recalcitrant among the three xylene isomers which are released into the atmosphere as a kind of fugitive emission from industrial production with healthy risk and environmental pollutions. Biotechnologies show off their advantages of completely degrading low-concentration contaminants into end-products like carbon dioxide, water, biomass, salts even biofuels since the 1920s. However, the practically insolubility of o-xylene may cause obvious deterioration in the mass transfer, thus reducing the performance of bioreactors.

Recently, an emerging biotechnology for removal of hydrophobic volatile organic compounds (VOCs) is the two-liquid phase bioreactor (TLPB). In a TLPB, an organic liquid-phase, which is always immiscible, non-biodegradable and biocompatible, is added to for an improvement of the VOC dissolution. A large variety of bioreactor configurations have been utilized in the research of TLPB so far, of which the continuous stirred tank bioreactor (CSTB) is the most popular system. Nevertheless, as a traditional bioreactor, the biotrickling filter (BTF) also showed an elimination capacity improvement of more than 100% relative to the one without silicone oil as non-aqueous phase. Although the airlift bioreactors (ALBR) didn't have a significant enhancement after adding silicone oil, it would be different for the abatement of gas-phase o-xylene. Actually, the literatures about the removal of o-xylene from off-gas in the BTF and ALBR using silicone oil as non-aqueous phase are sparse.

The main aim of this research was to make a comparison observation on the performance of a two-liquid phase (TLP) BTF and ALBR for the biodegradation of o-xylene existing in contaminated air under different empty bed residence times (30-60 s), inlet loading rates (0-100 g m^{-3} h^{-1}) as well as the response to shock-loads. The pilot plant consisted of two identical bioreactors with or without silicone oil to water phase ratio of 5% operating in parallel. Removal efficiency increased as the EBRTs were improved and inlet concentrations of o-xylene were decreased, exhibiting higher elimination capacities at higher inlet loading rate. Under shock-load condition, adding 5% silicone oil, removal efficiency of the biotrickling filter was dropped 23%, while that of the airlift bioreactors was essentially maintained at approximately 90% when the inlet concentration was increased from 500 mg m^{-3} to 1500 mg m^{-3}. When inlet concentration was recovered, the two-liquid phase airlift bioreactor could restore its removal efficiency to 100%, but the two-liquid phase biotrickling filter could not, which reveals that the former was more stable than the latter because of fully stirring and protection from silicone oil.

Effect of Various Fibrous Filter Media and Newly Designed Packaging Configurations on Sea Salt Particles Loading Performance

AUBURN UNIVERSITY

SAMUEL GINN COLLEGE OF ENGINEERING

2014 AICHE Conference

Nov 16-21, 2014

Pengfei Zhao

Center of Microfibrous Material Manufacturing (CM[3])

Department of Chemical Engineering

Auburn University, Auburn, AL

1

AUBURN UNIVERSITY

SAMUEL GINN COLLEGE OF ENGINEERING

Outline

- ❑Motivation
- ❑Objective
- ❑Introduction
- ❑Estimation of particles loading
- ❑Effect of various filter media on particles loading
- ❑Effect of filters packaging configurations on particles loading
- ❑Summary

2

AUBURN UNIVERSITY
SAMUEL GINN COLLEGE OF ENGINEERING

Motivation

- ❑ Removal of sea salt particles is a critical need for a number of current and future air breathing ocean platforms
- ❑ Sources
  - ✓ *Naturally from wind stresses over whitecaps*
  - ✓ *Wake or the bow spray by ship itself*

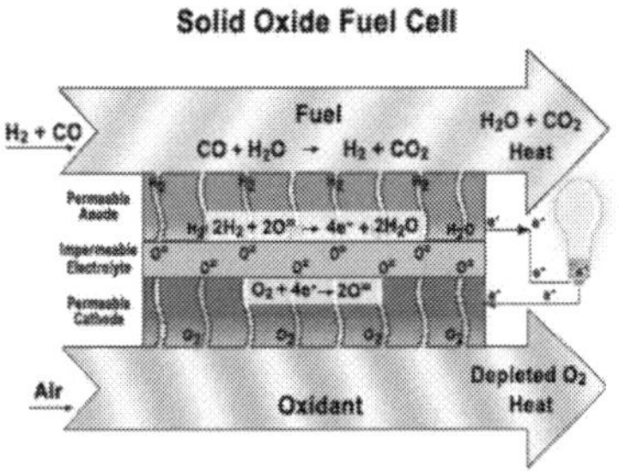

- ❑ Harmful effects of sea salt particles on fuel cell (SOFC) systems
  - ✓ *Clog the finely divided and highly tortuous flow channels in the fuel cells*
  - ✓ *Significantly degrade the performance of fuel cells (SOFCs)*

3

Sources: Raymond Majchrzak, Toby Cole, U.S. Navy Sea Salt Particle Removal Systems, Carderock Division Naval Surface Warfare Center

AUBURN UNIVERSITY
SAMUEL GINN COLLEGE OF ENGINEERING

Objective

- ❑ Significantly enhance sea salt particles loading capacity
  - ✓ *Longer filter lifetime*
  - ✓ *Lower cost for changing filters*
- ❑ Using pleated filters in novel designed packaging configurations- Approach I
- ❑ Using new type of filter media- Approach II

4

AUBURN UNIVERSITY
SAMUEL GINN COLLEGE OF ENGINEERING

Filter Design

- Two main aspects of filter design
  - ✓ *Filter pleats and configuration*

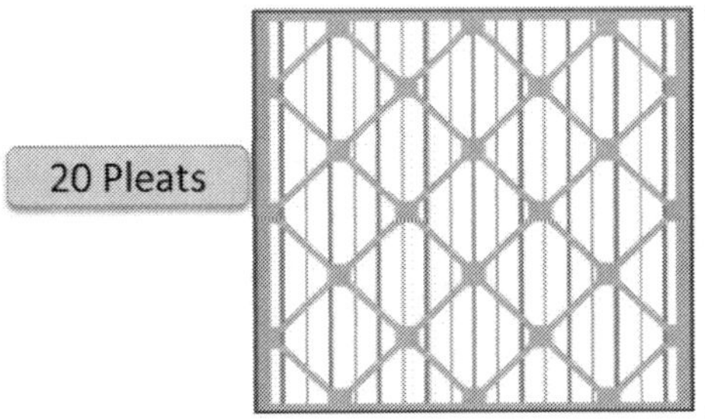

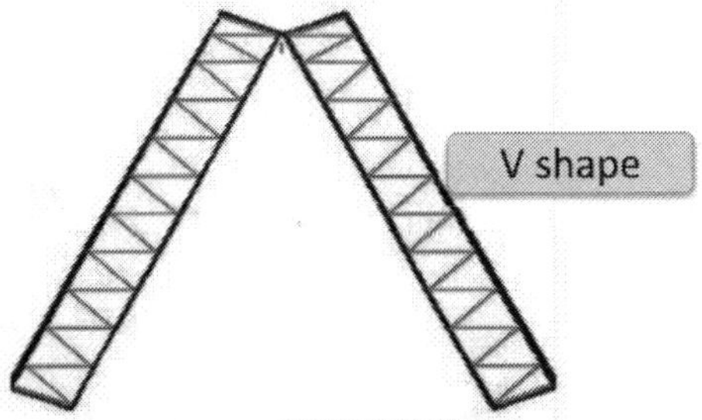

  - ✓ *Filter Media design*
    - ➢ *Woven*
    - ➢ *Nonwoven*
    - ➢ *Nanofiber entrapped*
    - ➢ *Nanofiber coated*

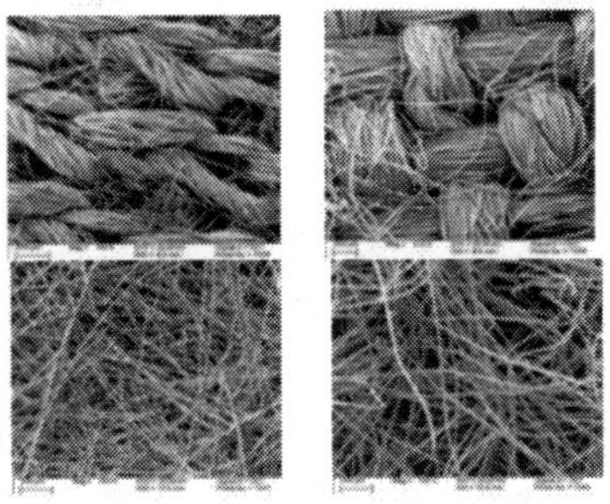

5

AUBURN UNIVERSITY
SAMUEL GINN COLLEGE OF ENGINEERING

Particles loading capacity

- In addition to pressure drop and filtration efficiency, particles loading capacity is another significant parameter that should be considered

- Particles loading capacity is defined as the quantity of contaminant a filter element can trap and hold before the maximum allowable back pressure or pressure drop is reached

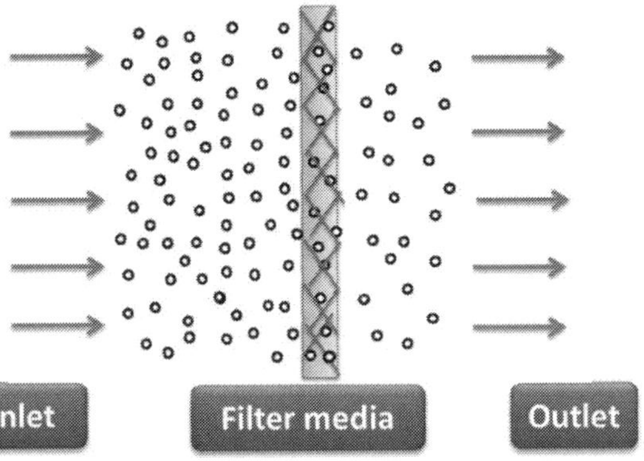

- Particle loading capacity is not only dependent on filter configuration (flow rate and media area), but also on filter media structure (depth filter and surface filter)

6

Estimation of Particles Loading

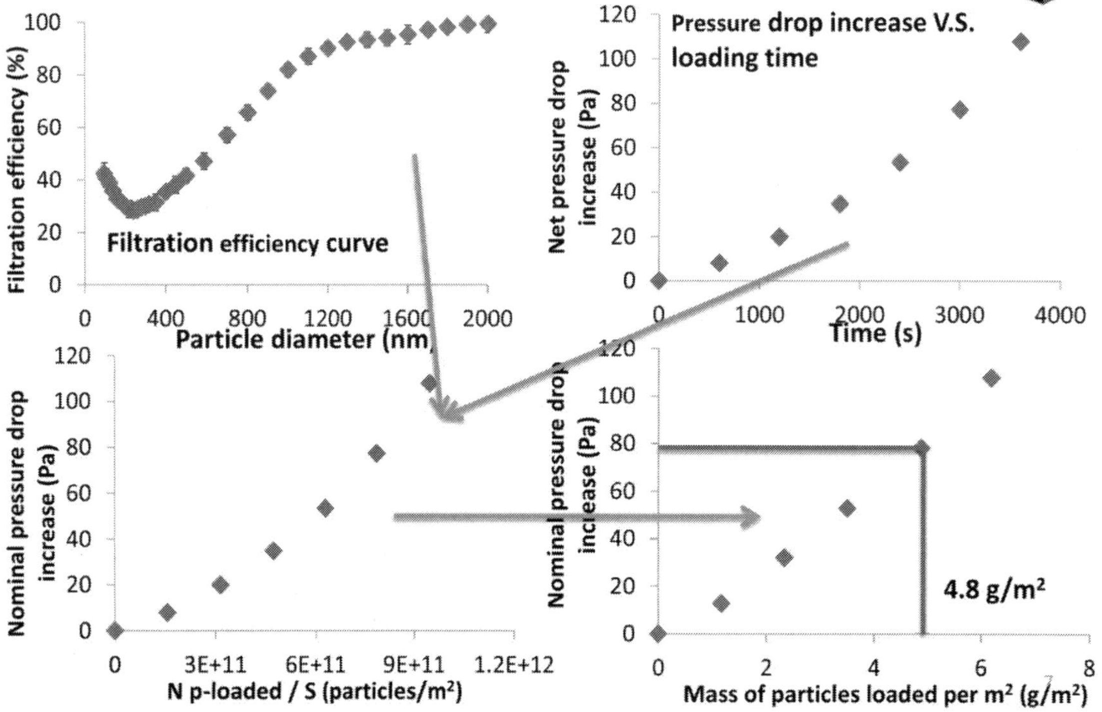

7

Estimation of Particles Loading (Cont)

- $Mass_{<200nm} = N_{<200nm} \times \pi/6 \times (D)^3 \times \rho$ — Spheres
- $Mass_{200-800nm} = N_{200-800nm} \times (D)^3 \times \rho$ — Cubes
- $Mass_{>800nm} = N_{>800nm} \times \pi/6 \times (1.4D)^3 \times \rho$ — Agglomerates
- $Mass_{Total} = Mass_{<200nm} + Mass_{200-800nm} + Mass_{>800nm}$

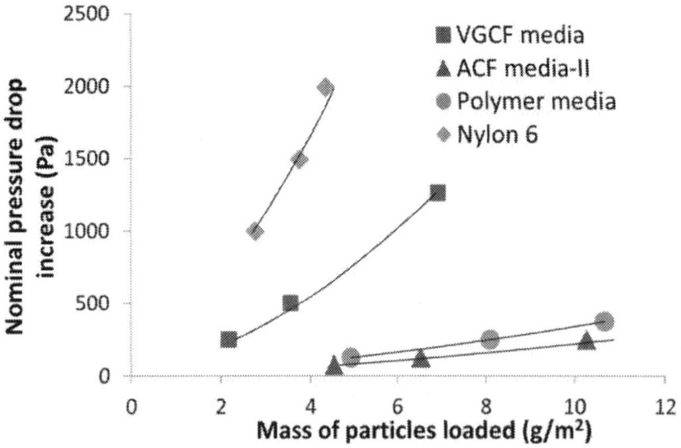

$$Error_{RMS}(\%) = \sqrt{\frac{\sum\left(\frac{\Delta P_{model} - \Delta P_{experimental}}{\Delta P_{model}}\right)^2}{n}} \times 100$$

	RMS
ACF media	4.37%
VGCF media	4.81%
Polymer media	2.45%
Nylon 6 media	3.03%

8

Experimental Setup (Bench scale)

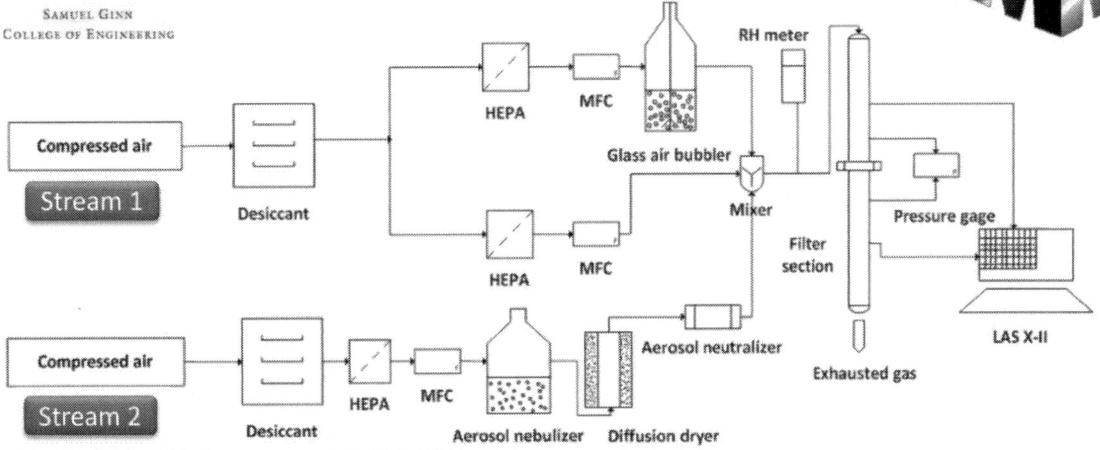

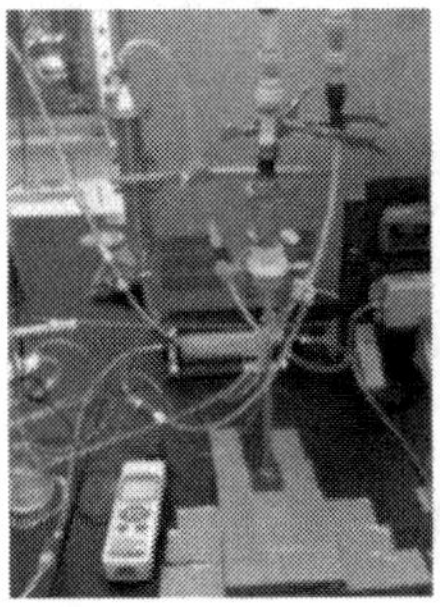

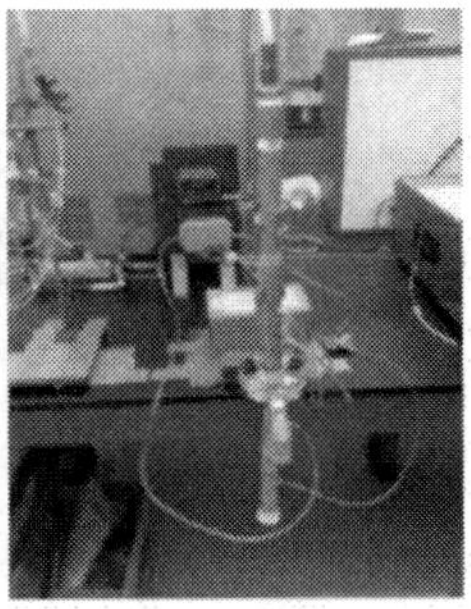

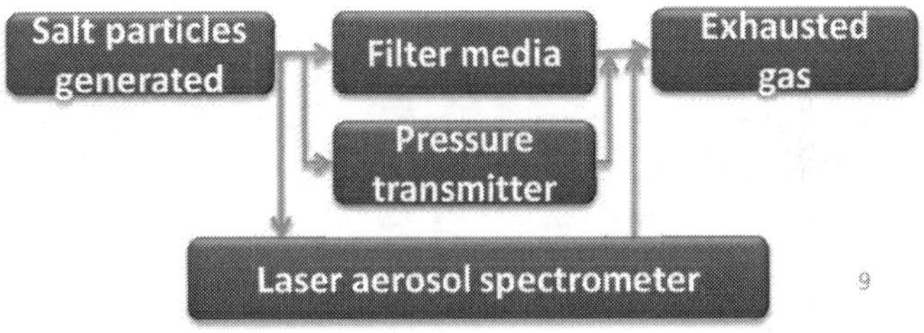

9

Effect of Various Filter Media on Particles Loading Performance

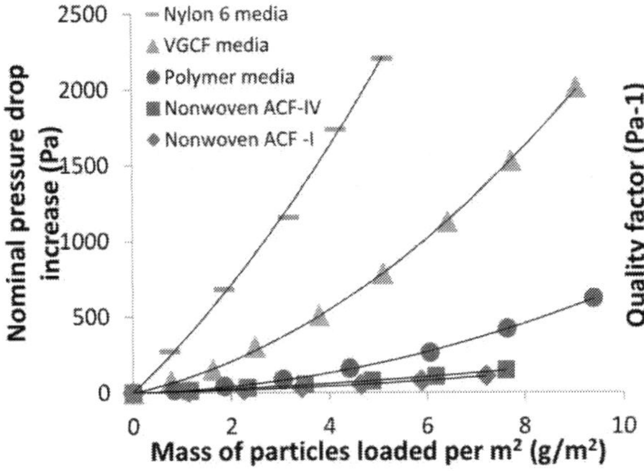

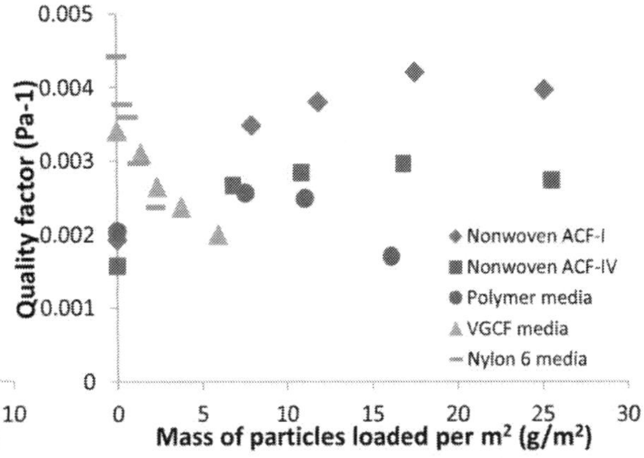

	Initial pressure drop (Pa)	Final pressure drop (Pa)	Particles loading capacity (g/m²)	Filter media life time (Days)
Nonwoven ACF-I	280	996	21.52	40
Nonwoven ACF-IV	471	996	17.47	31
Polymer filter media	119	996	11.36	17
VGCF filter media	362	996	4.43	5
Nylon 6 filter media	425	996	1.65	2

10

AUBURN UNIVERSITY
SAMUEL GINN COLLEGE OF ENGINEERING

Comparison of Particles Loading between Woven and Nonwoven Media

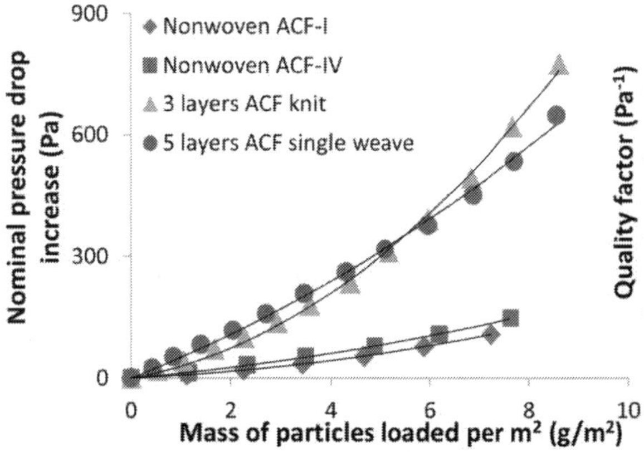

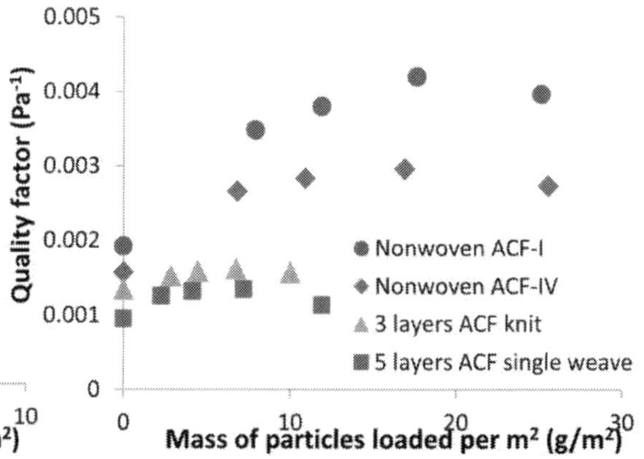

	Initial pressure drop (Pa)	Final pressure drop (Pa)	Particles loading capacity (g/m^2)	Filter media life time (Days)
Nonwoven ACF-I	280.0	996	21.52	40
Nonwoven ACF-IV	471.7	996	17.47	31
3 layers ACF knit	247	996	8.52	14
5 layers ACF single weave	293	996	9.30	15

11

AUBURN UNIVERSITY
SAMUEL GINN COLLEGE OF ENGINEERING

Schematic of Full Scale Test Rig

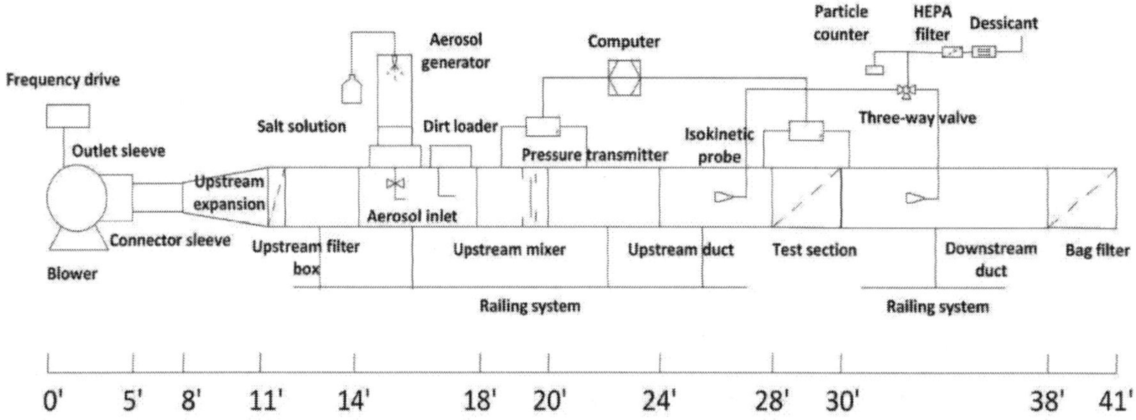

12

AUBURN UNIVERSITY
SAMUEL GINN COLLEGE OF ENGINEERING

Experimental Results of Single Filters

Merv 13 2 inch 15 pleats

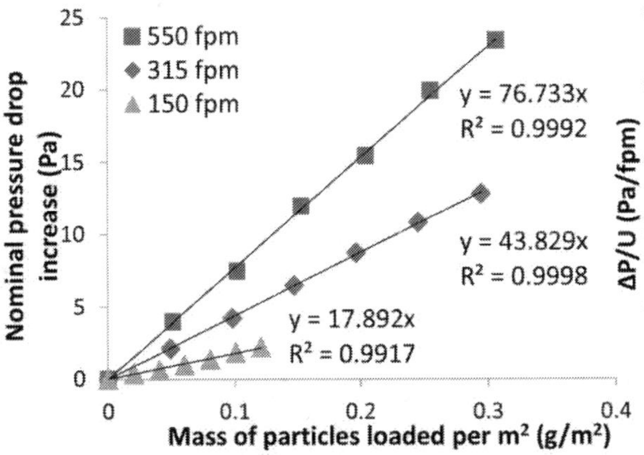

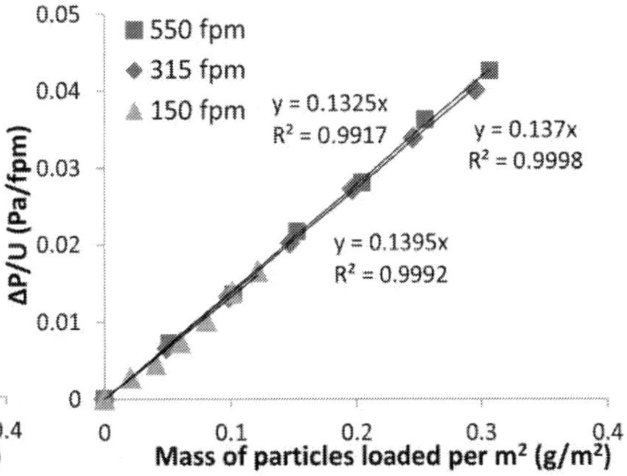

	Maximum mass of salt particles (g)	Maximum time (h)	Life time (Days)
550 fpm	1.4	8.9	4
315 fpm	4.5	27.1	14
150 fpm	12.2	72.8	36

$$\frac{\Delta P_1}{U_1} = K\frac{M_1}{A_1}, \frac{\Delta P_2}{U_2} = K\frac{M_2}{A_2}$$

$$M_1 = \frac{\Delta P_1 A_1}{U_1 K}, M_2 = \frac{\Delta P_2 A_2}{U_2 K}$$

$$\frac{M_1}{M_2} = \frac{A_1}{A_2}\frac{U_2}{U_1}$$

$$A_1 U_1 = A_2 U_2$$

$$\frac{M_1}{M_2} = (\frac{A_1}{A_2})^2$$

13

AUBURN UNIVERSITY
SAMUEL GINN COLLEGE OF ENGINEERING

Experimental Results of Single Filters

Merv 13 2 inch 550 fpm

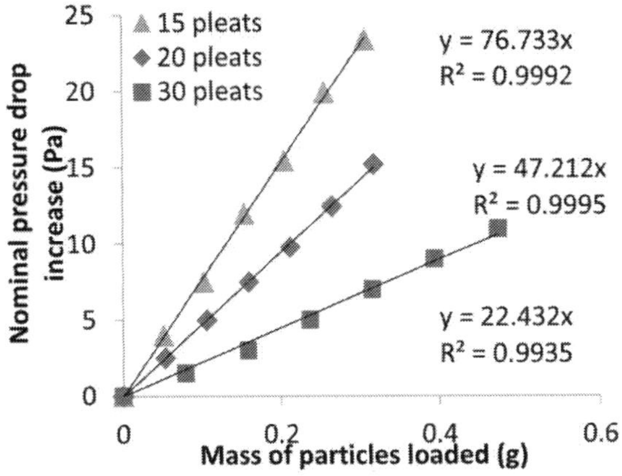

Merv 8 20 pleats 315 fpm

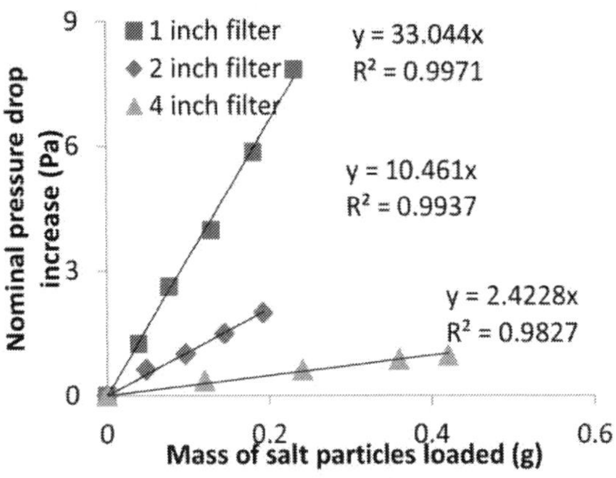

	Maximum mass of salt particles (g)	Maximum time (h)	Life time (Days)
15 pleats	1.4	8.9	4
20 pleats	2.6	16.4	8
30 pleats	5.7	35.6	18

	Maximum mass of salt particles (g)	Maximum time (h)	Life time (Days)
1 inch	6.0	33.1	17
2 inch	20.5	119.0	60
4 inch	91.4	510.5	255

14

AUBURN UNIVERSITY
SAMUEL GINN COLLEGE OF ENGINEERING

Multi-element Structured Array (MESA)

Merv 13 2 inch 15 pleats 315 fpm

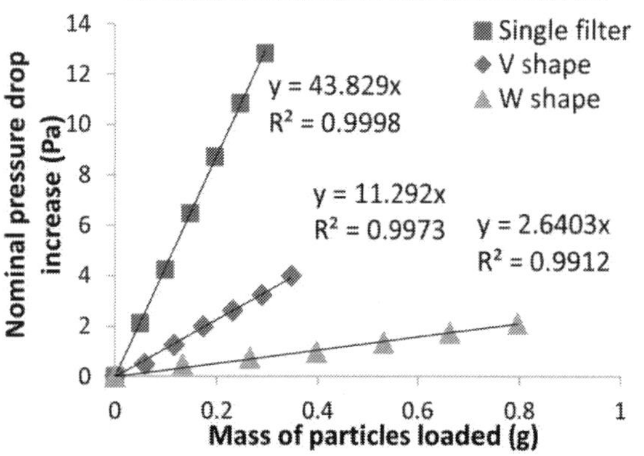

	Maximum mass of salt particles (g)	Maximum time (h)	Life time (Days)
Single filter	4.5	12.6	6
V shape	18.5	49.8	25
W shape	83.1	219.7	110

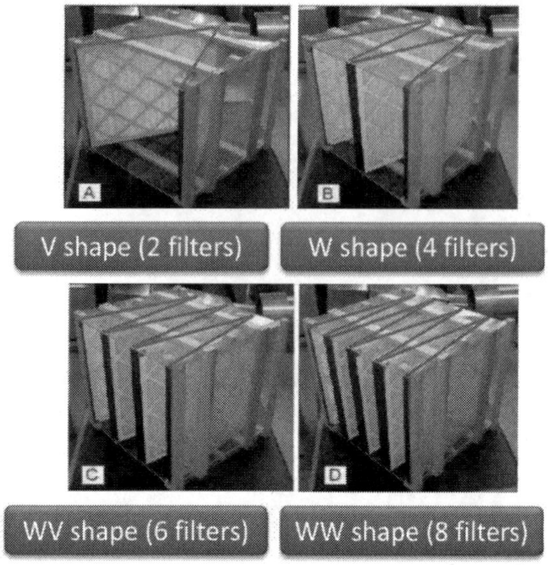

15

AUBURN UNIVERSITY
SAMUEL GINN COLLEGE OF ENGINEERING

Acknowledgement

- Advisor: Dr. Bruce J. Tatarchuk
- Staff members: Dr. Donald Cahela, Dwight Cahela, Kimberly Dennis, Sue Ellen Abner, Elaine Jimmerson, Karen Cochran and Brain Scweiker
- Members of CM^3 Group:
  - Especially filtration group
- Funding:
  - Office of Naval Research (ONR) Contract No. Navy-N00014- 11-1-0388

16

AUBURN UNIVERSITY
SAMUEL GINN COLLEGE OF ENGINEERING

Summary

- Mass of salt particles deposited can be estimated and calculated and the results fit very well with experimental data
- Nanofiber entrapped and coated filter media could lower particles loading capacity and filter lifetime
- Compared to woven filter media, nonwoven filter media have much better performance on particles loading due to different structure
- Filter media area and filtration velocity through the media could affect particles loading performance of filters simultaneously
- Larger filtration area and lower filtration velocity could lead to higher particles loading capacity and filter lifetime
- MESAs (Multi-element structured arrays) could significantly enhance particles loading performance and filter lifetime, which is able to be applied in real applications, such as SOFCs during ocean platforms

17

AUBURN UNIVERSITY
SAMUEL GINN COLLEGE OF ENGINEERING

Thank you for your attention!!

Questions?

18

387818 Removal of Heavy Metals from Wastewater Using Immobilized Biochar

Abstract:

Biochar, a product of pyrolysis, is receiving increased attention recently due to its unusual physical and chemical characteristics. In addition to its long-term carbon sequestration capabilities, biochar has been proposed to be used from anywhere from soil augmentation to removal of toxic metals from wastewaters. This paper is about developing a system that optimizes the removal of heavy metals, particularly lead, from wastewater using immobilized biochar in alginate beads. The advantages of immobilization in a polymeric bead is that biochar can be used as a fine powder, increasing its surface area/mass ratio, and by incorporating the powder into an open polymeric matrix the exposed biochar is more accessible to flow and, for packed bed operations, enables higher flow rates to be used with less pressure drop. Preliminary experiments in our lab indicate that a bead can be formed that has improved properties per unit mass over the non-immobilized biochar. The technical significance of developing this new process is important because it could help make remediation of waters contaminated with toxic metals less expensive.

375805 UV-Assisted Stabilization of Modified Softwood Kraft Lignin Fibers

Abstract:

Production of high strength carbon fibers from bio-derived precursors is of topical interest. Carbon fibers were produced from a partially acetylated softwood kraft lignin, but the thermo-oxidative stabilization required a low heating rate below 0.2°C/min to prevent the fibers from being heated too rapidly and becoming soft / tacky. Therefore, in this study, stretched as-spun fibers were UV irradiated for 15 min. The fibers underwent sufficient crosslinking on the surface such that they could be subsequently thermally stabilized at heating rate as high as 1.4°C/min. The extent of reaction was studied using FTIR, and the glass transition temperature evolution was assessed by DSC. Stabilized fibers were successfully carbonized at 1000°C, and resulting carbon fibers displayed a tensile strength as high as 1 GPa. These results establish that UV irradiation can effectively shorten the total stabilization time for lignin-based carbon fibers.

384364 Effect of a Natural Cactus Based-Mucilage Dispersant on the Surface Tension and Droplet Size of Dispersed Crude Oil

Abstract:

The purpose of this work is to evaluate the effect of the cactus plant-based mucilage on the surface tension and droplet size of dispersed crude oil in oil in water emulsions. The mucilage is extracted from the cactus plant, Opuntia ficus-indica, commonly known as Nopal or Prickly pear cactus. The extraction process yields two types of mucilage: the non-gelling extract (NE) and the gelling extract (GE). Cactus mucilage extracts and conventional dispersants were evaluated under different conditions and concentrations, including two concentrations of oil (3, 6% v/v) and several dispersant to oil ratios (1:1,1:3, 1:20, 1:40) at room temperature. Synthetic seawater and surrogate oil from Britsh Petroleum (properties similar to the Macondo rig) were used. The results of the dispersion obtaned using mucilage extracts were compared with results using conventional dispersants (Corexit EC9500A). The average droplet size was smaller in the systems with cactus mucilage (1.5% NE generated 5 µm droplets) when compared with the systems using the conventional dispersant Corexit EC9500A (1.5% EC9500A formed 6.2 µm droplets). The surface tensions of Oil/Water emulsion (3, 6% v/v) were reduced from around 50 mN/m to 30 mN/m by adding the cactus mucilage extracts. A standard EPA toxicity test using Daphnia magna colonies was exposed to both NE and GE mucilage extracts in concentrations ranging from 0 to 5000 mg/L showed that mucilage can be considered non-toxic to the evaluated species with a median lethal concentration (LC_{50}) above 2000 mg/L. Cactus mucilage is an alternative technology to mitigate the damage that oil may cause to the aquatic ecosystem and minimize undesired effects associated with the use of synthetic dispersants in oil spills.

388378 Artificially Engineered Protein Gels Derived from Nucleoporins

Abstract:

The nuclear pore, a tiny channel 45 nm in diameter, is filled with a polypeptide-based hydrogel capable of selectively filtering < 0.1% of all proteins in a eukaryotic cell while translocating over 1,000 molecules per second. The responsible polypeptides are nucleoporins, commonly containing "FG" repeat sequences, which are known to associate with one another to construct physical gels and give rise to the hydrogel's selective filtering property. Major obstacles in utilizing nucleoporin-based hydrogels for separation technology include low biosynthetic yield and long gel processing time, which typically takes several hours to days.

In order to overcome these drawbacks, we developed artificially engineered protein building blocks that attempt to mimic the selective permeability of nuclear pore hydrogels, are produced in high yields, and possess reduced gelation times. A well-investigated nucleoporin, Nsp1, was used as a template sequence for its well-defined repeat domains. To mimic Nsp1's transport properties, nucleoporin-like polypeptides (NLPs) were designed from consensus Nsp1 repeat sequences. To facilitate hydrogel formation, NLP sequences were flanked with coiled-coil domains which create three dimensional polymer networks. Here, we demonstrate that artificially engineered biopolymer building blocks can be expressed in high yield and can construct a physically crosslinked hydrogel within a few seconds. Using a combination of selective permeability fluorescence assays, shear rheology, and Raman spectroscopy, the mechanism of artificially engineered NLP hydrogel function is studied.

379734 Enhancing Oxygen Permeability in Hydrogel Wound Dressing By Cyanobacterial Gas Vesicles

Abstract:

Hydrogel wound dressing is commonly used nowadays. However, for some wounds the use of hydrogel wound dressing may incur insufficient oxygen availability for proper healing. Oxygen has an important role in angiogenesis, cell motility, and extracellular matrix formation. Oxygen limitation during the wound-healing process is one of the major causes that can lead to formation of chronic wounds. Cyanobacterial gas vesicles are hollow, protein-walled, gas-filled structures in cylindrical shapes with conical caps. Gas vesicle wall is impermeable to liquid water but permeable to gases with very low mass transfer resistance. Because oxygen diffusivity in a gas phase is much higher than that in liquid and solid phases, introducing gas vesicles into the hydrogel has the possibility of effectively enhancing oxygen permeation through the gel. Nonetheless, we have shown earlier that if dispersed randomly, gas vesicles need to be provided in high volume fractions to be effective in enhancing oxygen transfer, because the oxygen transferring rapidly through the gas vesicles is bottlenecked by the slow transfer across the liquid/gel continuous phase that separates the dispersed vesicles. We have thus taken the approach of attaching gas vesicles as continuous layers on the surface of fibers that extend across the whole hydrogel thickness. For example, PMMA coated, highly porous, and large pored polyurethane foam can be introduced as a core porous structure in the hydrogel wound dressing. After the coated PMMA surface is gently modified by acid hydrolysis, gas vesicles collected from the *Anabaena flos-aqua* culture are chemically attached onto the modified surface. This foam structure creates connected gas vesicle layers or channels that allow fast oxygen transfer through the hydrogel wound dressing. The PMMA surface modification has been optimized and tested. AFM pictures confirmed the attachment of gas vesicles on the modified surface. Much faster oxygen permeation through attached gas vesicle layers has been visualized by the localized color change in resazurin solution immediately adjacent to the gas vesicle-attached surface of a PMMA strip placed in the originally colorless anaerobic solution. We are currently measuring the oxygen permeability to provide more quantitative information on the oxygen transfer enhancement offered by different methods and amounts of gas vesicle attachment on surfaces. Results of cyanobacterial gas vesicle production and harvesting as well as the oxygen transfer enhancement achieved through the use of gas vesicle will be reported and discussed in the presentation.

386251 Adhesive Elastin-Based Proteins As Soft Tissue Glues

Abstract:

Each year, millions of major surgeries are performed worldwide. Many of these major surgeries involve suturing of tissue, which results in additional damage to the tissue and slows the healing process. Surgical adhesives are a promising alternative to sutures. To be a successful surgical adhesive, a product must be: biocompatible, set in a wet environment, and have proper adhesive and cohesive properties. In addition, a soft tissue surgical adhesive must also be flexible to prevent excessive stress on surrounding tissue. Unfortunately, most modern adhesives lack at least one of these characteristics.

The goal of this project is to utilize protein engineering to develop and characterize a biomimetic soft tissue surgical adhesive. The protein-based adhesive will contain two domains: a structural domain that provides flexibility and an adhesive domain that provides the ability to adhere in wet environments. The structural domain is inspired by elastin, a protein that provides elastic properties to tissues such as skin, blood vessels, and ligaments. The adhesive domain is inspired by mussel adhesive proteins (MAPs), which have the ability to bond underwater to nearly any material. MAPs contain numerous 3,4-dihydroxyphenyl-L-alanine (DOPA) residues, a chemical moiety that provides adhesion and cohesion through a number of mechanisms. Together, the elastin-based domain and the mussel-inspired domain will provide the characteristics necessary for an effective surgical adhesive.

Using standard molecular biology techniques, a family of proteins has been designed. Each protein contains elastin-like pentapeptides in combination with numerous tyrosine residues. Following expression in *E. coli* and subsequent purification, the proteins will undergo enzymatic conversion to produce the adhesive DOPA residues from tyrosines. To be successful soft-tissue surgical adhesives, the proteins must be biocompatible, have strong bulk adhesive strength, and a low elastic modulus. A LIVE/DEAD fluorescence assay will be utilized to assess protein cytotoxicity. Lap shear adhesion testing will be used to examine bulk adhesive strength. Finally, uniaxial tensile testing of crosslinked protein hydrogels will be performed to assess the elastic modulus of the adhesive. Initial results show high cytocompatibility (>98% viability) and significant bulk adhesive strength on aluminum (approximately 1 MPa). In summary, this new family of flexible adhesive proteins combines the mechanical properties of elastin and the adhesive properties of MAPs. Its unique properties are ideal for application as a soft tissue surgical adhesive, as well as coatings for cell attachment or scaffolds for tissue engineering.

Simulation and experimental investigation of osteogenic activity of hydrogel-conjugated BMP-2 peptide

Seyedsina Moeinzadeh, Esmaiel Jabbari

Department of Chemical Engineering, University of South Carolina, Columbia, SC, USA 29208

Introduction: The Peptide derived from the active domain of osteoinductive rhBMP-2 protein increased osteogenic differentiation and mineralization of the bone marrow stromal cells. Peptides can be conjugated to a macromer and crosslinked into a gel to immobilize the peptide at the site of regeneration. However, peptide aggregation and nanoscale structure formation in the conjugated state in aqueous solution can influence morphogenesis of the encapsulated mesenchymal stem cells (MSCs). In this work, we simulated nanostructure formation in the precursor solutions of polyethylene glycol (PEG) based hydrogels containing BMP2 peptide, BMP2 pre-conjugated to PEG macromers and BMP2 pre-conjugated to amphiphilic PEG macromers.

Materials and methods: The atomistic structure of the peptide and macromers were coarse-grained into different beads (set of atoms). There were totally 16 bead types for peptide back bone and amino acid side chains, ethylene oxide repeating unit (EO), acrylate group (Ac) and lactide repeating unit (L) (Fig 1a). Simulations were performed under NVT ensemble dynamics using the MARTINI force filed.[3] A Lennard-Jones (12-6) function was used for Van der Waals interactions. A Coulombic field with a cutoff distance of 12 Å was employed for electrostatic interactions. A harmonic function for bond length, a cosine harmonic function for bond angles and a dihedral function for torsions were used to account for interaction between bonded beads. A time step of 10-fs was employed and total simulation time was 1-μs. The temperature was held at 37°C using the Nose thermostat. All of the simulations were performed in 150×150×150 Å boxes with 3D periodic boundary conditions. The Mesocite module of Materials Studio (v5.5, Accelrys) was used for the mesoscale simulations. The osteogenic activity of the conjugated peptide was assessed with human mesenchymal stem cells by measuring alkaline phosphatase activity (ALP) and the extent of mineralization with incubation time.

Results: The simulation results showed that acrylated BMP2 peptides in the PEG aqueous solution formed aggregates with a distribution of aggregation numbers nagg (number of peptides per aggregate) ranging from 1 to 6. The average Nagg did not change significantly with conjugation of BMP2 peptide to the PEG macromer but the distribution of Nagg was narrower. The number of aggregates decreased and the average size of each aggregate increased when the peptide was pre-conjugated to the macromers containing degradable lactide segments. The average Nagg increased from 2.2 to 3.1, 4.2 and 5.8 and the number of aggregates decreased from 11 to 8, 5 and 4 when the number of lactide monomers per macromer increased from 0 to 2, 4 and 6, respectively. This was attributed to the contribution of hydrophobic lactide units to the formation of micelle like aggregates. Moreover, the hydrophobic side chains of the peptide (I, V, L) formed the center of the aggregate and the charged residues (K and E) located on the surface of the aggregates. Furthermore, the Ac-Ac integration number (INAc-Ac) which indicates the proximity of reactive acrylate groups in PEG solutions increased from 1.9 to 2.3, 2.6 and 3.3 with increasing the number of lactide monomers per peptide-macromer conjugate from 0 to 2, 4 and 6. The increase in INAc-Ac resulted in an increase in the conjugation efficiency of BMP2 peptide. Conjugation of the BMP-2 peptide to PEG increased critical micelle concentration (CMC) and osteoinductive potential of the peptide whereas conjugation to lactide-capped PEG reduced CMC and osteoinductive potential of the peptide. Experimental and simulation results revealed that osteoinductive potential of the BMP-2 peptide is correlated with its CMC and the free peptide concentration in aqueous medium and not the total concentration.

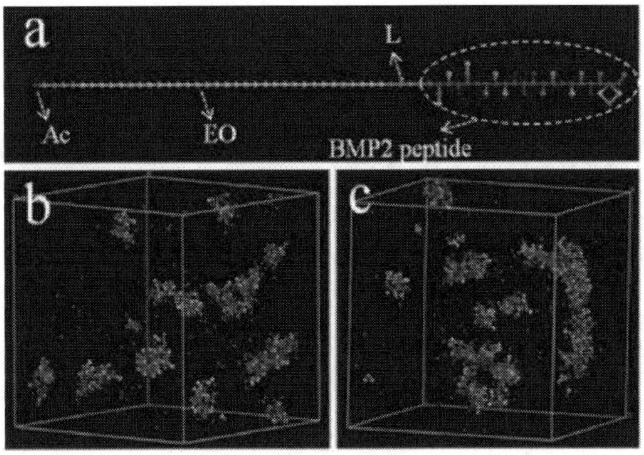

Figure 1. (a) Coarse-grained structure of BMP2 peptide conjugates; aggregate formation in BMP2-PEG (b) and BMP2-PEG-Lactide conjugates (c).

370147 Designing Multi-Component Nanostructured Soft Biomaterials Interacting with Charged Nanoparticles

Abstract:

We develop a computational model to design and characterize multi-component nanostructured soft biomaterials and their interactions with functionalized and charged nanoparticles using the Molecular Dynamics simulation technique. Our objective is to generate a stable vesicle through the self-assembly of amphiphilic lipid molecules using implicit solvent coarse-grained molecular models in order to investigate biologically important physiological processes occurring on the mesoscopic spatio-temporal scales. The amphiphilic lipid molecules are composed of a hydrophilic head group and two hydrophobic tails. In addition, we use two component lipid mixtures to study the formation of hybrid lipid vesicles through the self-assembly. We find that the degree of phase segregation between the two lipid species is tunable via their distinct effective chemical specificity and molecular architecture. Furthermore, we have extended the model to introduce the charged lipid molecules into the vesicle by using long-range electrostatic interactions between charged molecules. We investigate the interactions of various functionalized nanoparticles with the charged lipid species and effects of these interactions on the dynamics and morphologies of the lipid vesicles. Our findings can be used to design new soft biomaterials for various applications in medicine, sensing and energy.

390724 Novel 3D-Printed Oleophillic Absorbents

Abstract:

Here we describe efforts to design and produce absorbent materials containing internal capillary networks capable of selectively extracting oil dispersed in water. Unlike conventional methods, tour approach leverages commercial 3D printing technology a vehicle to rapidly produce absorbent materials with a high capacity for oil uptake. The 3D printing approach enables internal porosities to be precisely tailored to match the properties of different petroleum fractions and maximize the selectivity against water. The internal pore architectures function in much the same way as capillary networks in trees which are capable of generating enormous pressure driving forces. The polymer used to construct the absorbent structures is biodegradable and environmentally friendly, and the collected oil can be straightforwardly recovered. We report results of experiments aimed at determining the optimal pore network size and processing conditions to maximize recovery of different petroleum samples dispersed in water. We also evaluate the extent to which the rate of uptake can be used as a parameter to estimate the fluid viscosity.

390807 Fate and Removal Behaviour of Antibiotics in an Osmotic Membrane Bioreactor for Municipal Wastewater Treatment

Abstract:

Recently, there has been growing interest in the development of the osmotic membrane bioreactor (OMBR) process due to its low membrane fouling propensity, reduced energy demand and high rejection capability of various contaminants by the membrane. In this work, an OMBR was developed and operated with raw municipal wastewater as feed and seawater brine as draw solution (DS). The fate and removal behaviour of 21 types of antibiotics present in the municipal wastewater, including four β-Lactams, five fluoroquinolones, four macrolides, four tetracyclines, three sulphonamides and trimethoprim, were investigated in the OMBR system. With an influent concentration of 2.3-875.0 ng/L (the most prevalent antibiotics being enrofloxacin, sulfamethazine and cefalexin, followed by amoxicillin, lomefloxacin and ampicillin), the system showed 47%-99% overall removal of all the antibiotics. Almost all the antibiotics were efficiently rejected (87.3-99.7%) by the FO membrane, except sulphonamides (44.8-57.6%) due to their low molecular weights (250.1-278.1 g/mol). Relatively low biodegradation (10.7%-69.0%) was observed for all the antibiotics except macrolides (78.6-84.2%), which resulted in their accumulation (by 1.9-7.1 fold, with a maximum concentration of 2653.6, 2347.4 and 1081.1 ng/L for enrofloxacin, sulfamethazine and cefalexin, respectively) in the mixed liquor. Different antibiotics showed distinct sorption affinity to the activated sludge. The dominant antibiotics found in the activated sludge were ofloxacin (533.9 μg/gMLSS), amoxicillin (514.1 μg/gMLSS) and ciprofloxacin (476.7 μg/gMLSS) due to their high sorption constants (K_d). Ampicillin, lomefloxacin, enrofloxacin and oxytetracycline showed the poorest sorption to the activated sludge. All the antibiotics were detected in the DS (with an average concentration of 0.8-205.1 ng/L) except macrolides (with enrofloxacin, amoxicillin and sulfamethazine being the most predominant), which highlights the need for the enhancement of their biodegradation in the bioreactor. Accumulation of antibiotics in the bioreactor did not significantly affect pollutant removal efficiencies. High and constant removal of TOC (95%) and NH_4^+-N (99%) were achieved during the 98-day operation, with the DS TOC and NH_4^+ –N concentrations being less than 5.0 mg/L and 0.1 mg/L, respectively.

383292 Removal of Trihalomethanes (THMs) By Electro-Coagulation Process Using Aluminium Plate

Abstract:

Nowadays, chlorine is commonly used as the disinfectants in water treatment process. However, it would react with nature organic matter (NOM) to form disinfection by-products (DBPs) such as trihalomethanes (THMs) and Haloacetic acids (HAAs). The above THMs presented in drinking water have been considered as toxic compounds due to the risks Bladder cancer, rectal cancer, colon cancer and brain cancer according to the laboratory animals. The standard regulation for the concentration of THMs in drinking water is 80 ug/L in European Union and Taiwan. On the other hand, the USEPA formulated the standard concentration for THMs of 80 ug/L and would be further restricted to 40 ug/L in the future. Therefore, the removal of DBPs in drinking water be an important issue in environment engineering.

Electrocoagulation (EC) is an electrochemical process for treating wastewater using sacrificial anodes to generate coagulant precursors into solution. Aluminum ions would be released from the sacrificial anode instead of adding traditional coagulants, e.g., aluminum salts, thereby increasing the treatment efficiency of chemical precipitation and reducing the consumption of chemicals. As the result, the objectives of this investigation are (1) to evaluate the removal efficiency of THMs by applying the EC process in a batch reactor, and (2) investigate the effects of different operating conditions including initial pH, surfactant dose, charge loading and electrical conductivity on the removal efficiency of THMs, and (3) determine the optimized conditions for EC operation.

Abstract: Sulfur-containing radicals including sulfate radical ($SO_4^{\bullet-}$) and sulfite radical ($SO_3^{\bullet-}$) are strong but short-lived oxidant and reductant that can react with a verity of emerging contaminants, respectively. Therefore, they can be applied in either advanced oxidation or reduction process for water treatment and reuse. This study is aimed to investigate the generation of these two radicals by UV activation. In oxidation treatment process, production of oxidative $SO_4^{\bullet-}$ radical was achieved through UV/persulfate activation. Using benzene as a probe compound, it was determined that $SO_4^{\bullet-}$ radical was more efficient than traditionally used hydroxyl radical ($OH^{\bullet}$) when generated by persulfate and is less likely to be scavenged by common water constituents. Results from our recent study showed that under chemical conditions encountered in treated effluent, persulfate was converted by UV to $SO_4^{\bullet-}$ radical with a yield that was two orders of magnitude higher than $OH^{\bullet}$ generated from UV/hydrogen peroxide. Sulfate radicals then initiated the oxidation of recalcitrant organics (*e.g.*, benzene) through a complex chain reaction involving other intermediate species, suggesting an effective delivery of oxidants during water reuse application.

In reduction treatment process, production of $SO_3^{\bullet-}$ radical was achieved through UV/sulfite irradiation. Using hexavalent chromium Cr(VI) as a probe compound, results showed that the yield of $SO_3^{\bullet-}$ radical depended on solution pH, with a maximal yield at circumneutral pH range. It was also observed that the reduction reaction proceeded at a fast rate, with all Cr(VI) removal within 20 minutes of UV/sulfite treatment. In addition, the generation of strongly reductive $SO_3^{\bullet-}$ radical can also be converted to other intermediate species, including hydrogen atom ($H^{\bullet}$) and hydrated electron (e_{aq}^{-}) at different alkalinity and pH levels. This conversion impacted the efficiency of $SO_3^{\bullet-}$ radical treatment.

By studying the detailed chemistry of sulfate radical ($SO_4^{\bullet-}$) and sulfite radical ($SO_3^{\bullet-}$) through UV activation system, respectively, we can make the applications of sulfur-containing radical based system in water reuse application. For example, existing UV disinfection unit process can be retrofitted to remove pharmaceuticals through sulfate radical based oxidation process, and disinfection by-products, chromate and bromate can be removed through sulfite radical based reduction process.

Copper (II) Removal Using Activated Neem Bark from Waste Water as a Low Cost Adsorbent: Column Studies

Utkarsh Maheshwari[1], Suresh Gupta[2*]
[1]Lecturer, Department of Chemical Engineering, Birla Institute of Technology and Science (BITS), Pilani- 333031, Rajasthan, India, Email: utkarsh@pilani.bits-pilani.ac.in
[2]Assistant Professor, Department of Chemical Engineering, Birla Institute of Technology and Science (BITS), Pilani - 333031, Rajasthan, India.Phone: +91-1596-515224; Email: sureshg@pilani.bits-pilani.ac.in
*Corresponding Author

Abstract

With an rise in industrialization and water usage in the industries, there is an increase in the water pollution activities. There are a number of industries using heavy metals in their processing. These heavy metals are also available in their effluent streams in a very vast quantity. These heavy metals will make the water source impure and not suited for day-to-day use. Thus there is a prerequisite for obtaining a suitable method for the handling of the effluent streams. There are a number of procedures available for heavy metal removal out of which adsorption is proven to be the most desirable method[1-5].

In the present study, a fixed-bed adsorption column is used to perform continuous adsorption experiments in order to assess the performance of low-cost adsorbent (activated neem bark) developed for the removal of Copper (II) from aqueous solutions. Flow rate is among one of the most important parameter which is affecting the adsorption rate in the fixed bed column. Here, the effect of flow rate on the removal of copper (II) in the continuous adsorption column is studied. The corresponding breakthrough curves are also obtained. Using the developed breakthrough curve, various parameters such as breakthrough time, total percentage removal of Cr(VI), adsorption exhaustion rate and fraction of unused bed-length were obtained[4-6].

Keywords: Heavy Metal; Adsorption; Copper; Continuous Studies;

References:

1. Al-Rashdi B, Tizaoui C, Hilal N. Copper removal from aqueous solutions using nano-scale diboron trioxide/titanium dioxide (B2O3/TiO2) adsorbent. *Chem. Eng. J.* 2012;183(0):294-302.
2. Maheshwari U, Gupta S. Kinetic and Equilibrium Studies of Cr (VI) Removal from Aqueous Solutions using Activated Neem. *Research Journal of Chemistry and Environment.* 2011;15(2):939-943.
3. Gangadhar G, Maheshwari U, Gupta S. Application of Nanomaterials for the Removal of Pollutants from Effluent Streams. *Nanosci. Nanotech. -Asia.* 2012;2:140-150.
4. Gupta S, Babu BV. Experimental investigations and theoretical modeling aspects in column studies for removal of Cr(VI) from aqueous solutions using activated tamarind seeds. *J. Water Resource Prot.* 2010;2:706-716.
5. Gupta S, Babu BV. Modeling, simulation, and experimental validation for continuous Cr(VI) removal from aqueous solutions using sawdust as an adsorbent. *Bioresour. Technol.* 2009;100(23):5633-5640.
6. Baral SS, Das N, Ramulu TS, Sahoo SK, Das SN, Chaudhury GR. Removal of Cr(VI) by thermally activated weed Salvinia cucullata in a fixed-bed column. *J. Hazard. Mater.* 2009;161(2–3):1427-1435.

377457 Preparation Characterization and Application of an Affinity Flat Sheet Membrane

Abstract:

Arsenic is one of the main toxic heavy metals which are listed in US EPA's 129 priority pollutants. On one hand, arsenic in the groundwater is of natural origin and being released from the sediment into the groundwater due to the anoxic conditions of the subsurface. It is unmanageable and an issue of major concern. On the other hand, industrial waste discharge of arsenic exacerbates the problem. The Arsenic pollution of ground water is found in many countries throughout the world. The more strict regulations also bring a big challenge to conventional treatment technologies, such as ion exchange chemical precipitation, adsorption and membrane. Thus, to overcome the disadvantages of the current treatment technologies, a combined technology of adsorption and membrane filtration is proposed. The objectives of this study were to obtain a highly performed affinity membrane and to understand the working mechanism.

PSF/zirconium based nanoparticle blend flat sheet membranes with different amount of nanoparticle loaded were synthesized. The morphology and structure of membranes were analyzed by scanning electron microscope (SEM) and atomic force microscope (AFM). The performance of membranes was evaluated in terms of pure water flux (PWF), porosity, contact angle, tensile strength and elongation at break respectively. The adsorption capacity of the membrane also was tested by adsorption isotherm experiment.

Results showed that the addition of zirconium based nanoparticle additive increased the ratio of large pore in the skin layer and weakened the tensile strength. PWF and porosity of membranes increased with the increase of zirconium based nanoparticle dosage. The tensile strength attained its maximum before the nanoparticle was added. The nanoparticle had high stability in the nanoparticle blend membranes. The adsorption isotherm experiments showed the maximum adsorption capacity of the affinity membrane was as high as 120 mg-As/g-membrane.

It was observed that there is a certain buffering effect from the membrane. The optimal pH for arsenic adsorption on to the affinity membrane is wider than the zirconium based nanoparticle. The kinetics study shows that the adsorption of oregano-arsenic onto the membrane is much slower than that of inorganic arsenate. The adsorption kinetics can be well described by pseudo second order rate model. The Langmuir model can be well used to describe the adsorption isotherms of arsenic onto the membrane. The presences of humic acid dose not significantly influence the adsorption of arsenic on the membrane. The existence of fluoride, silicate or bicarbonate slightly hinders the adsorption of arsenic onto the membrane. However, the presence of phosphate does not significantly affect the adsorption of arsenic onto the membrane. The adsorption of arsenic is less sensitive to the ion strength, which indicates the inner-sphere adsorption mechanism is involved in the adsorption of arsenic onto the membrane. The continuous filtration experiments shows the As(V) concentration in the permeate remains at a level of less than the maximum contaminant level (MCL = 10ppb) at the treated volume less than 10 L. The adsorption capacity before breakthrough (when the effluent concentration is above 10 ppb) is 12 mg/g. XPS spectroscopic analyses demonstrate that the arsenic was uptake onto the membrane.

AUTHOR INDEX

Abron, Jessica....228
Adewuyi, Yusuf G (Debo)....322
Agrawal, Rakesh....391
Aichele, Clint P....14
Akinmolayan, Folashade....95
Alanazi, Yousef....394
Albadarin, Ahmad....304
Albassam, Abdulrahman....384
Alcantar, Norma....585, 620
Alderink, Erin....93
Alkhamis, Lamees....630
Almendrala, Michelle....500
Al-Musleh, Easa I....391
Alves, Rita M. B....411
Amador Sr., Oscar E....64
Amar, Vinod S....385
Ammerman, Michelle....377
Ang, Gerald Aaron....500
Anggara, Trunojoyo....148
Anozie, Uchechukwu Chamberlin....622
Antle, Chad....377
Antonakopoulou, Mariela....16
Araújo, Ofélia....397
Araújo, Ofélia Q. F....411
Arce, Pedro E....393
Arifin, Darwin....382
Augenbroe, Godfried....176
Aurandt, Jennifer....377
Avij, Paria....56
Avila, Pedro....143
Aydin, Fikret....626
Badiee, Ali....174
Bajpai, Rakesh....499
Balakotaiah, Vemuri....128, 211, 323
Bañares, Miguel A....143
Bansal, Mukesh C....203
Bates, Shane A....148
Bava, Luciana....567
Bayles, Taryn....155
Beckman, Eric J....330
Behdani, Fattaneh Naderi....287
Bellona, Christopher....392
Belope, M. A. Abia-Biteo....38
Benavides, Pahola T....532
Bernadas, Edmar....500
Bhadbhade, Navdeep....374
Bharadwaj, Vivek....329
Bocanegra-Martinez, Andrea....283
Boguski, Terrie....192
Bohanan, Montgomery....521
Boningari, Thirupathi....60
Booshehri, Amin Yoosefi....275
Borgese, Christina M....596
Borole, Abhijeet P....229, 383
Boyd, Christopher....51
Brennan, M. Jane....623
Bricka, R. Mark....549
Brigham, Jill....520
Browe, Matthew A....495
Brown, Jason....176
Byl, Tom....310
Caballero, José Antonio....127
Cabezas, Heriberto....532
Camacho, Lucy Mar....599
Canter, Christina....25, 290, 293
Capón-García, Elisabet....125
Cardona, Carlos A....428, 437, 504
Carmichael, Gregory....55
Carrales-Alvarado, Damarys H....580
Casasus, Anna....567
Castro-Arellano, Jose J....373
Champion, Julie A....52
Chang, E-E....409
Chawla, Ramesh....548
Chen, Chun-Chi....330
Chen, Daniel H....311
Chen, Grace....486
Chen, Hai-Ying....59
Chen, Hao....143, 289, 291, 292
Chen, J. Paul....114, 565, 581, 598, 632
Chen, Long....566
Chen, Rachel Ruizhen....501
Chen, Rui....49
Chen, Wesley....621
Chen, Yi-Hung....409
Chen, Yudong....191
Cheng, Peng....131
Chiang, Andrew....409
Chiang, Pen-Chi....409
Chopra, Shauhrat....546
Choudhury, Hanif A....381
Christofides, Panagiotis D....48
Cogswell, Christopher....496
Cohen, Yoram....48
Coker, Eric N....382
Colombo, Mauricio....301
Colville, Alexander....244
Corbet, Tiphaine....294
Cordoba Sr., Luis F....64
Critzer, Faith....427
Cruz-Narváez, Yair....373
Cucek, Lidija....123
Cuesta, Rafael Serra....294
Currier, Neal W....59
Da Fonseca, Fabiana Valéria....470
Dai, Fei....392
Davidson, P. Michael....427
Davila Sr., Javier....428, 437
De, Sirshendu....600
Dean, Anthony M....312, 329

AUTHOR INDEX

Delgass, W. Nicholas ... 148
Deng, Shuguang ... 24, 82, 332, 378
Depaoli, David ... 484
Depcik, Christopher D. ... 154
Desai, Tanviben Pareshbhai ... 599
Di Maggio, Jimena A. ... 113
Diaz, María Soledad ... 113
Diwekar, Urmila ... 532
Dobashi, Ritsu ... 314
Doshi, Rohan ... 532
Downes, Cara Meghan ... 24
Dubinsky, Jonathan ... 97, 528, 564
Dufreche, Stephen ... 499
Dungan, Barry ... 332, 378
Dunn, Russell F. ... 276
Dunsworth, Hailey ... 70
Dutt, Meenakshi ... 626
Easterling, Vencon ... 134
Eleazar, Elisa G. ... 409
El-Halwagi, Mahmoud ... 115, 280, 283
Eljack, Fadwa T. ... 296
Epling, William ... 61
Epling, William S. ... 151
Ergas, Sarina ... 585
Espuña, Antonio ... 125
Esteller, Laureano Jiménez ... 531
Estrada, Vanina ... 113, 300
Evangelisti, Sara ... 121
Fagan, Melissa ... 96
Falahat, Rana ... 620
Fam, Yakub ... 294
Fang, Randy ... 73
Felix, Steve ... 149
Figueroa, Nancy ... 241
Fisher, Galen B. ... 457
Fontes, Fernanda M. ... 411
Fuentes-Cortes, Luis Fabian ... 210
Fuertes, Lady Andrea ... 299
Galeano, Sergio ... 98
Galloway, Benjamin ... 58, 485
Gan, Jing ... 503
Gao, Larry ... 48
Garcia-Montoya, Mariana ... 283
Garg, Anurag ... 472, 478
Gargalo, Carina ... 386
Gatica, Jorge E. ... 301
Gençer, Emre ... 391
Ghassemi, Abbas ... 287, 577
Ghosh, U K ... 325
Giorgio, Christopher ... 499
Glaspie, Rebecca M. ... 2
Glassman, Matthew J. ... 621
Gomez, Joshua ... 82
Gong, Bo ... 495
Gong, Jian ... 117
Gonzalez, Michael A. ... 529
Gotti, Alberto ... 16
Gounder, Rajamani ... 148
Graziano, Diane J. ... 1, 388
Greenlee, Lauren F. ... 277
Greer, Jessica ... 230
Grossmann, Ignacio E. ... 123
Gu, Han ... 48
Gu, Sai ... 381
Gude, Veera Gnaneswar ... 227, 333
Guillén-Gosálbez, Gonzalo ... 127, 531
Guo, Fei ... 585, 620
Guo, Zhanhu ... 597
Gupta, Suresh ... 631
Hadi, Atefe ... 577
Hadjkali, M. ... 450
Hamzehlouyan, Tayebeh ... 151
Han, Chonghun ... 408
Han, Duanduan ... 627
Han, Wei ... 143, 289, 294
Handakas, Evaggelos ... 16
Harinath, Arvind ... 150
Harold, Michael P ... 134, 211, 323
Harris, Justin ... 310
Hayden, Olivia ... 377
Hazlett, Melanie ... 61
He, Cheng-Chun ... 629
He, Jinsong ... 565, 598
He, Qiyang ... 503
He, Yi ... 498
Herceg, Zoran ... 449
Hernández, María Rosa ... 301
Hernandez, Rafael ... 499
Herrud, Joseph ... 521
Ho, Thomas C. ... 375
Hoch, Patricia M. ... 300
Holguin, Francisco ... 332, 378
Holmes, William ... 499
Holsen, Thomas ... 392
Hong, Pui-Kwan Andy ... 395
Honjoya, Karina ... 548
Honorato-Cervantes, Hever ... 373
Hu, Bo ... 503
Hu, Yongtao ... 53
Huang, He ... 56
Huang, Yinlun ... 374, 530
Hubler, Berlyn ... 389
Huey, Greg ... 51
Hung, Francisco R. ... 56
Hung, Pik Shuen ... 291
Hungerbühler, Konrad ... 125
Hutzel, Bill ... 175
Ighere, Jude ... 548
Iizuka, Atsushi ... 410

AUTHOR INDEX

Inoue, Motoki 410
Inturi, Siva Nagi Reddy 458
Irisawa, Jun 314
Jabbari, Esmaiel...624
Jaberi, Masoume 287
Jain, Sapna 228
Janho, Michel Kahwaji...301
Jemai, Abdelbasset Bessadok 450
Jensen, Cory 191
Jeong, Soyeon 63
Jeong, Yeong Su 408
Jimenez, Laureano 127
Jin, Jing..619
Jing, Guohua 278
Jobson, Megan...335
John, Vijay T. 578
Johnson, Darin 366
Johnson, Lauren...192
Johnson, Matthew 311
Jones, Christopher W. 486
Jones, Kelvin 228
Ju, Lu-Kwang...622
Jung, Jaeheum..408
Kaeli, Emma 91
Kail, Brian 426
Kamasamudram, Krishna.................................59, 211
Kan, Eunsung..396, 556, 579
Kang, Jeon Woong 621
Karakitsios, Spyros 16
Karunanithi, Arunprakash T......97, 120, 273, 528, 564
Kawi, Sibudjing..379
Kazi, Monzure-Khoda 296
Keith, Jason M. 155
Kelemen, Caroline 94
Khanna, Vikas..389, 546
Khasnavees, Sravya 92
Khezami, L. 450
Khorshed Sr., Alam Akm 598
Kim, Do Heui..63
Kim, Jihyun R. 396, 556, 579
Kim, Mi-Kyoung 501
Kim, Minkyu 621
Kim, Woonhoe..584
King, Pulipati 555
Kizil, Ramazan...502
Kokabian, Bahareh...333
Koros, William J. 486
Kota, Arun..128
Koutahzadeh, Negin...393
Koutinas, Michalis 550
Kravanja, Zdravko 123
Kreider, Peter...394
Krishnaiah, K. 575
Krishnan, Sundar Rajan 155
Kulkarni, Sarika 55
Kulkarni, Shilpa 203
Kumar, Amit...25, 290, 293
Kumar, Ashok 59
Kumar, Sandeep...227
Kumar, Vivek 203
Kuo, Chin-Sheng 395
Kwan, Joseph Kai Cho..291
Kwan, Siu Ming 291
Kwan, Thomas..330
Ladshaw, Austin 484
Lai, Yue Tak 289
Laínez, José Miguel 125
Lam, Ka Wo...291
Lammers, Pete 332
Lammers, Peter 24, 378
Lastoskie, Christian 497
Lauterbach, Jochen...58
Lee, Ilsoon 49
Lee, In-Beum 202
Lee, Jaime 567
Lee, Oi Wa 291
Lee, Soo Bin 202
Lemaire, Paul C. 495
Lettieri, Paola 15, 121
Lewis, Alex..383
Leyva-Ramos, Roberto 580
Li, Chun 284
Li, Hui 274, 302, 303
Li, Jiang-Feng...274, 302, 303
Li, Jihong 334
Li, Liping..294
Li, Lucy 630
Li, Mengmeng...134
Li, Mingheng...26
Li, Shizhong 334
Li, Su-Jing...224, 563, 608
Li, Wei..563, 608
Li, Xianchang..311
Li, Yan...289
Li, Ying..294
Liang, Bin..284
Liao, Wei 49
Lima, Bruna...397
Lin, Cheng-Fang..395
Lin, Hongjian..503
Lin, Ronghong 484
Liu, Haizhou 630
Liu, Julie C..623
Liu, Peng 53
Liu, Qing 498
Liu, Xiaoxi..51
Liu, Yi...150, 211
Liu, Zhiguo 49
Lo, Shang-Lien 629
Losego, Mark D. 495

AUTHOR INDEX

Lou, Helen H. 311
Loya, Jim 287
Lu, Houfang 284
Luebke, David 426
Lum, Jacob 394
Luo, Yangchao 427
Luss, Dan 57, 128, 211
Lyu, Yuan 451
Ma, Yue 114
Mahanta, Narahari 598
Maheshwari, Utkarsh 631
Mallapragada, Dharik S. 391
Maneerung, Thawatchai 379
Mangwandi, Chirangano 304
Manousiouthakis, Vasilios 384
Mantalaris, Athanasios 550
Martens, Joshua 273
Martín, Mariano 123
Martínez-Gomez, Juan 115
Martinez-Guerra, Edith 227
Masanet, Eric 1, 388
Mateo-Sanz, Josep Maria 531
Matute, Willy Giron 297
Maupin, C. Mark 329
Mavrikakis, Manos 132
McCullough, Jack 243
McDaniel, Anthony H. 382
McNeill, V. Faye 50
McPherson, Gary 578
Mededovic, Selma 392, 449
Medeiros, José L. 397, 411
Mele, Fernando Daniel 301
Menendez, Fernando Garcia 54
Meng, Binglu 274, 302, 303
Menicucci Jr., Joseph Anthony 15
Metkar, Pranit S. 211
Miao, Yong 149
Miller, Jeffrey T. 148
Mills, Patrick L. 139
Min, Yong 274, 302, 303
Moeinzadeh, Seyedsina 624
Mohammed, Fahd M. 296
Mohite, Rajendra 478
Moholkar, Vijay 381
Monier, Erwan 54
Morgado, Claudia 397
Mostafa, Mohamed 105
Mousavi, Bahador 120, 273
Muhich, Christopher L. 382
Mukherjee, Munmun 600
Mukherjee, Rajib 122
Muñoz, Edrisi 125
Muppaneni, Tapaswi 332
Muppaneni, Tapaswy 24, 378
Musgrave, Charles B. 382
Nagaraj, Anuradha 139
Nagarajan, Krishnan 575
Nahata, Mohit 457
Nápoles-Rivera, Fabricio 115, 280, 283
Narsimhan, Ganesan 451
Nartker, Steven 377
Nazem-Bokaee, Hadi 518
Nenes, Athanasios 53
Neoh, Koon Gee 380
Newkirk, Dan 175
Ng, Nga Lee 51, 52
Nimana, Balwinder 293
Nirmalakhandan, Nagamany 332, 378
Nogami, Tatsuhiro 314
Noguchi, Miyuki 410
Nordin, Nadiah 242
Noriega, Mario Andres 299
Nulwala, Hunaid 426
Nunn, William T. 495
Ocampo-Perez, Raul 580
Ochoa, María P. 300
Ogale, Amod A. 619
Okamoto, Hidekazu 314
Olasehinde, Owoseni 578
Olsen, Bradley D. 621
Olukoya, Ife 14
Oluleye, G. 335
Ong, Cedrick 500
Otsuka, Tetsuo 314
Owoseni, Oluwatosin 298
Padak, Bihter 58, 485
Painter, Roger 310
Pan, Fujun 278
Pan, Kang 427
Pan, Shu-Yuan 409
Panda, Swapna Rekha 600
Paolucci, Christopher 148
Papageorgiou, Lazaros G. 121
Pappas, Dimitrios 60
Parekh, Atish 148
Park, Kyoung Eun 229
Park, Young-Il 501
Parker, Amy M. 549
Parks, James E. 222
Parra, Abdón 470
Parsons, Gregory N. 495
Pascual-González, Janire 531
Paudel, Ghana 390
Peng, Tunan 620
Peters, Robert W. 105
Peterson, Gregory W. 495
Petty, Brian 96
Pierce, Jason 2
Pihl, Josh A. 222
Pirbazari, Massoud 584

AUTHOR INDEX

Pistikopoulos, Efstratios N. 550
Pommerenck, Jordan 394
Pommerenck, Justin 394
Ponce-Ortega, José María 115, 210, 280, 283
Ponnusamy, Sundaravadivelnathan 24, 332, 378
Portela, Raquel 143
Prikhodko, Vitaly Y. 222
Puigjaner, Luis 125
Puzsynski, J. 385
Qiu, Bin 597
Qiu, Guanglei 628
Radzi, M. R. Aliff Mohamad 38
Ragha, Divya Shankari Srinivasa 628
Rahardianto, Anditya 48
Rahman, Md. Mustafizur 25, 290, 293
Raihan, Mostafa S. 155
Raj, Richa 323
Rajendran, Aravindan 503
Ramachandriya, Karthikeyan 14
Ravindran, Varadarajan 584
Ray, Amiya Kumar 325
Reddy, Harvind 332
Reddy, Harvind Kumar 24, 378
Renganathan, T. 575
Renner, Julie N. 623
Rentz, Nikki S. 277
Ribbeck, Katharina 621
Ribeiro, Fabio H. 148
Rich, Stephanie 71
Richmond, Peyton C. 311
Rico-Arzate, Enrique 373
Riddle, Mathew 1, 388
Ristau, Jarrid 276
Ro, Kyoung S. 331
Rogers, Alexis 228
Rogers, Shane 449
Rojas, Miguel 437, 504
Roland, Michael 519
Roman, Jessica K. 623
Román-Leshkov, Yuriy 133
Rong, Le 380
Rosenberg, Moshe 428, 437, 504
Rubio-Maya, Carlos 210
Ruiz-Femenia, Ruben 127
Ruiz-Mercado, Gerardo J. 529
Rupp, Erik C. 483
Russell, Armistead 53
Ryu, Jun-Hyung 202
Saliyeva, Sayara 15
Sampara, Chaitanya S. 151
Sanchez, Javier 51
Sanchez-Bautista, Aurora De Fatima 280
Santibañez-Aguilar, José Ezequiel 115, 280
Sanyal, Oishi 49
Saraeian, Alireza 577
Sarigiannis, Dimosthenis 16, 358
Sasmaz, Erdem 58, 485
Sastri, Bhima 191
Schaub, Tanner 332, 378
Schneider, William F. 148
Schwank, Johannes W. 457
Selin, Noelle 54
Selvaratnam, Thinesh 378
Senger, Ryan S. 518
Sengupta, Debalina 122
Serna-González, Meadardo 115
Serna-González, Medardo 210, 280, 283
Shakya, Bijesh M. 211
Sharma, Hom 223
Sharma, Pooja 325
Sharpe, Lonnie 310
Shende, Rajesh V. 385
Sheshamma, Garika 555
Shi, Fan 426
Shi, Junjie 498
Shi, Yao 498
Shi, Yun 224
Shin, Hyun-Dong 501
Shrestha, Sachi 211
Sidi, Fahim I. 495
Siebenhofer, Matthaeus 459
Siefert, Nicholas 426
Sikdar, Subhas 122
Silva-Castro, Vanessa 373
Simakov, David 133
Simmons, Christopher W. 527
Simons, Stefaan J. R. 15
Sin, Gürkan 386
Singh, Sanjeev K. 203
Smirniotis, Panagiotis 60, 376, 458
Smith, Joseph D. 390
Smith, Raymond L. 529
Smith, Robin 335
Sobhani, Negin 55
Soh, Lindsay 330
Sommerfeld, Anna 49
Song, Hao 374, 530
Sorensen, Eva 95
Sotomayor, Francisco 497
Srinivasan, Kalyan K 155
Srivastava, Pulkit 381
Stagg-Williams, Susan M. 154
Stangland, Eric 132
Stebbins, Daniela M. L. 585, 620
Steiner, Daniel 390
Stevens, Eric C. 495
Stogsdill, Michael 377
Stratton, Gunnar 392
Strzelec, Andrea 221
Su, Changsheng 150

AUTHOR INDEX

Su, Renay S.-C. 623
Su, Yang 578
Subramaniam, Ramalingam 499
Suidan, Makram 458
Sun, Dezhi 597
Sunkara, Bhanukiran 578
Taborda Sr., Gonzalo 428
Tahir, Muhammad Suleman 557
Tanaka, Toshiyuki 314
Tanhaemami, Mohammad 288
Tatarchuk, Bruce 131
Tatarchuk, Bruce J. 226, 609
Tavlarides, Lawrence L. 484
Teymouri, Ali 227
Thai, Chris 273
Thayil, Shaj 114
Thomas, Sylvia 620
Thornhill, Nina F. 95
Thorpe, R. B. 38
Ting, Yen-Peng 628
Toghiani, Hossein 155
Tong, Yen Wah 380
Toomey, Ryan 620
Toops, Todd J. 222
Trinh, Tran D. 497
Trujillo, Edward 618
Tsipa, Argyro 550
Tsouris, Costas 229, 484
Tuet, Wing-Yin 51, 52
Ueno, Katsuya 314
Ugaz, Victor M. 627
Upadhyay, Devesh 212
Valsaraj, Kalliat T. 56
Van Nieuwstadt, M. 212
Van Wyk, James 452
Vargas, Julio C. 64
Verma, Anuj 148
Verma, Vishal 52
Villano, Stephanie M. 312
Vongpanish, Napaporn 622
Voorhies, Wayne 332, 378
Vorobiev, E. 450
Vukusic, Tomislava 449
Wade, Elijah 72
Walker, Gavin 304
Wang, Chao 58, 284
Wang, Chi-Hwa 379, 380
Wang, Kun 312
Wang, Qi 514
Wang, Xiang-Qian 563, 608
Wang, Yufei 284
Wang, Zixuan 231
Wang, Ziyuan 375
Webster, Thomas J. 514
Wei, Suying 597
Weimer, Alan W. 382
Wentworth, Travis 154
Weston, Kayla 382
Weston, Theresa 191
White, Sam 245
Wilcox, Jennifer 483
Wilhite, Benjamin A. 57
Wilker, Jonathan J. 623
Wilkins, Mark R. 14
Woisetschläger, Dieter 459
Wu, Chao 608
Wu, Ching-Yao 629
Xiang, Ning 451
Xie, Heping 284
Xu, Bai-Long 608
Xu, Cuixia 597
Xu, Lang 132
Xu, Lu 51
Xu, Qiang 375
Xu, Rong 275
Yadav, Bholu Ram 472
Yamasaki, Akihiro 410
Yan, Liwei 530
Yan, Mi 503
Yang, Wen-Bin 274, 302, 303
Yang, Yan 503
Yang, Zhanyu 379
Yao, Yuan 1, 388
Ye, Longpo 284
Yen, Jiun 518
Yenkie, Kirti Maheshkumar 532
Yeung, King Lun 143, 289, 291, 292, 294
Yezerets, Aleksey 59, 148, 211
Yiacoumi, Sotira 229, 484
Yokochi, Alexandre F. T. 394
Yokoyama, Lidia 470
You, Fengqi 117
Youn, Seunghee 63
Yu, Hiu Pang 292
Yu, Ling 632
Yu, Youhai 274, 302, 303
Yue, Hairong 284
Zaimes, George G. 389
Zanganeh, Navid 155
Zappi, Mark E. 499
Zhan, Jingjing 578
Zhang, Di 121
Zhang, Jian 375
Zhang, Lele 274, 302, 303
Zhang, Meng 619
Zhang, Yue 427
Zhang, Yueheng 578
Zhang, Zenghui 56
Zhao, Junjie 495
Zhao, Pengfei 609

AUTHOR INDEX

Zhao, Wen 585, 620
Zheng, Shudong 498
Zheng, Yang 57, 211
Zheng, Yu Ming 632
Zhong, Qixin...427
Zhou, Meng 82
Zhou, Zuoming...278
Zhu, Jiahua...284, 566
Zhu, Wenhua H..226
Zimmerman, Julie..330